IMPRINTING AND CORTICAL PLASTICITY

Wiley Series in Neuroscience

Editor: Josef P. Rauschecker

VOLUME 1 **Imprinting and Cortical Plasticity:**
Comparative Aspects of Sensitive Periods
by Josef P. Rauschecker and
Peter Marler

IMPRINTING AND CORTICAL PLASTICITY

COMPARATIVE ASPECTS OF SENSITIVE PERIODS

Josef P. Rauschecker
Max-Planck-Institut für Biologische Kybernetik

Peter Marler
Rockefeller University

A WILEY-INTERSCIENCE PUBLICATION

JOHN WILEY & SONS
NEW YORK   CHICHESTER   BRISBANE   TORONTO   SINGAPORE

Library of Congress Cataloging in Publication Data:

Imprinting and cortical plasticity.

  (Wiley series in neuroscience)
  1. Critical periods (Biology) 2. Imprinting
(Psychology) 3. Developmental neurology. 4. Neuro-
plasticity. I. Rauschecker, Josef P. II. Marler,
Peter Robert. III. Series.
QL763.2.I476   1987   591.51   87-2100
ISBN 0-471-84368-7

Printed in the United States of America

10 9 8 7 6 5 4 3 2 1

CONTRIBUTORS

Arthur P. Arnold*
Department of Psychology and
  Laboratory of
  Neuroendocrinology
Brain Research Institute
University of California, Los
  Angeles
Los Angeles, California

Patrick Bateson*
Sub-Department of Animal
  Behaviour
University of Cambridge
Cambridge, England

Sarah W. Bottjer
Department of Psychology and
  Laboratory of
  Neuroendocrinology
Brain Research Institute
University of California, Los
  Angeles
Los Angeles, California

* = Authors for correspondence

Philip Bradley*
Department of Anatomy
University of Newcastle
Newcastle upon Tyne, England

Leon N. Cooper*
Department of Physics and
  Center for Neural Science
Brown University
Providence, Rhode Island

Yves Frégnac*
Laboratoire de Neurobiologie du
  Développement
Université Paris XI
Orsay Cedex, France

Helmut V. B. Hirsch*
Neurobiology Research Center
State University of New York
Albany, New York

Klaus–Peter Hoffmann*
Abt. für Vergleichende
  Neurobiologie
Universität Ulm
Ulm, FR Germany

Gabriel Horn
Department of Zoology
University of Cambridge
Cambridge, England

Eric I. Knudsen*
Department of Neurobiology
Stanford University School of
  Medicine
Stanford, California

Masakazu Konishi*
Division of Biology
California Institute of Technology
Pasadena, California

Daniel Margoliash*
Department of Anatomy
University of Chicago
Chicago, Illinois

Peter Marler*
Rockefeller University
Field Research Center
Millbrook, New York

Ernest J. Nordeen
Department of Psychology
University of Rochester
Rochester, New York

Kathy W. Nordeen
Department of Psychology
University of Rochester
Rochester, New York

Josef P. Rauschecker*
Max-Planck-Institut für
  Biologische Kybernetik
Tübingen, FR Germany

Dale R. Sengelaub
Department of Psychology and
  Laboratory of
  Neuroendocrinology
Brain Research Institute
University of California, Los
  Angeles
Los Angeles, California

Wolf Singer*
Max-Planck-Institut für
  Hirnforschung
Frankfurt, FR Germany

David G. Tieman
Neurobiology Research Center
State University of New York
Albany, New York

Suzannah B. Tieman
Neurobiology Research Center
State University of New York
Albany, New York

Brian Timney*
Department of Psychology
University of Western Ontario
London, Ontario, Canada

Nina Tumosa
Department of Psychology
University of Wisconsin
Madison, Wisconsin

FOREWORD

It is of course flattering when observations about which rather tentative assumptions were made over 50 years ago, do actually turn out to be worth following up. I consider the conceptual framework of this book *Imprinting and Cortical Plasticity* exceptionally fruitful, and I am delighted with it. The only thing that is missing, it would seem to me, is an allusion to the *great plasticity* of the original prevailing release mechanisms and the suddenness with which their selectivity increases—perhaps a subject to address on some other occasion? A gosling begins by following anything in the size range between a hen and a small boat. In the morning it will still readily leave a goose in favor of a human or a plastic bucket. A few hours later (assuming that the gosling has been with its real mother from 9 AM until 3 PM), it is impossible to incite it to run after anything other than a goose. The phrase "plasticity declines in the course of early development" does not fully convey the dramatic speed with which this change takes place. It has never been denied that the result of imprinting can be modified *to some extent*. It is self-evident that the animal is still capable of learning a great deal after the process of imprinting is completed (for example, gaining an understanding of the individual characteristics of its parents over a longer period—about two daylight cycles). But this is obviously learning in the more traditional sense, *because it is so readily reversible*.

Interwoven with behavior that is modified by learning in the more typical way, there are imprinted components that remain unchanged. For example, Immelmann imprinted zebra finches onto Bengalese finches and subsequently got them to accept female mates of their own species, which is tantamount to learning to accept a substitute for the imprinted object. When such a male (which had mated with a conspecific female to produce purebred offspring several times) was given a choice between females of the two species, the bird was found to return regularly to the imprinted species, that is, the Bengalese finch, and would reject the conspecific mother of his offspring. The bond with this female had been acquired by "learning" rather than "imprinting" and the learning remained readily reversible!

Years ago, I speculated that changes in the direction of increased selectivity of afferent mechanisms were responsible, at least in part, for the subsequent "blocking" of the afferent pathways, stubbornly resisting attempts to unblock them. It is my belief that there are different critical imprinting phases, not only in the development of each instinct system, but possibly also in the development of the parts of each such system. The new data appear to indicate something very much along those lines.

As the years go by, I have become less able to keep abreast of current publications, and I am indeed glad and thankful to find that some of my original ideas have proved to be correct and have given rise to further research. It was clear from the start that these would extend into the field of neuroethology, as emerges very clearly from this book, leading in turn to further ethological questions.

I would like to see a more precise investigation of the stimulus configurations that facilitate imprinting most effectively in the first place. We know that imprinting to musical tones in the grey goose occurs in the gosling in precisely that psycho-physiological state in which it utters a whistle signifying that it has been abandoned, as I have indicated in my book on grey geese. Another feature of imprinting which, in my opinion, sets it apart from all other learning processes, is that imprinting apparently requires no reinforcement such as is needed by all other forms of "conditioning." The organism is exposed to a stimulus for an amazingly short period of time and—bang—it becomes imprinted. I compare this in my own mind with human "falling in love," "ever so quickly may one catch the plague," says Olivia in *Twelfth Night*, after speaking for a few minutes with a young man, who turns out to be a woman in disguise into the bargain. I hope you will forgive me the conscious anthropomorphism!

Konrad Lorenz

Altenberg, Austria

PREFACE

The discovery by Konrad Lorenz in the 1930's that there are sensitive or "critical" periods during postnatal development in birds, during which external stimuli can heavily influence later behavior, was one of the major findings in early ethology. Imprinting, the term that Lorenz used to describe this phenomenon, suggested that certain behavioral and corresponding neural patterns were irreversibly established during the sensitive period. In later research, it turned out that modifications during the sensitive period were not always totally permanent, but that plasticity nevertheless declined during the course of early development. About thirty years later, David Hubel and Torsten Wiesel rediscovered sensitive periods on the single unit level in the visual system of cats. The time scale of this sensitive period, which they described later also for the monkey, was expanded as compared to imprinting in birds, but the similarity as such suggested to them that they were dealing with a related phenomenon. More detailed similarities were discovered later between the sensitive periods in birds, for example in the development of their auditory-vocal communication system, and mammals, in particular their visual system.

This book is not the first attempt to treat comparative aspects of sensitive periods, and to present related findings in both areas side by side. Several conferences have recently made similar efforts, and our joint experience from participation in some of these had a profound influence on our selection of topics in this volume. We are grateful to all colleagues who contributed to this book. The opportunity for an overview of what have been considered, until now, as distant and unrelated fields will, we hope, provide the reader with the opportunity to form an independent judgment as to whether sensitive periods in birds and mammals are comparable phenomena, and whether the findings at the behavioral and the neural level are mutually applicable, or at least heuristically illuminating. In our concluding chapter we have tried to summarize some of the useful parallels, and some of the contrasts. It is self-evident that exchange of ideas and findings between neuroscientists and ethologists working behaviorally or neurophysiologically leads to increased understanding of the nervous system. The approach of "Verhaltensphysiologie" (behavioral physiology), in the tradition of Erich von Holst, may ultimately turn out to be the only way to achieve a full understanding of the way in which the brain develops its remarkable ability to sustain behavior and to render it both stable and predictable, while at the same time retaining the potential for adaptation.

We would like to thank the Max-Planck-Society and the Deutsche Forschungsgemeinschaft for various forms of support during this project. In particular, we are grateful to The Neurosciences Institute of the Neurosciences Research Program for the fellowship to J.P.R. enabling us to have intensive discussions during the final phase of manuscript preparation.

JOSEF P. RAUSCHECKER
PETER MARLER

Tübingen, FR Germany
Millbrook, New York
May 1987

CONTENTS

Part Three
EPILOGUE

IMPRINTING AND CORTICAL PLASTICITY

PART I

IMPRINTING

1

DEVELOPMENTAL PLASTICITY IN THE AUDITORY SYSTEM

Masakazu Konishi

Division of Biology, California Institute of Technology, Pasadena, California

The ability of the auditory system to reorganize in response to modi-fication in its input channels has been reported in several studies. The reorganization includes the restoration of binaural frequency matches (Zakon, 1983), the asymmetrical and compensatory growth of dendrites in the first binaural neurons receiving unbalanced left–right inputs (Feng and Rogowski, 1980; Parks, 1981; Smith et al., 1983), and some physi-ological changes in binaural neurons (Silverman and Clopton, 1977). However, the role of this ability in the normal development of auditory function or acoustic behavior is largely unknown. Like monaural deaf-ening input modifications that exceed the normal range of operation have limited use in revealing behavioral significance. In general, inap-propriate approaches are those that fail to satisfy one or both of the following prerequisites. One is to determine how an animal uses the auditory cues to be manipulated and the other is to identify neurons that are sensitive to the proposed manipulation, if plasticity is to be studied at the neuronal level.

The work on the barn owl reviewed here by Knudsen satisfies both prerequisites. The owl localizes sound by using binaural intensity and time differences, respectively, for elevation and azimuth. Space-specific neurons, which are exquisitely selective for a particular combination of the two binaural cues, form a map of auditory space in the external nucleus of the owl's inferior colliculus. Topographical projection of this map onto the optic tectum gives rise to a visual–auditory bimodal map of space. Partial plugging of one ear causes an owl to commit systematic errors in sound localization. A young owl can correct for the error, whereas an adult owl cannot. The bimodal map in such a young owl is normal, that is, the visual and auditory maps are in register, so long as the same ear remains partially plugged. Removal of the plug causes both the auditory map to deviate from the visual map and the owl to localize sound with errors consistent with the deviation. These findings clearly show that the owl's developing auditory system is plastic enough to allow for growth as well as to compensate for an inherent or accidental imbalance in binaural inputs. Although the site of plasticity is not yet identified, it can be traced back systematically from space-specific neu-rons to lower-order neurons.

Birdsong is another ideal subject that satisfies the prerequisites (Kon-ishi, 1985). A bird must hear himself sing in order to develop either an imitated or normal song. The feedback control of voice makes song de-velopment plastic. Plasticity is in both the sensory and motor aspects of song development as discussed here by Marler. The vocal motor system is obviously plastic enough to produce a wide range of outputs. A bird's auditory system appears to contain both fixed and variable filters, the latter being susceptible to early auditory exposure as well as to auditory feedback of the bird's own voice during song development.

The discrete nuclei and fiber tracts of the song control system are particularly attractive for the study of anatomical and physiological changes associated with singing behavior. The discovery of auditory inputs to the song control system has raised the hope of identifying the site of sensorimotor integration during song development as Margoliash discusses in this volume. One of the song control nuclei, HVc, contains both vocal premotor and auditory neurons. In the white-crowned sparrow, some of the HVc auditory neurons respond exclusively to particular acoustic cues contained in the individual bird's own song. Although other HVc auditory neurons are less specific, they respond better to the bird's own song than to any other stimuli. The stimulus specificity of HVc auditory neurons is perhaps due not to the selection of the neurons matched to the bird's own song signature from a library of neurons tuned to different acoustic cues but to epigenetic "shaping" by vocal–auditory interaction during song development. Thus, Margoliash's work provides by far the most convincing evidence for the neurophysiological correlates of a naturally occurring imprinting-like phenomenon.

The above works of Knudsen and Margoliash, besides satisfying the aforementioned prerequisites, show the importance of selecting animals appropriate for the aim of the investigation. Traditional auditory physiology ignores this obvious rule. Similarly, use of biologically irrelevant stimuli in this field has biased the choice of topics and retarded the discovery of neural mechanisms relevant to behavior. In this respect, the study of developmental plasticity in the auditory system is no exception. In integrative neurophysiology, natural behavior should be the primary guide for the selection of both topics and animals.

REFERENCES

Feng, A. S. and B. A. Rogowski (1980). Effects of monaural and binaural occlusion on the morphology of neurons in the medial superior olivary nucleus of the rat. *Brain Res.* **189**:530–534.

Konishi, M. (1985). Birdsong: From behavior to neuron. *Ann. Rev. Neurosci.* **8**:125–170.

Parks, T. N. (1981). Changes in the length and organization of nucleus laminaris dendrites after unilateral otocyst ablation in chick embryos. *J. Comp. Neurol.* **202**:47–57.

Silverman, M. and B. Clopton (1977). Plasticity of binaural interaction. I. Effects of early auditory deprivation. *J. Neurophysiol.* **40**:1266–1274.

Smith, Z. D. J., L. Gray, and E. W. Rubel, (1983). Afferent influences on brainstem auditory nuclei of the chicken: N. laminaris dendritic length following monaural conductive hearing loss. *J. Comp. Neurol.* **220**:199–205.

Zakon, H. H. (1983). Reorganization of connectivity in amphibian central auditory system following VIII nerve regeneration: Time course. *J. Neurophysiol.* **49**:1410–1427.

2

EARLY EXPERIENCE SHAPES AUDITORY LOCALIZATION BEHAVIOR AND THE SPATIAL TUNING OF AUDITORY UNITS IN THE BARN OWL

Eric I. Knudsen

Department of Neurobiology, Stanford University School of Medicine, Stanford, California

To determine the location of a sound source, the auditory system evaluates a variety of cues and associates a given combination of cues with the appropriate location in space. The most important cues for sound localization are the differences in the timing and amplitude of sound at the two ears and the spectral properties of sound at each ear. The correspondence between these cues and sound source locations depends on the size and shape of the head and external ears. For example, the larger the head, the greater the separation of the ears, and the greater the interaural differences in timing that an animal experiences. Also, the larger the head and ears, the more sound they reflect and collect, and the greater the interaural differences in amplitude that they create. This means that as an animal grows early in life, the auditory system is exposed to new localization cues and to continually changing correlations between cues and sound source locations.

Because localization cues change during growth, it is unlikely that the neural circuitry that underlies sound localization develops in a completely predetermined fashion. A more likely hypothesis is that sensory experience plays a role in shaping the correct associations. This latter hypothesis is supported by experimental results from barn owls that demonstrate that both sound localization behavior and a neurophysiological representation of auditory space are modified by auditory experience (Knudsen, 1983a, 1985; Knudsen et al., 1982, 1984a,b). These experiments will be reviewed in this chapter.

The barn owl has proven to be an excellent model system for studying the neural basis of sound localization. Although phylogenetically distant from humans, it is remarkably similar in the lack of mobility of its external ears, its frequency range of auditory sensitivity, and the integrating stations in the central system (Konishi, 1973; Knudsen, 1980; 1983b; Moiseff and Konishi, 1983). More importantly, however, the owl localizes sounds with an accuracy unsurpassed in the animal kingdom (Payne, 1971; Konishi, 1973; Knudsen et al., 1979). Associated with this highly developed behavioral ability, a large proportion of the owl's auditory system is devoted to analyzing spatial cues, and, at higher levels in the auditory system, the tuning of neurons for spatial cues is exceptionally sharp (Knudsen et al., 1977; Knudsen and Konishi, 1978; Moiseff and Konishi, 1981). This latter property makes the task of studying neural correlates of sound localization easier in this species.

In the experiments described below, auditory experience was manipulated by chronically occluding one ear with a foam rubber plug. An earplug attenuates and delays the sound reaching the eardrum in a frequency-dependent manner (Knudsen et al., 1984a) and, therefore, alters interaural amplitude and interaural timing cues for localization. As a result, the perceived location of a sound source shifts toward the side of the unplugged ear. Because of a vertical asymmetry in the directional sensitivities of the barn owl's ears (Payne, 1971), plugging the left ear causes a localization error to the right and above the sound source, while plugging the right ear causes an error to the left and below the source (Knudsen and Konishi, 1979; Knudsen et al., 1984a).

SENSITIVE AND CRITICAL PERIODS IN THE DEVELOPMENT OF SOUND LOCALIZATION

Sound localization was assessed behaviorally by comparing the way in which owls turn their heads toward auditory versus visual stimuli. Normal owls orient identically to sounds and lights. However, owls that were monaurally occluded oriented differently to these two kinds of stimuli, and this difference was interpreted as an error in sound localization. The stimuli were presented in a darkened anechoic chamber; the auditory stimulus consisted of repetitive noise bursts presented from a remotely controlled, movable speaker, and the visual stimulus was provided by the glow of a light-emitting diode that was centered in the speaker cone and that traveled with the speaker. The orientation of the owl's head in response to a stimulus was indicated by the position of a reflected infrared beam on a screen in front of the owl, and was quantified relative to the location of the stimulus. Using this paradigm, the mean accuracy of normal owls in localizing auditory targets was about 1.5° in azimuth and elevation. Errors greater than 3° in either dimension were considered abnormal.

Prolonging the period of monaural occlusion of an adult owl had essentially no effect on its sound localization performance. The data shown in Figure 1A are from an owl that was monaurally occluded at 186 days of age and remained occluded for 161 days. Before the earplug was inserted, the animal localized sounds with normal accuracy. As expected, when the plug was placed in the owl's right ear, the owl exhibited a large localization error to the left of and below the sound source. Although the magnitude of the error fluctuated substantially during the period of monaural occlusion, there was no indication that

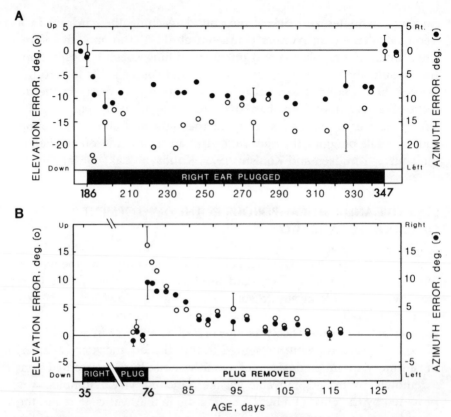

FIGURE 1. Sound localization errors plotted as a function of age for two owls, one monaurally occluded when fully grown (A), the other monaurally occluded while still immature (B). The azimuth (closed circles) and elevation (open circles) components of the localization errors are plotted separately in each graph. The acoustic stimulus was repetitive noise bursts; the behavioral measure was the orientation of the head to the source. Representative standard error bars are shown periodically. The owl in (A) shows no evidence of adjusting its sound localization in response to the abnormal cues induced by the earplug. The owl in (B) learned to localize accurately with the earplug in place. When the earplug was removed, it exhibited a large localization error, which it corrected over a period of weeks. The auditory histories of each bird are shown below. The data are from Knudsen et al. (1984a).

the owl was adjusting its error. On the day the earplug was removed, after more than 5 months of monaural occlusion, the owl immediately localized sounds with normal accuracy. Thus, continuous and prolonged exposure to abnormal auditory cues seemed to have no effect on sound localization in a bird of this age.

When performed on younger owls, the same experiment yielded entirely different results (Knudsen et al., 1982). The data in Figure 1B are from an owl that had its right ear plugged at 35 days of age. As soon as the owl was old enough to be trained and tested (72 days of age), it localized sounds with normal accuracy. This implies that the owl had altered its associations between auditory cues and locations in space and had learned to localize sounds accurately using the abnormal cues imposed by the earplug. This interpretation was confirmed when the earplug was removed and normal cues were restored: The owl exhibited a large sound localization error opposite in direction to the one originally induced by the insertion of the earplug.

These results, when considered together, indicate that the owl's auditory system can adjust its interpretation of localization cues based on experience, but only during a restricted period in early life. To determine more precisely the extent of this sensitive period, owls were monaurally occluded at various ages and their ability to recover accurate localization was evaluated (Knudsen et al., 1984a). As summarized in Figure 2B, owls occluded as late as 45 days of age recoverd accurate sound localization quickly (within 5 weeks in some cases); those occluded between 45 and 55 days of age recovered normal accuracy, but much more slowly; and owls occluded at 60 days of age or older never showed signs of adjusting their sound localization. Thus, abnormal auditory experience leads to changes in sound localization only during a brief period early in development.

The termination of this sensitive period at about 55 days of age might be triggered by an age-dependent mechanism. On the other hand, the time course of growth of the owl's head and ears is consistent with the alternative hypothesis that sensory experience brings the sensitive period to a close. The head and ears reach adult size at about 45 days of age (Fig. 2A). Since these structures determine the correspondence between auditory cues and sound locations, animals below this age experience continually changing sound localization cues while animals beyond this age experience constant localization cues. The coincidence between the maturation of the head and ears and the termination of the sensitive period could mean that exposure of the auditory system to stable adult cues, rather than age itself, brings the sensitive period to a close. Experiments are being conducted presently to determine whether age or experience triggers the termination of the sensitive period.

The adjustment of sound localization during monaural occlusion was

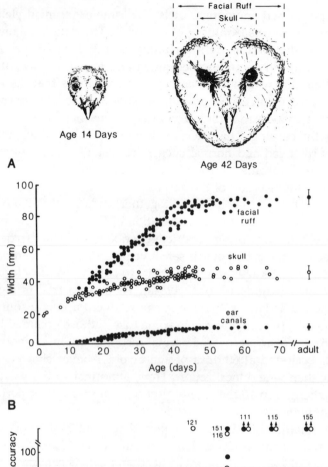

A

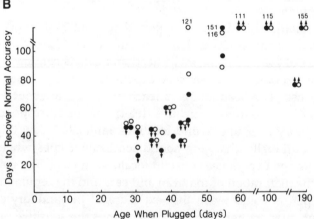

B

FIGURE 2. Growth of the barn owl and the ability of barn owls to recover accurate sound localization while monaurally occluded. The sketches are scaled drawings of the owl's face. In (A), the width of the skull and facial ruff (which functions as the pinnae for the owl) and the inner diameters of the ear canals are plotted as a function of age for eight

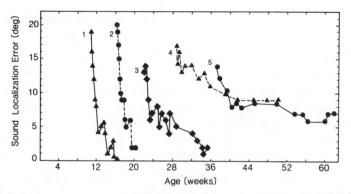

FIGURE 3. Adjustment of sound localization accuracy by barn owls after the removal of unilateral earplugs. Data are shown for five representative owls. All animals had one ear occluded at 4–6 weeks of age, that is, before they reached adult size (Fig. 2). The data are from Knudsen et al. (1984b).

gradual. Even the youngest animals took as long as 4–6 weeks to attain normal accuracy. The slow rate of adjustment probably reflects the fact that the animals were attempting to localize sounds using interaural amplitude differences that they never would have experienced normally. In other words, experience was dictating the formation of highly abnormal associations between binaural cues and locations. It is not surprising, therefore, that localization errors caused by *removing* an earplug from animals that had been plugged since early life were corrected more quickly, and could be corrected at much later ages than errors caused by earplug insertion (Knudsen et al., 1984b). Upon removal of the earplug, owls up to 20 weeks of age corrected their localization errors in 3–4 weeks (Fig. 3). However, this capacity to recover accurate localization following earplug removal also decreased with age: Animals that remained monaurally occluded beyond about 28 weeks of age continued to exhibit localization errors for more than 24 weeks after the earplug

---

owls. The sizes of these features determine the magnitudes of sound localization cues. In (B), the number of days required for an owl to regain normal localization accuracy (error less than 3°) is plotted against the age of the owl at the time one ear was plugged. Closed symbols represent the day that normal accuracy in azimuth was attained; open symbols signify the day that normal accuracy in elevation was attained. A downward arrow indicates that the owl exhibited normal accuracy in that dimension on the first day it was tested. Upward arrows signify that normal accuracy was never achieved through the day of testing indicated above (in fact, these birds made no adjustment of their localization accuracy at all). The data are from Knudsen et al. (1984a).

was removed. These data suggest that by about 28 weeks of age the neural connectivity that establishes associations between auditory cues and locations in space has become stable. Such a period in the development of the nervous system is called a critical period (Jacobson, 1978). It is interesting to note that the end of this critical period coincides approximately with the age at which these animals reach adulthood, as indicated by sexual maturity.

In summary, two periods in the development of sound localization in the barn owl have been identified: a very early sensitive period, ending at about 8 weeks of age during which owls can learn to interpret and make use of highly abnormal auditory cues, and a much longer critical period during which auditory cues become permanently associated with locations in space. It is possible that these periods are terminated by different neural mechanisms, one mechanism limiting the ability of the auditory system to establish unusual patterns of connectivity and the other causing the established pattern to become permanent. Another possibility is that both periods are ended by a single, experience-dependent, but time-limited, mechanism. The available data do not discriminate between these possibilities.

EXPERIENCE-DEPENDENT ALTERATIONS IN NEURONAL SPATIAL TUNING

In order to search for the sites and mechanisms that underlie these experience-dependent changes in sound localization, a neurophysiological measure was developed that assessed the associations between auditory cues and locations in space. The assay was provided by the spatial tuning of auditory–visual bimodal units in the optic tectum, the homologue of the mammalian superior colliculus. Units in the optic tectum are highly selective for the location of a sound source (Knudsen, 1982): They respond strongly to sounds only when the source is located in a particular "best area." The best area does not change with the amplitude or spectral content of the sound. Unit best areas vary systematically across the tectum to form a map of auditory space. The visual receptive fields of these bimodal units coincide with their auditory best areas, which means that the maps of auditory and visual space in the tectum are in register.

The alignment of auditory best areas and visual receptive fields rep-

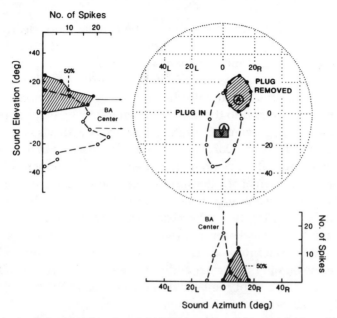

FIGURE 4. Auditory spatial tuning and visual receptive field of a single unit measured before and after removing the earplug. A left earplug was inserted at 41 days old and removed at 342 days old. This unit was recorded on the day the earplug was removed. The visual receptive field is dotted. The auditory receptive field with the earplug in place is dashed, and with the earplug removed is solid. Below and to the left are graphed the responses of the unit to eight noise bursts presented at various locations across the unit's receptive field. The noise bursts were 20 dB above unit threshold. The circled A's represent best area centers as determined from the spike-count data. These data are from Knudsen (1983a).

resents an association of auditory cues with appropriate locations in visual space. What happens to this alignment in owls that are raised monaurally occluded? Figure 4 shows the auditory and visual spatial tuning of a unit recorded in an animal that had its left ear occluded at 41 days of age (Knudsen, 1983a). The unit was recorded 301 days later, on the day the earplug was removed, and the spatial tuning of the unit was measured both before and after earplug removal. While the earplug was still in place, the auditory best area was aligned with the visual receptive field, indicating that the auditory spatial tuning of the unit was adjusted to select for the cues corresponding to a sound source at the location of its visual receptive field. This interpretation was confirmed by the observation that removal of the earplug (from the left ear) caused the auditory best area to shift to the right and up, clearly out of alignment with the visual receptive field. The magnitude and direction of this auditory–visual misalignment were those expected from sound localization errors measured behaviorally in birds treated in a similar way (Fig. 1). In birds raised with the opposite (right) ear occluded, auditory best areas shifted to the left and down following earplug removal (Fig. 5B). If the alignment of auditory best areas with visual receptive fields is a reliable neural correlate of sound localization behavior, then the sensitive and critical periods that affect the development of sound localization (Knudsen et al., 1984a,b) should also affect the development of this auditory–visual alignment. This predicts that auditory spatial tuning can be modified by monaural occlusion only if it is imposed before the owl reaches 8 weeks of age (the end of the behavioral sensitive period), and that auditory–visual alignment can be recovered following earplug removal only if normal auditory cues are restored by about 28 weeks of age (the end of the behavioral critical period).

Neurophysiological data from nine owls were consistent with these predictions (Knudsen, 1985). In animals that were monaurally occluded

---

FIGURE 5. Behavioral, physiological, and anatomical effects of early and prolonged monaural occlusion. These data were collected beginning 4 months after earplug removal from an owl that had its right ear occluded from 43 to 202 days of age, as indicated. On the left in (A), the orienting responses of this animal to noise bursts are plotted relative to the location of the speaker. The vector represents the mean localization error based on this sample. Data from a control bird are shown on the right for comparison. On the left in (B), auditory best area centers (open triangles) and visual receptive field centers (closed circles) are plotted on coordinates of space. The lines connect data from individual units and indicate the direction and magnitude of the misalignment. On the right in (B), best area centers of all bimodal units in the rostral tectum are plotted relative to the centers of their visual receptive fields. The vector indicates the median auditory–visual misalignment for this sample. In (C) are drawn horizontal sections through the left and right tecta after HRP injections and electrolytic lesions. The lesions were placed at the representations of 0° azimuth, 0° elevation in visual space (lesions 1 and 4) and in auditory space (lesions 2 and 3). The positions of these lesions indicate that the auditory, but not the visual, maps of space were shifted. The inset is a photograph of the lesions on the left side. The extents of the HRP injection sites (revealed by TMB processing) are shown by the stippled regions. The auditory and visual spatial tuning measured at the injection sites are indicated. If the altered auditory tuning were due to a reorganized projection from an unaltered map in the external nucleus of the inferior colliculus (ICX), then retrogradely filled cells (solid dots) should have been found at the sites marked with ×'s (Knudsen and Knudsen, 1983). However, as shown, the labeled cells were located at the same positions as in control animals, indicating that the ICX-to-optic tectum projection was unchanged. Abbreviations: ICC, central nucleus of the inferior colliculus; OT, optic tectum. The data are from Knudsen (1985).

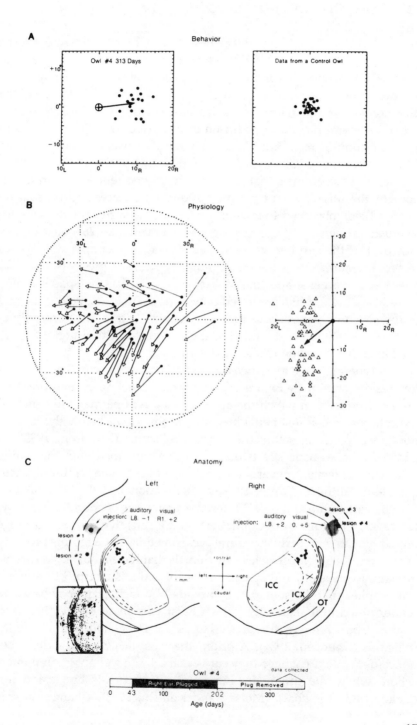

A   Behavior

Owl #4 313 Days   Data from a Control Owl

B   Physiology

30°L   0°   30°R

C   Anatomy

Left   Right

auditory visual   auditory visual
injection: L8 −1 R1 +2   injection: L8 +2 0 +5

lesion #1   lesion #3

lesion #2   lesion #4

rostral

left ⟷ right

caudal

1 mm

ICC   ICX   OT

#1

#2

data collected

Owl #4

Right Ear Plugged   Plug Removed

0   43   100   202   300

Age (days)

before 8 weeks of age, the auditory best areas of most units were aligned with their visual receptive fields when measured with the earplug still in place several months after earplug insertion. When the earplugs were removed from these animals, auditory best areas moved in the expected direction and out of alignment with visual receptive fields. In contrast, experience with monaural occlusion caused no change in the auditory spatial tuning of tectal units in an animal that was earplugged as an adult (at the age of 250 days) for more than one year.

Among those animals that were monaurally occluded early in development the misalignment between auditory best areas and visual receptive fields observed following earplug removal remained stable for as much as one year, if the earplug was removed after 28 weeks of age. However, if the earplug was removed earlier, for example at 24 weeks of age, the auditory–visual misalignment decreased progressively over a period of 3 weeks, and finally disappeared. Thus, the spatial tuning of auditory units could be altered only in birds occluded before the end of the sensitive period, and these alterations were permanent only in birds that remained occluded after the end of the critical period.

Ultimately, one would like to determine the site(s) in the auditory system where these experience-dependent changes occur. The anatomical results shown in Figure 5C (retrograde labeling of neurons that projected to tecta in which the auditory map was permanently shifted) indicate that the adjustments in spatial tuning occur below the level of optic tectum, at or before the inferior colliculus (Knudsen, 1985). In addition, the acoustic effects of the earplug combined with the tuning properties of tectal units argue that the changes in spatial tuning occur quite early in the auditory pathway. As mentioned in the introduction, earplugs attenuate and delay different frequencies of sound differently. However, the units in the optic tectum are sensitive to a broad range of frequencies, and early monaural occlusion does not alter this property (Knudsen, 1985). Nevertheless, in animals that have adjusted to chronic monaural occlusion, the spatial tuning of units is sharp, but shifted. This implies that the spatial cues provided by each frequency band are reinterpreted independently, which must take place before the information from different frequency bands are integrated in the inferior colliculus (Knudsen, 1984). A priori, the most likely site for the experience-dependent changes to occur is at the first level of binaural interaction, where the tuning for interaural differences is broad and frequency tuning is sharp (Moiseff and Konishi, 1983). Changes at this level probably could be accomplished by modulations in synaptic efficacy or changes in dendritic morphology rather than by the formation of new connections. In addition, if adjustments were made at this initial stage, all subsequent processing of spatial cues could be carried out normally, that is, the spatial information would be correct from the outset.

CONCLUSION

The perception of a sound's location is based on the association of sets of auditory cues with locations in space. Certainly, the basic neuronal circuitry necessary for extracting and interpreting localization cues develops according to genetic determinants. However, within the constraints imposed by these determinants there remains a substantial amount of flexibility that is subject to the shaping forces of experience. Early in life this plasticity is sufficiently great to allow the association of highly abnormal auditory cues with locations in space. In the course of normal development, this capacity for adjustment enables the auditory system to fine tune its localization circuitry and to respond to changes in auditory cues brought about by growth of the head and ears. However, once the owl reaches adult size, the plasticity of this system decreases substantially, and the calibration of sound localization and its neural substrates become extremely resistant to further change.

ACKNOWLEDGMENTS

I thank Phyllis Knudsen for technical assistance, and Drs. Steven Esterly and John Middlebrooks for reviewing the manuscript. This research was supported by the March of Dimes (1-863), Alfred P. Sloan Foundation (BR-2383), McKnight Foundation, and National Institutes of Health (2 R01 NS 16099-06).

REFERENCES

Jacobson, M. (1978). *Developmental Neurobiology*, 2nd ed. Plenum Press, New York.

Knudsen, E. I. (1980). Sound localization in birds. In: *Comparative Studies of Hearing in Vertebrates*, A. N. Popper and R. R. Fay, eds., pp. 287–322, Springer-Verlag, New York.

Knudsen, E. I. (1982). Auditory and visual maps of space in the optic tectum of the owl. *J. Neurosci.* **2**:1177–1194.

Knudsen, E. I. (1983a). Early auditory experience aligns the auditory map of space in the optic tectum of the barn owl. *Science* **222**:939–942.

Knudsen, E. I. (1983b). Subdivisions of the inferior colliculus in the barn owl (*Tyto alba*). *J. Comp. Neurol.* **218**:174–186.

Knudsen, E. I. (1984). Synthesis of a neural map of auditory space in the owl. In: *Dynamic Aspects of Neocortical Functions*, G. M. Edelman, W. M. Cowan and W. E. Gall, eds., pp. 375–396, Wiley & Sons, New York.

Knudsen, E. I. (1985). Experience alters spatial tuning of auditory units during a sensitive period in the barn owl. *J. Neurosci.* **5**:3094–3109.

Knudsen, E. I., G. G. Blasdel, and M. Konishi (1979). Sound localization by the barn owl measured with the search coil technique. *J. Comp. Physiol.* **133**:1–11.

Knudsen, E. I., S. D. Esterly, and P. F. Knudsen (1984a). Monaural occlusion alters sound localization during a sensitive period in the barn owl. *J. Neurosci.* **4**:1001–1011.

Knudsen, E. I., P. F. Knudsen, and S. D. Esterly (1982). Early auditory experience modifies sound localization in barn owls. *Nature* **295**:238–240.

Knudsen, E. I. and P. F. Knudsen (1983). Space-mapped auditory projections from the inferior colliculus to the optic tectum in the barn owl (*Tyto alba*). *J. Comp. Neurol.* **218**:187–196.

Knudsen, E. I., P. F. Knudsen, and S. D. Esterly (1984b). A critical period for the recovery of sound localization accuracy following monaural occlusion in the barn owl. *J. Neurosci.* **4**:1012–1020.

Knudsen, E. I. and M, Konishi (1978). A neural map of auditory space in the owl. *Science* **200**:795–797.

Knudsen, E. I. and M. Konishi (1979). Mechanisms of sound localization by the barn owl (*Tyto alba*). *J. Comp. Physiol.* **133**:13–21.

Knudsen, E. I., M. Konishi, and J. D. Pettigrew (1977). Receptive fields of auditory neurons in the owl. *Science* **198**:1278–1280.

Konishi, M. (1973). How the owl tracks its prey. *Am. Sci.* **61**:414–424.

Moiseff, A. and M. Konishi (1981). Neuronal and behavioral sensitivity to binaural time differences in the owl. *J. Neurosci.* **1**:40–48.

Moiseff, A. and M. Konishi (1983). Binaural characteristics of units in the owl's brainstem auditory pathway; Precursors of restricted spatial receptive fields. *J. Neurosci.* **3:**2553–2562.

Payne, R. S. (1971). Acoustic location of prey by barn owls (*Tyto alba*). *J. Exp. Biol.* **54:**535–573.

3

NEURAL PLASTICITY IN BIRDSONG LEARNING

Daniel Margoliash

Department of Anatomy, University of Chicago, Chicago, Illinois

Many of the attributes of birdsong—sensitive periods during development, interaction between sensory templates and motor programs, and individual recognition—are the behavioral manifestations of neuronal plasticity. Song thus serves as an attractive system for the study of the modifiable physiological processes underlying memory, learning, and perception. Investigations of the neural basis of song learning benefit from the focus provided by behavioral observations, and in turn may help to explicate the concepts developed to describe song.

The research on the song system nucleus hyperstriatum ventrale, pars caudale (HVc) is a good example of profitable interaction between behavior and neurophysiology. Behavioral observations have established that auditory feedback is important in the development of song, suggesting that sites of auditory–motor interaction are likely to be sites of plasticity during the development of the song motor pattern. These observations have focused attention on HVc because it exhibits both motor activity during singing and auditory activity during playback of song. Indeed, recent physiological data have demonstrated that adult auditory neurons in HVc are maximally responsive to each individual bird's own song, proving that response properties of HVc auditory neurons are modified during ontogeny. In turn, the physiological observations suggest several hypotheses: The response properties of HVc auditory neurons are established during development; during development HVc auditory neurons have a role in shaping the song motor program; and in adulthood the auditory representation of the bird's own song may contribute to song recognition. The existence of an auditory representation of "autogenous" (self-produced) song is an unpredicted result of the action of auditory feedback; its discovery may help to further organize behavioral investigations of song.

BEHAVIORAL OBSERVATIONS OF SONG LEARNING

Song development exhibits two distinct phases that require neuronal plasticity. Young birds pass through an impressionable phase during which they are sensitive to exposure to song (Thorpe, 1961; Marler, 1970). The duration of the impressionable phase is dependent on a variety of natural and experimental factors (Immelmann, 1969; Marler, 1970; Kroodsma and Pickert, 1980; Baptista and Petrinovich, 1984; see also Marler, this volume). The memory of song formed early in life—the

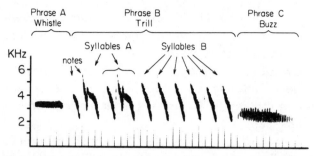

FIGURE 1. The terminology of song. Frequency on the *ordinate* is plotted as a function of time, on the *abscissa*. The time marks are 50 msec. Throughout this chapter, the smallest division of song, the note, is called an element (from Konishi, 1985).

"acquired song template"—guides vocal development later in life. Acoustic isolation disrupts the development of normal song, but only if imposed during the impressionable phase. For example, white-crowned sparrows (*Zonotrichia leucophrys*) exposed to song ("tutored") only before 10 days or after 50 to 100 days of age develop abnormal isolate songs, while birds tutored with tape-recorded songs of their species between 10 to 50 days of age but otherwise isolated develop normal songs (Marler, 1970). Thus, the development of song is insensitive to exposure to song outside of the impressionable phase, with the exception of those species that exhibit plasticity into adulthood (e.g., mockingbird—*Mimus polyglottos*). Song development in isolation is said to be guided by an "innate template" (Konishi, 1965b, 1978). A conspicuous feature of the song of isolated birds is the large variability among individuals. The functional organization of the template may help explain this phenomenon, a developmental process that has proven largely refractory to behavioral analysis.

Song emerges over an extended period of time. In many species, juveniles begin their first efforts at singing well past the close of the impressionable phase (Marler, 1970; Marler and Peters, 1982a). Song syllables gradually emerge from the amorphous structure of subsong (see Fig. 1 for song terminology). The production of an overabundance of syllables that are later subject to attrition precedes the establishment of the adult motor patterns (Marler and Peters, 1982b). While some of the syllables that survive into adulthood are accurate copies from the original tutor repertoire, others are new or are improvisations (Marler and Peters, 1981). An understanding of how sensory information is transferred to the motor system will presumably explain why new elements and imperfect copies form. The overproduction and attrition of elements suggests a developmental competition for survival among motor subprograms. An emerging rule, that species with large syllable repertoires copy song less faithfully than those species with simpler songs (Kroodsma and Pickert, 1984), perhaps reflects a limit on the memory size of the template.

The coordination of the motor system for song develops by reference to auditory feedback. Birds deafened early in life develop highly abnormal and unstable songs, the magnitude of the effect being somewhat dependent on species (Konishi, 1965a,b; Nottebohm, 1968; Marler and Waser, 1977; Marler and Sherman, 1982). In the majority of species, deafening adults has little or no effect on song maintenance (Konishi, 1965b, 1985). Thus, the crystallization of song releases central pattern generators from the effects of auditory feedback, although this need not be the case for those species that retain plasticity of song into adulthood, such as the canary (*Serinus canarius*—Nottebohm et al., 1976; Nottebohm, 1984). These results suggest that (1) there are connections, direct or indirect, between the auditory and motor (vocal) systems; (2) these are sites of sensorimotor plasticity during development; and (3) these interactions become fixed during song crystallization.

THE SITES OF SENSORIMOTOR INTERACTIONS

An important step in the investigation of the physiological basis of song learning was the identification of a discrete motor pathway controlling song production (Nottebohm et al., 1976, 1982; Gurney, 1981). Several nuclei in this pathway—NIF, HVc, RA, and nXIIts (see Fig. 2) are necessary for song production. Each of these nuclei exhibits temporally correlated neuronal activity during singing (McCasland, 1987; McCasland and Konishi, 1981), and their bilateral ablation eliminates singing (Nottebohm et al., 1976; McCasland, 1987). Other nuclei—area X, MAN, UVA, DM—are considered components of the song system by virtue of their connectivity, although the first three are not known to be active during song production (McCasland, 1987). The latency from neuronal activity to sound production is shorter as one proceeds toward the periphery along the "motor" pathway: NIF–HVc–RA–nXIIts

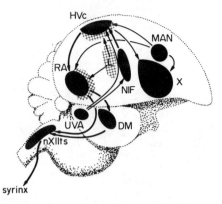

FIGURE 2. Anterograde connections in the song system. Bold *arrows* show connectivity of nuclei that are active during song production. Other nuclei do not exhibit time-locked neural activity during singing. Field L and its projections are shown as hatched areas. DM: nucleus dorsalis medialis; HVc: nucleus hyperstriatum ventrale, pars caudale; L: field L; MAN: nucleus magnocellularis of the anterior neostriatum; NIF: nucleus interfacialis; RA: nucleus robustus archistriatalis; UVA: nucleus uvaeformis; X: nucleus X; nXIIts: nucleus hypoglossus, pars tracheosyringealis. The projection from nXIIts innervates the syrinx, the vocal organ (from Konishi, 1985).

(McCasland, 1987). The axons of the nXIIts form the part of the hypoglossal nerve that innervates the syrinx, the avian vocal organ.

The results of the deafening experiments suggested that there ought be communication between the auditory and motor systems. Indeed, quantitative autoradiographic analysis of anterograde axonal transport of labeled amino acid has demonstrated that HVc and RA of the canary (*Serinus canarius*) adjoin zones of termination of fibers projecting from field L, the avian primary auditory telencephalic nucleus (Kelley and Nottebohm, 1979). The fibers projecting to HVc terminate in the "shelf," a cell-sparse zone ventral and primarily medial to the nucleus proper. The field L projection to RA also terminates adjacent to the nucleus. The somata of origin for the projection to HVc appear to be distinct from those projecting to RA. Field L neurons projecting to HVc are located in field L proper, the thalamic recipient zone (Karten, 1968), whereas the RA-projecting neurons are located ventral and caudal to field L. Although the physiological characteristics of those field L neurons that project to HVc and RA are not known, different patterns of activity have been observed for HVc and RA auditory neurons (Katz, 1982).

The projections from field L investigated by Kelley and Nottebohm did not invade the nuclear boundaries of HVc or RA. Recent studies of

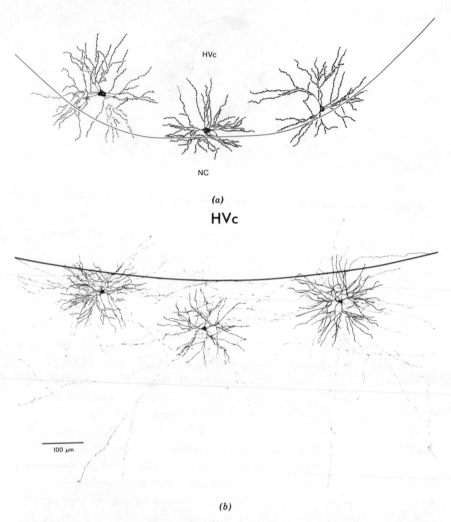

(a)

HVc

(b)

FIGURE 3. Morphology of HVc and shelf neurons. (A) A montage of HVc neurons with somata lying near the ventral border of HVc. Only the dendritic arborizations are shown. Note that the ventrally directed dendrites of all three cells ignore the HVc–shelf boundary (*dashed* line) and encroach into the shelf. (B) A montage of neostriatal shelf neurons. Note that the axons send few processes into HVc, but ramify extensively within the shelf and also project ventrally into the caudal neostriatum. All cells are camera-lucida drawings of lucifer yellow-filled cells, 100x oil-immersion objective (from L. Katz, unpublished results).

HVc have addressed the nature of the auditory pathway into these song system nuclei. In one study, a direct projection from the neostriatum onto HVc neurons with small somata has been observed (Rausch, 1985). In another study, using intracellular injections of the fluorescent dye lucifer yellow, Katz (unpublished results) visualized HVc and neostriatal "shelf" neurons maintained in vitro in zebra finch (*Poephila guttata*) brain slices. The axons of shelf neurons arborized extensively within the shelf, as well as ventrally into the caudal neostriatum, but sent few processes dorsally into HVc. As an example, the neurons in Figure 3B had axonal processes arborizing primarily within the shelf, as well as the caudal neostriatum (see also Fig. 2 of Katz and Gurney, 1981). Furthermore, axonal processes of shelf neurons exhibited relatively few terminal boutons, typically *en passant*, and were free of bushy terminal endings (Fig. 3B). The paucity of connections of shelf axons within the boundaries of HVc suggests these as unlikely as a major source of auditory input to HVc. On the other hand, Katz (unpublished results) also observed that the dendritic fields of HVc neurons do not respect the ventral border of the nucleus, so that the dendrites of somata located ventrally within HVc regularly invade the shelf (Fig. 3A). Thus, synaptic contact between either shelf fibers arising in field L or intrinsic shelf neurons, and HVc dendrites invading the shelf is perhaps a second pathway for auditory input to HVc, although a direct verification is lacking. In contrast, the dendritic arbors of RA neurons that approach the nuclear boundary tend to curve so as to remain within the nucleus proper (Gurney, 1980; DeVoogd and Nottebohm, 1981). In the zebra finch, HVc neurons projecting to RA do not appear to be auditory, as judged by lack of supra- or subthreshold response to noise bursts (Katz and Gurney, 1981; cf. Williams and Nottebohm, 1985). Hence, the details of the auditory pathway into RA remains unresolved.

THE NATURE OF SONG SELECTIVITY IN HVc

HVc is necessary for and active during song production (Nottebohm et al., 1976; McCasland and Konishi, 1981). The details of song are incorporated in the motor pattern of activity during singing, so an obvious question that arises is how HVc auditory neurons respond to the bird's own song. Analysis of HVc indicates that autogenous song elicits a stronger response than conspecific songs, even similar songs from the same dialect. The selectivity for autogenous song arises from the sensitivity of HVc auditory neurons to the details of the individual's song. Thus, auditory neurons in HVc are shaped by, and presumably help to shape, the motor pattern for song.

Using intracellular recordings in the zebra finch, Katz and Gurney (1981) were the first to unambiguously demonstrate auditory responses from HVc neurons. Reconstruction of electrode penetrations identified 59 neurons within the borders of HVc, of which 36 responded to noise bursts with time-locked subthreshold or spiking activity. The zebra finch neurons systematically preferred noise over tone bursts. The song of the zebra finch lacks tonal components, comprising a sequence of syllables rich in harmonic content. Thus, even these initial experiments provided a hint of correspondence between song and efficacy of stimulus. Furthermore, McCasland and Konishi (1981) presented preliminary evidence that HVc neurons in male canaries exhibited stronger responses to the bird's own song when compared with other conspecific songs or reversed canary song. That study was primarily directed toward demonstrating motor activity and inhibition of auditory responsiveness during singing, so that a detailed analysis of the auditory responses in HVc was not undertaken.

The principle analysis of auditory response properties of HVc neurons has been in the white-crowned sparrow (Margoliash, 1983a, 1986; Margoliash and Konishi, 1985). The incorporation of complex stimuli such as song into experimental paradigms in the absence of analytic techniques sufficiently powerful to describe and manipulate those stimuli inevitably results in limited or equivocal interpretations. Unfortunately, this has been a serious limitation of many physiological and ethological investigations.

These considerations in mind, a song analysis/synthesis system was developed (Margoliash, 1983b). As with many bird songs, the songs of the white-crowned sparrow are typically absent of harmonics and broadband noises, and are of the class of limited bandwidth signals that can be recovered from their zero crossings (Greenewalt, 1968; Staddon et al., 1978; Logan, 1977). Analysis by zero crossings does not suffer the $\Delta f/\Delta t$ limitation of Fourier analysis. The system that was developed could therefore reproduce the fine details of the frequency modulation in song while remaining faithful to the temporal parameters of song. Furthermore, the computational ease of reconstruction of a signal from its zero crossings permitted the rapid manipulation of songs in the spectral and temporal domain. An integral part of the approach was the development of graphic representations of the stimulus and a convenient "language" to describe manipulations of that representation (Margoliash, 1983b). To this end, successive zero crossings were displayed as a function of time in which the positions of phrases were defined, resulting in a natural, sonographlike representation. This system has facilitated the analytic description of the relationship between neuronal responses to simple and complex stimuli.

SINGLE UNIT BASIS OF SONG SELECTIVITY: SONG-SPECIFIC NEURONS

One class of HVc neurons are relatively rare, comprising perhaps 5% or less of the population, and are highly selective for autogenous song. These are the song-specific neurons (Margoliash, 1983a). Unlike other HVc neurons, song-specific neurons do not exhibit excitation to broadcast of an individual song phrase or element presented in isolation, nor to tone or noise bursts. Rather, song-specific neurons respond only to broadcast of a sequence of two single-note phrases or elements of song. This may be termed "temporal combination sensitivity" and has been reported in other systems (O'Neill and Suga, 1979; Sullivan, 1982). Neurons that require combinations of spectral components have also been described (Suga et al., 1979; Fuzessery and Feng, 1983). In these systems, the responses to single tone bursts do not form an adequate basis for describing the processing of behaviorally relevant stimuli (see also Leppelsack and Vogt, 1976; Scheich et al., 1979; Langner et al., 1981; Rose and Capranica, 1983).

Song-specific neurons are optimally stimulated by autogenous song. The sequence of sounds that elicits the maximal response from song-specific neurons is a naturally occurring sequence of two consecutive phrases or elements from autogenous song. Sequences of other phrases, or even sequences of artificial stimuli that mimic autogenous song, are less effective stimuli. Combined with temporal combination sensitivity, these are sufficient cues that song-specific neurons exhibit little response to most conspecific songs or to artificial song variants and simpler artificial stimuli. In contrast, these stimuli elicit a significant excitatory response for the majority of HVc neurons (Margoliash, 1986).

Many song-specific neurons are sensitive to the fine details of auto-

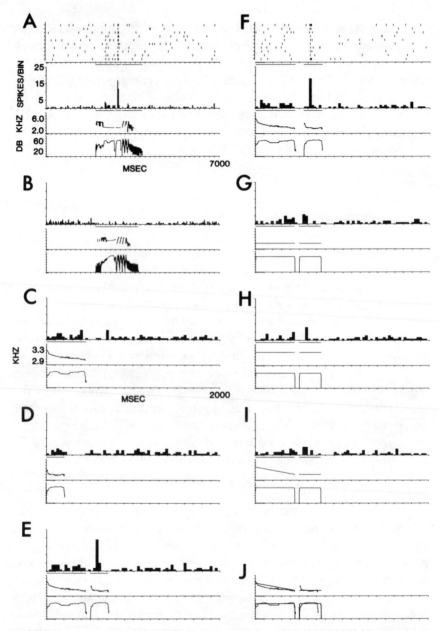

FIGURE 4. Response of whistle–whistle song-specific unit 120503. (A) The top section is a raster representation of the unit's response. Each vertical tick represents one spike; each row is marked on the ordinate by a horizontal tick and represents the response to one stimulus presentation. The middle section is a histogram of the raster; *ordinate*: number

genous song. For example, consider the whistle–whistle song-specific neuron of Figure 4. The neuron responded with a weak sustained response at the second phrase of autogenous song, and exhibited a strong response at the third phrase of song; both phrases were whistles (Fig. 4A). The second or third phrases presented in isolation were ineffective, eliciting only the weak sustained component (Fig. 4C and D), while the sequence of the two phrases isolated from the rest of the song (Fig. 4E) elicited a strong, facilitated response (i.e., enhanced response after the onset of the second element in the sequence). Thus, this neuron exhibited one of the identifying characteristics of song-specific neurons, namely strong temporal facilitation. Furthermore, as with all other song-specific neurons tested to date (Margoliash, 1983a, unpublished results), the duration of the interphrase interval (Fig. 4F) or the duration of the initial phrase of the sequence did not significantly affect the response. Interestingly, several sequences of artificial stimuli were presented (e.g., Fig. 4G–I). None of these elicited a strong facilitated response, although some matched the natural stimulus within 100 Hz and approximately 5 dB (see Fig. 4J). Thus, fine details of autogenous song were sufficient for this song-specific neuron to distinguish between autogenic and other stimulus features. The example illustrates that even some whistle–whistle song-specific units, as well as whistle–buzz, buzz–trill, trill, and buzz–buzz song-specific neurons, are "tuned" to fine features of the frequency or amplitude modulation of autogenous song, and cannot be stimulated with sequences of simple tones or frequency modulations (Margoliash, 1983a).

The facilitated response of whistle–whistle and whistle–buzz song-

---

of spikes per bin. The bottom section is a computer representation—frequency and amplitude versus time—of the stimulus. The stimulus is one of two songs the bird sang: note weak response at second phrase, strong response at third phrase. (B) Histogram and trill stimulus only. The other song in this bird's repertoire, with second phrase of higher frequency than song of (A) and missing the third phrase of (A): no response. (C–F) Phrases from song of (A). For (C–J) note expanded axes. (C) Second phrase: weak response. (D) Third phrase: no response. (E) Second and third phrase in sequence with naturally occurring 43-msec interphrase interval: strong facilitated response. (F) Same as (E) but 93 msec interphrase interval: strong facilitated response. (G) Two tone bursts of 3.0 kHz with 43-msec intertone interval: no response. (H) Same as (G) but frequency is 3.25 kHz: no response. (I) Frequency modulation from 3.25 to 3.0 kHz followed by tone burst of 3.0 kHz: no response. (J) Stimuli from (E) and (I) are superimposed. Note maximum difference in frequency at any point is ≤ 100 Hz; maximum amplitude difference is ≤ 8 dB. The time axes of all plots are aligned. All histograms 10 repetitions.

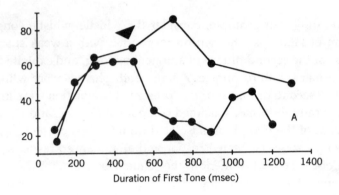

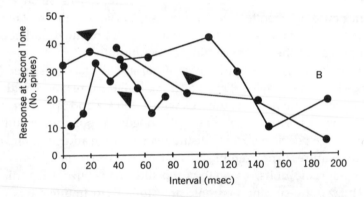

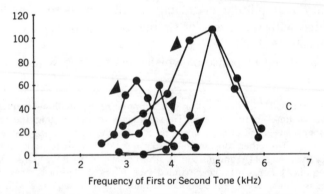

FIGURE 5. Response of whistle–whistle song-specific neurons (WWSSN) to sequences of tone bursts. Response is defined as number of spikes during the second tone burst. Arrow heads point to the naturally occurring value in autogenous song. (A) Response of two WWSSN as a function of duration of first tone burst. Note the response is affected only by large changes in duration. (B) Response of three WWSSN as a function of the interval between the tone bursts. Note the response occurs over a broad range of intervals. (C)

specific neurons is largely unaffected by either the duration of the first phrase of the sequence or the duration of the interphrase interval, and changes of 200–300% from the values occurring in the natural song typically have little effect on the magnitude of the response. This is most easily demonstrated for those whistle–whistle song-specific neurons that exhibit the facilitated response to simple sequences of tone bursts (Fig. 5A and B). Changes in the frequency of either tone burst, however, systematically reduces the response strength (Fig. 5C). These attributes contribute to the intradialect selectivity of whistle–whistle and whistle–buzz song-specific neurons. The process that elicits the facilitated response is gated by the offset of the first element in the sequence, so that consecutive identical tone bursts can elicit a facilitated response (Margoliash, 1983a). In contrast, the facilitated response of the FM/FM combination-sensitive neurons of the bat auditory cortex is gated by the onset of the first element of the pulse–echo pair (Suga et al., 1983). These differences presumably reflect different underlying neuronal mechanisms for combination sensitivity.

The selectivity of song-specific neurons can be explained on the basis of neuronal specificity to acoustic parameters of autogenous song. For example, a whistle–whistle song-specific neuron recorded in a wild-caught bird singing a Bodega Bay, CA, dialect song responded vigorously only to autogenous song and to a lesser degree one of nine other conspecific songs tested, including four songs that were from the same dialect (Fig. 6). This neuron responded to consecutive (but not single) tone bursts of frequencies matching the center frequencies of the corresponding whistles of autogenous song. The strength of response at the second tone burst was a function of the frequencies of the first and second tone bursts. The song selectivity of the neuron could be explained on the basis of the tuning to consecutive tone bursts. That is, the strength of response elicited by a song was dependent on the match between the frequencies of the whistles of that song and the whistles of autogenous song. Nevertheless, both whistles were necessary: One of the test songs that had an initial whistle of the same frequency as autogenous song

---

Response of two WWSSN as a function of frequency of first or second tone burst, with the other tone burst at the optimal frequency. Arrow heads denote the average frequency of the corresponding whistle. The strength of response is affected by the frequency of both tone bursts. Unlike for (A) and (B), the curves for (C) are approximately at maximum and centered on the values that occur in autogenous song.

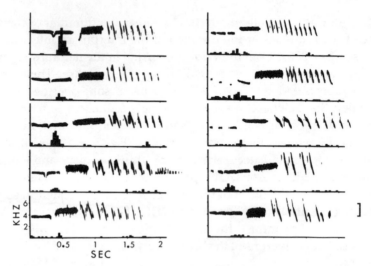

FIGURE 6. Song selectivity of whistle–whistle song-specific neuron 810102. Sonagrams (frequency versus time representations) of 10 white-crowned sparrow songs with response histograms superimposed. The five songs in the left column are all of the same Bodega Bay dialect; the top song is the bird's own song. Note that only one other song elicits strong excitation. The five songs in the right column are sample songs from five different dialects of the same race of white crowns (top to bottom: Berkeley, Tilden Park, Pacific Grove, Lompoc, Lake Merced, CA, dialects). Note that no song elicits strong excitation. All histograms 20 repetitions, 30-msec binwidth. The calibration mark in lower right represents 10 spikes (from Margoliash, 1983a).

but lacked a second whistle did not elicit a response. As suggested by the response to tone bursts, it was observed that similarities in test songs to the temporal pattern of autogenous song (i.e., duration of whistles and interwhistle interval) did not contribute to the efficacy of the test songs.

The parameters to which song-specific neurons in the white-crowned sparrow are sensitive suggest they may contribute to neighbor–stranger discrimination. In an investigation of the features of song used for neighbor–stranger discrimination in the congeneric white-throated sparrow (*Zonotrichia albicollis*), Brooks and Falls (1975) assessed which changes of a neighbor's song were sufficient so that a territorial male reclassified it as a stranger's song. With a 5% decrease in frequency in a neighbor's entire song or a 15% decrease in the frequency of the initial whistle only, white-throated sparrows responded aggressively when the song was broadcast, as if reacting to the songs of strangers. In contrast, changes of 5 or 15% in duration of the entire song or of the first phrase did not result in a statistically significant change in behavior. Analytic behavioral tests such as these have yet to be conducted with the white-crowned sparrow. Nevertheless, it is intriguing that song-specific neurons are unaffected by changes in duration but are rather sensitive to changes in frequency of the two note sequences. If white-crowned and white-throated sparrows rely on similar cues for neighbor–stranger discrimination, it suggests the possibility that the parameters used for individual recognition on the basis of song, and the parameters that song-specific neurons are tuned to, are established by a common mechanism.

SONG SELECTIVITY OF MULTIUNIT RESPONSES

Recent experiments have employed recordings from clusters of neurons to enable the characterization of the response properties of the population of HVc auditory neurons in each bird (Margoliash, 1986; Margoliash and Konishi, 1985). The characteristic physiology of HVc permits easy identification of appropriate multiunit recording sites, so that the issue of potential sampling bias introduced by use of autogenous song as a search stimulus can also be addressed. Multiunit response strength was quantified by measuring the time-varying neuronal signal (see Fig. 7). At each recording site, the response to autogenous song was compared with the response to other conspecific songs. All wild-type songs used in these experiments were of the same dialect, recorded from birds caught at the Bodega Marine Laboratory, Bodega Bay, CA.

Even at the multiunit level, autogenous song consistently proved to be the most effective stimulus. In one experiment (Margoliash and Konishi, 1985), 24 multiunit recording sites in two birds were presented with autogenous song and with a set of three conspecific "sample" songs. Of the 72 pairwise comparisons between autogenous song and a sample song, autogenous song elicited the stronger response in all but four. The responses to the sample songs were reasonably strong, on average 50% of the response to autogenous song. The phrases from the sample songs that elicited excitation always corresponded to those phrases from autogenous song that elicited excitation (e.g., Fig. 7). In another experiment, at two recording sites in each of three birds, a battery of 10 intradialect songs were presented along with the autogenous song (Figs. 8 and 9). These songs were very similar to each other. Furthermore,

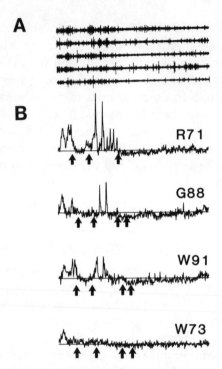

FIGURE 7. (A) Multineuronal activity during five repetitions of the bird's own song, R71. Note strong activity during whistle and trill (first and third phrases) of song, and reduced activity during the intervening buzz. (B) Summed responses (see Margoliash, 1986, for details) of the neuronal cluster of (A) to four songs, including R71. The baseline represents spontaneous activity. Arrows delineate the offset of each phrase (the fourth phrase of W73, G88, and W91 was a terminal buzz). Note response to R71 is strongest and that excitatory responses for all the songs are elicited by whistle and trill, but not buzz. Baseline represents 5 sec. Fifty repetitions per song, 1 song per 12 sec (from Margoliash and Konishi, 1985).

recording sites in two of the birds had been chosen without prior knowledge of the bird's own song. Nevertheless, for 50 of the 60 pairwise comparisons, autogenous song elicited a stronger response. As a control for sampling bias, in a third experiment four adult birds that presumably had prior singing experience but whose songs were not known to the experimenter were permanently implanted with electrodes in HVc. After song was induced with testosterone, all the recording sites in these birds were also most effectively stimulated by autogenous song over all the days they were tested. Five of the recording sites remained viable for extended periods of time, at least 25 days and up to 97 days after

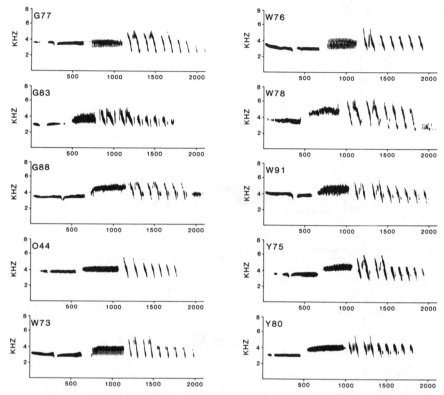

FIGURE 8. Sonograms of 10 songs from the Bodega Bay dialect. All songs share the same basic pattern: whistle, buzz, trill. All whistles comprise two parts except for W78. All trills comprise two syllable types. The presence of terminal buzz is variable. The frequencies of the whistle and buzz and the details of the frequency modulation of the phrases are the features that endow each song with uniqueness. HVc neurons are selective within this range of variability (from Margoliash, 1986).

the birds began to sing. Without fail, autogenous song elicited stronger responses than all the conspecific test songs for all days of experiment. At three of these recording sites, the pattern of activity elicited by autogenous song (e.g., strong response to whistle and trill with little response to buzz) did not vary significantly over time, whereas somewhat greater variability was observed at the other two sites (Margoliash, 1986). The technical difficulties in maintaining the stability of a recording over a period of 3 months suggests that these results are consistent with the hypothesis that HVc auditory neurons in the white-crowned sparrow do not exhibit plasticity in adulthood.

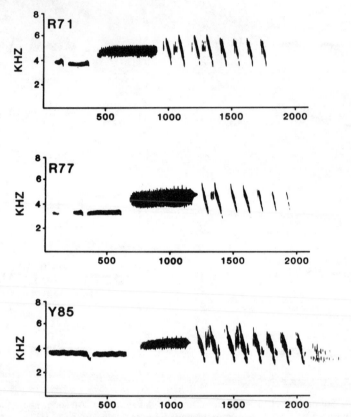

FIGURE 9. Sonograms of autogenous song of three birds. Note the songs are similar to each other and to the songs of Figure 8. Two recording sites in each bird were tested with autogenous song and the 10 songs of Figure 10. Autogenous song elicited a stronger response than did the other songs: for R71, 18–20 times; for R77, 15–20 times; for Y85, 17–20 times (from Margoliash, 1986).

It was also observed that HVc neuronal clusters preferred autogenous song over the song that birds were exposed to early in life (Margoliash, 1986). The effectiveness of the tutor song model was directly related to its similarity to autogenous song. In one bird, the tutor and adult (autogenous) song were quite similar, and the tutor song was a relatively effective stimulus. A second bird sang an approximate copy of the tutor song; in this bird the tutor song elicited moderate excitation from HVc neurons. A third bird entirely failed to copy his tutor song. In this bird autogenous song was a far more effective stimulus than the tutor song.

Clearly, in the white-crowned sparrow the response properties of adult HVc auditory neurons are established after the impressionable phase.

MULTIUNIT SPECIFICITY FOR ACOUSTIC PARAMETERS OF AUTOGENOUS SONG

The selectivity of HVc neuronal clusters for autogenous song suggests an underlying specificity for the set of acoustic parameters that define autogenous song. A total of 432 phrases from the sample songs was presented, of which only 10% elicited greater response than the corresponding phrases from the bird's own song (Margoliash, 1986). This suggests that the heightened response to autogenous song resulted from parametric differences at the level of individual phrases. It should be noted that all these songs were from the same dialect, and therefore shared the same sequence of phrases and similar temporal patterning.

A variety of parametric manipulations of song identified several of the relevant parameters of autogenous song (Margoliash, 1986). Increasing or decreasing the overall frequency of autogenous song by 500 or 1000 Hz reduced its efficacy. The response decreased monotonically with increasing frequency shift (Fig. 10A). This was observed for all songs, even when the shift in frequency did not create a song outside the normal intradialect range of variation in frequency. Another test involved reversing autogenous song, which alters the time-varying frequency and amplitude components but not the overall spectrum. Almost without exception, reversed song elicited a weaker response (Fig. 10B). Thus, HVc clusters were sensitive not only to static parameters of song such as frequency but also to the dynamic (time-varying) frequency and/or amplitude parameters. To distinguish between these, two song variants were created. For the first song variant, spectrally similar random frequency modulation was substituted for the frequency modulation of autogenous song. In the other song variant, essentially all frequency modulation was eliminated. Both song variants had amplitude modulation identical to autogenous song. Systematically, both these song variants also elicited weaker responses than autogenous song (Fig. 10C), with roughly the same decrement in response magnitude as observed for reversed song. Thus, in accordance with the single unit data, the time-varying frequency components of song were implicated in the response selectivity of HVc neuronal clusters.

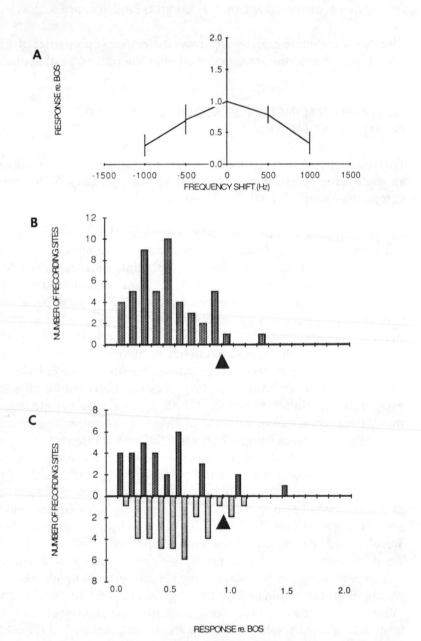

FIGURE 10. Effects of modification of autogenous song on HVc multineuronal responses. (A) Change in frequency. Note systematic decrease in response with increasing frequency shift (*n* = 17). (B) Reversed song. Only two sites exhibited stronger responses to reversed song than to forward song (*n* = 49). (C) Upper graph, response to the frequency ran-

THE SITE OF PLASTICITY

Since field L is a likely source of auditory input into HVc, it is of value to compare the physiology of these nuclei. Leppelsack (1981, 1983) recorded the response to synthetic song of neurons in field L and the mediocaudal neostriatum of the white-crowned sparrow. The birds in those experiments sang wild-type songs, or had been successfully tutored with computer-synthesized songs incorporating either downwards (natural) or upwards (unnatural) sweeping FMs in the trill phrase of song (Konishi, 1978). Although the majority of the field L neurons exhibited no directional selectivity for FMs (i.e., downwards/upwards discrimination), when directional selectivity was encountered it was positively correlated with the direction of FMs in the tutor and individual's song, and was statistically significant. Furthermore, a few specialized neurons were encountered that responded to FMs but not to tone bursts. For these neurons, the optimal starting and ending frequencies for FM in the synthetic song often approximated those in the tutor and individual's song.

In contrast to the response in HVc, however, multineuronal field L recordings exhibited no special selectivity for autogenous song (Margoliash, 1986). Frequency-shifted songs often elicited stronger responses than the unshifted song, largely dependent on the position of the recording site along the tonotopic axes (Fig. 11A). Field L neurons exhibited stronger responses to reversed song than did HVc neurons (Fig. 11B), and the two song variants described above often were more effective than autogenous song (Fig. 11C). A restricted portion within field L that exhibited HVc-like activity was not discovered, although the size of field L precluded a complete scan. Whereas in HVc autogenous song was often the most effective stimulus throughout an entire series of tests, this was essentially never observed in field L.

Although the multiunit technique could have easily missed a few selective units, these results demonstrate that field L neurons are not shaped during song development by the same processes that establish the response properties of HVc neurons. Since field L neurons are not

---

domized song variant (see text). Only three sites exhibited stronger response than to autogenous song ($n = 31$). Lower graph, response to the constant-frequency song variant. Three (different) sites exhibited stronger response than to autogenous song ($n = 36$). All graphs, 20 or 50 repetitions per song (adapted from Margoliash, 1986). BOS refers to bird's own song.

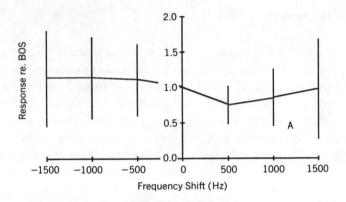

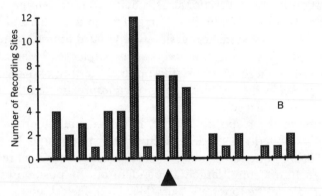

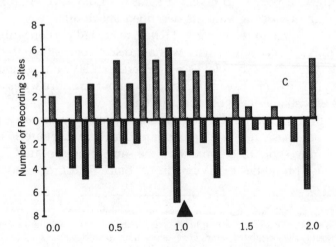

selective for autogenous song, an alternative hypothesis is that the correlation Leppelsack observed reflects exposure to song during the impressionable phase (Leppelsack, 1983). That conclusion is rather speculative, however, because the existence of the correlation per se does not distinguish between the role of autogenous and tutor songs, nor does ontogenetic modification necessarily reflect developmental processes: Are field L response properties established during the impressionable phase, song crystallization, or in adulthood? Furthermore, whether the FM-sensitive neurons in field L project into the motor system for song has yet to be explored.

THE EFFECT OF HORMONAL STATE ON HVc AUDITORY RESPONSES

HVc concentrates androgen (Arnold et al., 1976). The white crown's song is under hormonal control, although unlike the canary (Nottebohm, 1981) the volume of the white crown's HVc does not fluctuate seasonally (Baker et al., 1984). If auditory neurons in HVc contribute to song discrimination (see below), it is of interest to know if response properties are affected as hormone levels fluctuate throughout the year.

A thorough study of this issue has yet to be completed. In a fortuitous test, however, I recorded an adult male white-crowned sparrow in molt. Earlier, after the bird had received subcutaneous testosterone proprionate, he sang vigorously and loudly—thousands of times each day—and responded to conspecific song by countersinging. At this time blood hormone level was measured at 1695 pg/ml testosterone equivalents. Subsequently, the hormone was exhausted and the bird went into characteristic molt. Behavioral manifestations of molt included a total cessation of singing.

FIGURE 11. Effects of modification of autogenous song on field L multineuronal responses. (A) Change in frequency. Note normal song is not optimal. The large error bars are a consequence of presenting frequency-shifted songs in different parts of the tonotopic axis. This results in a flat averaged response ($n = 21$). (B) Reversed song. Of 60 recording sites, 23 exhibited stronger responses to reversed song than to forward song. Overall response to reversed song was quite strong. (C) Response to song variants. Upper graph, response to frequency-randomized song. Of 59 recording sites 26 exhibited stronger response than to autogenous song. Lower graph, response to constant frequency song. Of 55 sites 21 exhibited stronger response than to autogenous song. Note the difference of the field L data with the HVc data (see Fig. 10). All graphs, 10 or 20 repetitions per song (adapted from Margoliash, 1986). BOS refers to bird's own song.

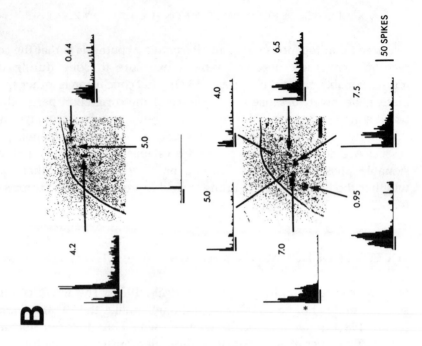

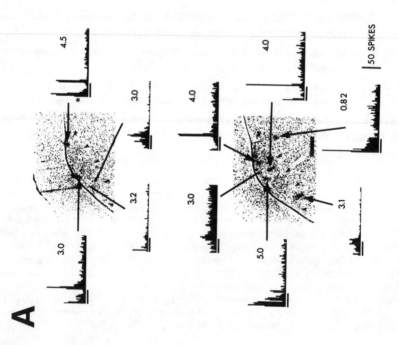

The first of two recording sessions was conducted two weeks into the molt. In 9 penetrations into the left hemisphere HVc, 26 single units were isolated that exhibited clear responses to auditory stimuli. Response properties included the gamut of previously reported classes, but particular attention was devoted to elucidating the response properties of the two whistle–whistle song-specific units isolated. One of these is the unit of Figure 4. The other unit had somewhat unusual response properties, perhaps because this bird sang two distinct songs—an unusual occurrence for white-crowned sparrows—and the unit responded to both songs. Note that the whistle–whistle song-specific neuron of Figure 4 responded only to one of the two songs (Fig. 4A and B). For this and a subsequent recording session 5 days later, while the bird was still in molt, a consistent observation was that rather than being diminished, auditory responses to tone bursts were unusually vigorous as compared to those I had previously observed for HVc.

The second recording session was devoted to sampling from units throughout the extent of HVc. Neurons were characterized by their robust responses to tone bursts. Response profiles included on, on–off, off, sustained, and with and without inhibition, as previously reported for HVc. The best frequency of these HVc neurons fell between 3.0 and 7.5 kHz, while neurons immediately outside of HVc responded best to very low frequencies (Fig. 12). As with hormonally primed, singing birds, the borders of HVc were easily discerned on the basis of HVc's characteristic "bursting" background activity, as demonstrated by the rows of electrolytic lesions placed at the dorsal (Fig. 12A) and ventral (Fig. 12B) borders of the nucleus. A blood sample taken one day later contained 253 pg/ml testosterone equivalents, a 6.7-fold reduction in hormone levels from premolt condition. Thus, with the possible excep-

---

FIGURE 12. HVc auditory responses in a molting bird. Each histological section is surrounded by histograms of HVc neuronal responses to tone bursts. Lines from the histograms point to the location where the unit was recorded. Lines with bold arrowheads point to sites of electrolytic lesions. Small arrowheads demark the ventral border of HVc. All histograms show 2 sec of response to 400-msec tone bursts. The time of occurrence and frequency of the tone bursts is shown for each histogram. For the histograms marked with a star, the calibration mark is 100 spikes. All histograms, 10 repetitions. Histological calibration bar: 200 μm. (A) Two sections in which the dorsal border of HVc is demarked by lesions. (B) Two sections in which the ventral border of HVc is demarked by lesions. Throughout HVc, note the variety of response patterns. Neurons within HVc have best frequencies of 3.0–7.5 kHz; neurons at the border or outside of HVc have best frequencies of 440–950 Hz.

tion of a greater magnitude of response, the 3 weeks of low hormone levels and no song had no apparent effect on auditory responses in HVc (see also Williams, 1985). The increased vigor in response to tone bursts may have been coincidental or potentially could reflect release of auditory activity from inhibition as motor neurons are depleted of hormone.

THE ROLE OF HVc AUDITORY NEURONS: AN HYPOTHESIS

The song selectivity and specificity for autogenous song of HVc auditory neurons results from ontogenetic modification, for many acoustic parameters of the white crown's song are learned. An individual cannot know the details of his song until he learns to sing, thus the crystallization of song is the earliest developmental process that can reasonably account for the properties observed in the adult HVc. Indeed, adult HVc neurons rather systematically prefer the adult form of song established during the crystallization phase over the tutor songs birds are exposed to during the impressionable phase (Margoliash, 1983a, 1986). The preponderance of evidence suggests that in the white-crowned sparrow, HVc auditory neurons do not exhibit plasticity in adulthood, although whether these neurons are modified by exposure to the songs of neighboring birds remains to be tested. Several lines of evidence support this contention. First and foremost, the stability of the adult white crown's song is a poor behavioral substrate for plasticity. Second, auditory responsiveness in HVc is actively inhibited during singing (McCasland and Konishi, 1981). Furthermore, HVc selectivity for song appears to be quite stable over a period of months and insensitive to changes in hormone levels, and even the shape of the recorded response waveform exhibits stability over time. Taken together, these results suggest that in the white-crowned sparrow HVc neurons do not exhibit plasticity into adulthood (Margoliash, 1986). Whether the response properties of HVc auditory neurons are established during the crystallization of song, and contribute to the development of song, has yet to be directly tested. Nevertheless, it is clearly parsimonious to suggest that plastic auditory neurons in the motor system contribute to auditory–motor plasticity during development. It will also be of interest to determine if in duetting and other birds that exhibit vocal learning throughout life, HVc auditory neurons continue to exhibit plasticity in adulthood.

I hypothesize that during development, auditory neurons in HVc participate in shaping the motor program for song. I assume HVc has direct access to the template—either intrinsic to HVc or residing in another nucleus (e.g., Leppelsack, 1983; Bottjer et al., 1984). This would result in excitation of HVc auditory neurons only by those motor patterns that produce a match between auditory feedback (via the shelf) and the memorized song. A bird would have to "search" for appropriate sounds, hence plastic song and the overproduction and improvisation of song syllables (Marler and Peters, 1981, 1982b). Activation of HVc auditory neurons by fortuitous motor patterns could help stabilize those patterns. In turn, the stabilization of specific motor patterns would commit HVc neurons, resulting in attrition of syllable types and concomitant emergence of the motor program for, and auditory representation of, autogenous song. Perhaps in later stages of song development, the interaction in HVc between auditory and motor activity would become prominent (McCasland and Konishi, 1981). The inhibition of HVc auditory neurons during singing may function to emancipate singing from susceptibility to auditory feedback in the adult. If so, in species that learn new syllables as adults, the inhibitory interaction should wax and wane with song learning. Thus, HVc auditory neurons could contribute to shaping the motor program during song crystallization, resulting in the formation of an auditory representation of autogenous song by the time song has crystallized.

In the adult, an auditory representation of autogenous song—an "autogenous reference"—could contribute to discrimination among conspecific songs. Auditory neurons that are shaped during song acquisition become sensitive to the parameters of communication, and are well suited to discriminate among the vocalizations of conspecifics. The linkage of production and perception is predetermined by the use of auditory feedback during vocal learning, and is well established, for example, for human speech. Thus, contribution of the uniquely appropriate autogenous reference to song recognition would be an efficient utilization of a functional structure in any case established by developmental processes. There is some evidence in other species relating the perceptual capabilities of adult birds with the form of autogenous song (McArthur, 1986; see also McGregor et al., 1983). For the white-crowned sparrow, playback experiments have established that the birds can distinguish intradialect songs from other conspecific songs and neighbors' songs from strangers' songs (Milligan and Verner, 1971;

Baker et al, 1981a,b), but have yet to investigate the potential role of autogenous song in song recognition.

It should be noted the "autogenous reference" hypothesis may not be generalizable to all songbirds. Some species possess an extensive repertoire—we do not know the rules for the representation of their songs. It is unlikely that the 2000-syllable repertoire of the brown thrasher (*Toxostoma rufum*—Kroodsma and Parker, 1977) can be maintained by the same rules as for the white-crowned sparrow, with fewer than 10 syllables. Furthermore, the auditory plasticity required by those species that continue to modify song into adulthood may preclude a stable auditory representation of song. Finally, if the female also relies on HVc to contribute to song recognition, what with those, such as the zebra finch, in whom HVc is greatly diminished? All hypotheses that suggest HVc plays a special role in song recognition (e.g., Nottebohm, 1984) must address these issues (see also Williams and Nottebohm, 1985).

CONCLUSION

Early in the study of song learning, the classical white-crowned sparrow deafening experiments suggested the importance of auditory feedback and central pattern generators for the establishment and maintenance of song (Konishi, 1965b). The recent studies of the auditory representation of autogenous song in HVc have confirmed and extended these results. The study of the interactions between HVc auditory and motor components during song crystallization is likely to further clarify how motor patterns are selected for and shaped during development. The behavioral observations that have provided a valuable logic for guiding these initial steps are beginning to be redefined in terms of specific mechanisms, as only an interplay between behavioral and physiological experimentation can hope to provide. It is reasonable to presume that other manifestations of song learning, including the innate predisposition to learn conspecific song, as well as the elusive acquired template, will also prove amenable to neuroethological analysis.

ACKNOWLEDGMENTS

These experiments were conducted in the laboratory of Dr. Mark Konishi, who provided support, encouragement, and insight. I thank Dr.

John Wingfield for performing the radioimmunoassays. Drs. Konishi, Peter Marler, and Larry Katz provided useful critiques of the manuscript.

REFERENCES

Arnold, A., F. Nottebohm, and D. W. Pfaff (1976). Hormone concentrating cells in vocal control and other areas of the brain of the zebra finch (*Poephila guttata*). *J. Comp. Neurol.* **165**:487–512.

Baker, M. C., Thompson, D. B., and G. L. Sherman (1981a). Neighbor/stranger song discrimination in white-crowned sparrows. *Condor* **83**:265–267.

Baker, M. C., K. J., Spitler-Nabors, and D. C. Bradley (1981b). Early experience determines song dialect responsiveness of female sparrows. *Science* **214**:819–821.

Baker, M. C., S. W. Bottjer, and A. P. Arnold (1984). Sexual dimorphism and lack of seasonal changes in vocal control regions of the white-crowned sparrow brain. *Brain Res.* **295**:85–89.

Baptista, L. F. and L. Petrinovich (1984). Social interaction, sensitive phrases and the song template hypothesis in the white-crowned sparrow. *Anim. Behav.* **32**:172–181.

Bottjer, S. W., E. A. Miesner, and A. P. Arnold (1984). Forebrain lesions disrupt development but not maintenance of song in passerine birds. *Science* **224**:901–903.

Brooks, R. J. and J. B. Falls (1975). Individual recognition by song in white-throated sparrows. III. Song features used in individual recognition. *Can. J. Zool.* **53**:1749–1761.

DeVoogd, T. J. and F. Nottebohm (1981). Sex differences in dendritic morphology of a song control nucleus in the canary: A quantitative Golgi study. *J. Comp. Neurol.* **196**:309–316.

Fuzessery, Z. M. and A. S. Feng (1983). Mating call selectivity in the thalamus and midbrain of the leopard frog (*Rana p. pipiens*): Single and multiunit analyses. *J. Comp. Physiol.* **150**:333–344.

Greenewalt, C. H. (1968). *Bird Song: Acoustics and Physiology.* Smithsonian Institute Press, Washington, D. C.

Gurney, M. E. (1980). Sexual differentiation of brain and behavior in the zebra finch (*Poephila guttata*): A cellular analysis. Ph.D. Thesis, California Institute of Technology, Pasadena, CA.

Gurney, M. E. (1981). Hormonal control of cell form and number in the zebra finch song system. *J. Neurosci.* **1**:658–673.

Immelmann, K. (1969). Song Development in the Zebra Finch and other Estrildid Finches. In: *Bird Vocalizations*, R. A. Hinde, ed., pp. 61–74, Cambridge University Press, Cambridge.

Karten, H. (1968). The ascending auditory pathway in the pigeon(*Columba livia*). II. Telencephalic projections of the nucleus ovoidalis thalami. *Brain Res.* 11:134–153.

Katz, L. C. (1982). The avian motor system for song has multiple sites and types of auditory input. *Soc. Neurosci. Abstr.* 8:1021.

Katz, L. C. and M. E. Gurney (1981). Auditory responses in the zebra finch's motor system for song. *Brain Res.* 211:192–197.

Kelley, D. B. and F. Nottebohm (1979). Projections of a telencephalic auditory nucleus–field L–in the canary. *J. Comp. Neurol.* 183:455–470.

Konishi, M. (1965a). Effects of deafening on song development in two species of juncos. *Condor* 66:85–102.

Konishi, M. (1965b). The role of auditory feedback in the control of vocalization in the white-crowned sparrow. *Z. Tierpsychol.* 22:770–783.

Konishi, M. (1978). Auditory Environment and Vocal Development in Birds. In: *Perception and Experience*, R. D. Walk and H. L. Pick, Jr., eds., pp. 105–118, Plenum Press, New York.

Konishi, M. (1985). Birdsong: From behavior to neuron. *Ann. Rev. Neurosci.* 8:125–170.

Kroodsma, D. E. and L. D. Parker (1977). Vocal virtuosity in the Brown Thrasher. *Auk* 94:783–784.

Kroodsma, D. E. and R. Pickert (1980). Environmentally dependent sensitive periods for avian vocal learning. *Nature* 288:477–478.

Kroodsma, D. E. and R. Pickert (1984). Repertoire size, auditory templates, and selective vocal learning in songbirds. *Anim. Behav.* 32:395–399.

Langner, G., D. Bonke, and H. Scheich (1981). Neuronal discrimination of natural and synthetic vowels in field L of trained mynah birds. *Exp. Brain Res.* 43:11–24.

Leppelsack, H.-J. (1981). Antwortverhalten auditorischer Vorderhirnneuronen eines Singvogels unter besonderer Berücksichtigung des Gesangslernens. Habilitationsschrift, Univ. Bochum.

Leppelsack, H.-J. (1983). Analysis of Song in the Auditory Pathway of Song Birds. In: *Advances in Vertebrate Neuroethology*, J. P. Ewert, R. R. Capranica, and D. J. Ingle, eds., pp. 783–799, Plenum Press, New York.

Leppelsack, H.-J. and M. Vogt (1976). Responses of auditory neurons in the forebrain of a songbird to stimulation with species-specific sounds. *J. Comp. Physiol.* 107:263–274.

Logan, B. F. (1977). Information in the zero crossings of bandpass signals. *Bell Sys. Tech. J.* **56:**487–510.

Margoliash, D. (1983a). Acoustic parameters underlying the responses of song-specific neurons in the white-crowned sparrow. *J. Neurosci.* **3:**1039–1057.

Margoliash, D. (1983b). Songbirds, grandmothers, and templates: A neuroethological approach. Ph.D. thesis, California Institute of Technology, Pasadena, CA.

Margoliash, D. (1986). Preference for autogenous song by auditory neurons in a song system nucleus of the white-crowned sparrow. *J. Neurosci.* **6:**1643–1661.

Margoliash, D. and M. Konishi (1985). Auditory representation of autogenous song in the song-system of white-crowned sparrows. *Proc. Natl. Acad. Sci. USA* **82:**5997–6000.

Marler, P. (1970). A comparative approach to vocal learning: Song development in white-crowned sparrows. *J. Comp. Physiol. Psychol.* Monograph **71:**1–25.

Marler, P. and M. S. Waser (1977). Role of auditory feedback of canary song development. *J. Comp. Physiol. Psychol.* **91:**8–16.

Marler, P. and S. Peters (1981). Sparrows learn adult song and more from memory. *Science* **213:**780–782.

Marler, P. and S. Peters (1982a). Structural changes in song ontogeny in the swamp sparrow, *Melospiza georgiana. Auk* **99:**446–458.

Marler, P. and S. Peters (1982b). Developmental overproduction and selective attrition: New processes in the epigenesis of birdsong. *Dev. Psychobiol.* **15:**369–78.

Marler, P. and V. Sherman (1982). Structure in sparrow song without auditory feedback: An emendation of the auditory template hypothesis. *J. Neurosci.* **3:**517–531.

McArthur, P. D. (1986). Similarity of playback songs to self song as a determinant of response strength in song sparrows (*Melospiza melodia*). *Anim. Behav.* **34:**199–207.

McCasland, J. S. (1987). Neuronal control of bird song production. *J. Neurosci.* **7:**23–39.

McCasland, J. S. and M. Konishi (1981). Interaction between auditory and motor activities in an avian song control nucleus. *Proc. Natl. Acad. Sci. USA* **78:**1815–1819.

McGregor, P. K., J. R. Krebs, and L. M. Ratcliffe (1983). The reaction of great tits (*Parus major*) to playback of degraded and undegraded songs: The effect of familiarity with the stimulus song type. *Auk* **100:**898–906.

Milligan, M. M. and Verner, J. (1971). Inter-populational song dialect discrimination in the white-crowned sparrow. *Condor* **73**:77–80.

Nottebohm, F. (1968). Auditory experience and song development in the chaffinch, *Fringilla coelebs. Ibis* **110**:549–568.

Nottebohm, F. (1981). A brain for all seasons: Cyclical anatomical changes in song control nuclei in the canary brain. *Science* **194**:211–213.

Nottebohm, F. (1984). Birdsong as a model in which to study brain processes related to learning. *Condor* **86**:227–236.

Nottebohm, F., T. M. Stokes, and C. M. Leonard (1976). Central control of song in the canary, *Serinus canarius. J. Comp. Neurol.* **165**:457–486.

Nottebohm, F., D. B. Kelley, and J. A. Paton (1982). Connections of vocal control nuclei in the canary telencephalon. *J. Comp. Neurol.* **207**:344–357.

O'Neill, W. E., and N. Suga (1979). Target range-sensitive neurons in the auditory cortex of the mustache bat. *Science* **203**:69–73.

Rausch, G. (1985). Ontogenese der Vokalisation und des vokalmotorischen Kernes Hyperstriatum Ventrale, pars Caudale (HVc) im Telencephalon des Beos (*Gracula religiosa intermedia*). Ph.D. thesis, Technical University, Darmstadt.

Rose, G. and R. R. Capranica (1983). Temporal selectivity in the central auditory system of the leopard frog. *Science* **219**:1087–1089.

Scheich, H., G. Langner, and D. Bonke (1979). Responsiveness of units in the auditory neostriatum of the Guinea fowl (*Numida meleagris*) to species-specific calls and synthetic stimuli. II. Discrimination of iambus-like calls. *J. Comp. Physiol.* **132**:257–276.

Staddon, J. E. R., L. W. McGeorge, R. A. Bruce, and F. F. Klein (1978). A simple method for rapid analysis of animal sounds. *Z. Tierpsychol.* **48**:306–330.

Suga, N., W. E. O'Neill, and T. Manabe (1979). Harmonic-sensitive neurons in the auditory cortex of the mustache bat. *Science* **203**:270–274.

Suga, N., W. E. O'Neill, K. Kujirai, and T. Manabe (1983). Specificity of combination-sensitive neurons for processing of complex biosonar signals in auditory cortex of the mustached bat. *J. Neurophysiol.* **49**:1573–1626.

Sullivan, W. E. III (1982). Neural representation of target distance in auditory cortex of the echolocating bat *Myotis lucifugus. J. Neurophysiol.* **48**:1011–1032.

Thorpe, W. H. (1961). *Bird Song: The Biology of Vocal Communication and Expression in Birds.* Cambridge University Press, Cambridge.

Williams, H. (1985). Sexual dimorphism of auditory activity in the zebra finch song system. *Behav. Neural Biol.* **44**:470–484.

Williams, H., and F. Nottebohm (1985). Auditory responses in avian vocal motor neurons: A motor theory for song perception in birds. *Science* **229**:279–282.

4

HORMONES AND CRITICAL PERIODS IN BEHAVIORAL AND NEURAL DEVELOPMENT

Arthur P. Arnold, Sarah W. Bottjer,* Ernest J. Nordeen,[†]
Kathy W. Nordeen,[†] and Dale R. Sengelaub**

**Department of Psychology and Laboratory of Neuroendocrinology,
Brain Research Institute, University of California, Los Angeles, California*

*[†]Department of Psychology, University of Rochester, Rochester, New
York*

Critical periods have been identified in a wide variety of behavioral and physiological systems. There are at least two general types of critical periods, those that occur early in development (Scott, 1978), and those that recur in adulthood, tied to circadian or other cycles (e.g., Everett, 1977). In this chapter, we will restrict our attention to developmental critical periods that occur early in ontogeny. One example is involved in the process of imprinting, in which a young nidifugous bird follows a moving object to which it has been exposed during a restricted ontogenetic period. Because of the attraction of the young bird to the object, its filial attachment to the mother is ensured, thus enhancing its survival. If the bird is isolated from potential imprinting stimuli for the entire duration of the critical period, it will never develop the following response when subsequently exposed to such objects, but instead will attempt to avoid them. The phenomenon of imprinting demonstrates a selective influence of the environment during a single, nonrecurrent time window in development.

In the present chapter, we discuss the relation between gonadal steroid hormones and critical periods in development of sexually dimorphic regions of the rat spinal cord and the zebra finch brain, and critical periods for zebra finch song learning. Our central question is what neural and hormonal events account for the occurrence of these critical periods. We begin by discussing the general characteristics of critical periods in neural and behavioral development, and pose several general questions that crop up in any discussion of the basis for critical periods. Thereafter, we will analyze the present state of understanding the development of the two systems.

CHARACTERISTICS OF DEVELOPMENTAL CRITICAL PERIODS

Imprinting offers us one of the archetypal examples of a critical period in behavioral development, since it has general characteristics similar to many other developmental critical periods, including those discussed in this chapter. In his famous paper on imprinting originally published in 1935, Lorenz (1970) drew a clear analogy between imprinting and morphological development of the embryo.

> These two facts, the determination of later behavior by an external influence . . . during *a specific ontogenetic period* and the *irreversibility* of this

determination process, provide a remarkable analogy between the developmental processes of instinctive behavioral systems and processes which have been identified in morphological development . . .

If, at a certain time in development, cell material is taken from the ectoderm of the posterior abdominal region of a frog embryo—where it would normally form a piece of abdominal skin during further embryonic development—and grafted onto the posterior end of the outer surface of the neural groove, it will form a constituent part of the spinal cord in accordance with its position. The cells are thus influenced by the organizational determinant of the local environment—an effect which Spemann terms "induction". . . . If [at a later stage of development] . . . the abdominal ectoderm cells (now determined as spinal cord cells) are transplanted back to their site of origin, . . . those cells which would have otherwise become spinal cord constituents do so even at the new [abdominal] site. . . . In other words, the cell material does not possess inherited "knowledge" of its fate—this is determined by the site it occupies. The local environment *imprints* the final organ character of the cells. (Lorenz, 1970, pp. 127–128, italics added).

In the same way, argued Lorenz, the environment of the young bird imprints on it the character of its filial and sexual preferences. In both morphological and behavioral imprinting, the process is limited to a critical period, and is irreversible.

The critical period represents a time when development is influenced by some environmental event. In making this generalization, we take a broad definition of environment. Thus, with regard to behavioral development, environment includes all forces outside the individual. With regard to neural or other morphological development, environment also includes forces coming from outside of specific cells, and includes hormones or other chemical signals within the body.

Lorenz emphasized three basic characteristics of critical period phenomena, both in behavioral and morphological development (see also Scott, 1978).

1. The effects of the environment during the critical period are *irreversible* and *stable*.

2. The environmental events all have the effect of *selecting* one avenue of development from a menu of developmental possibilities. The events of the critical period represent a developmental choice, the differential enhancement of some behavioral or functional quali-

ties. For example, the imprinting process involves enhancement of filial responses to one specific object, rather than to other objects in the environment.

3. There are temporal limits to the critical period.

Although more recent work has questioned whether the irreversibility, stability, and temporal limits of imprinting are absolute or relative (cf. Bateson, this volume), we have chosen to present these characteristics of critical periods in their original form. By doing so, we emphasize the analogy to morphological differentiation, and the greater importance of environmental events during (rather than outside of) the critical or sensitive period. This is not to deny the influence of later environmental events.

QUESTIONS REGARDING THE NEURAL AND HORMONAL BASES OF CRITICAL PERIODS

Each of the three main attributes of critical periods requires an explanation at the cellular level, leading us to ask the following three sets of questions.

Question 1: Why are the events of the critical period partially or wholly irreversible? What cellular processes account for the stability of the differentiating effects initiated during the critical periods? Why do the environmental influences that occur during critical periods differ from others that are less permanent?

Question 2: How is selectivity achieved during the critical period? What mechanisms allow specific enhancement of selected behavioral or neural capacities? In analyzing this question, we realize that neural and morphological development are not merely constructive or progressive, involving positive growth, but also destructive and regressive, involving death and atrophy. For example, research in the last two decades indicates that differentiation of neural circuits involves not only the constructive processes of neurogenesis, neuritic outgrowth, and synaptogenesis, but also the destructive or regressive processes of cell death and synapse elimination. In many brain regions, it is now well documented that there is an overproduction of neurons early in development, followed by loss of many neurons. This process may be a mechanism by which errors in development are corrected, or it may be adaptive in allowing an adjustment in numbers of pre- and postsynaptic elements in a neural circuit, thus ensuring that target regions will receive an adequate number of afferents (Oppenheim, 1981; Hamburger and Oppenheim, 1982). Similarly, evidence from many neural systems also suggests that early in ontogeny of the CNS, there may be an overproduction of synapses between neurons, followed by a regressive elimination of synapses to achieve the adult pattern (Purves and Lichtman, 1980). In considering how environment selectively enhances specific behavioral or neural capacities, both the progressive and regressive events represent processes that could be regulated to achieve specificity. For example, in the process of imprinting, the environment of the young duckling causes an approach response specific to a selected object. Does the development of this attachment involve enhancement of specific neural circuits via a facilitation of progressive events (e.g., by triggering the formation of certain synapses), or does this development involve the regulation of regressive events (e.g., by preventing the loss of selected neurons or synaptic connections)? In asking this question, we are following the lead of many previous workers in development. For example, this question has dominated work on the establishment of retinal-tectal specificity (Keating, 1976; Meyer and Sperry, 1976), and others have discussed these alternatives in behavioral development (Gottlieb, 1976). In the present chapter, we will consider this question as it pertains to studies of sexual differentiation and development of bird song.

Question 3: What cellular processes account for the temporal limits of the critical period? What events (neural, hormonal, etc.) explain the *onset* of the critical period? Since the critical period involves sensitivity to an environmental influence, what accounts for the initial change in sensitivity? What events explain the *end* of the critical period? By separating this second question from the first, we acknowledge that events controlling the beginning and end may be different. These questions have also been asked by previous authors, and one can discern several classes of processes that could explain the temporal limits of critical periods. One possibility is that a critical period exists because two (or more) *independent* developmental processes coincide temporally (or are temporally contiguous) in such a way as to facilitate sensitivity to the environment. The coincidence or contiguity of these processes lasts only for a certain time, thus delimiting the critical period. For example, it has been proposed that the end of the critical period for imprinting occurs when the duckling develops the capacity to fear novel stimuli, and hence will avoid rather than follow a new stimulus. This sort of explanation implies that the decline in imprinting is imposed by a process (development of fear) that may be *unrelated* to the imprinting process itself (reviewed by Bateson, 1966).

A second approach is to consider that the events of the critical period themselves play a role in terminating the critical period. In other words, the processes ending the critical period are not independent of those that begin the critical period, but are a natural consequence of them. There are two general ways in which this dependence may arise, which we will call negative feedback and positive feedback models. A negative feedback model envisions the environmental events of the critical period as terminating any further sensitivity to such events. For example in imprinting, one might imagine that exposure to an imprinting stimulus might facilitate an increase in fear of other objects. In this case, the process of imprinting triggers events that inhibit the imprinting process (i.e., negative feedback). Alternatively, as Bateson (1966, 1979) has argued, the attachment formed during imprinting by the young duckling for its mother also serves to keep the duckling close to the mother and away from other potential objects for imprinting. Thus, further imprinting is avoided through a positive feedback process that stabilizes the attachment to the mother and directs the duckling's attention toward her. Positive and negative feedback processes need not be mutually exclusive and might operate together.

CRITICAL PERIODS OF SEXUAL DIFFERENTIATION OF RAT SPINAL CORD

Perspectives on Sexual Differentiation

Critical periods have long been recognized in the sexual differentiation process (Burns, 1961). In the 1950s and 1960s, it became apparent that sex differences in behavior are controlled by gonadal steroid hormones secreted during a critical period of development (Young et al., 1964; Whalen, 1968; Gorski, 1974; Goy and McEwen, 1980). In mammals, androgens secreted by the male's testes act on the brain to *masculinize* the male (enhancing his capacity to show masculine behavior patterns as an adult) as well as *defeminize* the male (reducing his capacity to show feminine behavior patterns as an adult). The processes of masculinization and defeminization can be mimicked in genetic females. For example, a female exposed to an androgen such as testosterone early in development will be permanently masculinized and defeminized by the hormone. That is, when tested in standardized testing situations in adulthood, the behavior of an androgenized female is more masculine and less feminine by virtue of the prior androgen treatment.

Early research on the influence of gonadal steroids emphasized the presence of a critical period and the permanence of the androgenic effect. For example, to show the permanent effects of androgen, female rats had to be exposed to androgen either prenatally or just after birth. Comparable injections of androgen after the first postnatal week failed to masculinize or defeminize behavior. This result defines the critical period. For some reason, the targets of androgen action no longer respond in this permanent fashion after the first postnatal week.

The permanence of perinatal androgen actions contrasts with other more transient effects (Beach, 1971; Arnold and Breedlove, 1985). For example, the frequency of male copulatory patterns in rodents (mounting, intromission, ejaculation) is controlled by circulating levels of androgens in the adult male. If the male is castrated to reduce androgen titers, the masculine copulatory behaviors decline in frequency. Clearly, these adult effects of androgen are reversible and transient. In contrast, neonatal steroid actions are irreversible and permanent. Phoenix et al. (1959), making an analogy to the permanent androgenic effects on development of the genitalia, called the permanent, differentiating behavioral effects of steroids *organizational* effects. They envisioned gonadal steroids as organizing masculine neural pathways that were later activated by androgens. Hence, the adult, impermanent steroid effects were called *activational* effects. It is interesting that Phoenix et al. used an analogy to embryonic morphological development, as did Lorenz in the imprinting literature, to conceptualize the permanent effects of gonadal steroid effects on behavior.

Description of the SNB System

In recent years it has become apparent that neonatal steroid secretions produce permanent sex differences in behavior by creating sex differences in the central nervous system (CNS) (Arnold and Gorski, 1984). One dimorphic CNS region, the rat spinal nucleus of the bulbocaver-

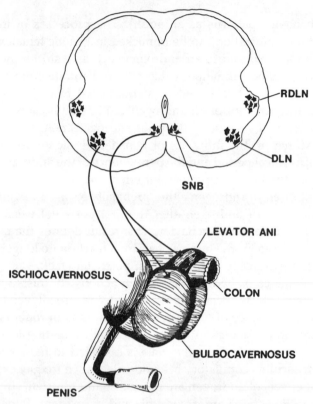

FIGURE 1. Drawing of major motor nuclei of the male rat fifth lumbar spinal cord. The spinal nucleus of the bulbocavernosus (SNB) innervates the levator ani and bulbocavernosus muscles which attach exclusively to the penis, as well as the anal sphincter (not shown). The dorsolateral nucleus (DLN) innervates the ischiocavernosus muscle. The SNB and DLN are both smaller in adult females, and the penile muscles are absent.

nosus (SNB), offers significant advantages for studying steroid regulation of neural development. The rat SNB consists of motoneurons that innervate sexually dimorphic striated muscles (the bulbocavernosus and levator ani, which attach exclusively to the penis in the male) and the nondimorphic anal sphincter muscle (Fig. 1) (Breedlove and Arnold, 1980; McKenna and Nadelhaft, 1984). The penile muscles play a role in penile erection and intromission, and in ejaculation (Hart and Melese-d'Hospital, 1983; Wallach and Hart, 1983; Sachs, 1982). Adult female rats lack both the bulbocavernosus and levator ani muscles, and the female SNB contains many fewer motoneurons, presumably only those that innervate the anal sphincter. Both the SNB motoneurons and their target muscles appear to be androgen targets in adulthood, since SNB motoneurons accumulate androgens selectively as shown in autoradiographic experiments, and the penile muscles contain androgen receptors (Breedlove and Arnold, 1980, 1983c; Dube et al., 1976). Moreover, both the muscles and the motoneurons respond to androgens in adulthood with growth (of soma volume, dendritic length, and muscle fiber size; Breedlove and Arnold, 1981; Kurz et al., 1986; Venable, 1966).

The permanent sex differences in the SNB are created by androgens during an early critical period. Breedlove and Arnold (1983a,b) exposed genetically female rats to testosterone propionate during one of three perinatal periods: in the last week of gestation (embryonic days 17–22, day 23 is day of birth), during the first five postnatal days, or during postnatal days 7–11. Females exposed during the first two of these periods possess more SNB neurons in adulthood than control females not exposed to androgens. The critical period is demonstrated because injections of androgens during the third period (postnatal days 7–11) fail to masculinize (i.e., increase) the number of SNB neurons (Fig. 2A). The beginning of the critical period has not been determined since we do not yet know when early embryonic injections of androgens would fail to masculinize the SNB system. However, the embryonic testes do not secrete androgens before embryonic day 14, and males do not secrete higher levels of androgens than females until day 18 of gestation (Picon, 1976; Weisz and Ward, 1980). These observations suggest that the endogenous period of androgen action begins around days 14–18 of gestation and ends around postnatal day 7 (with regard to androgenic effects on SNB neuron number).

Mechanism of Androgen Regulation: Explaining the Critical Period

How might androgens act to regulate SNB neuron number? In general, we know that spinal motoneurons proliferate by mitotic division of neuroblasts along the germinal epithelium lining the central canal. They then migrate into position in the ventral horn, sending out axons very early in development to contact their target muscles in the periphery. As a rule, more neurons contact the target muscles than are present in adulthood, and there is a period of massive motoneuronal death that brings neuron number into the adult range. Because the number of motoneurons surviving to adulthood depends on the availability of tar-

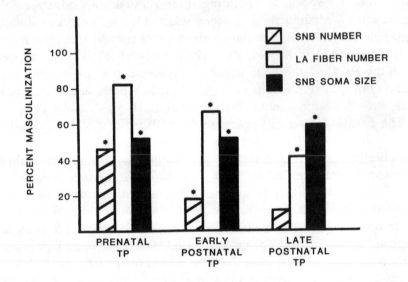

EFFECT OF TP ON SNB SYSTEM IN FEMALES

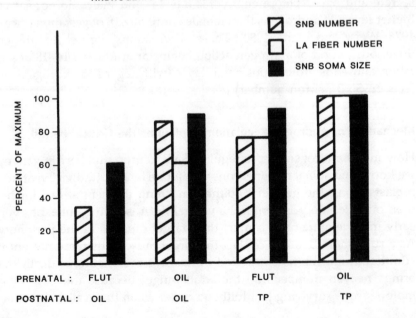

ANDROGENIC MANIPULATION OF SNB SYSTEM IN MALES

get musculature, it is thought that the overproduction and loss of motoneurons reflects a competitive process by which only some motoneurons survive and make stable synaptic contacts (Oppenheim, 1981). Producing an excess of motoneurons thus may ensure that all muscle fibers are properly innervated, both in amount and source of innervation.

This picture of early motoneuronal differentiation and growth suggests several major processes that androgens might modulate to regulate neuron number. Androgens might alter the mitotic activity of the germinal epithelium, producing more SNB motoneurons. Second, androgens might influence the migration of SNB motoneurons, directing them in greater numbers to the SNB region of the ventral horn, where local signals may determine their subsequent differentiation as SNB motoneurons. Third, androgens might influence the differentiation of SNB motoneurons. In the absence of androgens, fewer undifferentiated neuroblasts would differentiate into SNB motoneurons, but would become other types of neurons. This third possibility is similar to the second alternative in that an androgenic influence on migration of neurons may cause or be caused by an androgenic effect on differentiation. However, this need not be the case, and androgens could alter the process by which a neuroblast is specified as an SNB motoneuron without modi-

---

FIGURE 2. (A) Critical periods for masculinizing the SNB in female rats. Female rats were exposed to testosterone propionate (TP) for 6 days before birth (prenatal TP group), days 1–5 after birth (early postnatal TP group), or days 7–11 after birth (late postnatal TP group). In adulthood, three dependent variables were measured: number of SNB motoneurons, number of muscle fibers in the levator ani (LA), and size of SNB motoneuron somata. One hundred percent masculinization is equivalent to the control male value of each dependent variable, and 0% masculinization is equivalent to the control female value. The critical period for TP effects on SNB number seems to end around postnatal day 7, since TP had no significant masculinizing effect in the late postnatal group. * = significantly different from control females, $p \leq 0.05$. From Breedlove and Arnold (1983b) and Arnold (1984). (B) Critical periods for demasculinization of the SNB in males. Male rats were exposed prenatally (embryonic days 11–22) to the antiandrogen, flutamide, or to oil as a control. The prenatal oil group was thus exposed to endogenous androgen from their testes. Postnatally, they were castrated and given either testosterone propionate (TP) (days 1–11) or oil. The dependent variables were the same as in Figure 2A. One hundred percent indicates the most masculine values obtained in the group of males (extreme right) exposed both prenatally and postnatally to androgen. The group on the left, which was deprived of androgens both prenatally and postnatally, was completely feminine in number of SNB motoneurons. Exposure to androgens either just prenatally (second group from left) or just postnatally (third group from the left) caused nearly total masculinization of SNB number. From Breedlove and Arnold (1983a) and Arnold (1984).

fying the migration process. Fourth, androgens could prevent normally occurring death of SNB neurons. Note that this list suggests the possible androgenic regulation of both progressive processes (e.g., proliferation, migration) and of regressive processes (death).

Current evidence allows us to reject the first option, regulation of mitotic proliferation of SNB cells. As mentioned above, the time of androgenic regulation of SNB neuron number begins after embryonic day 14, the onset of androgen secretion, and probably not later than embryonic day 18, the earliest age at which fetal plasma levels of androgen are higher in males than in females (Picon, 1976; Weisz and Ward, 1980). This time of onset of androgen secretion rules out an effect on SNB motoneuronal proliferation since SNB cells undergo their last mitotic division before day 14 (Breedlove et al., 1983). Moreover, it has been shown that the SNB neurons present in masculinized females were "born" (i.e., underwent their last mitotic division) before the time of postnatal androgen injections that induced the masculinization (Breedlove, 1984). Clearly, androgens do not regulate SNB motoneuronal proliferation.

In contrast, there is strong evidence for the fourth alternative, androgenic prevention of SNB neuron death (Nordeen et al., 1985; Breedlove, 1984). Figure 3 shows the ontogenetic development of the number of motoneurons in the SNB region. On embryonic day 18, there are only about 50 motoneurons in this nucleus in each sex. This number rises rapidly over the next four days, so that on the day before birth (E22), males and androgenized females have nearly 300 motoneurons and untreated females have about 200 motoneurons. After this time, the number of motoneurons drops in all animals to adult levels by postnatal day 10 (males and androgenized females have about 210, females about 60). Throughout this period, cells are found in the SNB region whose pycnotic appearance is characteristic of dying cells. These degenerating cells are more prevalent in females, who lose more motoneurons, than in males or androgenized females, who lose fewer SNB motoneurons (Fig. 4). Moreover, the degenerating cells are more prevalent in all groups at those times (E22, P4) when SNB motoneurons are being lost than at P10 when SNB motoneuron number is stable. These results imply strongly that in males, androgens prevent normally occurring cell death, and that the sex difference in SNB motoneuron number is a result of more massive death in females.

At one level, this result explains the physiological events that close

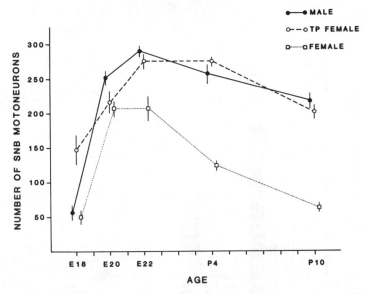

FIGURE 3. The ontogeny of the SNB nucleus. Graphs depict the number of SNB moto-neurons present at various ages in untreated males and females, and females injected daily with testosterone propionate (TP) from embryonic day 16 (E16) to birth, and on alternate days after birth. Except for day E18, there are no differences between masculinized females and males. The sex difference in SNB motoneuron number develops because motoneurons are lost from the SNB predominantly in the days just before and after birth. This loss is more pronounced in females and is prevented by androgens. From Nordeen et al. (1985).

one critical period in the development of the SNB. The critical period for androgenic masculinization of SNB neuron number ends around postnatal day 7, after which androgen injections into females fail to influence SNB neuron number. The reason for this failure is that the period of motoneuron death is virtually completed at this point. Androgens no longer save SNB motoneurons from dying because after day 7 most SNB neurons in females are already dead. Thus, these data provide us with answers to some of the questions posed above. They suggest that the selectivity (question 2) in the androgenic effect is achieved by a prevention of a degenerative event, and that the end of the critical period (question 3) occurs because of the time course of naturally occurring motoneuronal death. Because the time course of motoneuron death is probably independent of the action of androgens (i.e., androgens do not trigger the death process or shift it in time), we perceive the critical period as a coincidence of two *independent* processes, andro-

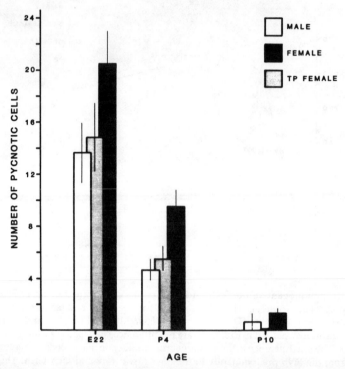

FIGURE 4. Pycnotic, degenerating cells occur in the greatest number from embryonic (E) day 22 to postnatal (P) day 4 at the time of the greatest loss of SNB motoneurons (Fig. 3). The incidence of degenerating cells is greater in females and is prevented by treatment with testosterone propionate (TP). The combined data of Figures 3 and 4 implicate SNB motoneuronal death as a major factor in the ontogeny of sex differences in this nucleus. From Nordeen et al. (1985).

gen secretion and motoneuronal death. The timing of the second process limits the timing of the effects of the first.

Although this explanation represents progress in our understanding of the critical period, a great deal has yet to be learned. We do not know for example, why the androgenic effect is permanent (question 1). Once having been saved from dying during the critical period, why do SNB neurons not require continuous exposure to androgens to survive to adulthood? Presumably, in the process of being saved, the SNB neurons underwent a permanent change (e.g., establishing stable synaptic connections) that maintains them in the absence of androgen. We do not yet understand the forces that maintain SNB neurons, which might involve androgenic regulation of progressive events, for example, muscle maturation and differentiation, synaptogenesis, and so on.

To approach the remaining questions in this system, we need to evaluate the cellular mechanisms by which androgens prevent SNB death. For example, since androgens also save the SNB target muscles from atrophying in neonatal females (Cihak et al., 1970; Breedlove and Arnold, 1983a,b), it is possible that androgens act directly on the target muscles only, thus providing the SNB motoneurons with a target necessary for their survival. Alternatively, androgens might act on SNB motoneurons to enhance their growth or health. Androgens might promote SNB axon outgrowth or stabilization of the neuromuscular junction, which would in turn promote motoneuron and muscle survival. At present, some evidence argues for a muscular site of action. SNB motoneurons initially send their axons to the periphery in approximately equal numbers in males and females, suggesting that the female SNB cells do not die because they fail to reach their targets (Sengelaub and Arnold, 1986). Moreover, Breedlove (1985b) has made genetic mosaic mice who have clusters of androgen-sensitive and androgen-insensitive tissue, and has demonstrated that SNB neurons need not be androgen sensitive in order to be saved from dying by androgens. This result implies that androgens act on a target other than the SNB to regulate the number of SNB neurons. However, androgens appear to influence survival of spinal ventral horn neurons in culture, even if no muscle cells are present, implying a direct action on the neurons (Hauser and Toran-Allerand, 1985). Thus, although present evidence suggests that the SNB neurons may not be targets for androgenic regulation of cell survival, the data do not uniformly imply this, and more work is needed.

Is cell death the only process influenced by androgens to regulate SNB neuron number? This question is not fully answered yet, but there are tantalizing bits of evidence that SNB migration may be influenced by androgens. In Figure 3, we see that between embryonic days 18 and 22, there is a remarkable increase in the number of motoneurons in the SNB region. This implies either that SNB motoneurons are migrating into the SNB at this stage, or that neurons in the SNB region are differentiating into recognizable motoneurons. The former explanation is more likely (Sengelaub and Arnold, 1986; Nordeen et al., 1985). Because androgen treatment of females starting on E16 increases the number of motoneurons in the SNB region on E18, it is conceivable that androgens facilitate this migration into the SNB. Alternatively, the elevated number on E18 could be explained if androgens merely prevent cell death at this early age. However, androgenic regulation of migration has not been disproved, and may play some role in determining the number of SNB neurons (see Breedlove, 1985a).

CRITICAL PERIODS FOR SEXUAL DIFFERENTIATION AND SONG LEARNING IN ZEBRA FINCHES

Introduction to Song Learning and Sexual Differentiation of the Vocal Control System

In song birds there are interesting critical periods, both for sexual differentiation of brain regions controlling song, and for the imprintinglike process of song learning. These two processes may well be related because they involve similar cellular events. Moreover, these events may be similar to those described above for sexual differentiation of the SNB system in the rat.

Song is a reproductive behavior in passerine birds, used for courtship and territorial display. In adult zebra finches, male song is stimulated by androgens and estrogens, and castration decreases the propensity to sing (Pröve, 1974; Arnold, 1975b; Harding et al., 1983). In many species, females sing less than males, and zebra finch females do not sing at all (Arnold, 1974).

Studies on the neural and hormonal bases of song have been facilitated by the discovery of brain regions controlling song (Nottebohm et al., 1976) and we now have an extensive map of the neural circuits that control song (Fig. 5) (Nottebohm, 1980b; Arnold, 1982; Bottjer and Arnold, 1986). The main descending path of motor vocal control regions is thought to include HVc to RA to the hypoglossal motor neurons innervating the vocal organ, the syrinx. Lesions of HVc and RA disrupt song in adult canaries (Nottebohm et al., 1976; Nottebohm, 1980b). Area X and MAN are also thought to be importantly involved in song control because they are connected to HVc or RA by heavy monosynaptic connections. Unilateral lesions of area X in adulthood have not been reported to have major effects on singing behavior (Nottebohm et al., 1976) whereas bilateral MAN lesions prevent song learning in juvenile zebra finches (Bottjer et al., 1984, see below). Many of the vocal control regions contain neurons that accumulate androgens (Fig. 5) (Arnold et al., 1976;

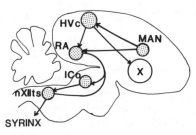

FIGURE 5. Schematic drawing showing connections of some major constituents of the vocal control neural network in the songbird brain. Shading represents presence of androgen accumulating cells in adult male zebra finches. HVc, hyperstriatum ventrale pars caudale; RA, robust nucleus of the archistriatum; MAN, magnocellular nucleus of the anterior neostriatum; X, area X of the lobus parlofactorius; ICo, nucleus intercollicularis; nXIIts, tracheosyringeal portion of the hypoglossal nucleus. (Nottebohm, 1980b; Arnold, 1982; Nottebohm et al. 1982).

Arnold, 1980a; Harding et al., 1984), suggesting that they are sites of action in androgenic control of song in adulthood.

The sex difference in song behavior reflects large sex differences in the size of vocal control regions in the brain (Nottebohm and Arnold, 1976). Compared with females, male zebra finches have more neurons in RA, larger neurons in HVc, RA, and MAN, and more androgen accumulating neurons in HVc and MAN (Gurney, 1981; Arnold and Saltiel, 1979; Arnold, 1980a). Other sex differences in neuron number, size, and androgen sensitivity have either been observed casually or are suspected to occur, but have not yet been reported (Table 1). These pervasive sex differences in brain and behavior are controlled by neonatal steroid secretions (Gurney and Konishi, 1980; Gurney, 1981, 1982; K. Nordeen et al., 1986). On days 3–10 after hatching, male zebra finches have much higher plasma levels of estradiol than do females, whereas androgen levels are equivalent in the two sexes during this period (Hutchison et al., 1984). Estrogen treatment of female zebra finches on the day of hatching causes profound masculinization of brain regions controlling song. The estrogen increases the number and size of neurons, and the number of androgen-accumulating neurons (Gurney, 1981; K. Nordeen et al., 1986) (Fig. 6).

There is likely to be a critical period for estrogenic masculinization of the song system, since adult females are not masculinized by exposure to estrogen (Gurney and Konishi, 1980; Gurney, 1981, 1982). However, the critical period for estrogen action has not been carefully defined. We do know that estrogen treatment of female zebra finches as late as day

TABLE 1. Sexual Dimorphisms in Zebra Finch Telencephalic Song System[a]

Attribute	Nucleus			
	HVc	RA	MAN[b]	Area X
Volume of nucleus	1, 2, 4, 5, 6, 7	1, 2, 4, 5, 6, 7, 8	—	1, 4, 6
Cell number	10	5, 8	—	S[c]
Cell size	3	3, 5, 8	3, 9	—
Dendritic size	—	5	—	—
Androgen accumulation	3,9	S[c] (3)	3, 9	—

[a] 1—Nottebohm and Arnold (1976); 2—Gurney and Konishi (1980); 3—Arnold and Saltiel (1979); 4—Arnold (1980a); 4—Arnold (1980b); 5—Gurney (1981); 6—Gurney (1982); 7—Bottjer et al. (1985); 8—Konishi and Akutagawa (1985); 9—K. Nordeen et al. (1986); 10—Arnold (unpublished).

[b] The borders of MAN are difficult to recognize in females, so that accurate measurements of volume and cell number have not yet been made.

[c] S = Sex difference suspected but not yet reported.

40 after hatching has some effect, so the critical period is not completely closed by that age (Gurney, 1980). However, females given estrogen on day 40 are less masculinized than those given estrogen on day 1. The decline in the influence of estrogen can be attributed either to a loss of sensitivity to estrogen or to the lower dose per body weight of estrogen given to day 40 females in the Gurney (1980) study. Although females

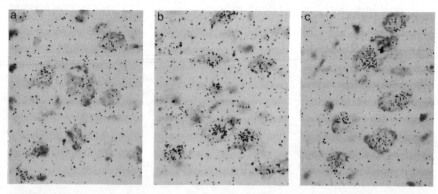

FIGURE 6. Photomicrographs demonstrating cellular accumulation of label in MAN after injection of tritiated dihydrotestosterone in adult zebra finches. There is very little accumulation by cells in control females treated at hatching with cholesterol (a). Adult males (c) and adult E2 females (treated at hatching with estradiol) (b) have many more androgen-labeled cells. Thus estrogens regulate the number of MAN neurons that are androgen targets. From K. Nordeen et al. (1986).

may remain responsive to estrogens until day 40, the period of masculinization of males may be much earlier. Because male zebra finches
secrete very high levels of estradiol on days 3–10 after hatching, this
time may be the main period of masculinization (Hutchison et al., 1984).

The Process of Song Learning

A male zebra finch learns his reproductive song by copying his father's
song early in life (Immelmann, 1969). Young males are fed by their
parents until about 30 days of age, at which point their initial song
vocalizations are produced. At first, the male utters only subsong, a
quiet and highly variable song. By around 50–60 days after hatching,
male song has progressed so that most features of the father's song are
clearly recognizable, and the notes are only slightly variable in form.
This relatively minor instability is cleared up by around 75–90 days of
age, when full, stereotyped song has crystallized and males are fully
reproductive.

Immelmann (1969) originally established that there is a critical period
for song learning in zebra finches. Isolation of males from their fathers
at 80 days of age has no effect on song learning, indicating that learning
is complete at this age. However, if males are isolated at earlier ages,
there are deficits in copying the father's song. Males isolated beginning
at 25–40 days of age form only a poor copy, but their adult songs do
contain some song elements heard from the father before isolation. This
finding indicates that song learning begins before 25–40 days of age, as
does the observation that the song of a 40-day old male already reflects
the influence of the father's song (Arnold, 1975a). Eales (1985) has recently suggested that songs heard between 35 and 65 days of age have
a greater effect than those heard at other times.

The process of song learning probably involves at least two learning
processes, often called "sensory" and "sensorimotor" learning (Marler,
1984; Konishi and Nottebohm, 1969). The young male must memorize
the acoustic features of the song to be copied in a period of sensory
learning. Then, the male must learn the neuromuscular coordinations
that allow him to produce an accurate copy of the song he previously
heard (sensorimotor learning). The separation of sensory and sensorimotor learning was originally made because motor rehearsal of song
typically lags behind the perceptual process of song memorization. For
example in white-crowned sparrows, males exposed to conspecific song

in the first 50 days of life will eventually copy the song accurately, even though rehearsal of learned songs does not begin until weeks or months later (Marler, 1970; see also Marler and Peters, 1982b). The period of motor development is known to be a learning process since it requires the ability to hear one's own song. Thus, Konishi (1965; see also Nottebohm, 1968; Marler and Waser, 1977; Price, 1979) demonstrated that white-crowned sparrows deafened at three months, after the period of sensory learning, were not able to complete motor development of song. Presumably the male needs to hear his own vocalizations to assess the accuracy with which they match the song model.

In zebra finches, the periods of sensory and sensorimotor learning appear to overlap. Zebra finches may be able to acquire song elements heard as late as 60 days, although most sensory learning has probably ended earlier (Immelmann, 1969; Böhner, 1983). Motor development progresses most rapidly between 30 and 60 days, but some further motor refinement occurs until about 90 days of age (Arnold, 1975a).

Neural Correlates of Song Learning in Males

In asking what is known about events responsible for critical periods in sexual differentiation and learning of song, we direct our attention to the four main telencephalic nuclei for vocal control: HVc, RA, MAN, and area X. As mentioned above, all of these nuclei are strikingly sexually dimorphic and their sexual differentiation is controlled by estrogens. HVc and RA have been suggested as sites for neural changes responsible for song learning in canaries, since growth of neurons in these regions correlates with some song learning phenomena (Nottebohm, 1984; Bottjer and Arnold, 1984; Arnold, 1985). For example, androgens induce dendritic growth of neurons in RA in adult female canaries, allowing production of new synapses, which correlates with the female's production of new song elements (DeVoogd and Nottebohm, 1981; DeVoogd et al., 1985). At the same time, both HVc and RA grow substantially in volume (Nottebohm, 1980a). Deafening, which prevents some aspects of song learning, partially blocks this effect of androgens on HVc volume (Bottjer et al., 1986b). In male canaries, there are seasonal changes in the volumes of HVc and RA that correlate with seasonal changes in the song repertoire (Nottebohm, 1981). Finally, HVc and RA volume both correlate positively with the size of song repertoires in canaries, implying a structural basis for learned individual differences in song (Nottebohm et al., 1981). These correlations between vocal plasticity and morphology of HVc and RA suggest that to understand critical periods for song learning in zebra finches, one must pay close attention to developmental changes in HVc and RA.

MAN also deserves close scrutiny. Lateral MAN, which projects to RA, seems to play a critical role in song learning (Bottjer et al., 1984). Lesions of lateral MAN in adult zebra finches have no effect on song behavior, suggesting that MAN is not on the main descending path for motor control of crystallized song. However, if lesions are placed in lateral MAN during song ontogeny (between 35 and 50 days of age), song development is profoundly disrupted. This result suggests that MAN neurons are required for song learning, but not maintenance of song, and that their role changes during the learning process. It is important to ask if the MAN lesions prevent sensory or sensorimotor learning. This question has not yet been answered, but it is unlikely that the MAN lesion is tantamount to deafening. Because deafening male zebra finches after 80 days of age can disrupt song development (Price, 1979), but MAN lesions after that age do not, it is unlikely that MAN represents a simple conduit for auditory information en route to the song system.

As an initial strategy in the analysis of the neural basis of critical periods in the zebra finch, it is important to describe the normal ontogenetic changes occurring in the song system at the time of song learning and sexual differentiation. This allows us to formulate specific hypotheses regarding which neural and hormonal events may be responsible for the timing of critical periods. In what follows, we will summarize these neural changes and then entertain specific hypotheses.

Changes in Neuronal Size and Number During Song Learning in Males

Bottjer et al. (1985, 1986a) compared development of the telencephalic vocal control areas at 12, 25, and 53 days of age in male and female zebra finches. These ages were chosen to represent (a) a very early stage in the song development process, when many nuclei are clearly discernable (day 12); (b) an age just before the onset of subsong, when MAN lesions are effective in preventing learning (day 25); and (c) an age when much song development has been completed, and MAN lesions no longer consistently disrupt song (53 days). During this period, HVc, RA, and area X all grow in volume in male zebra finches, much

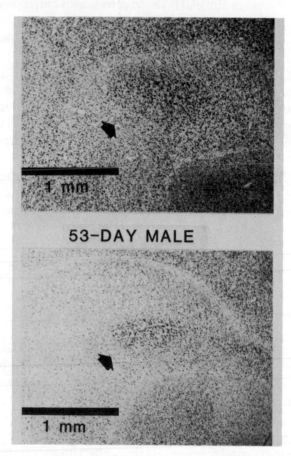

FIGURE 7. Photomicrographs of Nissl-stained sections of lateral MAN (arrows) in male zebra finches before (25-day male) and after (53-day male) the massive loss of neurons from this nucleus. The decrease in size of the nucleus is evident. From Bottjer et al. (1985).

more than nonsong nuclei. The volumetric increase in HVc leads that in RA somewhat, suggesting that the growth in RA may depend on growth in HVc, since HVc neurons innervate RA. The development of MAN contrasts markedly with the general growth of HVc, RA, and area X. Between 25 and 53 days of age, the volume of MAN drops to less than half of its 25-day volume (Figs. 7 and 8). This decline in MAN comes at a time when lesions of MAN decline in their impact on song,

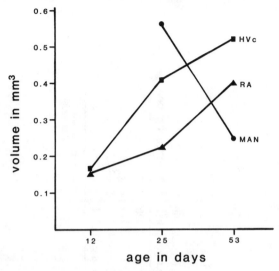

FIGURE 8. Graphic representations of development of the volumes of HVc, RA, and MAN in juvenile male zebra finches. From Bottjer et al. (1985).

suggesting that degenerative changes in MAN (e.g., cell death, see below) may play a role in the processes underlying song learning.

To evaluate what cellular processes might account for the volumetric changes in HVc, RA, and MAN during song learning, Bottjer et al. (1985, 1986a) measured neuron size, number, and density in these nuclei at 25 and 53 days of age (Table 2). Some of the same parameters were also measured by Konishi and Akutagawa (1985) and Herrmann and Bischoff (1986). In RA, neurons grow in size and become more widely spaced, but there is no change in the number of neurons. The increase in spacing of RA neurons suggests an increase in the volume of the neuropil, perhaps representing significant changes in the size or number of RA neuronal dendrites or their afferents. In HVc, neuron size does not change as this nucleus grows, but there is a large increase in the number and density of HVc neurons. In MAN, neuron size and spacing do not change, but half of the neurons are lost between 25 and 53 days of age. Thus, three striking changes occur in the song system at the time of most rapid motor acquisition of song: HVc increases in neuron number, MAN decreases in neuron number, and RA neurons grow in size and spacing.

Because we have come to expect neuronal death as a natural part of

TABLE 2. Values ± Standard Deviations for 25- and 53-day Male Zebra Finches

		Volume of Nucleus (mm^3)	Area of Neuronal Somata (μm^2)	Neuron Density (per $10^3 \mu m^3$)	Neuron Number
RA	25 days	0.228 ± 0.030	112.0 ± 24.3	0.129 ± 0.020	29,081 ± 2,472
	53 days	0.402 ± 0.101	157.0 ± 12.7	0.066 ± 0.010	26,085 ± 4,535
HVc	25 days	0.408 ± 0.085	98.6 ± 10.7	0.123 ± 0.011	49,698 ± 9,272
	53 days	0.521 ± 0.124	98.2 ± 7.3	0.148 ± 0.014	76,485 ± 17,928

From Bottjer et al. (1986a).

the genesis of neuronal circuits, this would seem to be the most likely explanation for the loss of MAN neurons. However, it is also conceivable that MAN neurons migrate out of this nucleus during this period, and further work is required to discriminate between these alternatives. In the case of HVc, the recent discovery of neurogenesis in adult canary HVc (Goldman and Nottebohm, 1983; Nottebohm, 1984) leads to the expectation that the increase in the number of HVc neurons in juvenile zebra finches represents a similar addition of newly generated neurons. Again, however, there is an alternative. HVc neurons could have been generated earlier, but only migrate into place between 25 and 53 days of age. No matter what the cellular mechanisms responsible for the change in neuron numbers, these data imply a major structuring of telencephalic circuitry at the time of song learning, including dramatic addition and deletion of functional neuronal units. Both regressive and progressive events seem to be occurring.

Konishi and Akutagawa (1985) have focused attention on the ontogeny of connections between HVc and RA during this juvenile period. They reported that the volume of male RA increases about threefold between 20 and 45 days of age. Furthermore, using tritiated amino acids injected into HVc, they studied the development of HVc-to-RA connections with autoradiography. At 15 days of age, they found little transport of labeled protein into male RA, implying few HVc-to-RA axons. The labeling increased in 25-day-old males, and RA was "completely labeled" by day 35. These results suggest that HVc innervation of RA is being established around days 25–35. This new ingrowth of HVc axons might come about because of the addition of new HVc neurons at this time, suggesting that some new HVc neurons project to RA. Moreover, the increase in spacing of RA neurons between days 25 and 53 (Konishi and Akutagawa, 1985; Bottjer et al., 1986a) may reflect in part this addition of HVc axons in RA. Most importantly, the implied ingrowth of HVc axons suggests that synaptic connections may begin to be organized around days 25–35, as the young male begins subsong.

The reciprocal addition of neurons to HVc, and loss of neurons from lateral MAN during the subsong period, suggest that these two processes may be causally related. Neurons in both lateral MAN and HVc project to RA, and indeed may compete with each other for synaptic contacts in RA. The addition of HVc-to-RA neurons might trigger a loss of MAN-to-RA neurons if the HVc neurons successfully compete against MAN neurons, depriving them of a target necessary for their survival.

Alternatively, MAN neurons may be lost for other reasons, but the decline in MAN neurons might leave open synaptic sites for newly generated HVc neurons, allowing them to survive.

Hormonal Correlates of Zebra Finch Development

Because androgens and estrogens are so strongly implicated in the development and maintenance of the song control system and vocal behavior in zebra finches (Pröve, 1974, 1978; Arnold, 1975a,b; Gurney and Konishi, 1980; Gurney, 1981, 1982; Harding et al., 1983, 1984), it is important to compare the cellular and behavioral changes described above with the ontogenetic pattern of steroid hormone secretion in zebra finches. Pröve (1983) has reported detailed measurements of plasma levels of testosterone, dihydrotestosterone, progesterone, and estradiol in male zebra finches between the ages of 8 and 75 days after hatching (Fig. 9). The levels of testosterone show three peaks, at 18–21 days, 34–37 days, and after 74 days. Dihydrotestosterone levels are relatively constant. Progesterone is secreted at higher levels than testosterone, and peaks occur at about 14–25 days and 70–73 days. Estradiol titers are relatively high after age 39 days, with occasionally lower values recorded at around 50, 60, and 70 days of age. These data are intriguing for several reasons. The high levels of progesterone suggest the need for further work on this previously ignored hormone in the song system. The peaks in testosterone secretion around 20 days of age probably precede and may trigger the first subsong around 25–30 days of age. The second peak of testosterone around 35 days of age corresponds to the time at which Konishi and Akutagawa (1985) find HVc fibers entering RA, so that testosterone may play some role in terminating or regulating that event. Finally, the 35-day peak of testosterone, together with the general increase in estradiol around 38 days of age could play some role in the motor acquisition of song, which is in full sway at that time, and may be responsible for some of the cellular events described above.

As part of the interpretation of the ability of steroids to influence neural development in the song system, it is important to measure the sensitivity of vocal control neurons to steroids. One recent study (Bottjer, 1987) reports that there is a lower percentage of androgen-accumulating cells in both MAN and HVc of male zebra finches at 25 than at 60 days of age. Because there is a loss of neurons from MAN during this period (Bottjer et al., 1985), the increase in the proportion

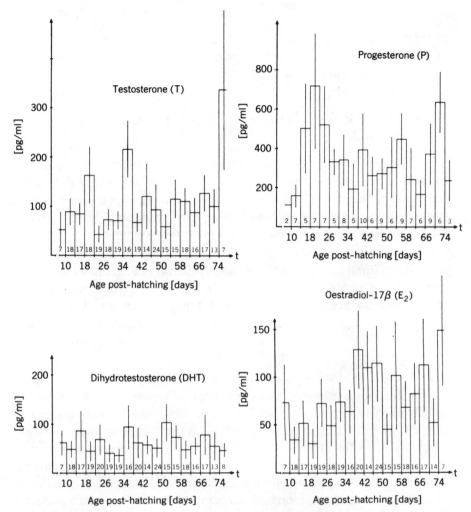

FIGURE 9. Ontogeny of plasma titers of four sex steroids in male zebra finches (means ± standard errors). Numbers under each column are sample sizes. From Pröve (1983).

of hormone-concentrating cells may indicate that the ability of these cells to accumulate androgen is somehow related to their selective retention in the nucleus. There is also an increase in the degree of labeling in individual cells in MAN. In HVc, the increase in the proportion of labeled cells coupled with the increase in neuronal number seen between 25 and 53 days (Bottjer et al., 1986a) suggests that there is a substantial increment in the absolute number of hormone-concentrating cells. These changes in accumulation may reflect ontogenetic alterations in the ability of MAN and HVc cells to react to androgen that are importantly related to neural development and to song learning.

Possible Causal Relations between Neural and Hormonal Events and Song Learning

These descriptions of cellular and hormonal events during subsong lead to specific hypotheses regarding the cellular events of the critical period for motor acquisition of song. Most of the following hypotheses are not mutually exclusive, and each addresses a different facet of the correlations between neural and behavioral song development.

1. The natural loss of MAN neurons by 53 days of age may close the sensorimotor phase, crystallize song, and allow no further changes in motor performance. This hypothesis is supported by the observation that MAN is large and has more neurons at day 25, when subsong is beginning (Bottjer et al., 1985), and because MAN neurons are critical for motor acquisition of song (Bottjer et al., 1984).

2. The loss of MAN neurons may be an important part of the selection of specific neural elements, but in contrast to hypothesis 1, MAN neuronal loss may not prevent further motor acquisition. If one imagines that different MAN neurons are involved in production of specific song elements, then the attrition in MAN neurons may reflect a selection of those circuit elements that control the song elements that are retained into adulthood. In swamp sparrows, Marler and Peters (1981, 1982a,b) found that the number of song elements in plastic song was much greater than the number retained in adult song. Assuming that a similar process occurs in zebra finches, this behavioral overproduction and then selective loss may parallel the apparent neuronal overproduction and loss in MAN. Could the two be causally related? Are the MAN neurons lost during ontogeny those that controlled the deleted song elements? In order for this hypothesis to be compatible with extant observations, one would have to assume that adult MAN neurons are important in the development, but not maintenance of song elements retained into adulthood, since lesions of adult MAN do not disrupt adult song elements.

3. The addition and retention of new androgen-accumulating (and

possibly other) neurons in HVc may be a necessary component of the processes of sensorimotor or sensory learning. Clearly, the massive increase in number of HVc neurons should allow for considerable change in vocal behavior. HVc occupies an important position in the neural circuit for song control. It is potentially only three synapses away from the vocal muscles of the syrinx, which affords it strong control of motor performance. It also receives patterned auditory input crucial for song learning (Kelley and Nottebohm, 1979; Katz and Gurney, 1981; Margoliash, 1983; Paton and Nottebohm, 1984; Margoliash and Konishi, 1985; Williams, 1985; Williams and Nottebohm, 1985). Size of HVc in canaries is more closely correlated with vocal complexity than is the size of RA or other brain regions (Nottebohm et al., 1981). Finally, growth of canary HVc in some contexts depends in part on proper auditory input, which is required for song learning (Bottjer et al., 1986b). Thus, the addition of HVc elements in juvenile male zebra finches could enhance auditory–motor integration required for song learning.

4. The neural changes in RA (growth of size of neurons, presumed growth of axonal and dendritic elements in RA neuropil) may represent important events underlying sensory or sensorimotor learning. This hypothesis is supported by the timing of these events during song learning (Bottjer et al., 1985; Konishi and Akutagawa, 1985). Furthermore, RA may receive auditory information (Kelley and Nottebohm, 1979), and it projects directly to the syringeal motoneurons (Nottebohm et al., 1976).

5. Peaks of testosterone and progesterone around 18 days of age may trigger neural events responsible for the onset of sensory or sensorimotor learning. Peaks of testosterone and estradiol around days 35–40 may regulate or begin the termination of neural events required for either of the two phases of song learning. There has been interest in the possible role of androgens in song learning for some years (Nottebohm, 1969; Arnold, 1975a; Nottebohm, 1981), and androgens and estrogens stimulate neural growth in zebra finches and canaries (Gurney and Konishi, 1980; Gurney, 1981, 1982; DeVoogd and Nottebohm, 1981). However, there is so far no unequivocal evidence that androgens are required for learning, and more needs to be learned about the roles of the various steroids in regulating song development (Pröve, 1983).

6. The maturation of androgen accumulation in male MAN may be required for song learning. The adult pattern of accumulation may limit vocal plasticity. These hypotheses reflect the realization that the com-

position of nucleus MAN changes in the course of song development. Although many neurons are lost between ages 25 and adulthood, the percentage of androgen-accumulating cells increases during this period (Bottjer, 1987). Perhaps this shift in the types of neural elements in MAN represents a process that regulates or is regulated by the song learning process.

Although further research is needed to evaluate these various hypotheses, there are already some relevant findings. Some stem from research on sexual differentiation of the song system in zebra finches, so we will turn first to a consideration of the neural changes leading to sex differences in the song system.

Ontogeny of Sex Differences in Neuronal Size, Density, Number, and Hormonal Sensitivity

In females, the volumes of HVc and RA increase between day 12 and 25, then decrease between day 25 and 53 (Bottjer et al., 1985). Bottjer et al. (1985) found sex differences in the volumes of HVc and RA as early as day 12, implying that the sex differences in plasma titers of estradiol, which occur as early as day 3 (Hutchison et al., 1984), have already begun to influence ontogeny of these nuclei. In contrast, Konishi and Akutagawa (1985) found that the volume of RA was not significantly sexually dimorphic until about day 25, after which the male RA increased greatly, and female RA decreased in volume. The decrease in volume was accompanied by a decrease in female RA neuron size, contrasting with an increase in male RA neuron size after day 30. Konishi and Akutagawa (1985) reported that between day 15 and adulthood, the number of RA neurons in males does not change, whereas female RA neuron number declines to about 30% of the day 15 value. This decline may represent either neuronal death or migration of neurons out of RA (more likely the former). Moreover, the ingrowth of HVc axons into RA, found in males between 15 and 35 days, fails to occur in females. These results imply that sex differences in RA neuron number may result because of a failure of HVc neurons to innervate the female RA. Thus, incoming HVc axons could prevent RA neuronal death (or migration) in males, and the failure of HVc ingrowth could allow the striking reduction in RA neuron numbers in females. Alternatively, the loss of RA neurons in females could be controlled via other mechanisms, but lead in turn to the failure of HVc axonal ingrowth.

Adult male and female zebra finches differ in the hormonal sensitivity of brain regions controlling song. A greater percentage of cells in male HVc and MAN accumulate androgens than homologous female cells (Arnold and Saltiel, 1979; Arnold, 1980a). Estradiol regulates the development of this masculine attribute, since adult females previously implanted with estradiol at birth have a greater percentage of androgen-labeled cells in HVc and MAN than do control females implanted with cholesterol (Fig. 6) (K. Nordeen et al., 1986). This estradiol treatment could increase the relative numbers of androgen target cells in one of two general ways. On the one hand, the ability of individual cells to accumulate androgen may be fixed, and estradiol might regulate processes of cellular addition or attrition to increase the proportion of androgen-labeled cells (the "cell number" hypothesis). Alternatively, the ability of individual cells to accumulate androgen might be labile, and estrogen might turn nonaccumulating cells into accumulators (the "cell conversion" hypothesis).

Mechanisms of Estradiol Action

To discriminate among these alternatives, E. Nordeen et al. (1987) studied changes in cell numbers induced by estradiol treatment of female zebra finches. The first finding was that estradiol implants into female zebra finches on day 20 were capable of masculinizing volume, cell size, and androgen accumulation in HVc and MAN. Because the brain of the day 20 female is nearly adult in size, yet still capable of masculinization, it is possible to measure cell numbers and androgen accumulation on day 20, before masculinization, and in adult females previously masculinized by estradiol implants on day 20 ("adult E2 females") (Table 3). In MAN, 20-day-old females have many more cells than do adult E2 females, indicating that many MAN cells are lost (through cell death or outward migration) between day 20 and adulthood. Although the percentage of androgen-labeled cells is lower on day 20 than in an adult E2 female, the number of androgen-labeled cells is about the same. In adult females not masculinized by estradiol, there are very few androgen-labeled cells, although their absolute number has not been assessed because of the difficulty in detecting the boundaries of MAN in such females. Thus, estradiol's effect is to maintain the number of androgen

TABLE 3. Developmental and E2-induced Changes in MAN and HVc Cell Number, Size, and Androgen Accumulation[a,b,c]

		20-day-old Female[d]	Adult Ch Female[e]	Adult E2 Female[f]
MAN	Volume (μm^3)	0.611 ± 0.046	NM	0.187 ± 0.029
	Cell size (μm^2)	170 ± 10	83 ± 4	93 ± 11
	Cell density	0.081 ± 0.003	NM	0.101 ± 0.008
	Total cells	50,301 ± 4,261	NM	18,499 ± 2,272
	Percent labeled	19 ± 2	17 ± 5	40 ± 7
	Total labeled	9,656 ± 1,091	NM	7,410 ± 1,487
HVc	Volume (μm^3)	0.174 ± 0.024	0.110 ± 0.013	0.271 ± 0.054
	Cell size (μm^2)	91 ± 8	53 ± 4	68 ± 2
	Cell density	0.163 ± 0.006	0.262 ± 0.013	0.187 ± 0.024
	Total cells	28,077 ± 4,326	29,036 ± 4,461	42,598 ± 5,170
	Percent labeled	21 ± 5	11 ± 2	45 ± 6
	Total labeled	6,217 ± 2,239	3,285 ± 1,899	20,099 ± 985

[a] Data shown (mean ± SEM) are the total volume (L + R), cell size and density (cells/1000 μm^3), estimated total cell number, percentage labeled following ^{3}H-DHT injection, and estimated total number of androgen target cells in MAN of 20-day-old females, adult control females, and adult females previously masculinized with E2 treatment at day 20 (E2 females).
[b] NM = not measurable because of ambiguity of MAN borders.
[c] From E. Nordeen et al. (1987).
[d] Untreated.
[e] Treated at hatching with cholesterol (Ch) serving as unmasculinized controls.
[f] Masculinized with an estradiol implant on day 20.

target cells in the face of massive cell loss from MAN. Estradiol might accomplish this either by selectively preventing the loss (death or outward migration) of androgen target cells (cell number hypothesis) or by facilitating androgen accumulation in individual cells to offset the loss of androgen target cells during the crash in MAN cell number (cell conversion). If the former mechanism is correct, then estrogenic regulation of the number of androgen target cells is closely comparable to androgenic regulation of neuron number, which we have documented in the SNB system (Nordeen et al., 1985).

The pattern of overall cell loss in MAN stands in diametric contrast to an overall addition of cells in HVc in response to estradiol (Table 3). Estradiol increases both the total number of cells and the number of androgen-accumulating cells in HVc between day 20 and adulthood. Because neurogenesis occurs in HVc even in adulthood in canaries and zebra finches (Goldman and Nottebohm, 1983; Nottebohm, 1984), it is likely that the neurons added to HVc in response to estradiol are newly generated. Thus estradiol may stimulate neurogenesis, or enhance the survival of newly generated cells. In doing so, estradiol seems to be selective, enhancing the proliferation or survival of androgen target cells over other HVc cells, thus increasing the proportion of HVc neurons that are accumulators (cell number hypothesis). Alternatively, estradiol may not be selective in regulating the numbers of specific cell types, but may instead increase the proportion of androgen target cells by conversion of nonaccumulators into accumulators.

Although we have considered a variety of cellular mechanisms by which estradiol enriches HVc and MAN with androgen target cells, it is worth noting that a single mechanism, estrogen prevention of cell death, could account for all changes observed. Because the decline in MAN cell numbers seems most likely to represent cell death, estrogen may selectively save androgen target cells in MAN from dying. In HVc, steroid regulation of neuronal proliferation in canaries has not been observed, although glial proliferation or survival is sensitive to androgens (Goldman and Nottebohm, 1983). Estradiol might prevent the death of HVc cells normally generated in zebra finches during juvenile life. Although more work is needed to establish a common mechanism for estradiol's diverse effects, this possibility is esthetically pleasing, especially considering the similar steroid prevention of cell death in the ontogeny of the SNB (Nordeen et al., 1985).

Evaluation of Hypotheses Linking Morphological Changes and Learning

The cellular attrition in MAN and cellular addition in HVc in masculinized females is comparable to that seen in males (Bottjer et al., 1986a). Presumably, these changes may be induced by the male's own estradiol, which is secreted during days 3–10 of life and thereafter (Hutchison et al., 1984; Pröve, 1983). This comparison of males and masculinized females allows a test of some of the hypotheses listed above concerning the causal relation among neural and hormonal events, and song development. Female zebra finches masculinized by estradiol sing very little, if at all (Gurney, 1982; Pohl-Apel and Sossinka, 1984), even though their song system is greatly masculinized. Such females begin to sing if given androgens in adulthood, and their song becomes gradually more complex, suggesting a process similar to song development in males

(Gurney, 1982; Pohl-Apel and Sossinka, 1984). This androgen treatment also induces growth in the song system, increasing the volume of HVc by about 40% (Gurney and Konishi, 1980; Gurney, 1982). These observations allow us to return to the specific hypotheses discussed above and summarize evidence for and against each.

Does cell attrition in MAN close the critical period for song learning? If "song learning" in this context means motor acquisition of song (i.e., sensorimotor learning), then this hypothesis seems unlikely. Estradiol-treated females lose neurons in MAN between day 20 and adulthood, as do males (E. Nordeen et al., 1987; Bottjer et al., 1985), yet adult E2 females are able to experience some sort of motor acquisition of song in response to androgen. Assuming that the adult E2 female's motor acquisition is a process comparable to that in juvenile males, motor acquisition can occur *after* MAN has lost many neurons. These data do not tell us whether cell loss from MAN constrains the perceptual (sensory) learning process.

Because E2 females do not practice song, it is clear that neither cellular attrition in MAN nor cellular addition in HVc lead to motor acquisition of song by themselves. This is not to say that if these processes occur, as in juvenile males, at the time of motor acquisition of song, that they do not play an important role in facilitating song development. But these processes are clearly not sufficient causes of normal song development.

Because adult E2 females seem to show some sort of song development in response to androgens, we must ask if the cellular changes that correlate with song development in juvenile males (cell loss in MAN, cell addition in HVc, innervation of RA, increase in androgen accumulation in MAN) also occur during this belated song development in androgen-treated adult E 2 females? If they do not, then we can conclude that to the extent that this late song development is homologous to juvenile song development of males, motor acquisition of song does not require synchronous changes in MAN or HVc cell number. At present we have little information on these questions, especially on the comparability of the behavioral changes in juvenile males versus adult E2 females. However, HVc does seem to increase in volume as E2 females sing in response to androgen (Gurney, 1982), suggesting the possibility of further addition of HVc cells at this time. Perhaps the strongest conclusion we can make is that further comparison of male and female song development promises to allow us to refine hypotheses concerning the roles of various cellular changes that may underlie song learning.

SUMMARY AND CONCLUSIONS

At this point, we can return to the three sets of general questions posed at the beginning of this chapter and summarize our current understanding of the neural and hormonal events of the critical periods for sexual differentiation and song learning in zebra finches. In neither case do we know what accounts for the irreversibility of the differentiating effects of estradiol or the stability of learned changes in song (question 1). We might speculate that both sexual differentiation and song learning selectively enhance the formation of specific synapses, and such synapses are self-maintaining. But this only begs the question of how the synapses operate to prevent their own demise.

With regard to question 2, the basis for *selective* neural and behavioral development, we can say that differential steroid-induced cell survival may be a predominant mechanism of achieving selective enhancement of masculine brain structure and behavioral function. Similarly, song learning may involve selective survival of certain neurons. We can state, for example, that androgen-concentrating cells are selectively favored over other cells during development of HVc and MAN. Although it is not clear whether progressive or regressive events account for the favored status of androgen target cells, a major hypothesis is that, directly or indirectly, the death of these neurons is selectively prevented by estradiol. These androgen target cells could conceivably play a particularly important role in song learning.

Finally, question 3 demands that we analyze the processes leading to the onset and offset of the critical periods of sexual differentiation and song learning. Again, there are no firm answers, but we are excited by the realization that we can now frame specific hypotheses to begin the task of answering these questions. Does the addition of neurons to HVc span the entire critical periods for sensory and sensorimotor song learning? Does the end of cellular accretion in HVc close either of these critical periods? Is cellular addition to HVc always a concomitant of sensory or sensorimotor song learning, or can it be dissociated from these learning phenomena? What role is played by the loss of MAN neurons, by steroid hormones, and by the increase in androgen target cells? Does the ingrowth of HVc axons into RA close the critical period for motor development when all RA synaptic sites are filled?

As we have stressed from the beginning, the analysis of behavioral critical periods in development has historically drawn heavily on analogies to morphological development of the embryo. This may be no accident, and the comparison may be stronger than a mere analogy. It is axiomatic that all physiological changes have a morphological or biochemical basis, since they are potentially explained by some alteration in the structure of molecules that aggregate to define morphological structure. Similarly, behavior is an external manifestation of neural morphology and function. Thus, the kinds of progressive and regressive developmental events that specify critical periods in embryonic morphological development may be essentially the same kinds of events that specify critical periods in neural and behavioral development. The similarity of these events may explain the similarity of many developmental critical periods.

REFERENCES

Arnold, A. P. (1974). Behavioral effects of androgen in male zebra finches (*Poephila guttata*) and a search for its sites of action. Ph.D. Dissertation, Rockefeller University, New York.

Arnold, A. P. (1975a). The effects of castration on song development in zebra finches (*Poephila guttata*). *J. Exp. Zool.* **191**:261–278.

Arnold, A. P. (1975b). The effects of castration and androgen replacement on song, courtship, and aggression in zebra finches (*Poephila guttata*). *J. Exp. Zool.* **191**:309–326.

Arnold, A. P. (1980a). Quantitative analysis of sex differences in hormone accumulation in the zebra finch brain: Methodological and theoretical issues. *J. Comp. Neurol.* **189**:421–436.

Arnold, A. P. (1980b). Effects of androgens on volumes of sexually dimorphic brain regions in the zebra finch. *Brain Res.* **185**:441–444.

Arnold, A. P. (1982). Neural Control of Passerine Bird Song. In: *Acoustic Communication in Birds*, D. E. Kroodsma and E. H. Miller, eds., Vol. 1, pp. 75–94, Academic Press, New York.

Arnold, A. P. (1984). Androgen regulation of motor neuron size and number. *Trends Neurosci.* **7**:239–242.

Arnold, A. P. (1985). Gonadal Steroid-Induced Organization and Reorganization of Neural Circuits Involved in Bird Song. In: *Synaptic Plasticity and Remodeling*, C. Cotman, ed., pp. 263–285, Guilford Press, New York.

Arnold, A. P. and S. M. Breedlove (1985). Organizational and activational effects of sex steroid hormones on vertebrate brain and behavior: A re-analysis. *Horm. Behav.* **19**:469–498.

Arnold, A. P. and R. A. Gorski (1984). Gonadal steroid induction of structural sex differences in the CNS. *Ann. Rev. Neurosci.* **7**:413–442.

Arnold, A. P. and A. Saltiel (1979). Sexual difference in pattern of hormone accumulation in the brain of a song bird. *Science* **205**:702–705.

Arnold, A. P., F. Nottebohm, and D. W. Pfaff (1976). Hormone concentrating cells in vocal control and other areas of the brain of the zebra finch (*Poephila guttata*). *J. Comp. Neurol.* **165**:487–512.

Bateson, P. P. G. (1966). The characteristics and context of imprinting. *Biol. Rev.* **41**:177–220.

Bateson, P. P. G. (1979). How do sensitive periods arise and what are they for? *Anim. Behav.* **27**:470–486.

Beach, F. A. (1971). Hormonal Factors Controlling the Differentiation, Development, and Display of Copulatory Behavior in the Ramstergig and Related Species. In: *Biopsychology of Development*, E. Toback, L. Aronson, and E. Shaw, eds., pp. 249–296, Academic Press, New York.

Böhner, J. (1983). Song learning in the zebra finch (*Taeniopygia guttata*): Selectivity in the choice of a tutor and accuracy of song copies. *Anim. Behav.* **31**:231–237.

Bottjer, S. W. (1987). Ontogenetic changes in the pattern of androgen accumulation in song-control nuclei of male zebra finches. *J. Neurobiol.*, in press.

Bottjer, S. W. and A. P. Arnold (1984). Hormones and structural plasticity in the adult brain. *Trends Neurosci.* **7**:168–171.

Bottjer, S. W. and A. P. Arnold (1986). The Ontogeny of Vocal Learning in Song Birds. In: *Developmental Processes in Psychobiology and Neurobiology*, E. Blass, ed., pp. 129–161, Plenum Press, New York.

Bottjer, S. W., E. A. Miesner, and A. P. Arnold (1984). Forebrain lesions disrupt development but not maintenance of song in passerine birds. *Science* **224**:901–903.

Bottjer, S. W., S. L. Glaessner, and A. P. Arnold (1985). Ontogeny of brain nuclei controlling song learning and behavior in zebra finches. *J. Neurosci.* **5**:1556–1562.

Bottjer, S. W., E. A. Miesner, and A. P. Arnold (1986a). Changes in neuronal number, density and size account for increases in volume of song-control nuclei during song development in zebra finches. *Neurosci. Lett.* **67**:263–268.

Bottjer, S. W., J. N. Schoonmaker, and A. P. Arnold (1986b). Auditory and hormonal stimulation interact to produce neural growth in adult canaries. *J. Neurobiol.* **17**:605–612.

Breedlove, S. M. (1984). Androgen forms sexually dimorphic spinal nucleus by saving motoneurons from programmed death. *Soc. Neurosci. Abstr.* **10**:927.

Breedlove, S. M. (1985a). Hormonal control of the anatomical specificity of motoneuron-to-muscle innervation in rats. *Science* **277**:1357–1359.

Breedlove, S. M. (1985b). Androgen-insensitive spinal nucleus of the bulbocavernosus motoneurons can be spared by androgen from death during development. *Soc. Neurosci. Abstr.* **11**:160.

Breedlove, S. M. and A. P. Arnold (1980). Hormone accumulation in a sexually dimorphic motor nucleus of the rat spinal cord. *Science* **210**:564–566.

Breedlove, S. M. and A. P. Arnold (1981). Sexually dimorphic motor nucleus in rat spinal cord: Response to adult hormone manipulation, absence in androgen insensitive rats. *Brain Res.* **225**:297–307.

Breedlove, S. M. and A. P. Arnold (1983a). Hormonal control of a developing neuromuscular system: I. Complete demasculinization of the male rat spinal nucleus of the bulbocavernosus using the antiandrogen flutamide. *J. Neurosci.* **3**:417–423.

Breedlove, S. M. and A. P. Arnold (1983b). Hormonal control of a developing neuromuscular system: II. Sensitive periods for the androgen induced masculinization of the rat spinal nucleus of the bulbocavernosus. *J. Neurosci.* **3**:424–432.

Breedlove, S. M. and A. P. Arnold (1983c). Sex differences in the pattern of steroid accumulation of motoneurons of the rat lumbar spinal cord. *J. Comp. Neurol.* **215**:211–216.

Breedlove, S. M., C. L. Jordan, and A. P. Arnold (1983). Neurogenesis in the sexually dimorphic spinal nucleus of the bulbocavernosus of the rat. *Dev. Brain Res.* **9**:39–43.

Burns, R. K. (1961). Role of Hormones in the Differentiation of Sex. In: *Sex and Internal Secretions*, W. C. Young, ed., Vol. I, pp. 76–160, Williams and Wilkins Co., Baltimore.

Cihak, R., E. Guttmann, and V. Hanzlikova (1970). Involution and hormone-induced persistence of the muscle sphincter (levator) ani in female rats. *J. Anat. (Lond.)* **106**:93–110.

DeVoogd, T. J. and F. Nottebohm (1981). Gonadal hormones induce dendritic growth in the adult avian brain. *Science* **214**:202–204.

DeVoogd, T. J., B. Nixdorf, and F. Nottebohm (1985). Synaptogenesis and changes in synaptic morphology related to acquisition of a new behavior. *Brain Res.* **329**:304–308.

Dube, J. Y., R. Lesage, and R. R. Tremblay (1976). Androgen and estrogen binding in rat skeletal and perineal muscles. *Can. J. Biochem.* **54**:50–55.

Eales, L. A. (1985). Song learning in zebra finches: Some effects of song model availability on what is learnt and when. *Anim. Behav.* **33**:1293–1300.

Everett, J. W. (1977). The timing of ovulation. *J. Endocrinol.* **75**:3P–13P.

Goldman, S. and F. Nottebohm (1983). Neuronal production, migration, and differentiation in a vocal control nucleus of the adult female canary brain. *Proc. Natl. Acad. Sci. USA* **80**:2390–2394.

Gorski, R. A. (1974). The Neuroendocrine Regulation of Sexual Behavior. In: *Advances in Psychobiology*, G. Newton and A. H. Riesen, eds. Vol. II, pp. 1–58, Wiley & Sons, New York.

Gottlieb, G. (1976). The Roles of Experience in the Development of Behavior and the Nervous System. In: *Neural and Behavioral Specificity*, G. Gottlieb, ed., pp. 25–54. Academic Press, New York.

Goy, R. W. and B. S. McEwen (1980). *Sexual Differentiation of the Brain*. MIT Press, Cambridge, MA.

Gurney, M. (1980). Sexual differentiation of brain and behavior in the zebra finch (*Poephila guttata*): A cellular analysis. Ph.D. Dissertation, California Institute of Technology, Pasadena, CA.

Gurney, M. (1981). Hormonal control of cell form and number in the zebra finch song system. *J. Neurosci.* **1**:658–673.

Gurney, M. (1982). Behavioral correlates of sexual differentiation in the zebra finch song system. *Brain Res.* **231**:153–172.

Gurney, M. and M. Konishi (1980). Hormone induced sexual differentiation of brain and behavior in zebra finches. *Science* **208**:1380–1382.

Hamburger, V. and R. W. Oppenheim (1982). Naturally occurring neuronal death in vertebrates. *Neurosci. Comment.* **1**:39–55.

Harding, C. F., K. Sheridan, and M. J. Walters (1983). Hormonal specificity and activation of sexual behavior in male zebra finches. *Horm. Behav.* **17**:111–133.

Harding, C. F., M. J. Walters, and B. Parsons (1984). Androgen receptor levels in hypothalamic and vocal control nuclei in the male zebra finch. *Brain Res.* **306**:333–339.

Hart, B. L. and P. Y. Melese-d'Hospital (1983). Penile mechanisms and the role of the striated penile muscles in penile reflexes. *Physiol. Behav.* **31**:807–813.

Hauser, K. F. and C. D. Toran-Allerand (1985). Testosterone increases cell numbers in organotypic cultures of fetal mouse spinal cord. *Soc. Neurosci. Abstr.* **11**:898.

Herrmann, K. and H-J. Bischoff (1986). Delayed development of song control nuclei in the zebra finch is related to behavioral development. *J. Comp. Neurol.* **245**:167–175.

Hutchison, J. B., J. C. Wingfield, and R. E. Hutchison (1984). Sex differences in plasma concentrations of steroids during the sensitive period for brain differentiation in the zebra finch. *J. Endocrinol.* **103**:363–369.

Immelmann, K. (1969). Song Development in the Zebra Finch and Other Estrildid Finches. In: *Bird Vocalizations*, R. E. Hinde, ed., pp. 61–77, Cambridge University Press, Cambridge.

Katz, L. C. and M. E. Gurney (1981). Auditory responses in the zebra finch's motor system for song. *Brain Res.* **211**:192–197.

Keating, M. J. (1976). The Formation of Visual Neuronal Connections: An Appraisal of the Present Status of the Theory of "Neuronal Specificity." In: *Neural and Behavioral Specificity*, G. Gottlieb, ed., pp. 59–110, Academic Press, New York.

Kelley, D. B. and F. Nottebohm (1979). Projections of a telencephalic auditory nucleus—field L—in the canary. *J. Comp. Neurol.* **183**:455–470.

Konishi, M. (1965). The role of auditory feedback in the control of vocalization in the white-crowned sparrow. *Z. Tierpsychol.* **22**:770–783.

Konishi, M. and E. Akutagawa (1985). Neuronal growth, atrophy, and death in a sexually dimorphic song nucleus in the zebra finch brain. *Nature* **315**:145–147.

Konishi, M. and F. Nottebohm (1969). Experimental Studies in the Ontogeny of Avian Vocalizations. In: *Bird Vocalizations*, R. A. Hinde, ed., Cambridge University Press, Cambridge.

Kurz, E. M., D. R. Sengelaub, and A. P. Arnold (1986). Androgens regulate dendritic length of sexually dimorphic mammalian motoneurons in adulthood. *Science* **232**:395–398.

Lorenz, K. (1970). Companions as Factors in the Bird's Environment. In: *Studies in Animal and Human Behavior*, K. Lorenz, pp. 101–258, Harvard University Press, Cambridge, MA.

Margoliash, D. (1983). Acoustic parameters underlying the responses of song-specific neurons in the white-crowned sparrow. *J. Neurosci.* **3**:1039–1057.

Margoliash, D. and M. Konishi (1985). Auditory representation of autogenous song in the song system of white-crowned sparrows. *Proc. Natl. Acad. Sci. USA* **82**:5997–6000.

Marler, P. (1970). A comparative approach to vocal learning: Song development in white-crowned sparrows. *J. Comp. Physiol. Psychol. Monograph* **71**:1–25.

Marler, P. (1984). Song learning: Innate Species Differences in the Learning Process. In: *The Biology of Learning*, P. Marler and H. S. Terrace, eds., pp. 289–309, Springer-Verlag, New York.

Marler, P. and S. Peters (1981). Sparrows learn adult song and more from memory. *Science* **213**:780–782.

Marler, P. and S. Peters (1982a). Developmental overproduction and selective attrition: New processes in the epigenesis of birdsong. *Dev. Psychobiol.* **15**:369–378.

Marler, P. and S. Peters (1982b). Subsong and Plastic Song: Their Role in the Vocal Learning Process. In: *Acoustic Communication in Birds*, D. E. Kroodsma and D. H. Miller, eds., Vol. II, pp. 25–50, Academic Press, New York.

Marler, P. and M. S. Waser (1977). Role of auditory feedback in Canary song development. *J. Comp. Physiol. Psychol.* **91**:8–16.

McKenna, K. E. and I. Nadelhaft (1984). Organization of the pudendal nerve in the male and female rat. *Soc. Neurosci. Abstr.* **10**:902.

Meyer, R. L. and R. W. Sperry (1976). Retinotectal Specificity: Chemoaffinity Theory. In: *Neural and Behavioral Specificity*, G. Gottlieb, ed., pp. 111–152, Academic Press, New York.

Nordeen, E. J., K. W. Nordeen, D. R. Sengelaub, and A. P. Arnold (1985). Androgens prevent normally occurring cell death in a sexually dimorphic spinal nucleus. *Science* **229**:671–673.

Nordeen, E. J., K. W. Nordeen, and A. P. Arnold (1987). Sexual differentiation of androgen accumulation in the zebra finch brain through selective cell loss and addition. *J. Comp. Neurol.* (in press).

Nordeen, K. W., E. J. Nordeen, and A. P. Arnold (1986). Estrogen establishes sex differences in androgen accumulation in zebra finch brain. *J. Neurosci.* **6**:734–738.

Nottebohm, F. (1968). Auditory experience and song development in the chaffinch (*Fringilla coelebs*). *Ibis* **110**:549–568.

Nottebohm, F. (1969). The "critical period" for song learning. *Ibis* **111**:386–387.

Nottebohm, F. (1980a). Testosterone triggers growth of brain vocal control nuclei in adult female canaries. *Brain Res.* **189**:429–436.

Nottebohm, F. (1980b). Brain pathways for vocal learning in birds: A review of the first 10 years. *Prog. Psychobiol. Physiol. Psychol.* **9**:85–125.

Nottebohm, F. (1981). A brain for all seasons: Cyclic anatomical changes in song control nuclei of the canary brain. *Science* **214**:1368–1370.

Nottebohm, F. (1984). Birdsong as a model in which to study brain processes related to learning. *Condor* **86**:227–236.

Nottebohm, F. and A. P. Arnold (1976). Sexual dimorphism in vocal control areas of the song bird brain. *Science* **194**:211–213.

Nottebohm, F., T. M. Stokes, and C. M. Leonard (1976). Central control of song in the canary (*Serinus canarius*). *J. Comp. Neurol.* **165**:457–486.

Nottebohm, F., S. Kasparian and C. Pandazis (1981). Brain space for a learned task. *Brain Res.* **213**:99–109.

Nottebohm, F., D. B. Kelley, and J. A. Paton (1982). Connections of vocal control nuclei in the canary telencephalon. *J. Comp. Neurol.* **207**:344–357.

Oppenheim, R. W. (1981). Neuronal Cell Death and Some Related Regressive Phenomena during Neurogenesis: A Selective Historical Review and Progress Report. In: *Studies in Developmental Neurobiology: Essays in Honor of Victor Hamburger*, W. M. Cowan, ed., pp. 74–133, Oxford University Press, New York.

Paton, J. A. and F. Nottebohm (1984). Neurons generated in the adult brain are recruited into functional circuits. *Science* **225**:1046–1048.

Phoenix, C. H., R. W. Goy, A. A. Gerall and W. C. Young (1959). Organizing action of prenatally administered testosterone propionate on the tissues mediating mating behavior in the female guinea pig. *Endocrinol.* **65**:369–382.

Picon, R. (1976). Testosterone secretion of foetal rat testes in vitro. *J. Endocrinol.* **71**:231–238.

Pohl-Apel, G. and R. Sossinka (1984). Hormonal determination of song capacity in females of the zebra finch: Critical phase of treatment. *Z. Tierpsychol.* **64**:330–336.

Price, P. (1979). Developmental determinants of structure in zebra finch song. *J. Comp. Physiol. Psychol.* **93**:260–277.

Pröve, E. (1974). Der Einfluss von Kastration und Testosteronsubstitution auf das Sexualverhalten männlicher Zebrafinken (*Taeniopygia guttata castanotis* Gould). *J. Ornithologie* **115**:338–347.

Pröve, E. (1978). Quantitative Untersuchungen zu Wechselbeziehungen zwischen Balzaktivität und Testosterontitern bei männlichen Zebrafinken (*Taeniopygia guttata castanotis* Gould). *Z. Tierpsychol.* **48**:47–67.

Pröve, E. (1983). Hormonal Correlates of Behavioral Development in Male Zebra Finches. In: *Hormones and Behaviour in Higher Vertebrates*, J. Balthazart, E. Pröve, and R. Gilles, eds., pp. 368–374, Springer-Verlag, Berlin.

Purves, D. and J. W. Lichtman (1980). Elimination of synapses in the developing nervous system. *Science* **210**:153–157.

Sachs, B. D. (1982). Role of penile muscles in penile reflexes, copulation, and induction of pregnancy in rats. *J. Reprod. Fertil.* **66**:433–443.

Scott, J. P. (1978). *Critical Periods.* Dowden, Hutchison, and Ross, Stroudsberg, PA.

Sengelaub, D. R. and A. P. Arnold (1986). Development and loss of early projections in a sexually dimorphic rat spinal nucleus. *J. Neurosci.* **6**:1613–1620.

Venable, J. H. (1966) Morphology of the cells of normal, testosterone-deprived and testosterone-stimulated levator ani muscles. *Amer. J. Anat.* **119**:271–302.

Wallach, S. J. R. and B. L. Hart (1983). The role of striated penile muscles of the male rat in seminal plug dislodgement and deposition. *Physiol. Behav.* **31**:815–821.

Weisz, J. and I. L. Ward (1980). Plasma testosterone and progesterone titers of pregnant rats, their male and female fetuses and neonatal offspring. *Endocrinology* **106**:306–316.

Whalen, R. E. (1968). Differentiation of the Neural Mechanisms which Control Gonadotrophin Secretion and Sexual Behavior. In: *Reproduction and Sexual Behavior*, M. Diamond, ed., pp. 303–340, Indiana University Press, Bloomington, IN.

Williams, H. (1985). Sexual dimorphism of auditory activity in the zebra finch song system. *Behav. Neurol. Biol.* **44**:470–484.

Williams, H. and F. Nottebohm (1985). Auditory responses in avian vocal motor neurons: A motor theory for song perception in birds. *Science* **229**:279–282.

Young, W. C., R. W. Goy, and C. H. Phoenix (1964). Hormones and sexual behavior. *Science* **143**:212–218.

5

SENSITIVE PERIODS AND THE ROLES OF SPECIFIC AND GENERAL SENSORY STIMULATION IN BIRDSONG LEARNING

Peter Marler

Rockefeller University Field Research Center, Millbrook, New York

Variation in the effects of a given stimulus event experienced at different phases of an organism's development is a virtually universal phenomenon (Immelmann and Suomi, 1981). The term *sensitive period* does not imply that the phase of maximal effectiveness is necessarily completely tied to chronological age, nor that equivalent effects cannot be obtained later (Hinde, 1983), but the fact remains that temporal constraints on the acquisition of responsiveness to stimuli can impose strong and in some cases virtually irreversible limitations on the readiness to learn. Analogous principles are to be found in other biological domains, including not only the neurobiology of visual and auditory development, as attested to in this volume, but also in embryology (Oppenheim and Haverkamp, 1985), neural development and differentiation of sexual mechanisms (Adkins-Regan, 1983; Arnold et al., this volume; Goy and McEwen, 1980; Hutchison and Hutchison, 1985; McEwen, 1983), development of recognition processes (Colgan, 1983), language and speech development (Lenneberg, 1967), the ontogeny of motor and sensorimotor integration (Fentress, 1981; Fentress and McLeod, 1986), and the development of social attachments in both nonhuman primates and man (Ainsworth et al., 1978; Hinde, 1974; Hofer, 1981; Murray Parkes and Stevenson-Hinde, 1982; Suomi, 1985). Sensitive periods for vocal learning are thus a special case of a widespread phenomenon.

Two areas of comparative behavioral and neuroethological research on birds that have revealed a great deal of basic information about sensitive periods for learning are filial and sexual imprinting (Lorenz, 1965, 1981; Bateson, 1976, 1978a,b, 1979; Gottlieb, 1971; Hess, 1973; Horn, 1985; Horn et al., 1973; Rajecki et al., 1978) and vocal learning. There is no lack of controversy, one bone of contention being the relative roles of generalized and specialized stimulation (Bateson, 1979; Hess, 1973; Johnston and Gottlieb, 1981). Some argue that stimuli influence the prospects of acquisition primarily because of their arousal effects. There is also evidence that access to or deprivation of very specific kinds of stimulation can have major effects in closing out or extending sensitive learning periods. Indications that the neural circuitry involved in imprinting on generalized and on specific stimuli may be distinct (Horn, 1985) heighten the importance of clarifying this issue.

Birdsong studies have been a rich source of insights into many aspects of the neural control of behavior, including sexual dimorphisms in brain mechanisms for song control (Arnold, 1982; Arnold and Gorski, 1984; Bottjer et al., 1984, 1985; Gurney and Konishi, 1980; Konishi, 1985; Konishi and Gurney, 1982; Nottebohm and Arnold, 1976; Nottebohm et al., 1976), their neurophysiological properties (Konishi, 1985; Leppelsack, 1983; Margoliash, this volume; Margoliash and Konishi, 1985; Mc-Casland and Konishi, 1981; Williams and Nottebohm, 1985), relationships between learning, hormones, and volume change in brain nuclei (Nottebohm, 1981, 1984a,b), correlations between neural ontogeny and song development (Bottjer and Arnold, 1986), and the roles of neurogenesis and synaptogenesis in the plasticity of avian song (DeVoogd et al., 1985; Goldman and Nottebohm, 1983; Nottebohm, 1984a,b).

These neuroethological investigations of song learning and vocal development have evolved directly from ethological studies, with special reference to the issue of sensitive periods (Immelmann, 1969; Baptista and Petrinovich, 1984, 1986; Cunningham and Baker, 1983; Kroodsma, 1978, 1982; Marler, 1970; Nottebohm, 1969; Slater, 1983; Slater and Ince, 1982), going back to the classic work of Thorpe (1958, 1961). In some respects research on the neural and hormonal mechanisms involved in song learning has out-paced behavioral work, particularly with regard to systematic and precise temporal characterization of sensitive periods for song acquisition. It is well established that some bird species have "age-independent" plasticity, for example, the canary (Nottebohm, 1984a; Nottebohm and Nottebohm, 1978), and others are what may be called "age-limited learners." There has yet to be a full accounting for any "life-long plasticity" species of variations through the life cycle in the readiness to acquire new songs, however. This review will focus mainly on data for "age-limited" learners, the main illustrations being drawn from emberizine sparrows.

"AGE-LIMITED LEARNERS": SONG DEVELOPMENT IN SPARROWS

If a young male song or swamp sparrow is reared out of hearing of adult sounds of its own kind, the major part of its vocal repertoire will develop normally. The most elaborate of its vocalizations, however, is invariably abnormal. This is the song, which is learned, unique to males for the most part, and serving functions of aggression and sexual stimulation. The song of such an isolate is delivered vigorously and frequently, as in a normal bird, but it is structurally and communicatively deficient both as a territorial signal to other males and as a sexual signal to females (Searcy et al., 1985, Searcy and Marler, in press).

By the time the song of an isolated male sparrow has fully matured at an age of about 200–300 days, the developmental stages of subsong and plastic song, which vary greatly in structure, have crystallized into stereotyped songs that display no further significant plasticity. Beyond this stage the songs typically remain fixed for life, recurring in the same form each spring and summer singing season for as long as the bird survives, irrespective of whether or not it has access to normal songs of its species at any later time. There is thus a more or less irreversible loss of motor plasticity at around the time of song crystallization.

If, instead, the young male is allowed to hear adult songs of his species in the first weeks of life, his own song will develop normally, incorporating imitations of the models to which he has been exposed. Thus normal development of the species-specific song depends on stimulation early in life by sounds of the bird's own species. As a consequence of this vocal learning local dialects are a common occurrence (Baker, 1982; Baker and Cunningham, 1985; Baptista, 1975). Once a male's song themes have crystallized there is again, typically, no further plasticity, either in the laboratory or in the field.

There are thus sensitive periods for song learning in sparrows. The aims of this chapter are (1) to present new data characterizing the temporal features of sensitive periods for song acquisition, (2) to review evidence that sensitive periods are not fixed but labile, within limits, depending on environmental circumstances, (3) to explore the role of specific and nonspecific external stimulation in the behavioral plasticity shown during sensitive periods, (4) to show that innate responsiveness to key stimuli embodied in natural songs plays an integral part in the process of learning to sing, and (5) to comment on hormonal correlates of sensitive learning periods.

The approach will be that of a comparative ethologist, exploring some of the ramifications of species differences in reactivity to the same environments (Lorenz, 1965, 1981; Tinbergen, 1951).

TEMPORAL FEATURES OF SENSITIVE PERIODS FOR SONG LEARNING

Birds vary greatly in the length and timing of the phase of the life cycle in which vocal plasticity occurs. In some species vocal plasticity, manifest as the ability to change the motor patterns of song, persists throughout life (e.g., the canary; Nottebohm and Nottebohm, 1978). This ex-

1. ACQUISITION = THE SENSORY PHASE

2. STORAGE

3. RETRIEVAL & PRODUCTION ⎤
   ⎬ = THE SENSORIMOTOR
4. MOTOR STABILIZATION ⎦   PHASE

FIGURE 1.

tended plasticity probably occurs during annually recurrent sensitive phases in the canary, although as yet there has been little systematic investigation of the precise timing of the readiness to acquire new songs in such life-long learners. For example, it has not yet been demonstrated that canaries more than a year old can actually acquire and imitate new songs, as they do in their first year of life (Waser and Marler, 1977; Marler et al., 1973). In sparrows learning is restricted to the first weeks or months of life, with temporal constraints that are sufficiently well defined to lend themselves to experimental investigation and manipulation in the laboratory. In turn, this makes them potentially amenable to exploration of their neural and hormonal basis.

Another advantage of sparrows as experimental subjects is that they do not require a live bird as a song tutor, as some species do (e.g., treecreepers; Thielcke, 1984; review in Slater, 1983). Live tutors are intractable as experimental sources of learning stimuli. Tape-recorded songs, on the other hand, are readily standardized and quantified, lending themselves to synthesis and independent control of their acoustic properties.

It is important to distinguish the different phases of song development (Fig. 1). These include stimulus acquisition, storage and retrieval, motor rehearsal and stabilization, and song crystallization. If a male sparrow is brought into the laboratory soon after hatching, before the onset of song learning at about 10 days of age, and exposed to tape-recorded songs of its own species for 30 days, it will learn them. Acquisition occurs in the *sensory* phase of song learning (Fig. 1). Learned songs are then retained in memory, in some cases for weeks or months, without rehearsal (the *storage* phase) (Marler and Peters, 1982a). Then it will begin to sing (the *sensorimotor* phase). First, the highly variable and amorphous patterns of subsong are produced. These involve no retrieval and rehearsal of acquired material, and are not involved in song learning in the narrow sense. Subsong may be important in learning how to use

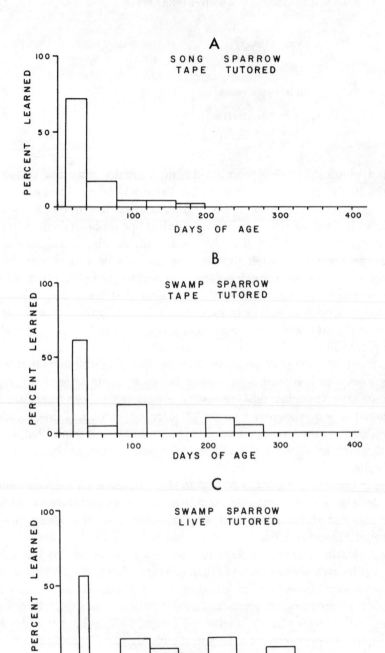

the vocal tract as an instrument for sound production (Nottebohm, 1972). Next in the developmental progression is plastic song, in which motor rehearsal of learned songs begins. This is the first evidence the male provides of retrieval of song material from memory. Finally, his song stabilizes into the stereotyped patterns that will comprise his crystallized song repertoire for the rest of his life.

In a recent experiment designed to delineate the timing of song acquisition, young song sparrows were individually isolated in the laboratory, and exposed to a continuously changing program of tape-recorded songs for the first year of life. The learning that was subsequently revealed by the imitations produced in the sensorimotor phase was overwhelmingly concentrated in the first 50 days (Fig. 2A). Defined in this fashion, the acquisition phase for learning in the song sparrow extends from about 20 to 50 days of age (Marler and Peters, in press).

A few subjects also learned in later stages of stimulus exposure, revealing that, as with many other aspects of behavior, there are appreciable individual differences in the precise timing of sensitive periods for song learning. When manifest in subjects raised under controlled conditions, such variations perhaps imply biologically significant individual differences in the underlying control mechanisms (Kroodsma and Pickert, 1984). In the song sparrow case the variation is manifest as an extension of the sensitive period, rather than a delay since every subject that engaged in late learning also acquired some songs early in the stimulation period. The significance of this variation may lie in its potential impact on patterns of dispersal of young birds from their birthplace and their settlement in new areas, opening up alternative options for different individuals (Kroodsma et al., 1984; Baker, 1982; Baker and Cunningham, 1985; Slater 1983).

In addition to learning from purely auditory stimuli, there is growing evidence that exposing a bird to a live tutor can affect the course of song learning. Interactions between tutor and pupil may exert an influence

---

FIGURE 2. The timing of song learning in male sparrows reared in the laboratory and exposed to programs of song stimulation that changed on a weekly or six-weekly basis, for the first year of life. In song sparrows trained with tape-recorded song (A) nearly 80% of the learning occurs in the first 40 days of stimulation. In swamp sparrows trained with tape recordings (B) the early learning peak was less prominent, and there was more learning at later ages. With the audiovisual stimulation provided by live singing tutors (C) the result was very similar. Note that the first two sessions of live tutoring each lasted three weeks rather than six.

on learning preferences, as when young males learn from the father in preference to other males within earshot, or favor as tutors males with high rank in the local community (e.g., Payne, 1973, 1983). Live tutoring may also extend sensitive periods beyond the limits defined by auditory stimulation alone (Kroodsma, 1982; Baptista and Petrinovich, 1984). Thus it is appropriate to ask whether the acquisition phase for song learning in sparrows would differ if, instead of tape recordings, males were exposed to a changing program of live tutors.

This comparison has been made with the swamp sparrow (Marler and Peters, unpublished work). Figures 2B and C summarize data on the timing of song acquisition by two groups of male swamp sparrows, one trained for the first year of life with tape-recorded songs, the other with a comparable program of live tutoring. Both tapes and tutors were changed on a six-weekly timetable, beginning at 10 days of age. Analyses of the imitations produced in crystallized song, at an age of 300–400 days, provided the evidence of retrieval from storage of songs acquired earlier. The results for tape- and live-tutored groups are very similar. Thus in these sparrows, auditory stimulation provides a satisfactory characterization of temporal features of the sensitive period. Tape tutoring and live tutoring do not, of course, provide identical patterns of stimulation, and the differences may have a significant bearing on other aspects of song development, as will be discussed in the next section.

In the swamp sparrow, both with tape-recorded stimulation and live tutoring, song acquisition was strongly focused in an early period, from 20 to 50 days of age. The concentration in this early phase was less extreme than in the song sparrow, however, amounting to 60% instead of 80% of the total number of songs acquired. In the swamp sparrow, appreciable learning took place later, between 50 and 300 days of age. Again there were individual differences, and once more they involved variation in the overall duration of the learning phase, rather than postponement of the entire sensitive period. The timing of closure clearly varied to a significant degree.

The distribution of learning through the first year of life differs in the swamp sparrow and the song sparrow, and a comparison between the two species is illuminating. The sensory phase for song learning is more extended in the swamp sparrow, and may prove to be multiphasic, in contrast with the short, uniphasic sensitive period of the song sparrow (Fig. 4). There is a precedent for extended or multiphasic sensitive periods in other species. Kroodsma (1974) found hints of a biphasic sensitive period in the Bewick's wren, one phase coming before dispersal from the birthplace and the second after emigration and establishment of a territory. The trend is clearer in the marsh wren, with a hiatus between the pre- and postdispersal learning periods (Kroodsma, 1978). The crystallized repertoire of some wrens thus consists of a mixture of songs learned at more than one locality.

Song and swamp sparrows differ in their life cycles. Song sparrows tend to be year-round residents, whereas swamp sparrows are typically migrants. It will be instructive to explore whether swamp sparrows sing on migration, and whether they establish and defend territories on their wintering grounds, behavior recorded in some other migrant song birds.

White-crowned sparrows illustrate another dimension of song ontogeny in which there are species and subspecific differences, namely, the interval between song acquisition and the start of production (DeWolfe, Baptista, and Petrinovich, unpublished work). In migrant subspecies, such as the montane white-crowned sparrow (*Zonotrichia leucophrys oriantha*), the interval is long, as in the swamp sparrow. In Nuttall's whitecrown (*Z. l. nuttalli*), a year-round resident, the interval is shorter so that crystallized song may occur as early as 100 days of age (DeWolfe et al., loc. cit.). In New York, the song sparrow is also a resident species, and again song crystallization occurs significantly earlier than in the migratory swamp sparrow (Marler and Peters, unpublished work).

Such species differences in programs for motor development and in sensitive periods for song acquisition are clearly innate, opening up opportunities for comparative study of underlying hormonal and neural mechanisms. The existence of both open-ended and one-time learners is another indication of strong genomic influences on the physiological mechanisms underlying sensitive periods.

SENSITIVE PERIODS ARE NOT FIXED, BUT LABILE

What are we to conclude from the lack of contrast between sensitive periods in the swamp sparrow defined by tape recordings and by live tutoring? Does this mean that the two classes of stimulation are functionally equivalent in all respects? We know that this is not the case. In fact interplay between tape and live tutoring has been a prime source of evidence of lability of sensitive period learning.

Some lines of evidence suggest that a live, socially interactive tutor is not only different but in some sense a stronger learning stimulus than a tape recording. In the marsh wren, and the white-crowned sparrow, in experiments in which tape stimulation was followed by live tutoring, learning from live tutors was demonstrated after closure of the sensory phase for acquisition of auditory stimuli alone (Marler, 1970; Cunningham and Baker, 1983). Males that were exposed to tape recordings when young, and no longer amenable to further training by that means, were induced to acquire new songs by exposing them to singing, interactive tutors (Kroodsma, 1982; Kroodsma and Pickert, 1984; Baptista and Petrinovich, 1984, 1986). In this way the sensitive period was lengthened for several weeks or even months.

One could interpret this result as implying that a weaker stimulus (song alone) was presented in the first part of the sensitive phase, followed by a stronger stimulus (a conspecific live tutor producing song), thus extending the sensitive period. A prediction from this interpretation would be that the sensitive phase for acquisition from the "strong" stimulus should be longer than that for the "weaker" stimulus. The swamp sparrow data show, however, that closure of the sensitive period occurs at the same time and with the same finality, whether the "strong" or the "weaker" stimulus is used.

Weak and strong may not, in fact, be completely appropriate terms to use in comparing tape recordings and live tutors as learning stimuli. They imply that the two types of stimulation exert quantitatively different degrees of influence on the learning process. As stimuli they differ in quality as well as quantity. Specific kinds of stimulation are obviously provided by a live tutor but not by a tape recording, and deprivation of those stimuli, such as inevitably occurs in an experiment in which tape tutoring precedes live tutoring, may delay closure of the sensitive period for song acquisition. This point will be taken up again later.

Serious consideration must also be given to the possibility that optimal requirements for learning stimuli are qualitatively distinct at different stages of development, especially in species with extended sensitive periods. In nature there is considerable variation in the stimulus situations that birds encounter in the course of song learning, both within and between species. In many birds, including sparrows, personal contact between young males and their fathers is brief and perfunctory and field studies have revealed no significant tendency for father–son learning in sparrows (Nice, 1943; Baptista, 1985; Baptista and Morton, unpublished work). In other species personal bonds between young and adult males may be sustained and intense, and the focus of special ritualized interactions, continuing after weaning. In such cases as the zebra finch and the bullfinch, a young male learns song, not in his relationships with adult males in general but specifically within the father–son relationship (Nicolai, 1959; Immelmann, 1967, 1969; Price, 1979; Böhner, 1983). Father–son song learning has been well documented under both field and laboratory conditions in Darwin's finches (*Geospiza* spp.) (Bowman, 1983; Grant, 1984; Millington and Price, 1985).

In sparrows intense, sustained interactions, whether filial or agonistic, between young males and particular adults are unusual in early life, once weaning is completed. Intense interactions are more likely to occur in late adolescence, toward the end of the sensitive period for song learning or after it has usually closed, as young males begin to engage in agonistic interactions and establish their first territory. This is when interaction with a live tutor might have effects on song acquisition. Territorial confrontations may lead to levels of arousal and degrees of attentiveness greater than those previously encountered by sparrows, prior to dispersal.

There is also the potential for variation in the factors shaping song acquisition within a species, depending on the date of hatching. On occasion young are hatched after the singing season has ended. The experience of late-hatched young is abnormal in two senses: They are deprived of song stimulation, and they are exposed to a different photoperiodic regime from early-hatched young. Kroodsma and Pickert (1980) demonstrated that both of these factors delay closure of the sensitive period during which marsh wrens learn songs from auditory stimulation.

The possibility that optimal requirements for learning stimuli are qualitatively distinct at different stages of development, especially in species with extended periods of song acquisition that span a range of developmental changes and different stages of the life history, is a serious complication for studies of the causal basis of sensitive periods. Auditory stimulation with tape recordings may be adequate or optimal in the first phase, in the white-crowned sparrow, for example (Marler, 1970; Baptista and Petrinovich, 1986, Petrinovich 1985), with other kinds of social stimulation assuming maximal effectiveness in later phases, as an accompaniment to song. I will return to this theme later.

OPTIMAL LEARNING STIMULI AND EFFECTS OF DEPRIVATION

At the height of the sensitive period for song acquisition, some birds behave as though they are innately attuned to highly specific auditory stimuli, to which they are responsive during the sensitive period and whose absence will delay its closure.

If the first experience marsh wrens have of conspecific songs is postponed, closure of the learning period can be delayed by several months (Kroodsma and Pickert, 1980). In the zebra finch, closure of the sensitive period for song learning is delayed for several weeks if, instead of being allowed to learn song from a conspecific father, young are fostered out to an alien species pair and thus deprived of access to conspecific song. The foster father's song will be learned, and the ability to learn new songs of the foster species persists until about 60 days of age. In males reared normally, with full access to zebra finch song, this ability is lost earlier, at about 35 days of age (Immelmann, 1967, 1969; Böhner, 1983, unpublished work). Several factors may contribute to the delay of sensitive period closure (e.g., Cate, 1984), but deprivation of optimal auditory stimuli is almost certainly one of them.

Thus there are several indications that while there is a general drive to acquire auditory stimuli of many kinds during the sensitive period, certain specific stimuli are learned more rapidly and accurately than others. This class of optimal learning stimuli typically includes songs of the subject's own species. Learning is not confined to such optimal stimuli, and if they are withheld, songs of other species may be learned. Even when this occurs, specific predispositions may be involved, in the sense that, as indicated below, the acceptability of alien songs is affected by whether they have acoustic features in common with the bird's own species' song.

INNATE PREFERENCES FOR SPECIFIC STIMULATION DURING THE SENSITIVE PERIOD

One way to explore the role of specific stimulation in song learning is to give subjects a choice of stimuli during the sensitive period. This has been done with several songbirds, providing a test of whether or not different sounds are equipotential. If preferences emerge, they provide us with another approach to questions about the relative importance of quantitative and qualitative stimulus properties.

The comparative approach of the classical ethology of Lorenz and Tinbergen is of value here. Birds that must learn their song are confronted with natural problems of choice when stimulated by songs of different species. Contrasting song patterns can be used as a basis for investigating the existence of preferences and for designing synthetic stimuli to explore the acoustic features on which preferences are based.

Young of two closely related species, swamp and song sparrows, that often live together within earshot and on occasion sing at the same time and place have been reared in isolation from adults and exposed to tape-recorded songs at the height of the sensitive period. Analysis of their crystallized songs revealed that they had acquired models experienced when they were young, committed them to memory, and used them later to guide their own song development (Marler and Peters, 1977, 1981b). When the birds were allowed to choose from an array of natural conspecific and heterospecific songs, a clear and unequivocal bias was revealed in both species in favor of conspecific song.

Alien songs were occasionally learned and reproduced as well. On rare occasions this also occurs in nature (e.g., Baptista et al., 1981; Helb et al., 1985). These occurrences are of interest for they show that the vocal tract of one species is capable of producing reasonable renditions of another's songs. Preferences are thus more likely to be based on sensory and perceptual mechanisms than on species biases built into the sound-producing apparatus, although the possibility of subtle differences must always be born in mind.

The two major types of acoustic cues on which such preferences might be based are the overall temporal organization of song—its syntax—and the acoustic morphology of the minimal units or notes from which the song is constructed—its phonology. Even species as closely related as the song and the swamp sparrow differ in the cues on which learning preferences are based.

Male swamp sparrows base their acceptance of conspecific songs and their rejection of alien song sparrow songs as models for learning entirely on phonology. Song sparrows, on the other hand, use both phonological and syntactical cues. These interact in interesting ways. Given the appropriate syntax song sparrows will overlook deficiencies in phonology that would otherwise lead them to reject an alien song. Thus a synthetic song with swamp sparrow phonology is more likely to be accepted by a young male song sparrow as a model for learning if it has multipartite song sparrow syntax (Marler and Peters, unpublished work).

This species difference in responsiveness to what are evidently key stimuli or, as Tinbergen (1951) and Lorenz (1981) would call them, sign stimuli, incorporated in their songs, is apparently innate. Natural song and swamp sparrow songs differ consistently in both syntax and phonology. The differences in phonology have only recently been discovered, in studies provoked by the realization that they must be there if the birds can make such unerring phonological discriminations. The differences have yet to be fully characterized, but each species appears to have its own distinctive set of note types. All species-typical songs are constructed by assembling in different combinations note types and subtypes drawn from a common pool that constitutes the species-specific repertoire (Marler and Pickert, 1984; Marler and Seaman, unpublished work).

Preferences for conspecific songs as models for vocal learning have been demonstrated in other bird species (Thorpe, 1958; Marler, 1970; Immelmann, 1969; Konishi, 1985). Furthermore, in cases where alien songs are accepted as models, they often have features in common with conspecific song (Konishi, 1985), as with the song of the tree pipit that Thorpe (1958, 1961) taught to the European chaffinch, chosen because its songs are somewhat chaffinchlike to the human ear.

Systematic attempts to independently manipulate phonology and syntax confirm that the presence of particular conspecific cues can render otherwise unacceptable songs more acceptable as learning stimuli (Fig. 3). Thus male swamp sparrows, which display a strong innate preference for conspecific phonology over that of the song sparrow, can be persuaded to accept song sparrow notes if they are incorporated into a note cluster to form a so-called syllable, the rest of which is made of swamp sparrow notes. Swamp sparrows will also learn parts of entire synthetic song sparrow songs if they contain one or more swamp sparrow syllables.

Birds evidently hear the songs of other species and can be taught to discriminate between them with precision (Sinnot, 1980; Dooling, 1982, unpublished work). Young males behave as though any song presented as a stimulus is subjected to normal sensory processing, but is then quickly lost from short-term memory in the normal course of events. If conspecific sign stimuli are present, however, it is as though a brief time

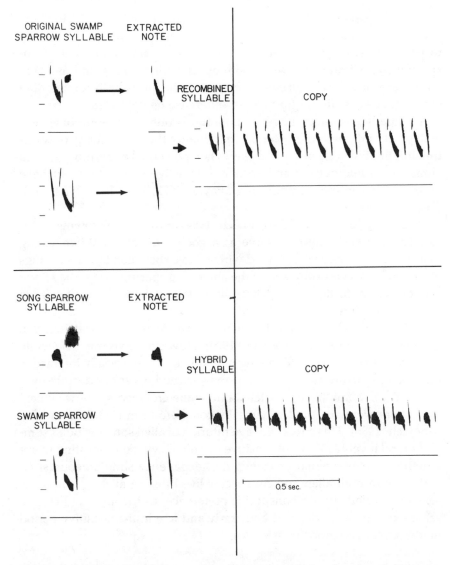

FIGURE 3. An illustration of the use of synthetic songs to induce learning of alien notes by embedding them in conspecific notes to create a hybrid syllable. On the right are copies by male swamp sparrows of these synthetic songs. The upper one is a control stimulus, created with swamp sparrow notes. The lower one shows a copy of a note extracted from a song sparrow syllable normally rejected by swamp sparrows as a model for song learning, rendered acceptable by placing it in a conspecific phonological context.

window is opened during which the stimulus cluster in view at the time becomes the focus of closer attention, and is correspondingly more likely to pass into storage for later use in guiding vocal development. Conspecific sign stimuli thus serve both as cues for learning and as "enabling" signals, whose presence increases the probability of learning other associated stimuli that would otherwise be neglected.

As already indicated, a live, singing tutor contributes additional "enabling" stimuli, with the potential to influence the acceptability of songs that might otherwise be passed over when presented as auditory stimuli alone. Such audiovisual and socially interactive stimulation can have potent effects on song learning (e.g., Payne, 1973, 1976, 1981, 1982, 1983; Todt et al., 1979).

Studies by Baptista and Petrinovich (1984) of the white-crowned sparrow (*Zonotrichia leucophrys*) serve as a good illustration. When young male white crowns are presented with a choice between tape recordings of their own species song and of the alien song sparrow song, they favor the former. With tape-recorded auditory stimulation of the alien song alone, male white crowns behaved as though they had received no song stimulation at all, and developed like birds reared in social isolation (Marler and Tamura, 1964; Marler, 1970). However, if young male white crowns were confronted for a long period with a live, socially interactive male song sparrow as a song tutor, they learned his songs despite their alien nature. Some were translated into white-crowned sparrow syntax, but others were accepted in virtually complete form (Baptista and Petrinovich, 1986). This successful live tutoring of alien song included stimulation relatively late in the sensitive period. It will be interesting to see whether white-crowned sparrows that experience simultaneous stimulation with both alien and conspecific live tutoring at the peak of the sensitive period show a conspecific preference, as I would predict. The same experiment performed both early and late in the sensitive period might reveal variations in selectivity.

ROLE OF AUDITORY FEEDBACK IN SONG DEVELOPMENT

A different line of evidence implicates specific stimulation in the process of song development, in this case involving stimuli generated by the bird's own vocal behavior. We know from the now classic experiments on the effects of deafening on song development that auditory feedback

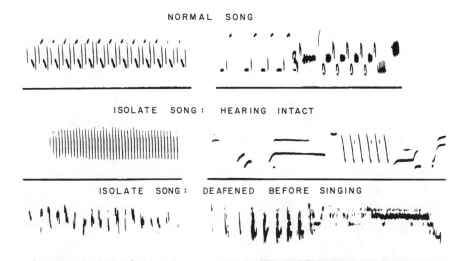

FIGURE 4. Typical examples from swamp and song sparrows of normal song, songs of birds reared in isolation with their hearing intact, and birds deafened prior to the onset of any singing behavior. The extreme degradation of song in an early-deafened male is the basis for invoking innate auditory templates presumed to guide an isolated bird in generating a degree of normal syntactical song structure, as long as the auditory feedback loop is intact.

plays a crucial role in the process of song ontogeny (Konishi, 1965; Nottebohm, 1968). If a male sparrow is deafened after song motor patterns have crystallized, the effects are minimal, suggesting that at that stage the motor program is internalized. If a sparrow is deafened prior to singing, however, it develops a very basic, amorphous song, having much in common with the earliest, subsong phase of development (Fig. 4). Unlike songs of an isolation-reared sparrow, which have some residual effectiveness as communicative signals, songs of deaf males evoke no responses from other birds at all (Searcy and Marler, in press). This highly degraded form of song results both if a male is deafened before song stimulation and also if deafening is delayed until after song acquisition during the sensitive period, but during storage, prior to any motor development of song. Thus there appears to be no internal circuitry that makes memorized songs directly available to guide motor development. To transform a memorized song into a produced song, the bird must be able to hear its own voice (Konishi, 1965).

In contrast with the minimal degree of species-specific structure in the song of an early-deafened sparrow, more normal song structure develops if a male is reared in isolation with hearing intact. This is the basis for invoking innate auditory templates for song development (Konishi, 1965; Marler and Sherman, 1983).

Detailed comparison of songs of male swamp and song sparrows reared in complete isolation reveals that, despite the abnormalities, many syntactical song features develop in species-typical form (Marler and Sherman, 1985). A comparison of these isolate songs with those of early deafened males shows that many species-specific syntactical features depend on auditory feedback for their development, thus implying innate auditory responsiveness to them. This is a third line of evidence that specific stimulation has a special role to play in song development. It may not be coincidental that some of the same aspects of song structure lacking in deaf males also appear to guide song learning preferences, although these have yet to be fully specified (Marler and Peters, unpublished work).

EMERGENCE OF SONG SYNTAX IN MOTOR DEVELOPMENT

In the swamp sparrow species-specific song syntax emerges late in development, during crystallization. The mature song is typically unisegmental, consisting of a string of identical syllabic repetitions. If young males are exposed in the sensory phase only to multisegmental songs, they produce more two-segmented songs after crystallization than is normal, but there is nevertheless an overwhelming tendency to produce songs consisting of a single segment. Yet, in plastic song, multisegmental songs are the rule rather than the exception. This is true even of birds trained with unisegmental songs; normal syntax only emerges as songs crystallize (Fig. 5).

At present we can only speculate about why these aspects of song structure emerge so late in development in the swamp sparrow. Elaboration and perfection of syllabic structure, a feature of plastic song, might have to progress to a certain degree of completion before a central motor program for normal swamp sparrow syntax is accessed. We cannot exclude the possibility of central motor programs capable of gen-

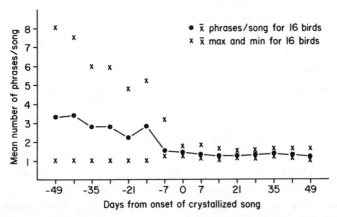

FIGURE 5. A demonstration of the late emergence of species-specific song syntax in male swamp sparrows, with one phrase per song, at the time of song crystallization. Mean numbers of phrases per song, and average maxima and minima, are plotted in relation to the day of song crystallization.

erating more structure than is evident in the song of deaf swamp sparrows, but withheld from manifestation because a deaf bird never achieves sufficient syllabic stability to initiate the last stages of normal motor development.

Turning from song production to song perception, it may be relevant that male swamp sparrows are unresponsive to the syntax of songs presented as stimuli, both during the sensitive period for acquisition and as mature territorial adults. This is not the case with song sparrows, whose song syntax, it will be remembered, is more complex. There is evidence of responsiveness to syntax both in infancy and in adulthood in this species (Marler and Peters, unpublished work). Thus it is instructive to examine the emergence of syntax in the produced song.

There is a contrast with the swamp sparrow in that signs of normal syntax emerge much earlier in song sparrow motor development. In certain cases entire renditions of imitated songs with normal syntax can be discerned early in plastic song, albeit with imperfect and highly variable phonology (Marler and Peters, unpublished work). Detailed analyses have yet to be completed, but guidance by auditory feedback during plastic song development appears to involve phonology and syntax more or less simultaneously. This is not to say that there is no evidence of a species-specific central motor program for song syntax in the song sparrow. The existence of such a program seems to be implied by the contrasts that can be extracted, albeit of a statistical nature, from data on the syntactical structure of songs of swamp and song sparrows deafened early in life (Marler and Sherman, 1983). Nevertheless in intact birds it appears that selective auditory mechanisms responsive to syntax are operative at earlier stages of motor development in the song sparrow than in the swamp sparrow.

Thus there is a degree of concordance between species differences in responsiveness to the syntax of songs as stimuli and differences in the role of auditory feedback mechanisms in the emergence of syntax in produced song. It is reasonable, although speculative, to postulate that the same auditory mechanisms are involved in shaping vocal acquisition preferences and in guiding the production of song motor patterns by auditory feedback. It remains to be seen whether the "autogenic" song units discovered by Margoliash (this volume) in the brain of white-crowned sparrows that are uniquely sensitive to an individual's own song play an active role in shaping the motor development of song by auditory feedback.

Evidently there are significant species differences in programs for motor development, and in the timing with which information in the central nervous system is made available.

Another indication of an association between sensitive period phenomena and the role of auditory feedback is to be seen in Nottebohm's discovery that effects of adult deafening on song are more drastic and disruptive in the canary (*Serinus canarius*) (Nottebohm et al., 1976), a species which, unlike sparrows, retains a degree of plasticity in its song throughout life (Nottebohm and Nottebohm, 1978). In age-limited learners, such as sparrows, deafening of mature birds, after song patterns have fully crystallized, has minimal effects on song structure and stability (Konishi, 1965).

Thus for the present it appears parsimonious to assume that shared sensory mechanisms are involved in guidance of song development by auditory feedback and in the manifestation of song learning preferences. In both cases birds display innate responsiveness to specific auditory stimuli.

HORMONAL CORRELATES OF SENSITIVE LEARNING PERIODS

In most birds, males typically sing and females do not. In some species female singing can be induced by androgen therapy, as in canaries, but even this may be difficult or impossible, as in the zebra finch. This vocal sexual dimorphism is a reflection of neuroanatomical differences between the sexes, found in the brain nuclei known to be involved in the control of song (Nottebohm and Arnold, 1976). The dimorphism is more extreme in some species, such as the zebra finch, than in others, such as the canary. These contrasts in brain structure are known to result, at least in part, from differences in the hormonal regimens prevailing in males and females in the pre- and immediately postnatal period. High estradiol production, peaking at about four days posthatching (Hutchison et al., 1984) is one factor responsible for the organizational effects that prepare the male brain for learning to sing (Arnold and Gorski, 1984; Gurney, 1981, 1982; Gurney and Konishi, 1980; Hutchison et al., 1984; Konishi and Akutagawa, 1985; Konishi and Gurney, 1982; Nordeen et al., 1986; Pohl-Apel, 1985; Pohl-Apel and Sossinka, 1984). It appears to be clear that the estrogen present in such high concentration in male songbird blood serum is in fact estradiol (Harding et al., unpublished work). Although the source has yet to be established, the tendency for a reciprocal relationship between levels of testosterone and estradiol in male swamp sparrows (see below) suggests the possibility of aromatization of testosterone as a mechanism, either in the brain (Callard et al., 1978; Hutchison and Steimer, 1983) or by some peripheral mechanism.

A week or two later, at about twenty days posthatching, the sensitive period for song acquisition begins, and again unusual hormonal profiles occur, perhaps hinting at a causal role. Once more, estradiol is implicated, as well as testosterone, in male zebra finches (Pröve, 1983) and canaries (Güttinger et al., 1984). In the chaffinch, Nottebohm (1969) has evidence of a significant role for testosterone. Castration delays closure of the sensitive period in this species, and testosterone therapy first induces learning later than is normal, and then brings the sensitive period to a close.

In the swamp sparrow the correlations between different phases of song development and unusual sex steroid hormone profiles in the blood plasma are striking, although their causal significance remains to be tested. It will be recalled that the sensitive period for song acquisition is extended, with a curious hiatus around 40 days of age. Song production occurs weeks later, and testosterone levels, which are at baseline throughout the winter (Fig. 6), rise in synchrony with the onset of singing (Marler, Peters, and Wingfield, unpublished work). There is ample

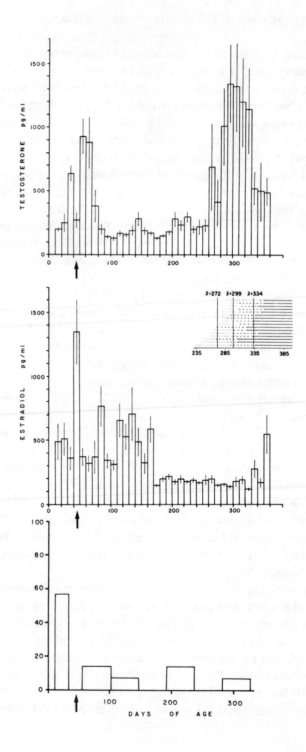

evidence from sparrows and other birds that the relationship between singing and androgens is a causal one, testosterone inducing song (Arnold, 1975a,b; Heid et al., 1985; Hutchison and Hutchison, 1983).

In swamp sparrows an earlier period of raised testosterone levels was also found from about 10–100 days, coinciding generally with the song acquisition phase, and in particular with the first part of the sensitive period. Estradiol levels were also high from about 10–170 days of age, again overlapping with the sensitive period for song learning. In addition, a remarkable estradiol peak was found between 40 and 50 days of age, coinciding with a trough in testosterone levels. In general testosterone levels tended to fall as those of estradiol rose. Especially intriguing is an indication that, in male swamp sparrows, there is a break in the song acquisition process that coincides with this sharp estradiol peak. The detail with which hormonal variations correlate with the changing readiness of male swamp sparrows to acquire new songs makes this species an ideal subject for direct investigation of the causal contributions of sex steroids to sensitive periods for song acquisition.

It is instructive to compare data for the zebra finch. When Pröve (1983) first described the remarkable peak in estradiol levels in the plasma of male zebra finches between the ages of about 35–50 days of age, this was thought to *coincide* with a major part of the sensitive period for song learning. However, Böhner (unpublished work) has shown that the sensitive period in male zebra finches for learning from a conspecific tutor is even shorter than previously estimated, beginning between 15 and 20 days of age and ending at about 35 days. Thus, as with the first phase of swamp sparrow song learning, an estradiol peak in early life may mark the termination of the normal song learning sensitive period.

The timing of the estradiol peak should be examined in zebra finches in which the sensitive period has been extended as a result of having been fostered onto an alien species and thus deprived of access to conspecific stimuli (Immelmann, 1967, 1969). It will also be illuminating to

---

FIGURE 6. Plasma levels of testosterone and estradiol in male swamp sparrows stimulated with live tutors from 10 to 300 days of age. Learning from the tutors is indicated in percentages in the lower histogram. The arrow marks a trough in testosterone levels, a peak in estradiol, and a hiatus in song learning. Relationships between hormone levels and later phases of the sensitive period for song learning are less clear. The insert, lined up with the same age scale, indicates the timing of the motor development of song in 16 males from another experiment. Plastic song (broken lines) and crystallized song (solid lines) follow subsong (dotted lines), the first stage, in a predictable fashion.

see whether a precise delineation of the recurrent sensitive phases for song learning in the canary, a life-long learner (Nottebohm and Nottebohm, 1978), reveals a decline or an increase in learning during the periods of high estradiol levels described in males by Guttinger et al., (1984), and during the seasonal variation in size of song-controlling nuclei in the adult canary brain, known to be under testosterone control (Nottebohm, 1981).

DOES OPTIMAL STIMULATION VARY AT DIFFERENT PHASES OF A SENSITIVE PERIOD?

It is possible that behavioral, neural, and hormonal mechanisms for song learning vary in different phases of a sensitive period. The suggestion has been advanced that social stimulation may be a more salient reinforcer of song learning in late stages of the sensitive period than at the time of onset. Experiments are needed to address this question directly.

In addition to those already considered, the Bengalese finch is another suggestive case. Dietrich (1980) defined some aspects of the sensitive period by switching young from one set of parents to another, or from their parents into isolation. Her data indicate that there are two phases. In the first, prior to independence, young males learn whatever songs they hear both from neighboring adults and the father. If a male with such a history is isolated at independence, the result is a song that incorporates imitations from both sources. If a male is left with the father after independence, an exclusive focus on his songs emerges, overriding the earlier learning from other males.

The wrens studied by Kroodsma and Pickert (1980) displayed late learning associated with social stimulation, provided in this case by territorial rivals and not by the father. Other possible examples are the indigo bunting (Rice and Thompson, 1968; Payne, 1983), the parasitic indigo bird (Payne, 1973), the chaffinch (Slater and Ince, 1982), and the white-crowned sparrow (Baptista and Petrinovich, 1984).

In some of the illustrations of relatively late learning, live tutoring overlaps with the motor development of song, raising the possibility of yet another form of behavioral plasticity. Stimulation may influence song development by affecting retrieval and retention, rather than stimulus acquisition. To appreciate the potential complications, we have to review the course of motor development in a little more detail.

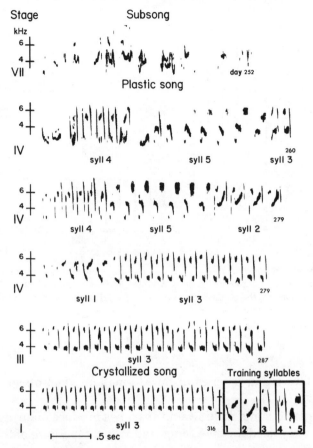

FIGURE 7. Samples of developing song in a male swamp sparrow progressing from subsong, through plastic song to crystallization. Age in days is indicated on the right. The box shows five training syllables presented as stimuli from 20 to 50 days of age. Roman numerals on the left indicate the numbered stage of development from I to VII. The overproduction during plastic song is clearly evident.

Study of the progression from early subsong to song crystallization in the swamp sparrow has revealed a strong tendency toward overproduction. Only subsong is produced up to about 300 days of age and little or no developmental progression is evident during this time. This is a period without rehearsal of any memorized material above the level of single notes. Storage of acquired stimuli without rehearsal lasts about 240 days (Marler and Peters, 1982a). Plastic song then appears for the first time, and song development proceeds rapidly to completion in

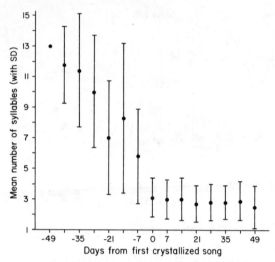

FIGURE 8. Syllable repertoire sizes of a group of young male swamp sparrows plotted in relation to the day of song crystallization. Mature repertoires average about three song types per male. During the 50 days or so prior to crystallization, male swamp sparrows overproduce songs in abundance, reducing the repertoire size as the time for crystallization approaches.

about 50 days. After a series of transformations, mature song finally crystallizes (Fig. 7).

In early plastic song, male swamp sparrows produce many more song types than they need for the average individual repertoire of three song types (Marler and Peters, 1982b). As the time for crystallization approaches, the 10 or so excess songs are discarded at a rate of one every 5 days on average (Fig. 8). Those that remain—three on average—become the permanent repertoire of that male. We also know that swamp sparrows persuaded by one means or another to learn alien song are more likely to discard alien components than conspecific ones in the attrition process.

The question of present interest is whether the attrition of conspecific songs is random, or whether stimulation at this time exerts an influence on choice of the final repertoire. Experimental data are lacking, but field observations on the song sparrow (Nice, 1943) are consistent with the assumption that song stimulation can affect the final choice of themes. As a male sparrow in plastic song engages in his first territorial disputes, he may be more likely to choose as songs for his crystallized repertoire the subset of his plastic song types that most closely match themes of the rivals with whom he is interacting. Laboratory experiments indicate that male sparrows do not normally acquire new songs at this late stage of development. On the other hand, there is still a limited range of potential plasticity in the process of retrieval and choice of previously learned song material.

In any case it is clear that a full understanding of the processes underlying late song learning induced by intensive social stimulation requires the analysis of all developmental stages. In the swamp sparrow, mature crystallized song reveals only a fraction of what a male has learned in early life (Marler and Peters, 1981a).

A final comment on hormonal and neural substrates for song learning is in order. Although correlations between sex steroid titers and learning appear to be close early in the sensitive phase, later on the correlations become looser, especially toward the closure of the sensitive period. It is intriguing to contemplate the possibility that hormones such as testosterone have an influence in later stages, not so much by direct neural effects but by indirectly affecting the intensity of social interactions. Wingfield (1985) has shown that social confrontations in sparrows elevate androgen titers. If these changes feed back on the intensity of social interactions and countersinging becomes intense, the stimulation may be increasingly likely to influence the process of selection from the plastic song repertoire.

There are also hints that the neural substrates for early and late song learning may be different, the latter perhaps associated with the curious process of shrinkage of one song control brain nucleus (MAN) in late stages of song development in the zebra finch (Bottjer et al., 1985; Arnold et al., this volume).

CONCLUSIONS

The evidence indicates that there are well-defined sensitive periods for song learning in birds. In some species they are extended, and perhaps even life-long, and in others they are restricted to a short period in infancy. Sensitive periods are not fixed, however, but labile. One way they can be extended is by withholding certain kinds of stimulation. Auditory stimulation with conspecific song is a significant and perhaps a dominant factor, but there are others. Conspecific visual stimuli may play a role, as well as the generalized arousal generated in the course of social interaction.

Another relevant factor is variation in photoperiod, presumably influencing song learning by way of its effects on the hormonal milieu. There are correlations between readiness to learn song and plasma levels of sex steroids, the causal significance of which has yet to be established. Early peaks of testosterone and estradiol correlate with maximal readiness of males to acquire new songs in some species. An exceptionally strong peak of estradiol, together with a drastic reduction in testosterone level, appears to correlate with a hiatus in learning in one species. Relationships between steroid levels and learning are less regular in later phases of the sensitive period, perhaps hinting at changes in the kind of influence exerted by physiological factors on the readiness to learn. One should not assume that the same exogenous and endogenous factors control vocal plasticity in different stages of the sensitive period.

Given a choice, conspecific song tends to be favored over alien songs for learning. Intensive social interaction imposed in a laboratory situation with a live singing tutor of another species can, however, override innate preferences for conspecific song. Responsiveness to sign stimuli contained in conspecific song is based both on phonological and on syntactical cues. There is some tradeoff between the two, deficiencies in one criterion being balanced by the presence of others. Certain cases of learning of alien songs become intelligible because their structure includes some of these innately specified criteria. In the laboratory, learning of alien songs is readily induced by embedding conspecific notes or syllables within them. Responsiveness to sign stimuli is implicated in the acquisition of song stimuli and their commitment to memory during the sensitive period.

A recurring difficulty in studies of learning including those on birdsong is how to derive a complete account of what has been acquired and placed in storage. There is evidence that more songs are learned and transferred into memory storage than are finally retrieved for use in producing the mature song repertoire. Auditory and audiovisual stimulation may influence which of the songs acquired in the sensitive period are finally chosen for crystallized song. Such an influence would be exerted, not in the "sensory phase," during song acquisition, but during retrieval, in the later "sensorimotor phase" of song development. This process of environmentally influenced selection from previously learned material may constitute a novel form of behavioral plasticity that is distinct from song acquisition in the customary sense. All stages of motor development must be analyzed to establish with certainty the kind of processes that underlie any given apparent case of song learning.

ACKNOWLEDGMENTS

I am grateful to many for discussion of the data and ideas presented here, including Drs. Josef Rauschecker, Patrick Bateson, Mark Konishi, Stephen Nowicki, Gregory Ball, Donald Kroodsma, Luis Baptista, and John Wingfield. My colleague Susan Peters is a coproducer both of the new data and many of the hypotheses presented. Esther Arruza prepared the figures and typed the manuscript. Research was supported in part by BRSG SO7 RR07065 awarded by the Biomedical Research Support Grant Program, Division of Research Resources, National Institutes of Health, and by grant number MH 14651 to Peter Marler.

REFERENCES

Adkins-Regan, E. (1983). Sex Steroids and the Differentiation and Activation of Avian Reproductive Behavior. In: *Hormones and Behaviour in Higher Vertebrates*, J. Balthazart, E. Pröve, and R. Gilles, eds., pp. 218–230, Springer-Verlag, Berlin.

Ainsworth, M. D. S., M. C. Blehar, E. Waters, and S. Wall (1978). *Patterns of Attachment*. Lawrence Erlbaum, Hillsdale, NJ.

Arnold, A. P. (1975a). The effects of castration and androgen replacement on song, courtship and aggression in zebra finchs (*Poephila guttata*). *J. Exp. Zool.* **191**:309–326.

Arnold, A. P. (1975b). The effects of castration on song development in zebra finches (*Poephila guttata*). *J. Exp. Zool.* **191**:261–277.

Arnold, A. P. (1982). Neural Control of Passerine Bird Song. In: *Acoustic Communication in Birds*, D. E. Kroodsma and E. H. Miller, eds., Vol. 1, pp. 75–94, Academic Press, New York.

Arnold, A. P. and R. A. Gorski (1984). Gonadal steroid induction of structural sex differences in the central nervous system. *Ann. Rev. Neurosci.* **7**:413–442.

Baker, M. C. (1982). Genetic Population Structure and Vocal Dialects in *Zono-*

trichia (Emberizidae). In: *Acoustic Communication in Birds,* D. E. Kroodsma & E. H. Miller, eds., Vol. 2, pp. 209–235, Academic Press, New York.

Baker, M. C., and M. A. Cunningham (1985). The biology of birdsong dialects. *Beh. Brain Sci.* **8**:85–133.

Baptista, L. (1975). Song dialects and demes in sedentary populations of the white-crowned sparrow (*Zonotrichia leucophrys nuttalli*). *Univ. Calif. Publ. Zool.* **105**:1–52.

Baptista, L. F. (1985). The functional significance of song sharing in the white-crowned sparrow. *Can. J. Zool.* **63**:1741–1752.

Baptista, L. F., M. L. Morton, and M. E. Pereyra (1981). Interspecific song mimesis by a Lincoln Sparrow. *Wilson Bull.* **93**:265–267.

Baptista, L. F. and L. Petrinovich (1984). Social interaction, sensitive phases and the song template hypothesis in the white-crowned sparrow. *Anim. Behav.* **32**:172–181.

Baptista, L. F. and L. Petrinovitch (In press). Song development in the white-crowned sparrow: social factors and sex differences. *Anim. Behav.*

Bateson, P. P. G. (1976). Specificity and the origins of behavior. *Adv. Study Behav.* **6**:1–20.

Bateson, P. P. G. (1978a). Early Experience and Sexual Preferences. In: *Biological Determinants of Sexual Behavior,* J. B. Hutchison, ed., pp. 29–53, Wiley & Sons, Chichester.

Bateson, P. P. G. (1978b). Sexual imprinting and optimal outbreeding. *Nature* **273**:649.

Bateson, P. P. G. (1979). How do sensitive periods arise and what are they for? *Anim. Behav.* **27**:470–486.

Böhner, J. (1983). Song learning in the zebra finch (*Taeniopygia guttata*): selectivity in the choice of a tutor and accuracy of song copies. *Anim. Behav.* **31**:231–237.

Bottjer, S. W. and A. P. Arnold (1986). The Ontogeny of Vocal Learning in Songbirds. In: *Handbook of Behavioral Neurobiology,* E. M. Blass, ed., Vol. 8, pp. 129–161, Plenum Press, New York.

Bottjer, S. W., S. L. Glaessner, and A. P. Arnold (1985). Ontogeny of brain nuclei controlling song learning and behavior in zebra finches. *J. Neurosci.* **5**:1556–1562.

Bottjer, S. W., E. A. Meisner, and A. P. Arnold (1984).. Forebrain lesions disrupt development but not maintenance of song in passerine birds. *Science* **224**:901–903.

Bowman, R. I. (1983). The Evolution of Song in Darwin's Finches. In: *Patterns of Evolution in Galapagos Organisms*, R. I. Bowman, M. Berson, and A. E. Levinton, eds., pp. 237–538, A.A.A.S., San Francisco.

Callard, G. V., Z. Petro, and K. J. Ryan (1978). Phylogenetic distribution of aromatase and other androgen-converting enzymes in the central nervous system. *Amer. Zool.* **18**:511–523.

Cate, C. (1984). The influence of social relations on the development of species recognition in zebra finch males. *Behaviour* **91**:263–285.

Colgan, P. (1983). *Comparative Social Recognition*. Wiley & Sons, New York.

Cunningham, M. A. and M. C. Baker (1983). Vocal learning in white-crowned sparrows: sensitive phase and song dialects. *Behav. Ecol. Sociobiol.* **13**:259–269.

DeVoogd, T. J., B. Nixdorf, and F. Nottebohm (1985). Synaptogenesis and changes in synaptic morphology related to acquisition of a new behavior. *Brain Res.* **329**:304–308.

Dietrich, K. (1980). Vorbildwahl in der Gesangsentwicklung bein Japanischen Mövchen (*Lonchura striata* var. *domestica*, Estrildidae). *Z. Tierpsychol.* **52**:57–76.

Dooling, R. J. (1982). Auditory Perception in Birds. In: *Acoustic Communication in Birds*, D. E. Kroodsma, and E. H. Miller, eds., Vol. 1, pp. 95–130, Academic Press, New York.

Fentress, J. C. (1981). Sensorimotor Development. In: *The Development of Perception: Psychological Perspectives*, R. N. Aslin, J. R. Alberts, and M. R. Petersen, eds., Vol. 1, pp. 293–318, Academic Press, New York.

Fentress, J. C. and P. J. McLeod (1986). Motor Patterns in Development. In: *Handbook of Behavioral Neurobiology*, E. M. Blass, ed., Vol. 8, pp. 35–97, Plenum Press, New York.

Goldman, S. A. and F. Nottebohm (1983). Neuronal production, migration, and differentiation in a vocal control nucleus of the adult female canary brain. *Proc. Natl. Acad. Sci. USA* **80**:2390–2394.

Gottlieb, G. (1971). *Development of Species Identification in Birds*. University of Chicago Press, Chicago.

Goy, R. W. and B. S. McEwen (1980). *Sexual Differentiation of the Brain*. MIT Press, Cambridge, MA

Grant, B. R. (1984). The significance of song variation in a population of Darwin's finches. *Behaviour* **89**:90–116.

Gurney, M. E. (1981). Hormonal control of cell form and number in the zebra finch song system. *J. Neurosci.* **1**:658–673.

Gurney, M. E. (1982). Behavioral correlates of sexual differentiation in the zebra finch song system. *Brain Res.* **231**:153–172.

Gurney, M. E. and M. Konishi (1980). Hormone induced sexual differentiation of brain and behavior in zebra finches. *Science* **208**:1380–1383.

Güttinger, H. R., E. Pröve, K. Weichel, and A. Pesch (1984). Hormonelle Korrelate zur Gesangsentwicklung der Kanarienvögel. *J. Ornithol.* **125**:245–247.

Heid, P., H. R. Güttinger, and E. Pröve (1985). The influence of castration and testosterone replacement on the song architecture of canaries (*Serinus canaria*). *Z. Tierpsychol.* **69**:224–236.

Helb, H.-W., F. Dowsett-Lemaire, H.-H. Bergmann, and K. Conrads (1985). Mixed singing in European songbirds—A review. *Z. Tierpsychol* **69**:27–41.

Hess, E. H. (1973). *Imprinting.* Van Nostrand Reinhold Co., New York.

Hinde, R. A. (1974). *Biological Bases of Human Social Behaviour.* McGraw-Hill Book Co., New York.

Hinde, R. A. (1983). Ethology and Child Development. In: *Handbook of Child Psychology*, M. M. Haith and J. J. Campos, eds., Vol. II, pp. 27–93, Wiley & Sons, New York.

Hofer, M. A. (1981). *The Roots of Human Behavior*, W. H. Freeman & Co., San Francisco.

Horn, G. (1985). *Memory, Imprinting and the Brain*, Clarendon Press & Co., Oxford.

Horn, G., S. P. R. Rose, and P. P. G. Bateson (1973). Experience and plasticity in the central nervous system. *Science* **181**:506–514.

Hutchison, J. B., and R. E. Hutchison (1983). Hormonal Mechanisms of Mate Choice in Birds. In: *Mate Choice*, P. Bateson, ed., pp. 389–405, Cambridge University Press, Cambridge.

Hutchison, J. B. and R. E. Hutchison (1985). Phasic Effects of Hormones in the Avian Brain during Behavioral Development. In: *Neurobiology*, R. Gilles and J. Balthazart, eds., pp. 105–120, Springer-Verlag, Berlin.

Hutchison, J. B. and T. Steimer (1983). Hormone-mediated Behavioural Transitions: A Role for Brain Aromatase. In: *Hormones and Behaviour in Higher Vertebrates*, J. Balthazart, E. Pröve, and R. Gilles, eds., pp. 161–174, Springer-Verlag, Berlin.

Hutchison, J. B., J. C. Wingfield, and R. E. Hutchison (1984). Sex differences in plasma concentrations of steroids during the sensitive period for brain differentiation in the zebra finch. *J. Endocrinol.* **103**:363–369.

Immelmann, K. (1967). Zur ontogenetischen Gesangsentwicklung bei Prachtfinken. *Verh. Dtsch. Zool. Ges.* **60**:320–332.

Immelmann, K. (1969). Song Development in the Zebra Finch and Other Estrildid Finches. In: *Bird Vocalizations*, R. Hinde, ed., pp. 61–74, Cambridge University Press, Cambridge.

Immelmann, K. and S. J. Suomi (1981). Sensitive Phases in Development. In: *Behavioral Development*, K. Immelmann, G. W. Barlow, L. Petrinovich, and M. Main, eds., pp. 395–431, Cambridge University Press, Cambridge.

Johnston, T. D. and G. Gottlieb (1981). Development of visual species identification in ducklings. *Anim. Behav.* **29**:1082–1099.

Konishi, M. (1965). The role of auditory feedback in the control of vocalization in the white-crowned sparrow. *Z. Tierpsychol.* **22**:770–783.

Konishi, M. (1985). Birdsong: from behavior to neuron. *Ann. Rev. Neurosci.* **8**:125–170.

Konishi, M. and E. Akutagawa (1985). Neuronal growth, atrophy and death in a sexually dimorphic song nucleus in the zebra finch brain. *Nature* **315**:145–147.

Konishi, M. and M. E. Gurney (1982). Sexual differentiation of brain and behavior. *Trends Neurosci.* **5**:20–23.

Kroodsma, D. E. (1974). Song learning, dialects, and dispersal in the Bewick's wren. *Z. Tierpsychol.* **35**:352–380.

Kroodsma, D. E. (1978). Aspects of Learning in the Ontogeny of Bird Song: Where, from Whom, When, How Many, Which, and How Accurately. In: *Ontogeny of Behavior*, G. Burghardt and M. Beckoff, eds., pp. 215–230. Garland Press, New York.

Kroodsma, D. E. (1982). Ontogeny of Birdsong. In: *Behavioural Development*, K. Immelmann, G. W. Barlow, L. Petrinovich, and M. Main, eds., pp. 518–532, Cambridge University Press, Cambridge.

Kroodsma, D. E., M. C. Baker, L. F. Baptista, and L. Petrinovich (1984). Vocal "Dialects" in Nuttall's White-crowned Sparrow. In: *Current Ornithology*, R. F. Johnston, ed., Vol. 2, pp. 103–133, Plenum Press, New York.

Kroodsma, D. and R. Pickert (1980). Environmentally dependent sensitive periods for avian vocal learning. *Nature* **288**:477–479.

Kroodsma, D. E. and R. Pickert (1984). Sensitive phases for song learning: effects of social interaction and individual variation. *Anim. Behav.* **32**:389–394.

Lenneberg, E. H. (1967). *Biological Foundations of Language*. Wiley & Sons, New York.

Leppelsack, H.-J. (1983). Analysis of Song in the Auditory Pathway of Song Birds. In: *Advances in Vertebrate Neuroethology*, J. P. Ewert, R. Capranica, and D. J. Ingle, eds., pp. 783–799, Plenum Press, New York.

Lorenz, K. (1965). *Evolution and Modification of Behavior*. University of Chicago Press, Chicago.

Lorenz, K. Z. (1981). *The Foundations of Ethology*. Springer-Verlag, New York.

Margoliash, D. and M. Konishi (1985). Auditory representation of autogenous song in the song system of white-crowned sparrows. *Proc. Natl. Acad. Sci. USA* **82**:5997–6000.

Marler, P. (1970). A comparative approach to vocal learning: song development in white-crowned sparrows. *J. Comp. Physiol. Psychol.* **71**:1–25.

Marler, P., M. Konishi, A. Lutjen, and M. S. Waser (1973). Effects of continuous noise on avian hearing and vocal development. *Proc. Natl. Acad. Sci. USA* **70**:1393–1396.

Marler, P. and S. Peters (1977). Selective vocal learning in a sparrow. *Science* **198**:519–521.

Marler, P. and S. Peters (1981a). Sparrows learn adult song and more from memory. *Science* **213**:780–782.

Marler, P. and S. Peters (1981b). Birdsong and Speech: Evidence for Special Processing. In: *Perspectives in the Study of Speech*, P. D. Eimas and J. L. Miller, eds., pp. 75–112, Lawrence Erlbaum, Hillsdale, NJ.

Marler, P., and S. Peters (1982a). Long-term storage of learned birdsongs prior to production. *Anim. Behav.* **30**:479–482.

Marler, P. and S. Peters (1982b). Developmental overproduction and selective attrition: new processes in the epigenesis of birdsong. *Dev. Psychobiol.* **15**:369–378.

Marler, P., and S. Peters (in press). A sensitive period for song acquisition in the song sparrow, *Melospiza melodia*, a case of age-limited learning. *Ethology*.

Marler, P., and R. Pickert (1984). Species-universal microstructure in the learned song of the swamp sparrow, *Melospiza georgiana*. *Anim. Behav.*, **32**:673–689.

Marler, P., and V. Sherman (1983). Song structure without auditory feedback: Emendations of the auditory template hypothesis. *J. Neurosci.* **3**:517–531.

Marler, P., and V. Sherman (1985). Innate differences in singing behavior of sparrows reared in isolation from adult conspecific song. *Anim. Behav.* **33**:57–71.

Marler, P., and M. Tamura (1964). Culturally transmitted patterns of vocal behavior in sparrows. *Science* **146**:1483–1486.

McCasland, J. S. and M. Konishi (1981). Interaction between auditory and motor activities in an avian song control nucleus. *Proc. Natl. Acad. Sci. USA* **78**:7815–7819.

McEwen, B. A. (1983). Gonadal Steroid Influence on Brain Development and Sexual Differentiation. In: *Reproductive Physiology IV. International Review of Physiology*, R. O. Greep, ed., pp. 99–145, University Park Press, Baltimore.

Millington, S. J., and T. D. Price (1985). Song inheritance and mating patterns in Darwin's finches. *Auk* **102**:342–346.

Murray Parkes, C. and J. Stevenson-Hinde (1982). *The Place of Attachment in Human Behavior*. Basic Books, New York.

Nice, M. (1943). Studies in the life history of the song sparrow. II. *Trans. Linn. Soc. N.Y.* **6**:1–328.

Nicolai, J. (1959). Familientradition in der Gesangsentwicklung des Gimpels (*Pyrrhula pyrrhula* L.). *J. Ornithologie* **100**:39–46.

Nordeen, K. W., E. J. Nordeen, and A. P. Arnold (1986). Estrogen establishes sex differences in androgen accumulation in zebra finch brain. *J. Neurosci.* **6**:734–738.

Nottebohm, F. (1968). Auditory experience and song development in the chaffinch *Fringilla coelebs*. *Ibis* **110**:549–568.

Nottebohm, F. (1969). The "critical period" for song learning. *Ibis* **111**:386–387.

Nottebohm, F. (1972). Neural lateralization of vocal control in a passerine bird. II. Subsong, calls and a theory of vocal learning. *J. Exp. Zool.* **1979**:35–49.

Nottebohm, F. (1981). Testosterone triggers growth of brain vocal control nuclei in adult female canaries. *Brain Res.* **289**:429–436.

Nottebohm, F. (1984a). Vocal Learning and Its Possible Relation to Replaceable Synapses and Neurons. In: *Biological Perspectives on Language*, D. Caplan, A. R. Lecours, and A. Smith, eds., pp. 65–95, MIT Press, Cambridge, MA.

Nottebohm, F. (1984b). Birdsong as a model in which to study brain processes related to learning. *Condor* **86**:227–236.

Nottebohm, F. and A. P. Arnold (1976). Sexual dimorphism in vocal control areas of the songbird brain. *Science* **194**:211–213.

Nottebohm, F. and M. E. Nottebohm (1978). Relationship between song repertoire and age in the canary, *Serinus canarius*. *Z. Tierpsychol.* **46**:298–305.

Nottebohm, F., T. M. Stokes, and C. M. Leonard (1976). Central control of song in the canary, *Serinus canarius*. *J. Comp. Neurol.* **165**:457–486.

Nowicki, S. 1987. Vocal tract resonances in oscine bird sound production: evidence from birdsongs in a helium atmosphere. *Nature* **325**:53–55.

Oppenheim, R. W. and L. Haverkamp (1985). Early Development of Behavior and the Nervous System. In: *Handbook of Behavioral Neurobiology*, E. M. Blass, ed., Vol. 8, pp. 1–33, Plenum Press, New York.

Payne, R. B. (1973). Behavior, mimetic songs and song dialects, and relationships of the parasitic indigobirds (*Vidua*) of Africa. *Ornithol. Monog.* **11**:1–333.

Payne, R. B. (1976). Song mimicry and species relationships among the West African pale-winged indigobirds. *Auk* **93**:25–38.

Payne, R. B. (1981). Song learning and social interaction in indigo buntings. *Anim. Behav.* **29**:688–697.

Payne, R. B. (1982). Ecological consequences of song matching: breeding success and intraspecific song mimicry by indigo buntings. *Ecology* **63**:401–411.

Payne, R. B. (1983). The social context of song mimicry: songmatching dialects in indigo buntings (*Passerina cyanea*). *Anim. Behav.* **31**:788–805.

Petrinovich, L. (1985). Factors influencing song development in the white-crowned sparrow (*Zonotrichia leucophrys*). *J. Comp. Psychol.* **99**:15–29.

Pohl-Apel, G. (1985). The correlation between the degree of brain masculinization and song quality in estradiol treated female zebra finches. *Brain Res.* **336**:381–383.

Pohl-Apel, G., and R. Sossinka (1984). Hormonal determination of song capacity in females of the zebra finch: Critical phase of treatment. *Z. Tierpsychol.* **64**:330–336.

Price, P. H. (1979). Developmental determinants of structure in zebra finch song. *J. Comp. Physiol. Psychol.* **93**:260–277.

Pröve, E. (1983). Hormonal Correlates of Behavioural Development in Male Zebra finches. In: *Hormones and Behaviour in Higher Vertebrates*, J. Balthazart, E. Pröve and R. Gilles, eds., pp. 368–374, Springer-Verlag, Berlin.

Rajecki, D. W., M. E. Lamb, and P. Obmascher (1978). Toward a general theory of infantile attachment: a comparative review of aspects of the social bond. *Behav. Brain Sci.* **3**:417–464.

Rice, J. O., and W. L. Thompson (1968). Song development in the indigo bunting. *Anim. Behav.* **16**:462–469.

Searcy, W. A., and P. Marler (in press). Response of sparrows to songs of isolation-reared and deafened males: further evidence for innate auditory templates. *Dev. Psychobiol.*

Searcy, W. A., P. Marler, and S. S. Peters (1985). Songs of isolation-reared sparrows function in communication, but are significantly less effective than learned songs. *Behav. Ecol. Sociobiol.* **17**:223–229.

Sinnott, J. M. (1980). Species-specific coding in bird song. *J. Acoust. Soc. Amer.* **68**:494–497.

Slater, P. J. B. (1983). Bird Song Learning: Theme and Variations. In: *Perspectives in Ornithology*, pp. 475–499, Cambridge University Press, Cambridge.

Slater, P. J. B. and S. A. Ince. 1982. Song development in chaffinches: what is learnt and when? *Ibis* **124**:21–26.

Suomi, S. J. (1985). Ethology: Animal Models. In: *Comprehensive Textbooks of Psychiatry*, H. I. Kaplan & B. J. Sadock, eds., Vol. IV. Williams & Wilkins Co., Baltimore.

Thielcke, G. (1984). Gesangslernen beim Gartenbaumläufer (*Certhia brachydactyla*). *Die Vogelwarte* **32**:282–297.

Thorpe, W. H. (1958). The learning of song patterns by birds, with especial reference to the song of the chaffinch, *Fringilla coelebs*. *Ibis* **100**:535–570.

Thorpe, W. H. (1961). *Birdsong: the Biology of Vocal Communication and Expression in Birds*. Cambridge University Press, Cambridge.

Tinbergen, N. (1951). *The Study of Instinct*. Oxford University Press, London.

Todt, D., H. Hultsch, and D. Heike (1979). Conditions affecting song acquisition in nightingales (*Luscinia megarhynchos* L.). *Z. Tierpsychol.* **51**:23–35.

Waser, M. S., and P. Marler (1977). Song learning in canaries. *J. Comp. Physiol. Psychol.* **91**:1–7.

Williams, H. and F. Nottebohm (1985). Auditory responses in avian vocal motor neurons: a motor theory for song perception in birds. *Science* **229**:279–282.

Wingfield, J. C. (1985). Short term changes in plasma levels of hormones during establishment and defense of a breeding territory in male song sparrows, *Melospiza melodia*. *Horm. Behav.* **19**:174–187.

6

NEURAL CONSEQUENCES OF IMPRINTING

Philip Bradley

Department of Anatomy, University of Newcastle upon Tyne, Newcastle upon Tyne, England

Gabriel Horn

Department of Zoology, University of Cambridge, Cambridge, England

IMPRINTING AS A LEARNING PROCESS

Filial imprinting occurs in the domestic chick, as it does in some other animals that show coordinated locomotor activity shortly after birth. Imprinting in the domestic chick (*Gallus domesticus*) involves an interaction between learning processes by which young birds come to recognize objects to which they are exposed and a predisposition to approach objects possessing some features of conspecifics. Some of the neural networks and mechanisms underlying the predisposition and the learning processes appear to be different (see Horn, 1985).

A visually naive young chick will approach a wide range of objects. If the chick is exposed to one object for some time, the chick may learn its characteristics and avoid other objects. This change in behavior suggests that the chick recognizes the object to which it was exposed and that information about the object is stored in the chick's central nervous system. What is the basis of this storage process?

It has long been considered that the storage of information acquired through learning involves changes in the synapse (Tanzi, 1893; Cajal, 1911; Hebb, 1949). There is much evidence that synapses are modifiable structures. However, with the exception of some aspects of habituation, the evidence that synapses are critically involved in information storage has been tenuous. In this chapter we discuss imprinting, provide evidence that a particular brain region is critical for this learning process, and show that specific changes in the morphology of synapses occur in this region as a consequence of training.

CHARACTERIZATION OF NEURAL CHANGES IN IMPRINTING

Biochemical Studies

If the storage process that underlies imprinting involves changes in the connections between neurons, changes in protein and RNA metabolism may be expected to occur in those brain regions in which storage takes place. To examine this possibility, dark-reared, visually naive chicks were exposed to a rotating illuminated box, which is an effective imprinting object. Dark-reared birds were used as controls. There was an increase in the incorporation of [^{3}H]lysine into protein and also of [^{3}H]uracil into RNA in a restricted brain region of the imprinted chicks

(Bateson et al., 1972). The region was the dorsal part of the cerebral hemispheres, which had been divided into two parts, the forebrain roof and the forebrain base. No effects of training were observed in the forebrain base or in the midbrain.

There are, of course, many ways in which the trained chicks differed from the dark-reared controls, and the biochemical changes may have been related to some or all of these differences. A number of control experiments were therefore undertaken in order to determine whether the biochemical changes were specifically related to the learning process, or whether they reflected some nonspecific differences in the behavior of the trained and dark-reared chicks, such as differences in arousal, sensory stimulation, and motor activity. The control procedures involved training chicks for different periods of time (Bateson et al., 1973) or for a fixed period of time (Bateson et al., 1975) and training one side of the brain of "split-brain" chicks (Horn et al., 1973a). The results of all these procedures together suggested that the changes in the incorporation of radioactive uracil into RNA in the forebrain roof of trained chicks were closely related to the imprinting process.

By using an autoradiographic technique with [^{14}C]uracil as the probe molecule, an increased incorporation into RNA was found in a restricted part of the medial hyperstriatum ventrale (Horn et al., 1979; see also Kohsaka et al., 1979). The region lies approximately halfway between the anterior and posterior poles of the cerebral hemispheres. Accordingly, this intermediate and medial part of the hyperstriatum ventrale was referred to as IMHV.

Lesion Studies

If the IMHV is involved in the storage of information, destruction of the region should prevent the acquisition of a preference through imprinting and impair the retention of an acquired preference. Both of these predictions were met by the results of two lesion studies that involved the bilateral destruction of IMHV (McCabe et al., 1981, 1982). Lesions of two other brain regions, the Wulst and the lateral cerebral area, were without effect on retention.

Two imprinting stimuli were used in the lesion studies of McCabe et al. (1981, 1982). Half of the chicks were trained by exposing them to a rotating flashing red box; the other half were trained by exposing the chicks to a rotating stuffed jungle fowl (*Gallus gallus*). Preferences were tested by placing the chicks individually in running wheels in front of which stood either the rotating box or the fowl. The wheel rotated if the chick attempted to run toward an object. The number of rotations made in a set period of time provided a measure of approach activity. Each chick's approach activity toward the training object and, separately, to a novel object (the box or fowl, whichever the chick had not previously seen) were recorded. The chicks were considered to prefer one object to the other if they approached that object relatively more vigorously than the other in the preference test. The chicks were considered to have no preference if they approached the two objects with equal vigor.

In each of four lesion studies (see Horn and McCabe, 1984) an analysis of variance had been performed on the preference measures contributed by the lesioned and control groups of chicks. In any one study the preferences of chicks for the object on which they had been trained was not significantly affected by the nature of the training object (box or jungle fowl). However, when data for all the lesion studies were combined, a clear effect appeared: lesions to IMHV profoundly impaired the preferences of chicks exposed to the box but had a much smaller, though significant, effect on chicks trained on the jungle fowl (Horn and McCabe, 1984).

PREDISPOSITIONS AND ACQUIRED PREFERENCES

The results of two further studies suggest that the neural consequences of exposing visually naive chicks to the rotating red box or to the jungle fowl are different. The neurotoxin N-(2-chlorethyl)-N-ethyl-2-bromo-benzylamine hydrochloride (DSP4), which lowers brain noradrenaline levels, impairs the acquisition of a preference for the box (Davies et al., 1985). In contrast, exogenous testosterone enhances the acquisition of a preference for the jungle fowl but is without effect on the acquisition of a preference for the box. Furthermore, there is a positive correlation between plasma testosterone concentration and the preference of fowl-trained chicks for the training object; there is no such correlation for chicks trained on the red box (Bolhuis et al., 1985).

In order to interpret these results, it is necessary to consider again the chicks' behavior. Visually naive chicks will approach a wide range of objects. If a chick remains exposed to a particular object for an adequate length of time, it learns the characteristics of that object. The evidence reviewed above suggests that when the stimuli are artificial, IMHV is involved in the learning process, probably in storage. These stimuli include red and orange rotating boxes, red cylinders, and flashing bars of light. What may be happening when chicks are exposed to the jungle fowl? They run toward the fowl, and if exposed to it for long enough, they learn the characteristics of the individual fowl. Chicks with bilateral lesions of IMHV also run toward the jungle fowl during the training period. However, these chicks do not selectively approach that fowl when given a choice between it and another stuffed jungle fowl (Johnson and Horn, 1985). The lesioned chicks show no evidence, therefore, of recognizing the individual to which they had been exposed. These results suggest that IMHV may be implicated in the recognition of individual jungle fowls, a process that involves learning. More generally, it seems likely that IMHV is involved in the storage of information acquired through exposure learning, whether the object to which a chick is exposed is an artificial imprinting stimulus or an individual conspecific (Horn and McCabe, 1984; Johnson et al., 1985).

THE MORPHOLOGICAL BASIS OF IMPRINTING

If, as the evidence suggests, IMHV is critically implicated in the storage of information acquired through imprinting, IMHV is an appropriate place to investigate the putative morphological changes that form the basis of the storage process. The results of such investigations are described below. It is, however, important to bear in mind that, in the chick, any neural change that occurs as a consequence of training is imposed on a developing brain. For this reason, the developmental status of IMHV is described below. It is also necessary to study the internal structure of IMHV and to examine the connections it makes with other areas of the brain. All this information may need to be taken into account by those who wish to draw comparisons between imprinting and plasticity in the visual cortex.

Structure and Connections of IMHV

The avian telencephalon appears, histologically, to be a laminated structure. The dorsal laminae are collectively known as the hyperstriatum and comprise the hyperstriatum accessorium, intercalatum, dorsale, and ventrale. Of these, the first three are most dorsal and comprise the visual Wulst. The Wulst together with the ectostriatum are major primary telencephalic visual projection areas. Of the more ventral structures of the telencephalon, the paleostriatal complex is considered to be homologous with components of the mammalian basal ganglia (Reiner et al., 1984) while the archistriatum contributes to descending motor pathways as well as to the avian "limbic" system (Zeier and Karten, 1971). Parts of the neostriatum receive inputs from auditory (Karten, 1968), somatic sensory (Delius and Bennetto, 1972), and visual (Parker and Delius, 1972) pathways.

The connections of IMHV in the chick brain have been investigated (Bradley et al., 1985). IMHV has reciprocal connections with the visual Wulst, the paleostriatum augmentatum, and the dorsal part of the archistriatum. It receives additional afferents from the hippocampus, area septalis, neostriatum, and the caudal part of the hyperstriatum ventrale. IMHV sends efferents to the posterior part of the archistriatum and to the dorsolateral part of the cerebral hemispheres. A number of these connections are bilateral. IMHV is thus ideally placed to act as an integrative center in the telencephalon, linking sensory and motor systems and subject to influences from regions of the brain thought to be concerned with motivational aspects of behavior.

A Golgi study of the internal organization of IMHV revealed no evidence of any clear lamination but showed that there are at least four distinct neuronal types present within this region (Bradley and Horn, 1982). The two major cell types were classified as either large multipolar neurons or as pyramidal-type neurons. The remainder of the neurons present were either small and bipolar or multipolar. The axons of the pyramidal-type neurons were traced beyond the boundaries of IMHV, and it is likely that these neurons form a major source of the efferent projection fibers. The multipolar neurons, in contrast, possessed axons that arborized extensively within IMHV. It is not yet known how the afferents to IMHV are distributed on the various intrinsic neuronal types.

It is useful to search for homologies between structures in avian and mammalian brains (see Karten, 1969; Reiner et al., 1984), and it is possible that IMHV corresponds to certain areas of association cortex of the mammalian brain. However, all such comparisons must be treated with caution (for further discussion see Horn, 1985).

Development of IMHV

Changes in synapses that occur in IMHV as a result of imprinting (see the next section) are not the only changes that occur in this region in the early posthatch period. During the first 30 days after hatching there are significant developmental changes within IMHV. Neuronal cell density decreases significantly with age, while between the time of hatching and 7 days of age there is a significant rise in the density within the neuropil of synapses on dendritic spines (Bradley, 1985). This rise is matched by an increase in the synapse–neuron ratio over the same period. The change in this ratio is partially accounted for by an increase in the length of the terminal dendritic branches of pyramidal-type neurons (Bradley, 1983b). These changes are most rapid in the first few days of life so that IMHV is in a labile state during this time. IMHV is not unique in this respect since the hyperstriatum accessorium is also undergoing developmental changes over the same period. Thus, a change in synaptic morphology associated with imprinting may easily be confounded with developmental changes unless the ages of experimental and control chicks are carefully matched.

Experimental Analysis of Synaptic Changes Associated with Imprinting

In the first experiment designed to investigate morphological consequences of imprinting two groups of chicks were used (Bradley et al., 1981). Both groups were hatched and reared in darkness. When the chicks were approximately 21 hours old, they were exposed to a situation known to be effective in inducing imprinting. The imprinting stimulus consisted of a horizontal red bar of light 2 cm high and moving upward at 47 cm/sec in a window 10 cm wide and 15 cm deep (Horn et al., 1979). The window was approximately 40 cm in front of the chick. During the training period chicks attempted to approach the stimulus. One group of chicks was exposed to the stimulus for 20 min (undertrained chicks). Chicks in the other group were exposed to the stimulus for a total of 140 min (overtrained birds).

After training the chicks were returned to the dark and 3 hr later, approximately 30 hr after hatching, all the chicks were killed by perfusion under anaesthesia with 0.1 M phosphate buffer, pH 7.4, followed by 2% glutaraldehyde/2% paraformaldehyde fixative. Blocks were removed from the right and left IMHV and processed for electron microscopy. Standard stereological sampling procedures were used, and for each block a total of 25 micrographs printed at a final magnification of 22,200 were examined. A number of features of synapses was measured. These features included the mean percentages of micrograph area occupied per synaptic bouton and dendritic spine and the length of the postsynaptic density. In this, and in the other experiments described below, all measurements were made without knowledge of the treatment each chick had received or the side of the brain from which a block had been cut.

Of the various measures of synapse structure, only the length of the postsynaptic density was significantly affected by further training. The effect was found in synapses of the left IMHV, not the right. The length of the postsynaptic density was shorter in the left IMHV than the right in undertrained chicks. Further training eliminated this asymmetry, primarily through an effect on the left IMHV.

These studies were extended by including a dark-reared group of birds and by increasing the number of birds in the undertrained group (Horn et al., 1986). Furthermore, synapses were subdivided into two groups, those on the shafts and those on the spines of dendrites.

The only significant effect of training was in the left IMHV. The mean lengths of the postsynaptic density of spine synapses ($\overline{Sp}_L$) in this region of dark-reared chicks was $286.00 + 12.24$ nm. This mean was significantly less, by $49.2 + 17.8$, than that of overtrained birds ($P < 0.01$) and not significantly different from that of undertrained birds. The $\overline{Sp}_L$ was significantly greater in overtrained than in undertrained chicks ($P < 0.01$). No other measure of synapse morphology in the left IMHV was significantly affected by training.

A second experiment (Horn et al., 1986) was conducted to allow comparisons to be made with a study that had been designed to investigate some electrophysiological consequences of imprinting (Payne and Horn, 1984). Two groups of chicks were used. One group of 15 chicks were dark reared throughout. The other group of 15 chicks were reared in the same way until the chicks were 21 hr old. They were then exposed to a rotating flashing red box (McCabe et al., 1982) for 1 hr. The chicks were trained for further 1-hr periods at 24 and 42 hr posthatch. Between training periods and before tissue removal the chicks were replaced in the dark-holding incubators. Three hours after the end of training the chicks were killed and blocks of tissue were removed from the right and left hyperstriatum accessorium as well as from the right and left IMHV. The blocks were processed for electron microscopy as described above.

There were no significant effects of training on any measure of synapse morphology in the hyperstriatum accessorium or in the right IMHV. There was, however, an effect of training on spine synapses in the left IMHV. The $\overline{Sp}_L$ in the left IMHV of trained chicks was significantly greater than that in the dark-reared birds ($P < 0.05$).

The relationship between approach activity during training and the length of the postsynaptic density of spine synapses was examined. For the left IMHV the correlation between $\overline{Sp}_L$ and approach counts recorded during the first hour of training was significant ($r = 0.52$, d.f. 13, $p <$ 0.05). The correlation was not significant when approach counts during the second and third hour of training were used as variables. Nor was there a significant correlation between the $\overline{Sp}_L$ of the right IMHV and approach activity or between the $\overline{Sp}_L$ in the hyperstriatum accessorium and approach activity.

DISCUSSION

The most striking feature of the results presented here is the consistent involvement in imprinting of one set of structures in the left IMHV. Thus, in two separate studies using different training protocols, different stimuli, and chicks of different ages, the only morphological measure to be significantly altered by training was the length of the postsynaptic density of synapses on dendritic spines in the left IMHV. This measure was also significantly correlated with a measure of the chick's behavior, namely their approach activity during the first hour of training. Previous studies have shown that this approach activity is correlated with the strength of imprinting (Bateson and Jaeckel, 1974).

The evidence of a hemispheric asymmetry in the imprinting process has been followed up by lesion studies (Cipolla-Neto et al., 1982; Horn et al., 1983). The results of these studies have shown that the right and left IMHV do indeed play different roles in this process. Briefly, the left IMHV appears to be critical for information storage, but the right IMHV may serve as a "buffer store," passing information on to some other brain region, or regions, collectively referred to as S' (Cipolla-Neto et al., 1982; Horn, 1981, 1985).

Using an entirely different learning task, Rose and his collaborators

(Stewart et al., 1984) have found changes in synaptic morphology in the medial part of the hyperstriatum ventrale, a region that overlaps with the IMHV, in the chick telencephalon. It would seem, therefore, that the medial part of the hyperstriatum ventrale can act as a modifiable, integrative structure in the brain of the chick and possibly of other birds as well (Maier and Scheich, 1983). This is not to say that it is the only such region in the avian brain, nor that it is the only region involved in learning. Indeed, as stated above, there is evidence that in the process of imprinting other brain regions are also involved in the storage of information (Cipolla-Neto et al., 1982; Horn, 1985).

It is a widely held belief that long-term memory is dependent on the structural modification of particular sets of synapses, and the evidence presented here supports this view. Such a modification could be achieved, for example, by increasing receptor number (see Horn, 1962) or receptor affinity in the postsynaptic density. These changes may be associated with changes in the levels of protein phosphorylation. The present study provides evidence that the overall size of the postsynaptic density is increased, but it is not yet known in what ways, if any, this change is associated with a change in the number of receptors in this membrane. Nor is it known whether the synapses affected by training are excitatory or inhibitory. However, indirect evidence suggests that spine synapses are excitatory (Gray, 1959; Eccles, 1964).

Because of the methods used for quantifying the synaptic changes, it is not possible to state whether they represent small changes in synaptic length distributed among a large population of dendritic spines in the left IMHV or whether they represent large changes in a small subset of these spine synapses. It is, however, clear that a short period of training leads to changes in a specific population of synapses within a region of the chick brain that plays a critical part in the storage of information acquired through a learning process. It is possible that there are continuities between these neural changes and those that occur in the visual cortex of kittens reared in a visually restricted environment (see Horn et al., 1973b): both kinds of neural change reflect the modifiable properties of neurons. Whether these continuities have a deeper significance remains an open question (see Horn, 1985).

REFERENCES

Bateson, P. P. G. and J. B. Jaeckel (1974). Imprinting: Correlations between activities of chicks during training and testing. *Anim. Behav.* 22:899–906.

Bateson, P. P. G., G. Horn, and S. P. R. Rose (1972). Effects of early experience on regional incorporation of precursors into RNA and protein in the chick brain. *Brain Res.* **39**:449–465.

Bateson, P. P. G., G. Horn, and S. P. R. Rose (1975). Imprinting: Correlations between behaviour and incorporation of [^{14}C]-uracil into chick brain. *Brain Res.* **84**:207–220.

Bateson, P. P. G., S. P. R. Rose, and G. Horn (1973). Imprinting: Lasting effects on uracil incorporation into chick brain. *Science* **181**:576–578.

Bolhuis, J. J., B. J. McCabe, and G. Horn (1986). Androgens and imprinting. Differential effects of testosterone on filial preference in the domestic chick. *Behav. Neurosci.* **100**:51–56.

Bradley, P. (1983a). Development of two regions of the chick telecephalon. *Neurosci. Lett.* **10**:588.

Bradley, P. (1983b). Development of dendritic trees in a region of the chick brain involved in a learning process. *Fidia Research Series, Frontiers in Neuroscience, Abstract book no.* **1**:149–151.

Bradley, P. (1985). A light and electron microscopic study of the development of two regions of the forebrain. *Dev. Brain Res.* **20**:83–88.

Bradley, P. and G. Horn (1982). A Golgi analysis of the hyperstriatum ventrale in the chick. *J. Anat.* (Lond.) **134**:599–600.

Bradley, P., G. Horn, and P. P. G. Bateson (1981). Imprinting: An electron microscopic study of chick hyperstriatum ventrale. *Exp. Brain Res.* **41**:115–120.

Bradley, P., D. C. Davies, and G. Horn (1985). Connections of the hyperstriatum ventrale of the domestic chick (*Gallus domesticus*). *J. Anat.* (Lond.) **140**:577–590.

Cajal, S. R. (1911). In: *Histologie du système nerveux de l'homme et des vertèbres*, Vol. 2, pp. 886–890, Maloine, Paris. (Republished 1955, Instituto Ramon y Cajal, Madrid).

Cipolla-Neto, J., G. Horn, and B. J. McCabe (1982). Hemispheric asymmetry and imprinting: The effect of sequential lesions to the hyperstriatum ventrale. *Exp. Brain Res.* **48**:22–27.

Davies, D. C., G. Horn, and B. J. McCabe (1985). Noradrenaline and learning: The effects of the noradrenergic neurotoxin DSP4 on imprinting in the domestic chick. *Behav. Neurosci.* **99**:652–660.

Delius, J. D. and K. Bennetto (1972). Cutaneous sensory projections to the avian forebrain. *Brain Res.* **37**:205–221.

Eccles, J. C. (1964). *The Physiology of Synapses*, Springer-Verlag, Berlin.

Gray, E. G. (1959). Axo-somatic and axo-dendritic synapses of the cerebral cortex: An electron microscope study. *J. Anat.* (Lond.) **93**:420–433.

Hebb, D. O. (1949). *The Organization of Behavior*, Wiley & Sons, New York.

Horn, G. (1962). Some Neural Correlations of Perception. In: *Viewpoints in Biology*, J. D. Carthy and C. L. Duddington, eds., Vol. 1, pp. 242–285. Butterworths, London.

Horn, G. (1981). Neural mechanisms of learning: An analysis of imprinting in the domestic chick. *Proc. R. Soc. Lond. B* **213**:101–137.

Horn, G. (1985). *Memory, Imprinting and the Brain*. Clarendon Press, Oxford.

Horn, G. and B. J. McCabe (1984). Predispositions and preferences. Effects on imprinting of lesions to the chick brain. *Anim. Behav.* **32**:288–292.

Horn, G., B. J. McCabe, and P. P. G. Bateson (1979). An autoradiographic study of the chick brain after imprinting. *Brain Res.* **168**:361–373.

Horn, G., B. J. McCabe, and J. Cipolla-Neto (1983). Imprinting in the domestic chick: The role of each side of the hyperstriatum ventrale in acquisition and retention. *Exp. Brain Res.* **53**:91–98.

Horn, G., P. Bradley, and B. J. McCabe (1986). Changes in the structure of synapses associated with learning. *J. Neurosci.* **5**:3161–3168.

Horn, G., S. P. R. Rose, and P. P. G. Bateson (1973a). Monocular imprinting and regional incorporation of tritiated uracil into the brains of intact and "split-brain" chicks. *Brain Res.* **56**:227–237.

Horn, G., S. P. R. Rose, and P. P. G. Bateson (1973b). Experience and plasticity in the central nervous system. *Science* **181**:506–514.

Johnson, M. H. and G. Horn (1986). Is a restricted brain region of domestic chicks involved in the recognition of conspecifics? Proc. European Brain and Behaviour Society, Oxford. *Beh. Brain Res.* **20**:109–110.

Johnson, M. H., J. J. Bolhuis, and G. Horn (1985). Interaction between acquired preferences and developing predispositions during imprinting. *Anim. Behav.* **33**:1000–1006.

Karten, H. J. (1968). The ascending auditory pathway in the pigeon (*Columba livia*). II. Telencephalic projections of the nucleus ovoidalis thalami. *Brain Res.* **11**:134–153.

Karten, H. J. (1969). The organisation of the avian telencephalon and some speculations on the phylogeny of the amniote telencephalon. *Ann. N.Y. Acad. Sci.* **167**:164–179.

Kohsaka, S., K. Takamatsu, E. Aoki, and Y. Tsukada (1979). Metabolic mapping of chick brain after imprinting using [^{14}C]2-deoxyglucose technique. *Brain Res.* **172**:539–544.

Maier, V. and H. Scheich (1983). Acoustic imprinting leads to differential 2-deoxy-D-glucose uptake in the chick forebrain. *Proc. Natl. Acad. Sci. USA* **80**:3860–3864.

McCabe, B. J., G. Horn, and P. P. G. Bateson (1981). Effects of restricted lesions of the chick forebrain on the acquisition of filial preferences during imprinting. *Brain Res.* **205**:29–37.

McCabe, B. J., J. Cipolla-Neto, G. Horn, and P. P. G. Bateson (1982). Amnesic effects of bilateral lesions placed in the hyperstriatum ventrale of the chick after imprinting. *Exp. Brain Res.* **48**:13–21.

Parker, D. M. and J. D. Delius (1972). Visual evoked potentials in the forebrain region of the pigeon. *Exp. Brain Res.* **14**:198–209.

Payne, J. K. and G. Horn (1984). Long-term consequences of exposure to an imprinting stimulus on spontaneous impulse activity in the chick brain. *Beh. Brain Res.* **13**:155–162.

Reiner, A., S. E. Brauth, and H. J. Karten (1984). Evolution of the basal ganglia. *Trends Neurosci.* **7**:320–325.

Stewart, M. G., S. P. R. Rose, T. S. King, P. L. A. Gabbott, and R. Bourne (1984). Hemispheric asymmetry of synapses in chick medial hyperstriatum ventrale following passive avoidance training: A stereological investigation. *Dev. Brain Res.* **12**:261–269.

Tanzi, E. (1983). I fatti e le induzioni nell' odierna istologia del sistema nervosa. *Riv. sper. Freniat. Med. leg. Alien. ment.* **19**:419–472.

Zeier, H. and H. J. Karten (1971). The archistriatum of the pigeon: Organization of afferent and efferent connections. *Brain Res.* **31**:313–326.

7

IMPRINTING AS A PROCESS OF COMPETITIVE EXCLUSION

Patrick Bateson

Sub-Department of Animal Behaviour, University of Cambridge, Cambridge, England

Imprinting was first made famous by Konrad Lorenz (1935). The extraordinary phenomenon of recently hatched ducklings or chicks approaching humans and developing social preferences for them had been described long before (see reviews in Bateson, 1966; Hess, 1973; Sluckin, 1972), but Lorenz drew attention to its importance in biology. The word *imprinting* vividly suggested that a permanent irremovable image had been left by the impact of experience on the soft wax of the developing brain. The wax was only thought to be soft at a particular stage in development and no impression could be left either before or after that crucial stage. Without doubt, this metaphor has been persuasive and strongly influenced the thinking of those working in other fields of research. For instance, Hubel and Wiesel (1970), when developing their own theories of cortical plasticity, explicitly referred to imprinting and the critical period concept associated with it.

As the ethological ideas were being adopted by neurophysiologists and others, the hypothesis of an all-or-nothing process occurring within a sharply defined developmental period had been disintegrating in the hands of the experimentalists who were attempting to test its implications at the behavioral level. In this chapter I shall briefly review what happened. The change in ethological thought, brought about by the empirical evidence collected in the 1960s, was very similar to the changes in thinking about the development of cortical plasticity occurring among physiologists 20 years later. I believe that knowledge of the earlier debate among the ethologists and its outcome is relevant to the current discussions about what enables the occurrence of plastic changes in the cortex and what subsequently prevents them.

Changing ways of thinking is never easy, particularly when the change means that the old but attractive metaphors have to be discarded. When the nature of the evidence and the varieties of explanations are clarified, potentially misleading assumptions are more easily detected. It is also easier to see how rival interpretations for sensitive periods can be distinguished. For these reasons I shall focus initially on how the evidence is collected and how it is then explained. This part of the chapter draws extensively on a joint article with Robert Hinde (Bateson and Hinde, 1987) and the reader is referred to it for further discussion and examples. After suggesting that a maturational clock explanation for the end of sensitive periods is less adequate than a newer competitive exclusion model, I finally discuss the benefits of the theoretical exercise for empirical research.

DEFINING SENSITIVE PERIODS

Times of susceptibility to long-term effects of current experience were originally termed *critical periods* by ethologists. This term, previously used in embryological studies, was first applied in the English language to behavioral examples by Lorenz (1937), when he published a translation of his famous Kumpan paper. "Critical period" implies a sharply defined phase of susceptibility preceded and followed by lack of susceptibility; if the relevant experience is provided before or after the period, no long-term effects are supposedly detectable. Experimental work on imprinting in the 1950s and 1960s showed, however, that the period was not so sharply defined as had been previously supposed and the term *sensitive period* was substituted by many ethologists (e.g., Hinde, 1961; see also Bateson, 1979, and Immelmann and Suomi, 1981, for discussions of terminology). The sensitive period concept implies a phase of greater susceptibility preceded and followed by lower sensitivity, with relatively gradual transitions.

Even though the terminology has been softened and made more appropriate, two bodies of data continue to present considerable conceptual difficulties. First, considerable lability has been found in the chronology of the sensitive period for filial imprinting, and lots of influences seemed to impinge on the underlying processes. Second, under certain conditions, plasticity could be renewed long after the sensitive period was supposedly over [see Bateson (1979, 1983b) for discussion of these issues]. Briefly, the following list of factors are all important in influencing a bird's responsiveness to an object:

1. The characteristics of the object; some things are much more attractive than others.
2. The stage of the bird's development at which the object was first presented; the presumption is that endogenous changes are at least partly responsible.
3. The length of time for which it was first presented.
4. The animal's experience prior to first presentation.
5. The animal's experience subsequent to first presentation.

Many other conditions can be added to this list such as the precise form of presentation (continuous, fixed intervals, unpredictably intermittent),

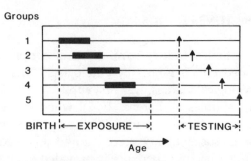

FIGURE 1. The operations required to demonstrate the existence of a sensitive period in development. Exposure is denoted by the thick bars. The number of groups is obviously arbitrary. Note that if the length of time from exposure to testing is kept constant (as shown by the arrows), the age of onset of exposure is confounded with the age of testing; but if the age of testing is kept constant, then the age of onset of exposure is confounded with the time from exposure to testing.

the bird's activity state (sleepy, alert, highly distressed), and whether or not other arousing or potentiating stimuli are used prior to or during the presentation (e.g., Johnston and Gottlieb 1981, 1985; Lickliter and Gottlieb, 1985). Furthermore, when the bird is developmentally ready, it seems to show a real hunger for learning and will actively work to present itself with appropriate forms of stimulation (Bateson and Reese, 1969).

The rich and complex data caused confusion because the concept of a sensitive period had been extended far beyond the operations that were used to define it. The sensitive period had been seen as the programmed opening of a window on the external world. After a predetermined time, the program supposedly shuts the window and ends possibilities for further modification of the system within. Of course, nobody could see the program—nor, indeed, the window. All that was observed was whether or not, at a given stage in development, the system could be modified by external manipulations. It is worth being clear, then, about the precise character of such manipulations.

The operations needed to establish the existence of a sensitive period are shown in Figure 1. The time when exposure to a given agent takes place is shown by the heavy lines. If the duration of exposure is merely started at different times and ended at the same time, then the evidence that is taken to support an age-specific effect is ambiguous; it could arise because the individuals exposed at an earlier age were also exposed for longer. Similarly, if the time from the end of exposure to testing is not kept constant, differences between the groups could arise because the effects of exposure could have attenuated more in group 1 than in group 5.

A more subtle point is that, if the length of time from exposure to testing is kept constant (as shown by the arrows), the age of onset of exposure is confounded with the age of testing; but if the age of testing is kept constant, then the age of onset of exposure is confounded with the time from exposure to testing. Strictly, the only way to be sure that neither of the confounded variables is responsible for the differences between the groups is to run two separate experiments. In one experiment the time from the end of exposure to testing is kept constant and in the other the age of testing is kept constant. However, in many cases the plausibility of one or other of the alternative explanations for the results would be low and the extra work unjustified.

Even when the evidence for a sensitive period satisfies the requirements outlined above, the conclusions that can be drawn from the findings are often limited in extent. Over the years ethologists have acquired some wisdom about these limitations. Their cautions, which should be remembered before generalizing too widely, are the following:

1. Even the most perfect demonstration of a sensitive period does not mean that the length of the period will be constant when the conditions are altered. In the case of filial imprinting in birds, social isolation lengthens the period (see Bateson, 1979).

2. Sensitive periods relate to specific types of input; other types of experience may exert effects at quite different times in the life cycle. While several types of behavior may be affected by experience during a given period of development, it does not follow that this is the only sensitive period in the life cycle (contrast this view with Scott, 1962).

3. Evidence for the existence of a sensitive period is not an explanation for the ineffectiveness of stimulation outside that period. Nothing new is added by stating that something has to occur "within the sensitive period" in order to be effective since the operations used to define the period are precisely those that generate the evidence to be explained.

4. The explanation for the onset of sensitivity need not necessarily be the same as the explanation for the end. The end of the sensitive

period for filial imprinting is almost certainly brought about by different processes from those responsible for the onset, a point I shall consider in greater detail later.

5. The grouping of phenomena under the descriptive heading of "sensitive periods" does not imply that they can all be explained in the same way.

6. Many descriptively defined sensitive periods may have no biological function whatsoever. They may merely represent a time of rapid reorganization when a given system is more easily disrupted by deprivation or insult. Even the implication that filial and sexual imprinting in a given individual have the same biological function is probably false (see Bateson, 1979, 1983a).

TYPES AND LEVELS OF EXPLANATION

Distinct types of explanations for sensitive periods are easily confused. This is partly because the difference between an immediate cause and one that operated further back in time is a matter of degree and, therefore, not always obvious. The difficulty is compounded because explanations are offered at a number of distinct levels of analysis ranging from the organismic to the molecular. In an attempt to bring some order into the tangle, Bateson and Hinde (1987) classified six types of explanation. The classification separated explanations in terms of events that accompany or immediately precede a change in sensitivity from those that postulate causes that lie further back in time. It also considered three levels of analysis at which the change might be explained: organismic, physiological, and molecular.

An example of an immediate organismic explanation for the end of sensitivity for filial imprinting is neophobia or the fear of novelty (Hess, 1959; Hinde et al., 1956). This in turn has been explained in terms of prior experience with another object leading to the establishment of a preference for the familiar and rejection of anything that the animal could detect as being different (Bindra, 1959; Sluckin and Salzen, 1961). At a lower level, when prior experience prevents or permits sensitivity to the long-term effects of novel input, it may preempt other types of experience from having the same impact. The phenomenon may be explained in terms of competitive exclusion (e.g., Bateson, 1981). The es-

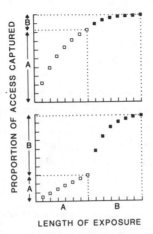

LENGTH OF EXPOSURE

FIGURE 2. Results of a simple competitive exclusion model when access to an executive system is captured at a particular stage in development. The amount of access captured by a stimulus equals $1 - e^{-ktr}$ where e is the base of the natural logarithm, k is a constant set at 0.5 in these simulations, t is the length of exposure to the stimulus, and r is the initial responsiveness to the stimulus. In the top simulation the initial responsiveness (r) is 0.25 to stimulus A and 0.5 to stimulus B. In the bottom simulation the values of r are 0.05 to A and 0.5 to B.

sence of this view is that a particular member of a class of inputs from the environment gains access to the systems responsible for executing the relevant patterns of behavior.

A number of possible physiological mechanisms could be suggested for competitive exclusion. For instance, suppose that gaining access involves growth of neural connections and that the area available for connections has finite size. When growth has proceeded beyond the halfway point and cannot be reversed easily, the input experienced first will be better able to control the behavior than other forms of output. The outcome of such a process is illustrated in Figure 2, which shows the proportion of access captured by first one (A) and then a second stimulus (B), assuming that the growth curves have an exponential character. This view of what happens was developed by Bateson (1981) and dubbed the "capacity model" by Boakes and Panter (1985). It also underlies the modeling attempts of Bischof (1985).

The alternative to the competitive exclusion model is the older notion of a physiological clock. Support for a clock model was seemingly provided by data that indicated that the beginning of the sensitive period for filial imprinting is better calculated in terms of age from the beginning of embryonic development rather than in terms of age from hatching (Gottlieb, 1961; but see Landsberg, 1976). Although preceding events at the behavioral level might influence a clock in nonspecific ways (see Aslin, 1981), the postulation of such mechanisms implies the proposal of a strong degree of autonomy from preceding external experience. Further, it was supposed that if such control operates on the onset of sensitivity, it also operates on the end.

Do the competitive exclusion and clock models make different predictions about what ends sensitivity? On the model of preemption by neural growth, once an animal is in a sensitive state, an influence of the input on the rate of growth will also affect whether or not the input experienced first in the sensitive state is subsequently more or less effective than input experienced later. For instance, a weak and impoverished stimulus might be reasonably expected to promote slower growth than a strong and rich one. This point is illustrated in Figure 2. When birds such as mallard ducklings or domestic chicks are reared in the dark or have their visual experience attenuated by being reared in patternless environments, the sensitive period for filial imprinting can be extended (e.g., Moltz and Stettner, 1961). This result is readily explained by the competitive exclusion model on the hypothesis that the rate of growth is thought to be slower when the intensity and quality of the input is low. However, the result can also be explained by an elaboration of the clock model if it is supposed that the running speed of the clock is partly dependent on external stimulation.

A different strand of evidence is needed to help us distinguish between the two rival explanations. Experience is portrayed as having a nonspecific stimulatory effect in the clock model, whereas it is thought to have a specific preemptive effect in the competitive exclusion model. In the second case some specific memory of the experience is required, otherwise it cannot have a preemptive effect. When domestic chicks or mallard ducklings are reared in isolation in a static but patterned environment, the sensitive period for imprinting is extended. However, when the sensitive period comes to an end, it can be shown that the birds have a memory for the environment they had experienced (e.g., Bateson, 1964a). The birds are more likely to respond socially to a moving object if it bears the same pattern as the walls of the cages in which they were previously isolated. This means that it would be incorrect to argue that rearing the animals in isolation merely delays a postulated clock bringing the sensitive period to a close. So, at least in the case of the end of the sensitive period for filial imprinting, the competitive exclusion model seems preferable to the clock one.

Of course, the competitive exclusion model only offers an explanation for the loss of sensitivity. It does not explain the onset of sensitivity. The onset is most satisfactorily accounted for in terms of programmed growth, even though it is necessary to postulate some influence on the timing processes of external conditions. The onset of sensitivity is clearly dependent on the sensory and motor development of the animals. In birds that are hatched naked and helpless, such as swallows, the onset occurs much later in relation to hatching than in the precocious ducklings and chicks (e.g., Burtt, 1977). It is possible to influence the onset of sensitivity by manipulating environmental stimulation. Birds exposed to patterned light are ready to learn earlier than those kept in darkness (Bateson and Wainwright, 1972). However, this merely seems to involve a speeding up of the increase in responsiveness, which, within limits, eventually happens anyway.

It is now worth taking a fresh look at the evidence that gave so much trouble when it merely seemed to create complication and confusion. In young precocial birds, the formation of a strong filial preference for a particular object seemed hopelessly variable. What the bird does depends critically on the conditions chosen by the experimenter and the length of a descriptively defined sensitive period always depends on those conditions. The theory strongly suggests that processes affecting the increase in sensitivity to the stimulation affecting social preferences are markedly different from those affecting the subsequent decline. Moreover, the birds are never in a vacuum, empty of experience. Once they are ready to be affected by particular types of stimulation from the environment, they are liable to be affected by the conditions around them. As this experience starts to leave its impact, the birds will reject objects they find noticeably different from the familiar conditions. The rate at which they become familiar clearly depends on the character of the stimulation. The process is fast if the birds are kept with members of their own species in the light; it is much slower if they are kept in isolation and slower still if they are kept in a patternless environment.

IMPLICATIONS FOR RESEARCH

The old notion of an internally determined sensitive period, within which imprinting had to occur, died hard. Since the data suggested that chicks and ducklings were most likely to respond to an experimenter's training object if they were presented with it on the first day after hatching, birds were usually trained on the first day. The mistaken assumption lay in supposing that initial responsiveness is closely linked to how much is learned and when imprinting naturally occurs. When I was a Ph.D. student I spent some time watching chicks that had been hatched out by live adult hens. The chicks spent nearly 80% of the daylight hours on the first day after hatching being brooded by the adult females (Bateson, 1963). I had expected to see something dramatic and was disappointed to find that when the chicks did occasionally emerge from under the adult hen, they hardly seemed to pay her any attention. Their activity around the hen did, however, increase substantially on the second day after hatching. Why did the timing of the peak initial responsiveness found in many laboratories not correspond to anything that was readily observable when the birds were reared with adult hens? It is possible to understand what happens when the old explanation for sensitive periods is cleared from our heads.

As soon as a dark-reared bird is brought into the light, it may attempt to escape from a moving object. This response becomes more and more pronounced as it gets older. Consequently, the time available for attending to the object and learning about its characteristics is reduced. When short training times of 15 min were used, the strength of the acquired preference unsurprisingly went down as the amount of escape went up (Hess, 1959). The mistake, though, was to assume that the amount of escape at a given age remains constant. It does not. Escape is much less obvious if the bird has been brought into the light prior to training (Polt and Hess, 1964; Bateson and Seaburne-May, 1973). Furthermore, escape from a particular moving object rapidly wanes (Bateson, 1964b; Ratner and Hoffman, 1974), and once that has happened the bird may approach and follow the object vigorously, learning about the object's characteristics as it does so. In older birds the approach and following of a chick is more intense than on the day after hatching (Bateson, 1964b; Smith and Nott, 1970). Consequently, once imprinting has started at 2 days of age, it may proceed more rapidly than it would have done at 1 day of age. Under natural conditions, even the initial avoidance of visual stimuli seen in the laboratory almost certainly does not happen. The natural mother's maternal calls, to which the chicks have been exposed and to which they have responded from before hatching, powerfully attract the chicks and can override any tendency to recoil from previously unfamiliar visual stimuli (see Gottlieb, 1971). Furthermore, even the brief periods of exposure to the natural mother on the first day after hatching would be quite enough to set the process of learning about her visual characteristics on the right course even though they probably operate rather slowly at that stage.

I agree with Gottlieb and his colleagues (e.g., Johnston and Gottlieb, 1981; Lickliter and Gottlieb, 1985) when they argue that the conditions under which imprinting is studied in the laboratory are so impoverished and artificial that the results can give a seriously misleading view of what happens in the wild. However, it does not follow that experimental analysis is, therefore, useless or that different neural systems are studied in laboratory and natural conditions. A car that is filled with low-grade gasoline and runs badly does not become another car on that account.

Even when opportunities to learn the characteristics of the natural mother are optimal, all that will be learned is not likely to be acquired at once. Apart from the time constraints involved in first detecting the most salient features and then the subtle ones that make an individual distinctive, the bird has to gather information about the front, side, and back views of its mother. All these views are physically distinct and they may also take on different appearances when viewed at different distances. The young bird is certainly capable of learning about a variety of different stimuli and, when they have been presented within the same time window, subsequently treats them all as though they were equivalent (Bateson, 1973). Therefore, the incisive, single-shot image conjured up by the term *imprinting* does not adequately represent what happens. Clearly, acquisition of the complex pattern recognition involved in detecting a particular individual from many different angles and distances takes some time. Precisely how long remains to be determined empirically.

In summary, it seems that the well-known sensitive period curves for chicks and ducklings, with their peaks within the first day after hatching, are misleading. Most processing and storage of information about the mother probably takes place at least a day later under natural conditions and, as we shall see, the system involved in analyzing the particular type of incoming information required for individual recognition may not be fully developed on the first day after hatching. These considerations require a more detailed look at the various developmental changes in the neural systems involved in imprinting.

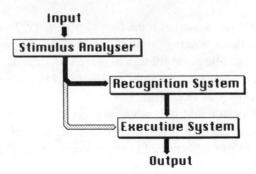

FIGURE 3. A scheme of the major processes believed to be involved in filial responses to a mother hen or her surrogate. In domestic chicks the stimulus analyzer is thought to continue developing until the second day after hatching. Plasticity resulting from imprinting occurs both in the recognition system and at the limited access to the executive system controlling filial behavior. Finally the direct input to the executive system from the stimulus analyzer may degenerate or become inhibited with increasing use of the recognition system.

CHANGES IN NEURAL MECHANISM

It seems likely that after hatching at least three changes are involved in the neural mechanisms concerned with filial imprinting in the domestic chick. These may be conveniently related to the simple scheme shown in Figure 3. The first change is relatively nonspecific, involving alterations in the stimulus analyzing systems—alterations that probably continue into the second day after hatching. A second change is specifically related to the characteristics of the imprinting stimulus in a system devoted to recognition. Part of this system has been localized in the intermediate and medial part of the hyperstriatum ventrale (see Bateson, 1984; Horn, 1985). The third involves changes in the access to the executive system from the neural mechanisms concerned with recognition of particular familiar objects.

The reasons for believing in the late maturation of parts of the stimulus analyzing system are discussed at length by Horn (1985). They were based initially on Horn and McCabe's (1984) discovery that on the first day after hatching a moving stuffed jungle fowl, the ancestral species of the domestic chick, is no more effective in eliciting approach than a conspicuous artificial stimulus, a red flashing and rotating box. Indeed, the two stimuli were carefully matched so that one was not preferred over the other in choice tests. However, by the second day after hatching, the fowl is markedly more attractive to the chicks. When a sensitive, simultaneous choice test was used, a novel fowl was even preferred to a familiar box (Johnson et al., 1985).

Subsequent experimental work has dissociated the neural analyzing mechanisms involved in enhanced responsiveness to jungle fowl from those involved in the recognition of familiar objects. The reader is referred to Horn (1985) for details. Bolhuis, Johnson, and Horn (1985) found that the maturation of the mechanisms that seem particularly sensitive to complex naturalistic stimuli can be facilitated by exposure to patterned light; Horn discusses how external factors as well as endogenous ones, such as testosterone, could influence the development of the system. From a functional standpoint it is significant that a system that must play a major part in enabling the bird to learn its mother's characteristics under natural conditions does not work with maximum efficiency until the age when the bird would normally start to learn about its mother.

The reasons for supposing that two kinds of plasticity are involved in imprinting are discussed elsewhere (Bateson, 1981, 1983b). I used to assume that imprinting only involved one type of plasticity (Bateson, 1978). The need for another emerged from extensive discussions with Gabriel Horn in the late 1970s. As it was originally conceived, the major neural effect of imprinting was the establishment of a representation of the familiar object. The behavioral consequence is that novel objects are responded to less strongly and eventually avoided. However, the behavioral effects can be reversed by exposing the bird to a novel object for a sufficient length of time. Moreover, old preferences can be eroded by the formation of new ones. While some evidence on the reversibility of preferences supported this point of view (Salzen and Mayer, 1968), other evidence suggested that the first preferences to be formed might be better protected from change than later ones (Immelmann and Suomi, 1981; Cherfas and Scott, 1981). These results led Horn and myself to the speculation that two forms of plasticity might be involved, one concerned with recognition of previously experienced inputs and another concerned with connecting up the mechanism involved in recognition of familiar inputs to the execution of filial or later in life sexual behavior (Bateson, 1981).

On the revised view of imprinting, the recognition mechanism has large capacity for storing representations of different objects. Once change has occurred here, further changes are made more difficult by escape from all novel objects. However, when escape from a particular object has waned, its characteristics can also be represented in the recognition system as a result of further exposure. By contrast the possibilities for accepting new connections into the executive system are limited. The capacity for plastic change here is believed to be severely restricted. Once this capacity is used up, the animal may become tame to a novel object, but will not respond to it socially. The suggestion is, then, that two different types of competitive exclusion take place as a result of social experience. I called the first type "acquired reduction in responsiveness to novel objects" (Bateson, 1978). The second type may be termed "acquired control of responsiveness to familiar objects."

It may prove helpful to postulate a fourth change in the neural mechanisms, namely a degeneration or inhibition of the direct control of highly effective stimuli on the executive system. The pathway that supposedly loses its function is depicted as a stippled arrow in Figure 3. Its existence as a pathway is inferred from the evidence that chicks with lesions in IMHV, the neural system that is required for the recognition of familiar objects, readily approach a wide range of moving objects (see Horn, 1985). Furthermore, in intact birds responsiveness to a given stimulus is an additive outcome of its naive responsiveness to such a stimulus and the experience it has had with that or other stimuli (Bateson, 1978). Degeneration or inhibition of such a pathway might be reasonably expected as experience with a familiar object increased. The idea is consistent with modern views that particular stimuli may play an important role in guiding development, but may cease to be important in the control of the fully assembled behavior (e.g., Marler, 1984). It might also be married with the old notion of the "law of transitoriness of instincts." In *The Principles of Psychology* William James (1981) described the evidence in the following terms:

> *Many instincts ripen at a certain age and then fade away.* A consequence of this law is that if, during the time of such an instinct's vivacity, objects adequate to arouse it are met with, a *habit* of acting on them is formed, which remains when the original instinct has passed away; but that if no such objects are met with, then no habit will be formed; and, later on in life, when the animal meets the objects, he will altogether fail to react, as at the earlier epoch he would instinctively have done. (p. 1017)

CONCLUSIONS

The major aim in this chapter is to show how imprinting has been profitably treated as a process of competitive exclusion. Its preemptive character explains much that would otherwise seem a mystery. Many if not all of the seeming contradictions in the experimental findings fall away and the surface complexity resolves itself into a simple principle. While the onset is dependent on general developmental processes, the end of sensitivity is caused by the specific effects of the imprinting process itself shutting off responsiveness to novel stimuli. This also seems to be a general message emerging from the current theorizing about cortical plasticity of the visual system. Similarities in the descriptive evidence for sensitive periods in the two areas of knowledge would flow simply from the common principle of competitive exclusion. No other common features are required.

Another general point that has come from analysis of the behavioral evidence is that several types of neural plasticity may be involved in the imprinting process. Apart from developmental changes that provide the enabling conditions for imprinting to take place, two specific consequences of imprinting are postulated. The first involves the neural representation of the familiar object, which would be the bird's mother in a natural environment. The second involves access from such a neural representation to the neural executive systems responsible for the control of behavior directed toward the mother. The behavioral evidence suggests that the access is limited and, once fully committed, a bird may become familiar with new objects but will not respond socially to them.

ACKNOWLEDGMENTS

I am very grateful to Gabriel Horn and Peter Marler for their comments on an earlier version of this chapter. I also thank Gabriel Horn for the pleasure and stimulation involved in many years of collaboration.

REFERENCES

Aslin, R. N. (1981). Experiential Influences and Sensitive Periods in Perceptual Development: A Unified Model. In: *Development of Perception: Psychobiological*

Perspectives, Vol. 2. *The Visual System*, R. N. Aslin, J. R. Alberts, and M. R. Petersen, eds., pp. 45–93, Academic Press, New York.

Bateson, P. P. G. (1963). The Development of Filial and Avoidance Behaviour in the Domestic Chicken. Ph.D. Dissertation, University of Cambridge.

Bateson, P. P. G. (1964a). Effect of similarity between rearing and testing conditions on chicks' following and avoidance responses. *J. Comp. Physiol. Psychol.* **57**:100–103.

Bateson, P. P. G. (1964b). Changes in chicks' responses to novel moving objects over the sensitive period for imprinting. *Anim. Behav.* **12**:479–489.

Bateson, P. P. G. (1966). The characteristics and context of imprinting. *Biol. Rev.* **41**:177–220.

Bateson, P. P. G. (1973). Internal Influences on Early Learning in Birds. In: *Constraints on Learning: Limitations and Predispositions*, R. A. Hinde and J. Stevenson-Hinde, eds., pp. 101–116, Academic Press, London.

Bateson, P. P. G. (1978). Early Experience and Sexual Preferences. In: *Biological Determinants of Sexual Behaviour*, J. B. Hutchison, ed., pp. 29–53, Wiley & Sons, Chichester.

Bateson, P. P. G. (1979). How do sensitive periods arise and what are they for? *Anim. Behav.* **27**:470–486.

Bateson, P. P. G. (1981). Control of Sensitivity to the Environment during Development. In: *Behavioral Development*, K. Immelmann, G. W. Barlow, L. Petrinovich, and M. Main, eds., pp. 432–453, Cambridge University Press, Cambridge.

Bateson, P. P. G., ed. (1983a). *Mate Choice*. Cambridge University Press, Cambridge.

Bateson, P. P. G. (1983b). The Interpretation of Sensitive Periods. In: *The Behavior of Human Infants*, A. Oliverio and M. Zappella, eds., pp. 57–70, Plenum Press, New York.

Bateson, P. P. G. (1984). The Neural Basis of Imprinting. In: *The Biology of Learning*, P. Marler and H. S. Terrace, eds., pp. 325–339, Dahlem–Konferenzen, Springer–Verlag, Berlin.

Bateson, P. P. G. and R. A. Hinde (1987). Developmental Changes in Sensitivity to Experience. In: *Sensitive Periods in Development*, M. H. Bornstein, ed., Lawrence Erlbaum, Hillsdale, NJ.

Bateson, P. P. G. and E. P. Reese (1969). Reinforcing properties of conspicuous objects in the imprinting situation. *Anim. Behav.* **17**:692–699.

Bateson, P. P. G. and G. Seaburne-May (1973). Effects of prior exposure to light on chicks' behaviour in the imprinting situation. *Anim. Behav.* **21**:720–725.

Bateson, P. P. G. and A. A. P. Wainwright (1972). The effects of prior exposure to light on the imprinting process in domestic chicks. *Behaviour* **52:**279–290.

Bindra, D. (1959). *Motivation: A Systematic Reinterpretation.* Ronald Press, New York.

Bischof, H.-J. (1985). Environmental Influences on Early Development: A Comparison of Imprinting and Cortical Plasticity. In: *Perspectives in Ethology,* Vol. 6, *Mechanisms,* P. P. G. Bateson and P. H. Klopfer, eds., pp. 169–217, Plenum Press, New York.

Boakes, R. and D. Panter (1985). Secondary imprinting in the domestic chick blocked by previous exposure to a live hen. *Anim. Behav.* **33:**353–365.

Bolhuis, J. J., M. H. Johnson, and G. Horn (1985). Effects of early experience on the development of filial preferences in the domestic chick. *Dev. Psychobiol.* **18:**299–308.

Burtt, E. H. (1977). Some factors in the timing of parent–chick recognition in swallows. *Anim. Behav.* **25:**231–239.

Cherfas, J. J. and A. M. Scott (1981). Impermanent reversal of filial imprinting. *Anim. Behav.* **29:**301.

Gottlieb, G. (1961). Developmental age as a baseline for determination of the critical period in imprinting. *J. Comp. Physiol. Psych.* **54:**422–427.

Gottlieb, G. (1971). *Development of Species Identification in Birds.* University of Chicago Press, Chicago.

Hess, E. H. (1959). Two conditions limiting critical age for imprinting. *J. Comp. Physiol. Psychol.* **52:**513–518.

Hess, E. H. (1973). *Imprinting.* Van Nostrand Reinhold Co., New York.

Hinde, R. A. (1961). The Establishment of the Parent–Offspring Relation in Birds with Some Mammalian Analogies. In: *Current Problems in Animal Behaviour,* W. H. Thorpe and O. L. Zangwill, eds., pp. 175–193, Cambridge University Press, Cambridge.

Hinde, R. A., W. H. Thorpe, and M. A. Vince (1956). The following response of young Coots and Moorhens. *Behaviour* **9:**214–242.

Horn, G. (1985). *Memory, Imprinting, and the Brain.* Oxford University Press, Oxford.

Horn, G. and B. J. McCabe (1984). Predispositions and preferences. Effects on imprinting of lesions to the chick brain. *Anim. Behav.* **32:**288–292.

Hubel, D. H. and T. N. Wiesel (1970). The period of susceptibility to the physiological effects of unilateral eye closure in kittens. *J. Physiol.* **206:**419–436.

Immelmann, K. and S. J. Suomi (1981). Sensitive Phases in Development. In: *Behavioral Development,* K. Immelmann, G. W. Barlow, L. Petrinovich, and M. Main, eds., pp. 395–431, Cambridge University Press, Cambridge.

James, W. (1981). *The Principles of Psychology, Vol. II.* Harvard University Press, Cambridge, MA.

Johnson, M. H., J. J. Bolhuis, and G. Horn (1985). Interaction between acquired preferences and developing predispositions during imprinting. *Anim. Behav.* 33:1000–1006.

Johnston, T. D. and G. Gottlieb (1981). Development of visual species identification in ducklings: What is the role of imprinting? *Anim. Behav.* 29:1082–1099.

Johnston, T. D. and G. Gottlieb (1985). Effects of social experience on visually imprinted maternal preferences in Peking ducklings. *Dev. Psychobiol.* 18:261–271.

Landsberg, J.-W. (1976). Posthatch age and developmental age as a baseline for determination of the sensitive period for imprinting. *J. Comp. Physiol. Psychol.* 90:47–52.

Lickliter, R. and G. Gottlieb (1985). Social interaction with siblings is necessary for the visual imprinting of species-specific maternal preferences in ducklings. *J. Comp. Psychol.* 99:371–379.

Lorenz, K. (1935). Der Kumpan in der Umwelt des Vogels. *J. Ornithologie* 83:137–213, 289–413.

Lorenz, K. (1937). The companion in the bird's world. *Auk* 54:245–273.

Marler, P. (1984). Song Learning: Innate Species Differences in the Learning Process. In *The Biology of Learning*, P. Marler and H. S. Terrace, eds., pp. 289–309, Dahlem–Konferenzen, Springer-Verlag, Berlin.

Moltz, H. and L. J. Stettner (1961). The influence of patterned light deprivation on the critical period for imprinting. *J. Comp. Physiol. Psychol.* 54:279–283.

Polt, J. M. and E. H. Hess (1964). Following and imprinting: Effects of light and social experience. *Science* 143:1185–1187.

Ratner, A. M. and H. S. Hoffman (1974). Evidence for a critical period for imprinting in khaki campbell ducklings (*Anas platyrhynchos domesticus*). *Anim. Behav.* 22:249–255.

Salzen, E. A. and C. C. Meyer (1968). Reversibility of imprinting. *J. Comp. Physiol. Psychol.* 66:269–275.

Scott, J. P. (1962). Critical periods in behavioral development. *Science* 138:949–958.

Sluckin, W. (1972). *Imprinting and Early Learning.* 2nd ed., Methuen, London.

Sluckin, W. and E. A. Salzen (1961). Imprinting and perceptual learning. *Quart. J. Exp. Psych.* 13:65–77.

Smith, F. V. and K. H. Nott (1970). The "critical period" in relation to the strength of the stimulus. *Z. Tierpsychol.* 27:108–115.

PART II

CORTICAL PLASTICITY

8

DEVELOPMENTAL PLASTICITY— SELF-ORGANIZATION OR LEARNING?

Wolf Singer

Max-Planck-Institut für Hirnforschung, Frankfurt, FR Germany

Because of its exceedingly high complexity, more than any other organ the brain depends on self-organizing processes for its development. By including epigenetic information these processes allow the creation of extremely complex structures that depart from rather global genetic instructions. The brain is particularly well adapted to use epigenetic information for its self-organization because it is designed as an information-processing system. It is now well established that neuronal activity, which serves as a vehicle for the transport and processing of information in the mature nervous system, also plays a crucial role in the self-organization of the developing system.

In all structures of the central nervous system investigated so far, the selectivity of connections appears to be achieved in a biphasic process. First, projection patterns are developed that are exuberant and rather imprecise. Second, this initial pattern is pruned by eliminating ectopic connections. This involves cell death and/or retraction of axon collaterals (for a recent review of the extensive literature see Singer, 1987). In a variety of structures neuronal activity has been shown to influence this pruning process. The blockade of neuronal activity can either lead to persistence of the imprecise and exuberant projection pattern or else it can lead to a complete removal of the silenced pathways (Stryker and Harris, 1986).

Of importance in the present context is that such activity-dependent shaping processes are not confined to embryogenesis but do extend into postnatal life. Here, neuronal activity becomes influenced by sensory signals. Environmental factors capable of eliciting or modulating sensory activity can thus affect the development of neuronal connections in that they influence the very activity patterns that promote the self-organizing process. Prominent examples for such experience-dependent modifications of neuronal circuitry come from the work performed in the visual system of vertebrates. Evidence is available that the responses to a structured visual environment are used for the refinement of projection maps. This has been shown both for the retino-tectal and for the retino-cortical projections. Experiments based on global or partial deprivation of vision have suggested that in the visual cortex of mammals neuronal activity can lead to both selective stabilization of synaptic connections and to their disruption.

These experience-dependent modifications are commonly thought to be confined to a critical period of postnatal development. Interestingly, however, recent data are suggesting that the timing of this sensitive period may itself depend on past sensory experience. This issue will be dealt with in the contributions to this volume by Timney and in part also by Hirsch. Moreover, evidence is accumulating that there is no unique critical period for the acquisition of visual functions but that critical periods exist with different time constants at the different processing levels (See Hirsch, this volume). As far as modifications of ocularity are concerned, the rules that describe these activity-dependent selection processes closely resemble the associative rules postulated by Hebb in 1949 for adaptive synaptic connections in a learning matrix (Rauschecker and Singer, 1979, 1981). These rules lead to selective stabilization of those pathways on a common target cell that have a high probability of conveying correlated activity patterns. Conversely, these rules imply that pathways that convey uncorrelated activity compete with each other, whereby the winner consolidates at the expense of the loser. Thus, by means of selective stabilization, these self-organizing processes associate neuronal connections with each other via a common postsynaptic target or effector cell whereby the selection criterion is the contingency in time and space of neuronal activity. Such Hebbian modifications have now been confirmed by the elegant demonstration of activity-dependent modifications of single-cell responses while recording from identified neurons in the visual cortex. These data will be discussed in the chapter by Frégnac.

The generality and power of such activity-dependent selection algorithms is emphasized in the contribution of Cooper by the demonstration that very similar processes can mediate the acquisition of a variety of different functional properties of cortical neurons. It comes as no surprise, therefore, that activity-dependent shaping processes occur also at other levels of visual processing than the striate cortex. Hoffmann will provide evidence in his contribution that the coordination of visual–motor reflexes requires experience during a critical period of early development. Again, functional criteria appear to be applied for selective stabilization of pathways that convey matching activity patterns.

Of particular relevance in our context is the recent evidence that such activity-dependent selection processes are not solely dependent on the local interactions between the modifiable afferents and their respective target neurons. Indications are now available that these local modifications do in turn depend on the activation state of central core modulatory systems such as the noradrenergic and the cholinergic projections to the neocortex (Bear and Singer, 1986; Kasamatsu and Pettigrew,

1979; Singer, 1982; Singer and Rauschecker, 1982). Neuronal activity per se, such as commonly recorded in anesthetized preparations, is not sufficient to promote changes in synaptic transmission and circuitry. A threshold that is different from the threshold of the sodium-dependent action potential—perhaps the threshold of dendritic calcium channels—has to be trespassed for the occurrence of activity-dependent long-term modifications. The involvement of modulatory systems such as the noradrenergic and the cholinergic projection suggests that state parameters such as arousal and attention play a role in the gating of neuronal plasticity since these modulatory systems have been implicated in the control of such states. Thus, during postnatal development, the activity-dependent autoshaping of circuits is extended to an experience-dependent self-organization process that is under the control of the central state of the brain. This notion has been extended by the recent demonstration that these early adaptation processes require brief consolidation for their expression during which brain functions must not be depressed. This important and hitherto controversial point will be discussed in some detail by Rauschecker.

It is in this context that the experience-dependent autoshaping processes in the developing visual system may have some bearing on processes occurring with imprinting. At least at the level of phenomenological description, the two processes have numerous features in common. Both are characteristic for early postnatal development; both are restricted to a sensitive period that can to some extent be prolonged if the relevant trigger stimuli are lacking; both depend on specific sensory experience; in both cases the sensory signals need to be processed by an awake and attentive brain in order to promote changes; in both cases changes are long-lasting and rather irreversible, the effects outlasting by far the time required for their induction; and in both cases deprivation ultimately leads to a deterioration of the respective function or behavior. Since nature is usually conservative with its inventions, it appears conceivable that the mechanisms that serve to adapt the building blocks of the brain to each other and those that serve to adapt the brain to its environment may be similar. This speculation together with the fact that the neuronal mechanisms of experience-dependent modifications in the visual system are probably easier to analyze than those underlying the complex behavioral changes following imprinting are the principal reason for the juxtaposition of imprinting on the one hand and visual system plasticity on the other. The investigation of both processes may have some bearing on the question of learning and engram formation.

In the central nervous system the program that determines responses to stimuli is stored in the architecture and the differential weighting of neuronal connections. Hence, modifications of circuitry can be considered as changes in the program of the central nervous system. In turn, if these occur in response to sensory stimulation, they can be considered the result of a learning process and the associated changes in the coupling of neurons as an engram.

REFERENCES

Bear, M. F. and W. Singer (1986). Modulation of visual cortical plasticity by acetylcholine and noradrenaline. *Nature* **320:**172–176.

Geiger, H. and W. Singer (1986). A possible role of Ca^{++}-currents in developmental plasticity. *Exp. Brain Res. Suppl.* **14:**256–270.

Kasamatsu, T. and J. D. Pettigrew (1979). Preservation of binocularity after monocular deprivation in the striate cortex of kittens treated with 6-hydroxydopamine. *J. Comp. Neurol.* **185:**163–182.

Rauschecker, J. P. and W. Singer (1979). Changes in the circuitry of the kitten's visual cortex are gated by postsynaptic activity. *Nature* **280:**58–60.

Rauschecker, J. P. and W. Singer (1981). The effects of early visual experience on the cat's visual cortex and their possible explanation by Hebb synapses. *J. Physiol. (Lond.)* **310:**215–239.

Singer, W. (1982). Central core control of developmental plasticity in the kitten visual cortex: I. Diencephalic lesions. *Exp. Brain Res.* **47:**209–222.

Singer, W. (1986). *Activity-dependent Self-organization of Synaptic Connections as a Substrate of Learning,* Dahlem-Konferenzen, Berlin. In press.

Singer, W. and J. Rauschecker (1982). Central core control of developmental plasticity in the kitten visual cortex: II. Electrical activation of mesencephalic and diencephalic projections. *Exp. Brain Res.* **47:**223–233.

Stryker, M. P. and W. A. Harris (1986). Binocular impulse blockade prevents the formation of ocular dominance columns in cat visual cortex. *J. Neurosci.* **6:**2117–2133.

9

CORTICAL PLASTICITY: THEORETICAL ANALYSIS, EXPERIMENTAL RESULTS

Leon N. Cooper

Department of Physics and Center for Neural Science, Brown University, Providence, Rhode Island

Do all animals learn? And do they learn in the same way? A theme that runs throughout this volume suggests how ubiquitous learning may be and further, possibly more exciting, that common factors may be involved in such seemingly diverse phenomena as the development of selectivity of cells in visual cortex and birdsong learning. We would all agree that some learning must depend on the details of information entering from the outside world. How else would the memory of a new face or the particular sequence that makes a learned birdsong be stored? In addition, there must exist some "global controller" or "enabling factor" that determines whether or not particular learning will take place.

For many years we have heard talk about possible modification of synapses between neurons as the physiological basis of learning and memory storage. These relatively vague ideas are becoming more precise. Insight into the molecular basis of synaptic modification is beginning to appear; the role of possible global controllers such as norepinephrine and acetylcholine is being clarified; and a mathematical structure for the network of neurons is rapidly evolving.

Presented here is a brief summary of some recent theoretical ideas as well as some experimental results related to plasticity in visual cortex, presumably related to the changes that take place when memory is stored and possibly also related to what is known about birdsong learning (see Konishi, Marler, and Margoliash, all this volume). More important than the details, what I hope to convince you is that what is being presented provides us with a language in which these questions can be discussed with clarity and precision.

WHY VISUAL CORTEX?

Experimental work of the last generation, from the path-breaking work of Hubel and Wiesel to some of the most recent results contained in this volume (see Frégnac, Hirsch et al., Rauschecker, Singer, and Timney, all this volume), has shown that there exist cells in the visual cortex of the adult cat that respond in a precise and highly tuned fashion to external patterns—in particular bars or edges of a given orientation and moving in a given direction—and that the number and response characteristics of such cortical cells can be modified. For example, the relative number of cortical cells that are highly specific in their response to visual patterns varies in a very striking way with the visual experience of the animal during the critical period.

Such results seem to provide evidence for the modifiability of the response of single cells in the cortex of a higher mammal according to its visual experience. Depending on whether or not patterned visual information is part of the animal's experience, the specificity of the response of cortical neurons varies widely. Specificity increases with normal patterned experience. Deprived of normal patterned information (e.g., dark reared or lid sutured at birth) specificity decreases. Further, even a short exposure to patterned information (during the critical period) after weeks of dark rearing can reverse the loss of specificity and produce an almost normal distribution of cells.

We do not claim and it is not necessary that all neurons in the visual cortex be so modifiable. Nor is it necessary that modifiable neurons are especially important in producing the architecture of the visual cortex. It is our hope that the general form of modifiability we require to construct interacting neural networks manifests itself for at least some cells of the visual cortex that are accessible to experiment. We thus make the conservative assumption that biological mechanisms, once established, will function in a more or less similar manner in different regions. If this is the case, modifiable individual neurons in the visual cortex can provide evidence for such modification more generally.

SUMMARY OF SINGLE-CELL THEORY

Cortical neurons receive afferents from many sources. In the visual cortex (e.g., layer IV) the principal afferents are those from the lateral geniculate nucleus and from other cortical neurons. This leads to a complex network that we have analyzed in several stages.

In the first stage we consider a single neuron with inputs from both eyes (Fig. 1). Here d^l, d^r, m^l, and m^r are inputs and synaptic junctions from left and right eyes. The output of this neuron (in the linear region) can be written as an inner product:

$$c = m^l \cdot d^l + m^r \cdot d^r \qquad (1)$$

This means that the neuron firing rate (in the linear region) is the sum of the inputs from the left eye multiplied by the appropriate left-eye

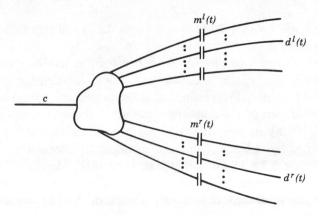

FIGURE 1. A model neuron.

synaptic weights plus the inputs from the right eye multiplied by the appropriate right-eye synaptic weights. Thus, the neuron integrates signals from the left and right eyes.

According to the theory presented by Bienenstock, Cooper, and Munro (1982) (BCM), these synaptic weights modify as a function of local and global variables. To illustrate, we consider the synaptic weight m_j (Fig. 2).

Its change in time, $\dot{m}_j$, is given by

$$\dot{m}_j = F(d_j \ldots m_j; d_k \ldots c; \bar{c} \ldots ; X, Y, Z) \tag{2}$$

Here variables such as $d_j \ldots m_j$ are designated local. These represent information (such as the incoming signal d_j and the strength of the synaptic junction m_j) available locally at the synaptic junction m_j. Variables

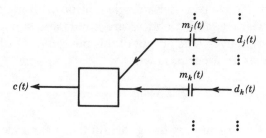

FIGURE 2. Local and quasi-local variables.

such as d_k . . . c are designated quasi-local. These represent information (such as c, the firing rate of the cell, or d_k, the incoming signal to another synaptic junction) that is not locally available to the junction m_j but is physically connected to the junction by the cell body itself—thus necessitating some form of internal communication between various parts of the cell and its synaptic junctions. Variables such as $\bar{c}$ (the time-averaged output of the cell) are averaged local or quasi-local variables. Global variables are designated X, Y, Z, and so on. These latter variables represent information (e.g., the presence or absence of neurotransmitters such as norepinephrine or the average activity of large numbers of cortical cells) that is present in a similar fashion for all or a large number of cortical neurons (distinguished from local or quasi-local variables presumably carrying detailed information that varies from synapse to synapse). These global variables are candidates for "enabling factors" discussed frequently in this volume (see chapters by Marler, Frégnac, Rauschecker).

In a form relevant to this discussion, BCM modification can be written as

$$\dot{m}_j = \phi(c, \bar{c}; X, Y, Z, \ldots)d_j, \tag{3}$$

so that the jth synaptic junction, m_j, changes its value in time as a function of quasi-local and time-averaged quasi-local variables c and $\bar{c}$ as well as global variables X, Y, and Z through the function ϕ and a function of the local variable d_j. The crucial function, ϕ, is shown in Figure 3.

What is of particular significance is the change of sign of ϕ at the

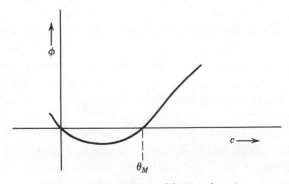

FIGURE 3. The BCM modification function.

modification threshold, θ_M, and the nonlinear variation of θ_M with the average output of the cell c. In a simple situation

$$\theta_M = (\bar{c})^2. \tag{4}$$

The occurrence of negative and positive regions for ϕ drives the cell to selectivity in a "normal" environment. This is so because the response of the cell is diminished to those patterns for which the output c is below threshold (ϕ negative) while the response is enhanced to those patterns for which the output c is above threshold (ϕ positive). The nonlinear variation of the threshold with the average output of the cell, $\bar{c}$, places the threshold so that it eventually separates one pattern from all of the rest. Furthermore, it provides the stability properties of the system.

A detailed analysis of the consequences of this form of modification is given in BCM. The results (as modified by the network analysis outlined next) are in general agreement with what we might call classical experiments of the last generation. Neurons in normal (patterned) environments become selective and binocular. In various deprived environments (MD, BD, etc.) the theoretical behavior follows the experimental results.

An unexpected consequence of this theory is a connection between selectivity and ocular dominance. The analysis given in BCM shows that in monocular deprivation, nonpreferred inputs presented to the open eye are a necessary part of the suppression of deprived eye responses. It follows that the more selective the cell is to the open eye (increasing the probability of nonpreferred inputs), the more the closed eye will be driven to zero, thus increasing the dominance of the open eye.

For an experimental test of these ideas it is important to determine what happens *during* the ocular dominance shift produced by monocular deprivation. Consider the experimental situation in which monocular experience follows a period of dark rearing. [Such experiments are presently being performed by Saul and Daniels (private communication)]. During dark rearing it is known that most area 17 cells become less responsive (sluggish) and lose their selectivity and that some (perhaps 20%) become visually nonresponsive.

Our theoretical analysis indicates that in the course of monocular experience, those cells that have become visually nonresponsive during the dark rearing will first show an increase in responsiveness to the open eye followed by the development of selectivity. Those cells that

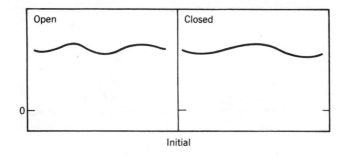

Initial

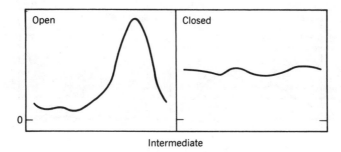

Intermediate

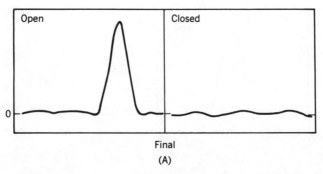

Final

(A)

FIGURE 4. (A) Progression of development of selectivity and ocular dominance. Note that selectivity develops for responsive binocular aspecific cells for the open eye *before* the response to the closed eye is driven to zero. (B) Progression of development of responsiveness and selectivity. Note that responsiveness to the open eye develops before, or along with, selectivity.

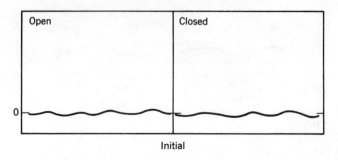

Initial

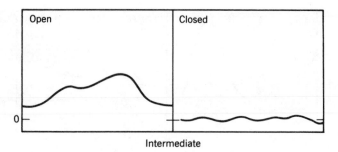

Intermediate

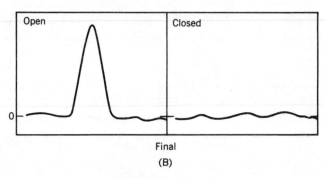

Final

(B)

FIGURE 4. (*continued*)

have survived the period of dark rearing as binocular and aspecific will exhibit a progression in which their selectivity to the open eye is increased while maintaining their response (often nonselective) to the closed eye. This should result in the two sequences shown in Figures 4A and 4B.

EXTENSION TO NETWORKS

To better confront these ideas with experiment, the single neuron discussed above must be placed in a network with the anatomical features of the region of interest. For visual cortex this suggests a network in which inhibitory and excitatory cells receive input from the lateral geniculate nucleus (LGN) and from each other. A simplified form of such a network, a first-order representation of the anatomy and physiology of layer IV of cat visual cortex (Fig. 5), has been studied by Scofield and Cooper (1985).

In a network generalization of Equation (1), we write

$$c_i = m_i^l \cdot d^l + m_i^r \cdot d^r + \sum_j L_{ij} c_j \tag{5}$$

where L_{ij} are the intracortical connections.

Analysis by Scofield and Cooper (1985) of the network along lines similar to that of the single-cell analysis described above shows that under proper conditions on the intracortical synapses, the cells converge to states of maximum selectivity with respect to the environment formed by the geniculate signals. Their conclusions are therefore similar to those of BCM with explicit further statements concerning the independent effects of excitatory and inhibitory neurons on selectivity and ocular dominance. For example, shutting off inhibitory cells lessens selectivity and alters ocular dominance. The inhibitory cells may be selective, but there is no theoretical necessity that they be so.

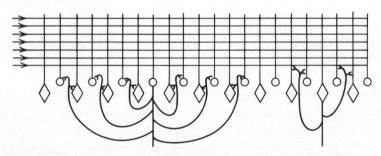

FIGURE 5. A simplified neural network. Shown are the two cell types: inhibitory, represented by circles, and excitatory, by diamonds. Geniculate afferents enter at the top of the figure and synapse with all cells in the network at the intersection of the horizontal and vertical fibers. Also shown are the intracortical fibers for each cell type. The exact ratio of inhibitory to excitatory cells is not important.

A mean field approximation to the above network (Cooper and Scofield, to be published) shows that if the average effect of intracortical connections results in inhibition of individual cells, then in monocular deprivation, the geniculocortical synapses to the cell will converge to nonzero states that give, as the result of stimulation of the closed eye, total responses that are zero. However, the fact that the geniculocortical states are nonzero means that the removal of cortical inhibition through the chemical blocking of inhibitory synapses would uncover responses from previously nonresponsive cells. This result is in accord with the experimental observation of "masked synapses" after the removal of the inhibitory effects of GABA with the blocking agent bicuculline (cf. Duffy et al., 1976; Sillito et al., 1980).

These results are obtained without assuming that the intracortical inhibitory synapses are very responsive to visual experience. Learning can occur entirely among the excitatory LGN–cortical synapses. Another point of view is espoused, for example, by Rauschecker and Singer (1982). They suggest that since cells lose their orientation specificity when binocularly deprived of pattern vision, and since it has been shown that inhibitory connections play a major role in determining orientation selectivity (Sillito, 1975), the cortical inhibitory synapses may suffer more than excitatory ones.

These conflicting ideas led us to perform experiments on changes in inhibitory activity due to visual experience (Bear et al., 1985) (described in the next section). They indicate that one measure of cortical inhibition is relatively constant even during dramatic manipulations of the visual environment.

EXPERIMENTAL TEST OF CHANGES IN INHIBITORY ACTIVITY DUE TO VISUAL EXPERIENCE

One of the consequences of the network theory discussed in the previous section is that experimental results that have been obtained in the visual cortex over the last generation can be explained primarily by modification of the LGN-to-cortex synapses with minimum changes among intracortical synapses. Thus, the possibility is opened that most learning takes place in the geniculocortical synapses. This somewhat surprising result has as one consequence the possibility of great simplification in the analysis of network modification.

An alternate hypothesis that has been considered for some time is that intracortical synapses bear heavy responsibility for modification in cortical circuitry during learning. In particular, it has been suggested that ocular dominance shifts in monocular deprivation are due to increased activity of GABAergic neurons, the open eye suppressing the closed. Sillito (1975) documented, in normal cats, that visually unresponsive cells may be "unmasked" by iontophoretic bicuculline. Thus, it is not unreasonable to speculate that many of the unresponsive cells in visually deprived kittens are being suppressed. Together, these data suggest as a possible hypothesis that *in kitten striate cortex the GABAergic neurons respond to sensory deprivation by forming new synapses.* This hypothesis implies that the density or strength of GABAergic synapses will increase in zones of cortex that are deprived of a normal thalamic input; in the case of monocular deprivation, these zones correspond to the closed-eye ocular dominance columns and to the monocular segment contralateral to the deprived eye. On the other hand, the theory described above suggests that there will be minimal response of GABAergic neurons to sensory deprivation. This hypothesis has been put to the test in a recent series of experiments in our laboratory (Bear et al., 1985) (described below).

To examine the distribution of GABAergic synapses, Bear et al. immunocytochemically localized GAD in sections of striate cortex. While immunocytochemistry is not a quantitative measure, they reasoned that changes restricted to deprived ocular dominance zones should be readily detected with this method. As a quantitative estimate of GABAergic synapse density, they biochemically measured GAD activity in homogenates of striate cortex.

They found no evidence for a change in the distribution of GAD-positive puncta in 12 unilaterally enucleated kittens. The band of layer IV puncta remained uniform even though the periods of monocular deprivation examined would all be sufficient to cause a physiological ocular dominance shift in the striate cortex (Hubel and Wiesel, 1970). GAD immunoreactivity was unchanged even under conditions that produced alterations in the level of the metabolic enzyme, cytochrome oxidase. Measurements of GAD activity showed no consistent or significant difference between either the binocular segments of enucleated and control kittens or the monocular segments of enucleated animals.

This conclusion is in striking agreement with network analysis, which, as mentioned above, suggests that inhibitory synapses are much less modified by experience than excitatory synapses. In addition to its implications for the "site of learning," such a hypothesis leads to important simplifications in the analysis of complex networks.

GLOBAL CONTROLS: THE ELUSIVE "ENABLING FACTOR"

Any theory of learning generally requires global as well as local controls. As discussed previously, local controls are those that determine detailed changes of individual synaptic junctions while global information would be expected to influence all or large numbers of synaptic junctions in the same way. From the point of view of a learning theory, there must be some way to distinguish more interesting from less interesting input. Experimentally it is known that certain areas of cortex (visual, auditory, and somatic sensory) exhibit plasticity during early critical periods but not during adulthood.

For example, most neurons in the visual cortex of newborn kittens (and normal adult cats as well) are activated equally well by both eyes. During the critical period (3 weeks to 3 months in cats), monocular lid suture or misalignment of the eyes (called strabismus) leads to the domination of cortical cells by one eye. In contrast, adult cats that are given monocular visual experience when they are *older* than 6 months of age remain unaffected by the imbalance of visual inputs; adult cortical neurons remain binocularly activated.

A major question has been formulated as follows: What are the global factors acting singly or in combination that affect the development of synapses in cortex?

In recent years it has been suggested that catecholamines (CA) are required for neuronal plasticity in the neocortex. One test of this hypothesis has been made in a series of experiments performed by Kasamatsu and Pettigrew (1976, 1979), who used the monocular deprivation paradigm with kitten visual cortex as a test system. In their control kittens they found the usual effect of monocular deprivation during the critical period—within a week the majority of cells lost their normal binocular responsiveness and could be driven only by stimulation of the nondeprived eye. But in animals given the neurotoxin 6-hydroxydopamine (6-OHDA) to deplete cortical CAs, the ocular dominance shift failed to occur and cells remained binocularly driven.

In later experiments Kasamatsu et al. (1979, 1981) pioneered the use of miniature osmotic pumps to infuse 6-OHDA continuously to local regions of cortex in one hemisphere while they used the other hemisphere as a control by perfusing only the vehicle for the 6-OHDA. Following monocular deprivation, normal plasticity was again disrupted in CA-depleted cortex as indicated by the lack of a shift in ocular dominance of visual cells. Because both noradrenergic and dopaminergic fibers project to the visual cortex, Kasamatsu et al. also used the minipumps to add norepinephrine (NE) to the cortex previously depleted of CAs to demonstrate that NE itself can restore plasticity. This evidence suggests that catecholamines, especially NE, are necessary for the cortical changes observed in kittens that have restricted vision during the critical period.

However, the relationship between static NE levels and plasticity is not simple. In one experiment (Bear and Daniels, 1983; Bear et al., 1983) cortical catecholamines were permanently depleted in newborn kittens by intraperitoneal injections of 6-OHDA. Biochemical analysis demonstrated severe reduction of NE levels; but in this experiment plasticity, as evidenced by ocular dominance shift, remained intact. In the neonatal experiments kittens received 6-OHDA *at birth* and were monocularly deprived for 7 days at about 5 weeks of age. The minipump kittens received 6-OHDA continuously for the 7 days of monocular deprivation. Comparison of the two paradigms suggests the possibility that loss of plasticity is not caused by depletion of NE alone.

Thus, despite Kasamatsu's success in demonstrating support for the NE hypothesis, questions have arisen. The above experiments show that early depletion of cortical NE does not, by itself, prevent the later ocular dominance shift after monocular deprivation. Other experiments confirm this conclusion. Daw et al. (1984) depleted cortical NE by section of the locus coeruleus fiber bundle near lateral hypothalamus and found no diminution of the ocular dominance shift after monocular deprivation. Videen et al. (1984) recorded no difference in the reaction of kitten and adult cat visual cortex neurons to iontophoretically applied NE. Adrien et al. (1982) observed no lack of shift after lesion of the locus coeruleus itself; and that group was unable to reproduce Kasamatsu and Pettigrew's (1979) finding that intraventricular injection of 6-OHDA prevents ocular dominance shift. All of these results reinforce the idea that NE is not the only factor in the global control of learning.

In addition to the NE system, several lines of evidence suggest that the cholinergic (ACh) system may also serve as a global modulator of cortical function. Similar to the locus coeruleus–NE system, the basal forebrain cholinergic (ACh) system has a widespread input to cortex that stands in marked contrast to the highly organized thalamocortical systems that provide specific sensory input to the cortex. In addition, several findings link both the NE and ACh systems to acquisition and storage processes related to learning and memory.

While these global cortical inputs have been related to memory and learning, the specific cellular mechanisms of ACh and NE function are unclear. Present evidence indicates that both systems may modulate the response of cortical neurons to specific sensory inputs. NE appears to improve the signal-to-noise ratio of sensory responses both in somatic sensory and in the primary visual cortex, and NE may potentiate the action of both excitatory and inhibitory transmitters (Waterhouse et al., 1980). These effects of NE could be mediated through α-adrenergic receptors that appear to be concentrated in the deeper layers (IV–VI) of the cortex.

The cholinergic system may function in a manner similar to NE. Application of low levels of ACh enhances the excitatory response of cortical neurons to glutamate and modifies the task-related discharge of cortical neurons during behavior. ACh has also been shown to modify the membrane input resistance of cortical neurons. The slow onset, long duration of action, and sensitizing effects of ACh are all consistent with the conclusion that the cholinergic system acts as a modulator of cortical activity. All of this suggests that both NE and ACh may play a modulatory role in cortex.

This view has been reinforced by recent work of Bear and Singer (1986), which indicates that 6-OHDA, in addition to destroying NE neurons, also interferes with ACh effects on cortical neurons. In addition, Bear and Singer have shown that the simultaneous diminution of both ACh and NE appears to prevent the ocular dominance shift in monocular deprivation while the diminution of either ACh or NE (and not both) does not prevent this shift.

These results may enable a resolution of the apparent contradictions in previous NE experiments discussed above and, in addition, suggest the fascinating possibility that ACh and NE act together to provide a global modulator for plasticity.

ACKNOWLEDGMENT

This work was supported by the Office of Naval Research and the U.S. Army Research Office.

REFERENCES

Adrien, J., P. Buisseret, Y. Frégnac, E. Gary-Bobo, M. Imbert, J. Tassin, and Y. Trotter. (1982). Noradrenaline et plasticité du cortex visuel du chaton: Un reexamen. *C. R. Acad. Sci. Paris Serie III* **295**:745–750.

Bear, M. F., and J. D. Daniels (1983). The plastic response to monocular deprivation persists in kitten visual cortex after chronic depletion of norepinephrine. *J. Neurosci.* **3**:407–416.

Bear, M. F., D. E. Schmechel, and F. F. Ebner (1985). Glutamic acid decarboxylase in the striate cortex of normal and monocularly deprived kittens. *J. Neurosci.* **5**:1262–1275.

Bear, M. F., M. A. Paradiso, M. Schwartz, S. B. Nelson, K. M. Carnes, and J. D. Daniels (1983). Two methods of catecholamine depletion in kitten visual cortex yield different effects on plasticity. *Nature* **302**:245–247.

Bear, M. F. and W. Singer (1986). Modulation of visual cortical plasticity by acetylcholine and noradrenaline. *Nature* **320**:172–176.

Bienenstock, E. L., L. N. Cooper, and P. W. Munro (1982). Theory for the development of neuron selectivity: Orientation specificity and binocular interaction in visual cortex. *J. Neurosci.* **2**:32–48.

Cooper, L. N. and C. L. Scofield (in press). Mean field approximation for a neural network.

Daw, N. W., T. W. Robertson, R. K. Rader, T. O. Videen, and C. J. Coscia (1984). Substantial reduction of cortical noradrenaline by lesions of adrenergic pathway does not prevent effects of monocular deprivation. *J. Neurosci.* **4**:1354–1360.

Duffy, F. H., S. R. Snodgrass, J. L. Burchfiel, and J. L. Conway (1976). Bicuculline reversal of deprivation amblyopis in the cat. *Nature* **260**:256–257.

Hubel, D. H. and T. N. Wiesel (1970). The period of susceptibility to the physiological effects of unilateral eye closure in kittens. *J. Physiol. (Lond.)* **206**:419–436.

Kasamatsu, T. and J. D. Pettigrew (1976). Depletion of brain catecholamines: Failure of ocular dominance shift after monocular occlusion in kittens. *Science* **194**:206–209.

Kasamatsu, T. and J. D. Pettigrew (1979). Preservation of binocularity after monocular deprivation in the striate cortex of kittens treated with 6-hydroxydopamine. *J. Comp. Neurol.* **185**:139–162.

Kasamatsu, T., J. D. Pettigrew, and M. Ary (1979). Restoration of visual cortical plasticity by local microperfusion of norepinephrine. *J. Comp. Neurol.* **185**:163–182.

Kasamatsu, T., J. D. Pettigrew, and M. Ary (1981). Cortical recovery from effects of monocular deprivation: Acceleration with norepinephrine and suppression with 6-hydroxydopamine. *J. Neurophysiol.* **45**:254–266.

Rauschecker, J. P. and W. Singer (1982). Binocular deprivation can erase the effects of preceding monocular or binocular vision in kitten cortex. *Dev. Brain Res.* **4**:495–498.

Sillito, A. M. (1975). The contribution of inhibitory mechanisms to the receptive field properties of neurons in the cat's striate cortex. *J. Physiol.* **250**:304–330.

Sillito, A. M., J. A. Kemp and H. Patel (1980). Inhibitory interactions contributing to the ocular dominance of monocularly dominated cells in the normal cat striate cortex. *Exp. Brain Res.* **41**:1–10.

Scofield, C. L. and L. N. Cooper (1985). Development and properties of neural networks. *Cont. Phys.* **26**(2):125–145.

Videen, T. O., N. W. Daw, and R. K. Rader (1984). The effect of norepinephrine on visual cortical neurons in kitten and adult cats. *J. Neurosci.* **4**:1607–1617.

Waterhouse, B. D., H. C. Moises, and D. J. Woodward (1980). Noradrenergic modulation of somatosensory cortical neuronal responses to iontophoretically applied putative neurotransmitters. *Exp. Neurol.* **69**:30–49.

10

WHAT SIGNALS ARE RESPONSIBLE FOR SYNAPTIC CHANGES IN VISUAL CORTICAL PLASTICITY?

Josef P. Rauschecker

Max-Planck-Institut für Biologische Kybernetik, Tübingen, FR Germany

The first demonstrations of plasticity in the visual cortex on the single-cell level were given by Hubel and Wiesel more than 20 years ago. Since then innumerable studies have shown in detail that during a sensitive period after birth the connectivity in the mammalian visual system can be changed as a consequence of specific experience.

Again, already in their early papers Hubel and Wiesel drew the parallel to imprinting in birds, which was first discovered and studied by Konrad Lorenz in the 1930s. There, dramatic behavioral changes are observed as a consequence of specific sensory stimulation. The idea that these behavioral changes might be caused by neuronal changes similar to those observed in cortical plasticity was not farfetched.

The most conspicuous similarity between imprinting and cortical plasticity is the existence of a critical or sensitive period, during which changes in the behavioral or neuronal pattern of an individual are induced most easily (Lorenz, 1935; Hubel and Wiesel, 1970). The existence of a temporal window, however, with a rising and a falling phase and a clear peak for susceptibility to external influences, represents only a superficial descriptive similarity and does not necessarily imply that there are similarities on more profound levels.

It is necessary, therefore, to look in both areas for the neuronal mechanisms underlying the observed changes. Only if similar mechanisms can be defined may one start to wonder whether imprinting in birds and developmental plasticity in mammals really represent homologous features in evolution rather than mere coincidence. On this mechanistic level it is not so important whether the biochemical and biophysical carriers of information are totally identical. It is of much more interest if the signals causing changes of neuronal circuitry, the "software" of the system, is the same. The beginning and end of the sensitive period could, for instance, be caused by rather different pharmacological agents in birds and mammals. Nevertheless, one could speak of a striking parallel if the release of these effectors were triggered by the same event during the course of development.

Both fields, developmental ethology and developmental neuroscience, have shown a growing mutual interest recently, which was also documented by a number of conferences (Marler and Terrace: The Biology of Learning, Dahlem Konferenzen, 1983; Rauschecker and Bateson: Parallels between Imprinting and Cortical Plasticity, Schloß Ringberg, 1984; Bateson: Selection in Neural and Behavioral Development, Neurosciences Institute, 1985; Immelmann: Sensitive Phases in Development, ZIF-Zentrum Bielefeld, 1985). Certainly many neuroscientists have become aware of the fact that compared to the wealth of neurophysiological and neuroanatomical literature on the consequences of restricted vision relatively little is known about the resulting behavioral modifications. That only gross changes of visual performance have been measured seems not too surprising. The lack of more sophisticated behavioral testing is partly due to a parallel lack in the exposure paradigms. The usefulness of behavior as a guiding tool is demonstrated most convincingly by the new discipline of neuroethology, some outstanding examples of which are contained in this volume. This area may well prove to be the natural and strongest bridge between the neurosciences and classical ethology.

ACTIVITY DEPENDENCE AND POSTSYNAPTIC GATING OF CHANGES IN SYNAPTIC EFFICACY

The main observation both in imprinting and in cortical plasticity is that changes of behavior and neuronal circuitry are caused by specific sensory stimulation. It is clear that these stimuli evoke a specific pattern of neural activity in corresponding brain structures. It is further quite logical that this activity must in one way or another be responsible for the observed changes. Less compelling is the conclusion that the primary modifications in the neural structure should occur in the synapses connecting different neural elements. In principle it could be possible that all plastic changes are due to a selection among neurons involving cell death or reversible inactivation. Many recent studies have made it very likely, however, that synaptic modification is indeed the main process at least in cortical plasticity. Gross changes in the structure of neural elements, dendrites or axon terminals, may only be a secondary consequence of the changes on the synaptic level.

The most interesting question in this context concerns the problem of whether neural activity in a presynaptic element alone is sufficient to change the efficacy of synapses formed by it, or whether some form of interaction (correlation) between pre- and postsynaptic elements or between different presynaptic elements is required. Only if this is the case, the receiving structure itself, for example the striate cortex, by way of its response properties could play a role in influencing the outcome of specific stimulation.

Changes of Binocular Connectivity in the Visual Cortex

The Classical Deprivation Studies. Classically, in the Hubel and Wiesel experiments, three different paradigms have been used to study the effect of visual experience on the innervation of area 17 neurons from the two eyes: monocular lid suture, binocular lid suture or dark rearing, and surgically induced squint of one or both eyes (Wiesel and Hubel, 1963, 1965; Hubel and Wiesel, 1965). It is well established from these and subsequent studies by various authors that monocular deprivation, depending on its duration, can lead to an almost complete loss of excitability of cortical neurons through the deprived eye. On the other hand, binocular deprivation does not lead to total abolition of excitability in all cortical cells. Thus, one can draw the conclusion that competition must be taking place between the input from the two eyes (Wiesel and Hubel, 1965; Guillery and Stelzner, 1970).

Monocular Exposure to Various Stimuli. In deprivation studies on cats or monkeys the most frequent technique for withholding visual stimuli to an eye during the sensitive period is to suture the lids of the eye shut. One has to realize, however, that this only prevents vision of patterned stimuli through this eye, while light–dark changes may still be visible. Some authors have therefore compared monocular deprivation by lid suture with asymmetric activation of the two eyes using various forms of contact lenses.

Blakemore (1976) showed that a diffusing contact lens covering one eye of a kitten has similar effects as monocular lid suture. By contrast, neutral density filters in front of one eye do not lead to a substantial reduction of excitability of cortical cells through this eye. This demonstrates clearly that it is not the difference in luminance between the two eyes, that leads to the shift of ocular dominance but the lack of patterned stimulation of one eye.

In similar experiments, Singer et al. (1977) and Wilson et al. (1977) reared kittens in temporally modulated light, while one eye was covered with a black contact lens, the other with a diffusor. Both studies gave congruent results: No change in the binocular connectivity of cortical units was observed, although the mean activity levels in the pathways from the two eyes up to the visual cortex were highly asymmetric. Cortical neurons require stimulation by bars or edges and do not respond very well to diffuse light flashes even in very young kittens (Hubel and

Wiesel, 1962, 1963; Rauschecker, unpublished). Thus, these experiments were the first hint that the activity of the postsynaptic target cells plays an important role for changes in cortical circuitry.

The next step was a study by Rauschecker et al. (1981) in which kittens were reared in an environment lit by stroboscopic light (brief flashes of light with a frequency of 2 Hz), while one eye was covered with a black contact lens preventing all vision through this eye. This kind of stimulation did produce a rather clear shift of ocular dominance (with some dependence on laminar distribution). Again this correlates well with the fact that most cortical units (about 70% in adult cats, Eschweiler et al., 1984) respond quite well to flashing bars or edges of light.

An even more stringent approach to the question is provided by restricting the monocular stimulation to contours of a single orientation (Rauschecker and Singer, 1981), since orientation specificity is the response property most typical for cortical cells. These experiments will be described and discussed later, once the classical stripe-rearing studies have been presented.

Changes of Cortical Orientation Selectivity

Is Orientational Modification Possible or Not? Historical and Methodological Aspects. The first attempts to influence the orientation preference of cortical neurons by specific visual exposure during the sensitive period were made in the same year by Hirsch and Spinelli (1970) and by Blakemore and Cooper (1970). Both studies, though using different methodological approaches, reported positive results. Blakemore and Cooper (1970) (and later Blakemore and Mitchell, 1973) reared their kittens freely moving in a stripe drum. The neurons they recorded from in area 17 after only a few hours of exposure to the stripes mostly responded best to the same orientation. Hirsch and Spinelli (1970, 1971) used a more rigorous exposure method: They projected the stripes onto the kittens' retinas with the help of biconvex lenses fitted into goggles together with the exposure pattern. The two eyes received different stimuli, one eye horizontal, the other vertical stripes. Upon recording, cortical units could be activated only through one eye, cells driven through the eye with horizontal experience preferring horizontal orientations and vice versa.

Hirsch and Spinelli's exposure method was used again later in studies by Stryker and Sherk (1975), Stryker et al. (1978), and Gordon et al.

(1979). The results of these investigations were less clear and suggested a number of conditions, which have to be fulfilled for the distribution of orientation preferences to be changed. First, and most important, they pointed out that the way of restricting the visual input has to be reliably controlled if the outcome is not supposed to be subject to chance. Thus, goggles would be superior to stripe drums. Second, these authors suggested that an asymmetry in the input to the two eyes may be required. However, other and later studies finally demonstrated that the same orientation presented to both eyes may also be effective in altering the distribution of orientation preferences (Blasdel et al., 1977; Rauschecker, 1979; Rauschecker and Singer, 1981). The main factor seems to be the control of visual input, asymmetric input may simply accelerate the rate of change.

Tretter et al., (1975), Singer (1976), and Blasdel et al., (1977) tried to overcome the uncertainty of rearing in stripe drums by fixing the animals and their heads in the center of the drum. Rauschecker and Singer (1979, 1981) used individually fitted goggles from polystyrole foam, which contained strong astigmatic lenses blurring all orientations but one.

Nature–Nurture or the Selective versus Instructive Issue. Apart from the methodological problems, there was another more interpretational issue that was discussed extensively in the context of stripe-rearing studies: the nature–nurture question or, transferred to the neuronal level, the selective versus instructive issue. This is one of the examples where modern neuroscience has touched on an old philosophical problem, which was already raised by Kant, Descartes, and many others (see Ganz, 1978).

The question of whether an a priori machinery is given to us for perception or whether we have to learn how to see starting with a tabula rasa, boils down to the question whether the cells in our visual cortex show orientation preferences prior to any visual experience of the individual or not. Closely related to this is the question of whether a bias in orientation preference induced by stripe rearing is due to a loss of neurons responding to other orientations or to an actual change in orientation preference, including the case of acquisition of a preference starting from a nonoriented origin.

All these questions seem to be finally settled now, as far as the specific effects of stripe rearing on the visual cortex are concerned. As to the existence of oriented cells in the visual cortex of very young, visually inexperienced kittens, Hubel and Wiesel's (1963) original finding turned out to be more or less correct. Undoubtedly, orientation selectivity can be generated without postnatal experience, as is found at least in some cortical cells of inexperienced animals (Imbert and Buisseret, 1975; Bonds, 1979; Vital-Durand and Blakemore, 1982; Albus and Wolf, 1984). Certainly not all neurons are selective, as Sherk and Stryker (1976) had assumed, but a substantial grid of selective cells is provided by "nature" that "nurture," that is, visual experience, can fill.

The issue of modifiability of orientation selectivity in individual neurons is less easy to settle. *Direct* evidence could only be gained if it were possible to observe a change in orientation preference during an acute experiment (Frégnac et al., 1984; see Frégnac, this volume).This is difficult for reasons having to do with the lack of enabling signals under anaesthesia and muscle paralysis (see section below on "Enabling Signals"). However, ample *indirect* evidence is available suggesting that cortical neurons can indeed change their orientation selectivity, at least within certain limits. One possibility is to compare quantitatively the magnitude of the orientation bias in a sample of striate cortex cells after rearing in a one-orientation environment with the expected bias, if it resulted merely from selective dropout of cells with other orientations (Rauschecker, 1979; Rauschecker and Singer, 1981). This assessment forces one to conclude that part of the cells must have changed their original orientation preference and some have newly acquired their selectivity from a state of unresponsiveness or nonorientedness.

In a final key experiment the two eyes of kittens were exposed to the same orientation, say vertical, for different amounts of time and (in order to get decoupling of binocular interactions in the cortex) on alternating days (Rauschecker, 1982). In this case, due to the selective deprivation process, both eyes would drive an equally small number of cells with a preference for horizontal. However, the eye with more experience (e.g., 100 hr) clearly activated not only more cells with a preference for vertical but also fewer cells with a preference for oblique orientations than the eye with less experience (e.g., 50 hr). This indicates that, as visual experience for vertical progresses, more and more cells with a preference for oblique orientations become tuned to vertical. The fact that both eyes drive the same small number of horizontally tuned cells demonstrates that this change of orientation preference is limited. The borders may be given by the original width of orientation tuning of the neurons, which is known to be broader at birth (Sherk and Stryker, 1976; Bonds, 1979; Albus and Wolf, 1984).

Orientation Selectivity as a Gating Parameter for Changes in Ocular Dominance: Rules for Synaptic Modification. So far we have looked at the phenomenology and at some interpretational features of the changes in ocular dominance and orientation preference of visual cortex, which can be induced by specific early experience during the sensitive period. Now we can go ahead and bring the two parameters together in experiments that have been thought to change one as a function of the other. This is done in studies rearing kittens with one eye seeing a single orientation, while the other eye is covered completely or is allowed to see all orientations unobstructedly (Rauschecker, 1979, 1983; Rauschecker and Singer, 1981). I do not want to describe all these experiments again here, rather I would like to go straight ahead to the conclusions that can be drawn from these data with regard to the signals governing cortical plasticity.

It follows from the results of these studies that pre- and postsynaptic activity have to be correlated for increases in synaptic efficacy to occur. Lack of correlation or anticorrelation results in no change or in a decrease of synaptic gain. If one assumes in addition that different sources of presynaptic activity have to be correlated for the postsynaptic threshold to be exceeded, it follows further that positive correlation is required between a certain number of presynaptic input fibers. These conclusions are depicted schematically in two different ways in Figure 1. Similar statements have been made on theoretical grounds in other studies (Stent, 1973; Malsburg, 1973; Palm, 1982; Bienenstock et al., 1982). The essence of all this in its original form can be traced back to Hebb (1949).

From this model we can deduce that a certain preference can be imprinted on a cortical cell or an assembly of cells if the corresponding array of input fibers is concomitantly activated with the postsynaptic targets. It is obvious that the acquisition of such a preference is immensely facilitated if the postsynaptic network already possesses properties that allow it to respond to that stimulus. Thus the model predicts that in cases where the imprinting stimulus is contained in the innate repertoire of stimuli to which the target structure responds preferentially, the speed of acquisition of a certain preference is much higher than for other stimuli. It is in these terms that a comparison between

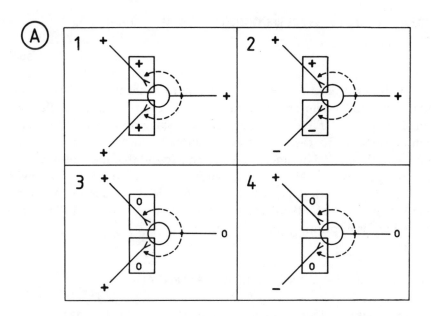

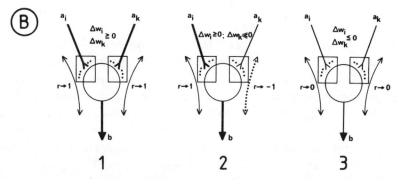

FIGURE 1. Rules for synaptic modification dependent on the correlation of pre- and post-synaptic activity. (A) Four different cases of activation of a cortex neuron by two different sets of inputs are drawn. Solid lines represent axonal pathways, dashed lines represent signal pathways feeding back into the vicinity of synaptic sites (framed by rectangular boxes). The efficacy of these synapses is changed as a function of pre- and postsynaptic activity. In all cases the average activity of the upper set of inputs is assumed to be above the mean resting discharge level ($+$), while the activity of the lower set can either be above (cases 1 and 3) or below (cases 2 and 4) this level. The discharge rate of the postsynaptic neuron can either be positive (1 and 2) or nil (3 and 4), since cortical neurons usually show very low spontaneous activity. The change of synaptic efficacy then simply results from the multiplication of pre- and postsynaptic activity. (B) This scheme is similar to (A), but the correlation r between input a and output b is directly used to change the synaptic weight w. Cases 1 and 2 are identical to cases 1 and 2 in (A), case 3 comprises both cases 3 and 4 from (A), because in both cases the correlation between pre- and postsynaptic activity goes to zero.

different forms of developmental plasticity, for example, in the visual cortex of mammals or in the telencephalon of birds, should be possible.

One other prediction that can be made from these considerations is that a neural system should remain plastic even beyond the classical sensitive period, if it is artificially kept in a state of low specificity. This can be done in the case of the visual cortex by dark rearing an animal for a lengthy period of time. Indeed, it was found that the critical period for changes in cortical ocular dominance can be prolonged if the animals are kept in the dark (Cynader et al., 1976, 1980; Cynader and Mitchell, 1980; Timney et al., 1980; Timney, this volume). Similar findings are available for imprinting (see Bateson, 1979).

ENABLING SIGNALS IN CORTICAL PLASTICITY

Several lines of evidence suggest that apart from specific sensory signals other, nonspecific sources of information have to be active for synaptic changes to take place. Since these signals do not by themselves determine the quality of specific preferences to be acquired or contents to be learned, but only allow them to occur, they are often referred to as gating or enabling signals (see Cooper, Marler, Singer, this volume).

At least two qualitatively different kinds of enabling factors may be discriminated. The first would be a global signal switching synapses into a state of increased plasticity, as is the case during the sensitive period. Other signals might have to do with what is usually referred to as "arousal" or "attention". The latter kind of signals would have to be available together with the specific activity that is generated by sensory signals and would thus enable local changes of neural circuitry.

Signals Enabling Global Plasticity during the Sensitive Period

This section will very briefly review some of the suggestions made for the control of the sensitive or critical period. The topic will be discussed more profoundly by other contributions in this volume (see, for example, Frégnac, this volume).

The Norepinephrine Hypothesis. The original norepinephrine theory of Kasamatsu and Pettigrew states that cortical plasticity can be abolished by depletion of catecholamines (Kasamatsu and Pettigrew, 1976) and that

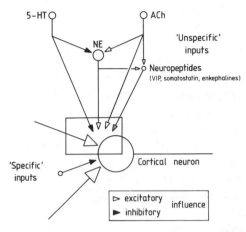

FIGURE 2. Substances possibly acting as enabling or disabling factors.

it can be reinstated, even in adult animals, by local perfusion with noradrenaline (Pettigrew and Kasamatsu, 1978). These findings have caused a large number of subsequent studies by various groups that have tried to prove or disprove the theory. Although for some years evidence seemed to accumulate against the theory (Adrien et al., 1982; Bear and Daniels, 1983; Daw et al., 1983, 1984), the current trend seems to attribute at least partial responsibility for the existence of critical period plasticity to norepinephrine (Bear and Singer, 1986).

As a possible alternative to norepinephrine, some authors suggested a similar role for acetylcholine (Sillito, 1983; Singer, 1984). In the meantime, experimental evidence argues also against an exclusive control of sensitive phase plasticity by acetylcholine (Bear and Singer, 1986).

It is also conceivable that disabling signals may be delivered by neurons in the dorsal raphe nuclei of the brain stem, which secrete 5-HT (serotonin) diffusely into most cortical regions including the occipital ones (Foote and Morrison, 1985). Such a system would thus act antagonistically to the noradrenergic locus coeruleus system.

Finally, some of the neuropeptides (possibly by an interaction with some of the classical transmitters) may play a role as enabling substances. Figure 2 shows some of these unspecific modulator candidates in a summary diagram.

Other Factors. As discussed in the section on activity dependence and postsynaptic gating, a self-termination of the sensitive period might be reached by increasing specificity of the neuronal network, thus reducing its degrees of freedom for fast synaptic changes. This would reduce the plasticity of single neurons to the degree still apparent in adult animals, which is then referred to ordinarily as "learning."

On the other hand, such a simple concept cannot explain the gradual onset of the sensitive period in kittens after birth, since susceptibility should be maximal in the totally inexperienced animal. In this context, comparative studies in different animal species, altricial and precocial, might be revealing, as they have been in neuroanatomy for the understanding of dendritic spine development (Schüz, 1981). Such studies suggest that the functional state of the neuronal machinery at birth depends on embryological factors governed by totally different programs.

Signals Enabling Local Changes of Synaptic Efficacy

The main line of evidence for additional, nonspecific signals required for synaptic changes to occur has come from studies of cortical plasticity in acute preparations: Attempts to induce reliable changes of cortical response properties under anaesthesia and muscle paralysis have failed almost invariably.

Eye Movements. One possible explanation for the lack of plasticity in anaesthetized, paralyzed animals is the lack of eye movements in these preparations. The very necessity of suppressing eye movements for the mapping of visual receptive fields may thus be responsible for the abolition of modifiability of their connectivity.

The most direct approach to this question was taken by Freeman and Bonds (1979). They stimulated an eye muscle in an anaesthetized paralyzed kitten by mechanically stretching it with a clock motor and at the same time visually stimulated one eye with optical patterns. Under these circumstances they were able to reduce the binocularity of cortical units, but not to induce an ocular dominance shift.

For eye muscle activity to play a role in enabling plasticity of the visual cortex, it has to reach this structure by way of direct or indirect pathways. Several studies have claimed a direct activation of visual cortical neurons by mechanical or electrical stimulation of extraocular muscles (Buisseret and Maffei, 1977; Ashton et al., 1984).

Other authors have pointed out that eye movement activity may reach the visual cortex by entering the pool of unspecific activity via the "reticular core" (see Singer, 1984). Such "central core control" of cortical plasticity has been associated with attention. It was demonstrated that electrical stimulation of the midbrain reticular formation or the medial thalamus can indeed restore part of the plasticity in acute preparations (Singer and Rauschecker, 1982). Nevertheless, the concept of central core control, attractive as it is, may simply defer the problem of gating to even less accessible structures.

Effects of Anaesthesia. The second factor that could be responsible for the impairment of cortical plasticity in acute preparations, alternatively or in addition to the lack of eye movements, is anaesthesia. Various kinds of anaesthetics have been used by different authors, including barbiturates, halothane, and ketamine, all of them mostly in combination with nitrous oxide. So far there have been no systematic studies as to whether there are differential effects of these drugs on cortical plasticity. Furthermore, there is not even a marginal theory about how general anaesthetics affect the normal functional activity of the brain. Certainly, most anaesthetics do not seem to alter responses in the primary visual cortex substantially (Wurtz, 1969). Only minor changes in initial transient components have been observed (Noda et al., 1971). Thus, an impairment of cortical plasticity cannot arise from a reduction of postsynaptic responsiveness in an activity-dependent Hebbian process.

There have been a number of reports that barbiturates and other anaesthetics modulate the activity of noradrenergic neurons in the locus coeruleus (Mason et al, 1983) and of cells in the midbrain reticular formation (Shimoji et al., 1984). This would make it possible that anaesthesia abolishes gating signals from this common pool of nonspecific activity, as mentioned above.

Recent experiments (Rauschecker, 1985; Rauschecker and Hahn, 1987) suggest a mechanism by which how anaesthesia might impair cortical plasticity and which role gating signals in general may play in the long-term storage of visual information. This view arose from the surprising finding that anaesthesia can impair plasticity even if it is given *after* visual stimulation. The exact paradigm was as follows: Young kittens were kept in the dark and taken out into the light for 20 min each day with one eye covered. The open eye was stimulated with high-contrast visual stimuli for a 30-day period. This procedure of 10 hr total monocular exposure led to a clear shift of ocular dominance in control kittens. For the experimental animals the same protocol was followed except that

the kittens were anaesthetized with an intramuscular injection of ke-
tamine/xylazine immediately after each visual stimulation period. Anaesthesia lasted for 1 hr. In none of these kittens was a change of cortical
binocularity observed. If, however, a break of 1 to 6 hr was allowed
between the end of each visual stimulation period and the onset of
anaesthesia, an ocular dominance shift did occur consistently.

The main conclusion that has to be drawn from this study is that
nonspecific enabling signals may not influence cortical plasticity by modulating the activity of cortical neurons, because the activity-dependent
process was already over by the time of interference. Rather, these experiments suggest that what is enabled by these signals is the structural
"consolidation" of the synaptic changes, which follows the initial activity-dependent step. The duration of this "transcription" or "printing"
process seems to be in the order of 1 hr or less and is thus much shorter
than had been assumed by other studies on consolidation in visual cortex
(Pettigrew and Garey, 1974; Freeman and Olson, 1979). It would be
interesting to see whether this process can also be impaired by other

Implementation of Enabling Factors

a) Inhibition/Disinhibition of Neural Activity

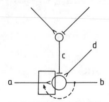

b) Impairment/Promotion of Synaptic Consolidation

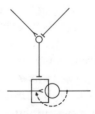

FIGURE 3. Two principles of implementing enabling factors for synaptic modifications: (A)
as a gating signal modulating the activity of the postsynaptic neuron, (B) as a signal enabling the "printing" or consolidation process, that is, the transfer of information from a
neural activity pattern into a structural change.

completely different anaesthetics or even by procedures such as elec-
troconvulsive shock or the application of specific protein synthesis
inhibitors.

Implementation of Enabling Signals in Cortical Plasticity. Figure 3 shows the
two conceivable ways of implementing unspecific signals in synaptic
modification. As argued above, the present experimental evidence forces
us to favor a role for enabling factors in synaptic consolidation processes
(Fig. 3B) rather than in the modulation of neuronal activity (Fig. 3A).

A complete model for activity-dependent synaptic modification, also
involving unspecific factors, is depicted in Figure 4. The structural
change may either take place in postsynaptic receptors (Fig. 4A) or in
the presynaptic mechanisms for transmitter release (Fig. 4B). In both
cases, a synaptic reward factor (SRF; Barlow, 1982) is assumed to be
released as a function of postsynaptic depolarization. In conjunction
with the secreted transmitter substance (Fig. 4A) or with presynaptic
depolarization (Fig. 4B) a chain of biochemical processes is initiated,

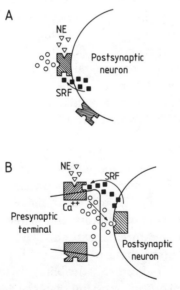

FIGURE 4. Scheme for activity-dependent synaptic modification under the presence of an
enabling factor such as norepinephrine (NE). A postsynaptic reward factor (SRF) is released
as a function of postsynaptic activity. The structural change itself can alternatively take
place in postsynaptic (A) or presynaptic (B) receptors. The first possibility would result in
a change in the amplitude of EPSPs evoked by a certain quantity of transmitter. In the
second case, the transmitter release after a presynaptic depolarization is influenced.

which is only completed if an enabling factor, such as norepinephrine (NE), is present in addition.

INTERACTIONS OF VISUAL AND AUDITORY SIGNALS DURING DEVELOPMENT

Behavioral Compensation of Visual Impairment

The Risk of the Critical Period. A major difference between imprinting in birds and visual plasticity in mammals seems to be that imprinting provides the animal with new behavioral features, whereas visual plasticity, despite possible modifications of prespecified response properties, often appears more like a selective deprivation process in which specific features are stabilized and others are lost. More research on the neurophysiological basis of imprinting may eventually reveal that both processes are in fact not so far apart in that respect.

One may wonder, therefore, what the survival value of a sensitive or critical period in the development of the visual cortex is, if its main effect is to put the individual in danger of becoming blind in one eye (Dews and Wiesel, 1970; Sherman, 1972) or losing visual resolution for nonexperienced orientations (Hirsch, 1972; Blasdel et al., 1977; Rauschecker, 1984). Certainly, most experimental situations are only caricatures of natural kinds of environments, but the fact remains that supernormal improvements of visual performance in nondeprived functions are difficult to demonstrate behaviorally or physiologically.

Nonvisual Capacities of the Blind. One may speculate therefore whether the survival value of a critical period in visual development could lie in the possibility of providing more information capacity for other sensory channels, if vision is not fully stretched. Casual reports about such compensatory improvements of auditory and tactile functions in blind persons exist in earlier literature (Griesbach, 1899; Woelfflin, 1909), though not without later contradictions (e.g., Axelrod, 1959). More recent quantitative investigations indicate that the blind are indeed significantly better than sighted persons in assessing distance and structure of surfaces by ear (Kellogg, 1962). Niemeyer and Starlinger (1981) found a clear superiority in functions they attribute to "higher levels of the auditory pathways." No comparison was made between early-acquired or inborn blindness and blindness acquired in later life.

Behavioral Compensation in Deprived Kittens. If one observes locomotor and orienting behavior in binocularly lid-sutured kittens, it is indeed very surprising how secure and unimpaired these animals appear. Only on closer inspection does it become apparent that they make use of hearing and touch much more intensely than their seeing mates. Quantitative measurements of auditory localization in blind and normal cats are being done in our laboratory, and indeed it seems that the visually deprived cats are better in localizing and discriminating sound sources (tone bursts), which are presented at different locations on the perimeter of a sound-attenuated chamber (Rauschecker and Dehnen, 1986).

Physiological and Anatomical Basis of Compensation Mechanisms

Auditory Compensation of Visual Deprivation in the Superior Colliculus. From the studies of Jenkins and Merzenich (1984) it appears that the auditory cortex is largely responsible for auditory localization. Changes of neural circuitry might therefore be expected to take place in the auditory cortex of cats. Nevertheless, we chose the midbrain superior colliculus (SC) to look at possible changes physiologically because this nucleus receives convergent input from auditory and visual cortex and may thus reflect changes in both structures (see Fig. 5). Furthermore, the superior colliculus, as part of the tectum, is known to play a major role in orienting behavior (Sprague et al., 1973; Harris, 1980).

The results of our study (Rauschecker and Harris, 1983) were rather straightforward: More than 40% of the cells in SC of binocularly deprived cats responded to our auditory stimuli (clicks and pure tone bursts up to 50 kHz delivered in free-field stimulation), whereas only about 10% of the cells in SC of normal cats responded to these stimuli. There was of course the danger of a sampling bias, since only the deeper layers of SC contain auditory neurons in normal cats. From track reconstruction it turned out, however, that our samples from the two groups of animals were actually opposed to this trend: We had relatively more cells from deep layers in normal than in deprived cats; in addition, some auditory cells were recorded even from superficial SC layers in the deprived animals.

The distribution of auditory response latencies of SC cells in the visually deprived cats as compared to normal cats indicates that the increase of auditory-responsive units is caused by a strengthened projection from auditory cortex rather than from subcortical sources, such as the inferior

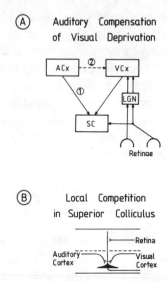

FIGURE 5. Neural pathways and mechanisms involved in auditory compensation of visual deprivation. (A) Two alternative possibilities are suggested explaining the increase of auditory-responsive neurons in superior colliculus (SC) after visual deprivation in kittens. In both cases a projection from cortical areas, either auditory (ACx) or visual (VCx), is involved. In addition, the strengthening of a transitory projection from auditory to visual cortex is discussed. (LGN = lateral geniculate nucleus). (B) Synaptic changes in superior colliculus can be explained in terms of local competition between visual and auditory inputs, if the spatial separation of retinal and cortical input fibers is taken into account. Thus, only cortical inputs would compete with each other, while retinal terminals remain firmly established.

colliculus (Rauschecker, 1984). This again nourishes the suspicion that the primary changes may be going on in auditory cortex. We are currently testing this possibility with a neuroanatomical approach using the enzyme horseradish peroxidase as a retrograde tracer (Rauschecker and Aschoff, 1987).

Possible Interactions between Visual and Auditory Signals in the Visual Cortex. An alternative hypothesis for explaining the compensatory increase of auditory-responsive cells in superior colliculus after dark rearing would be the following: Neurons projecting from visual cortex to SC could become responsive to auditory stimulation as a result of visual deprivation and could thus relay auditory signals to SC (see Fig. 5A). The corticotectal projection being bimodal in the first place could explain why most of the auditory-responsive cells in SC of deprived cats were found to be bimodal, retaining a visual input.

This second hypothesis is not completely farfetched since there have been casual reports about auditory-responsive units in area 17 of normal cats in the past (Jung et al., 1963; Morrell, 1972; Fishman and Michael, 1973). None of these reports have been taken really seriously, because the auditory responses were usually attributed to some nonspecific projection. This view has changed slightly recently because three groups using modern anatomical tracer techniques have independently discovered transitory projections in young animals (Innocenti and Clarke, 1984; Dehay et al., 1984; Cowan et al., 1984). Some of these projections are from auditory to visual cortex and although these projections seem to be only transitorily present during early development, the possibility exists that some remnants are there in the adult cortex. In particular, applied to the present question, it might be possible that the projection from auditory to visual cortex is maintained or even strengthened compensatorily as a result of visual deprivation.

Visual—Auditory Competition during Development. Both possible explanations for the compensatory increase of auditory-responsive cells in superior colliculus require an interaction between visual and auditory signals (Fig. 5A), which takes place either in the superior colliculus (hypothesis 1) or in the visual cortex (hypothesis 2). In both cases a competition between visual and auditory fibers seems to be going on.

Contrary to our original conclusion that assumed a noncompetitive ("live-and-let-live") interaction between visual and auditory inputs (Rauschecker and Harris, 1983), possible synaptic changes in SC (hypothesis 1) can be fully explained on the basis of competition (Guillery and Stelzner, 1970; Cynader, 1979). The only precondition is a separation of competing from noncompeting elements in time or space.

It is known that during normal development of the tectum the phylogenetically older, retinal input grows in earlier than the cortical input and establishes a retinotopic map. The cortical innervation of the tectum is diffuse at first and forms a precise map only later, probably under the influence of external experience and using the preestablished retinal map as a template (Anker and Cragg, 1974). It could thus be that competitive interactions are going on only between cortical fibers, while the retinal input is already firmly established. Indeed, in binocularly deprived cats the direct retinal input is practically unaffected, whereas the indirect input from striate cortex is highly impaired in its function (Hoffmann and Sherman, 1975; Rauschecker and Harris, 1983).

Competition restricted to cortical input fibers could also be facilitated by a local criterion (Fig. 5B): Retinal and cortical input fibers are spatially separated in different sublaminae of SC—retinal fibers terminating on the apical dendrites, cortical fibers on basal dendrites of cells in intermediate and deep layers (Huerta and Harting, 1984). If competition is a biophysical mechanism acting locally on the postsynaptic membrane, this could lead to an expansion of auditory cortical terminals at the expense of visual cortical ones, leaving the retinal input unaffected.

In the case of a preservation of the transitory projection from auditory to visual cortex (hypothesis 2), the interactions between visual and auditory signals would have to go on in the visual cortex. This could again be competition between the two types of input, similar to the possible interaction described above for SC or described earlier for competition in cortex between the two eyes (Wiesel and Hubel, 1965; Guillery and Stelzner, 1970) or between fibers encoding different orientations (Rauschecker, 1983).

CONCLUSIONS

I have presented three examples of conceptual and experimental frameworks of how neuronal signals can be responsible for synaptic changes during visual development.

The first part covered some of the work on monocular eye closure and rearing in uniorientational environments. I have tried to show that by varying the input to the open eye and by combining the two classical approaches of selective exposure one can deduce the existence of a very general mechanism for synaptic plasticity: The main parameter for increases in synaptic efficacy seems to be the correlation between pre- and postsynaptic activity and, as a consequence, between congruous presynaptic elements.

The second part dealt with signals necessary to enable developmental plasticity. Although these signals can be assumed to be present during a normal state of arousal, several situations can be demonstrated in which these enabling signals are obviously missing, thereby impairing or abolishing cortical plasticity. The most dramatic case is probably the state of anaesthesia and muscle paralysis, which is used under experimental electrophysiological recording conditions. In this situation no reliable changes of cortical circuitry seem to be possible. The finding that anaesthesia can impair plasticity even if it is given following visual stimulation suggests that enabling signals do not modulate the activity-dependent process. Rather they seem to participate in the transfer of activity patterns into structural synaptic changes, often referred to as printing or consolidation process.

In the third part I have demonstrated that rearrangement of neural connections in the visual system resulting from visual deprivation may at the same time cause changes in other sensory systems. In structures like the superior colliculus of the midbrain tectum, where convergent inputs from different sensory fibers are present, competitive interactions between signals from the auditory and the visual cortex lead to a compensatory increase in the number of auditory-responsive units. Similar cross-modality effects might be found in visual cortex via a stabilization of a normally transitory projection from auditory cortex.

Possible future advances in the understanding of neural plasticity during sensitive phases of postnatal development are quite obvious: One strategy is certainly to try and understand the synaptic mechanisms on the molecular level. More detailed information is necessary to understand the chain of biochemical processes triggered by neural activity and leading to changes of synaptic efficacy. An understanding of these processes would also facilitate answering the question of which steps are impaired by the lack of enabling signals and which steps discriminate the highly plastic state during the sensitive period from the less plastic one in later life.

A second way of progression, however, seems equally fruitful. In order to be able to appreciate the relevance of the neurophysiological and neuroanatomical changes observed during the sensitive period, one must relate these changes to behavior. This means that more refined measurements of behavioral consequences of specific experience have to be undertaken. A promising example might be the case of auditory compensation resulting from visual deprivation. On the other hand, taking into account behavioral aspects may also lead to the design of new paradigms for neurophysiological testing and to a more general understanding of neural plasticity in terms of learning (see Marler and Terrace, 1984). A whole number of lessons can be learned in that respect by neuroscientists from classical and modern ethology.

ACKNOWLEDGMENTS

I would like to thank Peter Marler and Ulrich Egert for comments on this chapter, Sabine Hahn for drawing the figures, and Shirley Würth for editing the manuscript.

REFERENCES

Adrien, J., P. Buisseret, Y. Frégnac, E. Gary-Bobo, J.-P. Tassin, Y. Trotter, and M. Imbert (1982). Noradrenaline et plasticité du cortex visuel du chaton. *C. R. Acad. Sc. Paris. Ser. III* **295**:745–750.

Albus, K. and W. Wolf (1984). Early postnatal development of neuronal function in the kitten's visual cortex: A laminar analysis. *J. Physiol. (Lond.)* **348**:153–185.

Anker, R. C. and B. G. Cragg (1974). Development of the extrinsic connections of the visual cortex in the cat. *J. Comp. Neurol.* **154**:29–42.

Ashton, J. A., A. Boddy, and I. H. L. Donaldson (1984). Directional selectivity in the responses of units in cat primary visual cortex to passive eye-movements. *Neuroscience* **13**:653–662.

Axelrod, S. (1959). *Effects of Early Blindness*. American Foundation for the Blind, New York.

Barlow, H. B. (1982). David Hubel and Torsten Wiesel: Their contributions towards understanding the primary visual cortex. *Trends Neurosci.* **5**:145–152.

Bateson, P. (1979). How do sensitive periods arise and what are they for? *Anim. Behav.* **27**:470–486.

Bear, M. F. and J. D. Daniels (1983). The plastic response to monocular deprivation persists in kitten visual cortex after chronic depletion of norepinephrine. *J. Neurosci.* **3**:407–416.

Bear, M. F. and W. Singer (1986). Modulation of visual cortical plasticity by acetylcholine and noradrenaline. *Nature* **320**:172–176.

Bienenstock, E. L., L. N. Cooper, and P. Munro (1982). Theory for the development of neuron selectivity: Orientation specificity and binocular interaction in visual cortex. *J. Neurosci.* **2**:32–48.

Blakemore, C. (1976). The conditions required for the maintenance of binocularity in the kitten's visual cortex. *J. Physiol. (Lond.)* **261**:423–444.

Blakemore, C. and G. F. Cooper (1970). Development of the brain depends on the visual environment. *Nature* **228**:477–478.

Blakemore, D. and D. E. Mitchell (1973). Environmental modification of the visual cortex and the neural basis of learning and memory. *Nature* **241**:467–468.

Blasdel, G. G., D. E. Mitchell, D. W. Muir, and J. D. Pettigrew (1977). A physiological and behavioural study in cats of the effects of early visual experience with contours of a single orientation. *J. Physiol. (Lond.)* **265**:615–636.

Bonds, A. B. (1979). Development of Orientation Tuning in the Visual Cortex of Kittens. In: *Developmental Neurobiology of Vision*, R. D. Freeman, ed., pp. 31–41, Plenum Press, New York.

Buisseret, P. and L. Maffei (1977). Extraocular proprioceptive projections to the visual cortex. *Exp. Brain Res.* **28**:421–425.

Cowan, W. M., J. W. Fawcett, D. D. M. O'Leary, and B. B. Stanfield (1984). Regressive events in neurogenesis. *Science* **225**:1258–1265.

Cynader, M. (1979). Competitive Interactions in Postnatal Development. In: *Developmental Neurobiology of Vision*, R. D. Freeman, ed., pp. 109–120, Plenum Press, New York.

Cynader, M. and D. E. Mitchell (1980). Prolonged sensitivity to monocular deprivation in dark-reared cats. *J. Neurophysiol.* **43**:1026–1054.

Cynader, M., N. Berman, and A. Hein (1976). Recovery of function in cat visual cortex following prolonged deprivation. *Exp. Brain Res.* **25**:139–156.

Cynader, M., B. N. Timney, and D. E. Mitchell (1980). Period of susceptibility of kitten visual cortex to the effects of monocular deprivation extends beyond six months of age. *Brain Res.* **11**:545–550.

Daw, N. W., R. K. Rader, T. W. Robertson, and T. O. Videen (1983). Effects of 6-hydroxydopamine on visual deprivation in the kitten striate cortex. *J. Neurosci.* **3**:907–914.

Daw, N. W., T. W. Robertson, R. K. Rader, T. O. Videen and C. J. Coscia (1984). Substantial reduction of cortical norepinephrine by lesion of adrenergic pathway does not prevent effects of monocular deprivation. *J. Neurosci.* **4**:1354–1360.

Dehay, C., J. Bullier, and H. Kennedy (1984). Transient projections from the fronto-parietal and temporal cortex to area 17, 18 and 19 in the kitten. *Exp. Brain Res.* **57**:208–212.

Dews, P. B. and T. N. Wiesel (1970). Consequences of monocular deprivation on visual behaviour in kittens. *J. Physiol. (Lond.)* **206**:437–455.

Eschweiler, G., M. Popp, J. P. Rauschecker, and W. Schrader (1984). Timing of flash responses in visual cortex of normal and strabismic cats. *Soc. Neurosci. Abstr.* **10**:469.

Fishman, M. C. and C. R. Michael (1973). Integration of auditory information in the cat's visual cortex. *Vision Res.* **13**:1415–1419.

Foote, S. and J. Morrison (1985). Noradrenergic and serotonergic innervation of cortical, thalamic and tectal structures in monkey. *Neurosci. Lett. Suppl.* **22**:S545.

Freeman, R. D. and A. B. Bonds (1979). Cortical plasticity in monocularly deprived immobilized kittens depends on eye movement. *Science* **206**:1093–1095.

Freeman, R. D. and C. R. Olson (1979). Is there a "consolidation" effect for monocular deprivation? *Nature* **282**:404–406.

Frégnac, Y., S. Thorpe, D. Schulz, and E. L. Bienenstock, (1984). Modification of the function in the cat visual cortical neurones induced by control of the correlation between postsynaptic activity and visual input. *Soc. Neurosci. Abstr.* **10**:1078.

Ganz, L. (1978). Innate and Environmental Factors in the Development of Visual Form Perception. In: *Handbook of Sensory Physiology*, Vol. VIII, R. Held, H. W. Leibowitz, and H.-L. Teuber, eds., pp. 437–488, Springer-Verlag, New York.

Gordon, B., J. Presson, J. Packwood, and R. Scheer (1979). Alteration of cortical orientation selectivity: Importance of asymmetric input. *Science* **204**:1109–1111.

Griesbach, H. (1899). Vergleichende Untersuchungen über die Sinnesschärfe Blinder und Sehender. *Pflügers Arch.* **74**:577–638.

Guillery, R. W. and D. J. Stelzner (1970). The differential effects of unilateral lid closure upon the monocular and binocular segments of the dorsal lateral geniculate nucleus in the cat. *J. Comp. Neurol.* **139**:413–422.

Harris, L. R. (1980). The superior colliculus and movements of the head and eyes in cats. *J. Physiol. (Lond.)* **300**:367–391.

Hebb, D. O. (1949). *The Organization of Behavior.* Wiley & Sons, New York.

Hirsch, H. V. B. (1972). Visual perception in cats after environmental surgery. *Exp. Brain Res.* **15**:405–423.

Hirsch, H. V. B. and D. N. Spinelli (1970). Visual experience modifies distribution of horizontally and vertically oriented receptive fields in cats. *Science* **168**:869–871.

Hirsch, H. V. B. and D. N. Spinelli (1971). Modification of the distribution of receptive field orientation in cats by selective visual exposure during development. *Exp. Brain Res.* **12**:509–537.

Hoffmann, K.-P. and S. M. Sherman (1975). Effects of early binocular deprivation on visual input to cat superior colliculus. *J. Neurophysiol.* **38**:1049–1059.

Hubel, D. H. and T. N. Wiesel (1962). Receptive fields, binocular interaction and functional architecture in the cat's visual cortex. *J. Physiol. (Lond.)* **160**:106–154.

Hubel, D. H. and T. N. Wiesel (1963). Receptive fields of cells in striate cortex of very young, visually inexperienced kittens. *J. Neurophysiol.* **26**:994–1002.

Hubel, D. H. and T. N. Wiesel (1965). Binocular interaction in striate cortex of kittens reared with artificial squint. *J. Neurophysiol.* **28**:1041–1059.

Hubel, D. H. and T. N. Wiesel (1970). The period of susceptibility to the physiological effects of unilateral eye closure in kittens. *J. Physiol. (Lond.)* **206**:419–436.

Huerta, M. F. and J. K. Harting (1984). The Mammalian Superior Colliculus: Studies of Its Morphology and Connections. In: *Comparative Neurology of the Optic Tectum*, H. Vanegas, ed., pp. 687–773, Plenum Press, New York.

Imbert, M. and P. Buisseret (1975). Receptive field characteristics and plastic properties of visual cortical cells in kittens reared with or without visual experience. *Exp. Brain Res.* **22**:25–36.

Innocenti, G. M. and S. Clarke (1984). Bilateral transitory projection to visual areas from auditory cortex in kittens. *Dev. Brain Res.* **14**:143–148.

Jenkins, W. M. and M. M. Merzenich (1984). Role of cat primary auditory cortex for sound-localization behavior. *J. Neurophysiol.* **52**:819–847.

Jung, R., H. H. Kornhuber, and J. S. Da Fonseca (1963). Multisensory Convergence on Cortical Neurons: Neuronal Effects of Visual, Acoustic and Vestibular Stimuli in the Superior Convolutions of the Cat's Cortex. In: *Progress in Brain Research. Brain Mechanisms: Specific and Unspecific Mechanisms of Sensory Motor Integration*, G. Moruzzi, A. Fessard, and H. H. Jasper, eds., Vol. 1, pp. 207–240, Elsevier, Amsterdam.

Kasamatsu, T. and J. D. Pettigrew (1976). Depletion of brain catecholamines: Failure of ocular dominance shift after monocular occlusion in kittens. *Science* **194**:206–209.

Kellogg, W. N. (1962). Sonar system of the blind. *Science* **137**:399–404.

Lorenz, K. (1935). Der Kumpan in der Umwelt des Vogels. *J. Ornithologie.* **83**:137–214, 289–413.

Malsburg, C. von der (1973). Self-organization of orientation sensitive cells in the striate cortex. *Kybernetik* **14**:85–100.

Marler, P. and H. S. Terrace (1984). *The Biology of Learning*. Dahlem Konferenzen, Springer-Verlag, Berlin, New York.

Mason S. T., R. A. J. King, P. Banks, and A. Angel (1983). Brain noradrenaline and anaesthesia: Behavioural and electrophysiological evidence. *Neuroscience* **10**:177–185.

Morrell, F. (1972). Visual system's view of acoustic space. *Nature* **238**:44–45.

Niemeyer, W. and I. Starlinger (1981). Do the blind hear better? Investigations on auditory processing in congenital or early acquired blindness. II. Central functions. *Audiology* **20**:510–515.

Noda, H., R. B. Freeman, B. Gies, and D. D. Creutzfeldt (1971). Neuronal responses in the visual cortex of awake cats to stationary and moving targets. *Exp. Brain Res.* **12**:389–405.

Palm, G. (1982). *Neural Assemblies. An Alternative Approach to Artificial Intelligence.* Springer-Verlag, New York.

Pettigrew, J. D. and L. J. Garey (1974). Selective modification of single neuron properties in the visual cortex of kittens. *Brain Res.* **66**:160–164.

Pettigrew, J. D. and T. Kasamatsu (1978). Local perfusion of norepinephrine maintains visual cortical plasticity. *Nature* **271**:761–763.

Rauschecker, J. P. (1979). Orientation Dependent Changes in Response Properties of Neurons in the Kitten's Visual Cortex. In: *Developmental Neurobiology of Vision*, R. D. Freeman, ed., pp. 121–133, Plenum Press, New York.

Rauschecker, J. P. (1982). Instructive changes in the kitten's visual cortex and their limitation. *Exp. Brain Res.* **48**:301–305.

Rauschecker, J. P. (1983). Competition and orientation-dependent recovery from monocular deprivation in the kitten's striate cortex. *Dev. Brain Res.* **10**:305–308.

Rauschecker, J. P. (1984). Neuronal mechanisms of developmental plasticity in the cat's visual system. *Human Neurobiol.* **3**:109–114.

Rauschecker, J. P. (1985). Retrograde impairment of visual cortical plasticity by anaesthesia. *Beh. Brain Res.* **16**:221–222.

Rauschecker, J. P. and A. Aschoff (1987). Changes in corticotectal projections of cats after visual deprivation. In press.

Rauschecker, J. P. and U. Dehnen (1986). Visual deprivation improves the ability for auditory localization in cats. *Perception* **15**:A34.

Rauschecker, J. P. and S. Hahn (1987). Ketamine-xylazine anaesthesia blocks consolidation of ocular dominance changes in kitten visual cortex. *Nature* **326**:183–185.

Rauschecker, J. P. and L. R. Harris (1983). Auditory compensation of the effects of visual deprivation in the cat's superior colliculus. *Exp. Brain Res.* **50**:69–83.

Rauschecker, J. P. and W. Singer (1979). Changes in the circuitry of the kitten visual cortex are gated by postsynaptic activity. *Nature* **280**:58–60.

Rauschecker, J. P. and W. Singer (1981). The effects of early visual experience on the cat's visual cortex and their possible explanation by Hebb synapses. *J. Physiol. (Lond.)* **310**:215–239.

Rauschecker, J. P., W. Singer, and M. von Grünau (1981). Effects of monocular stroboscopic experience on the kitten's visual cortex. *Adv. Physiol. Sci.* Vol. 2. *Regulatory Functions of the CNS. Subsystems,* J. Szentagothai, J. Hamori, N. Palkovits, eds., pp. 31–39, Akademiai Kiado, Budapest.

Schüz, A. (1981). Pränatale Reifung und postnatale Veränderungen im Cortex des Meerschweinchens: Mikroskopische Auswertung eines natürlichen Deprivationsexperiments. *J. Hirnforsch.* **22**:93–127.

Sherk, H. and M. P. Stryker (1976). Quantitative study of cortical orientation selectivity in visually inexperienced kitten. *J. Neurophysiol.* **29**:63–70.

Sherman, S. M. (1972). Visual development in cats. *Invest. Ophthalmol. Vis. Sci.* **11**:394–401.

Shimoji, K., H. Fujioka, and T. Ebata (1984). Anesthetics block excitation with various effects on inhibition in MRF neurons. *Brain Res.* **295**:190–193.

Sillito, A. M. (1983). Plasticity in the visual cortex. *Nature* **303**:477–478.

Singer, W. (1976). Modification of orientation and direction selectivity of cortical cells in kittens with monocular vision. *Brain Res.* **118**:460–468.

Singer, W. (1984). Learning to See: Mechanisms in Experience-dependent Development. In: *The Biology of Learning,* P. Marler and H. S. Terrace, eds., pp. 461–477, Dahlem Konferenzen, Springer-Verlag, Berlin, New York.

Singer, W. and J. P. Rauschecker (1982). Central core control of developmental plasticity in the kitten visual cortex. Electrical activation of mesencephalic and diencephalic projections. *Exp. Brain Res.* **47**:223–233.

Singer, W., J. Rauschecker, and R. Werth (1977). The effect of monocular exposure to temporal contrasts on ocular dominance in kittens. *Brain Res.* **134**:568–572.

Sprague, J. M., G. Berlucchi, and G. Rizzolatti (1973). The Role of the Superior Colliculus and Pretectum in Vision and Visually Guided Behavior. In: *Handbook of Sensory Physiology,* Vol VII/3B, R. Jung, ed., pp. 27–102, Springer-Verlag, New York.

Stent, G. (1973). A physiological mechanism for Hebb's postulate of learning. *Proc. Natl. Acad. Sci. USA* **70**:997–1001.

Stryker, M. P. and H. Sherk (1975). Modification of cortical orientation selectivity in the cat by restricted visual experience: a reexamination. *Science* **190**:904–906.

Stryker, M. P., H. Sherk, A. G. Leventhal, and H. V. B. Hirsch (1978). Physiological consequences for the cat's visual cortex of effectively restricting early visual experience with oriented contours. *J. Neurophysiol.* **41**:896–909.

Timney, B., D. E. Mitchell, and M. Cynader (1980). Behavioral evidence for prolonged sensitivity to the effects of monocular deprivation in dark-reared cats. *J. Neurophysiol.* **43**:1041–1054.

Tretter, F., M. Cynader, and W. Singer (1975). Modification of direction selectivity of neurons in the visual cortex of kittens. *Brain Res.* **84**:143–149.

Vital-Durand, F. and C. Blakemore (1982). Ontogénèse de la spécificité d'orientation dans le cortex visuel du singe. *J. Physiol. (Paris)* **78**:8B.

Wiesel, T. N. and D. H. Hubel (1963). Single-cell responses in striate cortex of kittens deprived of vision in one eye. *J. Neurophysiol.* **26**:1003–1017.

Wiesel, T. N. and D. H. Hubel (1965). Comparison of the effects of unilateral and bilateral eye closure on cortical unit responses in kittens. *J. Neurophysiol.* **28**:1029–1040.

Wilson, J. R., S. V. Webb, and S. M. Sherman (1977). Conditions for dominance of one eye during competitive development of central connections in visually deprived cats. *Brain Res.* **136**:277–287.

Woelfflin, E. (1909). Untersuchungen über den Fernsinn der Blinden. *Z. Sinnesphysiol.* **43**:187–198.

Wurtz, R. H. (1969). Visual receptive fields of striate cortex neurons in awake monkeys. *J. Neurophysiol.* **32**:727–742.

11

CELLULAR MECHANISMS OF EPIGENESIS IN CAT VISUAL CORTEX

Yves Frégnac

Laboratoire de Neurobiologie du Développement, Université Paris XI, Orsay Cedex, France

Propagation of nervous activity within the visual pathways may have a twofold influence on the development of the visual system. First, in an early phase of postnatal maturation, tonic afferent activity may participate in transcribing at higher levels of integration a functional specialization present at the periphery. Retinal discharge has been shown to be a key factor in the regulation of layer segregation and in maintenance of cellular integrity in the lateral geniculate body. A high degree of correlation in spontaneous activity has been found to exist between neighboring ganglion cells sharing similar receptive field organization. Such mechanisms could ensure the sharpening of precise topographical maps from one relay to the next and even produce differentiation of separate on and off channels in the geniculocortical pathway. The corresponding phase of development during which these processes occur can be considered to be independent of visual experience. The major role of spontaneous activity would be to generalize within the central nervous system endogenous constraints already expressed at the retinal level.

A second role of activity in the development of the visual pathway appears later during a postnatal period when the phasic component of the retinal message reflects interaction with the visual environment. This phase is usually referred to as the "critical period" during which evoked activity has great influence on the functional development of sensory nervous structures. Interestingly, the most experience-sensitive neurons are found in relays where levels of spontaneous activity are low.

This chapter will treat the role of sensory functioning during this latter phase and particularly the epigenetic mechanisms operating in kitten visual cortex. As summarized above, visual cortex development can be considered as a two-phase sequence: during the first three postnatal weeks a vision-independent period of maturation occurs under the influence of genetic constraints; this is followed by a period of modifiability during which functioning itself attunes or stabilizes this genetic "envelope." Several authors (Stryker, 1977; Bischof, 1983) have suggested that this sequence shows a strong analogy between cortical development and the phenomenon of imprinting in birds. Visual cortical cells may select their final selectivity state within a predetermined repertoire of response properties, but the exact characteristics of the features they must encode are not genetically fixed. Visual experience might help to refine or bias the specification process in the same way it helps a newborn duck to recognize its "mother" as the largest and noisiest object found outside the shell. However, cortical plasticity offers a limited analogy with imprinting, mainly because functional changes in neurons are generally not understood in terms of (nor correlated with) behavioral changes.

Consequently, this chapter will be more a catalog of experimentally induced modifications than a synthesis of their possible behavioral relevance. Its principal aim is to describe mechanisms linked with activity that could explain *when* and *how* visual cortical plasticity occurs.

The first point to be stressed is that, although there is no doubt about the vision-related nature of functional plasticity observed in mammalian visual cortex, certain conditions expressed by nonvisual signals must be met in order for such plasticity to occur. Both the onset and consolidation of visual functional changes may depend on attention processes or other kinds of evaluation of the behavioral relevance of the evoked activity patterns (Singer, 1980). The first two sections review the possible extraretinal signals that have been proposed to gate functional modifications of cortical receptive fields. Evidence is based on "population analysis," the statistical comparison of distribution of receptive field properties pooled from groups of kittens of different ages and rearing conditions that have been submitted to various types of surgery or chemical lesions.

The experimental approach presented in the final section is more "cellular" and explores putative mechanisms linked with activity that could regulate cortical plasticity. It is shown that, by artificially imposing temporal correlation between postsynaptic activity and certain characteristics of the visual afferent message, it is possible to record changes in the integration power of neurons considered individually.

EXTRAOCULAR PROPRIOCEPTION AND CORTICAL PLASTICITY

In order to describe the degree of selectivity of cortical organization as a function of age and visual experience, most experimenters since the sixties have compared receptive field properties of populations of single-units recorded in anaesthetized and paralyzed kittens. This "acute" preparation had the obvious advantage of a fixed isomorphism between retina and the stimulation screen and allowed recording from the same neuron long enough to characterize several functional properties of the studied receptive field. But its success in giving a coherent image of visual system development has been understood only recently. Although a few experimenters claim to have changed single-cell properties during the time of recording (review in Frégnac and Imbert, 1984), there is general agreement that anaesthesia and paralysis result in the *freezing* of the integrative properties of cortical neurons. Under these conditions, retinal input alone does not change receptive field properties (Frégnac and Bienenstock, 1981). Consequently, the state of cortical selectivity recorded in the "acute" preparation may be considered as resulting solely from the visuomotor experience given before the electrophysiological experiment and not of the visual stimulation given throughout the recording session.

The finding that *extraretinal signals* related to oculomotor activity are involved in gating modification processes of both orientation selectivity and ocular dominance is based on two types of electrophysiological studies: the cortical effects of delayed visual experience in deprived kittens, or of anomalous binocular experience in normally reared kittens, are compared for various conditions of ocular motility (under paralysis, following surgical akinesis, following passive or active visuomotor experience).

The first type of paradigm has been developed by Imbert and colleagues and concerns the restoration of orientation selectivity produced by delayed exposure to a normal visual environment. A brief flash of visual experience (6 hr) following 6 weeks of dark rearing restores a level of orientation selectivity in freely moving kittens comparable to that found in normally reared kittens of the same age (Imbert and Buisseret, 1975). This restoration process is abolished if kittens are given delayed visual experience under paralysis (and artifical ventilation) (Buisseret et al., 1978). A similar blockade is observed if visual experience is given after bilateral surgical intracranial section of oculomotor nerves (III, IV, VI), or after bilateral disinsertion of all extraocular muscles (e.o.m.) (Gary-Bobo et al., 1986).

The second type of experimental approach concerns shifts in ocular dominance produced by asynchronous binocular input. Freeman and Bonds (1979) showed that sensitivity to monocular deprivation (MD) was restored in the paralyzed kitten when eye occlusion was associated with passive eye movements of the experienced eye. In the same type of "acute" preparation, modifications of ocular dominance could be recorded in the same neuron following asynchronous binocular stimulation (produced by variable interocular disparity) associated with passive unilateral eye movements (Frégnac and Bienenstock, 1981). In this latter experiment long-term ocular dominance shifts were observed either for the initially passively moved eye or for the eye that remained fixed during the time of association (Fig. 1). For most neurons the main effect was a disruption of ocular dominance, but as in Freeman and Bonds's experiment, no systematic capture of the visual response was observed in favor of the mobile eye. This suggests that the gating signal does not have an ocular dominance itself, or does not give a strong competitive advantage to the eye that has been moved. In agreement with this inference is the report that akinesis of the experienced eye by surgical section of all extraocular muscles does not give protection against monocular deprivation effects as long as the deprived eye can still move freely (Singer et al., 1982).

Schematically, two factors related to ocular motility might intervene in triggering such plastic cortical changes. The first, the existence of which remains to be proven, is the *efferent copy* of the oculomotor out-flow; however, this does not explain the unfreezing of cortical plasticity in paralyzed kittens following passive eye movements. The second is the *proprioceptive reafference* of eye movements. Interestingly, suppression of the first factor appears experimentally unfeasible since the pre-motor or motor level from which the efferent copy could originate is unknown. Moreover, severing of oculomotor nerves inevitably affects the extraocular proprioceptive inflow. The only appropriate paradigm in which the two factors may be dissociated is to cut the contingent of extraocular proprioceptive afferent fibers that do not run intracranially together with the motor fibers of the oculomotor nerves.

Section of Ophthalmic Branch of Trigeminal Nerve as Model of Extraocular Inflow Deafferentation

Extraocular proprioceptive afferents in the cat run through oculomotor nerves in the intraorbital region, but before entering the cranial cavity, the majority decussates via anastomoses and joins the ophthalmic branch of the trigeminal nerve (V_1); only a small contingent remains within the oculomotor nerves. The localization of the extraocular pro-prioceptive neurons has been described in the mesencephalic nucleus of the V nerve (Alvarado-Mallart et al., 1975; Sivanandasingham, 1977) or, in a more recent study, only in the ipsilateral semilunar ganglion (Porter and Spencer, 1982). This latter finding is in agreement with most studies from other species (rat: Daunicht, 1983; lamb: Manni et al., 1966;

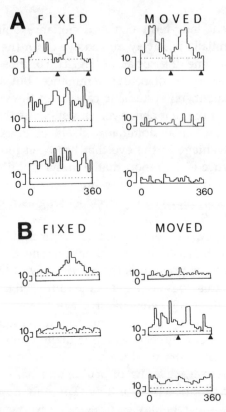

FIGURE 1. Ocular dominance shifts recorded from individual neurons. The histograms are orientation-tuning curves (number of spikes per sweep as a function of direction of the test stimulus on a 360° scale) established monocularly at various times. These curves are averaged over 3–5 runs using an interleaved multihistogram technique. Responses to stimulation through the eye, which is passively moved during the conditioning exposure, are presented in the right column. Responses to stimulation through the fixed eye are shown in the left column. Dotted lines indicate levels of spontaneous activity averaged over 30 acquisition periods. *Cell A* (reproduced from Frégnac and Bienenstock, 1981, with permission): the two orientation-tuning curves, shown in the upper row, established over a 2-hr preliminary observation period indicate that this cell was initially orientation biased and binocularly activated. The "conditioning" exposure for a 50-min period consisted of the association of horizontal movement of one eye with the binocular presentation of a moving grating of vertical bars (whose directions of displacement are indicated by filled triangles). The orientation-tuning profiles, shown in the middle and lower rows, represent, respectively, the averaged visual responses through each eye during the first and second hour of recording following the end of the conditioning exposure. The cell has lost its initial orientation preference and responds only through the previously fixed eye. *Cell B* (data from Frégnac, 1982): the upper row gives the orientation-tuning curves recorded during a 1-hr preliminary observation period. After a first period of 50 min of "conditioning" exposure, during which a rotating grating of concentric bars was associated with horizontal passive movement of one eye, a reversal of ocular dominance was observed

monkey: Porter et al., 1983). However, all these neuroanatomical studies conclude that the afferents corresponding to the chromatolyzed, or HRP-injected cells, whatever their somata location, run through the ophthalmic branch of the trigeminal nerve. Therefore, the section of the V_1 nerve appears to be an appropriate experimental model for the interruption of a major contingent of extraocular proprioceptive afferents.

Three questions must be answered in order to validate this nerve section as a possible model for extraocular deafferentation.

1. What is the real extent of the deafferentation produced by the V_1 section? In the intact animal there is growing evidence of responses to stretch of extraocular muscles or to passive eye movements at different levels of integration in the visual system (review in Frégnac and Imbert, 1984), but the effect of a unilateral or bilateral section of the V_1 nerve on stretch responses observed in the visual pathway is unfortunately still unknown.

2. What kind of nonproprioceptive information is running through the V_1 branches? In addition to the extraocular inflow, this sensory nerve conveys cutaneous afferents (nociceptive or not) from the face, mostly from the superior part of the orbit, the influence of which on visual cortical neurons is also almost unknown (but see footnote 1, p. 465, in Buisseret and Maffei, 1983). In order to find a control situation where one could look for the nonextraocular-inflow-related effect of the V_1 section, Trotter and colleagues (Trotter et al., 1981b; Graves et al., 1984, in press) have studied the effects on cortical plasticity and on perceptual performances produced by the section of the maxillary branch of the trigeminal nerve (V_2), which contains a similar type of sensory afferents innervating the inferior orbital region (Fig. 2).

3. What are the possible side effects of section of the V_1 nerve on ocular integrity? In addition to the loss of corneal sensitivity in the eye ipsilateral to the section, this surgery could result in clouding of the

---

for more than 30 min, as shown in the middle row. A second 50-min period of "conditioning" exposure to binocular vision of a vertical grating (filled triangles) associated with horizontal passive eye movements was then given. At the end of the exposure, a slight, but significant, orientation preference for the experienced orientation builds up when testing during the following 30 min through the mobile eye (lower histogram). The cell appeared at that time exclusively activated through this eye but was lost during the final completion of the control recording through the fixed eye (data not shown).

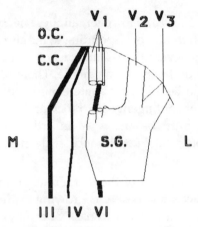

FIGURE 2. Schematic representation of section of the ophthalmic branch of trigeminal nerve (adapted from Trotter, 1981). The different branches of the trigeminal nerve, leaving the orbital cavity (OC) are shown in the cranial cavity (CC) before entering the semilunar ganglion (SG). They are represented as seen from above following a temporal approach. From a lateral position (L) to a more medial one (M) are the mandibular branch (V_3), the maxillary branch (V_2), and the ophthalmic branch (V_1) (usually 2 or 3 trunks). The section of the V_1 is indicated as practiced in Trotter et al. (1981a). The oculomotor nerves are represented by solid lines and displaced more medially for clarity.

optical apparatus of the eye. This phenomenon, while currently observed in the rat, is quite rare in the cat: a few cases of corneal opacity following bilateral section of the V_1 were observed by Trotter and colleagues (Graves et al., in press). The possible existence of an oculomotor disturbance is more difficult to assess. Bilateral section in kittens produces sensory neglect and the occurrence of eye movements may be severely reduced (Buisseret and Singer, 1983). After similar surgery in the adult monkey intermittent slow adductions of the nondominant eye are observed during monocular or binocular fixation, and vergence movements are also impaired (Guthrie et al., 1982). However, conjugate saccadic and pursuit-tracking performances appear normal. The effects of a unilateral section are less documented. Apart from a slight modification of visually guided behavior (Fiorentini et al., 1982), ocular motility appears unimpaired in unilaterally lesioned cats. Using the photographic method of Sherman (1972), no variation of interocular alignment could be detected in kittens after a unilateral section (Graves et al., 1984, in press). However, in adult cats (in a fixed head situation) submitted to a unilateral section of the V_1, the ipsilateral eye develops pendular nystagmus but only during periods of complete darkness. These abnormalities are not present in sham-operated cats (Fiorentini and Maffei, 1977). In summary, we can rule out at least major visuomotor disturbance such as observed in the case of strabismus, since if any oculomotor abnormality exists, it has been reported to be limited to periods spent in the absence of vision. Consequently, the lesioned animal would have a normal binocular visual experience for the rest of the time spent in light. However, since most of these observations are based on rather gross measurements, one should not completely discard a possible modification of oculomotor outflow signals secondary to the section of the V_1 nerve.

Evidence for an Extraretinal Critical Period

The participation of extraocular proprioceptive afferents in the maintenance or developmental processes of area 17 binocular integration was first proposed by Maffei and Bisti (1976). These authors observed a disruption of binocularity in surgically strabismic kittens maintained in the dark after the surgery; however, this effect could not be reproduced by others (Van Sluyters and Levitt, 1980) with somewhat shorter postsurgical delays. A second source of evidence comes from the description by several authors of a period of sensibility to surgical strabismus (Yinon, 1976; Levitt and Van Sluyters, 1982; Berman and Murphy, 1982). These described a critical period whose end coincides with that of the period of sensitivity to monocular deprivation. But since these animals were normally reared, the results were interpreted by these authors as being solely the result of binocular visual asynchrony.

We have reinvestigated the role of extraocular inflow by studying the cortical effects following a *unilateral section* of the V_1 nerve performed at different ages in both visually experienced (NR) and deprived (DR) kittens (Trotter et al., 1981a, in press). After unilateral extraocular proprioceptive deafferentation performed at *6 weeks of age*, disruption of binocular integration appears more than 1 week postoperatively and is still present 7 weeks after the section. This long-term effect corresponds to an abnormally high proportion of neurons dominated by the same eye. Depending on the animal, the control of visual response is taken either by the ipsilateral eye or the eye contralateral to the section. Similar results were observed in NR and DR kittens, which supports the nonretinal origin of the effect. Monocular cells were found principally in

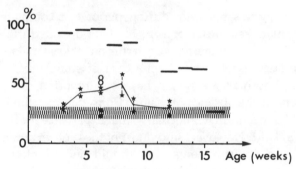

FIGURE 3. Critical period for retinal and extraretinal manipulation in kitten area 17 (reproduced from Frégnac and Imbert, 1984, with permission). Proportion of monocular cells among visual neurons is plotted with onset of manipulation period expressed in weeks. *Upper* (steplike) profile represents time course of period of susceptibility to unilateral eye closure (data adapted from Fig. 1 in Olson and Freeman, 1980, calculated from bins 1 and 7 of their histograms). Kittens were subjected to 10–12 days of monocular deprivation spaced regularly through the first 4 postnatal months. The period of monocular occlusion for each kitten is symbolized by a dark line ending with date of recording. *Lower* continuous profile represents time course of period of susceptibility to unilateral section of ophthalmic branch of trigeminal nerve (data from Trotter et al., 1981a). Kittens were reared either normally (★) or in total darkness (○) until recording session. Age at which section was performed corresponds to abscissa value of each experimental point. Electrophysiological recordings were made 4–7 weeks later, except in two dark-reared kittens where postoperative delay was reduced to 4–8 days (■). The same number of cells was recorded in each hemisphere. Shaded area indicates average level of monocular cells found in normally and dark-reared kittens of the same age when receptive fields are located in the 10° surrounding area centralis.

layers IV and VI. The period of sensitivity to unilateral section of V_1 extends from the fourth to the eighth postnatal week, although in kittens operated at 12 weeks, an abnormally high proportion of monocular cells was recorded in layer VI. In contrast, *bilateral section* of V_1 performed in 6-week-old NR kittens does not affect (later on) the proportion of binocular cortical cells; this indicates that the main factor responsible for disruption of binocularity is the *imbalance* between the extraocular proprioceptive inflow coming from each eye.

These data suggest that the period of sensitivity to the unilateral section of the ophthalmic branch of the trigeminal nerve has the same time course as the period of sensitivity to the unilateral occlusion of one eye (Fig. 3): at the same time during postnatal development, both extraretinal and retinal inputs might participate in the maintenance of the ocular dominance network.

Gating of Cortical Effects of Delayed Visual Experience

The major breakthrough in the implication of extraocular inflow authorizing epigenetic modifications at the cortical level was obtained by Buisseret and colleagues, who performed bilateral section of V_1 before delayed visual experience (Buisseret and Gary-Bobo, 1979; Trotter et al., 1981b). While freezing of the process of orientation selectivity restoration is observed in the freely behaving animal if no sensory extraocular feedback (section of V_1) is available during the time of visuomotor experience, sectioning of the maxillary branch of the trigeminal nerve (V_2) does not block the restoration kinetics (Trotter et al., 1981b). This last result reinforces the hypothesis according to which the observed functional kinetic depends on the integrity of the contingent of sensitive fibers coming from extraocular muscles and not from the somatosensory fibers innervating the face and the periorbital region, which travel along the different branches of the trigeminal nerve.

In a more recent series of experiments we could show that this gating could be generalized to both ocular dominance shift and orientation selectivity changes (Trotter et al., 1983). We used a paradigm of delayed *monocular* visual experience (ME) at 6 weeks of age, the onset of which was preceded by a few days, by a *unilateral* or *bilateral* section of the V_1 nerve. As expected, a bilateral section blocked both orientation selectivity restoration and capture of the visual response by the experienced eye. Most cells were nonoriented and binocularly activated as is the case in visually deprived kittens of the same age (i.e., the initial cortical state before the onset of visual experience). Monocular vision following unilateral section of V_1 induced an intermediate cortical selectivity state where both the orientation selectivity level and ocular dominance reflected the influence of the open eye. However, the amplitude of the epigenetic modifications appeared less pronounced than in the intact monocularly experienced kittens. In addition, the functional changes observed were slightly more pronounced when the unilateral section of V_1 was contralateral to the experienced eye, that is, when proprioceptive input of the exposed eye was available. But this effect of the laterality of the section relative to the open eye appears to be of secondary importance.

The general aspect of the blockade of cortical plasticity following the bilateral section of the V_1 has been confirmed for longer periods of monocular exposure (up to 1 year) by Buisseret and Singer (1983). In addition, these authors showed that it also gave protection from the effects of surgical strabismus associated with binocular vision.

From these results it may be concluded that extraocular proprioception might play a nonspecific role in the development of the functional properties of visual cortical cells: the sensory afferents coming from extraocular muscles would provide a signal of global context ("gate factor" in Frégnac, 1979) during an extraretinal critical period (see the previous section). They would allow the validation at the cortical level of functional modifications produced by specific retinal afferent signals. The amplitude of the epigenetic changes would depend more on the absolute quantity of extraocular proprioceptive inflow available during visual experience than on its ocularity.

Orientation-specific Gating or Global Gating?

Up to now, the experimental evidence reviewed here pleads for a control of the kinetic rate of cortical epigenetic changes by extraocular inflow. As a first approximation, the level of plasticity of the visual cortex is given by the relative quantity of extraocular inflow available during visual experience. This hypothesis allows the reinterpretation of a certain number of paradigms where injury of the extrinsic musculature such as cyclotorsion or strabismus might have protected visual cortical neurons from uncorrelated binocular input (review in Frégnac and Imbert, 1984).

However, two lines of recent evidence suggest a relationship between the response of a cortical visual cell to eye movements and its orientation preference. The first group of data has been obtained in the adult paralyzed and anaesthetized cat and is based on a single experimental paradigm: one eye is mechanically stimulated (stretch of disinserted extraocular muscle or passive eye movement imposed along certain radial directions) while the other eye is visually stimulated. Trotter and colleagues (1980) could elicit responses to passive eye movements (without visual stimulation) for a limited sample of cells; for most of these neurons they found that their preferred direction of eye movement was in fact orthogonal to the preferred orientation (when the cell's response was tested through the other fixed eye). For a larger sample of cells for which they could only describe a proprioceptive–visual interaction, they found a marked *facilitation* when the direction axis of the passive eye movement was *orthogonal* to the orientation of the preferred grating used for visual stimulation (Trotter et al., 1980). A remarkable corollary comes from the work of Enomoto et al. (1983), who showed a complementary suppressive interaction: in certain "transient" cortical cells, a significant reduction or *suppression* of the best visual response was obtained when visual activation was combined with stretch of particular extraocular muscles; but these muscles for which the interaction was observed were precisely the ones that would have been involved for exerting an eye movement in a direction *parallel* to the orientation preference of the neuron. The generality of these data remains to be confirmed, and a recent study of Ashton et al. (1984) reported much more complex responses; unfortunately, these authors did not look for the possible existence of a reciprocal organization of stretch responses and orientation preference, although they reported a significant proportion of cortical cells selective to the direction of passive eye movement.

The second group of data concerns development of visual cortex and the effects of delayed visual experience in kittens with biased ocular motility. Buisseret and Gary-Bodo (1981) suppressed four out of six extraocular muscles to restrict eye movements to one axis (horizontal or vertical). The cortical effects of binocular vision following bilateral disinsertion are comparable to those produced by directional rearing, and such an experiment does not permit by itself the dissociation of the relative influences of axial ocular motility and axially biased retinal flux. This study was recently completed with an ingenious paradigm where the exposed eye was left intact but without proprioceptive reafference (following V_1 section), while the closed eye had its motility restricted to one axis (Milleret et al., 1984). The striking result was that most cortical cells after monocular exposure to an unbiased visual environment were preferentially activated by an orientation orthogonal to the direction of the ocular movements experienced by the closed eye. One is tempted to conclude that the direction of eye movements gates only the orthogonal orientation preference. But in view of the high degree of synergy that normally exists between movements of each eye, one should ascertain the unbiased mobility of the intact eye. A proper control could be to leave intact the extraocular proprioceptive reafference of the experienced eye and section the V_1 branches ipsilateral to the closed (e.o.m. disinserted) eye and check whether this permits the gating of all orientation preferences. This test would help to decide whether or not the bias in orientation preference found by Milleret et al. (1984) could result from a biased directional retinal flux. A *negative* result (i.e., an overrepresentation of orientation preferences orthogonal to the direction of movement of the closed eye) would imply that disinsertion of extra-ocular muscles on one side (closed eye) impairs the motility of the other intact (open) eye. A *positive result* (i.e., a uniform distribution of orientation preference) would validate their paradigm as a new experimental tool in order to dissociate visual and extraocular proprioceptive factors.

In summary, eye motility has been shown to gate orientation selectivity and ocular dominance changes in response to modifications of retinal input during a critical period. This gating appears to be rather global since its ocular dependency (or "ocularity") is very weak, and it probably concerns all receptive field properties: bilateral extraocular proprioceptive deafferentation results in the blockade of the effects of a large variety of manipulation of visual input such as produced by delayed monocular or binocular vision and strabismus. However, it cannot be excluded that the direction itself of eye movement might influence in a more specific way the orientation preference development of certain cortical cells through proprioceptive afferents.

NORADRENALINE AND CORTICAL PLASTICITY

Models of learning and memory have often postulated the existence of global gating signals in the form of chemicals released throughout the brain, which might help to consolidate recently activated synapses (Crow, 1968; Kety, 1970; Gilbert, 1975). Among putative "rewarding" systems, noradrenergic projections appeared for a long time as the most eligible candidate to be involved in cortical plasticity. Most noradrenergic cell bodies are grouped in a restricted region of the brainstem formed in the cat by the locus coeruleus and subcoeruleus complex (Maeda et al., 1973; Moore and Bloom, 1979). They emit a diffuse ascending projection throughout the neocortical mantle, and one coeruleocortical axon forms thousands of monoamine-containing boutons, only a fraction of which appear to be junctional to neuronal elements (review in Foote et al., 1983). These morphological characteristics suggest an almost "humoral" action of NA-ascending projections, which could permit a global modulation of the transmission of sensory information within the neocortex.

More innovative is the idea that such a system plays a crucial role in kitten visual cortex during a limited postnatal period (review in Kasamatsu, 1983). The cortical NA level would locally control the degree of cortical plasticity by gating functional modifications in response to visual input. An impressive series of experimental evidence gives strong support concerning the role of catecholamines (CA) in modulating the functional cortical effects of monocular deprivation (Kasamatsu and Pettigrew, 1976, 1979; Kasamatsu et al., 1979, 1981a). Closure of one eye in 4–6-week-old normally reared kittens, even for a few hours, disrupts binocular integration with most neurons responding only through the remaining exposed eye as the ultimate result of the deprivation (review in Movshon and Van Sluyters, 1981). Kasamatsu and Pettigrew have shown that this dramatic effect of environmental manipulation can be completely prevented if the kitten has been submitted shortly before the start of the visual deprivation to two types of 6-OHDA lesions (intraventricular or intracortical injections). Their original interpretation was that both chemical lesions produced the same protective effect because they both depleted endogenous cortical CA content below a certain threshold.

The implication of the intracortical β-adrenoreceptor-adenosine-monophosphate (cAMP) system was furthermore suggested by the restoration of cortical sensitivity to monocular deprivation in 6-OHDA pretreated kittens, following local perfusion of exogenous NA or cAMP in the cortical tissue itself (Pettigrew and Kasamatsu, 1978; Kasamatsu et al., 1979; Kasamatsu, 1982; Kupperman and Kasamatsu, 1984). Indirect estimates of threshold values above which a significant action on cortical binocularity was found for both 6-OHDA and NA appeared to be in the range of values of specific uptake by NA terminals (Kasamatsu et al., 1981b). Final convincing evidence was the claim that sensitivity to monocular occlusion is blocked in untreated cortex after injection of β-adrenergic blockers (*d,l*-propranolol or sotalol) but was still present after injection of an α-adrenergic antagonist (phentolamine and phenoxybenzamine in Kasamatsu, 1979, 1983; Kasamatsu and Shirokawa, 1985; Shirokawa and Kasamatsu, 1986).

However, recent experiments, including our own (Adrien et al., 1982, 1985), have shed doubt on the conclusions of Kasamatsu and Pettigrew. The blockade of cortical plasticity following intraventricular injections of 6-OHDA could not be replicated (Adrien et al., 1982, 1985; Daw et al., 1985a; Allen et al., 1984). More refined techniques of NA deafferentation did not confirm the dependency of cortical plasticity on the NA level alone (Adrien et al., 1982, 1985; Daw et al., 1984, 1985b, Bear and

Singer, 1986). The following sections review several factors that have been invoked to explain this experimental controversy.

Relative Efficacy of NA Depletion Modes

In order to produce significant depletion at the visual cortex level, two neurotoxins have been used. 6-OHDA is known to be taken up by neurons transporting CA and to block CA uptake storage mechanisms. DSP-4 has been reported to be equally effective, although the mode of action is less known, and could affect both NA and 5-HT neurons (review in Jonsson, 1980).

Five modes of injection of these neurotoxins have been used by different experimental teams and differ in their degree of selectivity in lesioning the NA system.

1. *Systemic injection:* the neurotoxin is injected either intraperitoneally (IP) or subcutaneously (SC). In contrast to 6-OHDA, which has to be injected neonatally at an age when the blood brain barrier is still permeable (Bear and Daniels, 1983), DSP-4 passes the blood brain barrier in 4–6-week-old kittens (Daw et al., 1985b).

2. *Intraventricular injection* limited usually to the lateral ventricle (IV): see Fig. 4A.

3. *Coeruleus complex injection (LC):* see Figure 4B.

4. *Lesion of the noradrenergic ascending bundle* in the lateral hypothalamus region (LH): see Figure 4C.

5. *Intracortical injection (VC):* the neurotoxin is perfused by an osmotic minipump in the region where recording will take place.

One should note from Figure 4 the differences in the doses required to produce similar levels of NA depletion (50–95%) depending on the mode of lesion. Only injections of neurotoxin in the lateral hypothalamus region (LH) or directly in the complex coeruleus region (LC) avoid either damage of cortical homeostasis (VC lesion) or impairment of general behavior (IV injection). Of the five lesion modes, only intracortical injections (VC) appear to be successful in producing a replicable blockade of the cortical effects of monocular deprivation (see Table 1). This comparative survey leads to the conclusion that, more than the cortical

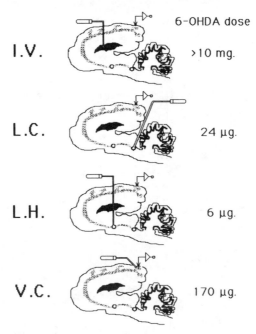

FIGURE 4. NA Depletion modes. The sites of lesion (minipump injection symbol) and of recording (electrode and amplifier symbol) are shown in a sagittal representation of the cat's brain. For each section both locus coeruleus (LC) and lateral hypothalamus (LH) locations are indicated by an open circle. The lateral ventricular zone (IV) is filled in. The shaded area represents the NA ascending projection to the neocortex. From top to bottom, each row corresponds to a different mode of injection of the neurotoxin (6-OHDA): in the lateral ventricle (IV), in the coeruleus complex (LC), in the lateral hypothalamus (LH), and in the visual cortex (VC). Varying with the mode of lesion, the dose of 6-OHDA is adjusted to produce a significant NA depletion of visual cortex (40–95%).

NA level, the injection mode of the neurotoxin appears crucial in the freezing of cortical function.

Specialized or Global Gating?

A major question raised from Kasamatsu and Pettigrew's pioneering work was to know if the blockade produced by NA depletion was found only for monocular deprivation (MD) or whether it would apply to other kinds of sensory manipulation. Protocols used in the literature are normal rearing (NR), monocular rearing (MR), monocular deprivation following normal rearing (MD), reverse suture following monocular oc-

TABLE 1. Effects of NA Depletion on Visual Cortical Plasticity[a]

Lesion Mode	Sensory Manipulation Mode						
	NR	MR	MD	RS	DR	ME	DR-Dir
Neonatal							
6-OHDA							
IP			Bear and Daniels (1983)				
LC	Adrien et al. (1985)	Adrien et al. (1985)			Adrien et al. (1985)		
Late							
DSP-4 IP } IV }			Daw et al. (1985b)				
6-OHDA IV			Kasamatsu and Pettigrew (1976, 1979); Adrien et al. (1982, 1985); Daw et al. (1985b), Allen et al. (1984)	Ary et al. (1979)			

Functional property	OS	OS, OD	OD	OD	OS	OS, OD	Dir.
LC							Adrien et al. (1985)
LH		Adrien et al. (1982, 1985) Daw et al. (1984), Bear and Singer (1986)					
VC			Kasamatsu et al. (1979) Bear et al. (1983) Paradiso et al. (1983)	Ary et al. (1979)		Daniels et al. (1981) Daw et al. (1983)	Daw et al. (1983)

^a This table summarizes the different studies (authors' names) of various modes of NA depletion (see Fig. 4 legend) on the cortical effects of various sensory manipulations (see text). Depending on the study, the functional properties looked for were orientation selectivity (OS), ocular dominance (OD), or directional selectivity (Dir.). For negative reports, only the name of the study is given. Reports of partial protection are outlined by a dashed-line rectangle. Reports of total blockade of visual cortical plasticity are shown inside a solid-line rectangle.

clusion (RS), dark rearing (DR), monocular exposure following dark rearing (ME), and restricted visual exposure to a unidirectional environment (DR-dir). These two latter paradigms are known to affect both orientation selectivity and ocular dominance in intact kittens. Furthermore, directional-deprivation- and monocular-deprivation-sensitive periods have different time courses (Daw et al., 1978) and could possibly concern different types of synapses (Sillito, 1977).

From inspection of Table 1, the blockade of plasticity appears confirmed following 6-OHDA intracortical injections for most types of sensory manipulations, which is in favor of a global gating role of NA. The only puzzling observation was made by Ary et al. (1979), who reported that neither intraventricular nor intracortical 6-OHDA injections prevented recapture of cortical cells by the previously deprived eye.

Influence of Respective Timings of Neurotoxin Lesion and of Sensory Manipulation

One of the reasons why involvement of coeruleus projections in postnatal plasticity of the visual system is attractive is the fact that the coeruleus complex is also part of a region where from 2 weeks of age a strong activation of the lateral geniculate body is generated (Bowe-Anders et al., 1974). The integrity of this generator of phasic activity—described in the adult as the ponto-geniculo-occipital wave (PGO)—could be crucial during an early phase of development (Davenne and Adrien, 1984, 1985). Since it has been reported that LC lesion has different consequences on paradoxical sleep maturation according to whether it is performed neonatally or after 3 weeks of age (Adrien, 1978), we compared the relative effects of such lesions on cortical plasticity. Our results show that LC lesions were ineffective in protecting cortical neurons from the gross epigenetic effects of visual experience whatever the age at which they were performed (Adrien et al., 1982, 1985).

A second related hypothesis was introduced by Bear and colleagues on the basis of similar paradigms of neonatal or late lesions (Bear and Daniels, 1983; Bear et al., 1983). They suggested that receptor compensation, which might occur with long periods of NA depletion, could explain an *acute* but *not chronic* blockade of cortical plasticity following CA depletion. However, this hypothesis was introduced by comparing the effects of similar periods of monocular deprivation in kittens lesioned with different techniques (IP vs. VC) neonatally or concurrently with the monocular occlusion period. Their conclusion does not hold if one restricts comparison to a given mode of neurotoxin injection. Following systemic DSP-4 injection (Daw et al., 1985b) or LC lesions (Adrien et al., 1985), no relation is found between the binocularity index and the delay between the chemical lesion and the onset of the monocular deprivation period. This observation might not apply to the case of intracortical NA deafferentation, where regrowth of CA terminals has also been observed in the area surrounding the perfusion site (Nakai et al., 1981).

Reexamination of "Noradrenaline Hypothesis"

The only remaining convincing arguments in favor of a noradrenergic gating of cortical plasticity are the following: (1) intracortical injection of NA restores plasticity following previous 6-OHDA intracortical pretreatment (Kasamatsu et al., 1981a); (2) blockade of cortical plasticity is obtained following injection of β-adrenoreceptor blockers in intact kitten cortex (Kasamatsu, 1979; Kasamatsu and Shirokawa, 1985; Shirokawa and Kasamatsu, 1986). However, blockade of plasticity following exogenous injection of chemicals is not a proof in itself: indeed, intracortical injections of glutamate (Shaw and Cynader, 1984) or of scopolamine (Singer, personal communication) are among other procedures producing qualitatively similar freezing of the cortical function.

To reconcile this field of experimental controversy, four further hypotheses will be presented that, without being exclusive, might lead to very different conceptions of NA involvement in cortical plasticity.

Proposal 1:   The first hypothesis is that in situ exogenous injection of neurotoxin might lead to a dysfunctioning of local cortical dynamics. Consequently, the loss of plasticity in visual cortex following intracortical 6-OHDA injection might be the early sign of a profound impairment of cortical homeostasis before any more drastic disorganization of receptive field properties takes place. In order to test such a hypothesis, one should look carefully and separately at nonspecific side effects of the injection technique and the use of the neurotoxin. Indeed, data from Bear et al. (1983) suggest that injection of the vehicle solution itself might produce protective effects. Other control experiments could consist of blocking NA (and 6-OHDA) uptake with desipramine or protriptyline

(Jonsson, 1980) or protecting NA terminals with nomifensin (Sievers et al., 1981). In contrast, other techniques of lesion should be developed that reduce the nonspecific effects of 6-OHDA metabolites, such as the combined use of 6-OHDA and bovine serum albumin (Allen et al., 1981). Finally, one should look for the specificity of the neurotoxin on noradrenergic neurotransmission itself and test possible interference with the action of other cortical neurotransmitters. Recent evidence from Bear and Singer (1986) indicates that concurrent iontophoresis of 6-OHDA significantly attenuates the enhancement of cortical visual responses by acetylcholine.

Proposal 2: Proponents of the noradrenaline hypothesis have more recently introduced the idea that blockade of cortical plasticity occurs where NA depletion reaches a certain threshold (Kasamatsu et al., 1985a). The level of CA depletion is not known for the intraventricular injections made originally by Kasamatsu and Pettigrew (1976, 1979), but it is hard to imagine that it differed greatly from that which we found (62–90%) when replicating their experimental procedure (Adrien et al., 1982). The situation appears different for intracortical injections. One cannot be sure that the relative positions of the injection needle and the recording electrode were comparable in each attempt: the differences in the degree of protection of binocularity found, for instance, by Bear et al. (1983) (or Paradiso et al., 1983) and Daw et al. (1983) might just reflect differences in the actual NA depletion level at the recording site. To test such a hypothesis, one could use electrochemical detection barrels assembled with the recording electrode (Armstrong-James et al., 1981); but it is clear from data of Daw and colleagues and of our own group that this hypothetical threshold value should exceed 97% of depletion: this value corresponds to the largest cortical NA depletion observed in some experimental cases where no functional protection effect was found. This nonlinear gating hypothesis is rather far from the initial suggestion of Kasamatsu and Pettigrew (1976, 1979), who proposed a graded control of plasticity by NA endogenous level.

However, NA level should not be expected to be a reliable indicator of the effectiveness of noradrenergic transmission. Intraperitoneal injection of an α-2-adrenergic agonist (clonidine) has been reported to block the physiological effects of monocular deprivation without affecting the cortical NA level (Nelson et al., 1985). This treatment is known to inhibit locus coeruleus cells' firing, and the amplitude of the protective effect was found to be correlated with the relative decrease in the cerebrospinal fluid of the NA metabolite MHPG (3-methoxy-4-hydroxy-phenylethylene glyolol). Other paradigms based on intracortical injection of β-adrenergic blockers indicate other possible biochemical correlates to be looked for, such as the relative availability of activatable β-adrenergic receptors (Kasamatsu and Shirokawa, 1985; Shirokawa and Kasamatsu, 1986).

Proposal 3:   Another interpretation has been suggested on a more speculative basis (Frégnac and Imbert, 1984; Videen et al., 1984; Adrien et al., 1985).

On the one hand, all procedures that increase the temporal correlation between the afferent geniculate message and the cortical response would improve synaptic transmission (see Hebb's principle, below). Local ionophoresis of NA or cAMP has been claimed to improve signal–noise ratio (Kasamatsu and Heggelund, 1982), but experimental evidence appears rather contradictory up to now (Videen et al., 1984).

On the other hand, all procedures that decrease correlation between sensory input and cortical activity would damp cortical plasticity. This could be the case when postsynaptic activity is driven to an abnormally low level (following anesthesia, paralysis, or intracortical scopolamine injection) or an abnormally high level (following intracortical glutamate). In this respect, it would be important to know at the single-unit level the effect of 6-OHDA injection on cortical activity.

However, an explanation attributing an uncorrelating influence of the neurotoxin on nervous activity suffers from its simplicity and from possible counter examples: in particular, it does not predict the absence of blockade of plasticity when high levels of activity are produced by penicillin application on the cortical surface (Videen et al., 1986). Similarly, how can we interpret the restoration of plasticity in adult cat cortex after LSD administration (McCall et al., 1982), a substance that is known to both enhance sensitivity to sensory input and inhibit responsiveness of cortical systems (Freedman and Halaris, 1978)?

Proposal 4:   The last hypothesis results simply from the observation of a fundamental distinction among procedures used to lesion NA terminals: in most protocols (systemic, IV, LH, LC) the morphological integrity of the NA neurons is affected throughout the brain. In the intracortical injection case, the NA lesion is confined to a remote terminal region (occipital cortex). Consequently, it could be assumed that all lesion procedures that affect the morphological integrity of coeruleus complex neurons put compensatory phenomena into play; therefore, this lesion-induced reactive process could modulate cortical plasticity and consequently would mask in lesioned kittens the otherwise predominant gating role of NA during cortical development (Adrien et al., 1985).

Data from Bear and Singer (1986) show that the combined destruction of the cholinergic and noradrenergic ascending pathways reliably reduces the sensitivity to monocular deprivation even though massive lesions of either system are ineffective. Although this finding has been interpreted as evidence that cortical plasticity is mediated through coactivation of several "gate factors" (ACh and NA), it does not contradict the view that the cholinergic system could play as well a palliative role following the lesion of the NA pathway. If that was indeed the case, combined lesions of NA and ACh systems would block NA control of cortical plasticity and the induction of compensatory phenomena.

This hypothesis allows two predictions:

1. A very limited NA terminal deafferentation should not trigger lesion-induced compensatory phenomena. This could be the case of intracortical 6-OHDA injection limited to a remote terminal zone (visual cortex); it could explain why this latter paradigm is up to now the only lesion mode that consistently blocks visual cortical plasticity.

2. Modulation of cortical plasticity should be observed when exogenous action is limited to an external control of the level of afferent activity arising in the ascending noradrenergic pathway, without affecting its cellular integrity. Plasticity of ocular dominance has been shown to be blocked during the critical period when silencing coeruleus complex cells (Nelson et al., 1985) and is restored at the adult age following pairing of monocular occlusion and electrical stimulation of the locus coeruleus (Kasamatsu et al., 1985b).

In view of the diversity of these proposals, and in spite of the coherence of the data accumulated by Kasamatsu and colleagues, the role of NA in relation to visual cortical plasticity does not appear as "unique" (Sillito, 1983; Bear and Singer, 1986) or necessary during development. In any case, the noradrenergic nature of blockade of plasticity induced by 6-OHDA or DSP-4 lesions must stand open to reexamination.

ROLE OF PROPAGATED ACTIVITY IN EPIGENESIS OF NEURONS CONSIDERED INDIVIDUALLY

Rules of Synaptic Modifications

Modifications in functional integration of cortical neurons may involve different classes of hypothetical mechanisms that are not exclusive.

1. They could result from changes in the geometry of the afferent connectivity and/or changes in the dendritic geometry of the target cell. Postsynaptic integration would consequently be profoundly transformed by geometric modifications in the connectivity graph even if the efficacy of transmission of each synapse remains unchanged. Such a process might intervene when there is sufficient time for pruning of contacts or reoriented growth to occur.

2. They could reflect local modifications in synaptic efficiencies while the geometry of both afferent fibers and target cell remains unaltered.

This second hypothesis has the advantage of being most economical and explains reversible and short-term effects. Its success in modeling visual cortical plasticity comes from the observation that effects of visual input are highly nonlinear during the critical period. In particular, delayed vision seems to permit the expression of a maturation process that until then was masked by the absence of visuomotor experience (Frégnac, 1979). A few hours of visual experience restores in 6-week-old visually deprived kittens an orientation selectivity level comparable to that observed after 5 weeks of normal rearing (Imbert and Buisseret, 1975). The most likely assumption used in most paradigms of short-term selective visual exposure is to suppose that modification of transmission efficiency of already present synapses has been gated by visuomotor interaction and is produced or "verified" by visual input.

The logical capacities of neuronal networks with synapses that are modifiable as a function of reverberating activity have retained the attention of numerous psychophysiologists and neuronal network modelers. The best known principle for synaptic modification was introduced by Hebb (1949). Although the original formulation of this hypothesis did not refer explicitly to synapses, it is used by most experimenters and theoreticians under the following form: the efficacy of transmission of a given synapse increases following a maintained *positive* temporal correlation between pre- and postsynaptic activities. This adaptive rule, where output has an influence on the transmittance of the system, has been widely applied in pattern recognition theory. Assemblies of formal neurons with appropriate cross-connections and Hebbian synapses can yield cybernetic structures possessing at least short-term memory properties. Apart from the conditional aspect of the input signal, the use of a positive feedback of the output signal in the modification rule is more often chosen by theoreticians of the visual system than negative retroaction. It seems particularly powerful in simulating cellular learning, but other hypotheses based on *negative* correlation be-

TABLE 2. Role of Temporal Correlation between Presynaptic and Postsynaptic Activities[a]

Rules of Synaptic Modifications	Experimental Tests
Correlation between pre- and post-synaptic activities	
1. **modifies firing threshold** of postsynaptic neuron (Shimbel, 1950) →	**Motor cortex** of cat: Woody et al. (1978), Baranyi and Feher (1981)
2. **decreases synaptic efficiency** of activated synapses (Albus, 1971) →	**Cerebellum**: Gilbert and Thach (1977), Ito et al. (1982)
3. **increases synaptic efficiency** of activated synapses (Hebb, 1949). Normalization rules:	**Motor cortex** of cat: Baranyi and Feher (1981)
3.1. Synaptic efficiency cannot increase beyond a maximum value (Marr, 1969)	**Hippocampus** (LTP): McNaughton et al. (1978)
	Hippocampus: Wigström et al. (1986)
3.2. Sum of efficiencies of all synapses afferent to the postsynaptic neuron is constant (Von der Malsburg, 1973)	**Aplysia**: Wurtz et al. (1967), Carew et al. (1984)
3.3. Efficiency of nonactivated synapses (while postsynaptic neuron is active) decreases (Stent, 1973)	

TABLE 2. (*continued*)

Rules of Synaptic Modifications	Experimental Tests
4. imposes sign and amplitude of synaptic modification	Kitten **visual cortex**: Fregnac et al. (1984)

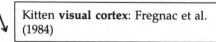

Synaptic modification is proportional to covariance between two presynaptic or between pre- and postsynaptic activities (Sejnowski, 1977)

Synaptic change depends on a floating threshold linked with mean activity of postsynaptic neuron (Bienenstock et al., 1982)

[a] Reports of functional modifications in agreement with predictions by a given hypothesis (arrow) are represented inside a solid-line rectangle. Only experiments based on recordings of neurons considered individually are listed.

tween input and output or on firing threshold modifications have been invoked as well (see Table 2).

In view of the high degree of correlation of firing that seems to exist between neighboring neurons or between neurons that share the same functional selectivity (Michalski et al., 1983; Krüger, 1983) in the adult cortex, it is not surprising that Hebb's hypothesis has become a panacea for experimenters as well. But interestingly, most theoreticians and experimenters use conditions of normalization of synaptic efficiencies that lead to the almost hidden implication of "anti-Hebbian" rules of modification. Later models have incorporated these additional rules in the Hebbian scheme (see Table 2). In particular, Bienenstock and colleagues (1982) have proposed a floating threshold of synaptic modification, the value of which is linked to the mean activity of the cell. This threshold determines the sign of the modification and avoids uncontrolled increases of the efficacy of the transmission at activated synapses after repeated presentation of a stimulus eliciting already postsynaptic activity. It also allows temporal competition among synapses and predicts both *increase* and *decrease* of efficiency of sets of synapses associated with different input patterns. In addition, it also simulates correctly a higher sensitivity to visual experience in previously silent neurons (as often recorded in deprived kittens) than in already active ones. A further interest to be found in this type of threshold hypothesis is that, although the key factor remains temporal correlation between pre- and postsynaptic activities, the evolutive power of the neuron (Changeux et al., 1973) also depends on parameters that may be associated with extraretinal input or with the internal state of the postsynaptic cell.

Functional Modifications Induced by Imposed Correlation between Postsynaptic Activity and Visual Input

As recalled above, a likely mechanism of epigenesis in visual cortex is modification in synaptic transmission dependent on temporal correlation between pre- and postsynaptic activities. Indirect evidence comes from recordings of populations of neurons in kittens submitted to selective visual exposure (Rauschecker and Singer, 1981). However, direct evidence recorded from a single neuron remains negative in the "acute" preparation when retinal manipulation is used alone.

In the first two sections, evidence for the existence of gating signals that would permit cortical neurons to switch from a passive *transmitting* state to an *adaptive* mode of functioning has been reviewed. Extraretinal activation such as produced by passive eye movements (Fig. 1), by intracortical injection of NA (Kasamatsu et al., 1979), or by electrical stimulation of intralaminar thalamic nuclei or of the mesencephalic reticular formation (Singer and Rauschecker, 1982) are the only three experimental situations where plasticity seems to be restored in the paralyzed and anesthetized kitten. It may be suggested that extraretinal activation is needed in order to change the internal state of cortical neurons and consequently abruptly enhance signal transmission or noise rejection. A possible mechanism by which a passive neuron could become a modifiable element is its membrane depolarization level. Singer and Rauschecker (1982) proposed more specifically that "Hebbian modifications have a threshold and occur only when dendritic depolarization of the postsynaptic target trespasses a critical level, the occurrence of soma spikes alone not being a sufficient condition."

In collaboration with Elie Bienenstock, Simon Thorpe, and Daniel Shulz, an experimental paradigm has been developed in the anesthetized and paralyzed kitten by which presentation of visual stimuli was paired with different levels of firing (and depolarization) of the recorded neuron (Bienenstock et al., 1983; Frégnac et al., 1984; Shulz et al., 1985; Frégnac and Shulz, 1985). This protocol has two advantages:

1. It relies on an exogenous control of the temporal correlation between postsynaptic activity and certain characteristics of the visual stimulus presented in the receptive field(s). In this respect, this situation corresponds to a test of Hebb's hypothesis, still indirect because of the unknown presynaptic activity.

2. In order to manipulate this correlation, the depolarization state of the recorded neuron was controlled by ionophoretic means. In this respect, this situation might mimic the pairing of extraretinal signals with visual stimulation. A more detailed presentation of these experiments follows.

Ionophoretic techniques were used to clamp neuronal firing at predetermined levels, which were made dependent on the orientation (or on the ocularity) of two stimuli shown in alternation (Fig. 5). One stimulus presentation was paired with a "high" level of discharge, while the other stimulus was present in the receptive field only while the cell's activity was clamped at a "low" level.

The choice of two stimuli paired with increased and decreased levels of activity is dictated by theoretical considerations. According to the "floating threshold" model (Bienenstock et al., 1982), competition between several stimuli should increase the speed of learning the Hebbian algorithm. Maintaining a mean visual activity during the pairing procedure of the same order as that observed without ionophoresis would correspond in the model to a fixed-modification threshold.

Clamp of neuronal activity was achieved by varying the retention/ ejection current of a potassium acetate or chloride electrode (3 M; impedance 2–15 MΩ). This technique has several advantages: (1) the recording channel is the ionophoretic channel, and we assume that the site of control of activity is close to the recording location; (2) continuous recording is possible even during the ionophoresis (applied current below 20 nA); and (3) modifications of firing rate could be obtained within 50–100 msec. In the optimal situation the current control resulted in enhancement (ejection) or complete suppression (retention) of the visual response. Care was taken to avoid potassium accumulation and depression effects.

Control studies showed that the preferred orientation of the recorded

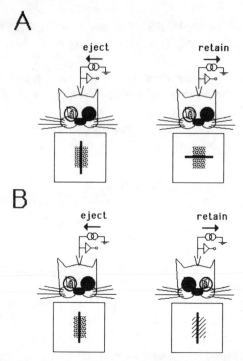

FIGURE 5. Protocols of imposed temporal correlation between postsynaptic activity and visual input. (A) This protocol was used in order to produce *orientation preference changes*. One cell is recorded from extracellularly in the visual cortex and tested through the dominant eye. Its receptive field is symbolized by a shaded rectangle on the stimulation screen. The presentation of a stimulus of a given orientation paired with positive current (left) alternates with presentation of a stimulus of a different orientation paired with negative current (right) applied through the recording 3 M KCl electrode. (B) This protocol was used in order to produce *ocular dominance changes*. A binocular cell is recorded from extracellularly in the visual cortex and tested separately through each eye. The corresponding receptive fields are symbolized by a shaded (right eye) or diagonally hatched (left eye) rectangle. Presentation of the preferred orientation stimulus through one eye paired with positive current (left) is alternated with presentation of the same stimulus through the other eye (right) paired with negative current.

neuron is unaffected by the level of a constant current applied through the microelectrode independently of the visual stimulation: the more positive the current, the higher is the visual response, the larger is the width of tuning, and the lower is the directional asymmetry; but the orientation for which the visual response is maximum remains unchanged and is independent of the value or of the polarity of the ionophoretic current.

In contrast, following the pairing procedure (performed monocularly) during which two different orientations were associated with two levels of firing (Fig. 5A), orientation-dependent changes were observed: they consisted in changes of the temporal profile of the responses (PSTH) and/or of the relative orientation preference and/or in modifications of the orientation tuning curve. These functional changes were generally consistent with Hebbian schemes: the orientation that had been paired with an increase of postsynaptic activity became relatively more effective in triggering visual response than the orientation that had been paired with a low level of activity. Significant results were obtained in normally reared kittens as old as 5 months of age. However, the clearest effects, such as change in orientation preference, were induced in deprived kittens recorded at the peak of sensitivity of the critical period, at 6 weeks of age (Fig. 6).

Concerning ocular dominance, control studies confirmed (as it has been already reported in the literature) a much higher intrinsic variability than for orientation preference. This variability seemed to be increased for low levels of discharge, that is, for high levels of retention. For some cells a slight positive dependency could be found between the binocularity index and the ionophoretic current, which indicates that a more positive current could unmask subliminally responses from the silent, or nondominant, eye. The effects of the pairing procedure associating different levels of firing with the ocularity of the visual stimulation (Fig. 5B) were much less convincing than for orientation preference, although significant changes could be found in 4-week-old normally reared kittens (Schulz et al., 1985; Frégnac and Shulz, 1985). An interesting result was also obtained in one cell recorded in an adult cat: the modification of ocular dominance was produced by selective suppression of spontaneous (but not visual) activity during stimulation through one eye. This last result appears compatible with a Hebbian hypothesis since the eye in favor of which the shift in ocular dominance occurred was the eye for which correlation between visual response and visual input was artificially increased. But this is up to now the only case where the improved response was obtained for the stimulation that had been paired with more negative current. This observation contrasts with increases of visual response induced by the first paradigm (orientation-dependent association), which were all produced by selective pairing with a more positive current (i.e., a more depolarized state).

The remarkable result is that all the provoked functional modifications

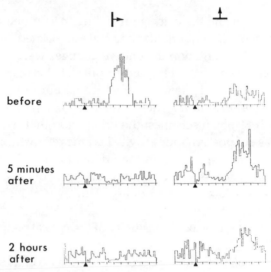

FIGURE 6. Orientation preference change (data from Frégnac et al., 1984). This cell was recorded in visual cortex of a 5-week-old kitten kept for the last 3 weeks in the dark. The cell was initially orientation biased. Poststimulus time histograms (PSTH) for the initially preferred stimulus (vertical bar) are shown on the left. PSTH for the initially nonpreferred stimulus (horizontal) are shown on the right. Filled triangles indicate onsets of the sweeps of bars across the receptive field (duration 4 sec). The *upper row* represents visual responses obtained in the control situation (retention −3 nA) *before* the pairing procedure. The pairing procedure was positive current (+3 nA) applied when presenting the horizontal bar, alternated with negative current (−7 nA) applied when presenting the vertical bar. This resulted during the pairing in an enhancement of the visual response for the initially nonpreferred stimulus and a total blockade of postsynaptic activity for the initially preferred stimulus. The two *lower rows* show the change in orientation preference observed 5 min and 2 hr *after* the end of the pairing procedure.

were consistent with Hebb's hypothesis, that is, the stimulus parameter (orientation or ocularity) associated with an increase (during current pairing) in the temporal correlation between postsynaptic activity and visual input, became more effective than that associated with a decrease in correlation. Some additional observations can be made. No response could be created de novo; the observed increases in visual response resulted from the enhancement of an already existing response, although additional zones of discharge could be revealed occasionally. Decreases in visual response could also be observed and eventually followed pairing with a low level or complete blockade of postsynaptic activity (Fig. 6). This could suggest that failure of an afferent fiber to repetitively trigger action potentials in the postsynaptic target could lead to a decrease in synaptic efficiency. This last rule, which is consistent with Sejnowski's covariance hypothesis (1977), is to be distinguished from the classical anti-Hebbian or pseudo-Hebbian modification rules proposed by Stent (1973) or Rauschecker and Singer (1981). Although these latter authors predicted a slow decrease (Rauschecker and Singer, 1981) or no synaptic change at all (Singer, 1983) when, irrespective of presynaptic activity, postsynaptic activity is absent, one should note that anti-Hebbian rules are usually introduced as normalization conditions that make total synaptic weight afferent to one cell constant. They predict a decrease in synaptic efficiency observed only when postsynaptic activity is present and presynaptic activity absent (i.e., for "silent" synapses). We tried in a few cases to simulate such a situation by unpairing (or mismatching) current pulses and visual stimulation (Frégnac and Shulz, 1985); but as already observed by Baranyi and Feher (1981), when randomizing the relative occurrences of action potentials and test excitatory postsynaptic potentials in the target cell, such "pseudoconditioning" (according to their terminology) did not induce any functional changes. The last comforting evidence of a nonartefactual effect is the absence of changes produced by a pairing procedure in which no effective control of postsynaptic activity was obtained in spite of high values of applied current.

Theoretical interpretation of these experiments calls for several remarks:

1. The main point is that the developed technique allows only an indirect test of Hebb's hypothesis. We avoid the difficulties of intracellular recording in the visual cortex but lose access to presynaptic information. Compared with the intracellular current injection technique as used by Baranyi and Feher (1981), extracellular potassium ionophoresis appears to have many side effects, such as changing the geometry or tortuosity of extracellular space and putting into play passive and active transport through a microcircuit of glia and neurons. But we have two reasons to assume that the main effect of the potassium leak is a modification of the membrane potential of the recorded postsynaptic neuron: (1) no action of current was found for positive monophasic short-duration action potentials (including mostly geniculate fibers), whereas positive current affected 80% of neurons recorded with biphasic spikes, some of them showing a somatodendritic inflexion; (2) receptive field

properties established for different levels of constant current, independent of visual stimulation, are predictable one from the other by simply assuming differences in the neuronal firing threshold for each current value. For more positive current, the neuron appears more depolarized, and subliminary responses are unmasked. This neuronal behavior is comparable to the "iceberg effect" describing spontaneous fluctuations in orientation tuning curves during long-duration recording of the same neuron (Frégnac and Bienenstock, 1981). However, these arguments do not preclude secondary modifications in presynaptic activity: since the time during which the level of firing was clamped extended in most cases over 1 sec, changes in the afferent message were certainly produced by activated intracortical loops. It is therefore illusory in this experimental situation to pretend to control postsynaptic activity without ulterior modification in presynaptic activity. Nevertheless, the key factor remains temporal change in the correlation between postsynaptic activity and visual input.

2. The Hebbian interpretation of such data relies at least on one "hidden" assumption: that distinct sets of synapses correspond to different orientation (or ocularity) stimuli. The first question one would like to ask is how different are the sets of synapses associated with two slits of light differing in orientation by a few degrees? This point has been given a minor value by most theoreticians who assume separability of the afferent messages under the form of nul scalar products (see Bienenstock et al., 1982). It could also explain the relative difficulty in producing ocular dominance changes. A reasonable guess concerning the experimental situation described in protocol B of Figure 5 is that most recorded neurons were second-order neurons, those that do not receive separate afferents from each eye. Consequently, the same set of synapses would have been associated to both "high" and "low" levels of activity. According to a Hebbian scheme, no functional modification is to be expected.

CONCLUSIONS

Experiments reviewed in this chapter give evidence that the functional expression of visual cortical plasticity is under the control of *extraretinal* signals. Activation of nonspecific afferents could distribute a global *gating* signal within sensory neocortex, during which modifications to specific sensory input is allowed. Among the possible sources of extraretinal input, experimental evidence favors a more decisive role for extraocular proprioception than for ascending noradrenergic projections. Nevertheless, it cannot be excluded that these two factors are somehow interdependent (Mc Bride and Suttin, 1984); projections from the locus coeruleus control the functioning of trigeminal neurons, some of which may relay extraocular inflow. It is possible that part of the diversity in the effects on cortical plasticity produced by distinct noradrenergic depletion protocols corresponds to differences in the degree of alteration of the extraocular pathway activation.

While the global nature of the gating by extraretinal input seems to be rather well established, almost no corresponding description has been given in terms of cellular events at the cortical level. Fragmental evidence suggests a control mechanism of the internal state of the target cell. In the adult visual cortex, visual response has been reported to be facilitated when eye movement direction is orthogonal to the orientation of the preferred stimulus (Trotter et al., 1980) and suppressed when eye movement is exerted in a direction parallel to it (Enomoto et al., 1983). The ionophoretic protocol we chose in order to change orientation preference (Bienenstock et al., 1983; Frégnac et al., 1984) strikingly recalls these findings on condition that one accepts the following assumption: the main effect of eye movement on the visual response of a cortical neuron is an orientation-dependent control of the membrane potential of the target cell.

A second indirect argument stresses the importance of *membrane potential* in gating cortical plasticity. Without denying the importance of temporal correlation between pre- and postsynaptic activities, it could be argued that changes induced by our pairing experiments (Frégnac et al., 1984) depend on the nature of the depolarizing agent, namely potassium. Modifications in receptive field properties induced by a constant leak of potassium in the extracellular medium are comparable to those following mesencephalic reticular formation (MRF) stimulation (Singer, 1979). This latter stimulation has been shown to increase extracellular potassium concentration (Lux and Singer, 1973). Exogenous control of extracellular potassium could thus mimic MRF stimulation, and both procedures have been reported to unblock cortical plasticity in the anesthetized and curarized preparation (Singer and Rauschecker, 1982; Frégnac et al., 1984).

The specificity of the epigenetic changes in response to visual manipulation are known to be correctly stimulated by *Hebbian schemes* (Rauschecker and Singer, 1981). Evidence presented in this chapter is based on an artificial control of activity localized around the site of recording (and not affecting the whole retinal message), and these experiments are among the very few to give a direct description of the evolutive power of a neuron considered individually. The functional changes observed in our pairing experiments suggest corresponding relative differential increases (for the stimulus whose correlation with postsynaptic activity was enhanced) and decreases (for the stimulus whose correlation with postsynaptic activity was reduced) in synaptic gain. This could indicate that Hebbian modifications could be associated with competitive inactivation of afferent synapses that suddenly have failed to elicit postsynaptic activity. Among the various schemes presented in Table 2, only the "floating threshold" model (Bienenstock et al., 1982) or the covariance hypothesis (Sejnowski, 1977) account correctly for such observations. But one should note that the basic hidden assumptions in these interpretations are that functional changes in the integrative power of a neuron result from modifications in synaptic efficiency and that modifiable synapses are excitatory. Nothing is said or predicted for local inhibitory connections, which are known to participate to a great extent in the genesis of orientation selectivity.

In conclusion, visual cortical plasticity may be thought of as a *threshold process*, which permits the integration of both extraretinal and visual activities. Extraretinal signals define the context in which modifications are authorized, while visually evoked activity could control distributed synaptic changes within the cortical network. The amplitude and sign of changes would depend on local modifications in the level of the temporal correlation (covariance) between pre- and postsynaptic activities. This twofold process could be realized at the cellular level by a sudden transition between two modes of functioning: a *passive* relay (or transmitting) mode and an *adaptive* (or modifiable) state. According to Singer (1980), the gating signal or "now print" command is identical with the sudden transition of an activated ensemble of interneurons from a passive relay mode to a highly resonant state. Changeux et al. (1984) proposed, in addition, a molecular scheme of how this transition could be reached; Interestingly, these authors assimilated the membrane potential of the postsynaptic cell to an allosteric effector intervening in the regulation of synaptic efficiency. Related experimental evidence comes from the recent work of Jahnsen and Llinas (1984), who showed that, depending on their membrane potential, individual neurons can show distinct modes of behavior.

Generalization of this latter observation to visual cortical neurons in the developing animal remains to be made. It is, however, conceivable that the main differences that could exist between imprinting, learning, and cortical epigenesis do not lie obligatorily in the activity-dependent mechanisms operating when the neuron has reached the adaptive mode; differences should be sought in the various types of enabling signals or genetic guidance that lead to the expression of functional plasticity.

ACKNOWLEDGMENTS

Research was supported by grants from C.N.R.S., D.G.R.S.T., M.I.R., and F.R.M. I am grateful to Joelle Adrien, Elie Bienenstock, Daniel Shulz, and Yves Trotter for their comments and to Kirsty Grant for help with the English. I wish to thank Professor Michel Imbert for continuous support.

REFERENCES

Adrien, J. (1978). Ontogenesis of some sleep regulations: Early postnatal impairment of the monoaminergic systems. *Prog. Brain Res.* **48**:393–403.

Adrien, J., G. Blanc, P. Buisseret, Y. Frégnac, E. Gary-Bobo, M. Imbert, J. P. Tassin, and Y. Trotter (1985). Noradrenaline and functional plasticity in kitten visual cortex: A reexamination. *J. Physiol. (Lond.)* **367**:73–98.

Adrien, J., P. Buisseret, Y. Frégnac, E. Gary-Bobo, J. P. Tassin, Y. Trotter and M. Imbert (1982). Noradrenaline et plasticité du cortex visuel du chaton. *C. R. Acad. Sci. Paris. Ser. III.* **295**:745–750.

Albus, J. G. (1971). A theory of cerebellar function. *Math. Biosci.* **10**:25–61.

Allen, C., J. Sievers, M. Berry, and S. Jenner (1981). Experimental studies on cerebellar foliation. II: A morphometric analysis of cerebellar fissuration defects and growth retardation after neonatal treatment with 6-OHDA in the rat. *J. Comp. Neurol.* **203**:771–783.

Allen, E. E., P. Trombey, J. Soyke, and B. Gordon (1984). Norepinephrine depletion: relation to visual cortical plasticity. *Soc. Neurosci. Abstr.* **10**:1079.

Alvarado-Mallart, R. M., C. Batini, C. Buisseret, J. P. Gueritaud, and G. Horcholle-Bossavit (1975). Mesencephalic projections of the "rectus lateralis" muscle afferents in the cat. *Arch. Ital. Biol.* **113**:1–20.

Armstrong-James, M., K. Fox, Z. L. Kruk, and J. Millar (1981). Quantitative ionophoresis of catecholamines using multibarrel carbon fibre microelectrodes. *J. Neurosci. Meth.* **4**:385–406.

Ary, M., J. D. Pettigrew, and T. Kasamatsu (1979). Manipulation of cortical catecholamines fail to affect suppression of deprived eye responses after reverse suture. *Invest. Ophthalmol. Vis. Sci. (ARVO Abstr.)* **18**:136.

Ashton, J. A., A. Boddy, and I. M. L. Donaldson (1984). Directional selectivity in the responses of units in cat primary visual cortex to passive eye-movements. *Neuroscience* **13**:653–662.

Baranyi, A. and O. Feher (1981). Long-term facilitation of excitatory synaptic transmission in the single cortical neurones of the cat produced by repetitive pairing of synaptic potentials following intracellular stimulation. *Neurosci. Lett.* **23**:303–308.

Bear, M. F. and J. D. Daniels (1983). The plastic response to monocular deprivation persists in kitten visual cortex after chronic depletion of norepinephrine. *J. Neurosci.* **3**:407–416.

Bear, M. F. and W. Singer (1986). Modulation of visual cortical plasticity by acetylcholine and noradrenaline. *Nature* **320**:172–175.

Bear, M. F., M. A. Paradiso, M. Schwartz, S. B. Nelson, K. M. Carnes and J. D. Daniels (1983). Two methods of catecholamine depletion in kitten visual cortex yield different effects on plasticity. *Nature* **302**:245–247.

Berman, N. and E. H. Murphy (1982). The critical period for alteration in cortical binocularity resulting from divergent and convergent stabismus. *Dev. Brain Res.* **2**:181–202.

Bienenstock, E. L., L. N. Cooper, and P. Munro (1982). Theory for the development of neuron selectivity: Orientation specificity and binocular interaction in visual cortex. *J. Neurosci.* **2**:32–48.

Bienenstock, E. L., Y. Frégnac and S. Thorpe (1983). Ionophoretic clamp of activity in visual cortical neurons in the cat: A test of Hebb's hypothesis. *J. Physiol. (Lond.)* **345**:123P.

Bischof, H. J. (1983). Imprinting and cortical plasticity: A comparative review. *Neurosci. Biobehav. Rev.* **7**:213–226.

Bowe-Anders, C., J. Adrien, and H. P. Roffwarg (1974). Ontogenesis of ponto-geniculo-occipital activity in the lateral geniculate nucleus of the kitten. *Exp. Neurol.* **43**:242–260.

Buisseret, P. and E. Gary-Bobo (1979). Development of visual cortical orientation specificity after dark rearing: Role of extraocular proprioception. *Neurosci. Lett.* **13**:259–263.

Buisseret, P. and E. Gary-Bobo (1981). L'orientation codée par un neurone visuel cortical dépend-elle de la direction des mouvements oculaires au cours du développement? *J. Physiol. (Paris)* **77**:50A.

Buisseret, P. and L. Maffei (1983). Suppression of visual cortical activity following tactile periorbital stimulation; its role during eye blinks. *Exp. Brain Res.* **51**:463–466.

Buisseret, P. and W. Singer (1983). Proprioceptive signals from extraocular muscles gate experience dependent modifications of receptive fields in the kitten visual cortex. *Exp. Brain Res.* **51**:443–450.

Buisseret, P., E. Gary-Bobo and M. Imbert (1978). Evidence that ocular motility is involved in the recovery of the orientational properties of visual cortical neurones in dark reared kittens. *Nature* **272**:816–817.

Carew, T. J., R. D. Hawkins, T. W. Abrams and E. R. Kandel (1984). A test of Hebb's postulate at identified synapses which mediate classical conditioning in Aplysia. *J. Neurosci.* **4**:1217–1224.

Changeux, J. P., P. Courrege and A. Danchin (1973). A theory of the epigenesis of neuronal networks by selective stabilization of synapses. *Proc. Natl. Acad. Sci. USA* **70**:2974–2978.

Changeux, J. P., T. Heidmann and P. Patte (1984). Learning by Selection. In: *The Biology of Learning*, P. Marler and H. S. Terrace, eds. pp. 115–133, Dahlem Konferenzen, Springer-Verlag, Berlin.

Crow, T. J. (1968). Cortical synapses and reinforcement: a hypothesis. *Nature* **219**:736–737.

Daniels, J. D., M. K. Ellis, S. A. Bianco, M. Garrett, S. B. Nelson and M. Schwartz (1981). Catecholamine depletion, ocular dominance shift and direction selectivity in kitten visual cortex. *Soc. Neurosci. Abstr.* **9**:1217.

Daunicht, W. J. (1983). Proprioception in extraocular muscles of the rat. *Brain Res.* **278**:291–294.

Davenne, D. and J. Adrien (1984). Suppression of PGO waves in the kitten: Anatomical effects on the lateral geniculate nucleus. *Neurosci. Lett.* **45**:33–38.

Davenne, D. and J. Adrien (1985). Maturation électrophysiologique des neurones du noyau géniculé latéral après lésion des voies ponto-géniculo-occipitales chez le chaton. *C.R. Acad. Sci. Paris. Ser. III.* **2**:59–64.

Daw, N. W., N. E. J. Berman, and M. Ariel (1978). Interaction of critical periods in the visual cortex of kittens. *Science* **199**:565–567.

Daw, M. W., R. K. Rader, T. W. Robertson and M. Ariel (1983). Effects of 6-

hydroxydopamine on visual deprivation in the kitten striate cortex. *J. Neurosci.* **3**:907–914.

Daw, N. W., T. W. Robertson, R. K. Rader, T. O. Videen and C. J. Coscia (1984). Substantial reduction of noradrenaline by lesions of adrenergic pathway does not prevent effects of monocular deprivation. *J. Neurosci.* **4**:1354–1360.

Daw, N. W., T. O. Videen, R. K. Rader, T. W. Robertson, and C. J. Coscia (1985a). Substantial reduction of noradrenaline in kitten visual cortex by intraventricular injections of 6-hydroxydopamine does not always prevent ocular dominance shifts after monocular deprivation. *Exp. Brain Res.* **59**:30–35.

Daw, N. W., T. O. Videen, D. Parkinson, and R. K. Rader (1985b). DSP-4 depletes noradrenaline in kitten visual cortex without altering the effects of monocular deprivation. *J. Neurosci.* **5**:1925–1933.

Enomoto, H., M. Matsumura, and J. Tsutsui (1983). Projections of extraocular muscle afferents to the visual cortex in the cat. *Neuroophthalmol.* **3**:49–57.

Fiorentini, A. and L. Maffei (1977). Instability of the eye in the dark and proprioception. *Nature* **269**:330–331.

Fiorentini, A., N. Berardi and L. Maffei (1982). Role of extraocular proprioception in the orienting behavior of cats. *Exp. Brain Res.* **48**:113–120.

Foote, S. L., F. E. Bloom, and G. Aston-Jones (1983). Nucleus locus coeruleus: new evidence of anatomical and physiological specificity. *Physiol. Rev.* **63**:844–914.

Freedman, D. X. and A. E. Halaris (1978). Monoamines and the Biochemical Mode of Action of L.S.D. at Synapses. In: *Psychopharmacology: A Generation of Progress*, M. A. Lipton, A. Dimascio, K. F. Killam, eds., pp. 347–359, Raven Press, New York.

Freeman, R. D. and A. B. Bonds (1979). Cortical plasticity in monocularly deprived immobilized kittens depends on eye movement. *Science* **206**:1093–1095.

Frégnac, Y. (1979). Development of orientation selectivity in the primary visual cortex of normally and dark reared kittens. II. Models. *Biol. Cyb.* **34**:195–203.

Frégnac, Y. (1982). Développement de la sélectivité neuronale dans le cortex visuel primaire du chat. Thèse d'Etat és-Science Naturelles, Paris VI.

Frégnac, Y. and E. Bienenstock (1981). Specific Functional Modifications of Individual Cortical Neurones, Triggered by Vision and Passive Eye-movement, in Immobilized Kittens. In: *Pathophysiology of the Visual System*, L. Maffei, ed., pp. 100–108. Dr. W. Junk Publisher, The Hague.

Frégnac, Y. and M. Imbert (1984). Development of neuronal selectivity in the primary visual cortex of the cat. *Physiol. Rev.* **64**:325–434.

Frégnac, Y. and D. Shulz (1985). Locally imposed temporal correlation of activities and visual cortical plasticity. Second Biennal Symposium on Neural Mechanisms of Conditioning (Abstr.), Cambridge, England.

Frégnac, Y., S. Thorpe, D. Shulz and E. L. Bienenstock (1984). Modification of the function in the cat visual cortical neurones induced by control of the correlation between postsynaptic activity and visual input. *Soc. Neurosci. Abstr.* **10**:1078.

Gary-Bobo, E., C. Milleret, and P. Buisseret (1986). Role of eye movements in developmental processes of orientation selectivity in the kitten visual cortex. *Vision Res.* **26**:557–567.

Gilbert, P. (1975). How the cerebellum could memorise movements. *Nature* **254**:688–689.

Gilbert, P. F. C. and W. T. Thach (1977). Purkinje cell activity during motor learning. *Brain Res.* **128**:309–328.

Graves, A., Y. Trotter and Y. Frégnac (1984). Influence of unilateral section of extraocular proprioceptive afferents on binocular depth perception in kittens. *Invest. Ophthalmol. Vis. Sci. (ARVO Abstr.)* **25**:293.

Graves, A. L., Y. Trotter, and Y. Frégnac (in press). Role of extraocular muscle proprioception in the development of depth perception in cats. *J. Neurophysiol.*

Guthrie, B. L., J. D. Porter, and D. L. Sparks (1982). Role of extraocular muscle proprioception in eye movements studied by chronic deafferentation of intra-orbital structures. *Soc. Neurosci. Abstr.* **8**:156.

Hebb, D. O. (1949). *The Organization of Behavior.* Wiley & Sons, New York.

Imbert, M. and P. Buisseret (1975). Receptive field characteristics and plastic properties of visual cortical cells in kittens reared with or without visual experience. *Exp. Brain Res.* **22**:25–36.

Ito, M., M. Sakurai and P. Tongroach (1982). Climbing fibre induced depression of both mossy fibre responsiveness and glutamate sensitivity of cerebellar Purkinje cells. *J. Physiol. (Lond.)* **324**:113–134.

Jahnsen, H. and R. Llinas (1984). Electrophysiological properties of guinea pig thalamic neurones: an in vitro study. *J. Physiol. (Lond.)* **349**:205–226.

Jonsson, G. (1980). Chemical neurotoxins as denervation tools in neurobiology. *Ann. Rev. Neurosci.* **3**:169–187.

Kasamatsu, T. (1979). Involvement of the beta-adrenergic receptor in cortical plasticity. *Invest. Ophthalmol. Vis. Sci. (ARVO Abstr.)* **18**:135.

Kasamatsu, T. (1982). A Role of the Central Norepinephrine System in Regulation of Neuronal Plasticity in the Cat Visual Cortex. In: *Neurotransmitters in the Retina and the Visual Centers (Biomedical Research Suppl.)* A. Kaneko, N. Tsukahara and K. Uchinozo, eds. pp. 88–93, Biomedical Research Found., Tokyo.

Kasamatsu, T. (1983). Neuronal Plasticity Maintained by the Central Norepinephrine System in the Cat Visual Cortex. In: *Progress in Psychobiology and Physiological Psychology, Vol. 10,* J. M. Sprague and A. N. Epstein, eds., pp. 1–112, Academic Press, New York.

Kasamatsu, T. and P. Heggelund (1982). Single cell response in cat visual cortex to visual stimulation during iontophoresis of noradrenaline. *Exp. Brain Res.* 45:317–327.

Kasamatsu, T. and J. D. Pettigrew (1976). Depletion of brain catecholamines: Failure of ocular dominance shift after monocular occlusion in kittens. *Science* 194:206–209.

Kasamatsu, T. and J. D. Pettigrew (1979). Preservation of binocularity after monocular deprivation in the striate cortex of kittens treated with 6-hydroxydopamine. *J. Comp. Neurol.* 185:139–161.

Kasamatsu, T., J. D. Pettigrew, and M. Ary (1979). Restoration of visual cortical plasticity by local microperfusion of norepinephrine. *J. Comp. Neurol.* 185:163–182.

Kasamatsu, T., J. D. Pettigrew, and M. Ary (1981a). Cortical recovery from effects of monocular deprivation: Acceleration with norepinephrine and suppression with 6-hydroxydopamine. *J. Neurophysiol.* 45:254–266.

Kasamatsu, T. and T. Shirokawa (1985). Involvement of beta-adrenoreceptors in the shift of ocular dominance after monocular deprivation. *Exp. Brain Res.* 59:507–514.

Kasamatsu, T., T. Itakura, and G. Jonsson (1981b). Intracortical spread of exogenous catecholamines: Effective concentration for modifying cortical plasticity. *J. Pharmacol. Exp. Therap.* 217:841–850.

Kasamatsu, T., T. Itakura, G. Jonsson, P. Heggelund, J. D. Pettigrew, K. Nakai, K. Watabe, B. D. Kupperman, and M. Ary (1985a). Neuronal Plasticity in Cat Visual Cortex: A Proposed Role for the Central Norepinephrine System. In: *Monoamine Innervation of Cerebral Cortex,* L. Descarries, T. A. Reader, and H. H. Jasper, eds., Alan Liss, New York.

Kasamatsu, T., K. Watabe, P. Heggelund, and E. Scholler (1985b). Plasticity in cat visual cortex restored by electrical stimulation of the locus coeruleus. *Neurosci. Res.* 2:365–386.

Kety, S. S. (1970). The Biogenic Amines in the Central Nervous System: Their

Possible Roles in Arousal, Emotion and Learning. In: *The Neurosciences: Second Study Program*, F. O. Schmitt, ed., pp. 324–336, Rockefeller University Press, New York.

Krüger, J. (1983). Simultaneous Individual Recordings from Many Cerebral Neurons: Techniques and Results. In: *Reviews of Physiology, Biochemistry and Pharmacology, Vol. 98*, R. H. Adrian et al., eds., pp. 177–233. Springer-Verlag, Berlin.

Kuppermann, B. D. and T. Kasamatsu, (1984). Enhanced binocular interaction in the visual cortex of normal kittens subjected to intracortical norepinephrine. *Brain Res.* **302:**91–99.

Levitt, F. B. and R. C. van Sluyters (1982). Recovery of binocular function in kitten visual cortex. *J. Neurophysiol.* **48:**1336–1346.

Lux, H. D. and W. Singer (1973). Changes of extracellular potassium concentration in cat visual cortex after stimulation of specific and unspecific afferents. *Pflügers Arch.* **S343:**137R.

Maeda, T., C. Pinc, D. Salvert, M. Ligier and M. Jouvet (1973). Les neurones contenant des catécholamines du tegmentum pontique et leurs voies de projection chez le chat. *Brain Res.* **57:**119–152.

Maffei, L. and S. Bisti (1976). Binocular interaction in strabismic kittens deprived of vision. *Science* **191:**579–580.

Manni, E., R. Bortolami and C. Desole (1966). Eye muscle proprioception and the semilunar ganglion. *Exp. Neurol.* **16:**226–236.

Marr, D. C. (1969). A theory of cerebral cortex. *J. Physiol. (Lond.)* **202:**437–470.

McBride, R. L. and J. Suttin. (1984). Noradrenergic hyperinnervation of the trigeminal sensory nuclei. *Brain Res.* **324:**211–221.

McCall, M. A., D. G. Tieman and H. V. B. Hirsch (1982). Chronic intraventricular administration of Lysergic acid diethylamid affects the sensitivity of cortical cells to monocular deprivation. *Brain Res.* **250:**301–308.

McNaughton, B. L., R. M. Douglas, and G. V. Goddard (1978). Synaptic enhancement in fascia dentata: cooperativity among coactive afferents. *Brain Res.* **157:**277–293.

Michalski, A., G. L. Gerstein, J. Czarkowska and R. Tarnecki (1983). Interactions between cat striate neurons. *Exp. Brain Res.* **51:**97–107.

Milleret, C., E. Gary-Bobo and P. Buisseret (1984). The preferred orientation acquired by the kitten's visual cortical cells depends on the direction of eye-movements during visual experience. *Neurosci. Lett. Suppl.* **18:**S74.

Moore, R. Y. and F. E. Bloom (1979). Central catecholamine neuron system: Anatomy and physiology of the norepinephrine and epinephrine systems. *Ann. Rev. Neurosci.* **2:**113–168.

Movshon, J. A. and R. C. van Sluyters (1981).Visual neural development. *Ann. Rev. Psychol.* **32:**477–522.

Nakai, K., G. Jonsson, and T. Kasamatsu (1981). Regrowth of central catecholaminergic fibers in cat visual cortex following localized lesion with 6-hydroxydopamine. *Soc. Neurosci. Abstr.* **7:**675.

Nelson, S. B., M. A. Schwartz, and J. D. Daniels (1985). Clonidine and cortical plasticity: possible evidence for noradrenergic involvement. *Dev. Brain Res.* **23:**39–50.

Olson, C. R. and R. D. Freeman (1980). Profile to the sensitive period for monocular deprivation in kitten. *Exp. Brain Res.* **39:**17–21.

Paradiso, M. A., M. F. Bear, and J. D. Daniels (1983). Effects of intracortical infusion of 6-hydroxydopamine on response of kitten visual cortex to monocular deprivation. *Exp. Brain Res.* **51:**413–422.

Pettigrew, J. D. and T. Kasamatsu (1978). Local perfusion of noradrenaline maintains visual cortical plasticity. *Nature* **271:**761–763.

Porter, J. D. and R. F. Spencer (1982). Localization and morphology of cat extraocular muscle afferent neurones identified by retrograde transport of horseradish peroxydase. *J. Comp. Neurol.* **204:**56–64.

Porter, J. D., B. L. Guthrie, and D. L. Sparks (1983). Innervation of monkey extraocular muscles: Localization of sensory and motor neurons by retrograde transport of horseradish peroxydase. *J. Comp. Neurol.* **218:**208–219.

Rauschecker, J. P. and W. Singer (1981). The effects of early visual experience on the cat's visual cortex and their possible explanation by Hebb synapses. *J. Physiol. (Lond.)* **310:**215–239.

Sejnowski, T. J. (1977). Storing covariance with nonlinearly interacting neurons. *J. Math. Biol.* **4:**303–321.

Shaw, C. and M. Cynader (1984). Disruption of cortical activity prevents ocular dominance changes in monocularly deprived kittens. *Nature* **308:**731–733.

Sherman, S. M. (1972). Development of interocular alignment in cats. *Brain Res.* **37:**187–203.

Shimbel, A. (1950). Contributions to the mathematical biophysics of the central nervous system with special reference to learning. *Bull. Math. Biophys.* **12:**241–275.

Shirokawa, T. and T. Kasamatsu (1986). Concentration-dependent suppression by beta-adrenergic antagonists of the shift in ocular dominance following monocular deprivation in kitten visual cortex. *Neuroscience* **18:**1035–1046.

Shulz, D., Y. Frégnac, S. Thorpe, and E. L. Bienenstock (1985). Correlation of Pre- and Postsynaptic Activities: Its Role in Visual Cortical Plasticity. In:

Brain Plasticity, Learning and Memory, P. Schmitt., J. Dalrymple-Alford and B. Will, eds., p. 579, Plenum Press, New York.

Sievers, J., M. Berry and H. Baumgarten (1981). The role of noradrenergic fibers in the control of post-natal cerebellar development. *Brain Res.* **207**:200–208.

Sillito, A. M. (1977). Inhibitory processes underlying the directional specificity of simple, complex and hypercomplex cells in the cat's visual cortex. *J. Physiol. (Lond.)* **271**:699–720.

Sillito, A. M. (1983). Plasticity in the visual cortex. *Nature* **303**:477–478.

Singer, W. (1979). Central Core Control of Visual Cortex Functions. In: *The Neurosciences: 4th Study Program*, pp. 1093–1110, MIT Press, Cambridge, MA.

Singer, W. (1980). Central gating of developmental plasticity in the cat striate cortex. *Verh. Dtsch. Zool. Ges.* **1980**:268–274.

Singer, W. (1983). Neuronal Activity as a Shaping Factor in the Self-Organization of Neuron Assemblies. In: *Synergetics of the Brain*, E. Basar, H. Flohr, H. Haken and A. J. Mandell, eds., pp. 89–101, Springer-Verlag, Heidelberg.

Singer, W. and J. P. Rauschecker (1982). Central core control of developmental plasticity in the kitten visual cortex. II Electrical activation of mesencephalic and diencephalic projections. *Exp. Brain Res.* **47**:223–233.

Singer, W., F. Tretter, and U. Yinon (1982). Central gating of developmental plasticity in kitten visual cortex. *J. Physiol. (Lond.)* **324**:221–237.

Sivanandasingham, P. (1977). Somata of feline extra-ocular muscle proprioceptors. *Acta Anat.* **99**:152–159.

Stent, G. (1973). A physiological mechanism for Hebb's postulate of learning. *Proc. Natl. Acad. Sci. USA* **70**:997–1001.

Stryker, M. P. (1977). The role of early experience in the development and maintenance of orientation selectivity in the cat's visual cortex. *Neurosci. Res. Prog. Bull.* **15**:455–462.

Trotter, Y. (1981). Rôle de la proprioception extraoculaire dans le développement fonctionnel du cortex visuel primaire du chat. Thèse de 3ème Cycle, Paris VI.

Trotter, Y., Y. Frégnac, and P. Buisseret (1981a). Période de sensibilité du cortex visuel primaire du chat à la suppression unilaterale des afférences proprioceptives extraoculaires. *C. R. Acad. Sci. Paris. Ser. III.* **293**:245–248.

Trotter, Y., Y. Frégnac and P. Buisseret (1983). Synergy between vision and extraocular proprioception in gating functional plasticity of kitten primary visual cortex. *C.R. Acad. Sci. Paris. Ser. III.* **296**:665–668.

Trotter, Y., Y. Frégnac and P. Buisseret (in press). The period of susceptibility of visual cortical binocularity to unilateral proprioceptive deafferentation of extraocular muscles. *J. Neurophysiol.*

Trotter, Y., E. Gary-Bobo and P. Buisseret (1981b). Recovery of orientation selectivity in kitten primary visual cortex is slowed down by bilateral section of ophthalmic trigeminal afferents. *Dev. Brain Res.* **1**:450–454.

Trotter, Y., L. Maffei and P. Buisseret (1980). Effets de mouvements oculaires passifs sur l'activité des neurones visuels corticaux chez le chat. *J. Physiol. (Paris)* **76**:62A.

van Sluyters, R. C. and F. B. Levitt (1980). Experimental strabismus in the kitten. *J. Neurophysiol.* **43**:686–699.

Videen, T. O., N. W. Daw, and R. C. Collins (1986). Penicillin-induced epileptiform activity does not prevent ocular dominance shifts in monocularly deprived kittens. *Brain Res.* **371**:1–8.

Videen, T. O., N. W. Daw, and R. K. Rader (1984). The effect of norepinephrine on visual cortex neurons in kittens and adult cats. *J. Neurosci.* **4**:1607–1617.

von der Malsburg, C. (1973). Self-organization of orientation sensitive cells in the striate cortex. *Kybernetik* **14**:85–100.

Wigström, H., B. Gustafsson, Y. Y. Huang, and W. C. Abraham (1986). Hippocampal long-term potentiation is induced by pairing single afferent volleys with intracellularly injected depolarizing current pulses. *Acta Physiol. Scand.* **126**:317–318.

Woody, C. D., B. E. Swartz, and E. Gruen (1978). Effects of acetylcholine and cyclic GMP on input resistance of cortical neurons in awake cats. *Brain Res.* **158**:373–395.

Wurtz, R. H., V. F. Castellucci, and J. M. Nusrala (1967). Synaptic plasticity: the effect of the action potential in the postsynaptic neuron. *Exp. Neurol.* **18**:350–368.

Yinon, U. (1976). Age dependence of the effect of squint on cells in kitten's visual cortex. *Exp. Brain Res.* **26**:151–158.

12

INFLUENCE OF VISUAL EXPERIENCE ON ONTOGENY OF OPTOKINETIC REFLEX IN MAMMALS

Klaus Peter Hoffmann

Abt. für Vergleichende Neurobiologie, Universität Ulm Ulm. FR Germany

The optokinetic reflex (OKR) in higher mammals is organized in a much more complex way than originally supposed. It is mainly by the development of the neocortical projections to the midbrain structures involved in the OKR that the quality of this simple reflex is changed. The basic function of the OKR is to stabilize the image on the retina during global movements of the visual environment or to counteract drifts of the eyes. This is achieved by direction-selective signals sent from the retina of each eye to the nuclei of the accessory optic tract and to the nucleus of optic tract (NOT) in the pretectum either directly or through the visual cortex (Fig. 1).

Horizontal optokinetic nystagmus (OKN) elicited through one eye can be subdivided into two components, temporal OKN, when the visual stimulus moves from nasal to temporal in the visual field, and nasal OKN, when the stimulus moves from temporal to nasal. According to a model based on neurophysiological and neuroanatomical data, nasal OKN is mediated through the nucleus of the optic tract (NOT) contra-

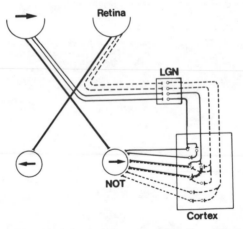

FIGURE 1. Wiring diagram showing the connections from the retina to the nucleus of the optic tract (NOT) either by direct retinopretectal W fibers or by Y axons relayed through the lateral geniculate nucleus (LGN) and visual cortex. Projections from the contralateral eye are shown by continuous lines, from the ipsilateral eye by broken lines, and binocular projections by a combination of both. The arrow in front of the left eye indicates the horizontal preferred direction, i.e., from temporal to nasal in the visual field, for on-center direction-specific ganglion cells, and for the contralateral NOT. Corticofugal axons can be direction specific for all directions and can be activated by either eye alone or binocularly. The axons contributing to the NOT response have to be selected by an activity-dependent process (see Fig. 7).

lateral to the stimulated eye, whereas temporal ORN is mediated through the ipsilateral NOT (Hoffmann, 1981; Precht and Strata, 1980). Independent of which eye is stimulated, an increase in activity in the left NOT over that in the right NOT leads to leftward OKN (slow phases directed to the left), whereas higher activity in the right NOT leads to rightward OKN. As is typical for subcortical visual centers, each retina has a strong projection to the contralateral NOT. It follows that nasal OKN could be triggered by this direct retinopretectal pathway (Collewijn, 1981). In addition, the NOT in the cat receives a strong cortical projection. The corticopretectal axons carry binocular information about retinal image slip, which can trigger OKN (Schoppmann, 1981). Each retina can therefore reach the contralateral NOT as well as the ipsilateral NOT via its cortical representations. Thus, in addition to the crossing retinopretectal fibers, nasal OKN could be triggered by this cortical pathway. Temporal OKN, on the other hand, would be strongly dependent on the uncrossed projection through the visual cortex because there are very few uncrossed direct retinal fibers to NOT (Ballas et al., 1981). This wiring diagram for OKN was tested by a variety of lesions in the visual pathway (Harris et al., 1980; Hoffmann, 1981; Precht et al., 1980; Wood et al., 1973). Bilateral cortical lesions nearly abolish temporal OKN and reduce the gain for nasal OKN. Transection of one optic tract leads to characteristic deficits for each eye, temporal OKN is decreased when the eye ipsilateral to the lesion is tested, and nasal OKN is weakened when the contralateral eye is stimulated.

In young kittens monocular OKN is asymmetric (van Hof-van Duin, 1978). Nasal OKN is much stronger than temporal OKN. During a sensitive period from 4 weeks to about 25 weeks after birth OKN becomes more and more symmetric until the adultlike pattern is reached. Early visual deprivation, monocular or binocular, as well as strabismus prevents the development of symmetric OKN (Cynader and Harris, 1980; van Hof-van Duin, 1976). Nasal OKN elicited through a deprived or strabismic eye is much stronger than temporal OKN. But even nasal OKN drops out at stimulus velocities above 20°/sec (Malach et al., 1984). OKN in the nondeprived eye in monocularly deprived cats or in the unoperated eye in strabismic cats also becomes abnormal due to the manipulation of the other eye. The asymmetry remains larger than after normal development (Cynader and Harris, 1980; Markner and Hoffmann, 1985).

RECEPTIVE FIELD PROPERTIES

Before discussing our ideas about the developmental mechanisms involved in the generation of the specificity of this system, the properties of the visual receptive fields in the NOT, in the retina, and in the visual cortex have to be described with respect to their relevance for the pathway of OKN.

Nucleus of Optic Tract

The visual receptive fields of neurons in the NOT of all mammals tested so far (for a review see Simpson, 1984) are very large (up to 90° horizontal width), and they respond direction-specifically over a very broad velocity spectrum (0.1–100°/sec). In the cat almost all cells in the left nucleus prefer movements to the left, and those in the right nucleus prefer movements to the right in the visual world (Hoffmann and Schoppmann, 1975). The output of these neurons goes to (or at least near to) three sites, the dorsal cap of the inferior olive, the nucleus prepositus hypoglossi, and the area of the nucleus reticularis tegmenti pontis (Precht et al., 1980; Lannou et al., 1984). All these areas are involved in stabilizing reflexes of eyes, head, or body. A lot of the evidence that the output of neurons in the NOT generates OKN is indirect. This type of evidence derives from the analysis of single-cell responses to visual stimulation in the anesthetized preparation and from the analysis of OKN in normal animals or in animals with pretectal or cortical lesions (Hoffmann, 1983a; Precht et al., 1980; Strong et al., 1984; Wood et al., 1973). However, our recordings from NOT neurons in awake cats with search coils implanted around their eyes to measure eye movements in a magnetic field during optokinetic stimulation showed similar results as recordings in the NOT of anesthetized cats (Hoffmann and Huber, 1983). The cells preferred large-area stimuli, responded well to very low velocities (<1°/sec), were direction specific for horizontal movement, and were mostly binocularly activated (see Figs. 2, 5A and 6). All cells responded to retinal slip over a range from less than 1°/sec to more than 100°/sec (best response at about 10°/sec). The discharge rate (spontaneous: 5–50 spikes/sec) was not modulated during optokinetic afternystagmus or during the vestibuloocular reflex in the dark. Electrical stimulation through the recording electrode (pulse width 1 msec, frequency 60 Hz, maximal amplitude 0.5 mA) elicited eye movements in

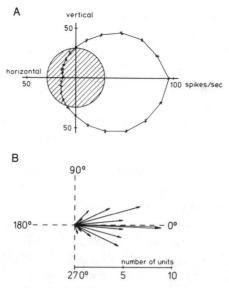

FIGURE 2. Population tuning curve (A) and individually preferred directions (B) for cells in the right nucleus of the optic tract. The stimulus consisted of a large-area random-dot pattern moving on a circular path. In A the activity (spikes/sec) of 36 cells is averaged and presented in polar coordinates. The shaded disc represents the mean spontaneous activity in the NOT. Response strength in different directions is represented by the vector from the origin to the crosses. In B the relative frequency of the preferred directions of NOT cells is given by the length of the arrows.

a clear OKN pattern as if the preferred stimulus was presented to the cells.

Therefore, NOT cells provide a direction-selective signal about retinal slip, which can lead to OKN. In a normal cat slow-phase eye velocity during OKN was always slower than the stimulus velocity, and cells in the left NOT would discharge at a rate above spontaneous activity as long as the stimulus moved leftward and vice versa (Hoffmann and Huber, 1983).

Retinal Output Relevant for Optokinetic Reflex

In a recent study (Hoffmann and Stone, 1985) we sought to identify the ganglion cells that project to the NOT by recording individual ganglion cells in the retina and by determining which could be activated antidromically and with weak stimulus currents by a stimulating electrode in the NOT. In a population of 558 retinal ganglion cells 98 cells were activated antidromically from the NOT, including 43 Y cells (35% out of 122), 21 X cells (7% out of 310), and 34 W cells (27% out of 126) (Fig. 3). An earlier study (Hoffmann and Schoppmann, 1981) indicated that the conduction time along retinal axons for the length from the chiasma to the NOT is 3–7 msec.

Thus, only W cells meet two criteria of putative retinal afferents to the NOT, appropriate conduction velocity and low threshold to antidromic stimulation. This group of W cells includes one example or more of the following subclasses of W cells distinguished by Stone and Fukuda (1974): phasic (3 cells), tonic (2 cells), suppressed by contrast (1 cell), and on-center direction selective (5 cells). Of interest is the presence of 5 on-center direction selective cells among the 11 putative NOT afferent cells. This is much higher than the proportion of these cells among W cells generally. The preferred directions of 2 of the 5 cells were horizontal (temporal to nasal) and of 3 were oblique (1 up and temporonasal, 1 up and nasotemporal, 1 down and nasotemporal). The directional tuning curves of two of these cells are depicted in Figures 4A,B.

Do these results provide evidence that direction-selective on-center W cells form the major visual input to the NOT and control OKN (Oyster et al., 1972)? The large size of NOT-receptive fields can be made up by a lot of convergence of retinal ganglion cells. In decorticated cats NOT cells hardly respond to velocities higher than 20°/sec, which is in the same low range as found in the direction-selective on-center ganglion cells. However, NOT cells in the adult cat respond to stimulus velocities up to 100°/sec. Also, many NOT cells can be activated from both eyes, whereas the retinal input seems to come almost exclusively from the

---

FIGURE 3. Relationship between receptive field types of retinal ganglion cells, their threshold for antidromic stimulation from the nucleus of the optic tract (NOT), and conduction time of their axons between the NOT and the optic chiasm (OX). We chose to present the latency difference between NOT and OX to eliminate different conduction times in the retina due to varying eccentricities and to allow the comparison with orthodromic latencies between OX and NOT (Hoffmann and Schoppmann, 1981). Different ganglion cell classes are plotted by the following symbols: Y = Y cells with the fastest conducting axons, X = X cells with axons of medium conduction velocity. The symbols representing W cells, which have the slowest conducting axons, are explained in the inset. Of interest are the on-center direction-selective W cells, which have a very low threshold and therefore probably form the major retinal input to the NOT. Many Y cells also have low thresholds, but their latencies are too short to qualify as input to the NOT. They have to be considered as fibers of passage. [From Hoffmann and Stone (1985).]

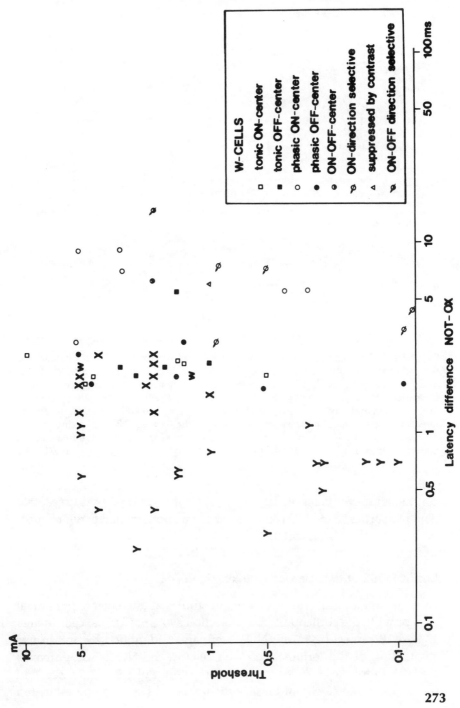

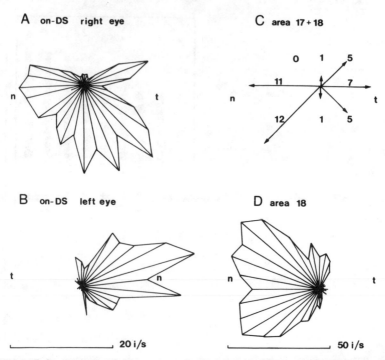

FIGURE 4. Polar diagrams to demonstrate the direction specificity of retinal ganglion cells with on-center receptive fields (A,B) and a cortical cell from area 18 (D) projecting to the nucleus of the optic tract (NOT). Response strength to movement of a large-area random-dot pattern in different directions is represented by the length of the lines from the origin to the border of the polar diagram. The symbols t (temporal) and n (nasal) indicate the direction in visual space for the eye contralateral to the recorded NOT. Movements from temporal to nasal are preferred by cells in the NOT. In C is presented the frequency of preferred directions for cortical cells projecting to the NOT. [From Schoppmann (1981).]

contralateral eye (Ballas et al., 1981; Hoffmann, 1983a). These receptive field properties in the NOT must come from another source, which most probably is the visual cortex.

Cortical Output Relevant for Optokinetic Reflex

As many anatomical and physiological studies show, layer V pyramidal cells provide the corticopretectal projection system. Cortical cells in area 17 and 18 projecting to the NOT were again identified by antidromic stimulation of the terminals of such cells in the NOT (Schoppmann, 1981). Antidromic latencies show a mean latency of 2.8 + 1.5 msec, and the conduction velocity of these corticopretectal fibers can be estimated to be approximately 13 m/sec. Response properties of antidromically activated cortical units were very similar to the so-called corticotectal layer V pyramidal cells. All units have (for the visual cortex) quite large receptive fields with up to 5° in diameter and could be equally well activated by oriented light bars as well as by large-area random dot patterns moved across their receptive field. By contrast, within the entire cortical population tested, only 48% of the recorded units could be activated by either of the two types of pattern. All the antidromically activated units preferred certain movement directions to others or were strongly direction selective with a preference for horizontal movements on the average (see Figs. 4C,D). All cells but one were binocular, and the responses enhanced by binocular stimulation as compared to stimulation of either eye alone. In our study these corticopretectal cells could rarely be activated by electrical stimulation of the geniculocortical pathway. Nevertheless, from other studies on the input to layer V pyramidal cells projecting to the mesencephalon, we can argue that such cells are most likely activated by Y fibers from the geniculate either mono- or disynaptically (Ferster and Lindstrom, 1983; Hoffmann and Stone, 1971; Singer et al., 1975). The comparison of retinal and cortical input to the NOT very clearly shows that the typical response profile of the cells in the adult cat's NOT reflects a strong cortical input. As has been elaborated in a previous review (Hoffmann, 1983a), following lesions of the visual cortex in both hemispheres, the remaining retinal input to the NOT leads to a much reduced modulation of the activity of these neurons and to the failure of these neurons to respond to velocities above 20°/sec. All cells are then exclusively driven by the contralateral eye. Concomitant changes can be seen in the optokinetic nystagmus, which reaches only much lower slow-phase velocities and becomes asymmetric when tested monocularly (Strong et al., 1984). Clearly, the functional role of the cortical visual input on top of the retinal input consists in adding binocularity and response strength, particularly at high stimulus velocities, to the visual input relevant for OKN.

ROLE OF EARLY VISUAL EXPERIENCE

Effects of Monocular Deprivation on Receptive Field Properties in Nucleus of Optic Tract

Monocular deprivation was achieved by closing one eye by lid suture within the first week after birth over a period of 6–24 months. After

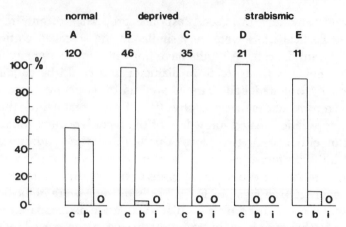

FIGURE 5. Relative frequency of binocular cells in the nucleus of the optic tract (NOT) in normal (A), monocularly deprived (B,C), and strabismic (D,E) cats. The number of analyzed neurons is given above the histograms. Symbols: c activated exclusively from the contralateral eye; b, binocularly activated; i, activated exclusively by the ipsilateral eye; 0, no cell. B, cells ipsilateral to experienced eye. C, cells ipsilateral to deprived eye. D, cells ipsilateral to unoperated eye. E, cells ipsilateral to deviated eye.

opening the deprived eye, OKN was tested (Markner and Hoffmann, 1985) and then the receptive fields in the NOT were analyzed. In monocularly deprived cats binocularity in both NOTs can be totally abolished (Fig. 5). In most cats all NOT neurons lack an ipsilateral input whether the deprived or the nondeprived eye is the ipsilateral one. The neurons driven by the deprived as well as by the nondeprived eye exhibit a clear direction specificity for temporonasal movements of optokinetic stimuli presented to the contralateral eye. They differ, however, significantly in their capacity to respond to high stimulus velocities (Hoffmann, 1979; 1983b). Only cells that can be stimulated through the nondeprived eye can respond when the stimulus moves faster than 20°/sec (Fig. 6). Of course, the same differences were found when we compared OKN in the deprived eye to OKN in the nondeprived eye (see introduction and Markner and Hoffmann, 1985).

If we compare these NOT cell properties to those in decorticated animals, it becomes evident again that direction specificity in NOT cells is independent of the input from visual cortex, whereas responses to high velocities and binocularity are mediated through visual cortex. Direction selectivity is thus not dependent on visual experience, whereas responses to high velocities and binocularity are.

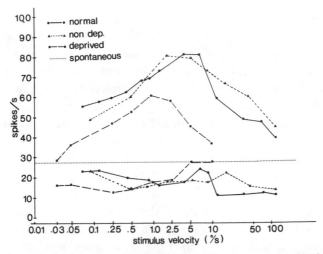

FIGURE 6. Relationship between stimulus velocity (degrees per second) (abscissa) and response strength (spikes/sec) (ordinate) for cells in the nucleus of the optic tract (NOT) driven through the deprived eye (squares, 10 cells) or through the nondeprived eye (triangles, 8 cells). Responses for 12 cells in the NOT of normal cats are given by the dots. Dotted line represents spontaneous activity of all cells. Curves above this line represent activation in the preferred direction, curves below the dotted line indicate the decrease of activity in the zero direction. [From Hoffmann (1983b).]

It was argued by Schoppmann (1981) that the same corticofugal axons may provide the input to NOT and to the superior colliculus (SC) by axon collaterals. Despite this, decortication and monocular deprivation have quite different effects in NOT and SC. Cells in SC driven by the deprived eye lose their specificity for the direction of stimulus movement just as after decortication. There is only very little change in the velocity preference, and most cells in both colliculi are driven by the nondeprived eye (Hoffmann and Sherman, 1974). Nevertheless, the changes in both structures, NOT and SC, have to be explained by modifications in the cortical influence. Ipsilateral to the deprived eye a simple explanation can be given for the changes. The deprived eye lost its cortical connections, and the retinal influence is predominantly crossed. All neurons in the NOT as well as SC are controlled by the nondeprived eye and, as a consequence, show normal properties. Why are the changes contralateral to the deprived eye so different in NOT and SC? In the colliculus contralateral to the deprived eye most cells are controlled by the nondeprived (ipsilateral) eye. These cells show normal properties. Cells connected to the deprived (contralateral) eye are abnormal. In the NOT of most monocularly deprived cats all cells are connected to the deprived (contralateral) eye. The ipsilateral cortical input, which should be able to mediate the influence of the nondeprived (ipsilateral) eye, has not developed its normal function. The anatomical substrate for this corticopretectal projection is still present and can be demonstrated with electrical stimulation of the corticofugal fibers that leads to postsynaptic action potentials in NOT cells. But this cortical projection to NOT cannot change the rate of action potentials in response to visual stimulation. At least two hypotheses to explain these results can be suggested:

1. The deficit is to be found in the cells of the visual cortex. This explanation is difficult to reconcile with the different results in NOT and SC of monocularly deprived cats if the same axons innervate both structures. The cortical cells must be functioning properly to control the SC ipsilateral to the nondeprived eye. In addition, in a preliminary experiment in one cat we tested cells in the visual cortex ipsilateral to the nondeprived eye with random-dot patterns, that is, the same stimulus that failed to drive the ipsilateral NOT cells. We found a normal percentage of cells in cortex responding to this pattern in the usual direction-specific way. If we assume that this population contains the corticopretectal cells of Schoppmann (1981), another hypothesis must be considered.

2. If normally functioning cortical cells send their axons to NOT, the disruption must be in the corticopretectal synapse. There is an innate predominantly crossed retinal projection to the NOT cells in the newborn cat. Following Hebb's paradigm (Hebb, 1949), the corticopretectal synapses can only increase their efficacy if the postsynaptic activity in NOT cells is modulated by this retinal input. Then, during the sensitive period, the cortical input is tuned to the direction specificity prespecified in the NOT by the retina. If, however, the retinal influence is blocked by a closed lid during the sensitive period and NOT cells discharge all the time at their spontaneous rate, the cortical axon terminals make wrong or only subthreshold connections in the NOT. The nondeprived ipsilateral eye remains unable to drive the NOT cells under the condition of natural visual stimulation. We have not even found a modulatory influence when monocular contralateral and binocular stimulation are compared.

Effects of Strabismus

Rendering kittens strabismic by surgical section of the medial rectus results again in severe disruptions of binocular connectivity in the NOT (Cynader and Hoffmann, 1981). Figure 5 indicates the proportion of monocularly versus binocularly influenced cells in the NOT of normal and strabismic cats. In the normal cat, described in Figure 5A, nearly one-half of the units encountered could be driven by visual stimuli presented through either eye (Hoffmann and Schoppmann, 1981) with the remaining units influenced only by visual stimuli presented through the contralateral eye. In the strabismic animals (Figs. 5D,E) virtually all the units encountered could be driven only via stimulation of one eye, namely, the eye contralateral to the NOT under study. Binocular input onto single cells was virtually abolished. Each eye became the sole source of input to the NOT on the opposite side of the brain. In the left NOT the dominant eye was the deviating eye, and in the other NOT the unoperated eye provided the sole route for visual stimulation.

Apart from the disruptions of binocular connectivity described above, response properties of NOT cells in normal and strabismic cats appeared similar. As in normal cats, single cells were characterized by large receptive fields and a preference for large textured stimuli. All cells recorded in the NOT of the strabismic cats responded best with horizontal movement from temporal to nasal and showed inhibition when the stimulus moved in the opposite direction. Maximum excitatory response occurs at lower stimulus velocities (near 1°/sec) than normally, and the units' ability to distinguish the two directions of motion declines steadily once stimulus velocity increases beyond a few degrees per second.

The loss of input from the ipsilateral eye in strabismic cats appears to be attributable again to a partial or complete functional suppression of the pathway from the visual cortex to the NOT. In general, suppression of input from one pathway during development occurs when that pathway is placed at a competitive disadvantage relative to another input pathway, for example, when one eye only is deprived of vision (Cynader, 1979; Hubel and Wiesel, 1970). However, in these experiments we have shown that input from one eye is suppressed on the ipsilateral side of the brain and enhanced relative to that on the opposite side of the brain even though both eyes are used for vision (Cynader and Harris, 1980) and no functional suppression of inputs from either eye is observed in the striate cortex (Hubel and Wiesel, 1965).

The functional suppression of ipsilateral eye responses in the NOT of the strabismic cats may be a consequence of unequal strength of input from the two eyes in normal cats. Input from the contralateral eye reaches the NOT via a direct pathway from the retina while that of the ipsilateral eye takes a route passing through the visual cortex and the strength of these two functional pathways is unequal. In the strabismic cats the different directions of view of the two eyes must cause a lack of spatial and temporal congruity of inputs reaching the NOT via these two pathways. It appears as though incongruity of input via two pathways of unequal strength is a sufficient condition for the functional suppression of the weaker pathway. Thus, even though each eye reaches adequate stimulation, the relatively weaker pathway from the ipsilateral eye to the NOT via the visual cortex becomes nonfunctional in the strabismic animals.

SPECULATIONS ABOUT ONTOGENY OF RECEPTIVE FIELD PROPERTIES IN THE NOT

We can now incorporate these findings in a general model trying to explain the formation of a direction-specific pathway (see Fig. 7). This model follows closely the summary made recently by Fawcett and O'Leary (1985) of the role of electrical activity in the formation of topographic maps in the nervous system (Changeux and Danchin, 1976; Cowan and O'Leary, 1984). Basically, our assumption is that survival of retinal or cortical projections or consolidation of their synapses is dependent on the availability of a critical amount of a trophic survival factor. As a model one could take some of the known actions of nerve growth factor (NGF), which is a trophic agent essential for the survival of sympathetic neurons (Levi-Montalcini, 1964; Thoenen et al., 1972; Yankner and Shooter, 1982). This survival factor is located in the target cells, and the amount released is strongly dependent on the degree of depolarization of the target cells. Many inputs with the same properties releasing their transmitter at the same time will depolarize the target cell more strongly and as a consequence receive more trophic survival factor than a few inputs with different properties, especially when they discharge asynchronously.

During early development axons of direction-selective retinal ganglion cells that carry signals about horizontal temporonasal movements ter-

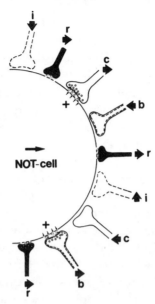

FIGURE 7. A possible mechanism for the formation of direction-specific connections in the nucleus of the optic tract (NOT) based on the assumption that survival of retinal or cortical projections or consolidation of their synapses is dependent on the availability of a critical amount of a trophic survival factor. This factor is released from the target cell upon strong depolarization. The NOT cell prefers movements in the direction of the arrow due to a prespecified input from direction-selective retinal ganglion cells preferring the same direction in visual space and mainly originating from the contralateral eye (solid terminals, r). Terminals from cortical axons (open terminals), which increase their discharge frequency to movements in the preferred direction of the NOT cell and can be activated through the eye contralateral to the NOT, will receive enough survival factor. c, i, b: corticopretectal axon terminals that can be activated exclusively through the contralateral, ipsilateral, or both eyes. Only synapses of c and b terminals that code rightward movement will be consolidated in the right NOT (+).

minate in the dorsal terminal nucleus and NOT either due to genetic prespecification or an as yet unidentified selection process. The postsynaptic cell is subsequently strongly direction-selective for global movement or retinal slip in temporonasal direction; that is, the cell will be strongly depolarized with movements in temporonasal direction presented to the contralateral eye and remain uninfluenced or become hyperpolarized with movements in the nasotemporal direction. At birth direct retinal axons from the ipsilateral eye have a much weaker influence on NOT cells. Also, as has been shown for the rabbit retina (Oyster and Barlow, 1967), most on-center direction-selective ganglion cells prefer the opposite horizontal direction as the ipsilateral NOT (the temporonasal direction for the ipsilateral eye is equivalent to the nasotemporal in the contralateral eye). Thus, retinal terminals from the ipsilateral eye will mostly discharge when the target cell is not depolarized by its major input from the contralateral eye, and the ipsilateral retinal input present at birth may even be eliminated during this shaping process. The retinal connection from the contralateral eye is not experience dependent because rotation of that eye will also rotate the preferred direction of the NOT cells by the same amount (Hoffmann and Cynader, unpublished). Thus, a genetic retinopretectal specificity is present when later, during development (about 4–6 weeks after birth), the corticopretectal projection matures (Schoppmann, 1985). The selection of the correct cortical axon can occur after the following rule (see Figs. 1 and 7): Cortical cells strongly activated by the contralateral eye and projecting to the ipsilateral NOT will discharge their terminals at a higher probability in synchrony with retinal axons than cells strongly activated by the ipsilateral eye. In addition, only those cortical cells qualify whose preferred direction is identical to the preferred direction of the retinal axons terminating on the target cells. These corticopretectal terminals receive enough survival factor because this substance is released by the target cell only upon strong depolarization due to the retinal input. The synapse formed by this arbor will be consolidated. The other synapses remain unsupported and thus subthreshold. This process will select cortical axons carrying the same direction-specific signal from the contralateral eye as the retinal axon. Other cells will be less successful depending on how much their preferred direction deviates from that of the retinal axons or how strongly they are influenced by the ipsilateral eye. Under the assumption that the activity of axons with information from the same retina is more correlated than the activity from different retinas, the ipsilateral retina will be connected to NOT cells only as an accompanist of the contralateral retina, that is, through binocular cortical cells.

CONCLUSIONS

This model could explain why in the normal cat we observe a dominance of contralateral input to NOT even through cortex. All NOT cells receive cortical input, but the contralaterally dominated corticopretectal axons are selected by their higher correlation with the retinal axons driving the NOT cells. This model could explain why in monocularly deprived cats the nondeprived eye can make no connections to the ipsilateral NOT. This structure is never depolarized enough by the retinal input in synchrony with the cortical activity because the specific contralateral retinal input is blocked by the lid suture. This model could also explain why in strabismic cats only contralateral input controls the NOT responses. The ipsilateral eye is connected to the NOT only through binocular cortical cells. The incongruity and almost complete separation of the pathways from the two eyes prevents the necessary correlated activity between the contralateral and ipsilateral eye to create this binocularity.

REFERENCES

Ballas, I., K.-P. Hoffmann, and H. J. Wagner (1981). Retinal projection to the nucleus of the optic tract in the cat as revealed by retrograde transport of horseradish peroxidase. *Neurosci. Lett.* **26**:197–202.

Changeux, J.-P., and A. Danchin (1976). Selective stabilization of developing synapses as a mechanism for the specification of neuronal networks. *Nature* **264**:705–711.

Collewijn, H. (1981). The Oculomotor System of the Rabbit and Its Plasticity. In: *Studies of Brain Function*, H. B. Barlow, H. Bullock, E. Florey, O. J. Grüsser, H. van der Loos, eds., Vol. 5, pp. 75–106, Springer-Verlag, New York.

Cowan, W. H. and D. D. M. O'Leary (1984). Cell Death and Process Elimination: The Role of Regressive Phenomena in the Development of Vertebrate Nervous System. In: *Medicine, Science and Society*, K. J. Isselbacher, ed., pp. 643–668, Wiley & Sons, New York.

Cynader, M. (1979). Competitive Interactions in Postnatal Development of the Kitten's Visual System. In: *Developmental Neurobiology of Vision*, R. D. Freeman, ed., pp. 109–120, Plenum Press, New York.

Cynader, M. and L. R. Harris (1980). Eye movement in strabismic cats. *Nature* **286**:64–65.

Cynader, M. and K.-P. Hoffmann (1981). Strabismus disrupts binocular convergence in cat nucleus of the optic tract. *Dev. Brain Res.* **1**:132–136.

Fawcett, J. W. and D. D. M. O'Leary, (1985). The role of electrical activity in the formation of topographic maps in the nervous system. *TINS* **8**:201–206.

Ferster, D. and S. Lindström (1983). An intracellular analysis of geniculo-cortical connectivity in area 17 of the cat. *J. Physiol. (Lond.)* 342:181–215.

Harris, L. R., F. Leporé, J. P. Guillemot, and M. Cynader (1980). Abolition of optokinetic nystagmus in the cat. *Science* 210:91–92.

Hebb, D. O. (1949). *Organization of Behavior.* Wiley & Sons, New York.

Hoffmann, K.-P. (1979). Optokinetic Nystagmus and Single-cell Responses in the Nucleus Tractus Opticus after Early Monocular Deprivation in the Cat. In: *Developmental Neurobiology of Vision*, R. D. Freeman, ed., pp. 63–72, Plenum Press, New York.

Hoffmann, K.-P. (1981). Neuronal Responses Related to Optokinetic Nystagmus in the Cat's Nucleus of the Optic Tract. In: *Progress in Oculomotor Research*, A. Fuchs and W. Becker, eds., pp. 443–454, Elsevier, North Holland.

Hoffmann, K.-P. (1983a). Control of the Optokinetic Reflex by the Nucleus of the Optic Tract in the Cat. In: *Spatially Oriented Behavior*, A. Hein and M. Jeannerod, eds., pp. 135–153, Springer-Verlag, New York.

Hoffmann, K.-P. (1983b). Effects of early monocular deprivation on visual input to cat nucleus of the optic tract. *Exp. Brain Res.* 51:236–246.

Hoffmann, K.-P. and H. P. Huber (1983). Responses to visual stimulation in single cells in the nucleus of the optic tract (NOT) during optokinetic nystagmus (OKN) in the awake cat. *Soc. Neurosci. Abstr.* 9:1048.

Hoffmann, K.-P. and A. Schoppmann (1975). Retinal input to direction selective cells in the nucleus tractus opticus of the cat. *Brain Res.* 99:359–366.

Hoffmann, K.-P. and A. Schoppmann (1981). A quantitative analysis of the direction-specific response of neurons in the cat's nucleus of the optic tract. *Exp. Brain Res.* 42:146–157.

Hoffmann, K.-P. and S. M. Sherman (1974). Effects of early monocular deprivation on visual input to cat superior colliculus. *J. Neurophysiol.* 37:1276–1286.

Hoffmann, K.-P. and J. Stone (1971). Conduction velocity of afferents to cat visual cortex: A correlation with cortical receptive field properties. *Brain Res.* 32:460–466.

Hoffmann, K.-P. and J. Stone (1985). Retinal input to the nucleus of the optic tract of the cat assessed by antidromic activation of ganglion cells. *Exp. Brain Res.* 59:395–403.

Hubel, D. H. and T. N. Wiesel (1965). Binocular interaction in striate cortex of kittens reared with artificial squint. *J. Neurophysiol.* 28:1041–1059.

Hubel, D. H. and T. N. Wiesel (1970). The period of susceptibility to the physiological effects of unilateral eye closure in kittens. *J. Physiol. (Lond.)* 206:419–436.

Lannou, J., L. Cazin, W. Precht, and M. LeTaillanter (1984). Responses of prepositus hypoglossi neurons to optokinetic and vestibular stimulations in the rat. *Brain Res.* **301**:39–45.

Levi-Montalcini, R. (1964). The nerve growth factor. *Ann. N. Y. Acad. Sci.* **118**:149–168.

Malach, R., N. P. Strong, and R. C. van Sluyters (1984). Horizontal optokinetic nystagmus in the cat: Effects of longterm monocular deprivation. *Dev. Brain Res.* **13**:193–205.

Markner, C. and K.-P. Hoffmann (1985). Variability in the effects of monocular deprivation on the optokinetic reflex of the non-deprived eye in the cat. *Exp. Brain Res.* **61**:117–127.

Oyster, C. W. and H. B. Barlow (1967). Direction-selective units in rabbit retina: Distribution of preferred directions. *Science* **155**:841–842.

Oyster, C. W., E. Takahashi, and H. Collewijn (1972). Direction-selective retinal ganglion cells and control of optokinetic nystagmus in the rabbit. *Vision Res.* **12**: 183–193.

Precht, W. and P. Strata (1980). On the pathway mediating optokinetic responses in vestibular nuclear neurons. *Neuroscience* **5**:777–787.

Precht, W., P. G. Montarolo, and P. Strata (1980). The role of the crossed and uncrossed retinal fibers in mediating the horizontal optokinetic nystagmus in the cat. *Neurosci. Lett.* **17**:39–42.

Schoppmann, A. (1981). Projections from areas 17 and 18 of the visual cortex to the nucleus of the optic tract. *Brain Res.* **223**:1–17.

Schoppmann, A. (1985). Functional and developmental analysis of a visual corticopretectal pathway in the cat: A neuroanatomical and electrophysiological study. *Exp. Brain Res.* **60**:363–374.

Simpson, J. I. (1984). The accessory optic system. *Ann. Rev. Neurosci.* **7**:13–41.

Singer, W., F. Tretter, and M. Cynader (1975). Organization of cat striate cortex, a correlation of receptive field properties with afferent and efferent connections. *J. Neurophysiol.* **38**:1080–1098.

Stone, J., and Y. Fukuda (1974). Properties of cat retinal ganglion cells: A comparison of W-cells with X- and Y-cells. *J. Neurophysiol.* **37**:722–748.

Strong, N. P., R. Malach, P. Lee, and R. C. van Sluyters (1984). Horizontal optokinetic nystagmus in the cat: Recovery from cortical lesions. *Dev. Brain Res.* **13**:179–192.

Thoenen, H., A. Saner, R. Kettler, and P. U. Angeletti (1972). Nerve growth factor and preganglionic cholinergic nerves: Their relative importance to the development of the terminal adrenergic neuron. *Brain Res.* **44**:593–603.

van Hof-van Duin, J. (1976). Early and permanent effects of monocular deprivation on pattern discrimination and visuomotor behavior in cats. *Brain Res.*
111:261–276.

van Hof-van Duin, J. (1978). Direction preference of optokinetic responses in
monocularly tested normal kittens and light deprived cats. *Arch. Ital. Biol.*
116:471–477.

Wood, C. C., R. D. Spear, and J. J. Braun (1973). Direction-specific deficits in
horizontal optokinetic nystagmus following removal of visual cortex in the
cat. *Brain Res.* **60**:231–237.

Yankner, B. A. and E. M. Shooter (1982). The biology and mechanism of nerve
growth factor. *Ann. Rev. Biochem.* **51**:845–868.

13

UNEQUAL ALTERNATING EXPOSURE: EFFECTS DURING AND AFTER THE CLASSICAL CRITICAL PERIOD

Helmut V. B. Hirsch, David G. Tieman, Suzannah B. Tieman, and Nina Tumosa

Neurobiology Research Center, State University of New York, Albany, New York

In long-lived species, normal behavioral development requires appropriate sensory stimulation. Such experience-dependent development introduces a degree of vulnerability. The young organism is dependent on certain types of experience in order to successfully complete neuronal and behavioral development. For example, rearing birds in isolation (see Bateson, this volume) or withholding song stimulation (see Marler, this volume) affects imprinting and song learning.

In general, stimulation is most likely to exert a long-term effect if it occurs at a particular stage in development—the so-called sensitive or critical period. Often, however, some of the behaviors affected by the experience do not develop until long after the necessary stimulation is no longer present; the animal's early experience may thus influence the course of subsequent behavioral development.

It has been suggested that plasticity of the mammalian visual system may provide a good model for studying mechanisms involved in experience-dependent development (e.g., Bischof, 1983); much of the evidence supporting this involves experience-dependent changes during the critical period. Less is known about the effects of visual experience during the critical period on behavioral and neuronal development *after* the critical period. In this chapter we describe some aspects of experience-dependent development in the mammalian nervous system, focusing on changes occurring during the critical period that have significant effects on later development.

Development of the mammalian visual system is influenced by environmental stimulation (for reviews see Movshon and Van Sluyters, 1981; Sherman and Spear, 1982; Frégnac and Imbert, 1984; Hirsch, 1985a,b), especially during a relatively brief postnatal period, which, in the cat, begins around the end of the third week and ends gradually by the twelfth to sixteenth week (e.g., Hubel and Wiesel, 1970; Olson and Freeman, 1980). Before or after this time, abnormal experience, such as monocular deprivation, has little or no effect, whereas at the height of the critical period, as little as six hours of monocular deprivation can be effective (Peck and Blakemore, 1975). The critical period coincides with a period of normal maturation. At the onset, many cells in the visual cortex are sluggish and immature; by its end or earlier, the majority have acquired their adult response properties (Pettigrew, 1974; Blakemore and Van Sluyters, 1975; Frégnac and Imbert, 1978).

Among these response properties is the ability to detect small differences in the relative position of the images on the two retinas (Pettigrew et al., 1968). These differences, called retinal disparity, provide a basis for perceiving the distribution of surfaces and their borders in three dimensions. For a cell to be able to detect retinal disparity, it must be binocular, that is, able to respond to stimulation of either eye. Most cells in the visual cortex of the adult cat are binocular, although many respond more strongly to one eye or the other; each eye dominates roughly equal numbers of cells so that the "ocular dominance distribution" is symmetric (Hubel and Wiesel, 1959, 1962).

Development of disparity selectivity requires normal visual experience (Pettigrew, 1974), and abnormal early visual experience can drastically alter both the binocularity of cortical cells and the distribution of their ocular dominance. A loss of binocularity occurs if the two eyes do not receive simultaneous, concordant stimulation (e.g., Hubel and Wiesel, 1965; Blake and Hirsch, 1975; Van Sluyters and Levitt, 1980); a bias in the proportion of cells activated by each eye occurs if the exposure of the two eyes is asymmetric or unequal (Wiesel and Hubel, 1963b, 1965; Blakemore and Eggers, 1978; Ikeda and Tremain, 1978; Maguire et al., 1982; D. Tieman et al., 1979, 1983a,b). By studying the course of development under conditions of abnormal stimulation, it is possible to discover some of the mechanisms by which visual experience influences normal development. One such mechanism appears to involve a competitive interaction between afferents from the two eyes; this "binocular competition" was inferred from the effects on the developing visual system of an imbalance in stimulation of the two eyes (Wiesel and Hubel, 1965; Guillery, 1972).

An imbalance in stimulation has most often been produced by suturing together the lids of one eye (e.g., Wiesel and Hubel, 1963b, 1965). This results in an imbalance in the amount of stimulation of the two eyes, but it also entails continuous deprivation of one eye. Some consequences of monocular lid suture result from the imbalance in stimulation, which is thought to bias the outcome of binocular competition. Other consequences, however, result because one eye is deprived of normal patterned visual stimulation; such stimulation is required for normal development since cells in visual cortex remain immature or develop abnormally if animals are reared in total darkness or have both eyelids sutured closed (Hubel and Wiesel, 1965; Pettigrew, 1974; Blakemore and Van Sluyters, 1975; Leventhal and Hirsch, 1977, 1980; Frégnac and Imbert, 1978; Watkins et al., 1978; Bonds, 1979; Derrington, 1984).

Although monocular lid suture confounds an imbalance in stimulation with continuous deprivation of one eye, it has been possible to separate out some of the consequences of an imbalance in stimulation using two criteria for identifying consequences of binocular competition (discussed by Sherman and Spear, 1982). First, changes that are observed only in those portions of the visual pathways that normally receive input from the binocular segment of the visual field (see Fig. 1) most likely result from competition (Guillery and Stelzner, 1970). For example, when tested with the deprived eye, monocularly lid-sutured cats orient to targets in the monocular portion of the visual field but not to targets in the binocular portion of the visual field (Sherman, 1973, 1974a; Tumosa et al., 1982). This is true even for an artificial monocular segment, created by placing a lesion in the retina of the experienced eye (Sherman et al., 1974). Second, changes that cannot be predicted by the amount and quality of stimulation received by each eye probably also result from competition. For example, monocular lid suture, in which the experienced eye has an advantage, disrupts vision through the deprived eye more than binocular lid suture, in which neither eye has an advantage (Wiesel and Hubel, 1965; Ganz et al., 1972; Lehmkuhle et al., 1982). These two criteria help to separate those effects of monocular lid suture that reflect binocular competition from those that reflect deprivation.

We have developed a procedure for producing an imbalance in the activity along the pathways from the two eyes that has two important advantages over monocular lid suture (Tumosa et al., 1980a, 1982): (1) It eliminates the confounding between continuous pattern deprivation of one eye and the imbalance in stimulation of the two eyes, and (2) it allows the magnitude of the imbalance to be systematically varied.

Kittens are reared by allowing each eye patterned visual input, but on alternate days and for different periods of time (unequal alternating monocular exposure or unequal AME). To vary the magnitude of the imbalance in stimulation, the relative duration of exposure is changed: the more experienced eye receives 8 hr of exposure every other day, and the less experienced eye receives 1 hr (AME 8/1), 4 hr (AME 8/4), or 7 hr (AME 8/7) of exposure on the intervening days. As controls, animals receive equal periods of exposure with each eye, again on alternate days (1 hr/day, AME 1/1; or 8 hr/day, AME 8/8). The rearing procedures are described in detail in Tumosa et al. (1982).

In this chapter we describe the effects that unequal AME has on the developing visual system. We show that these effects satisfy both of the criteria used to identify consequences of binocular competition, and we

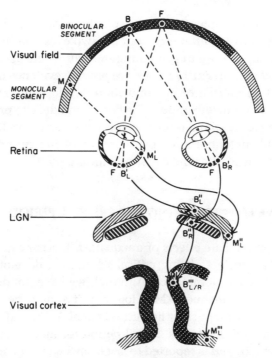

FIGURE 1. Diagram of the pathway from retina to visual cortex in the cat. The pathways for the left and right eyes are indicated by lines slanting right and left, respectively. The heavy lines indicate the ipsilateral pathways, and the light lines indicate the contralateral pathways. The two eyes are converged on a fixation point F in the visual field, the image of which falls at F on each retina. Information from the visual field is inverted by the lens onto the retinas such that a point M in lateral, monocular visual field maps onto the medial nasal retina of one eye (at M'_L), and a point B, in the central binocular visual field maps onto more central retina in both eyes (at B'_L and B'_R). The information is then relayed to the LGN, where it forms a topographic map of the contralateral visual field, so that points in the left visual field are mapped onto the right LGN, and vice versa. The inputs to the two eyes project to separate laminas. Lamina A receives input from the nasal retina of the contralateral eye; lamina A_1 receives input from the temporal retina of the ipsilateral eye. From medial to lateral in the LGN corresponds to points in visual space from the midline (0°) to 90° lateral. Thus, the monocular part of the visual field is represented at the lateral edge of lamina A (at M''_L).

Axons from cells in the different layers of the LGN (B''_L and B''_R) then converge onto a single region of visual cortex ($B'''_{L/R}$) so that each hemisphere contains only one representation of the contralateral visual field. Because the inputs from the two eyes converge, most cells in cortex are binocular.

describe the implications that our results have concerning the competitive struggle between afferents from the two eyes and the role of competition and deprivation in generating the acuity deficits seen following monocular lid suture. Furthermore, we provide evidence that some consequences of an imbalance in stimulation develop only after both eyes are allowed simultaneous input: The behavioral deficits produced by an early imbalance in stimulation become worse as a result of binocular exposure. This deterioration, apparently involving suppression of one eye, can occur well after the classical critical period.

COMPETITIVE EFFECTS DURING THE CRITICAL PERIOD

Many of the immediate effects of unequal AME occur during the period identified previously as being critical for normal development of binocularity and ocular dominance. Although we have not determined the critical period for unequal AME, logically it should be related to the critical periods for monocular lid suture, equal AME, and, perhaps also, for strabismus. Of these, the critical period for monocular lid suture is best documented, and is reported to last from 3 weeks of age to 16 weeks of age, with the maximum sensitivity occurring between 4 and 6 weeks postnatal (Hubel and Wiesel, 1970; Blakemore and Van Sluyters, 1974; Movshon, 1976; Olson and Freeman, 1980). Although not as well characterized, the critical periods for both AME (Presson and Gordon, 1979) and strabismus (Yinon, 1976; Ikeda et al., 1978; Levitt and Van Sluyters, 1982) are reported to be very similar to those for monocular lid suture. We are thus quite confident that the critical period for unequal AME follows a similar course.

Visual Fields

A normal cat, when tested monocularly for its ability to orient to targets in space (Fig. 2A), will respond to stimuli between 30° nasal and 90° temporal; thus the central 60° can be seen with either eye (Fig. 2B; also Sherman, 1973; Tumosa et al., 1980a, 1982). A cat reared with equal AME, no matter what the duration of the daily periods of exposure (24, 8, or 1 hr) has normal visual fields (Fig. 2B; also Tumosa et al., 1980a, 1982). A cat reared with an imbalance in stimulation, however, shows visual field deficits when tested with its less experienced eye: The greatest deficits are within the binocular portion of the visual field, and the size of the deficit is related to the degree of the imbalance in exposure. The deficits in monocularly lid-sutured cats are the most severe; when tested with the deprived eye alone, these cats fail to respond to targets presented anywhere within the binocular portion of the field (Fig. 2B; also Sherman, 1973; Tumosa et al., 1982, D. Smith et al., 1982a; D. Tieman et al., 1983b; but cf. van Hof-van Duin, 1977; Heitländer and Hoffmann, 1978, who observed some responses in the temporal half of the binocular field). The deficit is smaller in unequal AME cats, involving only the nasal portion of the binocular field. The deficit in animals reared with a large or moderate imbalance in stimulation (AME 8/1 or AME 8/4) encompasses the entire nasal field as early as 8 weeks of age (Fig. 2B). This deficit could result from a failure of development; Sireteanu and Maurer (1982) have shown that very young kittens respond poorly to targets in the nasal hemifield. However, our results in AME 8/7 cats show that unequal AME can cause a progressive loss of responsiveness in the nasal hemifield. When tested at 8–10 weeks, AME 8/7 cats show a deficit in only part of the nasal field (Fig. 2B). When the rearing is continued to 14–15 weeks, however, the nasal field deficit becomes complete and equal to that seen in AME 8/1 and AME 8/4 cats (Tumosa et al., 1983). Thus, the effects of unequal AME appear to be cumulative, and, at least in AME 8/7, the nasal field deficit appears to involve something more than a simple failure of development.

Competition, not deprivation, causes the nasal field deficit in unequal AME cats. The deficits are restricted to the binocular part of the visual field, and they are not seen in cats that receive equal AME, even when each eye receives as little exposure as does the less experienced eye of the AME 8/1 cats (e.g., in AME 1/1 cats; see Fig. 2B). The nasal field losses are therefore caused neither by insufficient exposure nor by a lack of concordant inputs, but by the imbalance in stimulation of the two eyes and, thus, by binocular competition. The fact that the deficits are restricted to the nasal portion of the binocular visual field suggests that the ipsilateral pathway from retina to visual cortex, which subserves the nasal visual field, is at a competitive disadvantage. This interpretation is consistent with our physiological and anatomical observations (see below).

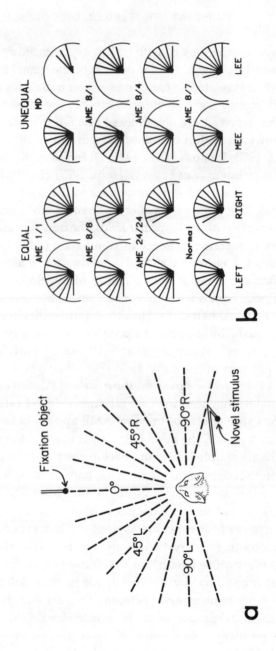

Primary Visual Cortex

Following a prolonged period of unequal AME, most cells in visual cortex can be activated by stimulation of only one eye, usually the more experienced eye (Fig. 3; D. Tieman et al., 1979, 1983a; D. Smith et al., 1982b). The Ame 8/1 cats show a greater shift in ocular dominance than the AME 8/4 cats (Fig. 3; D. Tieman et al., 1983a). Further, neither group shows as large a shift as is commonly seen in monocularly lid-sutured cats (Wiesel and Hubel, 1963b, 1965). Thus, the size of the shift in ocular dominance is a function of the size of the imbalance in stimulation of the two eyes imposed during rearing.

The size of the imbalance in stimulation required to produce a shift in ocular dominance may be a function of the relative timing of the exposure of the two eyes. Christen et al. (1983) report that they observe a shift in ocular dominance with an imbalance of 7 to 1 only if the two eyes are exposed on separate days (as they are in our experiments and those of D. Smith et al. (1982b; Reeves et al., 1982)), and not if both eyes are exposed on the same day, although still separately. They succeeded in producing a shift in ocular dominance by exposing both eyes separately on the same day, but only by increasing the imbalance to 50 to 1

FIGURE 2. Visual perimetry in cats. (A) Method of testing visual fields. The cat was restrained with its lateral canthi aligned along the 90° guidelines and its nose pointed along the 0° guideline toward the fixation object (a piece of food on a wire). A novel stimulus (a piece of food on another wire) was then introduced along one of the guidelines. A positive response was recorded, when, upon being released, the cat turned and immediately approached the novel stimulus. A negative response was recorded if the cat approached the fixation object or if it scanned the field before approaching the novel stimulus. If the animal turned toward the novel stimulus but then turned back toward and approached the fixation object, the response was recorded as plus/minus since the animal apparently saw the novel stimulus but chose not to respond normally to it. [Reprinted with permission from Tumosa et al. (1982).] (B) Visual fields for each eye of cats reared with equal (left) or unequal (right) exposure to the two eyes. The fields are represented by polar plots showing the responses to stimuli presented at every 15° of the visual field. The semicircle represents a level of 12 positive responses in 12 trials. Averaged data are shown for 5 normal cats, 1 AME 24/24 cat, 4 AME 8/8 cats, 9 AME 1/1 cats, 2 monocularly deprived cats, 20 AME 8/1 cats, 14 AME 8/4 cats, and 5 AME 8/7 cats (tested at 10 weeks). The data are plotted as if the less experienced eye were always the right eye, although for many animals it was the left. For all the cats reared with equal exposure, the fields of the two eyes are normal. For all cats reared with unequal exposure, the field of the more experienced eye is normal, but the field of the less experienced eye is restricted, and the degree of this restriction is a function of the size of the imbalance in exposure. Data from Tumosa et al. (1980, 1982, 1983), S. Tieman and Hirsch (1983), and unpublished observations.

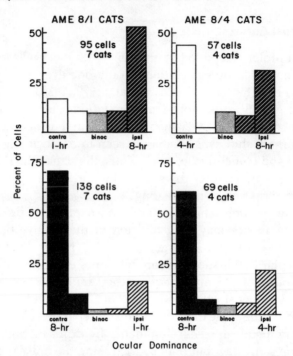

FIGURE 3. Histograms showing the proportion of cells in primary visual cortex (area 17) activated by each eye of AME 8/1 cats and AME 8/4 cats. Cells in the bin labeled contra were activated only by the contralateral eye; cells in the bin labeled ipsi were activated only by the ipsilateral eye; cells in the bin labeled binoc were activated equally by both eyes; cells in the remaining two bins were activated by both eyes, but in one case the contralateral eye was dominant and in the other case the ipsilateral eye was dominant. Cells recorded from the cortical hemisphere contralateral to the less experienced eye are shown above, and those from the hemisphere ipsilateral to the less experienced eye below. Darker bars represent cells dominated by the 8-hr eye and lighter bars represent cells dominated by the 1-hr eye. Solid bars represent cells driven predominantly by the contralateral pathway and striped bars represent cells driven predominantly by the ipsilateral pathway. In both hemispheres of the AME 8/1 cats and in the hemisphere ipsilateral to the 4-hr eye of the AME 8/4 cats, most cells are dominated by the 8-hr eye. Reprinted with permission from D. Tieman et al. (1983a).

(Christen and Mower, 1985b). We have not reared any unequal AME cats where each eye receives exposure separately but on the same day, but we have reared cats that received 1 hr of binocular exposure followed by 7 hr of monocular exposure each day. This imbalance produces a bias similar to that seen in AME 8/4 cats (S. Tieman et al., unpublished observations). Thus, our form of within-day 8/1 imbalance does produce an ocular dominance shift, although it is not as large as the shift that we see after our usual between-day 8/1 imbalance.

The shift in ocular dominance produced by unequal AME is presumably caused by a decreased spread of the geniculocortical afferents representing the less experienced eye. Tumosa et al. (1980b; Tumosa, 1982) used transneuronal transport after intraocular injection of [^{3}H]proline to label the regions of visual cortex that receive input from one eye in cats reared with unequal AME. They found that the zones of afferent input corresponding to the less experienced eye are smaller than those corresponding to the more experienced eye (Fig. 4A). Thus, afferents from the more experienced eye have access to more of visual cortex than do those from the less experienced eye.

When recording responses of single cells in visual cortex, the effects of unequal AME are more apparent ipsilateral to the less experienced eye than contralateral to it, suggesting that the contralateral pathway from retina to visual cortex has some advantage in the competition for access to cortical cells. If only the imbalance in inputs determined which pathways are at an advantage, then one would expect to find the same relative difference in the numbers of cells activated by the two eyes in each of the cortical hemispheres. However, the relative dominance of the more experienced eye is greater in the hemisphere contralateral to the more experienced eye. Both a large (AME 8/1) and a moderate (AME 8/4) imbalance in stimulation produce a dominance by the more experienced eye in the hemisphere contralateral to that eye; only a large imbalance in stimulation (AME 8/1) produces dominance of the more experienced eye in both hemispheres (Fig. 3). The contralateral pathway from retina to visual cortex must have an intrinsic advantage (e.g., earlier development) that is sufficiently strong to overcome a moderate but not a large imbalance in stimulation.

Cells dominated by the less experienced eye are distributed throughout the depths of the visual cortex in both hemispheres of unequal AME cats: During physiological recordings, cells activated by the less experienced eye were no more likely to be encountered in layer IV, where most geniculocortical afferents end, than in other layers (D. Tieman et al., 1983a). This is in sharp contrast to the laminar distribution of cells activated by the deprived eye in monocularly lid-sutured cats; such cells are more likely to be recorded in layer IV than in other layers (Shatz and Stryker, 1978).

Comparable results have been found by Tumosa and Tieman (1981;

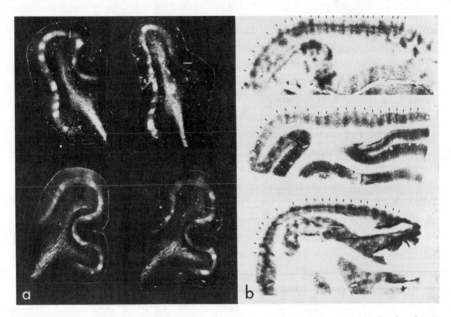

FIGURE 4. Anatomical correlates of the changes in ocular dominance. (A) Ocular dominance patches in layer IV of visual cortex in AME 8/1 cats. Tritiated proline was injected into one eye of AME 8/1 cats in order to label transneuronally the geniculocortical projection corresponding to that eye. These are dark-field photographs of autoradiograms of coronal sections through the visual cortex of one animal in which the 8-hr eye was injected (above), and one in which the 1-hr eye was injected (below). In each case, the sections are from the hemisphere ipsilateral to the injected eye. The patches for the 8-hr eye are larger than those for the 1-hr eye. [Reprinted with permission from Tumosa (1982).] (B) Distribution of cells activated by stimulation of one eye in the primary visual cortex (area 17) of two AME 8/1 cats and one normal cat as demonstrated with the metabolic marker [^{14}C]-2-deoxyglucose. [^{14}C]-2-deoxyglucose was administered intravenously to cats in which one eye had previously been removed. After 45 min, the cat was deeply anesthetized, and perfused. The brains were frozen, sectioned in a cryostat, and the sections applied to X-ray film to obtain the images shown here. For details of the method, see S. Tieman and Tumosa (1983). Columns of cells activated by one eye are indicated by tic marks. Sections are from the hemisphere ipsilateral to the stimulated eye. Top: Columns of cells labeled by stimulation of the 1-hr eye. Middle: Columns of cells labeled by stimulation of one eye of a normal cat. Bottom: Columns of cells labeled by stimulation of the 8-hr eye. Note that the columns of cells for the 1-hr eye are narrower than normal, whereas those for the 8-hr eye are wider than normal; in all cases columns extend throughout the layers of the visual cortex. [Reprinted with permission from Tumosa (1982).]

Tumosa, 1982) using the [^{14}C]-2-deoxyglucose technique to label ocular dominance columns. They found that the size of ocular dominance columns was affected by both monocular lid suture and unequal AME, but that only monocular lid suture changed the shape of the columns. The columns for the less experienced eye are smaller than normal in both kinds of cats; in monocularly lid-sutured cats, but not in AME 8/1 cats, these columns are also often restricted to layer IV (Bonds et al., 1980; Tumosa and Tieman, 1981; Tumosa, 1982; Kossut et al., 1983; S. Tieman, 1985). Some of these results are illustrated in Figure 4B.

Unlike the cortical cells dominated by the deprived eye in a monocularly lid-sutured cat (Wiesel and Hubel, 1963b, 1965), the cortical cells that are dominated by the less experienced eye in an unequal AME cat respond relatively normally to visual stimulation (D. Tieman et al., 1979, 1983a; Reeves et al., 1982; D. Smith et al., 1982b). Even after a large imbalance, that is, AME 8/1, the cells activated by the less experienced eye are orientation selective; only subtle differences can be detected between these cells and those that are activated by the more experienced eye (D. Tieman et al., 1983a).

The cortical effects of unequal AME differ, therefore, from those of monocular lid suture: (1) Significant numbers of orientation-selective cells can be activated by the less experienced eye, and (2) cells activated by the less experienced eye are present in all cortical layers. This suggests that an imbalance in stimulation of the two eyes does not, by itself, interfere with the development and/or maintenance of orientation selectivity of cells; it is only when that imbalance is coupled with continuous deprivation of one eye that few if any cells activated by the less experienced eye retain normal orientation selectivity. Furthermore, although monocular lid suture decreases the ability of cells in layer IV to activate cells outside layer IV, this is not due to the imbalance in stimulation alone since unequal AME does not have this effect. This disruption in interlaminar connectivity also cannot be explained by the deprivation alone since visually responsive cells are likely to be present throughout all layers of the visual cortex in binocularly lid-sutured cats. An imbalance in stimulation must therefore exacerbate effects of continuous deprivation such that the combination affects (1) the selectivity of cells in visual cortex and (2) the spread of activity from layer IV to the other layers of the visual cortex.

Lateral Geniculate Nucleus

Although monocular lid suture causes changes in the physiology of cells in the lateral geniculate nucleus (LGN; Sherman et al., 1972, 1975; Lehmkuhle et al., 1978, 1980; Sireteanu and Hoffmann, 1979), analogous changes have not been observed following unequal AME. Monocular lid suture causes changes in both the spatial acuity of individual cells and in the proportions of different cell types that can be recorded from the LGN: Unequal AME causes no detectable changes in either of these (Christen et al., 1983).

The spatial resolution of one cell type, the "X cells", is reduced in those layers of the LGN that receive input from the deprived eye of a monocularly lid-sutured cat (Lehmkuhle et al., 1978, 1980; Sireteanu and Hoffmann, 1979). In contrast, in those layers of the LGN that receive input from the less experienced eye of an unequal AME cat, the X cells are physiologically normal (Christen et al., 1983). The fact that X-cell acuity is normal in unequal AME cats is perhaps not surprising, given that neither dark rearing (Mower et al., 1981; Kratz, 1982) nor monocular occlusion (Christen and Mower, 1985a) affect X-cell acuity: As suggested by Ikeda and Tremain (1978), a reduction in X-cell acuity appears to require diffuse stimulation of one eye, such as is produced by light coming through the closed eyelids or through a diffusing lens (Christen and Mower, 1985a,b).

In addition to the change in spatial resolution of the X cells, there is a change in the number of "Y cells" that can be recorded in the layers of the LGN that receive input from the deprived eye of monocularly lid-sutured cats (Sherman et al., 1972, 1975). In contrast, normal percentages of Y cells are recorded from those layers of the LGN that receive their input from the less experienced eye of an unequal AME cat (Christen et al., 1983). This normality suggests that a decrease in the number of Y cells recorded is caused not by competition but by deprivation. This interpretation is consistent with the observation that depriving both eyes, whether by dark rearing or binocular lid suture, decreases the percentage of Y cells recorded in the LGN (Sherman et al., 1972; Kratz et al., 1979; Sherman and Spear, 1982). These deprivation effects may be exacerbated by competition since fewer Y cells are recorded from binocular portions than from monocular portions of those LGN layers getting input from the deprived eye of a monocularly lid-sutured cat (Sherman et al., 1972; Sherman and Spear, 1982).

Although no physiological changes have been seen in the LGN of unequal AME cats, there are anatomical changes (S. Tieman et al., 1984). Unequal AME causes the cells in the layers that receive their input from the less experienced eye to be smaller than normal. This shrinkage is presumably secondary to the changes in the geniculocortical projection described above.

Measurements of the soma size of LGN cells provide the clearest evidence that unequal AME affects binocular competition between afferents from the two eyes. First, unequal AME has an effect on cells in regions where competition can logically occur: In the binocular segments of the nucleus, cells in layers receiving input from the less experienced eye are smaller than those in layers receiving input from the more experienced eye. In contrast, unequal AME has no effect on cells in regions that are protected from competition: In the monocular segments of the LGN, cells that receive input from the less experienced eye do not differ in size from those that receive input from the more experienced eye. Second, the effects of an imbalance in stimulation of the two eyes cannot be explained by the stimulation received by each eye separately. In AME 8/1 cats, the cells in the layers of the LGN receiving input from the 1-hr eye are smaller than those in the layers receiving input from the 8-hr eye, yet cells in AME 1/1 cats are no smaller than those in AME 8/8 cats. The effects of unequal AME are therefore greater than could be predicted from the total duration of exposure given to the two eyes, and so must reflect the imbalance in the exposure. Thus, the effects of unequal AME on the sizes of cells in the LGN satisfy both criteria for identifying the effects of binocular competition.

Measurements of cell size in the LGN also confirm that the effects of the unequal AME are greater in the hemisphere ipsilateral to the less experienced eye than in the hemisphere contralateral to it. A similar, but less marked, difference between the ipsilateral and contralateral pathways is sometimes observed after monocular lid suture (Hickey et al., 1977; Hickey, 1980). It should be noted that although the effects of unequal AME are largely confined to the hemisphere ipsilateral to the less experienced eye, both the contralateral and ipsilateral pathways to that hemisphere are affected. The cells in lamina A1, which receives ipsilateral input from the less experienced eye, are smaller than normal, and the cells in the binocular segment of lamina A, which receives contralateral input from the more experienced eye, are larger than normal. Thus, in the LGN ipsilateral to the less experienced eye, the cells in the binocular, but not the monocular, segment of lamina A appear to hypertrophy, while the cells in lamina A1 shrink.

The pattern of results described here is true for both AME 8/1 cats and AME 8/4 cats, but the changes are smaller for the AME 8/4 cats than for the AME 8/1 cats (14% vs. 20% in the hemisphere ipsilateral to the less experienced eye), and in neither case were the changes as large as those previously reported for monocularly deprived cats (28–44%; Wiesel and Hubel 1963a; Guillery and Stelzner, 1970; Hickey et al., 1977). These results provide further evidence that the size of the effect is directly correlated with the size of the imbalance imposed during rearing and are consistent with the physiological and 2-deoxyglucose data on ocular dominance of cells in visual cortex (Wiesel and Hubel, 1965; Tumosa and Tieman, 1981; Tumosa, 1982; D. Tieman et al., 1983a; Christen and Mower, 1985b).

In summary, behavioral, anatomical, and physiological studies provide a coherent set of results that demonstrate that (1) unequal AME affects the developing visual system by altering the outcome of the competitive interaction between the afferents from the left and right eyes, (2) the contralateral pathway has an inherent advantage in this competition, and (3) the size of the effect is correlated with the size of the imbalance imposed during rearing.

We have used two manipulations to attempt to reverse these changes after the critical period. (1) Removing the more experienced eye is, as described below, effective; (2) providing unrestricted binocular exposure is not.

EFFECTS OF REMOVING THE MORE EXPERIENCED EYE

In monocularly lid-sutured cats, later removal of the experienced eye is accompanied by an expansion of the visual field of the deprived eye (D. Smith et al., 1982a) and by an increase in the ability of that eye to activate cells in both the cortex and the superior colliculus (Kratz et al., 1976; Hoffmann and Cynader, 1977). This increase may result from removal of tonic inhibition by inputs from the normal eye (Kratz et al., 1976; Burchfiel and Duffy, 1981) or from a form of "disuse supersensitivity" (Sharpless, 1975; Blakemore and Hawken, 1982).

We have found that, in AME 8/1 cats, as in monocularly lid-sutured cats, the visual field of the less experienced eye expands following removal of the more experienced eye (S. Tieman and Hirsch, 1983). Unlike monocularly lid-sutured cats (Heitländer and Hoffmann, 1978; Smith et al., 1982a), however, the AME 8/1 cats recover the ability to respond to targets in the nasal field (S. Tieman and Hirsch, 1983). This recovery presumably reflects changes in the retinogeniculocortical pathway or the corticocollicular pathway, since responses to nasal targets require an intact visual cortex (Sherman, 1974b; Sherman and Sprague 1979).

D. Smith et al. (1982b) report that the proportion of cells in visual cortex responding to activation of the 1-hr eye increases from 17 to 85% following removal of the 8-hr eye. This improved ability of the 1-hr eye to activate cortical cells could explain the improvement in the visual field observed by Tieman and Hirsch (1983).

It is unlikely that all of the changes produced by unequal AME are caused by tonic inhibition by afferents from the more experienced eye. First, the retinogeniculocortical pathway from the less experienced eye terminates over a smaller percentage of visual cortex (Fig. 4A; Tumosa et al., 1980b). Further, Tumosa and Tieman (1981; Tumosa, 1982) have shown in AME 8/1 cats that ocular dominance columns activated by the 1-hr eye are smaller than normal even after removal of the 8-hr eye (Fig. 4B). Thus, at least part of the changes produced by unequal AME involve decreases in the excitatory connections from the less experienced eye.

EFFECTS OF SUBSEQUENT BINOCULAR EXPOSURE

Information from the two eyes cannot be integrated into a single, three-dimensional representation of the visual world if the eyes are misaligned; disturbances of binocular visual perception such as diplopia and visual confusion are likely to result. Disruption of normal, concordant visual input to the two eyes during early postnatal life results in such misalignment of the eyes (e.g., Sherman, 1972; Blake et al., 1974; D. Tieman et al., 1983b). Alternating monocular exposure, both equal and unequal, produces a convergent strabismus (esotropia) (Blake et al., 1974; Packwood and Gordon, 1975; Tumosa et al., 1983), whose severity is not related to the presence or size of any imbalance in stimulation (Tumosa et al., 1983). This esotropia is probably a contributing factor to the changes described below.

The human clinical literature suggests that a patient's attempts to cope with disturbances in eye alignment can take one or two forms: (1) using first one eye and then the other (alternating fixation), or (2) using one eye to the exclusion of the other (uniocular fixation). In the latter case the eye that is not used becomes nonfunctional, that is, amblyopic (Duke-Elder and Wybar, 1973). Alternating monocular exposure forces an animal to alternate fixation. For unequal AME, the animal fixates with one eye more than the other and, thus, is forced to "favor" fixation with the more experienced eye. Two lines of evidence suggest that a cat reared with AME and a moderate or large imbalance in exposure will, when given unrestricted binocular exposure, continue to favor the more experienced eye to the extent of developing uniocular fixation.

Visual Fields

Animals were reared with AME, either equal or unequal, until they were between 3 and 9 months old, when they were given unrestricted binocular exposure. Visual fields were determined repeatedly for each eye separately and for both eyes together (Tumosa et al., 1983).

For cats given little or no imbalance during the initial rearing (AME 1/1 and AME 8/7), binocular exposure had no effect on the visual fields: Both the binocular and monocular fields remained unchanged throughout 4 months of binocular exposure. The visual fields of the AME 1/1 cat, which were normal after the initial rearing, remained normal (Fig. 5). For the AME 8/7 cats, the field of the 7-hr eye remained restricted after rearing, and the field of the 8-hr eye and of both eyes together remained normal.

FIGURE 5. Effects of subsequent binocular exposure on the visual fields of an AME 1/1 cat, an AME 8/1 cat, and an AME 8/4 cat. The exposure history of the animal is indicated above each visual field plot. The first number gives the animal's age in weeks at the end of AME, and the second number indicates the length of binocular exposure, also in weeks. Plus/minus responses are plotted as dashed lines. For the AME 1/1 cat, the fields are normal both immediately after the alternating exposure and throughout the 4 months of binocular exposure. For the AME 8/1 cat, both the binocular field and the monocular field of the 8-hr eye were normal in extent at the onset of binocular exposure, whereas that of the 1-hr eye was restricted to the temporal hemifield. With continued binocular exposure, both the binocular field and the field of the 1-hr eye became more restricted until eventually the binocular field was equivalent to the monocular field of the 8-hr eye, and the field of the 1-hr eye was restricted to the central 30° of the hemifield. Similar changes occurred in the fields of the AME 8/4 cat, whose binocular exposure did not start until the animal was 9 months old. The monocular fields of this cat were not tested at 5 and 7 weeks of binocular exposure, nor was the binocular field tested at 0 weeks. [Reprinted with permission from Tumosa et al. (1983).]

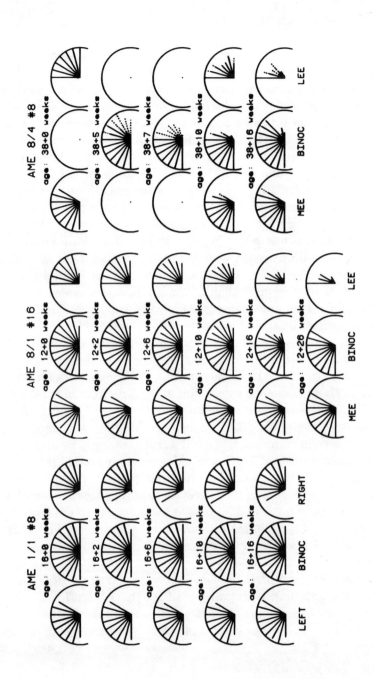

The results were very different for animals that received a larger imbalance in stimulation of the two eyes (i.e., AME 8/4 and AME 8/1). At the end of the alternating monocular exposure, the visual field of the less experienced eye was restricted to the temporal hemifield, and subsequent binocular exposure resulted in a further restriction of both the binocular field and the monocular field of the less experienced eye. This is illustrated in Figure 5 for one AME 8/1 cat that was reared with unequal AME for 12 weeks and one AME 8/4 cat that was reared with unequal AME for 9 months. After 26 weeks of subsequent binocular exposure, the binocular visual field of the AME 8/1 cat became equivalent to the monocular field of the 8-hr eye, suggesting that, when both eyes were open, this animal was not using its 1-hr eye at all. Furthermore, even when the animal was tested with the 1-hr eye alone, there was evidence of suppression: The field was restricted to the central 30° of the temporal hemifield. Similar, though slightly less dramatic, results were observed for the AME 8/4 cat after 16 weeks of binocular exposure. Thus, binocular exposure created an almost complete suppression of the less experienced eye, even when the onset of binocular exposure was delayed until the cat was 9 months old.

The time course of the development of suppression is illustrated in Figure 6. There was some suggestion in the data from binocular testing of an initial (4–6 week) period of improvement in visual fields, followed by a steady deterioration that proceeds at different rates depending on the imbalance in stimulation present during rearing. The animal's poor performance at the time it is first tested binocularly and shortly thereafter may reflect an initial confusion produced by its first experience with simultaneous exposure of the two eyes, which, since the animal is esotropic, is likely to result in diplopia. As the animal becomes familiar with this experience its performance may show a temporary improvement. Subsequent deterioration in performance is presumably due to suppression of the less experienced eye.

The amount of suppression seen after binocular exposure is a function of the imbalance imposed during rearing. The AME 8/1 cats show the greatest suppression, the AME 8/4 cats show intermediate levels of suppression, and the AME 8/7 and AME 1/1 cats show none. This is illustrated in Figure 7.

Optokinetic Nystagmus

The optokinetic responses of unequal AME cats given subsequent binocular exposure are consistent with suppression of one eye. We tested

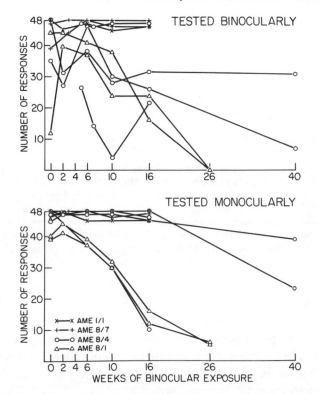

FIGURE 6. Suppression as a function of binocular exposure. The number of responses to targets presented 45°, 60°, 75°, and 90° ipsilateral to the less experienced eye is plotted on the ordinate (0° is straight ahead). Plus/minus responses are given half weight. Note that the number of responses decreased with time. This trend is clearer for the data obtained under monocular testing conditions (lower graph) than for the data obtained under binocular testing conditions (upper graph). [Reprinted with permission from Tumosa et al. (1983).]

optokinetic responses in a series of four AME 8/1 cats that had been given prolonged periods of binocular exposure after the end of the alternating monocular exposure (S. Tieman et al., 1985). Normal adult cats tested monocularly respond equally to nasal-to-temporal and temporal-to-nasal movement, whereas the AME 8/1 cats, like other cats with reduced binocularity (van Hof-van Duin, 1978; Cynader and Harris, 1980; Hoffmann, 1981; Malach et al., 1984), respond well to temporal-to-nasal movement and poorly or not at all to nasal-to-temporal movement. Normal cats tested binocularly respond equally to left-to-right movement and to right-to-left movement. Unlike the normal cats, or other cats reported in the literature (van Hof-van Duin, 1978; Cynader and Harris,

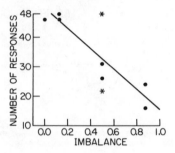

FIGURE 7. Suppression as a function of imbalance. Imbalance is defined as the difference in the duration of stimulation of the two eyes divided by the longer of the two durations. That is, $I = (A - B)/A$, where A is the longer duration (usually 8 hr) and B is the shorter (7, 4, or 1 hr). The ordinate is the same as in Figure 6. The data plotted are those obtained under binocular testing conditions after 16 weeks of binocular exposure. The data for the two late-exposed cats are plotted as asterisks. The line shown is the best-fitting straight line through all 9 data points. [Reprinted with permission from Tumosa et al. (1983).]

1980; Harris and Cynader, 1981; Hoffmann, 1981; Malach et al., 1984), the unequal AME cats show a marked asymmetry when tested binocularly: They respond better to the direction that corresponds to temporal-to-nasal movement for the more experienced eye (in this case, left to right) than to the opposite direction, which corresponds to temporal-to-nasal movement for the less experienced eye (Fig. 8). Thus, when both eyes are open, it is primarily the more experienced eye that controls the animal's behavior. This result fits very well with the idea that the less experienced eye is suppressed, and, like suppression of visual fields caused by subsequent binocular exposure, is not reported for monocularly lid-sutured cats. When tested with both eyes open, monocularly lid-sutured cats show little or no asymmetry in their optokinetic nystagmus (van Hof-van Duin, 1978; Hoffmann, 1981; Malach et al., 1984).

Although it is clear that the competitive changes described in the previous section involve modifications of the geniculocortical pathway, it is not clear in what part of the visual pathway suppression occurs. Unlike competition, suppression affects both the monocular and binocular segments of the visual pathway. The suppression is therefore unlikely to involve only local interactions between afferents from the more experienced eye and the less experienced eye. The suppression is also unlikely to involve a disconnection of the 1-hr eye from cortex. That would not explain why the direct retinal pathway, which is clearly functional under monocular testing conditions, fails to drive the optokinetic

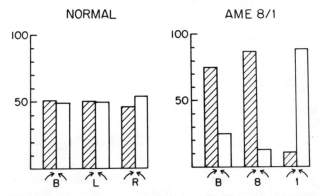

FIGURE 8. Percentage of optokinetic responses to clockwise and counterclockwise rotation for both eyes together (B) and for the left (L) and right (R) eyes separately for two 17-month-old normal cats and four 18-month-old AME 8/1 cats. The normal animals responded to either rotation about equally. In contrast, the AME 8/1 cats all showed a striking nasotemporal asymmetry when tested monocularly with either eye; they responded well only to motion in a temporal-to-nasal direction. When tested with both eyes open, they all showed a preference for clockwise motion (temporal-to-nasal for the 8-hr eye). [From S. Tieman et al. (1985).]

reflex under binocular conditions. The suppression could be caused by efferent control of the retina, as has been suggested for human amblyopes by van Hasselt (1972/3). An efferent pathway has been demonstrated to travel with the optic nerve and terminate along the border of the inner nuclear and inner plexiform layers of the retina (e.g., Ramon y Cajal, 1892/3; Brooke et al., 1965; Honrubia and Elliott, 1970; Hoogland et al., 1985). In birds this efferent pathway is well documented and is known to arise from the isthmooptic nucleus (Cowan, 1970). In mammals, however, the source and even the existence of this pathway remain controversial (e.g., Brindley and Hamasaki, 1966; Itaya, 1980; Schnyder and Kunzle, 1984; Hoogland et al., 1985), and its function, although possibly inhibitory (e.g., Ohno, 1980), is unknown. Finally, the suppression could involve changes in the corticotectal or corticopretectal pathways, since the two visuomotor tasks with which we have demonstrated suppression both involve cerebral cortical and midbrain pathways (Wood et al., 1973; Sherman, 1974b; Collewijn, 1975; Sherman and Sprague, 1979; Hoffmann, 1981; Strong et al., 1984).

In conclusion, unequal AME during the first months of postnatal life must create conditions that result in suppression of the less experienced eye once the animal is given simultaneous input to the two eyes. The resulting changes in visuomotor behavior occur well after the end of the sensitive period during which effects of an imbalance in stimulation of the two eyes have previously been demonstrated.

VISUAL ACUITY

Monocular lid suture produces severe deficits in visual acuity. The ability to detect a gap (Dews and Wiesel, 1970), grating acuity (Mitchell et al., 1977b, 1984), and contrast sensitivity (Lehmkuhle et al., 1982) are all sharply reduced. Because monocular lid suture confounds an imbalance in stimulation with continuous deprivation of one eye, it is not possible to determine which factor is responsible for these acuity deficits.

Our studies of unequal AME cats suggest that an imbalance alone does not affect visual acuity, at least for high-contrast stimuli. Even with a fairly large imbalance in stimulation, for example, AME 8/1, there are no significant differences in either grating acuity (Reeves et al., 1982) or vernier acuity (Hirsch et al., in preparation) between the two eyes. For example, we have tested AME 8/1 cats with a Mitchell jumping stand (Mitchell et al., 1977a) for their ability to detect an offset in a vertical line. There was no evidence for any difference in performance between the 8-hr eye and the 1-hr eye (Table 1). An imbalance in stimulation of the two eyes does not appear to produce any significant deficit either

TABLE 1. Performance on a Vernier Acuity Task for Each Eye of Four AME 8/1 Cats Tested before and after Binocular Exposure[a]

Cat	Gap Tested	Before (%)		After (%)	
		8-hr	1-hr	8-hr	1-hr
MRG	2.54	58	68	75	75
GRG	0.13	77	72	73	73
HRR	0.13	80	75	71	75
TBT	0.38	65	64	Not tested	
Mean	0.79	70	70	73	74

[a] Mean percent correct performance on a test series. The gap tested is given in degrees of visual angle and was determined by the animal's asymptotic performance on a staircase series. MRG had a torsional nystagmus or tremor, which may have contributed to his relatively poor acuity.

in grating acuity or vernier acuity for the less experienced eye. These results are consistent with the comparatively normal response properties of the cells driven by the less experienced eye in LGN (Christen et al., 1983; Christen and Mower, 1985b) and visual cortex (D. Tieman et al., 1979, 1983a; Smith et al., 1982b), and with Ikeda's (1979) suggestion that acuity deficits result from deprivation rather than competition.

Not only does AME 8/1 not result in acuity deficits, the suppression resulting from subsequent binocular exposure also does not produce such deficits. Hirsch et al. (in preparation) retested vernier acuity after 3–10 months of binocular exposure. All of the animals tested reached the same threshold levels as they had at the end of the alternating monocular exposure, and, as Table 1 shows, there was no evidence for any difference in performance between the two eyes. We can conclude that the suppression of visual field responses for the less experienced eye does not involve an acuity loss, at least for high-contrast stimuli.

If neither an imbalance nor suppression alone produces acuity deficits, then, in order to explain the acuity loss in strabismic animals, it is necessary to invoke the argument that, during the critical period, the image on the retina of the nonfixating eye will be blurred because the eye will be directed toward objects for which it is not properly accommodated (Ikeda, 1979). Rearing with blurred images does produce acuity deficits (Maguire et al., 1982; E. Smith et al., 1983; Deering and Peck, 1984). It is likely that the effects of an imbalance in stimulation of the two eyes exacerbate those of blurring since monocular blurring (either by lid suture or by atropinization) results in greater acuity deficits than binocular blurring (Lehmkuhle et al., 1982; E. Smith et al., 1983; but cf. Deering and Peck, 1984).

CONCLUSIONS

We have shown that unequal AME, like monocular lid suture, alters the outcome of the competitive interaction between the afferents from the two eyes during development. It is therefore not surprising that it has many of the same effects as monocular lid suture. However, unequal AME differs from monocular lid suture in that neither eye is totally deprived of patterned vision. This difference probably accounts for the additional effects seen in monocularly deprived cats. That is, we conclude that deprivation plays an important role in (1) restricting the laminar location of cells in visual cortex dominated by the deprived eye, (2) preventing the development or maintenance of normal response properties by these cells, (3) controlling the frequency with which Y cells can be recorded in the LGN, (4) determining the spatial resolution for X cells in the LGN, and (5) generating the acuity deficits seen in monocularly deprived cats.

Another way in which unequal AME differs from monocular deprivation is the manner in which the animal is affected by subsequent binocular exposure. Although we routinely observe suppression in our AME 8/1 cats, most monocularly deprived cats do not show suppression (Sherman, 1974a; D. Tieman et al., 1983b; Tumosa et al., 1983). In contrast, monocularly deprived cats usually show improvement with binocular exposure, especially in visual acuity (cf. Mitchell et al., 1977b, 1984). We believe that this exposure is helping them to overcome the effects of the deprivation rather than those of the altered competition. The unequal AME cats, since they are not deprived of pattern vision, have no such effects to overcome.

Our data show that experience continues to modify the mammalian visual system after the classical critical period; unequal AME imposed during the first few months of life alters the way in which the system responds to subsequent binocular exposure. The changes that experience produces during the critical period therefore help determine how the visual system will respond to, and in what ways it will be changed by, experience in the future. This property of continued plasticity, and its dependence on the organism's early experience, is shared by the neuronal mechanism involved in imprinting and song learning (see chapters by Marler and Bateson, this volume).

ACKNOWLEDGMENTS

We thank P. Caruccio, E. A. O'Neill, and D. Zimmerman for technical assistance, H. Ghiradella for comments on the manuscript, and R. Loos and B. Speck for helping to prepare the figures. Support provided by PHS Grants EY01268, EY02609, and BRSGRR0712215, NSF Grants BNS8217479 and BNS8217540, and the State University of New York at Albany.

REFERENCES

Bischof, H.-J. (1983). Imprinting and cortical plasticity: A comparative review. *Neurosci. Biobehav. Rev.* **7**:213–225.

Blake, R. and H. V. B. Hirsch (1975). Deficits in binocular depth perception in cats after alternating monocular deprivation. *Science* **190**:1114–1116.

Blake, R., M. L. J. Crawford, and H. V. B. Hirsch (1974). Consequences of alternating monocular deprivation on eye alignment and convergence in cats. *Invest. Ophthalmol. Vis. Sci.* **13**:121–126.

Blakemore, C. and H. M. Eggers (1978). Effects of artificial anisometropia and strabismus on the kitten's visual cortex. *Arch. Ital. Biol.* **116**:385–389.

Blakemore, C. and M. J. Hawken (1982). Rapid restoration of functional input to the visual cortex of the cat after brief monocular deprivation. *J. Physiol. (Lond.)* **327**:463–487.

Blakemore, C. and R. C. Van Sluyters (1974). Reversal of the physiological effects of monocular deprivation in kittens: Further evidence for a sensitive period. *J. Physiol. (Lond.)* **237**:195–216.

Blakemore, C. and R. C. Van Sluyters (1975). Innate and environmental factors in the development of the kitten's visual cortex. *J. Physiol. (Lond.)* **248**:663–716.

Bonds, A. B. (1979). Development of Orientation Tuning in the Visual Cortex of Kittens. In: *Developmental Neurobiology of Vision,* R. D. Freeman, ed., pp. 31–49, Plenum Press, New York.

Bonds, A. B., M. S. Silverman, G. Sclar, and R. B. Tootell (1980). Visually evoked potentials and deoxyglucose studies of monocularly deprived cats. *Invest. Ophthalmol. Vis. Sci. (ARVO Abstr.)* **21**, 225.

Brindley, G. S. and D. I. Hamasaki (1966). Histological evidence against the view that the cat's optic nerve contains centrifugal fibres. *J. Physiol. (Lond.)* **184**:444–449.

Brooke, N. L., J. de C. Downer, and T. P. S. Powell (1965). Centrifugal fibres to the retina in the monkey and cat. *Nature* **207**:1365–1367.

Burchfiel, J. L. and F. H. Duffy (1981). Role of intracortical inhibition in deprivation amblyopia: Reversal by microiontophoresis of bicuculline. *Brain Res.* **206**:479–484.

Christen, W. G. and G. D. Mower (1985a). Effects of monocular occlusion vs. diffusion on the cat visual system. *Invest. Ophthalmol. Vis. Sci. (ARVO Abstr.)* **26**, 251.

Christen, W. G. and G. D. Mower (1985b). Effects of unequal alternating monocular occlusion and diffusion on cat visual system. *Soc. Neurosci. Abstr.* **11**:462.

Christen, W. G., G. D. Mower, and F. H. Duffy (1983). LGN and cortical effects of alternating monocular experience. *Invest. Ophthalmol. Vis. Sci.* (*ARVO Abstr.*) **24**, 139.

Collewijn, H. (1975). Oculomotor areas in the rabbit's midbrain and pretectum. *J. Neurobiol.* **6**:3–22.

Cowan, W. M. (1970). Centrifugal fibers to the avian retina. *Brit. Med. Bull.* **26**:112–118.

Cynader, M. and L. Harris (1980). Eye movement in strabismic cats. *Nature* **286**:64–65.

Deering, K. R. and C. K. Peck (1984). Amblyopia results from chronic blurring of visual images during development. *Dev. Brain Res.* **15**:302–306.

Derrington, A. M. (1984). Development of spatial frequency selectivity in striate cortex of vision-deprived cats. *Exp. Brain Res.* **55**:431–437.

Dews, P. B. and T. N. Wiesel (1970). Consequences of monocular deprivation on visual behaviour in kittens. *J. Physiol. (Lond.)* **206**:437–455.

Duke-Elder, S. and K. Wybar (1973). Ocular Motility and Strabismus. In: *System of Ophthalmology*, S. Duke-Elder, ed., pp. 280–315, C. V. Mosby, St. Louis.

Frégnac, Y. and M. Imbert (1978). Early development of visual cortical cells in normal and dark-reared kittens: Relationship between orientation selectivity and ocular dominance. *J. Physiol. (Lond.)* **278**:27–44.

Frégnac, Y. and M. Imbert (1984). Development of neuronal selectivity in primary visual cortex of cat. *Physiol. Rev.* **64**:325–434.

Ganz, L., H. V. B. Hirsch, and S. B. Tieman (1972). The nature of perceptual deficits in visually deprived cats. *Brain Res.* **44**:547–568.

Guillery, R. W. (1972). Binocular competition in the control of geniculate cell growth. *J. Comp. Neurol.* **144**:117–130.

Guillery, R. W. and D. J. Stelzner (1970). The differential effects of unilateral lid closure on the monocular and binocular segments of the dorsal lateral geniculate nucleus in the cat. *J. Comp. Neurol.* **139**:413–422.

Harris, L. R. and M. Cynader (1981). The eye movements of the dark-reared cat. *Exp. Brain Res.* **44**:41–56.

Heitländer, H. and K.-P. Hoffmann (1978). The visual field of monocularly deprived cats after later closure or enucleation of the non-deprived eye. *Brain Res.* **145**:153–160.

Hickey, T. L. (1980). Development of the dorsal lateral geniculate nucleus in normal and visually deprived cats. *J. Comp. Neurol.* **189**:467–481.

Hickey, T. L., P. D. Spear, and K. E. Kratz (1977). Quantitative studies of cell size in the cat's dorsal lateral geniculate nucleus following visual deprivation. *J. Comp. Neurol.* **172**:265–282.

Hirsch, H. V. B. (1985a). The Tunable Seer: Activity-dependent Development of Vision. In: *Handbook of Behavioral Neurobiology*, Vol. 8, E. Blass, ed., pp. 237–295, Plenum Press, New York.

Hirsch, H. V. B. (1985b). The role of visual experience in the development of cat striate cortex. *Cell. Molec. Neurobiol.* **5**:103–121.

Hoffmann, K.-P. (1981). Neuronal Responses Related to Optokinetic Nystagmus in the Cat's Nucleus of the Optic Tract. In: *Progress in Oculomotor Research*, A. Fuchs and W. Becker, eds., pp. 443–454, Elsevier, North-Holland.

Hoffmann, K.-P. and M. Cynader (1977). Functional aspects of plasticity in the visual system of adult cats after early monocular deprivation. *Phil. Trans. Roy. Soc. B* **278**:411–424.

Honrubia, F. M. and J. H. Elliott (1970). Efferent innervation of the retina. II. Morphological study of the monkey retina. *Invest. Ophthalmol. Vis. Sci.* **9**:971–976.

Hoogland, P. V., A. Vanderkrans, F. D. Koole, and H. J. Groenewegen (1985). A direct projection from the nucleus oculomotorius to the retina in rats. *Neurosci. Lett.* **56**:323–328.

Hubel, D. H. and T. N. Wiesel (1959). Receptive fields of single neurones in the cat's striate cortex. *J. Physiol. (Lond.)* **148**:574–591.

Hubel, D. H. and T. N. Wiesel (1962). Receptive fields, binocular interaction and functional architecture in the cat's visual cortex. *J. Physiol. (Lond.)* **160**:106–154.

Hubel, D. H. and T. N. Wiesel (1965). Binocular interaction in striate cortex of kittens reared with artificial squint. *J. Neurophysiol.* **28**:1041–1059.

Hubel, D. H. and T. N. Wiesel (1970). The period of susceptibility to the physiological effects of unilateral eye closure in kittens. *J. Physiol. (Lond.)* **206**:419–436.

Ikeda, H. (1979). Physiological basis of amblyopia. *Trends Neurosci.* **2**:209–212.

Ikeda, H. and K. E. Tremain (1978). Amblyopic LGN cells in kittens raised with "penalisation" of one or both eyes. *Metabol. Ophthalmol.* **2**:127–129.

Ikeda, H., K. E. Tremain, and G. Einon (1978). Loss of spatial resolution of lateral geniculate nucleus neurones in kittens raised with convergent squint produced at different stages in development. *Exp. Brain Res.* **31**:207–220.

Itaya, S. K. (1980). Retinal efferents from the pretectal area in the rat. *Brain Res.* **201**:436–441.

Kossut, M., I. D. Thompson, and C. Blakemore (1983). Ocular dominance columns in cat striate cortex and effects of monocular deprivation: A 2-deoxyglucose study. *Acta Neurobiol. Exp.* **43**:273–282.

Kratz, K. E. (1982). Spatial and temporal sensitivity of lateral geniculate cells in dark-reared cats. *Brain Res.* **251**:55–63.

Kratz, K. E., S. M. Sherman, and R. Kalil (1979). Lateral geniculate nucleus of dark reared cats: Loss of Y cells without changes in cell size. *Science* **203**:1353–1355.

Kratz, K. E., P. D. Spear, and D. C. Smith (1976). Postcritical period reversal of effects of monocular deprivation on striate cortex cells in the cat. *J. Neurophysiol.* **39**:501–511.

Lehmkuhle, S. W., K. E. Kratz, and S. M. Sherman (1982). Spatial and temporal sensitivity of normal and amblyopic cats. *J. Neurophysiol.* **48**:372–387.

Lehmkuhle, S. W., K. E. Kratz, S. Mangel, and S. M. Sherman (1978). An effect of early monocular lid suture upon the development of X-cells in the cat's lateral geniculate nucleus. *Brain Res.* **157**:346–350.

Lehmkuhle, S. W., K. E. Kratz, S. Mangel, and S. M. Sherman (1980). Effects of early monocular lid suture on spatial and temporal sensitivity of neurons in dorsal lateral geniculate nucleus of the cat. *J. Neurophysiol.* **43**:542–556.

Leventhal, A. G. and H. V. B. Hirsch (1977). Effects of early experience upon orientation sensitivity and binocularity of neurons in visual cortex of cats. *Proc. Nat. Acad. Sci. USA* **74**:1272–1276.

Leventhal, A. G. and H. V. B. Hirsch (1980). Receptive-field properties of different classes of neurons in visual cortex of normal and dark-reared cats. *J. Neurophysiol.* **43**:1111–1132.

Levitt, F. B. and R. C. Van Sluyters (1982). The sensitive period for strabismus in the kitten. *Dev. Brain Res.* **3**:323–327.

Maguire, G. W., E. L. Smith III, R. S. Harwerth, and M. L. J. Crawford (1982). Optically induced anisometropia in kittens. *Invest. Ophthalmol. Vis. Sci.* **23**:253–264.

Malach, R., N. P. Strong, and R. C. Van Sluyters (1984). Horizontal optokinetic nystagmus in the cat: Effects of long-term monocular deprivation. *Dev. Brain Res.* **13**:193–205.

Mitchell, D. E., F. Giffin, and B. Timney (1977a). A behavioural technique for the rapid assessment of the visual capabilities of kittens. *Perception* **6**:181–193.

Mitchell, D. E., M. Cynader, and J. A. Movshon (1977b). Recovery from the effects of monocular deprivation in kittens. *J. Comp. Neurol.* **176**:53–64.

Mitchell, D. E., K. M. Murphy, and M. G. Kaye (1984). The permanence of the visual recovery that follows reverse occlusion of monocularly deprived kittens. *Invest. Ophthalmol. Vis. Sci.* **25**:908–917.

Mower, G. D., J. L. Burchfiel, and F. H. Duffy (1981). The effects of dark-rearing on the development and plasticity of the lateral geniculate nucleus. *Dev. Brain Res.* **1**:418–424.

Movshon, J. A. (1976). Reversal of the physiological effects of monocular deprivation in the kitten's visual cortex. *J. Physiol. (Lond.)* **261**:125–174.

Movshon, J. A. and R. C. Van Sluyters (1981). Visual neural development. *Ann. Rev. Psychol.* **21**:477–522.

Ohno, T. (1980). The possibility of centrifugal projections to the retina in the rat. *Experientia* **36**:1400–1401.

Olson, C. R. and R. D. Freeman (1980). Profile of the sensitive period for monocular deprivation in kittens. *Exp. Brain Res.* **39**:17–21.

Packwood, J. and B. Gordon (1975). Stereopsis in normal domestic cat, Siamese cat, and cat raised with alternating monocular occlusion. *J. Neurophysiol.* **38**:1485–1499.

Peck, C. K. and C. Blakemore (1975). Modification of single neurons in the kitten's visual cortex after brief periods of monocular visual exposure. *Exp. Brain Res.* **22**:57–68.

Pettigrew, J. D. (1974). The effect of visual experience on the development of stimulus specificity by kitten cortical neurones. *J. Physiol. (Lond.)* **237**:49–74.

Pettigrew, J. D., T. Nikara, and P. O. Bishop (1968). Binocular interaction on single units in cat striate cortex: Simultaneous stimulation by single moving slit with receptive fields in correspondence. *Exp. Brain Res.* **6**:391–410.

Presson, J. and B. Gordon (1979). Critical period and minimal exposure required for the effects of alternating monocular occlusion in cat visual cortex. *Vision Res.* **19**:807–811.

Ramon y Cajal, S. (1892/3). La rétine des vertébrés. *La Cellule* **9**:119–259.

Reeves, T. M., R. N. Holdefer, and D. C. Smith (1982). Unequal alternating monocular deprivation does not produce unequal visual acuity. *Soc. Neurosci. Abstr.* **8**:296.

Schnyder, H. and H. Künzle (1984). Is there a retinopetal system in the rat? *Exp. Brain Res.* **56**:502–508.

Sharpless, S. K. (1975). Disuse Supersensitivity. In: *The Developmental Neuropsychology of Sensory Deprivation*, A. H. Riesen, ed., pp. 125–152, Academic Press, New York.

Shatz, C. J. and M. P. Stryker (1978). Ocular dominance in layer IV of the cat's visual cortex and the effects of monocular deprivation. *J. Physiol. (Lond.)* **281**:267–283.

Sherman, S. M. (1972). Development of interocular alignment in cats. *Brain Res.* **37**:187–203.

Sherman, S. M. (1973). Visual field defects in monocularly and binocularly deprived cats. *Brain Res.* **49**:25–45.

Sherman, S. M. (1974a). Permanence of visual perimetry deficits in monocularly and binocularly deprived cats. *Brain Res.* **73**:491–501.

Sherman, S. M. (1974b). Visual fields of cats with cortical and tectal lesions. *Science* **185**:355–357.

Sherman, S. M. and P. D. Spear (1982). Organization of visual pathways in normal and visually deprived cats. *Physiol. Rev.* **62**:738–855.

Sherman, S. M. and J. M. Sprague (1979). Effects of visual cortex lesions upon the visual fields of monocularly deprived cats. *J. Comp. Neurol.* **188**:291–312.

Sherman, S. M., K.-P. Hoffmann, and J. Stone (1972). Loss of a specific cell type from the dorsal lateral geniculate nucleus in visually deprived cats. *J. Neurophysiol.* **35**:532–541.

Sherman, S. M., J. R. Wilson, and R. W. Guillery (1975). Evidence that binocular competition affects the postnatal development of Y-cells in the cat's lateral geniculate nucleus. *Brain Res.* **100**:441–444.

Sherman, S. M., R. W. Guillery, J. H. Kaas, and K. J. Sanderson (1974). Behavioral, electrophysiological and morphological studies of binocular competition in the development of the geniculocortical pathways of cats. *J. Comp. Neurol.* **188**:291–312.

Sireteanu, R. and K.-P. Hoffmann (1979). Relative frequency and visual resolution of X- and Y-cells in the LGN of normal and monocularly deprived cats: Interlaminar differences. *Exp. Brain Res.* **34**:591–603.

Sireteanu, R. and D. Maurer (1982). The development of the kitten's visual field. *Vision Res.* **22**:1105–1111.

Smith, D. C., R. N. Holdefer, T. M. Reeves (1982a). The visual field in monocularly deprived cats and its permanence. *Beh. Brain Res.* **5**:245–259.

Smith, D. C., T. M. Reeves, and R. N. Holdefer (1982b). Unmasking silent synapses in cats reared with unequal alternating monocular deprivation. *Soc. Neurosci. Abstr.* **8**:298.

Smith, E. L. III, R. S. Harwerth, and G. W. Maguire (1983). Effects of chronic atropinization on visual acuity in kittens. *Beh. Brain Res.* **7**:307–319.

Strong, N. P., R. Malach, P. Lee, and R. C. Van Sluyters (1984). Horizontal optokinetic nystagmus in the cat: Recovery from cortical lesions. *Dev. Brain Res.* **13**:179–192.

Tieman, D. G., M. A. McCall, and H. V. B. Hirsch (1979). Physiological effects of unequal alternating monocular deprivation. *Soc. Neurosci. Abstr.* **5**:631.

Tieman, D. G., M. A. McCall, and H. V. B. Hirsch (1983a). Physiological effects of unequal alternating monocular exposure. *J. Neurophysiol.* **49**:804–818.

Tieman, D. G., N. Tumosa, and S. B. Tieman (1983b). Behavioral and physiological effects of monocular deprivation: A comparison of rearing with occlusion and diffusion. *Brain Res.* **280**:41–50.

Tieman, S. B. (1985). The anatomy of geniculocortical connections in monocularly deprived cats. *Cell. Molec. Neurobiol.* **5**:35–45.

Tieman, S. B. and H. V. B. Hirsch (1983). Removal of the more-experienced eye decreases visual field deficits in cats reared with unequal alternating monocular exposure. *Brain Res.* **27**:170–173.

Tieman, S. B. and N. Tumosa (1983). [^{14}C]-2-deoxyglucose demonstration of the organization of ocular dominance in areas 17 and 18 of the normal cat. *Brain Res.* **267**:35–46.

Tieman, S. B., D. G. Tieman, and H. V. B. Hirsch (1985). Optokinetic nystagmus in cats reared with unequal alternating exposure. *Invest. Ophthalmol. Vis. Sci. (ARVO Abstr.)* **26**:256.

Tieman, S. B., D. L. Nickla, K. Gross, T. L. Hickey, and N. Tumosa (1984). Effects of unequal alternating monocular exposure on the sizes of cells in the cat's lateral geniculate nucleus. *J. Comp. Neurol.* **225**:119–128.

Tumosa, N. (1982). Binocular competition affects the development of the visual system in the cat. Ph.D. dissertation, State University of New York, Albany.

Tumosa, N. and S. B. Tieman (1981). Binocular competition determines the size and shape of ocular dominance columns in cats. *Soc. Neurosci. Abstr.* **7**:674.

Tumosa, N., S. B. Tieman, and H. V. B. Hirsch (1980a). Unequal alternating monocular deprivation causes asymmetric visual fields in cats. *Science* **208**:421–423.

Tumosa, N., S. B. Tieman, and H. V. B. Hirsch (1980b). Anatomical effect of unequal alternate monocular deprivation. *Invest. Ophthalmol. Vis. Sci. (ARVO Abstr.)* **21**:59.

Tumosa, N., S. B. Tieman, and H. V. B. Hirsch (1982). Visual field deficits in cats reared with unequal alternating monocular exposure. *Exp. Brain Res.* **47**:119–129.

Tumosa, N., S. Nunberg, H. V. B. Hirsch, and S. B. Tieman (1983). Binocular exposure causes suppression of the less experienced eye in cats previously reared with unequal alternating monocular exposure. *Invest. Ophthalmol. Vis. Sci.* **24**:496–506.

van Hasselt, P. (1972/3). The centrifugal control of retinal function. A review. *Ophthal. Res.* **4**:298–320.

van Hof-van Duin, J. (1977). Visual field measurements in monocularly deprived and normal cats. *Exp. Brain Res.* **30**:353–368.

van Hof-van Duin, J. (1978). Direction preference of optokinetic responses in monocularly tested normal kittens and light deprived cats. *Arch. Ital. Biol.* **116**:471–477.

Van Sluyters, R. C. and F. B. Levitt (1980). Experimental strabismus in the kitten. *J. Neurophysiol.* **43**:686–699.

Watkins, D. W., J. R. Wilson, and S. M. Sherman (1978). Receptive-field properties of neurons in binocular and monocular segments of striate cortex in cats raised with binocular lid suture. *J. Neurophysiol.* **41**:322–337.

Wiesel, T. N. and D. H. Hubel (1963a). Effects of visual deprivation on morphology and physiology of cells in the cat's lateral geniculate body. *J. Neurophysiol.* **26**:978–993.

Wiesel, T. N. and D. H. Hubel (1963b). Single-cell responses in striate cortex of kittens deprived of vision in one eye. *J. Neurophysiol.* **26**:1003–1017.

Wiesel, T. N. and D. H. Hubel (1965). Comparison of the effects of unilateral and bilateral eye closure on cortical unit responses in kitten. *J. Neurophysiol.* **28**:1029–1040.

Wood, C. C., P. D. Spear, and J. J. Braun (1973). Direction-specific deficits in horizontal optokinetic nystagmus following removal of visual cortex in the cat. *Brain Res.* **60**:231–237.

Yinon, U. (1976). Age dependence of the effect of squint on cells in kittens' visual cortex. *Exp. Brain Res.* **26**:151–157.

14

DARK REARING AND THE SENSITIVE PERIOD FOR MONOCULAR DEPRIVATION

Brian Timney

Department of Psychology, University of Western Ontario, London, Canada

Wiesel and Hubel's (1963a,b) demonstration that prolonged closure of the lids of one eye leads to an irreversible loss of binocularity in the kitten visual cortex was the beginning of an industry in visual neuroscience. One topic that has received a great deal of attention because of its obvious clinical implications concerns the period of susceptibility to the effects of monocular deprivation. In their initial papers Wiesel and Hubel (1963a,b) reported that if an adult cat is deprived, there are no perceptible anatomical or physiological deficits. The conclusion they drew from this finding was that there must be a period, early in an animal's life, when the visual system is susceptible to disruption by anomalous visual experience.

Although the concept of a period of susceptibility is well established, it has become apparent over the last several years that one cannot speak about a single sensitive period; there appear to be several, each with a different time course. For example, Daw and his colleagues (Berman and Daw, 1977; Daw and Wyatt, 1976) have shown that reversal of direction selectivity in cortical neurons can be accomplished only during the first 5 weeks of life. In my own laboratory we have found that the sensitive period for the loss of stereoscopic depth perception seems to be over before the end of the third month (Timney, 1985). Others have reported sensitive periods for certain of the effects of artificially produced strabismus (Berman and Murphy, 1982) and alternating monocular occlusion (Presson and Gordon, 1979) that do not seem to match exactly the time course for the effects of monocular deprivation on cortical ocular dominance. Although there is almost certainly some underlying common factor, one should be cautious in talking about "the" sensitive period.

In this chapter only susceptibility to monocular deprivation in kittens will be discussed and only a single issue addressed: To what extent can the period of susceptibility to monocular deprivation be modified through manipulation of the visual environment? Before doing that, however, the major consequences of monocular deprivation and the characteristics of the sensitive period for this form of visual deprivation will be reviewed.

EFFECTS OF MONOCULAR DEPRIVATION

Monocular deprivation imposed during the sensitive period makes its presence felt throughout the visual system. The major changes include the following:

Cortical Physiology

The deprived eye loses most of its functional connections with neurons in the visual cortex. In the normal kitten more than 80% of cortical neurons may be activated through either eye; this proportion may drop almost to zero following prolonged monocular deprivation (Wiesel and Hubel, 1963b).

Visual Acuity

Behavioral testing of the visual capacities of monocularly deprived cats reveals profound deficits of visual acuity (Giffin and Mitchell, 1978), stereoscopic depth perception (Timney, 1983), visual orienting (Sherman, 1973), and visuomotor skills (Dews and Wiesel, 1970).

Cortical Ocular Dominance Columns

In the adult cat, terminals from the lateral geniculate nucleus (LGN) within layer IV of the cortex are segregated into alternating segments known as ocular dominance columns. The presence of these columns may be demonstrated using autoradiographic techniques (Shatz et al., 1977). In a normal kitten, terminals from each eye occupy approximately the same amount of space. In a monocularly deprived kitten, however, the columns corresponding to the experienced eye expand to occupy the territory that would normally belong to the deprived eye (Shatz and Stryker, 1978).

Lateral Geniculate Nucleus

The encounter rate for Y cells in the LGN is reduced (Sherman, 1972). Also, most authors have reported a loss in the spatial resolution of X cells (e.g., Lehmkuhle et al., 1978), although others have not observed this (e.g., Shapley and So, 1980). Anatomically, there is a reduction of between 25 and 40% in the cross-sectional area of cells in the layers of the LGN connected to the deprived eye (Guillery and Stelzner, 1970; Wiesel and Hubel, 1963a). The cell shrinkage is much greater in the binocular segments of these layers.

THE SENSITIVE PERIOD FOR MONOCULAR DEPRIVATION

Susceptibility and Potential for Recovery

Although the sensitive period is most often thought of as a time of heightened susceptibility, it is important to emphasize that it is a time when there is also a great potential for the recovery of function. This has been demonstrated most clearly in studies of the effects of reverse occlusion (Blakemore and van Sluyters, 1974). Many of the deficits that follow a period of monocular deprivation may be reversed if the deprived eye is opened and the initially experienced eye is closed. Because of this recovery potential, Aslin (1982) has suggested that the sensitive period be regarded as a period of "elasticity", when the system may rebound back to its original state. Although this description is a little too simple, it does emphasize the fact that the sensitive period works both for and against the developing visual system.

Time Course of Sensitive Period

After showing that there must be a period of heightened susceptibility to monocular deprivation, Hubel and Wiesel (1970) attempted to obtain an estimate of its time course. They deprived kittens for different periods of time beginning at different ages, and noted the relative dominance of the experienced eye in ocular dominance histograms obtained from single-unit recordings. On the basis of a fairly limited sample, they concluded that the sensitive period "begins suddenly near the start of the fourth week, remains high until sometime between the sixth and eighth week, and then declines, disappearing finally around the end of the third month" (p. 419).

A number of studies since then have caused this original estimate to be refined. Blakemore and van Sluyters (1974) used reverse occlusion to study the changes in potential for recovery over the course of the sensitive period. Kittens were monocularly deprived from birth for various periods of time and then underwent reverse occlusion for 9 weeks before single-unit recordings were made. Blakemore and van Sluyters reported that, from a peak during the fifth week, there was a gradual decline in the capacity of the initially deprived eye to regain its functional connections. When the occlusion was reversed at 14 weeks, very little recovery occurred. Their data are summarized in Figure 1.

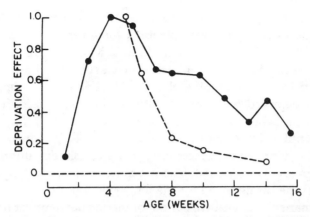

FIGURE 1. Profile of the sensitive period for monocular deprivation in light-reared kittens. Open circles: data from Blakemore and van Sluyters (1974) showing effect of 9 weeks of reverse suture in monocularly deprived kittens, beginning at the ages specified on the abscissa. The deprivation effect is defined as the proportion of cells dominated by the more recently experienced eye and is a measure of the reversal of ocular dominance. Closed circles: effect of 10–12 days of monocular deprivation, beginning at the age specified on the abscissa. Deprivation effect is defined as the proportion of cells dominated by the undeprived eye and is a measure of the effectiveness of the deprivation (from Olson and Freeman, 1980).

A different approach was taken by Olson and Freeman (1980), who examined susceptibility more directly. They deprived kittens for 10–12 days beginning at different ages, and then recorded from the visual cortex immediately following the deprivation period. They found that susceptibility to monocular occlusion was present at the age of 2 weeks, peaked during the fifth and sixth weeks, and then declined very slowly so that some residual susceptibility remained after 16 weeks. These data are presented also in Figure 1. The suggestion that the sensitive period was not over completely by the end of the fourth month was reinforced by Cynader et al. (1980). These investigators looked at both susceptibility and potential for recovery in older cats. In one set of studies cats were deprived of vision in one eye for 3 months beginning at different ages, and recordings were made immediately after the deprived eye was opened. If deprivation was imposed before the age of 7 months, there was a tendency for the cortex to be dominated by the eye with normal experience, suggesting residual susceptibility until at least 6 months.

Cynader et al. (1980) also examined the potential for recovery, both physiological and behavioral. Two cats were deprived from birth until

4 or 5 months of age, when their deprived eyes were opened. Behavioral estimates of visual acuity were obtained for several months following the end of the deprivation period. Although neither animal achieved the acuity levels of normal cats, both showed substantial recovery, with the cat deprived for 4 months obtaining higher acuity estimates than the animal deprived for 5 months. When single-unit recordings were made after all behavioral testing was complete, the cat that had been deprived until it was 4 months old had many cells that could be driven through the deprived eye, indicating that a certain amount of physiological recovery had occurred. There was no evidence for physiological recovery in the cat deprived for the longer period.

To summarize: the sensitive period for monocular deprivation appears to begin around the end of the second week, peaks during the fifth and sixth weeks, and then declines to a very low level by the end of the sixth month. It is possible that the degree of susceptibility and the potential for recovery may follow slightly different time courses.

EXTENDING THE SENSITIVE PERIOD BY DARK REARING

One feature of the periods of susceptibility to environmental exposure is that they seem to coincide with the periods of most rapid neural development within the visual system (De Courten and Garey, 1982; Cragg, 1975; Dobbing, 1968; Hickey, 1977). If this is true, then any manipulation that delays or prolongs the time course of neural development should have a similar influence on the timing of the sensitive period. One way in which the rate of development may be changed is to prevent an animal from having any visual experience. This may be done by dark rearing from birth.

Consequences of Early Dark Rearing

Several studies have shown that the majority of cortical neurons in a young kitten have immature response properties (Albus and Wolf, 1984; Barlow and Pettigrew, 1971; Frégnac and Imbert, 1978). If the kitten is permitted normal visual experience, virtually all the cells become adult-like by about the sixth week (Blakemore and van Sluyters, 1974; Frégnac and Imbert, 1978; Derrington and Fuchs, 1981; Pettigrew, 1974). In contrast, if the kitten is kept in darkness from birth, normal development does not proceed after the third week. For example, Frégnac and Imbert (1978) reported that there were no significant differences in the proportions of different classes of visually responsive cells in normal and dark-reared kittens before the age of 17 days. However, after this age there was a large increase in the proportion of orientation-specific cells in normal kittens. Among dark-reared kittens there was a comparable increase in nonselective cells. By the middle of the second month the cortex of dark-reared kittens appears less mature than that of kittens under 2 weeks old.

After the initial decline in responsiveness of cortical cells in young dark-reared kittens, there appears to be little change if the deprivation is prolonged. Although there are no direct comparisons available, descriptions of the properties of cortical neurons in older dark-reared kittens (e.g., Mower et al., 1981) are very similar to those of younger animals raised in darkness (e.g., Frégnac and Imbert, 1978). Thus, it seems that total visual deprivation may prevent development within the visual cortex. It should be stressed, though, that the cortex of a dark-reared cat cannot be considered as equivalent to that of a neonate; there are degenerative changes that begin very early in life.

Another characteristic of the dark-reared cortex is its potential for recovery of function. There is much more extensive recovery from prolonged dark rearing than from similar periods of monocular deprivation. Cynader et al. (1976) monitored the recovery of cortical function in dark-reared cats and cats with bilateral lid suture. They found that very substantial recovery of function could occur. With deprivation periods of 11–15 months followed by 6–12 months in normally lighted surroundings, about 50% of the neurons encountered were orientation specific. In a more recent study Cynader and Mitchell (1980) found that in cats dark reared until the age of 4 months, over 80% of the neurons they encountered were orientation selective after 4 weeks in the light.

The physiological recovery observed in dark-reared cats is accompanied by recovery of visual acuity (Mower et al., 1982; Timney and Mitchell, 1979; Timney et al., 1978). Cats deprived until the age of 6 months recover to essentially normal levels. Among animals deprived for longer periods, the recovery is not so extensive but is still much greater than that observed in monocularly deprived animals of comparable ages (Giffin and Mitchell, 1978; Cynader et al., 1980).

Monocular Deprivation Following Early Dark Rearing

The observation that developmental changes in the cortex may be delayed until normal visual experience is provided, led Cynader (1977) to suggest that perhaps the onset of the sensitive period might be delayed too. A series of studies carried out since then have supported this hypothesis in general terms, but they have also revealed differences between the "normal" and the delayed sensitive period.

In the following sections the effects of monocular deprivation following dark rearing on each of the characteristics that are known to be affected when otherwise normally reared kittens are monocularly deprived will be reviewed.

Cortical Physiology. Cynader and Mitchell (1980) raised kittens in the dark for between 4 and 10 months and then imposed a period of monocular deprivation lasting between 3 days and 3 months. Mower et al. (1980) used dark-rearing periods from 4 months to over a year and similar periods of monocular occlusion; in a more recent study Cynader (1983) extended the dark-rearing period to 2 years followed by 9–18 months of monocular deprivation. The results of all of these experiments may be summarized by saying that, in each case, the imposition of monocular deprivation caused a reduction in the proportion of neurons that could be driven through the deprived eye. Figure 2 illustrates the effects of three months of monocular deprivation in three light-reared and three dark-reared cats. It is quite clear that a very strong deprivation effect is present, even in the cat dark reared for the first 10 months of life, well beyond the end of the normal sensitive period. When the initial deprivation period was extended to 2 years, the shift in ocular dominance was less pronounced, even though the period of monocular closure was much longer (Cynader, 1983).

To quantify the extent to which the nondeprived eye dominated the visual cortex, Cynader (1983) calculated a *takeover index* and an *index of binocularity*. The takeover index was defined as the ratio of cells falling into ocular dominance classes 1 and 2 (deprived eye dominant) to the number falling into classes 6 and 7 (normal eye dominant). The lower the value of the index, the fewer neurons were dominated by the deprived eye. The binocularity index was defined as the percentage of cells falling into ocular dominance groups 3, 4, and 5. Figure 3 shows these indices for the cats studied by Cynader and Mitchell (Cynader, 1983;

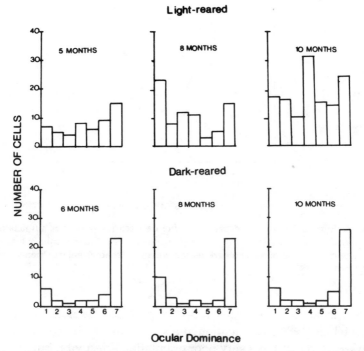

Ocular Dominance

FIGURE 2. Effect of 3 months of monocular occlusion in light-reared (top) and dark-reared (bottom) kittens, beginning at the ages shown in each panel. In the ocular dominance scale shown on the abscissa, 1–7 represent increasing dominance of the nondeprived eye, with cells in group 1 excitable only through the deprived eye and cells in group 7 excitable only through the experienced eye (redrawn from Cynader and Mitchell, 1980).

Cynader and Mitchell, 1980). The takeover index shows that the non-deprived eye dominates almost all of the neurons in the 4-month dark-reared animal, but the effect becomes somewhat less as the period of dark rearing is increased. The proportion of strongly binocular cells remains very low in all of the cats subjected to monocular deprivation, irrespective of the duration of initial deprivation.

Because the sensitive period is also a time when the cortex may recover from the deficits produced by anomalous experience, the potential for physiological recovery was examined by Timney et al. (1980). Two cats were dark reared and monocularly deprived and then permitted a recovery period following the opening of the deprived eye. In each case a large proportion of cells could be driven through the formerly deprived eye. As with the "normal" sensitive period, the extended one is a period of cortical elasticity (Aslin, 1982).

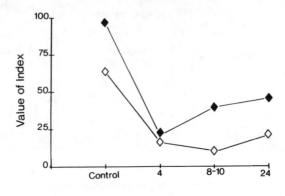

FIGURE 3. Effect of monocular deprivation after dark rearing plotted as a function of the initial period of deprivation. Filled symbols: takeover index. Open symbols: binocularity index. The control points refer to dark-reared kittens with no visual experience (redrawn from Cynader, 1983).

One curious feature of the recovery process was noted by Cynader and Mitchell (1980). As mentioned above, most of the cortical neurons in dark-reared cats have very poor orientation selectivity, but after a few weeks of normal experience they acquire this property. What Cynader and Mitchell observed was that there was also recovery of orientation selectivity in those neurons that could be driven by the *deprived* eye of their dark-reared animals even though they had received no visual experience through that eye. Figure 4 shows the increase in the percentage of orientation-specific units as a function of time in the light. Although the rate of recovery in the deprived eye is slower than that of the open eye, it is nevertheless substantial. This is quite different from what happens following monocular occlusion in light-reared cats, where there is no passive development of function.

Visual Acuity. The changes that occur in the dark-reared cortex are reflected in the cat's visual capacities. Timney et al. (1978) had observed that acuity developed to essentially normal levels in cats that had been dark reared until the age of 6 months. In a subsequent paper Timney et al. (1980) measured visual acuity in dark-reared cats that had been monocularly deprived. Data from two animals are shown in Figure 5. Although the periods of initial deprivation and monocular occlusion were somewhat different, the overall pattern of results was the same:

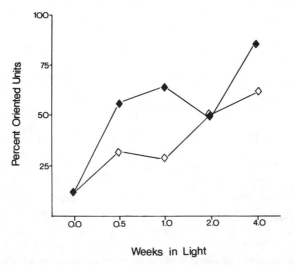

FIGURE 4. The time course of recovery of orientation selectivity in cats monocularly deprived following 4 months of dark rearing. Filled symbols: percentage of units dominated by the nondeprived eye (ocular dominance groups 5–7). Open symbols: percentage of units with a strong input from the deprived eye (ocular dominance groups 1–4). Note that the recovery of orientation selectivity in the deprived eye is occurring in the absence of any patterned visual experience (redrawn from Cynader and Mitchell, 1980).

after only a very brief period of apparent blindness, there was a rapid improvement in acuity to a level considerably below that achieved by the nondeprived eye. The very brief period of apparent blindness observed in these cats was much shorter than that seen in monocularly deprived cats raised in the light. It is possible that this finding is related to the passive physiological recovery noted by Cynader and Mitchell (1980).

Ocular Dominance Columns. In normal cats ocular dominance columns in the visual cortex are not present at birth; lateral geniculate terminals from each eye are freely mixed in layer IV. These terminals segregate into the adultlike arrangement by about the age of 6 weeks (LeVay et al., 1978). Among cats that have been denied visual experience, normal segregation does not occur and the terminals remain mixed (Swindale, 1981). However, if the cats are removed from the dark early enough and given time to recover, then the afferents will separate in the normal way. Swindale (1982) showed that a kitten removed from the dark at 6 weeks developed typical ocular dominance columns after several weeks in the light. But if the dark-rearing period was extended to seven

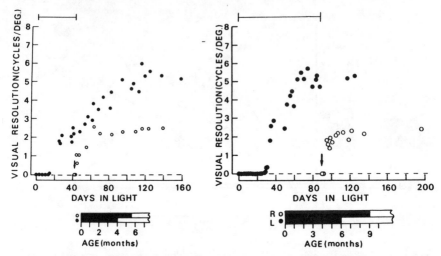

FIGURE 5. Development of visual acuity in dark-reared, monocularly deprived kittens. The kittens were raised in total darkness for either 4 or 6 months and then given a 6-week or 3-month period of monocular occlusion, as indicated in the lower panels. Recovery in the open eye (filled symbols) was monitored from the time the animals were brought into the light and in the deprived eye (open symbols) from the time that eye was opened. The horizontal bars at the top of the figure represent the monocular deprivation period. The arrows show results from the first test session after the deprived eye was opened (from Timney et al., 1980).

months, no such segregation took place. As part of a separate study (to be discussed below), Mower and his colleagues (1985) raised one cat in darkness for just over 4 months and then permitted a further 4½ months of normal exposure. There was no evidence for ocular dominance columns in this animal. Together, these results suggest that there is a sensitive period for the development of ocular dominance columns, but it is over sometime between the ages of 6 weeks and 4 months.

While dark rearing can prevent the normal segregation of terminals into ocular dominance columns, monocular deprivation permits the terminals from one eye to expand their territory at the expense of the other eye (Shatz and Stryker, 1978). Mower et al. (1985) examined the effects of monocular occlusion following 2½–4½ months of dark rearing. They found that the distribution of terminals was similar to that of their control dark-reared animals. There was no evidence for takeover by the non-deprived eye. Figure 6 permits a comparison of the results in light-reared and dark-reared cats following similar periods of monocular occlusion.

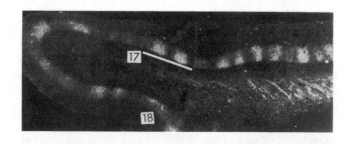

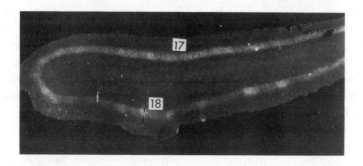

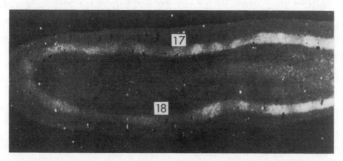

FIGURE 6. Autoradiograms of cortical areas 17 and 18 following intraocular injection of [³H]–proline. Upper panel: injection into the deprived eye of a normal cat given 7 months of monocular deprivation beginning at the age of 4 months. White marker bar indicates a 2-mm section of cortex that was selected for quantitative analysis. Middle panel: injection into the deprived eye of a cat dark reared until a little over 4 months followed by 9 months of monocular deprivation. Lower panel: injection into the nondeprived eye of a cat dark reared until 4 months followed by 3 months of monocular deprivation (from Mower et al., 1985).

Lateral Geniculate Nucleus. There are three major consequences of monocular deprivation in the LGN: a loss in the frequency with which Y cells are encountered, a reduction in the spatial resolving power of X cells, and a reduction in the average cross-sectional area of cells in the deprived layers. The effects of dark rearing on the LGN are much less severe. Kalil (1978) reported that the growth rate of cells in the LGN was slowed but that by 4 months of age cells are of normal size. Kratz and his associates (Kratz, 1982; Kratz et al., 1979) reported that except for a reduction in the encounter frequency of Y cells and an increase in the number of cells that could not be classified, all other physiological and anatomical properties of the lateral geniculate were normal in dark-reared cats.

Mower et al. (1981) confirmed these results and extended them by studying the influence of monocular occlusion on the lateral geniculate nucleus of dark-reared cats. Monocular deprivation for 3 months following 4 months of dark rearing had no effect. There were no differences between dark rearing and dark rearing followed by monocular deprivation.

In a more recent study Mower et al. (1985) measured the cross-sectional area of LGN neurons in dark-reared, monocularly deprived cats and, as shown in Figure 7, found no evidence for any selective effect of monocular deprivation. Together, these results suggest that the developmental plasticity of the LGN is not delayed by prolonged dark rearing.

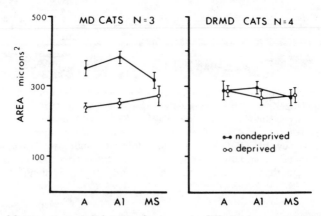

FIGURE 7. Mean cross-sectional areas of neurons in different layers of the LGN of light-reared and dark-reared cats following monocular deprivation. Vertical bars correspond to 1 standard error (from Mower et al., 1985).

Time Course of Extended Sensitive Period

The sensitive period for light-reared animals reaches a peak around the fifth week and then declines over the next 5 months or so. If dark rearing were to leave the visual system frozen at a very early stage of development, which then proceeded normally on exposure to light, one might expect that the delayed sensitive period would follow the same time course. But this does not seem to be the case. Cynader (1983) examined the decline in susceptibility following dark rearing by delaying the time at which monocular occlusion was imposed. After 4 months of dark rearing, kittens were permitted various periods of binocular experience before one eye was closed. He found that there was a decline in susceptibility as the deprivation period was delayed, so that after only about 6 weeks of normal exposure, monocular occlusion did not produce a shift in ocular dominance in favor of the open eye. However, even when the period of normal experience prior to eye closure was extended beyond 8 weeks, there was still a striking breakdown of binocularity, suggesting that the monocular occlusion was having some effect. These data are summarized in Figure 8. Behavioral measurements of visual acuity

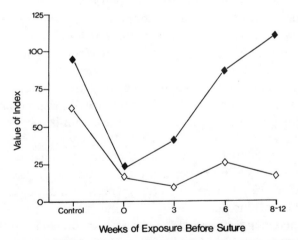

FIGURE 8. Profile of the sensitive period for monocular deprivation following 4 months of dark rearing. Closed symbols: value of takeover index (as defined in Fig. 3) for kittens monocularly deprived for 2 months, beginning at different times after being brought out from the dark. Open symbols: value of binocularity index (as defined in Fig. 3) for the same kittens. The control points refer to dark-reared cats with no visual experience (redrawn from Cynader, 1983).

in cats raised under similar conditions to those of Cynader (1983) were obtained by Timney et al. (1980). They found also that the monocular deprivation effect became less pronounced as the period of normal experience was extended.

NATURE OF EXTENDED SENSITIVE PERIOD

It is clear from the data reviewed so far that dark rearing modifies the age-dependent decline in visual plasticity observed in normal kittens. The simplest interpretation of this prolonged susceptibility is that dark rearing "freezes" the visual system in an immature state; then, when the animal is brought into the light, the normal developmental sequence is set in motion. According to this view, the changes that occur in the dark-reared cat brought out into the light should be very similar to those of a normally reared kitten with comparable visual experience and the delayed sensitive period should be indistinguishable from the normal one.

There are several arguments against such an interpretation. The first suggestion that the visual system is not simply frozen comes from studies of the development of cortical response properties in dark-reared kittens. As mentioned earlier, although normal development proceeds through the first few weeks of life, there is a decline in the response quality of cortical neurons beyond that age (Frégnac and Imbert, 1978). The dark-reared cortex is not a neonatal cortex.

Although there does not appear to be any marked decline in responsiveness in the cortex after about the first 6 weeks, there is an interaction between the period of deprivation and the extent of recovery. From the earliest studies (Cynader and Mitchell, 1980; Cynader et al., 1976; Timney et al. 1978, 1980) it was apparent that the rate and extent of recovery from early dark rearing depended on the duration of the initial deprivation period. In addition, the effectiveness of monocular deprivation in modifying ocular dominance diminishes as the dark-rearing period is lengthened, suggesting that the system is gradually losing its potential for recovery of any kind (see Fig. 3).

If we look at the change in susceptibility as a function of normal experience, a comparison of the data presented in Figures 1 and 8 suggests that the time course of the sensitive period after dark rearing is different from that in light-reared kittens. Although any conclusions should be drawn cautiously, because of procedural differences in the experiments and in the manner of calculating the deprivation effect and takeover index, it appears that dark-reared kittens are maximally susceptible immediately after being removed from the dark, and susceptibility declines rapidly with exposure to a normal environment. Together, these data argue strongly against the view that dark rearing simply "freezes" the system in an immature state. They show also that the sensitive period following dark rearing cannot be considered equivalent to the normal sensitive period.

A more moderate alternative suggested first by Cynader (1983) and elaborated upon by Mower and Christen (1985) is that the effect of dark rearing is somehow to slow down the whole developmental process. The data that are available are consistent with this view. One line of evidence, in addition to that cited above, comes from Cynader (1983). He raised kittens in normal illumination for 2 months before putting them in the dark for 3 months. After they were brought into the light, they were allowed 2 weeks of monocular vision before single-unit recording. A strong version of the freezing hypothesis would argue that the effect of total visual deprivation is to arrest development at a given level. Therefore, the cortex of the 5-month-old dark-reared animal should be equivalent to that of a normal 2-month-old animal. If dark rearing was having no effect, the kitten should be like a 5-month-old normal kitten. Cynader compared his experimental animals to control animals of both these age groups and found that the distribution of ocular dominance fell midway between the 2- and 5-month-old kittens. While this result is not definitive in isolation, it certainly supports the view that there is a gradual decline in plasticity as dark rearing is prolonged.

There is some limited evidence that the development of plasticity is slowed also. In normal kittens there is an increase in susceptibility up to a peak during the fifth week. Imbert and Buisseret (1975) have shown that changes in orientation selectivity are achieved more readily in 5-week-old light-reared kittens than in those that have been dark reared. Similarly, Schecter and Murphy (1976) reported that changes in ocular dominance required more time in dark-reared kittens than light-reared animals of the same age.

Role of Light in Developmental Process

When dark-reared kittens are brought into the light, initially they show a great deal of cortical plasticity, but this declines fairly rapidly. Obviously, the introduction of visual experience is the trigger that sets in motion the developmental changes. It is appropriate now to ask which aspect of visual experience is important. Dark rearing is the most extreme form of visual deprivation because it prohibits any kind of direct visual stimulation. But it is possible to deprive an animal of patterned visual experience without depriving it of light. This is usually accomplished by suturing closed the lids of both eyes and permitting the animal to live in a normally illuminated environment.

Although the two manipulations of dark rearing and binocular lid suture were considered at first to be equivalent, there is good evidence to suggest that they have quite different effects on the developing visual system. In contrast to the relatively high proportion of visually responsive, binocular, cells in dark-reared kittens, the majority of cells in the lid-sutured kittens tend to be unresponsive or have abnormal receptive fields. Of those cells that can be characterized, monocular units predominate (Cynader et al. 1976; Kratz and Spear, 1976; Mower et al., 1981). There are also differences with respect to recovery from deprivation. While Timney et al. (1978) obtained virtually complete recovery of visual acuity among dark-reared kittens, Smith et al. (1980) observed only partial recovery in lid-sutured animals. Although it is possible that procedural differences between the Timney et al. and the Smith et al. studies contributed to the observed differences between the experimental conditions, Mower et al. (1982) carried out a study in which dark rearing and lid suture were compared directly and confirmed the original findings.

Mower et al. (1981, 1982) extended their comparison of dark rearing and binocular occlusion by studying dark-reared kittens that had one eye closed after they were brought into the light and binocular lid-sutured kittens that were permitted a period of monocular vision after the initial deprivation period. Unlike the dark-reared animals, there was no indication that the period of monocular exposure had any effect on the previously lid-sutured kittens. Indeed, for these animals there was only limited evidence for any recovery from the initial period of binocular occlusion. Exactly the same pattern of results was obtained behaviorally: the lid-sutured kittens showed only limited recovery of acuity in the experienced eye, and there was no additional effect of monocular deprivation.

The major difference between the outcome of dark rearing and lid suturing is that while the cortex of the dark-reared animals appears normal but immature, the cortex following binocular lid suture is quite abnormal. This abnormal responsiveness and low susceptibility to monocular deprivation suggests that lid-sutured animals do pass through a developmental sequence but that it is atypical. The reason for this may be found by considering the type of visual stimulation received: dark rearing is total deprivation, whereas binocular lid suture degrades the quality of the retinal image but does not deprive the animal of visual experience. The extent of that experience has been documented by Spear et al. (1978), who reported that cortical neurons responded to stimulation through the closed lids. Also, Loop and Sherman (1977) have shown that simple brightness discriminations can be made by lid-sutured kittens.

It seems reasonable, therefore, to suggest that the anomalous visual stimulation experienced during lid suture is sufficient to allow development, but development that is abnormal. Such a proposal was made by Jampolsky (1978), who argued that a distinction should be made between total occlusion and diffusion, which he considered to be visual white noise. He proposed that under white-noise conditions the stimulation provided by the light permits development to progress at its normal rate, but because the quality of visual experience is inadequate, the developmental sequence is inappropriate.

It is known that if young dark-reared kittens are given very brief periods of visual exposure, there is a very rapid development of cortical selectivity (Buisseret et al., 1978, 1982). In addition, Rauschecker and Singer (1982) have reported that if kittens were given a brief period of normal monocular or binocular experience near the peak of the standard sensitive period and then binocularly deprived until they were well over a year old, cortical responsiveness was similar to that of kittens deprived from birth. This result indicates that large changes can occur in dark-reared cortex and suggest further the possibility that light may act as a trigger to set off a developmental sequence. Mower and his colleagues have explored this possibility (Mower et al., 1983, 1985). Kittens were reared in total darkness from birth until the age of 4 or 5 months except for 1 or 2 days during the second month when they were allowed normal visual exposure. After dark rearing single-unit recordings were made immediately and/or after a period of monocular occlusion. Compared to control dark-reared animals, cortical responsiveness was degraded in all of the kittens that had received the brief visual experience. When recordings were made after the monocular occlusion period, there was no evidence for a shift in cortical ocular dominance in favor of the experienced eye. Thus, it seems that brief exposure to light acts as a switch to activate a developmental mechanism that then continues to run in the absence of visual experience.

NEURAL DEVELOPMENT AND NEURAL PLASTICITY

Several authors have pointed out the relationship between periods of normal neural development and susceptibility to environmental manipulation. The peak of the normal sensitive period occurs at a time when synaptogenesis in the cortex is just being completed and cortical neurons are taking on their adultlike qualities (Buisseret and Imbert, 1976; Cragg, 1975; Pettigrew, 1974). The studies reviewed above demonstrate that there is a close relationship between the potential for developmental changes following dark rearing and susceptibility to monocular deprivation. This relationship is perhaps best illustrated by Cynader and Mitchell's (1980) report that recovery of cortical selectivity may take place over a few weeks following dark rearing and that the delayed sensitive period seems to follow a similar time course (Cynader and Mitchell, 1980).

A second line of evidence that the timing of the delayed sensitive period is associated with periods of neural development comes from Mower et al.'s (1982, 1985) demonstration that monocular occlusion following dark rearing has little effect on the anatomy or physiology of the LGN, a structure whose development is essentially independent of visual experience (Kalil, 1978; Kratz, 1982; Kratz et al., 1979). In a similar manner, although cortical ocular dominance columns do not develop normally in the absence of visual experience, there is no evidence of any recovery after 4 months of dark rearing and no effect of monocular deprivation (Mower et al., 1985). It would be of interest to determine if ocular dominance columns can be affected by monocular occlusion during the time when some recovery is possible (Swindale, 1982).

Most of the data discussed here are consistent with the hypothesis that the effect of dark rearing is to slow down development within the visual cortex and that the extension of the sensitive period for monocular deprivation is a direct consequence of this delay. The questions that remain to be answered are how dark rearing impedes cortical development and how a brief period of visual stimulation can activate the developmental process.

ACKNOWLEDGMENTS

Preparation of this chapter was supported by grants from the Medical Research Council and the Natural Sciences and Engineering Research Council of Canada. Vanessa Russell provided invaluable assistance in getting the manuscript into its final form.

REFERENCES

Albus, K. and W. Wolf (1984). Early post-natal development of neuronal function in the kitten's visual cortex: A laminar analysis. *J. Physiol.* **348**:153–185.

Aslin, R. N. (1982). Experiential Influences and Sensitive Periods in Perceptual Development: A Unified Model. In *Development of Perception: Psychobiological Perspectives*, R. N. Aslin, J. R. Alberts, and M. R. Petersen, eds., vol. 2, pp. 45–93, Academic Press, New York.

Barlow, H. B. and J. D. Pettigrew (1971). Lack of specificity of neurones in the visual cortex of young kittens. *J. Physiol. (Lond.)* **218**:98–100.

Berman, N. and N. W. Daw (1977). Comparisons of the critical periods for monocular and directional deprivation in cats. *J. Physiol.* (Lond.) **265**:249–259.

Berman, N. and H. Murphy (1982). The critical period for alteration in cortical binocularity resulting from divergent and convergent strabismus. *Dev. Brain Res.* **2**:181–202.

Blakemore, C. and R. C. van Sluyters (1974). Experimental analysis of amblyopia and strabismus. *Brit. J. Ophthalmol.* **58**:176–182.

Buisseret, P. and M. Imbert (1976). Visual cortical cells: Their developmental properties in normal and dark reared kittens. *J. Physiol. (Lond.)* **255**:511–525.

Buisseret, P., E. Gary-Bobo, and M. Imbert (1978). Ocular motility and recovery of orientational properties of visual cortical neurones in dark-reared kittens. *Nature* **272**:816–817.

Buisseret, P., E. Gary-Bobo, and M. Imbert (1982). Plasticity in the kitten's visual cortex: Effects of the suppression of visual experience upon the orientational properties of visual cortical cells. *Dev. Brain Res.* **4**:417–426.

Cragg, B. G. (1975). The development of synapses in the visual system of the cat. *J. Comp. Neurol.* **160**:147–166.

Cynader, M. (1977). Extension of the critical period in cat visual cortex. *Invest. Ophthalmol. Vis. Sci. (ARVO Abstr.)* **16**:163.

Cynader, M. (1983). Prolonged sensitivity to monocular deprivation in dark-reared cats: Effects of age and visual exposure. *Dev. Brain Res.* **8**:155–164.

Cynader, M. and D. E. Mitchell (1980). Prolonged sensitivity to monocular deprivation in dark-reared cats. *J. Neurophysiol.* **43**:1026–1039.

Cynader, M., N. Berman, and A. Hein (1976). Recovery of function in cat visual cortex following prolonged deprivation. *Exp. Brain Res.* **25**:139–156.

Cynader, M., B. N. Timney, and D. E. Mitchell (1980). Period of susceptibility of kitten visual cortex to the effects of monocular deprivation extends beyond six months of age. *Brain Res.* **191**:545–550.

Daw, N. W. and M. J. Wyatt (1976). Kittens reared in a unidirectional environment: Evidence for a critical period. *J. Physiol. (Lond.)* **257**:155–170.

De Courten, C. and L. J. Garey (1982). Morphology of the neurons in the human lateral geniculate nucleus and their normal development. *Exp. Brain Res.* **47**:159–171.

Derrington, A. M. and A. F. Fuchs (1981). The development of spatial-frequency selectivity in kitten striate cortex. *J. Physiol. (Lond.)* **316**:1–10.

Dews, P. B. and T. N. Wiesel (1970). Consequences of monocular deprivation on visual behaviour in kittens. *J. Physiol. (Lond.)* **206**:437–455.

Dobbing, J. (1968). Vulnerable Periods in Developing Brain. In: *Applied Neurochemistry*, A. N. Davison and J. S. Dobbing, eds., pp. 287–316, Blackwell, Oxford.

Frégnac, Y. and M. Imbert (1978). Early development of visual cortical cells in normal and dark-reared kittens: Relationship between orientation selectivity and ocular dominance. *J. Physiol. (Lond.)* **278**:27–44.

Giffin, F. and D. E. Mitchell (1978). The rate of recovery of vision after early monocular deprivation in kittens. *J. Physiol. (Lond.)* **274**:511–537.

Guillery, R. W. and J. Stelzner (1970). The differential effects of unilateral lid closure upon the monocular and binocular segments of the dorsal lateral geniculate nucleus in the cat. *J. Comp. Neurol.* **139**:413–442.

Hickey, T. L. (1977). Postnatal development of the human lateral geniculate nucleus: Relationship to a critical period for the visual system. *Science* **198**:836–838.

Hubel, D. H. and T. N. Wiesel (1970). The period of susceptibility to the physiological effects of unilateral eye closure in kittens. *J. Physiol.* **206:**419–436.

Imbert M. and P. Buisseret (1975). Receptive field characteristics and plastic properties of visual cortical cells in kittens reared with or without visual experience. *Exp. Brain Res.* **22:**25–36.

Jampolsky, A. (1978). Unequal Visual Inputs and Strabismus Management: A Comparison of Human and Animal Strabismus. In: *Symposium on Strabismus. Transactions of the New Orleans Academy on Ophthalmology*, pp. 358–492, C. V. Mosby, St. Louis.

Kalil, R. (1978). Dark rearing in the cat: Effects on visuomotor behavior and cell growth in the dorsal lateral geniculate nucleus. *J. Comp. Neurol.* **178:**451–467.

Kratz, K. E. (1982). Spatial and temporal sensitivity of lateral geniculate cells in dark-reared cats. *Brain Res.* **251:**55–63.

Kratz, K. E. and P. D. Spear (1976). Effects of visual deprivation and alterations in binocular competition on responses of striate cortex neurons in the cat. *J. Comp. Neurol.* **170:**141–152.

Kratz, K. E., S. M. Sherman, and R. Kalil (1979). Lateral geniculate nucleus in dark-reared cats: Loss of Y cells without changes in cell size. *Science* **203:**1353–1355.

Lehmkuhle, S., K. E. Kratz, S. C. Mangle, and S. Sherman (1978). The effect of early monocular lid suture upon the development of X-cells in the cat's lateral geniculate nucleus. *Brain Res.* **157:**346–350.

LeVay, S., M. P. Stryker, and C. J. Shatz (1978). Ocular dominance columns and their development in area IV of the cat's visual cortex: A quantitative study. *J. Comp. Neurol.* **179:**223–244.

Loop, M. S. and S. M. Sherman (1977). Visual discrimination during eyelid closure in the cat. *Brain Res.* **128:**329–339.

Mower, G. D. and W. D. Christen (1985). Role of visual experience in activating critical period in cat visual cortex. *J. Neurophysiol.* **53:**572–589.

Mower, G. D., J. L. Burchfiel, and F. H. Duffy (1981). The effects of dark-rearing on the development and plasticity of the lateral geniculate nucleus. *Dev. Brain Res.* **1:**418–424.

Mower, G. D., C. J. Caplan, and G. Letsou (1982). Behavioral recovery from binocular deprivation in the cat. *Beh. Brain Res.* **4:**209–215.

Mower, G. D., W. G. Christen, and C. J. Caplan (1983). Very brief visual experience eliminates plasticity in the cat visual cortex. *Science* **221:**178–180.

Mower, G. D., C. J. Caplan, W. G. Christen, and F. H. Duffy (1985). Dark-rearing prolongs physiology but not anatomical plasticity of the cat visual cortex. *J. Comp. Neurol.* **235**:448–466.

Olson, C. R. and R. D. Freeman (1980). Profile of the sensitive period for monocular deprivation in kittens. *Exp. Brain Res.* **39**:17–21.

Pettigrew, J. D. (1974). The effect of visual experience on the development of stimulus specificity by kitten cortical neurones. *J. Physiol. (Lond.)* **237**:49–74.

Presson, J. and B. Gordon (1979). Critical period and minimum exposure required for the effects of alternating monocular occlusion in cat visual cortex. *Vision Res.* **19**:807–811.

Rauschecker, J. P. and W. Singer (1982). Binocular deprivation can erase the effects of preceding monocular or binocular vision in kitten cortex. *Dev. Brain Res.* **4**:495–498.

Schechter, P. and E. H. Murphy (1976). Brief monocular visual experience and kitten cortical binocularity. *Brain Res.* **109**:165–168.

Shapley, R. and Y. T. So (1980). Is there an effect of monocular deprivation on the proportions of X and Y cells in the cat lateral geniculate nucleus. *Exp. Brain Res.* **39**:41–48.

Shatz, C. J. and M. P. Stryker (1978). Ocular dominance in layer IV of the cat's visual cortex and the effects of monocular deprivation. *J. Physiol. (Lond.)* **281**:267–283.

Shatz, C. J., S. Lindstrom, and T. N. Wiesel (1977). The distribution of afferents representing the right and left eyes in the cat's visual cortex. *Brain Res.* **39**:41–48.

Sherman, S. M. (1972). Development of interocular alignment in cats. *Brain Res.* **37**:187–203.

Sherman, S. M. (1973). Visual field defects in monocularly and binocularly deprived cats. *Brain Res.* **49**:25–45.

Smith, D. C., R. Lorber, L. R. Stanford, and M. S. Loop (1980). Visual acuity following binocular deprivation in the cat. *Brain Res.* **183**:1–11.

Spear, P. D., L. Tong, and A. Langsetmo (1978). Striate cortex neurons of binocularly deprived kittens respond to visual stimuli through the closed eyelids. *Brain Res.* **155**:141–146.

Swindale, N. V. (1981). Absence of ocular dominance patches in dark-reared cats. *Nature* **290**:332–333.

Swindale, N. V. (1982). Formation of ocular dominance patches in kitten visual cortex after periods of dark-rearing. *J. Physiol. (Lond.)* **332**:13P.

Timney, B. (1983). The effects of early and late monocular deprivation on binocular depth perception in cats. *Dev. Brain Res.* **7**:235–243.

Timney, B. (1985). Visual Experience and the Development of Depth Perception. In: *Brain Mechanisms and Spatial Vision*, D. J. Ingle, M. Jeannerod, and D. N. Lee, eds., pp. 147–174, Martinus Nijhoff, Netherlands.

Timney, B. and D. E. Mitchell (1979). Behavioural Recovery from Visual Deprivation: Comments on the Critical Period. In: *Developmental Neurobiology of Vision*, R. D. Freeman (ed.), pp. 149–160, Plenum Press, New York.

Timney, B., D. E. Mitchell, and M. Cynader (1980). Behavioral evidence for prolonged sensitivity to effects of monocular deprivation in dark-reared cats. *J. Neurophysiol.* **43**:1041–1054.

Timney, B., D. E. Mitchell, and F. Giffin (1978). The development of vision in cats after extended periods of dark rearing. *Exp. Brain Res.* **31**:547–560.

Wiesel, T. N. and D. H. Hubel (1963a). Effects of visual deprivation on morphology and physiology of cells in the cat's lateral geniculate nucleus. *J. Neurophysiol.* **26**:979–993.

Wiesel, T. N. and D. H. Hubel (1963b). Single cell responses in striate cortex of kittens deprived of vision in one eye. *J. Neurophysiol.* **26**:1003–1017.

PART III
EPILOGUE

15

CORTICAL PLASTICITY AND IMPRINTING: BEHAVIORAL AND PHYSIOLOGICAL CONTRASTS AND PARALLELS

Josef P. Rauschecker

Max-Planck-Institut für Biologische Kybernetik, Tübingen, FR Germany

Peter Marler

Rockefeller University Field Research Center, Millbrook, New York

NATURE–NURTURE AND THE ISSUE OF INSTRUCTIVE AND SELECTIVE PLASTICITY

One of the more enduring philosophical controversies is the problem of innate knowledge. On the one hand are the nativists exemplified by Kant and Descartes, who assert that the acquisition of human knowledge is only comprehensible if one postulates a range of specific, innate abilities. On the other hand are the empiricists, led by John Locke, who view the invocation of innate knowledge as redundant. An extreme manifestation of empiricist tenets is found in the behavioristic movement of J. B. Watson and his followers and their modeling of the mind of a human infant as a tabula rasa.

Modern equivalents to these antipodal philosophical positions on the acquisition of knowledge can be found in neurobiology. The presumption is held by some that aside from provision of the major physiological requisites for learning, the brain remains largely uncommitted as to the particulars of what kind of knowledge is acquired. According to this view, the processes of development and neural plasticity involved depend primarily on instructions derived from environmental interactions. This is essentially an empiricist position. An alternative stance is taken in theories of neuroselection, as espoused in recent years by Changeux and Edelman (Changeux and Danchin, 1974, 1976; Changeux et al., 1984; Edelman, 1978, 1981; Edelman and Finkel, 1984). Here the presumption is that developmental processes underlying the acquisition of some kinds of knowledge, if not all, are extensively and pervasively preordained. By this line of theorizing, effects of certain experiences, if not all, are thought to be acquired by selection from existing neuronal circuitry rather than by the instructive creation of circuitry de novo.

Neither of these positions is advocated by anyone in as extreme a form as characterized here, but the two trends can be discerned in contemporary research on the neurobiology of development. According to the nativist, innate predispositions loom large in interpretation of the plasticity as manifest in early development. External stimuli are by no means equipotential in their capacity to evoke enduring changes in neural organization and behavior. Some have no effect, even though there is every reason to suppose that they are processed normally by the organism's sensory apparatus. At the other extreme are forms of stimulation that result in rapid, incisive, and sometimes relatively irreversible changes. The threshold for potential plasticity in response to such stimulation is often lower during certain phases of development than at any other time, leading both neurobiologists and ethologists to speak of sensitive periods for experiential modifiability. Somewhere between these two types of stimulation, one ineffectual, the other uniquely potent, are forms of stimulation of many kinds that have the potential to engender neural and behavioral reorganization in the right circumstances. In other words, they can be learned. Here the thresholds for inducing durable changes are likely to be higher than in the previous case, and concomitant activation of state-dependent mechanisms may be required for learning to occur, such as are associated with motivational changes and noxious stimulation. The distinction between these two types of external stimulation, both effective, one with minimal thresholds for direct, specific, and relatively irreversible reorganizational effects, the other requiring longer periods of stimulation and other concomitant environmental and state-dependent events, can be discerned in both visual development and imprinting. It is a distinction that can be made even at the single-unit level in visual development.

WHICH SINGLE-UNIT RESPONSE PROPERTIES ARE PLASTIC DURING NEURAL DEVELOPMENT AND WHAT IS THE BEHAVIORAL SIGNIFICANCE OF THIS PLASTICITY?

The strength of classical ethology has always been to enquire into behavioral significance of natural events using teleonomical arguments. In more modern terms, one could refer to the same theme by stressing that biological systems are cooperatively organized or are governed by synergetic mechanisms (Haken, 1983; Edelman, 1981). Whatever terms are used, it is certainly important also for single-unit neurophysiologists to ask about the behavioral significance of cortical plasticity. An orthodox nativist could argue that cortical plasticity is not necessary because everything can be genetically predetermined. This is patently not the case, however. For example, if we think about the binocularity of cortical neurons in the growing organism, both interocular distance and diameter of the eyeballs are continually increasing during a certain period of postnatal development. This means that binocular correspondence cannot be fully preordained at birth, or it would necessarily be disrupted by growth of the head and the eyes. The only way to manage *without* experience-dependent plasticity of neural connections would be by means of a complicated genetic program taking into account growth rates of different organs at different times during ontogeny. Even the most sophisticated genetic program, however, could not foresee natural variations in rates of growth in different individuals. Completely pre-programmed changes in binocular wiring could never produce mapping as precise as that generated almost automatically by experience-dependent plasticity using the activation of nervous networks by external stimuli and internal feedback.

In the visual system these developmental interdependencies between genetic and epigenetic information have long been recognized (see, e.g., Barlow, 1975). Studies on the auditory system of owls have generalized this consideration to other sensory systems. Sound localization depends on interaural intensity differences and interaural delay. The correspondence between values in these two parameter domains and the spatial position of a sound source changes inevitably with head growth during ontogeny. Knudsen (this volume) showed that the nervous connections in the auditory system of the owl that participate in sound localization are highly plastic during a sensitive period of postnatal development and depend on experience for their stabilization.

Of course, nature has to pay a price for this elegant self-tuning process: it is highly susceptible to false or distorted information from an unusual environment or to errors in the peripheral sensory apparatus of the individual itself. This susceptibility to misinformation has been exploited in deprivation experiments on young animals to unravel the mechanisms of self-tuning and the time course of the sensitive periods involved. As long as, in a natural setting, unusual environments are only temporary during the sensitive period, their effects can still be corrected. In fact, it appears that restoration of the "normal" calibrated state is achieved more easily than induction of the distorted state. This is certainly the case for the owl's auditory localization system (Knudsen, this volume), and experiments on the orientation specificity of neurons in the cat's visual cortex can also be interpreted in this way: if young kittens are reared in the dark for 6 weeks until orientation specificity of cortical cells has been abolished, then only 6 hr of visual experience are sufficient to restore selectivity to normal levels (Imbert and Buisseret, 1975).

If misinformation from the environment persists until well after the end of the sensitive period, as is the case, for example, if the individual is born with an optical abnormality, it may be more beneficial to make the resulting neuronal changes permanent. If an animal is born with a squint, the following strategies are possible for the nervous system to prevent double vision: (1) an anomalous correspondence can be established between retinal loci that normally project to different cortical neurons; (2) the inputs from the two eyes to the cortex can be separated completely, leading to only monocular cells, alternating fixation, and a loss of stereopsis; and (3) one eye can be suppressed permanently ("amblyopia"). All three strategies are used by the brain both in the natural and experimental cases, but the circumstances under which one is chosen are not totally understood at present. Evidently, the highest degree of plasticity is required for establishing an anomalous correspondence, and this strategy is limited by the range over which afferent projections arborize (Gilbert and Wiesel, 1983). Alternating fixation is usually established with divergent squint in humans, while amblyopic suppression occurs most frequently in convergent squinters (Von Noorden, 1985). Obviously, loss of stereopsis and of vision in one eye are behaviorally less disturbing for an individual than conflicting information arising from double vision.

The neural correlate of deprivation amblyopia is a virtually complete loss of neurons in area 17 excitable from the deprived eye; the correlate of stereoblindness is an irreversible loss of binocular units in the striate cortex (Wiesel and Hubel, 1963; Hubel and Wiesel, 1965). In squint amblyopia suppression is correlated with an increase in the visual response latency of visual cortical units driven from the deviating eye (Eschweiler et al., 1984).

Thus, the neural consequences of visual deprivation can be quite severe, and some kind of compensation by other sensory modalities might be expected. Casual reports about blind people with supranormal tactile and auditory capacities find experimental confirmation in quantitative behavioral studies on visually deprived cats (Rauschecker and Dehnen, 1986; Rauschecker, this volume). The obvious next step is to explore the anatomical basis of these changes and to study the means by which the nervous system achieves this compensatory plasticity. The findings suggest that both cooperative *and* competitive interactions are indeed at work during ontogenetic development, having as their "goal" the optimization of nervous processing.

PHENOMENOLOGY OF SENSITIVE PERIODS

The threshold of responsiveness of organisms to stimulation is enormously variable, especially with regard to its long-lasting consequences for neural organization and behavior. Repeatedly one finds that a given amount and quality of stimulation results in profound organizational changes in certain phases of development and at other phases has little or no effect. Such phases, when the organism is uniquely susceptible to the influence of certain stimuli, are characterized as sensitive periods. They merit this designation irrespective of whether equivalent organizational changes are impossible at other times, or can be achieved, but only by dint of much more stimulation, or by means of other concomitant conditions not required at the peak of the sensitive period. It was the realization that imprinting is not totally irreversible, but only difficult to reverse outside this sensitive period, that inclined some ethologists to discard the term *critical* period, with its all-or-nothing implications in favor of the more neutral term *sensitive* period (Bateson, this volume).

It is also important to avoid deification of the sensitive period concept. It is self-evident from the material in this book that, within the developmental program of a single organism, there is not one sensitive period but many. Each has its own set of determinants and functional connotations. As Bateson (this volume) indicates, some may be no more than epiphenomena, reflecting stages of development when rapid and profound reorganization is taking place. There are many indications of principles of organization that are shared, but there is no *necessary* implication that different sensitive periods share common causal factors. On the contrary, the available evidence suggests a multiplicity of mechanisms. Some sensitive periods seem to be governed by endogenous "clocks" that are resistant to external influence, and others depend acutely on the prior and current experiential history of the organism.

Bearing in mind these hazards, the heuristic value of the concept of sensitive periods cannot be doubted. It serves to focus attention on the variability of responsiveness of organisms to certain kinds of stimulation and on the existence of nonuniformity in the consequences of activity for the subsequent state of the organism. It has profound implications for our understanding of the capacity of organisms to benefit from interactions with their environments. Recent research has revealed many illustrations of the diversity of sensitive period phenomena.

Different Sensitive Periods within One Visual Area

The classical work on visual cortical plasticity was performed in a single cortical area, the primary visual cortex (area 17). Moreover, for a long time only a single response property of visual cortical neurons was used for the study of the sensitive period in this cortical area: development of the ocular dominance of single neurons has been a primary focus, in part because manipulation of the binocular input by eye closure is by far the easiest way to restrict the visual input experimentally. Furthermore, monocular deprivation has proved to be very reliable in producing consistent results in different laboratories.

When attempts were first made to influence other response properties of cortical neurons, such as orientation or direction selectivity, it was automatically assumed that the sensitive period for these parameters would be the same as that for ocular dominance. Daw and co-workers first pointed out that there may be different sensitive periods for different response properties in area 17 neurons (Daw et al., 1978). It appears that the sensitive period for direction selectivity does indeed occur earlier than that for changes in binocularity.

Even more conclusive is the point that different critical periods may exist for the plasticity of different cell types, such as X and Y cells in the cat, or parvo- and magno-cellular layer inputs in the monkey (Sherman and Spear, 1982; Wiesel, 1982). Multiple sensitive periods have also been demonstrated at the behavioral level for different visual functions in primates (Harwerth et al., 1986). The sequence in which the sensitive periods come to a close seems to mirror the hierarchy of processing stages in the visual system with which the particular functions are associated.

Different Sensitive Periods within One Sensory System

Apart from areas 17, 18, and 19, the cerebral cortex of mammals is known to contain at least another dozen representations of the visual field (Woolsey, 1981). Neuroscientists are far from understanding the functional significance of these multiple representations, which exist for the auditory and somatosensory modalities as well. An obvious conjecture is, of course, that different maps emphasize different aspects of the visual world. Much of the processing of color, movement, and form may take place in different cortical areas (Zeki, 1978; Van Essen and Maunsell, 1983).

Research on the developmental plasticity of these different visual areas has only just begun. However, work by Spear and collaborators indicates that the sensitive period in at least one of the extrastriate areas is different from that in striate cortex (Tong et al., 1984; Jones et al., 1984).

Sensitive Periods May Vary from One Sensory Modality to Another

The next step in coming to a general understanding of sensitive periods in sensory development is to compare, in the same species, plasticity of sensory representations in different modalities. We have stated before that it was certainly not by chance that the visual system was chosen by experimenters for initial studies of plasticity. Perturbations of sensory input are more difficult to perform in other modalities.

Few developmental studies have been done on mammalian auditory cortex (see Konishi, this volume). In the somatosensory cortex the work of Merzenich and colleagues (1984b) indicates that plasticity of neural connections may extend well into adulthood. Whether this constitutes a real difference in the time course of sensitive periods or whether it is due to a difference in experimental procedure remains to be seen. Injury to sense organs, in this case, for instance, the amputation of a finger, is a much more drastic intervention than the manipulation of sensory input to an otherwise intact organ (see also Van der Loos and Woolsey, 1973 for plasticity of whisker representation in mice). Even increased stimulation of one finger in an adult monkey may cause an expansion of the corresponding representation in somatosensory cortex, however (W. Jenkins, personal communication; cited in Merzenich et al., 1984a). The question remains, then, whether adult plasticity in the form of "mobile maps" can usefully be viewed as a residual form of developmental modifiability or whether there are specific properties whose plasticity is indeed largely restricted to a sensitive period.

Species Differences in Sensitive Periods

One direct source of evidence of genomic influences on sensitive periods is the existence of species differences. Research on song learning has provided some of the most compelling illustrations (Marler, this volume). Some birds possess sharply defined sensitive periods restricted firmly to the first few weeks of life. Others are more extended, encompassing a period of up to 9 months. In some songbirds there is potential vocal plasticity throughout life, although it has yet to be determined whether the potential is present continuously or displays annually recurrent cyclicity, as seems probable in the canary (Nottebohm and Nottebohm, 1978). Here plasticity appears to coincide with an annual pattern of growth and shrinkage of some of the song control nuclei in the brain associated with changes in the extent of dendritic trees and even with indications of seasonal neurogenesis (Nottebohm, 1984; DeVoogd et al., 1985; Goldman and Nottebohm, 1983; Bottjer and Arnold, 1986). Even in close genetic relatives the precise timing of sensitive periods for song learning may differ in adaptive relationships with particulars of life style and patterns of dispersal.

For some developmental events optimal timing will differ from species to species, but for other events temporal requirements are general, widely shared across species and relatively conservative, as with the ontogeny of sensory processing. Such is the case with visual development, although even here there may be a potential for species differences (e.g., between animals with different degrees of binocularity). There has been little effort to place comparative investigation of visual development in functional or evolutionary contexts. The available data do suggest, however, a lack of plasticity in species with a low degree of binocular interaction, such as the rabbit (see Blakemore, 1978).

SENSITIVE PERIODS ARE MODIFIABLE, WITHIN LIMITS

An early objection to the term *critical* period was the implication of stereotypy and immutability. There is ample evidence that both the time course of sensitive periods and sensitivity within them can be changed, although there are usually limits to the extent of their modifiability. We propose that at least two kinds of factors participate in their modification. One involves generalized intense arousal mechanisms and the other involves deprivation of specific forms of stimulation and the resulting patterns of use of neural circuitry. Analogs of both are to be found in visual development and in imprinting.

Some of the most extreme modifications of the timing of song learning in birds have been achieved by substituting for tape recordings live, socially interactive males as tutors, presumably constituting a strong form of stimulation that generates a high degree of arousal. It is intuitively reasonable that one means for instating a readiness to learn and compensating for an increasing threshold for plasticity is to induce a strong state of arousal. Conversely a lack of arousal should reduce plasticity.

A possible pharmacological basis for the effects of arousal can be seen in the norepinephrine hypothesis of plasticity as it was formulated for visual development (Kasamatsu & Pettigrew, 1976). It has been suggested that alterations of visual plasticity can be achieved by exogenous manipulation of norepinephrine levels in the brain. Similar effects have been reported by interference with brain loci such as the locus coeruleus, known to be involved in arousal and to contain noradrenergic neurons (see Daw et al., 1985, and chapters by Cooper, Frégnac, Rauschecker, and Singer, all this volume, for a discussion of this still controversial hypothesis).

The withholding of certain kinds of stimulation is another effective means of modifying sensitive periods. Deprivation of general stimulation often has the effect of delaying closure of a sensitive period, as is the case with filial imprinting of chicks reared in either light or darkness (Bateson, this volume), perhaps attributable to a lack of arousal. There is also evidence, however, of an equivalent impact of the withholding of highly specific stimulation, the arousal potential of which is much less obvious. The possibility of special developmental roles for particular kinds of stimulation in both imprinting and visual development merits further consideration.

ROLE OF SPECIFIC STIMULATION

There is growing evidence that, far from being tabulae rasae, organisms are attuned to be especially responsive to particular types of stimulation at certain stages of development despite *potential* responsiveness to an enormously wide range of stimuli. The most effective environmental influences are often most appropriately defined in terms of their qualitative stimulus characteristics rather than quantitative ones such as overall stimulus strength. This is not to say that arousal may not be one of the relevant consequences of such stimulation, but the triggering events may be very particular and even species-specific stimuli and not just intensive stimulation of a generalized nature.

Horn's recent finding that in some circumstances a naturalistic model of a jungle fowl takes precedence as an imprinting stimulus over an arbitrary stimulus such as a colored box in chicks is a case in point (Horn

1985). In song learning some birds display innate preferences for certain sound patterns, the optimal set including songs of the bird's own species, even though they are capable of learning songs of other species as well (Marler, this volume). Withholding such specific stimulation may have especially potent effects in delaying closure of a sensitive period. Developmental changes resulting from such specific stimulation may be achieved more rapidly and with lower thresholds than with suboptimal stimulation.

Similar trends can be discerned in visual development. As measured by single-unit responsiveness in the visual cortex, preferences for certain orientations are instated more readily than others depending on the already prevailing tuning of these cells (Rauschecker, this volume). When placed in competition with different orientations, those orientations that are not adequately stimulated become suppressed. Fewer exposures may be required to establish a preference for certain orientations than to instate others, and already induced changes in the direction of less-preferred orientations may be reversible within limits.

Thus, both at the level of the complete organism and at the single-unit level of neuronal organization, there are predispositions to respond more readily to some stimuli than to others. Therefore, fewer exposures are required to establish permanent preferences. Knudsen (this volume) finds a similar phenomenon in the development of brain mechanisms in owls for sound localization. Owls are capable of developing accurate localization with one ear blocked. They display a much extended sensitive period for reversal under such conditions, however, as compared with a bird whose hearing in both ears is unimpaired. Again, allowing an owl with one ear blocked to experience sounds with the blockage removed, even for a short time, will reverse adjustments already instated in response to the unbalanced stimulation. Thus, specific patterns of stimulation have a degree of prepotency in their ability to gain control of development in the face of competing stimulation of other kinds.

The notion of a tabula rasa seems altogether inappropriate for characterizing such phenomena. The evidence suggests rather a relationship to the ethological concept of "sign stimuli," with its implication that organisms are typically capable of responsiveness to wide ranges of stimuli, but nevertheless display certain predispositions or preferences, often uniquely specific, in some cases even differing between closely related species. The major function of the concepts from classical ethology of sign stimuli and their complementary "innate release mechanisms" (Lorenz, 1950) may be not to generate organisms that are reflexive automata, but to aid them in adjusting in a biologically appropriate manner to the effects of individual experience. Far from being antithetical to learning, they often complement and sustain learning in a highly adaptive fashion (Marler, Dooling, and Zoloth, 1980). Such mechanisms may vary in their behavioral and neurodevelopmental significance at different stages of life. Many appear to play their primary role in infancy, becoming redundant as organisms mature, having served their purpose of encouraging certain trajectories during periods of plastic, experientially controlled development.

SPECIFIC AND GENERAL STIMULATION AS ENABLING FUNCTIONS

The critical importance of motivation and arousal in learning is well documented (Kety, 1970). The reinforcing effects of stimuli that elevate arousal above minimal levels and reduce arousal below unpleasantly high levels are well established. Arousal is implicated in the contributions of novelty and intense stimulation in imprinting and in the effects of social interaction on song learning. It also seems to be implicated in visual development, as can be inferred from the suggested effects of exogenous norepinephrine in reinstating plasticity of binocular connections in the visual cortex in adulthood, after closure of the sensitive period (Pettigrew and Kasamatsu, 1978).

There is an increased probability that stimulation experienced in an aroused state or accompanied by stimuli that induce arousal will exert an undue influence on neural and behavioral development. In this sense, exogenous and endogenous agents of arousal may be thought of as enabling stimuli, which facilitate the emergence of enduring effects of other stimuli that accompany them. External stimuli with such enabling effects are often characterized as generalized in nature, involving stimulation that is arousing, attention catching, and even noxious or repellent (Kety, 1976; McGaugh et al., 1984).

In addition to such generalized stimuli, highly specific stimulation can also have enabling effects. The clearest evidence comes from song-learning studies, where the provision of specific song stimuli to which a bird is innately responsive and that provide the basis for conspecific song-learning preferences, will increase the probability that songs of other species will be learned if presented in temporal proximity to them (Marler, this volume). There are indications of a similar phenomenon in imprinting, as when accompaniment of arbitrary visual stimuli by species-specific vocal signals increases the probability that imprinting of ducklings on them will occur (Gottlieb, 1971). It remains to be seen what kind of role specific enabling stimuli play in development of the visual system.

ROLE OF ENABLING FACTORS IN PLASTICITY

The distinction we have made between generalized and specific stimuli playing a role in various developmental processes is quite obvious if we consider the very qualities of the stimuli themselves and their distribution in the natural world, the former ubiquitous, the latter highly localized. The distinction becomes more difficult to make, however, if we look at the problem from a physiological viewpoint, that is, if we think about the effect that these stimuli may have on the underlying neuronal network. The common denominator of both kinds of stimulation is obviously that they greatly enhance the activity of certain parts of the brain, perhaps even at the very same locations. In that respect, one might even argue that some highly specific stimuli could also act like generalized stimuli in that they lead to an increased state of arousal. It could be, for example, that these stimuli evoke "emotional" reactions of fear or attachment or some subset of their neuropharmacological correlates, which then in turn initiate the enabling process.

The effect of these specialized stimuli (or "sign stimuli") may, however, be even simpler and more direct than that. Because of innate preferences, they may evoke a stronger response in the neuronal processing machinery than other stimuli. One could think of them as putting neuronal circuits in the equivalent of a resonant state. The subsequent reverberation would thus lead to exponential potentiation of the activation of the local network.

The common assumption of all modern neurobiological theories of learning, including developmental plasticity, is that stimulation leads to synaptic changes in the underlying neuronal machinery. These synaptic changes depend on neuronal activity. The enabling function of both generalized and highly specific stimuli would then consist of increasing the activity level of corresponding neurons, thus accelerating the synaptic changes.

There are problems with this simple concept of how enabling factors operate: one is that even if all conditions are fulfilled for optimal stimulation and for a normal state of arousal during the stimulation period, no permanent synaptic changes and consequently no learning occurs if arousal is dramatically lowered immediately after the stimulation. Clinically, this condition is referred to as retrograde amnesia, and a specific form of it has recently been demonstrated on the single-cell level for visual cortical plasticity (see Rauschecker, this volume; Rauschecker and Hahn, 1987). The case of retrograde amnesia, which can be caused, for example, by general anaesthesia, suggests that the original activity pattern caused by the stimulation first creates a kind of blueprint that is still labile. The blueprint then needs to be copied in order to be transferred into a more permanent state. Learning will occur only if a normal state of arousal is guaranteed during this copying process. One role of enabling factors would thus be to instate conditions for facilitating this transfer process.

ONSET AND CLOSURE OF SENSITIVE PERIODS MAY BE CONTROLLED BY DIFFERENT FACTORS

Just as sensitive periods for different developmental systems may have distinct physiological determinants, so the onset and closure of a given sensitive period may be differently determined. Bateson (this volume) argues that this is the case with imprinting, with onset a function of growth and closure a function of the accumulation of certain kinds of experiences. The onset is not independent of external stimulation, however, since it can be hastened or delayed by exogenous factors. The "capacity model" of Bateson (this volume) and Boakes and Panter (1985), as an alternative to a "clock" model, serves to epitomize the notion that cumulative effects of use and experience act to bring a sensitive period to a close, with onset viewed as a consequence of programmed growth. A similar interpretation may be applicable to aspects of visual development (see Timney, this volume; Rauschecker, this volume).

It is important not to exaggerate the irreversibility of closure of sensitive periods. Given sufficiently intense or sustained arousal, imprinting preferences can be reversed. There are cases where intensive social interaction appears to have induced song learning at times when auditory stimulation alone is no longer effective. The still controversial experiments on induction of visual plasticity by exogenous norepinephrine raise a similar prospect that closure of the sensitive period is not completely irreversible.

It is equally important not to overemphasize the significance of cases of reinstatement of plasticity after closure of a sensitive period by unusually stringent means. In the normal course of development such events are likely to be rare, and a sensitive period may be a quite adequate and reliable means of achieving a predictable timetable for completion of a phase of plasticity according to an optimal program.

REFERENCES

Barlow, H. B. (1975). Visual experience and cortical development. *Nature* **258**:199–204.

Blakemore, C. (1978). Maturation and Modification in the Developing Visual System. In: *Handbook of Sensory Physiology*, Vol. VIII *Perception*, R. Held, H. W. Leibowitz, and H.-L. Teuber, eds., pp. 377–436, Springer-Verlag, Berlin.

Boakes, R. and D. Panter (1985). Secondary imprinting in the domestic chick blocked by previous exposure to a live hen. *Anim. Behav.* **33**:353–365.

Bottjer, S. W. and A. P. Arnold (1986). The Ontogeny of Vocal Learning in Songbirds. In: *Handbook of Behavioral Neurobiology*, Vol. 8, E. M. Blass, ed., pp. 129–161, Plenum Press, New York.

Changeux, J.-P. and A. Danchin (1974). Apprendre par stabilisation sélective des synapses en cours de développement. In: *L'unité de l'homme*, E. Morin and M. Piatteli, eds., pp. 320–357, Le Seuil, Paris.

Changeux, J.-P. and A. Danchin (1976). Selective stabilization of developing synapses as a mechanism for the specification of neuronal networks. *Nature* **264**:705–712.

Changeux, J.-P., T. Heidmann, and P. Patte (1984). Learning by Selection. In: *The Biology of Learning*, P. Marler and H. S. Terrace, eds., pp. 115–133, Dahlem Konferenzen, Springer-Verlag, Berlin.

Daw, N. W., N. E. J. Berman, and M. Ariel (1978). Interaction of critical periods in the visual cortex of kittens. *Science* **199**:565–567.

Daw, N. W., T. O. Videen, T. Robertson, and R. K. Rader (1985). An Evaluation of the Hypothesis that Noradrenaline Affects Plasticity in the Developing Visual Cortex. In: *Lectures in Biology, Vol. 5, The Visual System*, A. Fein and J. S. Levine, eds., pp. 133–144, Alan Liss, New York.

DeVoogd, T. J., B. Nixdorf, and F. Nottebohm (1985). Synaptogenesis and changes in synaptic morphology related to acquisition of a new behavior. *Brain Res.* **329**:304–308.

Edelman, G. (1978). *The Mindful Brain. Cortical Organization and the Group-selective Theory of Higher Brain Functions*, MIT Press, Cambridge, MA.

Edelman, G. (1981). Group Selection as the Basis for Higher Brain Function. In: *The Organization of the Cerebral Cortex*, F. Schmitt et al., eds., pp. 535–563, MIT Press, Cambridge, MA.

Edelman, G. and L. Finkel (1984). Neuronal Group Selection in the Cerebral Cortex. In: *Dynamic Aspects of Neocortical Function*, G. M. Edelman, W. E. Gall, and W. M. Cowan, eds., Wiley & Sons, New York.

Eschweiler, G., M. Popp, J. P. Rauschecker, and W. Schrader (1984). Timing of flash responses in visual cortex of normal and strabismic cats. *Soc. Neurosci. Abstr.* **10**:469.

Gilbert, C. D. and T. N. Wiesel (1983). Clustered intrinsic connections in cat visual cortex. *J. Neurosci.* **3**:1116–1133.

Goldman, S. A. and F. Nottebohm (1983). Neuronal production, migration, and differentiation in a vocal control nucleus of the adult female canary brain. *Proc. Natl. Acad. Sci. USA* **80**:2390–2394.

Gottlieb, G. (1971). *Development of Species Identification in Birds*. University of Chicago Press, Chicago.

Haken, H. (1983). *Synergetics*. Springer-Verlag, Berlin.

Harwerth, R. S., E. L. Smith, G. C. Duncan, M. L. J. Crawford, and G. K. von Noorden (1986). Multiple sensitive periods in the development of the primate visual system. *Science* **232**:235–238.

Horn, G. (1985). *Memory, Imprinting and the Brain*. Clarendon Press, Oxford.

Hubel, D. H. and T. N. Wiesel (1965). Binocular interactions in striate cortex of kittens reared with artificial squint. *J. Neurophysiol.* **28**:1041–1059.

Imbert, M. and P. Buisseret (1975). Receptive field characteristics and plastic properties of visual cortical cells in kittens reared with or without visual experience. *Exp. Brain Res.* **22**:25–36.

Jones, K. R., P. D. Spear, and L. Tong (1984). Critical periods for effects of monocular deprivation: Differences between striate and extrastriate cortex. *J. Neurosci.* **4**:2543–2552.

Kasamatsu, T. and J. D. Pettigrew (1976). Depletion of brain catecholamines: Failure of ocular dominance shift after monocular occlusion in kittens. *Science* **194**:206–209.

Kety, S. S. (1970). The Biogenic Amines in the Central Nervous System: Their

Possible Roles in Arousal, Emotion, and Learning. In: *The Neurosciences, Second Study Program*, F. O. Schmitt, ed., pp. 324–336, Rockefeller University Press, New York.

Kety, S. S. (1976). Biological Concomitants of Affective States and their Possible Roles in Memory Processes. In: *Neural Mechanisms of Learning and Memory*, M. R. Rosenzweig and E. L. Bennett, eds., pp. 321–326, MIT Press, Cambridge, MA.

Lorenz, K. (1950). The comparative method in studying innate behaviour patterns. *Symp. Soc. Exp. Biol.*, IV. Cambridge University Press, Cambridge.

Marler, P., R. Dooling, and S. Zoloth (1980). Comparative Perspectives on Ethology and Behavioral Development. In: *The Comparative Method in Psychology*, M. Bornstein, ed., pp. 189–230, Lawrence Erlbaum, Hillsdale, NJ.

McGaugh, J. L., K. C. Liang, C. Bennett, and D. B. Sternberg (1984). Adrenergic Influences on Memory Storage: Interaction of Peripheral and Central Systems. In: *Neurobiology of Learning and Memory*, G. Lynch, J. L. McGaugh, and N. M. Weinberger, eds., pp. 313–332, Guilford Press, New York.

Merzenich, M. M., W. M. Jenkins, and J. C. Middlebrooks (1984a). Observations and Hypotheses on Special Organizational Features of the Central Auditory Nervous System. In: *Dynamic Aspects of Neocortical Function*, G. M. Edelman, W. E. Gall, and W. M. Cowan, eds., pp. 397–423, Wiley & Sons, New York.

Merzenich, M. M., R. J. Nelson, M. P. Stryker, M. S. Cynader, A. Schoppmann, and J. M. Zook (1984b). Somatosensory cortical map changes following digit amputation in adult monkeys. *J. Comp. Neurol.* **224**:591–605.

Nottebohm, F. (1984). Vocal Learning and Its Possible Relation to Replaceable Synapses and Neurons. In: *Biological Perspectives on Language*, D. Caplan, A. R. Lecours, and A. Smith, eds., pp. 65–95, MIT Press, Cambridge, MA.

Nottebohm, F. and M. E. Nottebohm (1978). Relationship between song repertoire and age in the canary, *Serinus canarius*. *Z. Tierpsychol.* **46**:298–305.

Tong, L., R. E. Kalil, and P. D. Spear (1984). Critical periods for functional and anatomical compensation in lateral suprasylvian visual area following removal of visual cortex in cats. *Science* **179**:395–398.

Pettigrew, J. D. and T. Kasamatsu (1978). Local perfusion of noradrenaline maintains visual cortical plasticity. *Nature* **271**:761–763.

Rauschecker, J. P. and U. Dehnen (1986). Visual deprivation improves the ability for auditory localization in cats. *Perception* **15**:A34.

Rauschecker, J. P. and S. Hahn (1987). Ketamine-xylazine anaesthesia blocks consolidation of ocular dominance changes in kitten visual cortex. *Nature* **326**:183–185.

Sherman, S. M. and P. D. Spear (1982). Organization of visual pathways in normal and visually deprived cats. *Physiol. Rev.* **62**:738–855.

Van der Loos, H. and T. A. Woolsey (1973). Somatosensory cortex: Structural alterations following early injury to sense organs. *Science* **179**:395–398.

Van Essen, D. C. and J. H. R. Maunsell (1983). Hierarchical organization and functional streams in the visual cortex. *Trends Neurosci.* **6**:370–375.

Von Noorden, G. K. (1985). Amblyopia: A multidisciplinary approach. *Invest. Ophthalmol. Vis. Sci.* **26**:1704–1716.

Wiesel, T. N. (1982). Postnatal development of the visual cortex and the influence of environment. *Nature* **299**:583–592.

Wiesel, T. N. and D. H. Hubel (1963). Simple-cell responses in striate cortex of kittens deprived of vision in one eye. *J. Neurophysiol.* **26**:1003–1017.

Woolsey, C. N. (1981). *Cortical Sensory Organization. Vol. 2: Multiple Visual Areas.* Humana Press, Clifton, NJ.

Zeki, S. M. (1978). Functional specialization in the visual cortex of the rhesus monkey. *Nature* **274**:423–428.

INDEX

369

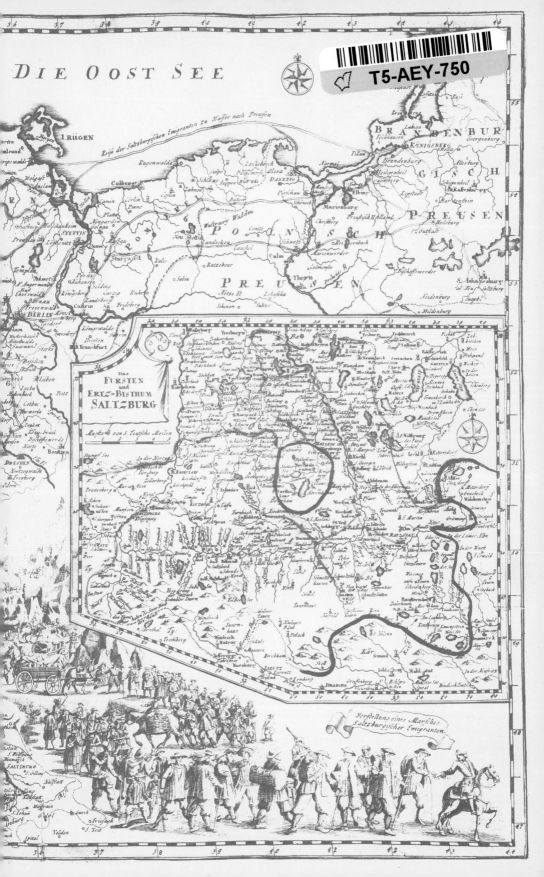

Henry Newman's
Salzburger Letterbooks

Hanns Klammer aus Bischoffshofen.

Ach Herr laß unsre Flucht im Winter nichtgeschehen,
dieswar sonst meine Bitt und meiner Wünsche Ziel.
doch nun bin ich getrost im Winter auch zu gehen,
weil Gottes warme Lieb uns selbst bedecken will.

O Lord, let our flight not occur in winter:
    this used to be my plea and the aim of my wishes,
but now I am content to go even in wintertime,
    because God's warm love will cover us.

Henry Newman's
Salzburger Letterbooks

Transcribed and edited by
GEORGE FENWICK JONES

WORMSLOE FOUNDATION PUBLICATIONS
NUMBER EIGHT

UNIVERSITY OF GEORGIA PRESS

ATHENS

Copyright © 1966

UNIVERSITY OF GEORGIA PRESS

Library of Congress Catalog Card Number: 66-25848

This edition is limited to 750 copies.

Printed in the United States by

FOOTE & DAVIES, DORAVILLE

Contents

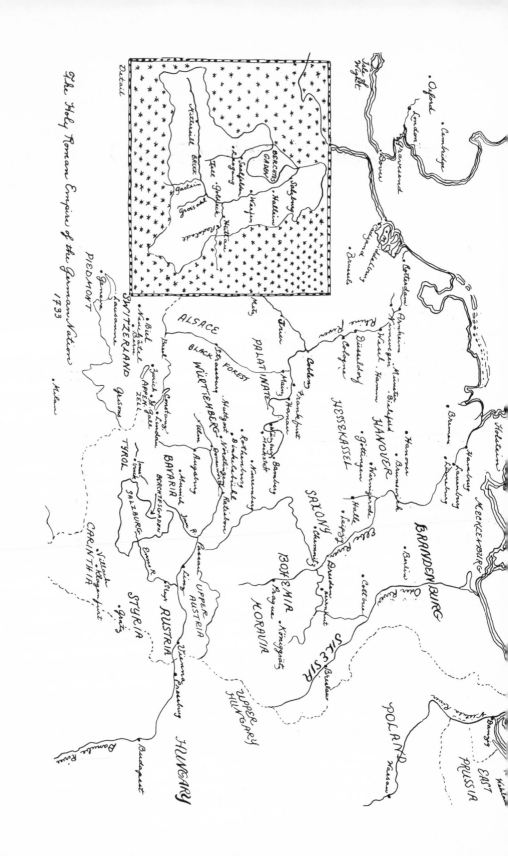

The Holy Roman Empire of the German Nation 1733

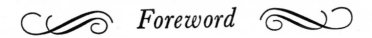

Foreword

THE Wormsloe Foundation is a non-profit organization chartered on December 18, 1951, by the Superior Court of Chatham County, Georgia. In the words of its charter, "The objects and purposes of this Foundation are the promotion of historical research and the publication of the results thereof; the restoration, preservation and maintenance of historical sites and documents and the conduct of an educational program in the study of history in the State of Georgia, and in states adjoining thereto."

As its first important activity, the Foundation has begun the publication of a series of historical works and documents under the title of "Wormsloe Foundation Publications." They will consist of important manuscripts, reprints of rare publications, and historical narratives relating to Georgia and the South. The first volume appeared in 1955, written by E. Merton Coulter, the General Editor of this series, and entitled *Wormsloe: Two Centuries of a Georgia Family.* This volume gives the historical background of the Wormsloe Estate and a history of the family which has owned it for more than two and a quarter centuries.

The second publication of the Foundation was *The Journal of William Stephens, 1741-1743* and the third volume was *The Journal of William Stephens, 1743-1745,* which is a continuation of the journal as far as any known copy is extant. However, there is evidence that Stephens kept up his journal for some years after 1745. Both of these volumes were edited by the General Editor of the Wormsloe Foundation series, and were published in 1958 and 1959 respectively.

The fourth volume of the series was the re-publication of the unique copy of Pat. Tailfer *et al., A True and Historical Narrative of the Colony of Georgia* . . . With Comments by the Earl of Egmont, in the John Carter Brown Library of Brown University. In this publication there appeared for the first time in print the comments of Egmont. With the permission of Brown University, this volume was edited by Clarence L. Ver Steeg of Northwestern University, Evanston, Illinois.

The fifth volume in the series of Wormsloe Foundation Publications was the long-missing first part of Egmont's three manuscript volumes of his journal. It was edited by Robert G. McPherson of the University of Georgia. This volume contains the journal from 1732 to 1738, inclusive, and is owned by the Gilcrease Institute of American History and Art, who gave permission for its publication.

In 1963, the Foundation published its sixth volume, *The Journal of Peter Gordon, 1732-1735,* which was edited by the General Editor of the series. Gordon came over with Oglethorpe on the first voyage, and he began his journal when he left England. The original manuscript was acquired by the Wormsloe Foundation in 1957, and was presented to the University of Georgia Library.

The seventh volume in the series was *Joseph Vallence Bevan, Georgia's First Official Historian.* It is a departure from the previous five volumes, which were documentary, and was written by the General Editor. It brings to light a historiographer who was appointed Georgia's first official historian by the state legislature.

The present volume, *Henry Newman's Salzburger Letterbooks,* begins a series which might be called a series within the Wormsloe Foundation series, for it is to be followed by several volumes of translations of *Urlsperger Reports (Ausführliche Nachrichten.* . . . edited by Samuel Urlsperger, Halle, 1735ff, and dealing with the Georgia Salzburgers). This volume is transcribed and edited by George Fenwick Jones of the University of Maryland, who, also, will edit future volumes of the Salzburger translations.

E. MERTON COULTER
General Editor

 *Preface*

This edition of Henry Newman's Salzburger Letterbooks is based on Xerox copies from microfilms of the originals, which are preserved in London in the archives of the Society for Promoting Christian Knowledge. Because its assistance to the Protestant exiles from Salzburg had necessitated considerable correspondence, the secretary of the S.P.C.K., Henry Newman, caused it to be entered into several letterbooks for greater safety and convenience. Consequently, this correspondence is not only preserved in its entirety but also with painstaking care. There are two volumes of outward correspondence, both in folio size: one of 136 pages from the years 1732 to 1734 and one of 117 pages from the years 1734-39. The inward correspondence is contained in three volumes, also in folio size: one of 168 pages from 1732-33, one of 174 pages from 1734-35, and one of 41 pages from 1735. In addition, but not included here, the S.P.C.K. owns various autographed items concerning the Salzburgers, some of which are copied in these letterbooks, as well as three books of account of 100 double pages recording all the Society's collections for and the disbursements to the Salzburgers.

This edition endeavors to follow the original syntax, spelling, and capitalization exactly; and it inserts correct forms, in brackets, only when the meaning might otherwise be obscure. The sole alteration in spelling is the occasional deletion of what appears to be an *s* at the end of certain words; for one of the copyists had the annoying habit of ending some words with a little flourish that happened to be identical with his *s*. Wherever the context indicates that the noun or verb in question is singular, the mark is assumed to have been merely a decorative flourish. Punctuation has been left largely intact, except that many commas have been omitted and a few added, since there seems to have been no rhyme or reason in their use. A collation of the few remaining original letters with their copies in the letterbooks indicates that the copyists paid no attention to the commas in the originals, which were in turn just as frequent and just as irrational as their own. Bracketed periods have been supplied where clarity demands.

Abbreviations have been resolved, and for very justifiable reasons. Most of them would have been resolved if the letters had been printed when written; and most of them were introduced by the copyists and therefore reveal nothing about the authors' intent. Although familiar enough in the 18th century, many of these abbreviations would confuse a modern reader: e.g., Sen. Urlsperger was not a Senator, but a Senior of the Consistory; and M. Vat did not have a first name beginning with M. but was merely being addressed as Monsieur. Being thoroughly familiar with scripture, the gentlemen of the S.P.C.K. would have immediately recognized abbreviations like Col. and Ps., even though they are not immediately recognized by everyone today. Some 18th-century abbreviations now seem quaint or comical; for example, in America today the word Gent. does not suggest gentlemen as much as it does a men's rest room. Some abbreviations have subsequently lost their reason for being: it is no longer necessary to write *would* as *wou'd* to indicate that the liquid consonant should be dropped, since it is now dropped regardless of spelling. Likewise, we no longer have to write *receiv'd* to show that the last *e* is silent, since it is silent now even when written. In the 18th century, on the other hand, the vowel sound of the weak verbal suffix could be pronounced or suppressed at will, as is still the case of the word *learned* in "The man learned" and "The learned man." Reproducing abbreviations like *yo'r* (with a raised *r*) would add more to the cost than the meaning, and it would give these letterbooks a quaint and archaic flavor that they do not deserve. The authors of these letters were clear-thinking, practical men, not quaint antiquarians; and the language they wrote was both that of the pulpit and that of the counting house.

The French letters in this collection are copied verbatim, together with their total disregard for accent marks. Some of the irregularities are perhaps due to archaisms, others to the fact that the Swiss and German correspondents seem to have been rather vague in matters of syntax and orthography. Peculiar is the rather consistent practice of ending the infinitives and second person plural verbs with *ès* or *es*. The errors in transcribing into the letterbooks suggest that the copyists were not familiar with French. This is even more obvious in the case of the Latin letters, the numerous errors of which indicate that the scribes copied them without any idea of their meaning.

Where possible, significant persons in these letterbooks are identified in the notes, and the location of the pertinent note may be ascertained in the appended index of proper names.

Because the S.P.C.K. discontinued its activities in the thirteen colonies when they won their independence, I had always thought of it as a relic of the past, as a venerable establishment somehow perpetuated by sheer force of tradition. Therefore, during my research in the Society's archives in Trinity Church, I was surprised to discover that the Society is still a vital and vigorous organization busily engaged in sending books and men into the four corners of the earth. Were Henry Newman to return today, he would certainly be gratified to see his beloved Society still performing its mission so diligently more than two centuries after his death. At this point I wish to thank the staff of the S.P.C.K., particularly their archivist, Mr. Arthur E. Barker, for the hospitality and cooperation I enjoyed during my research in their archives. I also wish to thank the American Philosophical Society, the Wormsloe Foundation, and the Georgia Salzburger Society for the financial support required for this undertaking, and I also wish to express my gratitude to the University of Maryland for its generous encouragement of scholarly research. Miss Bessie Lewis of Townsend, Georgia, deserves my gratitude for correcting various errors in transcription and for other valuable suggestions. To Dr. E. Merton Coulter, the General Editor of the Wormsloe Foundation, as well as to Mrs. Craig Barrow, its patroness, I owe the actual publication of these letters.

GEORGE FENWICK JONES

Department of Foreign Languages and Literature
University of Maryland

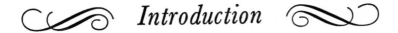 *Introduction*

ALTHOUGH he spent most of his life in London, Henry Newman was a New Englander, the grandson of a dissenting English clergyman who had emigrated to Massachusetts in the time of Cromwell.* Orphaned at the age of six, Newman was reared by his mother's people and educated at Harvard, where he later served for three years as librarian. Having been influenced by liberal tutors, he gradually dropped his Congregational faith and became an Anglican; yet he retained the stern Puritan virtues of his forebears, particularly those of diligence and devotion to duty. While chiefly concerned with theology and the humanities, he also occupied himself with astronomy and other natural sciences. As a youth he served as lay chaplain on several ocean voyages, during one of which he was wounded in an encounter with a French warship. When his ship reached England after his harrowing experience, Newman kissed British soil and never again left the land of his ancestors.

In England Newman first served in the household of the Duke of Somerset, a favorite of Queen Anne, in whose service he acquired considerable skill in diplomacy and administrative procedure and also a life-long loyalty to the newly established House of Hanover and to the policy of Protestant Succession. Because of his natural piety, good education, and diplomatic experience, to say nothing of his private means and celibate life, Newman was ideally qualified to be the secretary of a missionary society, and for such a call he did not have long to wait. During his residence in London, the leading English charitable enterprise there was the Society for Promoting Christian Knowledge (S. P. C. K.), which had its office in Bartlett's Buildings in Holborn, London. This pious organization had been founded by Dr. Thomas Bray in 1698 to carry the gospel to the underprivileged classes of Great Britain and of the British colonies, or "Plantations" as they were

*For an excellent account of Newman's life and works, see Leonard Cowie, *Henry Newman, An American in London, 1708-43* (London, 1956) .

1

then called. It did not restrict its membership to Englishmen but also invited certain prominent Protestant clergymen on the European Continent and in the "plantations" to become corresponding members.

Protestant supremacy was constantly threatened during the seventeenth century, and for three years England was ruled by a Catholic King, James II; but in 1691 the Protestants strengthened their position by crowning the Dutch prince William of Orange jointly with Queen Mary, and Protestant Succession became law by the Act of Settlement of 1701. After the reign of Queen Anne, the crown passed in 1714 to George IV, Duke of Hanover and Brunswick, who then became George I of England and founded the English House of Hanover. This personal union of England and Hanover under a common crown tightened England's natural ties with the Protestant states of Germany, since many of the new monarch's German officials followed him to London. As the letters of these letterbooks show, the Hanoverian Envoy to the Holy Roman Empire served George II both as Duke of Hanover and as King of England. As king, George II was the head of the Church of England, yet in private life he remained a German Lutheran of the Augsburg Confession; and, judged by royal standards of the time, he was an exceptionally pious ruler.

With the House of Hanover firmly established and Protestant Succession assured, many Englishmen looked upon England as the natural champion of Protestantism on the European continent as well as at home; and it was in this spirit that the S.P.C.K. solicited the services of Protestant divines, mostly Lutheran, in the various German states. Continental Protestantism was then in a sorry plight. Having been worsted in a series of wars, the Huguenots of France had been barely tolerated by the Edict of Nantes of 1598; and, after that safeguard was revoked in 1685, they were ruthlessly suppressed and those who did not escape or recant were sent to the galleys. Thus Protestantism quickly ceased being a major factor in France. Calvinism held its own in the Netherlands and Switzerland, but it was losing ground in Germany, where the Lutheran Church was fighting a rear-guard action too.

Popular concern for the Protestants on the Continent was shared by the members of the S.P.C.K., who wished to aid and assist all persecuted Protestants. Chief among these were, of course, the Huguenot refugees from France, who were scattered throughout the Protestant cities and countries of Europe or else living undercover in France and seeking any ways and means to escape. The S.P.C.K. also aided the Waldensians of Piedmont, who were being persecuted by the Duke of Savoy. This oldest Protestant sect in Europe had already endured some five centuries of persecution since the followers of Peter Waldo broke with the Church of Rome. Because the survival of their religion depended on the ability of their ministers to study abroad, the S.P.C.K. contributed funds to make such study possible.

The popular sympathy felt for the Huguenots and Waldensians was far surpassed by the surge of sympathy for the Salzburg Protestants in 1731, when the reigning Archbishop, Count Leopold Anton of Firmian, resolved to expel all Protestants from his domain. Nothing of this kind had yet happened in the 18th century, which considered itself an enlightened age. Whereas princes still had the legal right to regulate the religion of their subjects, most of them had found it more expedient to tolerate small and harmless sects.

To understand the Archbishop's motives, we must briefly review the religious situation then prevailing in the Holy Roman Empire of the German Nation, as the crazy-quilt of quarrelsome little states in Central Europe was called. Luther's defiance of the Pope at Worms in 1521 was popularly acclaimed as a blow for German freedom from Papal interference and oppression, and it was accepted and championed by the populace at large, from prince to peasant. The Lutheran, or Evangelical, faith was adopted by most secular princes and by nearly all Imperial Free Cities, except for some of those in the Southwest, which embraced the dogmas of Calvin or Zwingli. The only German states in which the Reformation made little headway were the ecclesiastical territories, especially those belonging to the archbishops and bishops along the Rhine and Main and to their south; for these clerics saw the Reformation as a threat to their secular power. After spreading rapidly at first, the reformed religion was slowed down by the Counter Reformation; and, with the aid of Spanish troops, the Catholic emperor came close to crushing it in the Schmalkaldic War. Nevertheless, the Peace of Augsburg of 1555 granted freedom of religion to those Protestant princes and free cities that had already subscribed to the Confession of Augsburg, the Lutheran articles of faith as formulated by Melanchthon, Luther's spokesman. This treaty put an end to further Evangelical expansion; and the initiative soon passed to the Counter Reformation, which was pressed relentlessly by the Vatican and Jesuits.

Except for minor dynastic skirmishes, the Treaty of Augsburg was followed by a long peace, albeit an uneasy one, during which the German princes aligned themselves in two opposing factions: the Protestant Union and the Catholic League. The long expected hostilities broke out in 1618, when Catholic forces undertook to suppress Protestantism in Bohemia; and what had begun as a war of religion swiftly degenerated into a thirty-year free-for-all in which nearly all the German states and most of their neighbors participated as expedience rather than conscience dictated. Few wars have been waged with such cynicism and perfidy or with such wanton destruction and cruelty; for religion served principally as a pretext for personal greed and ambition. Gustavus Adolfus of Sweden was one of the few participants to whom genuinely religious motives can be imputed, and it is ironic that he was subsidized by Cardinal Richelieu of Most Catholic France, who supported the German heretics, and also the Turkish infidels, as a means of weakening the Catholic emperor. After thirty violent years, peace was finally negotiated in 1648 at the Treaty of Westphalia, largely at the expense of the German Empire and to the advantage of France.

The great war was followed by a period of economic stagnation, caused in part by the destruction of life and property in the war and by the shift of trade routes to the Atlantic seaboard, but also by the greed and particularism of the petty princes, who had now won almost complete freedom from the Empire, an empire which was, as Voltaire later remarked, neither Holy nor Roman nor an Empire. Germany was a patch-work of principalities, duchies, counties, bishoprics, cantons, marches, free cities, and other jealous little entities, each fighting for itself and ready to sacrifice the common good for its own particular advantage. All the while France continued to nibble away at Germany's western frontier, while most of the German princelings welcomed any foreign encroachment that spared their territory and distracted the emperor from asserting his authority.

Through clever diplomacy and successful wars, France succeeded in keeping Germany a house divided, especially one divided by confession. The Calvinists of western Germany, who bore the brunt of French aggression, remained politically impotent; and the Zwinglians and Calvinists of Switzerland and the Netherlands had separated themselves from the affairs of the Empire. Consequently, the chief confessional division in Germany was between Roman Catholics and Lutherans, the latter of whom were on the defensive and feared Catholic more than French aggression. In reading the following letters about persecution in Austria and Hungary, it is easy to see that the Protestants of Central Europe could say "Rather the Sultan than the Pope" (*Lieber der Sultan als der Pabst*), for the Sultan was satisfied with tribute and did not try to enforce his religion by stake and sword.

Although most Protestants lived in the north of Germany and most Catholics lived in the south, this division was not clear-cut, since most of the Imperial Free Cities in the south were still Protestant, as were some large territories in Württemberg and the Palatinate, both of which had Catholic rulers. Augsburg, the spiritual home of the Georgia Salzburgers, was not unusual in being rather evenly divided between Protestants and Catholics, each group having its own mayor and council. With such a delicate balance of power, it is understandable that the Catholic part of the city hesitated to let so many alien Protestants lodge in the city.

There being literally hundreds of sovereign states and cities in the Empire, there were a corresponding number of representatives and envoys. In principle there was a representative from each city and state to each city and state, although in practice a single agent might represent several governments, as was the case of Baron von Reck, who represented George II in his capacity of King of England as well as of Duke of Hanover, Brunswick, and Zell. Some agents were subjects of the states in which they resided, rather than of those they represented, as seems to have been the case of Mr. Gullman, the British agent at Frankfurt, and of Mr. Wolters, the British agent at Amsterdam. Like von Reck, they probably represented the English king in all his capacities.

During the expulsion of the Salzburg Lutherans, the diet or parliament of the loosely-united Empire was accustomed to meet in the Danubian city of Regensburg, or Ratisbon as our letters call it, which was then a free city but is now in Bavaria. Being more concerned with maintaining their religious freedom than with defending the moribund empire, the Protestant representatives were united in a Protestant Body (Evangelick Body, Corpus Evangelicorum), in which they could take measures for their common defense; and it was this body that assumed the chief responsibility for defending the rights guaranteed to Protestant minorities by the Treaty of Westphalia. Baron von Reck, the English-Hanoverian envoy at the Diet, seems to have been an important agent in aiding the Salzburger emigrants, even though political considerations caused his name to be partially suppressed in our letters.

The fragmentation of the German realm into so many little parcels naturally hampered commerce, especially because of the fantastic number of tolls and other mercantile restrictions. John Vat's letter of 9 October 1734 from Rotterdam gives some idea of the tolls demanded on just a short section of the Rhine, tolls that were not entirely abolished until the Customs Union (Zollverein) of 1844. Despite such restrictions and tariffs, the diligent burghers and peasants of Germany did create wealth, but much of what they created was wasted by the countless little potentates, each of whom wished to imitate the court of Versailles.

Not the least extravagant of these rulers were the archbishops of Salzburg, who, although nominally men of God, maintained lavish courts and built handsome palaces at the expense of their tax-burdened subjects. The Protestants who resolved to emigrate from Salzburg had the right to take their property with them, yet their property was largely confiscated through all the real and pretended taxes and other claims of the archbishop. Moreover, with so many farms suddenly thrown on the market, the price of real estate dropped drastically, especially since few of the archbishop's subjects had much cash. By the time the King of Prussia sent commissioners to Salzburg to try to collect the moneys owed his new subjects from there, very little was forthcoming; and there is no evidence and little likelihood that the Georgia Salzburgers ever received any part of the claims they submitted.

Luther's revolt reached Salzburg quickly and two of his disciples actually preached his cause there, yet the archbishops succeeded in stamping out the movement in the capital and in most of their territory. Lutheran heretics continued to worship secretly in various remote valleys; but, as often as heresy was detected, the offenders were punished and usually dispossessed and expelled, and it was assumed that the trouble had been eradicated. In the year 1686 a Salzburger named Joseph Schaitberger publicly professed his Evangelical religion and was forthwith imprisoned. Exiled upon his release from prison, he settled in Nürnberg, where he lived a pious and industrious life as a wood-carver while remaining in communication with his co-religionists back home. It was he who wrote the "Exiles' Song", the hymn sung by all later groups of exiles who migrated through the countries of central Europe in search of freedom to practice their Protestant religion.

Partly as a result of Schaitberger's propagandistic and missionary activities, to say nothing of the good treatment he received from Protestant well-wishers, his co-religionists in Salzburg not only maintained but even increased their numbers despite constant persecution. Finding himself unable to stamp out this heresy, Archbishop Leopold Anton decreed that all Protestants would have to recant their faith or leave the country at once. He apparently believed the heresy to be limited to a few troublemakers, who would forsake their faith rather than their native land; and therefore it came as a surprise when nearly a fourth of his subjects declared themselves Protestant and announced that they were ready to abandon their homes rather than their religion. Historians differ concerning the total number of these exiles, which is usually set between twenty and thirty thousand; but there is evidence that Frederick William, the Elector of Brandenburg and King of Prussia, received no less than 18,000 of them into his territories in Prussia and Lithuania, which had been partially depopulated by plague. He also welcomed others into Brandenburg and elsewhere, as did the rulers of Denmark, Hanover, the Netherlands, and other Protestant states.

The earliest item in our letterbooks is an extract of a letter of 14 April 1732 from Samuel Urlsperger, the senior Lutheran pastor at Augsburg, acknowledging receipt of £ 125 contributed for the benefit of the Salzburg exiles. This sum came from England; and it may have been donated by King George II himself or by his family or court, it having been collected by the Rev. Frederick Michael Ziegenhagen, the king's German chaplain, a man of great influence at court because of his free access to the king. The next letter, Ziegenhagen's of 23 May to Newman, indicates that the S.P.C.K. had by then assumed responsibility for aiding the Salzburgers; and the next, Newman's of 13 June to Urlsperger, indicates that the S.P.C.K. accepted Ziegenhagen's suggestion to have their funds remitted by the German business firms of Johann Baptista Mayer of London and Christian Münch of Augsburg.

From this point on the letterbooks consist mainly of the correspondence between Newman and Urlsperger, the two prime movers of the Georgia Salzburger venture; and to this correspondence are added letters by all the more important participants and benefactors of the undertaking. Prominent among these benefactors were Sir John Phillips, Baronet, and James Vernon and William Tillard, Esquires, who were members of the S.P.C.K. entrusted with soliciting donations. Sir James Vernon was also a

Trustee for Establishing the Colony of Georgia and therefore able to assure close collaboration between the S.P.C.K. and the Georgia Company.

Because mail service was irregular, letters did not always arrive in the same sequence in which they were sent; and, because the mails were slow, Urlsperger, as the more prolific writer, often wrote successive letters before receiving an answer to his first. This was of course even more the case when he corresponded with the pastors in Georgia, from whom an answer took anywhere from six to twelve months. The length and frequency of Urlsperger's letters may owe something to a personal penchant for letter-writing, or one might say to a national inclination to this pastime; but they are also due to the need of keeping the S.P.C.K. and its donors abreast of the latest developments. Urlsperger supplied much of this information by enclosing extracts of letters from informants elsewhere, particularly from correspondents in Ratisbon. As is indicated in the incoming letter of 26 February, 1733, many of these letters were from Johann Göbel, the commissary appointed by the King of Prussia to deal with the exiles from Salzburg and Berchtolsgaden. Some seem to have been from Baron von Reck, the English king's ambassador to the Imperial Diet, who would have thought it diplomatic to keep his religious activities secret while dealing with the deputies of a Catholic emperor. Others may have come from a zealous Lutheran pastor named Esterlin, who was one of Urlsperger's correspondents there, as is shown in the incoming letter of 6 April 1733. Some of the atrocity stories transmitted by Urlsperger may have been more rhetorical than historical, for their veracity was questioned even by Pastor Heinrich Walther Guerdes of the so-called Swedish Church, which was actually German Lutheran and therefore keenly concerned with the persecutions. Like the American authorities in Berlin after World War II, the Protestant leaders in eighteenth-century Germany had to be constantly on guard against clever imposters posing as victims of oppression and all too ready to fabricate atrocity stories on order. But, whether accurate in their details or not, these accounts did serve to open the hearts and purses of potential benefactors.

It is interesting to observe how, despite the stilted epistolary commonplaces of the age and despite the hurdles of translation, Newman and Urlsperger soon came to a perfect understanding of each other through their letters. In view of so much correspondence with Germans, it is surprising that Newman never attempted to learn their language; but he was probably discouraged by the barbaric Gothic script then in use, which did much to isolate the Germans from their Western neighbors.

Newman's correspondence gives a play-by-play account of the entire Georgia Salzburg operation; but even more important, it gives an insight into the emotional environment of the time. Some of the persons and events of the early letters played no direct role in the Georgia operation, yet they do throw light on the thinking of the leading actors and benefactors. Several of these letters have been omitted because of irrelevance, but these are duly mentioned in the notes. The letters quickly show the changing motives for the S.P.C.K.'s activities. The first few remittances to Urlsperger were for the benefit of the exiles in general and without restrictions; but soon the purpose changed and money was solicited only for the exiles destined for Georgia. In other words, funds were solicited less for the succor of Salzburgers than for the recruitment of colonists, and Urlsperger spent more and more of his time convincing the exiles that they should migrate to Georgia. These mixed motives were typical of the age and of the whole colonization of Georgia, which was to be a haven for debtors and other impecunious people who had failed in Europe, a refuge for distressed Protestants, a military outpost, and a source of raw materials to supply the industrial needs of Great Britain. It is clear that no policy could have been developed to fit these conflicting purposes, each of which was very laudable in its own right.

Urlsperger's task of finding Salzburgers willing to settle in Georgia was made difficult by rumors and letters, many of them unfortunately true, concerning the hardships and cruel treatment awaiting most German colonists in America, even those who had been persuaded by promises just as grand and just as convincing as those made by or for the Trustees. Even more disturbing was the sad return of so many of the Tirnbergers, a group of Salzburg exiles who had settled on the island of Cadzand in the Netherlands at the invitation of the States General. Despite the sincere promises made by the Dutch ambassador at Ratisbon, these people were maltreated by the inhabitants of Cadzand, who looked upon them only as a cheap source of labor; and those who could escape did so and returned to Frankfurt, Ratisbon, and other Free Cities, where they became a burden and embarrassment for their hosts and a reason for their compatriots to trust no more promises.

Most of the German letters of this Salzburg correspondence

were done into English by J. C. Martini, Ziegenhagen's secretary. At first glance his translations appear bumpy and too German, yet a careful comparison with letters by Newman and others shows that the quaint vocabulary and awkward word-order of Martini's English conformed to common English usage of the period, and even his use of a singular verb with a multiple subject (the minister and the catechist was) agreed with English usage. When Martini writes "is dead" for "has died", that need not be the influence of German "ist gestorben", since Newman uses the same expression, which is still permissible in British English. In fact American readers should keep in mind that many of Martini's phrases that sound foreign or antiquated to our ears are still current in British speech. Despite a few Germanisms in unimportant passages, such as "upper servant" for *Oberknecht* (foreman) or "Country folks" for *Landsleute* (compatriots), Martini's translations, while very free, usually render the precise meaning of the originals. Martini may have been responsible for the anglicization of some of the German names, it having been customary at the time to naturalize foreign names, even family names. Thus Johann Heinrich Schmidt may become John Henry Schmidt, or even John Henry Smith. The same was true in reverse, and Henry Newman often appears as Heinrich Neumann in German translations of his letters. Martini's translations usually render German place-names in their normal eighteenth-century English forms, such as Cologne, Mayence, Munich, Frankfort, and Hall (Halle), but in a few cases he attempts to translate them, as in the case of Saltzkammergut, which he renders as Salt-Cammer-Good-or Estate. Sometimes the result is a hybrid, as for example when Leipzic becomes Leipswick, perhaps through analogy with Brunswick (Braunschweig). In some instances the German place-names in these letterbooks concur with neither normal German nor normal English eighteenth-century usage; yet this does not necessarily indicate carelessness on the part of the copyists, who were rendering the names as they were commonly spelled in South Germany. Whereas German orthography since Luther had been much better standardized than English, it still permitted some variations, particularly with regard to South German proper names. Upper German names like Lansperg, Innspruck, Goldeg, and Millstatt could be retained in standard German or they could be "corrected" to Landsberg, Innsbruck, Goldeck, and Mühlstadt.

Unfortunately, Martini's English script was sometimes illegible for Newman's clerks, who misread some names and words and gave up on others. This explains the occasional blank spaces in their copies, which they seem to have left just long enough to insert the missing word after consulting the translator. Since none of the existing blanks are of much significance, we can assume that the scribes neglected to follow the matter through in these cases. Most of the mistakes in spelling German names must have been due to the copyists, who apparently wrote from dictation. That the copyists were trusting their ears rather than their eyes is suggested by their complete disregard for the spelling, punctuation, or capitalization of the original letters, to say nothing of such phonetically correct but graphically wrong renditions as "the wholly Roman Empire." In the case of Newman's letter of 20 August, 1734, to J. von Reck, which was inadvertently entered twice; the punctuation and capitalization of the copies differ even though both entries were made by the same hand and within a short period of time.

Although Newman's letters would not pass a modern freshman composition test, they were acceptable by the standards of his day. Like his German contemporaries, he had little feeling for syntax or even for following a direct train of thought; and the most diverse grammatical structures can be juxtaposed in a single sentence. Not unusual is his letter of 21 November, 1732, to Urlsperger, the first paragraph of which is written as a single sentence comprising numerous subordinate clauses. Particularly annoying for a modern reader is the sometimes unclear relationship between pronouns or relative clauses and their antecedents. Newman writes, for example, that the officers at Gravesend are "to permit the Landing of all the Baggage of the Salzburgers and to show them all Civilities they can, but then they are to be locked up in some Warehouse. . . .", and then he writes that it will be more convenient for the Salzburgers "to come up to the Red House instead of stopping at Gravesend where they may embark on the prince of Wales at half the trouble and expense. . . ." It takes a second look to see that "they" refers to the baggage, not to the Salzburgers, and that "where" refers to the Red House and not to Gravesend.

Even though they do not meet present standards of syntax, Newman's letters are intelligible and informative; and they offer occasional passages of great charm, such as the description of Sir John Phillip's death. It should be noted that Newman's apparently casual style did not result from carelessness or lack of concern. He first composed his thoughts directly into a notebook and then edited the rough draft carefully, making numerous deletions, additions, and changes, before copying it, usually in his own hand, for sending. As some of his letters confess, ill health sometimes made him delegate this latter task to his clerk. Because the drafts of his Salzburg letters were scattered among many others in his innumerable notebooks, Newman caused them to be copied into his Salzburg Letterbooks, which thus furnished the S.P.C.K. with a compact record of the entire proceedings. Some, but not all of the letters in these letterbooks bear a code or file number which relates them to the overall correspondence of the S. P. C. K.

A comparison with the original drafts in the notebooks (or General Letterbooks as they are designated in the S. P. C. K. archives) show that the clerks copied them very carefully, at least in regard to the sound of the originals. One or two letters were entered into the Salzburg Letterbooks in Newman's own hand, but all the rest were in the clearer hands of his amanuenses. Being of legal and financial import, the correspondence of the Letterbooks is complete and is supplied with cross references between inward and outward correspondence. Very few letters were lost in the vast correspondence, and these were always replaced by copies from the originals. The importance of these letterbooks is indicated, for example, when they settled Urlsperger's and Newman's misunderstanding about the amount of salary promised to Bolzius and Gronau. Autographed originals of some of this correspondence by Newman, Urlsperger, and others are preserved in the Miscellaneous Letters of the S. P. C. K. Archives; and a collation of these with their copies in the letterbooks confirms the care taken by the scribes to retain the exact wording of the originals, even if not always the exact spelling, capitalization, or punctuation. In this regard, we should remember that the letters were written to be read aloud: when they were "laid before the Society", the members of the S. P. C. K. were interested in their meaning but not in their spelling or capitalization.

After Newman and Urlsperger, the next most frequent contributor to this correspondence was Baron von Reck, the English-Hanoverian envoy at Ratisbon, who must be distinguished from his nephew, the Baron von Reck who led the first transport of Salzburgers to Georgia. As is so often the case in this correspondence, the uncle's Christian names are never given, and only two of his signatures included the initial "J" (which could of course, have been an "I", since the two were not differentiated when capitalized). In first mentioning the nephew, Urlsperger gave his

Christian name as George; and consequently that is how he appears in many English records, even though the nephew himself almost always signed his name Philipp Georg Friedrich. The context often indicates whether the Baron von Reck under discussion is the envoy at Ratisbon or his nephew elsewhere, yet this is not always immediately apparent. Therefore, although anyone reading these letters consecutively can usually identify the von Reck in question, the initials (J) and (Ph) are always added for the benefit of those who might consult the letters out of sequence.

Needing a pastor for the first transport of Georgia Salzburgers, Urlsperger consulted his colleague, Professor Gotthilf August Francke of Halle, who was the son and successor of August Hermann Francke, the famous Pietist leader and founder of the Francke Foundations (Franckesche Stiftungen). Gotthilf August Francke, a corresponding member of the S. P. C. K., remained one of the three "Reverend Fathers" of the Georgia Salzburgers, together with Urlsperger and Ziegenhagen. Although few of his letters appear in this collection, he corresponded regularly for many years with John Martin Bolzius (sometimes written and *always* pronounced Boltzius!), the young Latin teacher he proposed to Urlsperger as chief minister of the Georgia Salzburgers. Bolzius and his assistant, Israel Christian Gronau, wrote many letters from Georgia, some of which appear in this volume, as well as a copious journal which should eventually appear in the Wormsloe Foundation Series.

In reading these letterbooks, the reader must keep in mind that all English letters are dated Old Style (O.S.), or according to the older Julian Calendar, whereas all letters from the Continent are dated New (N.S.), or according to the newer Gregorian Calendar. This fact explains, for example, how John Vat's letter of 4 October from Frankfurt was read in London on 1 October. Since 4 October O.S. equalled 22 September N.S., the letter took nine days to arrive and be read. The dates of letters acknowledged or referred to in the correspondence are always distinguished by the clues O.S. and N.S., or else by their Latin equivalents S.V. and S.N.

Although the Society tried to be prompt in attending to all correspondence, delays sometimes occurred in convening board meetings; and consequently some time might elapse between the arrival and the reading of a letter. Also, because of the slowness of the mails, there was an inevitable delay between the writing of a letter abroad and its arrival in London. On the average, letters from Dover took one or two days, those from Rotterdam a week (if not delayed by contrary winds), those from Augsburg, Ratisbon, and Halle two weeks or more, and those from Georgia anywhere from three to six months. Because of the irregularity of the mails, particularly by water, a letter from a distant correspondent might arrive long before another one sent much earlier by the same correspondent. For this reason it would be meaningless to try to rearrange these letters in any order based upon their date of writing; and it is clearly better to keep them in the order that Newman, in his wisdom, saw fit to collect them.

Since these letters furnish a journal or log of the whole Salzburg venture, they can best be read in the sequence in which they came to the attention of the board of the S. P. C. K., for only then can the reader grasp the board's motives, decisions, and policies. That means that Newman's letters should be read according to the date of writing, whereas incoming letters should be read according to the date they were read or "laid before the Society," this date being always given in the caption of the copy. When read in this order, these letters give an intimate insight into the whole undertaking.

First we see the reception of the Salzburger exiles at Augsburg, their sojourn there, and their trip by land to Marksteft and by boat down the Main and Rhine to Rotterdam and from thence by ship to Dover, Charleston, and Savannah. Little is heard from them after their arrival at Ebenezer, perhaps because it was hoped the initial hardships and sickness would be successfully surmounted. John Vat, leading a second transport a year later, furnished a vivid account of the trip down the Rhine, and we get a detailed report of the exiles' stay in England. Only when Vat reaches Ebenezer do we realize the dreadful conditions prevailing there as a result of the unhealthy climate, the barren soil, and the confusion caused by directives sent by a benign but misinformed rear-echelon headquarters.

Meanwhile, the letterbooks indicate that Philipp von Reck, elated by his first successful mission to Georgia and by his audience with the King, had set out without the authorization of the S. P. C. K. to recruit Bohemian Protestants for Georgia, an undertaking quickly stopped by stern letters from Newman to the young enthusiast and to the more mature Urlsperger in Augsburg. Nevertheless, our letters show that other Bohemians did emigrate to Georgia, even if they remained there only a short time. These were the Bohemian Brothers (Fratres Bohoemiae), also called the Moravian Brothers or Moravians, who had previously fled persecution in Bohemia and Moravia and had found refuge at

Herrnhut, the estate of Count Nicolaus Ludwig Zinzendorf, a lay minister of their brotherhood. The Trustees permitted Zinzendorf to send a contingent of his co-religionists to Georgia under August Gottlieb Spangenberg despite the objections and warnings of Urlsperger, von Reck, and Bolzius, who had witnessed Zinzendorf's radical views and innovative zeal at Halle. These letters also show that von Reck's enthusiasm encouraged certain citizens of Appenzell in Switzerland, under the leadership of John Tobler, to seek an asylum for their compatriots with the Salzburgers at Ebenezer. While the Georgia venture did not materialize, Tobler's group did pass through Ebenezer on their ill-starred journey to establish a city at New Windsor high up the Savannah River.

The second transport appearing to have been successful, the Trustees resolved to send another Salzburger transport, this time once again under Philipp von Reck. Because there were not enough Salzburgers willing to join the proposed third transport, the Trustees resolved to accept some exiles from Carinthia, the province bordering Salzburg to the southward and belonging to the hereditary possessions of the Emperor. Of the many Carinthians who had declared themselves Protestant, most were transported involuntarily to the Emperor's new domains in Hungary, where they were settled among the Saxons of Transylvania. Others had gathered at Ratisbon, where they hoped the Evangelical Body would persuade the Emperor to release the wives and children they had been forced to leave at home. Because the Emperor refused to let their dependents go, von Reck could recruit only a small number of Carinthians for his journey.

Supplementing these Carinthians with a few fellow-sufferers from Upper Austria, mostly from Salzkammergut adjacent to Salzburg, von Reck returned to Augsburg to pick up whatever Salzburgers might wish to join them. Despite the efforts of the Trustees and the S. P. C. K. to suppress the unfavorable reports coming from Georgia, people in Central Europe were becoming skeptical of all promises; yet von Reck succeeded in enlisting enough Salzburgers to complete his transport. In his hurry to reach England before Oglethorpe's departure, he does not seem to have allowed much time for writing, for we hear nothing about his journey from Augsburg to England. With his letter of 27 October 1735 to Newman, which he wrote aboard the London Merchant off the Downs, incoming letters ceased being copied into the letterbooks; and for the remainder of this narrative we must depend upon Newman's unilateral correspondence, which continued to be copied for four more years. Newman's correspondence gradually slacked off and finally limited itself almost entirely to practical matters concerning the ministers' salaries and the like, so the last four years are represented by fewer letters than were written in the single year 1733.

Nevertheless, Newman's diminishing correspondence did not imply lessened concern. His enthusiasm and optimism may have waned, but not his love or loyalty; and to his last day he remained the faithful champion of his children across the sea. It is perhaps significant that the last letter of this collection proposes the sending of a fourth transport of Salzburgers, a project strongly urged by the residents of Ebenezer despite the war that had just broken out with Spain and her possessions in nearby Florida. Newman was instrumental in organizing this fourth transport, which reached Georgia safely in 1741; and he continued giving his moral and material aid until his death two years later. After his death, the S. P. C. K. continued its benevolence toward the Salzburgers and paid the salaries of their ministers until the War of Independence severed Georgia's political ties with the Mother Country and thus removed it from the sphere of the Society's activities. Of all the settlements made in Georgia during the Trustee's administration, that of the Salzburgers was by far the most successful one, and its success was due in no small measure to the benefactions of the S. P. C. K. and to the wisdom and loyalty of its secretary, Henry Newman.

PART I

Outward Correspondence

A Monsieur

Monsieur Urlsperger Ministre de l'Evangile de Dieu à Augsbourg en Suabia.

London 13 June 1732   Per 50 £ Sterling or the value of it. At Sight pay this our first of Exchange to the Order of (Salv: Tit:) Mr. Urlsperger the Sum of Fifty Pounds Sterling or the Value of it, for the Value of Sir John Philipps Baronet Received by the Hands of Mr. Newman as per Advice from John Baptist Mayer and Company.
To Mr. Christian Munch at Augsburg.

[The following marginal note appears to the left of the above bill] This Bill drawn on Mr. Münch and enclosed in the following Letter to Mr. Urlsperger at Augsbourg.

Bartlet's Buildings 13 June 1732.

Reverend Sir: Sir John Phillips desires me to send his humble Service to You with the Bill of Exchange above on Mr. Münch for 50 £ Sterling drawn by Messrs. Mayer and Company which he desires may be laid out in the best manner You can towards the Relief of our distressed Brethren the Protestants who are banished from Saltzburg Arch-Bishoprick, and that You would be so kind as to signify when it comes to Your hands, with an Account of the Number now actually banished from the said Arch-Bishoprick, How many if not all of them are like to be received into His Prussian Majesty's Dominions, and to what places they that do not go into the Prussian Dominions intend to retire, and what further Number may be expected to follow those already Banished.

Our Good Friend Mr. Ziegenhagen is gone with His Majesty to Hanover in Order to take the Benefit of the Waters at Pyrmont for his health, but before he went he Communicated to the Society an Extract of Your Letter of the 14 April N.S. to him Signifying that the Number of Exiles was like to reach to 40,000, which is almost double the Number that Your former Letters mentioned; if this be a mistake the Society would be glad Your next may rectify it.

The same Letter acknowledged the	L	s	d	L	s	d
Receipt of the Remittance of _____	125.	—	—			
To which Mr. Ziegenhagen has since						
added 41 £ 1 S and 70 £ making_____	111.	1.	—			—
				236.	1.	—

And a few Guineas more passed through Mr. Ziegenhagen's hands, all which the Society will be glad to hear gets safe to Your hands, for the Seasonable Relief of our Saltzburg Brethren, whose Patience and Courage may God Almighty continue to Support by Succour from the Protestants far and near 'till they are all more happily provided for in this world than they could expect to be in their Native Country under a State of Persecution.

I hope you Received the Society's Packet sent to You in August, recommended to the Care of Mr. Ziegenhagen. I am sorry to tell You that our Friend Mr. Martini[1] very much wants his health, which he has many Years usefully employed in our German and East India Mission Correspondence[2]; may God restore Mr. Ziegenhagen and him in due time to the Society and long continue You with Your Protestant Brethren at Augsbourg the faithful distributors of Charity, as well as dispensers of Spiritual Consolation to all those of our Brethren of Saltzbourg whose distress may oblige them to come in the Way of Your Relief is the Wish of the Society as well as of

Reverend Sir   Your most Obedient humble Servant
Henry Newman

To the Reverend Mr. Urlsperger at Augsbourg.

[Copied here a bill of exchange for 40£, dated 23 June 1732, otherwise identical in content and wording with previous bill, and with same marginal note]

Bartlet's Buildings London, 23 June 1732.

Reverend Sir: My last was of the 13th Current accompanying a Bill of Exchange for 50 £ Sterling drawn by Messrs. Mayer and Company which I hope You have Received long before this, which accompanys another Bill for Forty Pounds Sterling drawn likewise by Messrs. Mayer and Company on Mr. Münch the same Person on whom the former Bill for 50 £ was drawn. This is a Benefaction from the Worthy Sir Richard Ellis Baronet which I Yesterday Received from Sir John Philipps towards relieving the most distressed Objects of Charity among the Protestant Exiles from Saltzburg who likewise paid to me the 50 £ which produced the last Bill, but who that was the Benefaction of, I cannot inform You; whoever it came from Sir John and Sir Richard both will be glad to hear You have Received the value of each Bill and what they produce in Florins or other Germany Money, there being as Mr. Mayer informs me no settled Exchange between London and Augsbourg, it is not doubted Mr. Münch will be as favourable as he can, considering how the Produce of it is to be applied nor need I to mention that it will be a particular Satisfaction to the Worthy Persons abovementioned and other Benefactors through their hands that the Protestant Sufferers are pleased to remember them in their Prayers to God, when they give thanks to the Fountain of all Goodness for the Effects of their Good Offices which You need not doubt will be still continued in Favour of our Suffering Brethren. May God reward Your Zeal to serve them on this Occasion and furnish You with the Supplies necessary to facilitate their Removal or Settlement in such Protestant Country, as the Providence of God may lead Them to is the wish of the Society here as well as of

Reverend and Dear Sir   Your most Obedient humble Servant
Henry Newman

I hope Mr. Ziegenhagen is got well to Pyrmont tho' I have not yet heard from him.

I am glad to tell you that Mr. Martini is much better.

To the Reverend Mr. Urlsperger at Augsbourg.

[Copies of two bills of exchange like the previous ones, one for 102 £ 2 s (money received of James Vernon, Esq.); the other for 37 £ 17 s, both dated 30 June 32.]

Bartlet's Buildings 30 June 1732.

Reverend Sir: I hope You have Received by this time my Letter of the 13th Current enclosing a Bill of Exchange for 50 £ Sterling by order of Sir John Philipps Baronet towards relieving the Distressed Protestants of Saltzbourg, and that You will ere this comes to Your hand receive mine of the 23rd Current enclosing another Bill of Sir John Philipps for 40 £ Sterling for the same Charitable Purpose. I am now by his Order to send the above Bill for 37 £ 17 s for the same Good Purpose this and the former Bills being all drawn by Messrs. Mayer and Company on Mr. Christian Münch at Augsbourg, of the Payment and Distribution of which Sir John will be glad to see some signification from You by Letter as soon as it may be with Your Convenience for the satisfaction of those Pious Persons who have sent these tokens of their Compassion to their Suffering Brethren. I am now Glad to add that the Honourable James Vernon Esq. one of the Clerks of his Majesty's most Honourable Privy Council has this Week paid to me 102 £ 2 s — d Sterling with a Desire that it might be remitted to You for for the Relief of the Saltzbourg Protestants in the same Manner as I had sent the Remittances from Sir John Philipps, and they both entirely approve of Your Application of part of these Remittances for relieving the Saltzburg Protestants who are admitted into the Hospitals at Augsbourg or elsewhere by reason of Sickness or other infirmities as being in some Respects the Great objects of Charity. The Society and They send their Compliments to You and thanks for all Your good Offices to these their persecuted Brethren and will be glad to be informed particularly the Numbers that have hitherto actually left the Saltzbourg Dominions and in what Places they are dispersed. I have here enclosed Messrs. Mayer's and Company Bill for the 102 £ 2 s Sterling abovementioned, and hope more Remittances may be made [hence] as God shall enable their Friends who are daily Soliciting Relief for the Saltzburg Protestants. May God Almighty prosper the Endeavours of the Evangelick Body[3] at Ratisbon for obtaining some Mitigation at least of the Oppressions of their and our Brethren, and may signify what Success they have had or are like to have in their Application to His Imperial Majesty in behalf of those who have not yet quitted the Arch Bishoprick of Saltzburg and what number there may be of them who are like to become Exiles hereafter and whether they are permitted to carry off their Effects. I am Reverend Sir etc.

To the Reverend Mr. Urlsperger at Augsbourg.

[Copied here a bill of exchange for 208 £ 16 s, dated 14 July 32, otherwise identical to that on p. 1 and with same marginal note]

Bartlet's Buildings 14 July 1732.

Reverend Sir: This covers a Bill of Exchange from Messrs. Mayer and Company on Mr. Münch for 208 £ 16 s Sterling towards relieving the Saltzburg Protestants at the Desire of Sir John Philipps Baronet and the other Gentlemen concerned as Receivers of the Benefactions for that Charitable use concerning the Distribution of which together with the former Remittances they will be glad to be favoured with some Account from You, because Your Account of the encrease of the Number of the Pious Confessors and the hardships they chearfully undergo for the sake of Religion when more known here may furnish their Friends in England with fresh Arguments to solicite the Charities of well disposed Persons.

This is the fifth Remittance which has passed through my hands, and it will be a Satisfaction to the Society, and the worthy Gentlemen who are pleased to be receivers of Charities for this Service to be informed particularly when each Remittance has come to Your hands as a Voucher for their Account hereafter if any Benefactor should be desirous to know that their Charity was Received and distributed by You.

The Society are sensible of the Great trouble which this Relief must draw upon You, but the Pleasure it must give You, as it would any Good man, to be the Instrument of Providence in assisting his distressed Brethren in the Case of the Saltzburg Exiles will I doubt not abundantly compensate that trouble.

Give me leave to refer You to my former Letters for such Enquiries as the Society and Your Friends here would be glad to receive some Answer to at the same time I assure You that
I am Reverend Sir Your most Obedient humble Servant

Henry Newman

London 25 July 1732. To the Reverend Mr. Urlsperger at Augsbourg.

Reverend and very Worthy Sir: The Gentlemen of the Society for Promoting Christian Knowledge having taken under their Consideration the Great and Good Work, which You and Your

fellow-labourers have been for some time employed in (Viz.) that
of Administring Spiritual as well as Bodily Food to the Numerous
Protestant Exiles from the Country of Saltzburg, think ourselves
obliged in the first Place to offer up our Thanks and Praises to
Almighty God for opening the Eyes of these distressed People to
see the Dangerous Errors and Superstitions of the Popish Persuasion and to incline their Hearts rather to quit and forsake their
All, than to sin against him by wounding and Violating their
Consiences.

Happy Sufferers, Glorious Resolution! the Genuine Fruit of a
true saving Faith, wrought in their Souls by the Influence of the
Holy Spirit.

In the next Place Sir we desire You to accept our Sincere Acknowledgements for the Account which You was pleased to send
to the Reverend Mr. Ziegenhagen of the Manifold hardships and
Sufferings of these poor persecuted Creatures, and of several other
Occurrences and Proceedings relating to them, which Relation is
drawn up by Yourself in a way so edifying and Affecting that 'tis
to be hoped few serious Christians will be able to read it without
feeling in their Hearts some tender Motions towards a People
whose Pious, meek, and resigned behaviour under all their Difficulties and Trials is very exemplary and amazing in so corrupt and
degenerate an Age as this is. And again Sir we are to bless and
magnify Almighty God for disposing the Minds of Protestant
Kings and Princes and other States, together with many of their
Good Subjects to Commiserate and relieve the Necessities of these
faithful Confessors. May it be the good Pleasure of God to
strengthen them more and more, and to establish them in the true
Faith, and to make them ever mindful of his Mercies to them.

As we have reason to be abundantly satisfied of Your unwearied
Labours for the inward and outward Advantage of these Exiles,
so we are in no doubt at all of Your discreet and Prudent Management in the Distribution of all such Charities as already have
been and (through the Goodness of God) may further be paid
into Your Hands, a particular Account of which Sums as have
been hitherto received, with the Disposition of them, when You
can find time to draw it out, will be very gratefull to us the Trustees, and a great satisfaction to all the Contributors. That God
Almighty would be graciously pleased to assist and prosper You
in this noble undertaking and continue You a happy Instrument
of Promoting his Glory and the further Benefit and Good of these

poor Exiles, and crown You hereafter with an everlasting Reward, is the unfeigned Desire and Prayer of

Reverend and Worthy Sir   Your very Affectionate and much obliged humble Servants   John Philipps[,] Ja. Vernon[,] Wm. Tillard

P.S. It is our Opinion Sir that if You could find leisure to acquaint Us with some other material Passages of the Sufferings and Behaviour of these Exiles it might (thro' the Divine Blessing) very much serve our present Design. Since writing of this, Yours of the 17th July 1732 to Mr. Newman is come to hand, and was read this Morning at the Society with great satisfaction, who return You their most hearty Thanks.

The Following P.S. was wrote by Mr. Newman to the foregoing Letter.

Bartlet's Buildings 25 July 1732.

Reverend Sir:

P.S.: Mr. Arch-Deacon Denne,[4] Mr. Hoare, and Mr. Ziegenhagen the other Receivers of Benefactions to the Saltzbourg Sufferers at the Desire of the Society, are all out of Town or they would also have joined in Signing this Letter. I hope You Received mine of the 30th of June with the Remittances of

$$\begin{array}{rlcl}
 & \pounds \quad s & & \\
 & 37.17 & & \\
\text{and} & 102.2 & \pounds \quad s & \\
 & & 139.19 & \\
\text{and of the 14th July with} & 208.16 & \pounds \quad s \quad d \\
 & & \underline{348.15 \; -}
\end{array}$$

of which I hope Your next will inform

Reverend Sir   Your most humble Servant   Henry Newman

To the Reverend Mr. Urlsperger at Augsbourg.

[Copied here a bill of exchange for 121 £ 8 s, dated 28 July 1732, otherwise identical to that on p. 1 and with same marginal note].

Bartlet's Buildings 28 July 1732.

Reverend Sir: My last was of the 25 Current accompanying a Letter from all our Trustees then in Town, to which I beg leave to refer You. I am glad now by Order of Sir John Philipps who is one of them to send the Bill above for 121 £ 8 s Sterling in

Compliance with Your Desire of the 17th Current N.S.[.] the Society and those Gentlemen who are pleased to be Trustees for the Benefactions collected for our Brethren the Saltzburgers, Exiles for the sake of the Protestant Religion, are so well pleased with the Account You give of them, that they have ordered an Extract of Your Letter to be printed for the Satisfaction of those who have already contributed, or shall hereafter Contribute to the Relief of those Pious Confessors. May God increase their Number, and may they never want Succours from their Protestant Brethren in other Countries. You mention a Benefaction of 2,000 Florins from two Foreign Protestant Congregations, but don't mention of what Country, the Society will be glad in Your next to know from whence that Remittance came, and that You would continue to signify what Numbers of Exiles leave the Country of Saltzburg, and into what parts of Germany or Europe they are dispersed or designed to be sent to. What is mentioned in a Letter You enclosed concerning a Commotion in Bohemia wants to be explained, and whether those that are oppressed there on Account of Religion may not be relieved by Application to the Evangelick Body at Ratisbonne as well as those in the Arch Bishoprick of Saltzbourg, they that are acquainted with the Constitution of the Empire and the Treaties under which the Protestants claim Protection are best Judges.[5] May God direct and prosper all Your good Offices for the Service of those who Suffer for the sake of Truth and the Christian Profession is the wish of

Reverend Sir   Your most Obedient humble Servant
Henry Newman

To the Reverend Mr. Urlsperger at Augsbourg.

[Copied here three bills of Exchange for 64 £ 11 s 6d; 63 £ 1 s (money received of James Vernon); 100 £ (money received of William Tillard), all three dated 8 Aug. 1732. Otherwise identical to that on p. 1 with same marginal note].

Bartlet's Buildings, London, 8 August 1732.

Reverend Sir: I Received by last Post Your Acceptable Letter of the 1st Current N.S. and upon Communicating it to the Society, the Receivers undermentioned have ordered me to procure Bills for the several Sums against their Names mentioned Respectively Viz.

	£ s d
Sir John Phillips Baronet for	64 . 11 . 6
The Honourable James Vernon Esq. for	63 . 1 . –
William Tillard Esq. for	100 . – . –

£ s d

making in all _____ 227 . 12 . 6

for which Sums there are herewith sent 3 Bills of Exchange drawn
by Good Mr. Mayer free of L'Agio as the worthy Mr. Münch is
pleased to allow the Payment on his part; May God reward them
for it. I congratulate Mr. Münch on the Honour His Imperial
Majesty has lately conferred him, may he long enjoy it to the
Glory of the true Fountain of Honour. A further Account of any
Motions in Bavaria and Carinthia in favour of Protestantism will
be very acceptable, and also of what is doing or done at Schemnitz
in Hungaria.

My last was of the 28 of July accompanying a Bill for 121 £
8 s – d which I doubt not but You will take Notice of when it
comes to Your hands together with those sent before it, which
I don't mention here, not doubting but they will all go safe to
Your hands; and it is desired that You will be pleased to return
separate Receipts for all these Sums as You have hitherto done
for the Satisfaction of my Masters that they are duly remitted to
You.

I hope in the midst of Your present Hurry You will not neglect
Your Health of which You make so good use for the Service of
our Common Brethren. Mr. Martini continues mending and
sends his humble Service and best wishes to You at the same time
I subscribe my Self.

Reverend and Dear Sir   Your most Obedient humble Servant
Henry Newman

To the Reverend Mr. Urlsperger at Augsbourg.

[Copied here a Bill for 100 £, dated 18 August 1732, otherwise
identical to that on p. 1 and with same marginal note].

Reverend Sir: I Received Your Letter on the 12th Current
N.S. on the Day of the last Meeting of the Society, to whom (Mr.
Martini being Present) the Contents were immediately Communicated, and were very acceptable for the Piety and Patience expressed by the Protestant Exiles under their hard usage. In Consideration of which Sir John Philipps has desired me to bespeak
a Bill of Exchange for One hundred Pounds Sterling to be sent

by this Post, which I have accordingly done and it is here enclosed being drawn by Messrs. Mayer and Company as usual on Mr. Munch, to whom the Society think themselves very much obliged for the Priviledge allowed them in these Remittances of an Exemption from L'Agio, for which may God reward him, and You also for Your Care and Zeal in making a faithful Distribution of the Charities remitted.

Your last mentions a more perfect Account to follow it, of the Numbers already come out of Saltzburg, and expected still to come out under the Character of Exiles on Account of Religion, which the Society will be glad to receive, together with an Account to what parts of Germany or other parts of Europe they intend to dispose of themselves, whether it be true that His Majesty of Prussia has given Directions for receiving all that will go into His Dominions, and to what parts thereof they are sent? what Trade or Profession the Exiles are generally bred to? and what proportion do the Children make, with the Grown Persons? with such other Observations as Your own Prudence may Suggest will be very acceptable.

I made Your Compliments to the Society, and Trustees, Collectors of Charity for the Saltzburgers, who desire me to return theirs, at the same time I assure You of being

Reverend Sir   Your most Obedient humble Servant

Henry Newman

Bartlet's Buildings London 8 September 1732.   To the Reverend Mr. Urlsperger at Augsbourg.

Reverend Sir: My last was of the 18th August enclosing a Bill of Exchange from Mr. Mayer and Company for One hundred Pounds drawn on Monsieur Munch by order of Sir John Philipps as a further Remittance towards relieving the Protestant Exiles from Saltzburg: since when I have received Your Letters of the 18th and 25th of August with the two enclosed extracts of Letters from Ratisbon, and also Your Favour of the 1st September N.S. with the 3 enclosed Receipts for Messrs. Philipps, Vernon, and Tillard, all which have been communicated to the Society, and they thank You for Your Care in applying the Remittances from hence for relieving the Confessors from Saltzburg.

I am also in the Name of the Society to desire You would in the most Respectful manner give their Thanks to good Monsieur

Munch for his ready Payment of all their Bills free of L'Agio, for which they hope God will recompence him, both in this and the next World.

In Answer to Your Enquiry about relieving Proselytes from the Church of Rome out of the Money remitted to You, the Society are of Opinion they cannot apply any of the Contributions paid to them for relieving the Saltzburg Protestants, toward relieving such Proselytes, except they come from the Arch-Bishoprick of Saltzburg professing themselves Protestants.

Care will be taken not to Print anything here which may expose You or Your Friends for the Good Offices they are continually doing to these Pious Exiles or to any others that are or may be inclined to embrace the Protestant Religion, which the Society hope God will establish in his Good Time through the World, to the utter Extirpation of all Anti Christianism and Idolatry, as well in the Romish as the Mahometan and Pagan Parts, and whenever this is done, no doubt it will be in such a Manner that both Jews and Infidels will be obliged to confess that the Finger of God has accomplished it.

In the mean time as we daily pray that the Kingdom of God may come so all Good Men will rejoice to be any wise Instrumental in the Advancement of it.

The Society doubt not of Your Prudent and Faithful Application of the Money remaining in Your hands to the purpose for which it was remitted, and tho' I have no Authority to desire You to send any of the Exiles (that may hereafter be obliged to leave Saltzburg for the sake of Religion) this way, yet there is reason to believe if it were not for the Greatness of the Expence that would attend their Transportation that they would be welcome to the worthy Gentlemen concerned in the Undertaking mentioned in the Enclosed Print,[6] which they themselves have published to make known their Design; His Majesty's Gracious Approbation so far as to incorporate them by Royal Charter, and a Grant of a considerable Tract of Land in a very happy Climate will in all likelihood soon enable them to make a good Provision for any Industrious People that shall transport themselves thither at their own Expence.

I made Your Compliments to the Society, who desire You would accept theirs while I Subscribe myself

Reverend Sir   Your most Obedient humble Servant
Henry Newman

N.B. 1 Printed Paper was enclosed in the Letter above containing some Account of the Designs of the Trustees for Settling the Colony of Georgia in America.

Bartlet's Buildings London 15 September 1732. To the Reverend Mr. Urlsperger at Augsbourg.

Reverend Sir: My last was of the 8th Current O.S. Since which I am favoured with Yours of the 8th Current N.S. the Contents of which have been Communicated to the Society and they thank You for them, as Sir John Philipps does particularly for the Receit of the last 100 £ remitted to You. In my last I Took the Liberty to cover to You a Printed Account published here by the Trustees for Settling the New Colony of Georgia on the South Borders of South Carolina; which as I then signified was not designed for any invitation of the Exiles from Saltzburg.

And as I have no Authority to give them any Invitation, I desire You will please not to construe it as such, it being represented here that His Majesty the King of Prussia has invited all that will go into His Dominions to a happy Settlement which all that wish well to the Distressed Exiles must rejoice at.

My most humble Service to Good Mr. Münch, and thanks for all his Good Offices to these Pious Confessors, I have as You desired communicated Your last Letters to Mr. Mayer who I doubt not will signify his Thanks to You for the Sight of them.

May God continue to enable You to assist these poor People, and long make You an Instrument for the Advancement of his Glory in the World is the sincere wish of

Reverend Sir   Your most Obedient humble Servant

Henry Newman

P.S. Please to continue Your Account of the Motions of the Saltzburgers and other Inhabitants in Your Parts and to what parts of the World or Europe they are Generally disposed.

[Copied here a Bill of Exchange for 50 £ dated 13th October 1732, otherwise identical to that on p. 1].

Bartlet's Buildings London 13th October 1732. To the Reverend Mr. Urlsperger at Augsbourg.

Reverend Sir: My last was of the 15th September which I hope You have received together with mine of the 8th of the same

Month, though I have had none from You since that of the 25 September N.S. I am now directed to inform You that there has been a Conference here between the Gentlemen [Trustees] empowered by His Majesty to make a new Settlement in Georgia in South Carolina, and the Gentlemen of this Society [S.P.C.K.] concerning the most effectual Method of relieving the Distresses of the Persecuted Protestants of Saltzburg, that the Gentlemen of both Societies are of Opinion that this would best be done by Settling them in Georgia because they will there be put immediately[7] into Possession of Land which will belong to themselves and their Posterity for ever, and will there enjoy all the Rights and Priviledges, Religious and Civil of English-born Subjects, and likewise that the Gentlemen have come to a Resolution to apply some of the Contributions which they shall receive to this Purpose if it shall be agreeable to the Poor People.

But before You make these their Resolutions known, they desire You would please to inform me whether You Judge such a a Proposal would be agreable to the poor Sufferers? When You think any of them may be ready to set forward on such a Design? and what numbers You apprehend may be willing to embrace these offers. The Gentlemen apprehend that they have great Reason to hope they shall be able to receive, transport, and Settle a Competent Number of Families by the next Spring; but of this You will be more fully informed when they shall have known from You how acceptable these Proposals in General may be and in what Manner received.

If they are well accepted You will then have time enough to consider of proper Persons to go with such as shall Settle in America as their Pastor and Catechists, and You are desired to inform the Society if You shall know of any whom You can recommend for those Offices. One Qualification which is Judged very necessary for those purposes is an Acquaintance with the English Tongue, since they will Settle there with many English Families, and it is to be wished that they may be incorporated with them as soon as possible.

The first Embarkation for this New Colony is now in such forwardness that it is expected to Sail the beginning of next Month, and will consist of 70 English Families.

With regard to the manner of Your making such Enquires as are necessary in order to give me the Informations beforementioned I need not Suggest that they should be done with all convenient Caution and Privacy. I am but too well acquainted with Your Discretion and Prudence to think You need such an Intimation.

Sir John Philipps has desired me to add to the Fund in Your Hands the Bill of 50 £ Sterling as a Benefaction from Mr. Hollis of this City,[8] drawn as usual on Mr. Munch, to whom I beg my humble Service may be acceptable, and that You may be assured that I am

Reverend Sir   Your most Obedient humble Servant
Henry Newman

You will please to return a Receit as usual for the enclosed Remittance when received. Mr. Ziegenhagen I am glad to tell You is returned from Germany very much improved in his Health.

Bartlet's Buildings 24th October 1732.   To the Reverend Mr. Urlsperger at Augsbourg.

Reverend and Dear Sir: I Received Your Favour of the 6th Current N.S. and the Translation of it being this Day communicated to the Society, they thank You for Your Letter and the other Papers accompanying it, and what You mention of the Depositions the Prisoners have made, though they may be too Bulky to come by the Post, yet if they contain things necessary to be known for the Service of the Exiles, they will be glad to receive the Information, and to make them less bulky You can get them transcribed in a small hand, which may perhaps reduce Your 12 or 15 Sheets to half or less, but let that be as it will, don't spare any Expence that You think necessary for Their Service who forsake all for the sake of Truth, and the Enjoyment of that Liberty of Conscience which every Man living has a Right to, in Proportion to the Light it Pleases God to give him.

If any of the Berchtolsgaden Subjects or Tirnbergers are Permitted to transport Themselves beyond Sea, and You signify beforehand their Inclination, and what Numbers of them are disposed to go, according to the Hints mentioned in my last of the 13th Current they need not doubt of being welcome to all the Assistance which their Friends here can give to facilitate their Transportation to a Country where it is hoped they will never meet with Persecution, but a quite Different Usage to what they have in their Native Country.

My humble Service always attends Good Mr. Münch, and I beg You to be assured that I am

Reverend and Dear Sir   Your most humble Servant
Henry Newman

Mr. Ziegenhagen I have just now parted with, who desires me to send You his Compliments with an Assurance that he has begun a long Letter in Answer to Your 3 Letters which he has Received from You. That his stay was so short in Germany, and always in Motion, that the Waters he went thither to Drink for his health, he is obliged to drink here which deprives him of the Pleasure of being so Punctual in his Correspondence with You as he wishes he could be, but hopes his long Letter when You receive it will make You Satisfaction. My last enclosed a Bill of 50 £ from Sir John Philipps which I hope You have Received.

Bartlet's Buildings. 21st November 1732. To the Reverend Mr. Urlsperger at Augsbourg.

Reverend Sir: My last was of the 24th of October which I hope You have Received, as I have Yours of the 3rd and 10th of November N.S. acknowledging the Receit of mine of the 13th of October wherein there was a Proposition made of so much Consequence for the Relief of our Persecuted Brethren, that when I laid this Day a Translation of Your said Letters and Papers, accompanying them, before the Society, they ordered me to recommend it to You in the most Pressing manner to favour me with as Speedy and Particular an Answer as You can, Every Day discovering more an earnest desire in the Benefactors to the Protestant Exiles from Saltzburg, that some of these Exiles may become His Majesty's in Georgia South of South Carolina, being a large Tract of fine Land between two large Rivers in a Temperate Climate, capable of a vast variety of Productions fit for the Benefit of Trade and Comfort of Life; whither a Ship Load of Artificers of all kinds necessary for establishing a New Colony are gone last Week; amply provided with all Materials to prepare the Country for the Reception of any indigent distressed Protestants that may be disposed to follow them from hence or any other Part of the World and to render the Establishment more effectual, a worthy Member of Parliament one of the Trustees for Settling that new Colony [Oglethorpe] is gone a Volunteer in the same ship to direct the first Scheme for Settling Towns and Villages therein with proper Instructions for building Churches, Schools etc. of what may make the Inhabitants florish to all Generations.

Some of the laborious Tirnbergers would have been exceeding welcome to that new Colony for the Service they might have done to themselves and their Neighbours by raking wealth out of the

Bowels of the Earth, but since they all seem to be disposed of in Germany the Exiles from Berchtolsgaden and their Neighbours will find infinite Scope to improve their Art of Turnery Ware etc. on the materials which that Country abounds with and when once You Favour me with a Particular Answer to the Proposition of the 13th of last Month, nothing will be wanting in the Power of the Society to facilitate the Transportation of a Competent Number by a Ship that may be provided for them, when the Number inclined to go thither shall be known. But as I hinted before, this is to be signified to them, and by You to me in the most Prudent manner to avoid giving umbrage of Offence to any of His Majesty's good Allies, who don't delight in Persecuting their Subjects for the sake of that Religion which their Consciences lead'em to Profess.

To avoid giving this Offence the Society have been as sparing as they could of Publishing in Print what You have Communicated to them, further than was absolutely necessary for the Satisfaction of those who have been and are still inclined to be further Benefactors of them, of which I have therefore covered to You the only Print which they have Published since that I sent to You of the 17th of July last; beside which there are weekly and almost daily Published in our News Papers concerning the Saltzburgers from the Foreign Papers, which are done without the Privity of the Society and are undoubtedly for the most Part known to You therefore I don't send them. However of these I send You two of our daily Papers of the 28 June last, and 15 November Current with remarkable Accounts from Ratisbonne to let You see how by one means or other the Publick here are daily informed of what Passes relating to the Saltzburgers from other hands than the Society, which makes it the less necessary for them to appear in it.

Your Worthy Correspondent at Ratisbon [J. von Reck] is I hope raised up by Providence to do many good Offices to these faithful Sufferers, may it please God still to bless his endeavours, and though I have not the honour to be known to him, let my humble Service be made acceptable to him from

Reverend and Dear Sir   Your most Obedient humble Servant

Henry Newman

Sir John Philipps sends his humble Service to You with Mr. Zeigenhagen, Mr. Martini and the rest of Your Friends here are well. My humble Service to Good Mr. Münch. I have not received your Packet in a Roll sent Franco to Amsterdam 28 October last.

Enclosed in the foregoing Letter to Mr. Urlsperger.

The Whitehall Evening Post of 16 September 1732. Daily Courant—28 June 1732. Daily Courant—15 November 1732.

Bartlet's Buildings. London 19 December 1732. To the Reverend Mr. Urlsperger at Augsbourg.

Reverend Sir: My last was of the 21st of November since which I have received Yours of the 27th November N.S. and acquainted the Society with the Contents of it, as soon as I could get it translated.

In answer to which the Society observe that You repeat what You formerly mentioned as Your own opinion of the aversion of the Saltzburgers to cross the Seas, but You don't signify whether You have confered with them upon it, and that what You now write is their Sentiments upon the offer mentioned in my Letter of the 13th October last and notwithstanding their supposed unwillingness to go into different Countries in separate Bodies, You now signify that the Tirnbergers have accepted the invitation by the States General to come into Holland, from whence they are informed that about 5000 Germans of the Palatinate were Shipped off for America last Summer.

The Society are glad to hear the Berchtolsgaders have an expectation of a Settlement to their Satisfaction. The Society do not think it proper to invite the Subjects of any Prince to leave their Country by the Prospect or Promise of Temporal Advantages untill they have declared their intention and obtained leave from their Prince to depart, but when that is done, they think it right to let them know that Great-Britain has opened an Azylum to receive People who are driven out of their Country for the sake of their Religion.

The Situation of Georgia and the Advantages they are to enjoy there I have already acquainted You with in my Letter of the 21st of November last, and other Preceeding Letters; Monsieur Pury a Swiss published last Summer a small Pamphlet of 4½ Sheets in French Sold at Neufchatel and St. Sulpy, under the Title of *Description Abrogée de l'Etat present de la Caroline Meridionale nouvelle Edition avec des Eclaircissemens* by which a Judgment may be made of the Climate and Soil of Georgia joining to that

Province, upon Publishing which Account several Swiss Families went last Summer to South Carolina[9] and

In Consequence of an Expectation that some of the Saltzburg Exiles would be inclined to go to Georgia, the Benefactions to them have been considerably augmented, and the Benefactors have generally appropriated their Charity to be applied to such of them as shall become Subjects of His Majesty of Great-Britain in Georgia, for which reason the Society don't think themselves at Liberty to apply the Contributions made to them, otherwise than they have been appropriated by the Charitable Benefactors.

The Society have ordered me to acquaint You that it is their opinion that whenever any of the Exiles shall be engaged or declare themselves ready to become the Subjects of any other State than Great-Britain it will from that time be the Concern of that State to provide for them.

In Consequence of which they desire that what remains in Your hands of the Contributions that have been remitted to You from the Trustees may not be otherwise employed than for the Relief of such distressed Protestant Emigrants as having left their Country for their Religion are willing to become Subjects of Great-Britain in the Colony of Georgia, except where any that are not engaged by any State have a Settlement yet to seek, in which Case they leave it to Your discretion to relieve their present Necessities.

In Answer to Your Enquiries, the Society have directed me to acquaint You that the Collections here are undertaken by several of their Members at the particular request of the Society, and carried on by them in a private manner being moved to it by their Christian Zeal and Compassion for Assisting their Brethren under Persecution.

There has been no General Collection throughout the Kingdom because nothing has been done in it by Publick Authority.

The Society were not a little surprized that a distinction should come from abroad between the Presbyterians and the Members of the Established Church concerning a Common Act of Christian Charity in promoting which both have concurred without any such distinction at home.

The Abstract of a Letter from Your Friend at Ratisbon is a Melancholy Proof of the cruel disposition of the Papists who will not let the Persecuted Protestants go 'till they expect from the Rigour of the Season many of them must Perish.

Your last Letter under cover to Mr. Mayer cost 5 s Postage and

Your former in Proportion, but if You please for the future to send Your Letters directly to me they will save the charge of the cover, and come a day or two sooner to my hands.

Dear Sir: You will excuse in regard to my hurry in other Business, the Liberty I have taken to write to You as above by my Clerks hand;[10] In answer to Your favour of the 27th November, I have now received Your favour of the 11th December N.S. and am ordered to send You the thanks of the Society for it, to which I hope soon to be able to send You a Particular Answer.

I am   Reverend Sir   Your most Obedient humble Servant
Henry Newman

Before Sealing I receive Your Role of Papers via Amsterdam and a worthy Member of the Society being present at the opening of it desires to know the price of two Such Screw Medals as were sent in it, and how they may be conveyed to me. I shall acquaint Sir John Philipps with Your kind present to him, and thank You for that You have sent to me which I shall carefully preserve (in memory of Your Zeal for the Saltzburgers) among the Curiosities[11] belonging to the Society. I shall be glad to know whether they are the Workmanship of any of the Protestant Emigrants or of the Berchtolsgaders who intend to be so as soon as they can.

Bartlet's Buildings London 29 December 1732.   To the Reverend Mr. Urlsperger at Augsbourg. Read at a Meeting of the Society 28th December 1732. Approved and ordered to be forwarded by next Post.

Reverend Sir: My last was of the 19th of this Month, wherein I acknowledged the Receit of Your favour of the 11th Current N.S. and referred You to a more particular Answer as soon as I could receive the Orders of the Society thereupon.

The Society again thank You for the Particular Advices of that Letter, and in order to make a proper Answer to it, it was necessary to consult the Honourable Trustees for the New Colony of Georgia, who being acquainted with so much of Your Letter as related to them, have for the present Signified to the Society their Resolutions thereupon in the following Order Viz.

"1stly. That the said Trustees for Georgia will defray all the Charges of the Protestant Emigrants to the Place of Embarkation and from thence to Georgia in South Carolina in America.

2ndly. On their Arrival in Georgia each Family will have Provision given them Gratis 'till they can take in their Harvest, and also Seed will be there given them Sufficient to Sow all the Lands they shall in the first Year make ready for Sowing, and likewise all Tools and Instruments of Agriculture, Building, and other necessary Uses.

3rdly. Each Man shall have Lands assigned, Sufficient for the comfortable Subsistence of himself and Family, which Lands are Granted *in Perpetuum* free from all Vassalage and Servitude, liable to no Rent for the first Ten Years, and then only to the small acknowledgement of Ten Shillings per Annum, for every hundred Acres and shall be treated in every Respect in the same manner with His Majesty's Natural Born Subjects.

4thly. They are to conform to such Orders and Regulations for the Maintenance of Property, Peace, and Good Government as the Trustees shall think necessary from time to time to establish; and on their Arrival are to assist each other in clearing their Lands, Building Houses, and such other works as shall be necessary for their mutual Safety, in Common with His Majesty's other Subjects there.

5thly. A Pious able Protestant Minister to preach to 'em in their own Language will be provided and Supported at the Charge of the Trustees for Georgia. They shall be protected in the free Exercise of their Religion and in the full enjoyment of all the Civil and Religious Rights of the Free Subjects of Great-Britain."

At the opening of the next Session of Parliament about the Middle of January the Society hope there will be some Provision made for enabling the Trustees for Georgia to provide for a considerable number of such Protestant Emigrants as shall be willing to go thither; In the mean while the Society will at present provide for Transporting 50 Families consisting of a Man and his Wife and Children under the Age of 7, and three Single Persons above that Age shall be esteemed as one Family. This last Article is only to help the Trustees to a Medium by which they may compute the Expence of a Greater Number. And as to Your Queries relating to the intervention of Publick Authority, You may expect an Answer after the Opening of the Parliament.

The Extracts of Your Friend's [J. von Reck's] Letters from Ratisbon were very acceptable to the Society, and discover him to be a Person of Prudence as well as Zeal for Serving the Protestant Interest.

The Society are very much pleased with the Effectual Interposition of the States General by their Minister [Gallieris] at Ratisbon for receiving so great a Number of the Emigrants under their Powerfull Protection.

The Role of Protocol etc. by way of Amsterdam is come safe to hand, and the Papers therein will be translated for the use of the Society as soon as may be. The two Medals in the same Parcel are also come to hand, and the choice of them presented to Sir John Philipps, who thanks You for Your Respect to him therein, though he desires me to tell You, he heartily wishes You had spared making him the Complement he being but one of several Gentlemen concerned as a Trustee in behalf of the Society for the Saltzburg Exiles. The other Medal will be laid up in the Repository of the Society as a Memorial of Your Zeal to serve them, for which Curiosity I beg You will accept of my most Respectful Thanks.

The Letters enclosed for Virginia and for Mr. Claüs in Covent Garden have been forwarded according to Direction.

I am again obliged to make use of my Clerk's hand which I hope You will excuse, and wishing You many happy new Years, I remain

Reverend and Dear Sir   Your most Obedient humble Servant

Henry Newman

Bartlet's Buildings London 2nd February 1732/33.   To the Reverend Mr. Urlsperger at Augsbourg.

Reverend Sir: I received the favour of yours of the 22nd January N.S. and have communicated the same to the Society, by whose direction I am to acquaint you for the present that our Parliament is now opened and that the Trustees for Georgia are in daily expectation of having a Sum of Money granted 'em by that Assembly towards defraying the expence of conducting, fitting out, and transporting a considerable number of the protestant Emigrants of Saltzburg to that Colony, of which you shall not fail to receive advice as soon as that resolution shall be passed, and at the same time care will be taken to send you according to your desire a short description or Account of the Country itself which appears so desirable a settlement to our own People that several of them have been already sent thither. In the mean while you will receive enclosed herein a German Translation of the Conditions or Terms proposed by the said Trustees to such of the Emigrants as shall be willing to go and settle there.[12]

The Society make no doubt but that the Georgian Trustees will give the same allowance to such of the Emigrants as shall accept of the above mentioned Conditions, as the Tirnbergers had from the States General: and they also believe that those Gentlemen when enabled so to do will very readily allow a proper maintenance for a Minister and also for a Catechist or School-Master for their Spiritual Comfort & Instruction, but this point will be more fully explained in the next Letter.

The Society likewise apprehend the Trustees have no objection to the extending this Liberality to the other Emigrants as well as the Saltzburgers: and they return you thanks for the promise of your good Offices in facilitating the Transportation of those Poor People to Georgia and in providing a fit and able Minister for them: as the Spiritual wants of this new Colony ought to be supplied no less than their Temporal, a Minister [Herbert] has been sent over with such of our people as have gone to Georgia and he being soon to return home, another [Quincy] is now going to relieve him.

As to the application of the Charity Money remaining in your hands, the Society desire their meaning may be understood to be, only, that no part of that money be sent after any of the Emigrants from the time they shall be engaged with other Powers, and shall be gone; but that you be at Liberty to relieve any fresh Comers as they pass, whether they be from Berchtolsgaden or other parts of Saltzburg, as your Charity and Prudence shall direct you; and when you advise the Society that your Stock is near Spent, they will remit you a farther supply.

Those Gentlemen have directed me to return you their sincere thanks for your remembrance of them in your prayers, and beg you will continue so to do.

The Society direct me to assure you Sir that they are well satisfied that you had no sinister intention in making the Query about the Contributions of those dissenters you call Presbyterians;[13] but the Society owe them the Justice to declare that many of them have been remarkably Charitable on this occasion, and therefore hope also that the Behaviour of the Calvinists who you were informed treated the Emigrants with Coldness and Indifference in a certain great City in Germany will prove not altogether so bad as has been represented to you as they shall always hope that every thing which looks like ill will or animosity among Protestants, who live in the mouth of danger from an enemy ready to devour them, will be wholly laid aside, this Testimony to the Behaviour

of the Dissenters is the more to be relied on, as it comes from a
Society whose Members are all of the established Church.

I am now to present you Sir John Philipps's most Affectionate
Service and to assure you from him that he is only concerned on
Account of the present received by him, he never desiring any
acknowledgement for carrying on a good work which brings its
own reward with it: and that in all Sir John's Letters to the Benefactors he has wrote to, he has made particular mention of your
most singular worth, and of your indefatigable pains and care
all along taken for the spiritual and Temporal relief and Benefit
of these faithful Confessors.

On reading that part of your Letter which relates to the Medals,
three Gentlemen of the Society, Mr. *Vernon,* Mr. *Zeigenhagen* and
Mr. *Copping* desired me to let you know they shall think themselves obliged to you if you will be pleased to procure for each
of them one of the new Medals of 5 Florins 30 Creuzers a piece,
the value of which shall be remitted to you by the first opportunity of a remittance on account of the exiles.

Dear Sir: My present great indisposition has obliged me to
make use of a friend's hand to signify the pleasure of the Society as
above. I am glad to find by your last that God grants you strength
still to pursue the labour you have so heartily engaged yourself
in for the service of so many thousand pious Confessors for which
you seem to be raised upon purpose to be the Instrument of providence. May your courage never fail and may you always rejoice
in the Effects of your laudable zeal for the Glory of God is the
sincere wish of

Reverend and Dear Sir   Yours etc.   H. Newman

*London 27th of March 1733. A Monsieur. Monsieur Urlsperger
Ministre de la Parole de Dieu a Augburgh en Suabia.*

Reverend Sir: I am very much Concerned that I have not been
able to write to you in Answer to your Severall Obliging Letters
to Mr. Ziegenhagen and my self Since mine of the second of February, all which have been translated and duly communicated to
the Society who are very thankfull for all the pains you take in
behalf of the Saltsburgh Exiles[.] But upon Reading your last letter
of the 16 Current N.S. to Mr. Ziegenhagen this day communicated
tho' I Can't yet write so fully as I Hope 'eer long to do, I am
ordered to let you know that the Society are every week taking

Measures in favour of the Saltsburgers notwithstanding the Effect of their Deliberations are not immediately seen.

The Parliament have been hitherto so taken up in affairs of a National concern that those who are to be the first movers in a thing of this nature as relieving the Saltsburgers have not yet had it in their power to follow their own inclinations in favour of them without hazarding the Success of their motion; therefore I must beg leave to Referr you still to Some future letter to inform you of What may be Expected from that August Body.

In the mean time your advices of all the motions of the Emigrants will be exceeding acceptable and the Society were this day very much Pleased at the Account you give of the good disposition of the Prussian Commissary [Goebel] and the Baron [J.] de Reck, the last of whom I should be glad to Know whether he be a Minister of our King or any other Prince and if so from whom. The Society desire particularly that you would Explain that passage of your letter wherein you say the Exiles are forbid to come to Ratisbone thro' the States they have hitherto permitted to pass. An Account of Georgia is now in the Press and as soon as Published will be sent to you which I shall be glad to do by the first opportunity.

Mr. Oglethorpe Member of Parliament and one of the Trustees for Settling the new Colony of Georgia I informed you in my Letter of the 21st of November last was then just Embarkt with a Ship Load of Artificers for Georgia by a letter from himself dated the 13th of January last is happily arrived at South Carolina and preparing to go to Georgia to facilitate the Comfortable reception of any Protestant Emigrants that may hereafter follow him. They gained their passage in 7 Weeks notwithstanding they went somewhat round about to avoid the cold northerly winds that only 2 young Children died in the Voyage the eldest being but a year and 1/2 old both of them very weak when they Embarked being half starved with want before they left London as many others were who are Recovered (he says) with food and Care upon the Sea, That the Governor [Robert Johnson] and Speaker of the Assembly [Jenys][14] of the Province wellcomed him on his landing and have promised all the Service in their power to farther the happiness of the new Establishment in Georgia concerning which I hope my next will give you a more Satisfactory account from

Reverend and Dear Sir   Your most humble Servant

H. Newman

P.S.: Dr. Herbert went over with Mr. Oglethorpe as Chaplain but with a design to return with him to England for which Reason the Trustees are Sending another [Quincy] over by the next Ship.

If you have not sent the 3 medals desired in my letter of the 2nd of February [Mr. Wilson and Mr. Belcher] another Member has desired me to Beg a 4th may be added to them[.] And a particular friend of mine desired a 5th may be added to be paid for in the next remittance made to you[.] My humble service to the good Chevalier de Munck [Münch].

London May 4th 1733. A Monsieur Monsieur Urlsperger Ministre de la parole de Dieu a Augsburg en Suabia.

Reverend and Dear Sir: I Received your Favour of the 27th of April N.S. and acquainted the Society with the contents of it and the other Papers accompanying it which were exceeding acceptable and tho' I cannot yet tell you what may be the Resolutions in Parliament I am ordered to acquaint you with the Contents of the Petition prepared by the Trustees for Georgia to the House of Commons which they are well Satisfied will meet with no opposition when offered at a proper juncture which is the only thing now waited for and there is Reason to believe it will be presented in a few days.

The Account of the Climate and Products etc. of Georgia is now also prepared by order of the said Trustees and Translating into high Dutch to be Sent to you as well as in English which I believe you will be desired to gett printed in order to make it more Known.

There is no Reason to doubt by the good disposition of some Members of Parliament but that a very handsome Summ will be Granted to the Trustees for Georgia according to the Prayer of the petition herewith sent.

I have also sent Copys of the last advices from Georgia as published in our news papers of last week which fully answer the Expectation of their friends here and if any of the Protestants drove out of Saltsburg should be inclinded to go thither I am persuaded nothing will be wanting to Encourage them and defray all the Expences of their passage of which I hope my next will be more particular from

Reverend Sir   Your most Obedient Humble Servant   H. N.

London May the 11th 1733. A Monsieur Monsieur Urlsperger Ministre de la parole de Dieu a Augsburg en Suabia.

Reverend Sir: I Received your Favour of the 7 Current N.S. and Acquainted the Society with the contents of it to whom they were very acceptable[.] I am now Glad to Acquaint you that yesterday the petition of the Trustees for Georgia in favour of his Majesties poor Subjects and the distressed Protestants abroad who are willing to become his subjects was presented to the House of Commons being recommended by his Majesty and that the same was well received and referred to a Committee to Consider what was proper to be done thereon So that by this day sennight I hope to be Able to inform you of their Resolution therein.

In the mean time I am ordered to send this Early Notice of it which gives me an opportunity of assuring you that I am
Reverend Sir   Your most Humble Servant   H. N.

London May 18, 1733. A Monsieur Monsieur Urlsperger Ministre de la parole de Dieu A Augsburg en Suabia.

Reverend Sir: According to the promise in my last of the 11th Current I Have the pleasure to inform you that the Parliament have come to a Resolution to grant a Sum of mony to the Trustees for Establishing the Colony of Georgia in America to be applyd towards the defraying the Charges of Carrying over and Settling foreign and other Protestants in the said Colony and that this Society will Endeavour to prevail with the Trustees to apply as much of that money as they can Obtain from them towards Settling your Saltsburgers or other of your persecuted Protestant Brethren as have yet found no other Settlement and are desirous to go to Georgia.

To facilitate this design You will find Enclosed an Account of Georgia which to prevent mistakes is translated into high Dutch and of such terms and priviledges as will be granted to Such foreign Protestants as the Trustees shall think fitt to Settle there which Account is left to your prudence to make a proper use of with this caution that the Trustees are not yet come to a Resolution of sending more at present than two ship-loads of passengers Consisting of 150 in Each ship to receive them on board at Rotterdam but the precise time of sending one or both those ships will depend on your Advice when a Sufficient number of passengers of one or both Ships may be ready at Rotterdam to Embark.

In order to this the Society recommend it to you to find out a proper person to Conduct them in their way from Germany to Rotterdam and that you would make it and transmitt to the Society as soon as you Can an Estimate of the whole charge of such a Conductor with such number of Protestants to be conducted.

The Society will in the mean time Endeavour to obtain from his Majesty an order to his Minister at Ratisbon for litteras requisitorias for providing them with leave to pass thro the severall Dominions in their way to Rotterdam.

The Society also desire you would Recommend a Person qualifyd to accompany these Protestants as their Minister to whom the Society are willing to allow a Sallary of 50 £ per Annum till the Settlement may be in a Condition to Support him.

The Society would be glad to know how the cash in your hands holds out, and to make a Remittance to you for Effectually Enabling you to answer all the Services now recommended to you

I am with great Respect Reverend and Dear Sir Your most Obedient Humble Servant H. N.

P.S.: The Society hope that with regard to pious books etc. the Evangelick body at Ratisbon will be as kind to those protestant Exiles that may come this way as they have been to others The account herewith sent is an Extract of which has been lately published here and was too large to be Sent to you by the Post.

The Account of Georgia Enclosed in the abovementioned Letter was as follows viz.

A Short Account of Georgia in America. Its Climate, Products, and the Priviledges there granted to the Protestant Inhabitants of what Nation soever.

Among the Various Methods of Assisting the Unfortunate those are the best which not only procure immediate Relief but provide for their future happiness.

With this View His Majesty King George the Second of Great Britain has given a large Tract of land in South Carolina (to be called Georgia) in Trust for providing a comfortable Subsistence for the poor of the said kingdom and such Protestants abroad as shall be driven from their own Country by Popish Persecution. The Disposal of this is in the hands of several Noblemen and Gentlemen who generously employ their time and Endeavours to the Improvement of the Undertaking.

Carolina (of which Georgia is a part) abounds with Provisions Deer, Hares, Rabbits, and Cattle in vast abundance Fish and

Fowls of Various kinds Fruits of the best sort Indian Corn, and Rice and European Grain of all sorts.

The Climate is known to be a good one and there are people ready to instruct in the Seasons proper to the Climate and the nature of Cultivating that Soil which is a very rich one. It lies in about 32 Degrees Northern Latitude, the Air is almost always clear, temperate and healthy.

It is Southward of our present Settlements in Carolina, and divided from them only by the river Savannah and bounded on the South by the River Alatamaha which are both large and Navigable[.] from one River to the other on the Sea Coasts is about 70 English Miles, and the Extent of the Country, from the Sea to the Apalachian Mountains is about 300 English Miles still widening in its Progress from the Sea.

The Country is at present filled with Oaks, Firr, Cyprus, Beach, Elm, Cedar, Chestnut, Walnut, Pines, Bay, Laurell, Apple, Peach, Mulberry, and many other valuable Trees, beside Vines which flourish there extremely well, and as it is a fruitful Soil, it may be soon made as beautiful and as profitable a Country by a number of hands as the best of our Foreign Plantations.

All who are sent thither by the Trustees will have free lands in Perpetuity given them. As their Children grow up and Marry, they will also have Lands granted them.

Those that are Poor have their Passage defrayed and all Conveniences allowed them in it. They are not crowded too many in a Ship for fear of Illness, They are Supplied on their Arrival in Georgia with arms for their defence Seeds of all kinds for their Lands and Provisions for a Year or till their Lands can yield them any.

Those who can bear their Expences will have a greater Quantity of Land proportioned to the Number of Servants they carry with them. These Servants however to be able working Men who will have at the Expiration of their Services Lands given them likewise These Lands to be Freehold to themselves and their Heirs male[15] for Ever. All Foreigners will have the same happy Priviledges as the English and those who shall be born in Georgia will Enjoy all the Libertys as Natural born Subjects of Great Britain where the Laws of the Land do firmly secure every Person from Persecution and all other Oppressions whatsoever and from the Insults of all officers and Soldiers.

Liberty of Conscience and a free Exercise of Religion are allowed to Protestants of all Denominations and all Encouragement will be given to Virtue and Religion & all Discouragement to Immorality and Profaneness.

The Trustees have come to a Resolution to give such Persons as are driven out of Germany by Popish Persecution and are not able to transport themselves 50 Acres of Land to each Family and then they will take care to have Provision for Protestant Ministers in their own Language as there shall be occasion to Cathechize, Instruct, Preach, and otherwise Edify 'Em in Religious Concerns.

That the People by the nearness of their situation may be of more use and better defence to one another, they will be Settled in Towns a hundred familys in Each, Their Lands will be devided into 3 Lotts, viz. one Lott for a House and Yard within the Town another for a Garden near the Town and a third for a Farm at a little Distance from it. Their Houses to be built and their Lands cleared by Generall Labour and mutual Assistance.

The People in this Colony will be in no danger from any Enemys. There are but Few Indian Familys within 400 English Miles and these in perfect amity with the English. Port Royall the Station of his Majesty's Ships is within 30 and Charles Town a great Mart is within 120 English miles. If the Colony should be attacked, it may be relieved by the Sea from Port Royal, or the Bahamas and the Militia of South Carolina will be Ready to Support it by Land.

The Raising of Raw Silk (which will be one of their Principall Employments) is so Easy a work that Every Person from Childhood to old Age can be Serviceable therein. And as it will be a usefull Commodity to Great Britain, It will be a profitable one to those who are Employed in it.

As the Success of the Colony and the Wellfare Support and Protection of the Inhabitants will be the chief Study of the Trustees they will allways be a free, and must in a few Years be a flourishing and happy People.

Bartlet's Buildings London 29 May, 1733. To the Reverend Mr. Urlsperger at Augsbourg in Suabia.

Reverend Sir: My Last was of the 18th Current O.S. since which I have the favour of Yours of the 27th Current N.S. and immediately Sent it to our Friend Mr. Ziegenhagen to be translated but he being at Kensington for the sake of the Country Air Cannot so Soon receive and Return it as he used to do.

In the mean time I am ordered by the Society to Acquaint you that if you have not according to my Last already Engaged a Conductor to come with the Emigrants from Ratisbone to Roterdam the Society have one well recommended to them as a person proper for the Service having languages etc. Viz. Mr. John Vat a Member of the Council at Bienne between Solethum [Solothurn] and Newfchattel [Neuchâtel] who is desired to attend such notice as he Shall receive from you for that Service.

This Gentleman lived many Years in England and is well known to Severall Gentlemen of the Society as a person both of Experience and Integrity; It is only to be feared that he may be Engaged in some Publick affairs as a Magistrate in his own Country to which he has been lately invited by the great Council of Bienne but it is to be hoped he may be spared to make an Excursion so farr as to do this piece of Service to the Protestant Cause in which he does not want [lack] zeal and I hope discretion.

I wish you all manner of prosperity and am

Reverend Sir   Your most humble Servant   Henry Newman

A Copy of a Letter from the Reverend Mr. Stephen Hales to Mr. Vat in Switzerland. London 1 June 1733.

A Monsieur Monsieur Jean Vat a Bienne en Suisse

Dear Sir: It is with great Pleasure that I have heard of your welfare from your Correspondents here.

The Society for Promoting Christian Knowledge having lately written to Mr. Urlsperger at Augsbourgh to send three hundred Emigrant Saltzburgers or other persecuted Protestants to Roterdam in order to their being thence transported to settle in Georgia, And I hearing that it was Requisite to have one to conduct them to Roterdam thought no one so proper as your self which made [sic] take the Liberty to Recommend you for that purpose to the Society who readily approving of You desired me to write to You immediately to make the proposal to You and to desire You if it suited Your Convenience to hold your Self in readiness to go upon notice from Mr. Urlsperger who has directions to write to You for that purpose if he is not already provided of a Conductor. As his Residence is at Augsburg letters directed thither to him will come to hand tho' he should be gone to Ratisbon.

You may depend on it that Your Expences to and fro' in this Service will be defrayed by the Society.

As I am going into Hampshire for two months whatever you shall have Occasion to write on this Subject please to write to Mr. Newman and not to me.

I have the Satisfaction to acquaint you that we hear that Mr. Pury and his People are well and go on with Success.

Mr. Oglethorpe and our Colony are also very well and go on with Success too. They are settled on an Eminence about 12 miles up the Savannah River. The Parliament has given us 10,000 £ and there are daily liberal Contributions coming in from private hands as also by Collections from house to house in several Parishes of London where the ministers of the respective Parishes do in their Sermons exhort their People to Liberality on the Occasion so that we have a good prospect that our Colony will flourish. We are in so perfect good Friendship with the native Indians that they are well pleased with their Neighbours.

I, am, Sir, Your Affectionate humble Servant   Stephen Hales

Bartlet's Buildings London 3rd July 1733. A Monsieur Monsieur Urlsperger Ministre de la Parole de Dieu à Augsbourg en Suabia.

Reverend & Dear Sir: I received your favour of the 29th of last month N.S. the Contents of which were this day laid before the Society and were so acceptable that they are very glad to find You are not weary of the good Office you have accepted and thank you for it.

And in answer to it they are glad to observe that Monsieur [J.] de Reck and Monsieur Hugo His Majesty's Ministers at Ratisbon are so fully instructed by Orders from hence to give You what Assistance they can to facilitate the Passage of any of the Saltzburg Emigrants that shall be disposed to accept of a Conveyance to Georgia on the terms mentioned in my former Letters.

The Trustees for the New Colony of Georgia act under the Royal Authority and all the Property of the Lands in Georgia is by His Majesty vested in them, in as strong terms as possible impowering them to dispose of the said lands to whom they please and therefore what they propose or recommend in favour of the Saltzburgers may be justly looked upon as done under that Authority.

His Majesty's Rescript to the Embassy at Ratisbon seems to authorize their granting any Assistance they can to further the designs of the Trustees for Georgia and of the Society in favour of the Protestant Emigrants but if any further Authority is wanting, and you please to signify in what manner that can be done the Society will endeavour to obtain it.

The Society are of Opinion that the small numbers that shall offer themselves to go to Georgia must be maintained at their Charge till those small numbers shall make up 150 sufficient for an Embarkation in one ship that will be ready to receive them at Rotterdam as you shall advise of their motion thither.

In answer to your desire for sending more than one Minister, the Society are inclined to send two Ministers and one School Master, with the Colony of Protestant Emigrants that shall be disposed to go to Georgia, and that those Persons receive their Ordination at Augsburg or some other protestant Church in Germany before they depart thence; but as the Charge of their Transportation will be considerable the Society leave it to your Prudence to settle the Salaries of those persons as frugally as you can, the Society relying on good Providence to enable them to comply with what you shall stipulate in their names.

If the Person recommended for a School Master be also be able to perform the Office of a Catechist or Clerk in assisting the Ministers it will be so much the better: in all which as I said the Society confide in Your prudence for making a proper Choice. Let me hear from you as often as any thing material occurs in this Service and you will oblige the Society as well as

Reverend & Dear Sir   Your most obedient humble Servant
Henry Newman

London 6 July 1733.   A Monsieur Monsieur Urlsperger Ministre de la Parole de Dieu à Augsburg en Suabia.

Reverend Sir: By the last Post the 3rd Current I acknowledged the Receipt of your Letter of the 29th of June N.S. and answered such particulars thereof as I had received the Orders of the Society upon.

I am now to inform you that whereas in my last Letter it was proposed to send 150 Emigrants in one ship, the Trustees for Georgia upon further Consideration at a Meeting they had the 4th Current are of Opinion for some good reasons that the first Embarkation should consist but of 70 Passengers, who will be sooner dispatched than a greater number, and therefore they will order a Ship to lie at Rotterdam well accommodated for the Reception of that number as soon as you signify that they are on their March thither.

They desire You would as near as you can signify their Quality Viz. How many Men and Women with their Age and Professions, the number of Children computing those under 15 Years of Age and above 2 years at the Allowance of 3 for 2 Passengers of grown People (it being usual to allow the Passage of 3 of that Age for the price of 2 Adult Persons) and all under 2 years of Age (being supposed to live at their Mother's breast) are with their respective Mothers estimated as one Passenger that accommodations may be made accordingly.

In case of Sickness the Captain will be provided with the usual necessarys of Medicines etc. for their relief that nothing may be wanting to make their Passage Comfortable.

One Minister 'tis hoped will be sufficient to attend this first Embarkation, and that they may be ready to depart from Rotterdam by the latter End of next month; but this will depend on the dispatch you make in sending them forward from Germany by easy Marches as the Weather will permit. Concerning all which it is desired you will be pleased to give me as early notice as you can.

I am ordered to get Bills of Exchange to remit to you Three hundred pounds Sterling which I hope to do by next Post towards defraying the Charges that will necessarily be occasioned in collecting and conveying these good People to Rotterdam

I am Reverend Sir   Your most humble Servant   H. Newman

A Copy of Mr. Mayer's Bill sent to Mr. Urlsperger 10 July 1733. London the 10th July 1733

At two days sight pay this my first Bill of Exchange to Senior Urlsperger or his Order Three Hundred pounds Sterling or the Value of it for the same received of Sir John Philipps Baronet and place it as per advice.

John Baptista Mayer

(First) To Mr. Christian Munch at Augsburg.

Bartlet's Building London 10 July 1733.   To the Reverend Mr. Urlsperger at Augsburg.

Reverend Sir: My two last letters were of the 3rd & 6th Current O.S. which I hope are gone safe to your hands and to which I shall be glad to receive an answer as occurrences may enable you to write. I am now at the desire of Sir John Philipps one of our Trustees for receiving Benefactions in favour of the Saltzburgers to cover to you a Bill for 300 £ Sterling to enable You to defray the present and future Charges that will necessarily attend one or more Transports of those good People to Rotterdam.

Since the Trustees for Georgia are willing to send a Ship to

Rotterdam for so small a number as 70 Passengers computed as I mentioned in my last, I hope it will be a great ease to your Care that you will not be obliged to keep so large a family on your hands as was at first proposed when they were to be made up 150.

The difficulty will be to find proper Conductors for so many little Transports to Roterdam, but if you can prevail with Mr. Vat at Bienne to undertake it, he will make a play of it, and return to Augsburg from Roterdam by that time you can be ready with a Transport for a second Embarkation.

As to printing the Account of Georgia formerly sent to you, it is a matter of discretion that must be left to His Majesty's Embassy at Ratisbon and You as the best Judges of things of that nature in your parts of the World. Nothing will be blamed here that ends well for the Protestant Interest, and the Glory of His Majesty of Great Brittain.

I am glad to find you are not weary in promoting the deliverance of these good People from their Bondage and that it pleases God to give you health to do what you do; may it long continue for accomplishing all the good Ends you wish is the hearty Prayers of

Reverend Sir   Your most obedient humble Servant   H. Newman

P.S. Mr. Ziegenhagen being at Kensington I have not heard from him since this day sennight but now I expect to see him less, since the Court is going to reside at Hampton Court whither he is to follow.

London 14th August 1733.   To the Reverend Mr. Urlsperger at Augsbourg.

Reverend & Dear Sir: I Received by last Post your very acceptable Letter of the 10th Current N.S. and this morning Acquainted the Society with the Contents of it who were very Glad to hear of the recovery of your health and safe Return from Saxony[.] May it long Continue for the Good use You make of it.

The Society think themselves obliged and desire you would Give their thanks to Mr. Münch for his ready Payment of all their Bills on him particularly the last for 300 £ which they Entirely approve of your reserving for such Emigrants as shall be inclined to Go to Georgia.

The Season of the Year being far advanced the Trustees for that Colony and the Society are Earnestly desirous that no time may be lost to dispose the Emigrants you have Engaged with others to make up a Sufficient Number for the first Embarkation which they hope may be easily be done if you can Engage the 50 Tirnbergers lateley Returned from Zeland to be of that Number.

The Society are absolute Strangers to the Complaints you mention of the Palatines in Pensylvania who resort thither in such Vast Numbers of their own Accord without invitation as must be attended with great inconvenience especially to such as are not able to Pay their Passage and have nothing to begin with but must perhaps sell themselves for 3 or 4 Years as Servants to Defray those Expences, and Acquire their daily Subsistence, These inconveniences must therefore lie at their own Doors by what appears to the Society; but the Case of those that go to Georgia upon Invitation whether they be of his Majesty's Subjects here or of Protestants from Saltzburg is Entirely different. The Trustees contract with Masters of Ships for their Transportation in the most Comfortable manner wherein they take all the Caution they can not to be disappointed by Employing only Persons of Integrity and Experience, and whatever they have Stipulated has hitherto been punctually observed nor can the Trustees be Supposed to have any interest in Suffering themselves or their Dependents to be deceived.

The Kings Embassy at Ratisbon have already His Majesty's Instructions in this Affair and the Trustees for Georgia can have no greater Authority than they have to give Sanction to the Proposals they have made and which to prevent all Ambiguity was therefore Sent to you in High Dutch.

The Society thank you for your early care in choosing two proper Persons to attend the first Embarkation as Religious Instructors; And the Trustees hope since they are willing to send a Ship on purpose for so small a Number as Seventy that you and your friends at Ratisbone will soon be able to dispatch them to Roterdam while the Season for travelling is so Agreeable as it is under a proper Conductor, which the Society Leave to your Prudence to make choice of as formerly Signified; Your next letter will be Exceedingly Welcome if it bring News of your having obtained a Sufficient Number for the first Embarkation there being a Ship now ready in the Thames to sail to Roterdam upon the first Notice you shall Give to

> Reverend Sir   Your most Obedient humble Servant
> H. Newman

Bartlet's Buildings London 28th August 1733.   To the Reverend Mr. Urlsperger at Augsburg.

Reverend Sir: I hope you received mine of the 14th August in answer to yours of 10th Current N.S. since which by the last Post I received your favours of 17th & 20th August N.S. Translations of both which together with the Papers and Letters accompanying them were this day laid before the Society, who thank you for the pains you take to inform them early and particularly of the motion of the Emigrants.

You will please to observe that I acquainted you in my former Letters that His Majesty has graciously invested the Trustees for settling the New Colony of Georgia with the whole Power necessary for that Service and therefore what you mention of Proper Powers from hence relating to those Emigrants which shall be disposed to go thither must come from that Corporation and you need not doubt but such a Power as shall be thought proper will be sent to you under their Common Seal after the matter has been submitted to their Consideration which will be at their next meeting to Morrow.

As to the Society for Promoting Christian Knowledge, I formerly acquainted you that they are a Voluntary Body of Men, who do not Act under a Royal Charter, as the Trustees for Georgia do, but in this affair have concerned themselves as private Men, out of Pure Pity and Compassion to the Sufferings of their Protestant Brethren, though they have had His Majesty's approbation for what good offices they intended to do in this affair.

The Society are glad to hear you have the approbation of the Kings Embassy at Ratisbon for publishing what was sent to you by way of Description of Georgia and desire their Excellencys may be assured of their Sincere thanks for all their good offices in promoting the Design in hand for making a Settlement of the Emigrants in that Province.

The Society are much concerned to hear of the Return of some of the Emigrants from the Island of Cadsan [Cadzand] & should be more so if any just occasion was given 'em of disaffection by the misbehaviour of those who had the conduct of that Settlement. but you may be assured that the utmost care under the good Providence of GOD will be taken here by the Trustees for performing and making good all that they have promised in behalf of such Emigrants as are willing to Settle in Georgia.

The Society heartily wish you all manner of Prosperity[.] at the same time I assure you that

I am Reverend and Dear Sir   Your most Obedient humble

Servant   Henry Newman

P.S. I acquainted Mr. Ziegenhagen to expect a Letter from you by next Post.

The Society will be glad to See a Copy of what you publish on the affair abovementioned.

Bartlet's Buildings London 7 September 1733.   To the Reverend Mr. Urlsperger at Augsburg.

Reverend & Dear Sir: I received both your Letters of the 31st of August & 3rd September N.S. in due time and immediately acquainted the Society with the Contents of them, who You may be sure were not a little pleased with the hopes of seeing some effectual fruit of all their concern & your toil towards relieving their persecuted Brethren; which your last seems to give them some prospect of.

To answer your Letters more fully I think I cannot do better than to place Your Queries in the Order you have transmitted them with the answers I am instructed to make, tho' I must observe to You that I have in my former Letters answered most of them already.

Your 1st Query is Whether these People their Children and Posterity shall continue to have Liberty of professing the Lutheran Religion & shall have the use of a Minister & Catechist continued to them?

Answer the Colony of Georgia is established upon the Foundation of Liberty and Toleration. With respect to a Minister the Protestant Saltzburgers will be in the same circumstance, and have the like priviledge as the Church of England Inhabitants whose Minister is supplied by the Trustees, and will be maintained till they shall be in a condition to support him in the Colony.

The 2nd Article seems to be a Request rather than an Enquiry viz. That they may be planted as near as possible to each other, in order to their mutual assistance & that they may meet with more ease as well as frequence at Divine Worship and to have their Children instructed.

Answer. The Trustees for Georgia have so much regard to this request that they have already given previous Instructions and will repeat them that the Saltzburgers may be placed as near as possible to each other being very desirous that Persons who have so good a sense of Religion as to leave their All for the sake of Jesus Christ should have all possible encouragement & assistance from their fellow Christians to keep them stedfast in the true Faith and to enable them to transmit the same to their Posterity.

Query 3rd. Whether care will be taken that they shall not be mocked and ridiculed upon Account of their Habit, Behaviour and Language as has been done in Zeeland?

Answer such Behaviour towards Strangers is very unbecoming and what is justly abhored by all good Men and particular Care will be taken that they shall not be so treated.

Q. 4. Whether the Saltzburgers may be secure of a Property in their Lands & Goods, & that they shall not be obliged to enter into Service?

Answer the Trustees for Georgia always designed that Lands should be given to them and that they should have the same Liberty as to their Persons & the same Property in their Lands & Goods as the rest of his Majesty's Subjects there.

To the Query about Numbers to be sent:

Answer. The Trustees would be very glad to have as many of the 300 as You can get sent away as soon as possible, and that You give me as early notice of their Coming as you can, upon which they will take immediate care for their reception & transport, but as there cannot go more than a Hundred in one ship, it may be convenient that such a Number should travel first & the other follow at some little Distance. You will be pleased to take notice that any number from 70 to 100 may be carried in one Ship, but rather than not make a beginning, if You can get no more, they will accept of 60 for the first Company and hope to hear by your next that you have secured that Number and engaged a Minister and Catechist to attend them. If after the first Number are departed a sufficient Number should be ready to make a Second Party, you are desired to send them with another Commissary and if You are assured of 2 Companies it would be prudent to send a Minister with one in one Ship, and the Catechist with the other, but whether to send one or both at the time time is left to Your discretion.

The Society take notice of a false Report raised about the great Mortality which might probably attend them; but by the Accounts of the Trustees for Georgia have received from their Settlement of the Persons that went over thither but one only & he an elderly Person died there since their arrival about half a Year ago.[16]

Your Powers for acting in behalf of the Trustees are drawn up, & would have been transmitted by this Post but the Season of the Year carrying so many Gentlemen out of Town there were not a sufficient number of Trustees to sign it, however you may expect it will be sent by the first Opportunity under cover to His Majesty's Embassy at Ratisbon.

If You please to signify how your money holds out care will be taken to supply you, or if on a sudden emergency you should want Two or Three Hundred pounds for serving the Saltzburgers you may take it up, & draw upon Sir John Philipps Baronet who is pleased to permit it upon your giving me proper advice thereof, to which if You please let your Bill refer. Upon mentioning which give me leave to put You in mind what I believe your hurry has made you forget viz. that You have acknowledged the receiving the 300 £ in your Letter to me, but have [not?] sent a Receipt in Form as usual to be laid before Sir John.

I wish all manner of Prosperity & am

    Reverend Sir Your most obedient humble Servant
                Henry Newman

Bartlet's Buildings 18th September 1733. To the Reverend Mr.
Urlsperger at Augsburg.

Reverend Sir: My last was of the 7th Current since which I received Your Favours of the 7th & 14th Current N.S. with Copies of several Papers and Letters which have been all translated and considered by the Society; The Powers and Instructions to Your self from the Trustees for Georgia are dispatched in due form, under their common Seal, and delivered to Baron Hattorff, who has been so kind as to undertake to forward them to his Majesty's Embassy at Ratisbon in order to be sent to You.

The Trustees will apply to his Majesty for such Instructions to His Ministers at Vienna as desired by Your Letter of the 7th Current Viz. That they Endeavour to obtain from His Imperial Majesty, proper Orders to the Magistracy at Augsburg for admitting the Emigrants to Quarters in that City.

The Trustees and Society both are glad to understand You have already retained 40 for the Colony of Georgia, & that You have an immediate prospect of having a number sufficient to compleat the first Transport out of the returned Emigrants. They hope by your next Letters to hear of the day they are to set forward, and how long (supposing common Accidents) they may be on their Journey to Roterdam.

They observe the Person recommended for a Conductor to the Emigrants [Ph. von Reck] is a very Young Man, And tho' they readily accept of the Young Man's Zeal in the good work, Yet apprehending he may not have sufficient Experience to discharge the part of a Conductor alone, it is again recommended to You to write immediately to Mr. Vat to wait upon You & take upon him the Conduct of the first Transport in conjunction with the Young Man recommended by the Baron von Recht [J. von Reck], because it is of the utmost Consequence especially in the first Embarkation that the same should be Conducted with all the Skill and prudence which can be made Use of to recommend the Design. And if the Young Gentleman and the Journeyman Apothecary [Zwiffler] You mention are desirous to proceed to Georgia, in regard that the First is recommended by the Baron von Recht and the other by Your Self, it is not at all to be doubted but the Trustees will make the same allowances and Grants to them as they do to the Emigrants.

I hope it will please God to continue Your Health to See happily accomplished the Emigration of all the Transports of our suffering Brethren who may embrace the invitation to Georgia and believe they that go first will make so good a Report of the Usage they will find in the Country they are going to as to make others very desirous of following them. I am

Reverend Sir   Your most obedient humble Servant
Henry Newman

P.S. I received last Saturday 5 Screw Medals from Mr. Wolter the King's Agent at Roterdam by Order of Mr. Gullman and the Gentleman for whom they are designed desire me to return their Thanks to You for Your Care in conveying them and that You would let me know the prime Cost and the Charge attending the Conveying of them with Your first Convenience in order to my remitting per Value.

Bartlet's Buildings London 18 September 1733.   To. Mr. J. Vat at Bienne in Switzerland.

Dear Sir: I received in due time Your most acceptable Letters, of the 30 March, 31 July, & the 4th September Current the last Enclosing a Letter to Dr. Hales at Teddington who being then at his Residence it was immediately forwarded to him; so I did not open it but shall I doubt not in due time know the Contents as Mr. Vernon does already.

I have not acknowledged the Receipt of these favours sooner being daily in expectation of writing to You by order of the Society as I now do to Beg you would be in readiness to proceed to Augsburg upon the first Notice you will receive from Mr. Urlsperger to go thither or to any other place he shall direct for Conducting a Transport of Emigrants from Saltzburg to Roterdam in order to their Embarking for Georgia.

The Society had (before you was wrote to) left it to Mr. Urlsperger to provide a proper Conductor but his last Letters signifying that he had Entertained a young Man of 21 Years of Age [Ph. von Reck] recommended by the Baron Van Recht [J. von Reck] one of his Majesties Ministers at Ratisbone in regard to his recommendation, The Society decline refusing him and have therefore ordered me to write to Mr. Urlsperger by this Post which I have done to desire he would immediately write to You and desire You would accept of acting in conjunction with him whose name he does not tell me but that he is willing to proceed even to Georgia with the first Transport which the Society accept of & had rather be at an Extraordinary Expence of Bearing your charges with his than that there should be want of a Gentleman of your Experience to conduct the first Embarkation. This Gentleman is recommended by Baron Van Recht as a person of Learning and ingenuity for his Years so that your Assistance may soon make him a man of Experience which seems to be the only thing he wants and Yet absolutely necessary.

Your Good inclinations to this Service will it is hoped prevail with You to accept of a partial Commission in Conjunction with a friend rather than none and it is very likely that Mr. Urlsperger may have You in his Eye for a succeeding Transport But the Society as things appear to them had rather You should attend this and that also.

I Pray to God to direct you in this and all Your affairs and give you health equal to Your Zeal for promoting any Service for the Protestant Interest[.] Mr. Urlsperger has mony and credit to supply You with what may be necessary for your self and those You may receive under your Care. Pray let me hear from You by all Opportunities how you Proceed and let me have directions how to write to You as you change your situation.

The University of Oxford in their late Publick Act Complimented Dr. Hales with his degree of Doctor of Divinity while he was on his Living in Hampshire not expecting any such honour in regard to his great merit and the Service he has done both to

Religion & Learning in the 2 Volumes he has published on Vegetable Staticks etc. I Lately received a very obliging Letter from Messrs. Polier and Seigneux at Lausania [Lausanne] which I hope soon to answer and Remain

Dear Sir   Your most humble Servant   Henry Newman

Bartlet's Building London 28th September 1733.   To the Reverend Mr. Urlsperger at Augsburg.

Reverend & Dear Sir: My last was of the 18th September which I hope You have received, since which I received Your favours of the 17th & 21st September Current N.S. with the Prints enclosed in them & communicated them to the Trustees for Georgia, who together with the Society are extreamly thankfull for the great prudence and care you have shewn in conducting this affair.

They are in hopes the opposition made by the Popish part of the Magistracy will be removed upon the application which is made to His Imperial Majesty and will be sent this day to the Kings Minister at Vienna.

As to the objections made by the Emigrants arising from their not knowing the Truth of the Case or the Misinformations they have received if the very wise answers which You have made cannot do it, there is no doubt but the hearty reception and good usage which You may assure them in the Strongest Terms they shall meet with will entirely remove them and therefore the Trustees and the Society are mighty Zealous to have some go tho' never so few & desire (agreable to the inclosed Minute) that you will immediately Send away what you can and that You will give the most early notice of their motion that a Ship may be ready to receive them at Rotterdam.

They hope You are provided with a good Minister to go along with them but as to a Catechist think the Person recommended in the inclosed Letter by Dr. Guerdes the Lutheran Minister of the Sweedish Church[17] here upon all Accounts most proper, as He is Master of the German and English Tongues and can therefore be more usefull than any Person You can send from Germany.

If after all, their Labour of Love proves unsuccessfull and You cannot engage any Saltzburgers to accept what they think are very good Offers and what are greedily sought after by our own People they must rest Content with having Shewn their good Will towards their persecuted Brethren having had no other Intention but the Benefit of these distressed Objects, no advantage being Possibly to be made to themselves.

Upon this Occasion I think proper to acquaint You from my self that the Trustees are much Sollicited in behalf of a great number of very worthy Confessors of the Truth who have been driven out of the Valleys of Piedmont[18] and have not yet obtained a Settlement but having made the first Offer to the Saltzburgers will receive no other till they have absolutely refused it.

I wish You all manner of Prosperity and am

Reverend Sir   Your most humble Servant   Henry Newman

Mr. Ziegenhagen is well and has seen your Complements to him. I congratulate you upon the prospect You have of a new alliance that I heartily wish may be prosperous to your Family and particularly to the Young Lady your Daughter and her intended Spouse.[Wenndrich].

Copy of the Minutes of the Trustees for Georgia.

Palace Court

At a Meeting of the Trustees for Establishing the Colony of Georgia September 26, 1733.

Mr. Vernon produced Letters from the Reverend Mr. Urlsperger to Mr. Newman acquainting him with Several Obstacles thrown in the way of the Saltzburgers coming over by the Popish part of the Magistracy of Augsburg; That the number ready to embark are at present but forty one Heads and that many of these object to their going to America, That at present there seems little Likelihood of increasing the Number.

Resolved

That Mr. Vernon be desired to procure Letters to be written to the Reverend Mr. Urlsperger to send immediately to Rotterdam the said Forty one Heads, or as many as are willing to come, Being assured that their Reception in Georgia will encourage others to follow them; and that Mr. Urlsperger be desired to dismiss such as are unwilling to embark. That an Application has been made to his Majesty to send Instruction to his Minister at Vienna to apply to the Imperial Court, that the Popish Magistrates at Augsburg may be induced to consent that the Georgian Colonists may be admitted into their City. By Order of the said Trustees Benjamin Martyn Secretary

Copy of a Letter from Dr. Henry Walther Guerdes Dated at New Broad Street London 11 September 1733.   To Mr. Newman.

Dear Sir: I must beg Leave to Ask; Whether? or no! I may apply to You for recommending a poor Family of my Congregation to the Honourable Trustees of Georgia for the settling of amongst the Saltzburghers and others Germans expelled as I hear from abroad?

You know perhaps Whether a good sober and diligent man that could keep School both in English and German is wanted or would be acceptable to the Colony?

For such a one it is that I could recommend to You in the Person of one Christopher Ortmann who has kept for these many Years a Charity School under my own Inspection and has always in his Office behaved to my entire Satisfaction. The man is about 50 Years of Age and is married to a Woman [Juliana] about 40. They have no Children. The Woman is likewise able to instruct Children in all manner of needle work, so that she may keep a School for Girls for that purpose and has done so before now.

Here they find it a difficult matter to Subsist and are therefore mighty desirous to go to Georgia.

I think 'em worthy of encouragement and recommendation and shall therefore be particularly obliged to You if You can put them in the way that they may obtain the end they desire.

I will in my turn endeavour to see (when I hear that the Saltzburgers come) Whether I can do some thing for them in my Congregation, being with due respect.

Sir   Your very humble Servant   Henry Walther Guerdes

Bartlet's Buildings 9 October 1733.   To the Reverend Mr. Urlsperger at Ausburg.

Reverend & Dear Sir: I this Day received Your very acceptable Letter of the 1st Current N.S. and having acquainted the Society with the contents of it they were glad to find good Providence continues to remove the difficulties that obstructed the Relief proposed to be given to the Protestant Saltzburgers by making them free Men of Georgia, and when you have received my Letters of the 18th & 28th of last month I doubt not but You will be enabled to answer all Objections that can with any shew of Reason be made to the Proposal on the part of the Trustees for the Colony of Georgia so as to confirm those Saltzburgers that have come to a resolution of embracing the Offer made to them that they will with the Blessing of God attending 'em see every day fresh Cause to rejoyce in the Track they have taken.

Sir John Philipps sends his humble Service and Thanks to You for Your Receipt for the 300 £ last remitted.

Mr. Ziegenhagen also giving his Company at the Society sends his humble Service and received from me Your Letter of the 28th of last month which Mr. Mayer to whom it was enclosed had sent to me.

The Society think themselves much obliged to the excellent Mr. von Münch and desire you would give their Thanks to him for his kind offer of advancing what money you may want if the supplies you have already received should fall short of what is necessary to dispatch the Emigrants, and defray their Expences on the Road to Rotterdam. Wishing You all manner of Prosperity, I remain

Reverend Sir   Your most humble Servant   Henry Newman

Bartlet's Buildings 16 October 1733.   To the Reverend Mr. Urlsperger at Augsburg.

Reverend Sir: I hope e'er this You have received mine of the 18th and 28th September and 9 Current O.S.

I am now to acknowledge the Receipt of Your Favour of 12th Current N.S. Yesterday in the Evening the Contents of which being this Day laid before the Society I am ordered to acquaint you that in Case you have not been able to find a proper Person to be a Minister to the Saltzburgers going to Georgia You may give your self no further trouble because they are in hopes to get one here a German Divine that understands English, has a good Heart and is well qualified for the work being recommended to the Society by our good Friend Mr. Ziegenhagen.

The Catechist [Ortmann] mentioned in mine of 28 of September has with his wife attended the Society and are accepted into such Services as they may be capable of beside Catechising the Children of the Saltzburgers, as soon as I receive Your advice of the Saltzburgers designed for Georgia that they are set out from Augsburg and about what time they may be Expected at Rotterdam. The Trustees for settling that new Colony have ordered a ship [Purysburg] now in the River for the Service to proceed to Rotterdam to receive them on Bord and the more Expedition there is used in hastening their March the less they will feel of the Cold weather which will e'er long come upon us in the Climate where we live.

Till this affair is in some forwardness of Dispatch the Society and all Your Friends here will be glad to hear from You once at least every week that nothing may be wanting in them to take proper measures for the Comfort of those Saltzburgers that shall Embark for Georgia.

I always make Your Compliments to the Society who as often return theirs with their best Wishes that God would prosper your Endeavours in Serving distressed Protestants wherein joins heartily

Reverend Sir   Your most humble Servant   Henry Newman

P.S.: I formerly mentioned your directing your Letters to me here in Bartlet's Buildings London as would occasion my receiving your Letters some hours sooner and at half the Postage I now pay for them beside saving Mr. Mayer the Trouble of sending a Servant on purpose with them which is attended with great inconvenience to him if it happen on a Post Day[.] I have been since thinking that perhaps you decline using that method for fear your Letters should be intercepted but if You direct 'em to me at Bartlet's Buildings in London without taking notice of my being *Secretary to the Society here* I believe they will never miscarry and as to any inconvenience that may accrue to your self by the Contents of them if they should fall into ill hands you are always at Liberty to sign them with the first or last Letters of Your name only which I shall understand tho' others may not. Being obliged to gett your Letters Translated before I can communicate them to the Society or Trustees for Georgia makes it necessary for me to save as much time as I can Especially now we are on a Crisis of Embarking the Emigrants that accept of the Offers made to them.

Bartlet's Buildings 23 October 1733.   To the Reverend Mr. Urlsperger at Augsburg.

Reverend & Dear Sir: I received Yesterday both Your Letters of the 15th and 22nd Current N.S. which were this Day Communicated to the Society and they Entirely approve of the measures You have taken for Providing a Minister and Catechist[.] But if they do not succeed so as to answer their and Your Expectations there is a worthy Divine here ready to supply the want of the former who has frequently assisted Mr. Ziegenhagen in the Kings German Chappel as mentioned in my last Letter of the 16th Current which I hope you have received 'eer this whose Resolution will be determined as Your next Letters shall advise, in the mean time Mr. Ortman the Schoolmaster is preparing himself to go over with the Transport ship to Rotterdam to attend the Saltzburgers there as an Interpreter to the English Officers and Sailors on bord the ship. It would have been a Pleasure to the Trustees as well as the Society if the honest Emigrant [George Buecher] with the Family of 11 Persons had continued in his Disposition of proceeding to Georgia because I have reason to believe that he would have been Distinguished by a Commission as a Magistrate among his Brethren if he had Gone. His not going with his numerous Family and others going in their Room has made a Considerable alteration in the List You transmitted for which reason it is Desired that You would send a Compleat List of the Emigrants designed for Georgia of the Men, Women, and Children with their respective Ages and Professions at the time of their setting out from Augsburg the Number of which may perhaps be encreased by that time they take their Departure.

I need not repeat to You that if You should want money to defray the Expences that will necessarily happen in conducting these Emigrants to Rotterdam upon Mr. Münch supplying you with it or with Credit where it may be wanting on the Road in Exchange for your Bills on Sir John Philipps as mentioned in my Letter of the 7th of September that your Bills will be punctually Honoured.

The Society are not a little pleased at Your having sent for Mr. Vat to conduct the first Transport because his Integrity and Experience gives them great hopes that nothing will be wanting to render the passage of those honest People to Rotterdam as comfortable as can be for persons under their Circumstances.

The Society have made no Bargain with Mr. Vat but leave him to make his own Demands relying on his being contented with what shall be reasonable for his Service which I mention that you may readily confide in him as the Society do for his faithfull Discharge of the Trust reposed in him and that you may give him Your Instructions with the same freedom and assurance as to a Person You had been long acquainted with.

I hope Your next will inform me of the Day of the Emigrants being set out from Augsburg as well as of their number etc. and the Route they are to observe and in the meantime you may be assured that Care will be taken by the Trustees for Georgia that a proper ship will be sent to Rotterdam to receive them as soon as they arrive there.

May You still Continue to Enjoy your health and make that good use of it as you do for the Glory of God and the good of mankind is the wish of the Society as well as of

Reverend Sir   Your most humble Servant   Henry Newman

Bartlet's Buildings London 26th October 1733.   To the Reverend Mr. Lowther at Rotterdam.

Reverend Sir: I hope this will find You happily returned to Rotterdam and thank you for the pleasure you very kindly gave to the Society while you was at London by attending their weekly meetings here.

They have Just now advice of 50 Saltzburg Emigrants setting out as supposed Friday the 19th Current our Stile from Augsburg under the Conduct of Mr. Vat an honest Swiss for Rotterdam in order to proceed to Georgia. Mr. Vat is furnished with *Literas Requisitorias* by the King's Embassy at Ratisbone and Money or Credit at Augsburg to defray all his Expences to and at Rotterdam till the Emigrants can Embark on bord the Purisburg—Capt. Fry Commander who is to sail next Thursday or Friday by order of the Trustees for Georgia to receive them on bord; but for fear it might happen by contrary winds or some other accident that the Ship should not arrive a week before the Emigrants do for their immediate reception, on their arrival I am directed by Mr. Philipps our present Chairman and Mr. Vernon a worthy Member of the Society as well as one of the Trustees for Georgia to request Your good Offices in bespeaking conditionally Quarters for these Emigrants till they can Embark to prevent any imposition which they & their Conductor may be liable to as Strangers, the Expence of which will be defrayed as I said by Mr. Vat and Your good Offices on this Occasion will be Esteemed as a favour to the Trustees for Georgia as well as to our Society, which I am sure both Body's will thankfully acknowledge.

I do not know Mr. Wolters the King's Agent at Rotterdam for foreign affairs but have upon this Occasion taken the liberty to write to him to ease You of any Trouble that he can in promoting this Service which by the Nature of it cannot be unacceptable to him as a good Protestant and servant to our Gracious King[.] I have therefore referred him to You to concert with You and he may think proper to be provisionally only for if they Embark immediately upon their arrival at Rotterdam, I hope neither of You will have any trouble but rather Pleasure, I am

Reverend Sir   Your most humble Servant   Henry Newman
P.S. The King's Barges are gone down the River to meet the Prince of Orange[19] hourly Expected and if his Highness lands at Tower Wharf his Majesty's Coaches are there to receive him.

Bartlet's Buildings 26 October 1733.   To Mr. D. Wolters Agent for his Britannick Majesty at Rotterdam.

Sir: I thank You for Your favour of the 15 of last month and for Your Care in forwarding from Mr. Gullman the Box of Screw Medals by Capt. Christian Master of the Hyam Sloop who delivered them safe to me.

I am now at the Desire of some Charitable Gentlemen to request your good Offices in behalf of 50 or more Protestant Emigrants from Saltzburg who have lately accepted of becoming Subjects to his Britannick Majesty and as such are supposed to have set out from Augsburg Fryday 19 Current O.S. in order to Embark at Rotterdam for the Province of Georgia[.] How soon they may arrive at Rotterdam is not Certain but the Purisburg Capt. Fry is ordered hence to be there for their reception as long before hand as he can Yet in Case contrary Winds or other accidents should prevent his being there time enough; The Reverend Mr. Lowther Minister of the English Church at Rotterdam is wrote to by this post to bespeak Quarters provisionally for them that Mr. Vat their Conductor may not be imposed upon in the Expence of their Lodging and Diet which as Strangers they may be liable to. And any assistance you are pleased to give in Concert with Mr. Lowther to facilitate this Service, I can assure You will be thankfully accepted by many worthy & Charitable Gentlemen here.

Mr. Vat or the Conductor of these People whoever he may be will be Enabled to Discharge all reasonable Expence on them, but the design of this recommendation is to prevent their being put to unreasonable Charges. And if they Embark immediately on their arrival You will have the Pleasure of seeing Your Trouble at an End. The Agreement for Quarters being provisional only.

I shall be glad to have it in my Power to return any Favour You shew to these Objects of Charity and remain

Sir   Your most humble Servant   Henry Newman

Bartlet's Buildings London 30 October 1733.   To the Reverend Mr. Urlsperger at Augsburg.

Reverend & Dear Sir: I have this Morning received Your Favour of the 29th Current and immediately procuring a Translation of it have acquainted the Society with the Contents of it, They are Concerned to find the Number of Emigrants a Second time reduced to 40 tho you don't mention the Reason.

My Last was of the 23rd Current which I hope You will have received e'er this, to inform You of the Purisburg's being the Ship appointed by the Trustees for Georgia to attend at Rotterdam to receive the Emigrants on their arrival there to which I am now to add that Capt. Fry master of the said Ship is ordered by the Trustees to Depart hence Friday 2nd of November Wind & Weather permitting so that in all likelihood she may arrive at Rotterdam in a Week before the Emigrants will.

I have wrote to the Reverend Mr. Lowther Minister of the English Church at Rotterdam and to Mr. Wolters the King's Agent there to assist with their good Offices the Commissary that shall attend the Emigrants in whatever Shall be thought necessary.

The Trustees for Georgia don't design the Purisburg shall come up to London in her return from Holland but to touch at Deal or Dover to be Visited by an Officer from them to see that all things be provided for the Saltzburgers which may be necessary to render their voyage Comfortable. Nor would it be for the advantage of the Passengers to Georgia to hinder their voyage so much as bringing the Ship into the River would do. I hope to hear often from Mr. [Ph.] Von Reck while they are on their March and shall be glad of the like Favour from You as occurrences may require & I Remain

Reverend Sir   Your most humble Servant   Henry Newman
P.S. I cannot account for Mr. Vat's Silence but that he intends to carry to You his Answer in Person.

Bartlet's Buildings 6 November 1733.   To the Reverend Mr. Urlsperger at Augsburg.

Reverend and Dear Sir: My last was of the 30 October. Since which I received none from You tho' I hear 3 Mails are come in since Your last of the 29 October N.S.

The Purisburg sailed for Rotterdam last Friday the 2 Current and as the Wind has been fair she must be arrived e'er this. Christopher Ortman the Schoolmaster went over in the Ship to assist the Saltzburgers as an Interpreter and if the Minister and Catechist does not attend them there as You Expect Mr. Ziegenhagen has a prospect of being furnished with a very good Minister here if he may be allowed a little time to Prepare himself for the Voyage.

Last Post I received from Mr. Vat a Letter at Bienne 30 October 1733 wherein he acknowledges the Receit of Your 2 Letters

of the 15 and 21 of the same Month and mine of 18 of September
And that according to your Desire he purposed the next Day viz.
31 October to set out for Augsburg by way of Zurich St. Gall and
Lindau.

I wish his Journey be not undertaken too late to arrive with You
time enough to attend the Saltzburgers to Rotterdam unless something unexpected should retard their departure from Augsburg
but perhaps he may find means to overtake them. A Letter from
Mr. Von Recht [von Reck] their Conductor will be very acceptable to inform me now and then of the Condition of the Emigrants
under his Conduct[.] I hope every thing will answer his and our
Wishes to him on the road[.] if Mr. Vat be with you when you
receive this please to let him know that I am concerned to find my
Letter should be 4 weeks going to him and his answer 17 Days
coming to me tho I sent it to his friend in London for more certain and Expeditious Conveyance.

I hope Your Letters for the future will come directly to me here
partly to avoid double Charges & more for Saving time not to say
also some trouble both to Mr. Mayer and my self for I send my
Clerk every Post almost to Mr. Mayer to save him trouble and if
he returns without a Letter from You he walks at least 2 English
Miles in vain. I am
    Reverend Sir   Your most humble Servant   Henry Newman

*Bartlet's Buildings 13th November 1733.   To the Reverend Mr.
Urlsperger at Augsburg in Suabia.*

Reverend Sir: My last was of the 6th Current since which I
have received your favours of the 2nd, 5th, 9th and 12th of this
month N.S. and the several Papers refered to in them a Translation of which were this day all laid before the Society and read[.]
They thank you for your very kind assistance and Care in dispatching the Protestant Emigrants and hope Success will attend
all your future good Offices in their behalf.

The Society took early care to engage the Reverend Mr. Lowther
Minister of the English Church at Rotterdam and Mr. Walters
[Wolters] Agent for His Brittish Majesty there to give all the
assistance they could to the Commissary that should attend the
Emigrants and accordingly Mr. Lowther has signifyed to me that
he has prepared the Magistrates to expect a Transport from
Augsburg for Georgia and that they need not doubt of a kind
Reception if they came in health and to be satisfyed of that they

will desire them to stay one night or two between Dort and Rotterdam, or so long as till the Magistrates are satisfyed that they are in health, when they will give orders for their proper reception.

The Society are glad to hear Mr. Vat is safely arrived at Augsburg, and that you have taken him under your roof in that kind manner you have done till it can be resolved how to employ him[.] there seems to be a probability of soon making up another Transport of Emigrants either at Augsburg or France [Frankfurt][20] but they think it very proper he should remain with you till it can be known what occasion there may be for his Service to conduct a future Transport to Rotterdam.

There is no advice yet of the arrival of the Purysburg at Rotterdam but She has had fair winds[.] tis very probable that she will get thither at least a week before Mr. [Ph.] Von Reck and his Troop arrive. The Society are well pleased with the Instructions you have given to Mr. Von Reck and approve of the allowance you have agreed to give him for his Service. I am now making Copies of your last Letters to be laid before the Trustees for Georgia at their next meeting and remain

Reverend Sir   Your most humble Servant   Henry Newman
P.S. I beg leave to refer you to the Enclosed Mr. Vat for other particulars relating to affairs here and shall now and then take leave to Subjoin a Postscript to you by way of Letter to him[.] Mr. Oglethorpe intended to return this fall to England but there is advice come that he in September last had a fall from his Horse in Georgia which gave him such a Contusion in his side as will go nigh to prevent his Return and if that be all the ill Consequence that will attend his fall some of his friends here will not be Sorry that it will detain him to see and take care of the Saltzburgers who it is wished may arrive in Georgia before he comes away[.] I mean the first Transport.

Bartlet's Buildings London 16 November 1733.   To the Reverend Mr. Richard Lowther at Roterdam.

Reverend Sir: I received your Letter of the 18 Current N.S. and acquainted the Society with the Contents of it, at their last meeting who thank you for your readiness to bestow your good offices in favour of the Saltzburgers and very much approve of your prudent caution in acquainting the Burgomasters of your expectation. I am glad to understand by a letter I received this Evening from Mr. Ortmann that the Purisbourg arrived at Rotterdam last Sunday safe and that She will be ready to receive the transport of Saltzburgers as soon as they arrive, and consequently give you a very little trouble, if any.

Mr. Vat who was designed for their Conductor did not arrive at Augsburg from Switzerland 'till a week after they were set out and therefore Mr. Philippe De Reck nephew of our King's minister at Ratisbone is so kind as to take upon him the charge of acting as Commissary to conduct them, from whom I have also a Letter dated at Franckfort on the Maine of the 15th Current N.S. where they had then rested two days all in good health, and that he had hired a Vessel to carry them all to Rotterdam so that in all likelihood they may be with You e'er this comes to Your hands and because I don't know how to direct to him better I have taken the liberty to cover a Letter for him to you imagining that he will apply himself to You as he is instructed immediately on his arrival.

A Minister [Bolzius] and a Catechist [Gronau] recommended by Professor Franck at Hall that could not join them on the Road will meet them at Rotterdam who are likewise recommended to your good offices as well as the minister [Schumacher] that comes with them from Augsburg who will return to Augsburg after seeing them embarked.

The news Papers will inform You of the Prince of Orange's illness which has occasioned a Suspension of the Nuptials, but I am glad to tell You that upon an Enquiry I have this day made at Somerset house, His Highness is much better and looked upon to be out of danger of any ill consequences from the fatigue of his voyage which occasioned his indisposition so that tis hoped he may be able in a few days to go abroad, and some say that orders are given to fit up an apartment for him at Kensington Palace and that the Wedding will be defered 'til the beginning of December.

I shall be glad to hear from You of the safe arrival of the Saltzburgers, and of any Circumstances which may serve to guide us in the future transport: and may perhaps e'er long follow these under the Conduct of Mr. Vat, out of the Eight hundred Emigrants expected from Saltzburg as Mr. [Ph.] De Reck informs me.

Sir James Lowther is just now come well to town having left Sir Thomas and Lady in good health at Holker who are not expected here 'till after Christmas.

All Your friends here particularly Sir John Philipps and his Sons are well and return their humble Service to You by

Reverend Sir   Your most humble Servant   Henry Newman

P.S. Whatever charges you are at in serving the Society on this

Occasion please to keep an account of it that I may take care to discharge it.

Bartlet's Buildings London 13th November 1733. To Mr. J. Vat at Mr. Urlspergers at Augsburg.

Dear Sir: It was with Pleasure I received by last Post the advice of your Arrival at Augsburg, both by Mr. Urlsperger's and Your Letters of 12th Current N.S. which were this day laid before the Society and since there is a Prospect of another Transport of the Saltzburg Emigrants for Georgia either at Augsburg or Franckfort wanting a Conductor to Embark at Rotterdam, they think it adviseable that you remain where you are till they or Mr. Urlsperger can point out the Service you are so kind as to undertake when occasion shall require.

We see but little into the ways of Providence and therefore I will not call it a disappointment that you did not sooner arrive at Augsburg, till I see how things may turn hereafter[.] your experience under the Prudent Conduct of Mr. Urlsperger will enable you dextrously to avoid the Artifices and Snares of those of the Romish Persuasion who are incessantly engaged with the Common Enemy of mankind to devour as many innocent Protestants as they can.

No doubt there is a Treasure of Wrath reserved for those who delight in persecuting the faithful followers of Jesus Christ and whether the time be now at Hand that Antichrist is to receive her doom from those of her own Party the Progress of the Wars in Italy and Poland may in a few months shew.[21]

Governor Shute and Colonel Valogne are now at Bath in good Health as the rest of your Friends here are but frequently Inquisitive of your Welfare.

I must not omit to acquaint you of the Prince of Oranges happy arrival here last Wednesday to be married to our Princess Royal, the Celebration of which was to have been as Yesterday at St. James's in a most Magnificent manner where accommodations are prepared for about 4000 Spectators in a Gallery that Issues from the Royal Apartments over the Garden into St. James's Park and thence by two long sides returns to the French and Dutch Chappel by Marlborough House but the Prince being taken ill on the day before which was Sunday while he attended Divine Service in the Dutch Church at Austin Fryers with Symptoms of a Pleuretick Fever the King immediately gave orders for suspending the Nuptials till His Highness's recovery, which 'tis said there is great likelihood of in a few days. His Highnesses weak Constitution not being accustomed to such fatigue as his late Voyage &c had necessarily subjected him to 'tis said occasioned his indisposition which 'tis hoped by the Blessing of God on the great Care that is taken of him at his Lodgings in Somerset House may soon be removed.

Pray let me hear from you now and then either in a Postscript to Mr. Urlsperger's letters or a Separate Letter as you find occasion to write to

Sir   Your most humble Servant   Henry Newman

Bartlet's Buildings London 16 November 1733.   To Mr. [Ph] De Reck at Rotterdam.

Sir: I received with great Pleasure your letter of the 15th Current N.S. from Franckfort and immediately communicated it to the Trustees for Georgia who were then sitting to whom it was very acceptable and shall take care to lay it before the Society at their next meeting the 20th Current and acquaint you with their orders thereupon.

In the meantime the Reverend Mr. Lowther Minister of the English Church at Rotterdam and a Member of the Society is desired to assist You with his advice and good Offices in every thing for the easy Embarkation of the Saltzburgers under your care on board the Purisbourg where I hope you and they will find all necessary accomodations for bringing them to England. Captain Fry the Master of the Purisburg has instructions from the Trustees for Georgia what Port to Stop at in England before he proceeds on his Voyage to Georgia from which Port I shall be glad to receive a Letter from you as soon as you arrive with an Account of any further necessaries that may possibly be wanting that Care may be taken to Supply them.

I received a Letter from Mr. Urlsperger of the 12th Current N.S. by which Mr. Vat arrived the 10 at Augsburg too late to follow you with any likelihood of overtaking you and I am very glad to find you have Personated an English Commissary so well as not to want his assistance. He lodges at present with Mr. Urlsperger in hopes of another Transport from the Prospect of 800 more Emigrants out of whom perhaps a number may be inclined to go to Georgia on the terms offered to those under your care.

I shall be glad to see you safe in England which I hope may be in a Short time assuring you that I am

Sir   Your most humble Servant   Henry Newman

P.S.: I take the Liberty of writing to you in English hoping you may understand it as well as I do french and if you dont Mr. Christopher Ortmann who is gone over in the Purisburg will be your Interpreter.

Please to let him know that I received his Letter of the 11th Current O.S. from Rotterdam and that his friends here are glad to hear of his safe arrival and hope to hear further from him as occasion offers.

Bartlet's Buildings London 20th November 1733. To the Reverend Mr. Urlsperger at Augsburg.

Reverend and Dear Sir: Since my last of the 13th Current I have received your favours of the 16th & 19th Current both which have been laid before the Society and the last of them being translated by hand by Mr. Vat was read, but the Subject of it being of great moment and too delicate to be answered hastily they have desired Mr. Ziegenhagen to answer it according to the sense of the Society at the reading of it. Your letter of the 16th Current being long the Translator could not finish the Translation of it, till the Society was risen, upon which I was ordered to give a Copy of it to be laid before them at their Meeting to morrow which I have taken care of.

We hourly expect advice of the arrival of Mr. de Reck with the Saltzburgers at Dover were [where] they are to Stop to receive their last Dispatches from the Trustees for Georgia and to be filled up with Passengers what may be wanting in the Saltzburgers to compleat the Number contracted for otherwise the Trustees would be at the same Expence for sending half a Ship as a whole Ship Load. For the same reason of frugality they will rather send the Passengers to the Ship than let the Ship come into the River where she will be liable to so many hindrances as cannot be foreseen or avoided, and yet the Trustees are engaged to pay demurage for all those Letts [hinderances] while the Ship is in their Service. The Society have ordered the books you desired to be sent by the first opportunity which I shall do as you advise[.] you forgot to send me the Price of the Screw Medals.

I hope I may be more particular in my next and in the mean time desire you may be assured that I am

Reverend Sir Your most humble Servant Henry Newman
P.S. In the meantime your direction to me here saves at least 6 Hours of the time your Letters are coming to my hands. Give me leave to subjoin what follows.

Bartlet's Buildings 20th November 1733.   To Mr. Vat at Mr. Urlsperger's at Augsburg.

Dear Sir: I was glad to see your hand writing to a long letter from Mr. Urlsperger and that you are so happily employed as to assist him till Providence shall direct you to some other service for our afflicted Brethren[.] I hope the pleasure you enjoy under Mr. Urlsperger's Roof will compensate for the trouble you have had of making an Excursion from Swisserland.

Colonel Pury is returned from South Carolina but I have not had the happiness yet to see him tho I hope it will not be long first.

I am Dear Sir   Your most humble Servant   Henry Newman

P.S. Dr. Hales I hear is well just now returned from his living in Hampshire[.] there is a Second Volume of his vegetable Staticks published, which will be sent with the first but I doubt I shall with difficulty get an Opportunity of Conveying them unless you instruct me.

Bartlet's Buildings London 24th November 1733.   To Mr. Philip George Fred. de Reck at Dover by Mr. Butienter.

Sir: I received your Letter of the 27th of November N.S. from Rotterdam; which I have this day laid before the Society, and they are glad to hear the Saltzburgers designed for Georgia are so far arrived in their way thither and if you are inclined to go with them to Georgia and to settle there, the Trustees for that Colony will give directions to Mr. Oglethorpe to grant you such encouragement in Land as you shall be fully satisfyed with.

In the meantime I am directed by the Society to acquaint you that they have agreed to defray your Expences to Rotterdam on the foot of [according to] Mr. Urlsperger's Stipulation as signified in his Letters to the Society and that upon the Embarkation of the Saltzburgers on the Purysburg they are under the direction of the Trustees for Georgia who will take care that every promise made to them by their direction as to free Passage Over Sea, & their Settlement in Georgia shall be punctually complied with on their arrival there. So that you will please to consider if you dont incline to settle in Georgia on the foot of the encouragement abovementioned whether it may be worth your while to proceed thither.

I am further directed by the Society to desire your acceptance of five Pounds to be paid to you by the Reverend Mr. Butienter as a gratuity from the Society over and above the allowance made to you for your Service to Rotterdam. And if you are not inclined to go to Georgia the Society will be glad to see you here in your passage back to Germany.

I am Sir   Your most humble Servant   Henry Newman
P.S. Sir: I should be glad [if] I had been able to confer with you and the Saltzburgers in High Dutch and that [if] I had been other wise qualified to wait on you at Dover[.] the Society have therefore desired the Reverend Mr. Butienter one of their Members and the worthy colleague of Mr. Ziegenhagen to wait on you at Dover with their complements to the Reverend Mr. Bolzius and Mr. Gronau the Minister and Catechist sent by Professor Franck at Hall to attend the Saltzburgers to Georgia, and at the same time to advance to them ½ a years Salary from the time they set out from Hall Viz. 25 £ to Mr. Bolzius and 15 £ to Mr. Gronau but Mr. Ortmann's Salary to Commence from the time of his Embarkation here for ½ a Year of which he will also receive 5 £, their Salaries to continue till Such time as they can be provided in Georgia. Capt. Coram one of the Trustees for Georgia will on their parts visit the Purysburg to see that all things necessary be supplied for the Comfort of the Passengers on bord & to fill up the ship with as many English Passengers as may be conveniently accommodated. If You come to London I shall be very glad to see You to have the Account of your Pilgrimage with the Emigrants referred to in your Letter of the 15 November from Franckfort. I hope You received mine of the 16th November O.S. at Rotterdam.

Bartlet's Buildings 29 November 1733.   To the Reverend Mr. Butienter at Dover or in his absence to Capt. Coram.

Reverend Sir: The Purysburg sailed from Rotterdam the 23rd Current but there is no News of her yet however the wind being contrary we are in no Pain for her Safety.

I have a Letter from Mr. Lowther Minister at Rotterdam of the 23 Current signifying that Capt. Fry refused to admit the two Clergymen attending the Saltzburgers into the great Cabin that being his own priviledge upon which I immediately complained of it to Mr. Symonds Capt Fry's owner who has wrote Yesterday and again by this Post in very strong Terms that they be immediately received and provided with the best accommodations in the Great Cabin and that whatever he Claims on Account of his priviledge must be charged to his Service and that no other Persons be admitted into the Cabin so as to hinder or incommode the two Clergymen and therefore I hope there will be no further occasion of Complaint on this Score.

A Chalice and Patten in Plate[22] are making and will be sent down as soon as made therefore pray let me know how they may be Directed at Dover in Case you should be come away from thence.

My humble Service to Capt. Coram who I hope is got well with you to Dover of which a Line from him or you will be very welcome.

My humble Service also to Mr. De Reck and Messrs. Boltzius[23] and Gronau of whose great merit I have received a long letter from Professor Franck which very much pleased the Society and therefore they will take it very well that You and Capt. Coram take care that nothing be wanting to their being accommodated in the best manner on bord during their voyage to Georgia and it is from these Gentlemen that the good usage [treatment] of the Saltzburgers will be made known to their friends in Germany so that 'tis wondered here that the Capt. should not for his own Interest give them the best accomodations without any direction from hence but Mr. Simonds has wrote so fully upon this head to the Capt. that I hope no more need be said about it.

I have 2 Letters from Mr. Urlsperger since You went by which there is no prospect of More Saltzburgers till next Spring but great Complaints of the Usage of Protestants in Hungary and Bohemia for whom tis hoped good Providence will raise up some relief from the Protestant powers in Europe.

The Prince of Orange continues on the mending hand so that Yesterday he received the Compliments of several of the Nobility etc. and my Lord Mayor in person waited on his Highness to congratulate his Recovery so much as he enjoys it.

Pray let me hear from You as occasion offers and believe me to be

Reverend Sir   Your most Obedient humble Servant
Henry Newman

Bartlet's Buildings 4 December 1733.   To the Reverend Mr. Lowther at Rotterdam.

Reverend Sir: I received the favour of the 4th Current N.S. which was this Day laid before the Society who thank you for all your good Offices to the Saltzburgers and the Gentlemen who attend them as Conductors.

I was in hopes e'er now to have heard of their Arrival at Dover but for above a week past the wind has been against them and this day the Society are informed that the Purisburg run aground in Your River at coming out where I hope your next may inform me that she received no damage but what may be easily repaired.

I am to thank You for the advice you give about the Captain's not admitting the Clergymen into the great Cabbin upon which 2 letters have been wrote to him in very strong Terms by the Captain's Owner to admit them into the great Cabbin preferably to any body else and to give them there the best accommodations which I doubt not will be complied with as soon as he receives those letters at Dover.

The Prince of Orange is much better in his health and is going to Kensington for the perfect recovery of it which every well wisher to Old England and the Protestant Interest must rejoice at.

I shall make Your Complements to Sir James Lowther as soon as I see him, and the Society desire Your acceptance of theirs at the same time I shall Subscribe my self

Reverend Sir   Your most obedient humble Servant
H. Newman

Bartlet's Buildings London 4 December 1733.   To the Reverend Mr. Urlsperger at Augsburg in Germany.

Reverend and Dear Sir: I have received Your Letters of the 23 and 26 November N.S. with the Extracts which accompanied them all which were this Day laid before the Society and they Expressed themselves not a little pleased at Your Zeal for procuring Relief to the Distressed Protestants in Saltzburg and other parts.

They were glad to understand by Mr. [J.] De Reck's Letter to you that the influence of the Protestant Powers in Europe is at present so great and hope God will incline their Hearts to make a right use of that Influence at this Juncture.

The Emigrants under the Conduct of Mr. [Ph.] De Reck embarked at Rotterdam the 4 Current N.S. with Messrs. Bolzius and Gronau but the Ship being rather too large for the Service run aground coming out of the River so that they were obliged to lighten her to get her off of the Bank but I hope without any other damage than hindering her a few Days of her Voyage to Georgia.

In the meantime one of the Trustees for Georgia and the Reverend Mr. Butienter a Member of our Society are gone to Dover to Expect [await] the Arrival of the Ship and to give her all the dispatch they can. If Mr. De Reck inclines to go with them the Trustees for Georgia have ordered the Capt. to afford him the best Accommodations the Ship will allow of thither and back to England.

I have received a Letter from Professor Franck giving an Extraordinary Character of the Minister and Catechist which he has recommended in regard to which the Society have taken Care that nothing may be wanting to make their Passage comfortable.

Pray give my humble Service to Mr. Vat, his friends here often enquire after his health, and will be glad of any good occasion of seeing him. I wish you all Manner of Prosperity and am
Reverend Sir   Your most Obedient humble Servant
Henry Newman

Bartlet's Buildings 4 December 1733.   To the Reverend Mr. Butienter at Dover.

Reverend & Dear Sir: I am directed by the Society to return You their Thanks for Your Letter and Charitable Designs for the Distressed Emigrants of Saltzburg. They beg of You to give them all the Comfort and Assistance in Your Power, and assure them that our Prayers and best Endeavours will be used that they shall meet with the most Christian Usage in their Passage to and after their arrival at Georgia. The Society are Extreamly concerned to hear that the Reverend and worthy Ministers Bolzius and Mr. Gronau were not treated at Rotterdam by Capt. Fry as they deserved. The Trustees of Georgia have sent repeated Orders by the owner Mr. Symonds to Capt. Fry that those Gentlemen shall be immediately received into the Cabbin and to the Captain's Table and that nothing may be wanting on our parts. You are desired by the Society to enquire into and supply those worthy Gentlemen with all necessaries, as bedding Clothes Refreshments etc. while at Dover and for their Passage and to assure them that we Esteem them greatly for their works Sake and that the Character Mr. [G. A.] Franck gives us of them is the highest Satisfaction and Pleasure we could receive. We are Sorry that the Poor Emigrants have met with Difficulties at their first Embarkation[.] God does it for a Tryal of their Constancy and Perseverance in the Truth[.] We are greatly pleased to hear from all hands that Monsieur [Ph.] Von Reck has approved himself worthy the Charge reposed in him, We Desire to be further assisting to him and if he wants any Necessaries for his Passage or at Dover You are desired to supply him, the Trustees for Georgia are very pressing that he should proceed with the Emigrants to Georgia and his Expences to and from thence will be defrayed; We hope to have the happiness of Seeing him here upon his return when we shall make our gratefull acknowledgements for his Care and Dexterity in managing this affair[.] And now we entreat You to Comfort the poor Emigrants and assure them that our best Endeavours have been used for their kind and Christian Treatment both here and abroad and that we may not be wanting upon our parts we desire You will present each grown person above 20 with 20 s Men as well as Women, To each Person under and above 12 Years of age 10 s and 5 s to all under 12 Years to be laid out as their Parents shall direct. The Society beggs that you will allow Your self every thing that is necessary and Convenient upon the Road and at Dover at our Charge[.] For the Payment of all these Expences, We desire You will take up Money from Mr. Minet, and his Bill drawn upon Wm. Tillard Esq. in Spittal Yard shall be paid at Sight. and now Dear Sir We recommend You and the poor Emigrants of Saltzburg to the Care and Protection of Almighty God, wishing their Prosperity and assuring them that no Care of ours shall be wanting to promote it. I am in the Society's Name

Your most affectionate Friend and Brother   Tho. Wilson

Bartlet's Buildings 4 December 1733. To the Reverend Mr. Butienter at Dover.

Reverend Sir: Mr. Wilson has wrote so fully the Sense of the Society upon reading Your Letter of the 1st Current to me that I have only to add that beside the Contrary winds having detained the Purysburg she had the Misfortune to run aground in coming out of the River from Rotterdam which obliged the Capt. to take out some of her Lading to Lighten her but it was hoped no other Damage would Ensue except detaining them a few days longer.

The Chalice and Patten are just now sent in and to save time that they may go tomorrow I have procured from a Trunkmaker the smallest Trunk I could get to Pack them in and shall send them by the first Opportunity if not tomorrow then on Thursday directed to Mr. Minet for You[.] my humble Service to Capt. Coram and tell him I hope he takes abundance of Care of You that You want nothing for Your Comfort that the Road affords but if he should be Sparing as a hardy Traveller in his way of living the Society desire you would allow your Self every thing that is Convenient for your own Person as well as for Making the presents to the Minister Catechist and the Saltzburgers as mentioned in Mr. Wilson's Letter, for which Service You are desired to draw Bills on William Tillard, Esq. in Spittal Square One of our Treasurers for any money that you shall take up from Mr. Minet or any other Gentleman at Dover Mr. Tillard being desired to give due honour to any Bills You shall draw upon him.

Pray let me hear from You as often as occasion requires. Mr. Ziegenhagen was here this day and very well, with the rest of Your Friends that usually attend here[.] I have had no Letters from Augsburg since my last to You of the 29th November but Expect some every Day.

I am   Dear Sir   Your most obedient Humble Servant
Henry Newman

Bartlet's Buildings 11 December 1733.   To Captain Coram at Dover, Kent.

Sir: Your Letter of Yesterday Concerning the Appearance of the Purisburg off of Dover was very welcome to the Society who were just breaking up when it came to hand. Mr. Vernon and Dr. Bundy had both the Satisfaction of reading it and I hope your next will confirm the Account of her Arrival at Deal or Dover with some Particulars which may Enable the Society and Trustees to judge of such further measures as may be necessary for them to take before the Saltzburgers leave England.

Pray give my humble service to Mr. Butjenter with my Thanks for his Letter of last Sunday and let him know that having communicated to the Society they agreed in the Construction he puts on Mr. Wilson's and my Letters of furnishing the Minister and Catechist with all necessaries.

I suppose You have Instructions from the Trustees to make a List of the Saltzburgers their Sex, Age and Professions respectively and if You please to favour Mr. Butjenter with a Copy of it he will send, or bring it at his return to Town to be laid before the Trustees.

You have also instructions about Mr. [Ph.] De Reck who by his last Letters seems much inclined to go on to Georgia and as I understand by Mr. Vernon, the Trustees are very well pleased he should go on the terms he himself has mentioned.

I hope God will Direct Mr. Butjenter and You to concert matters so as may result in his Glory and the Comfort of these pious

Confessors who tho' they are but few in number their good or ill report of the Treatment they meet with here will animate or discourage a considerable number to follow them next Spring.

I am   Sir Your most humble Servant   Henry Newman

Bartlet's Buildings London 13 December 1733.   To Mr. [Ph.] De Reck at Dover.

Sir: I received Your Favour of the 22nd Current N.S. giving an Account of the Trouble that attended your Passage from Holland which has been communicated to several of Your Friends here and given them not a little concern that the beginning of Your voyage should meet with so many rubbs and quietudes to You and your Fellow Travellers. I hope such Effectual Care is taken to remove them especially in those things which are in the Captains Power that You will have no Occasion to Complain hereafter and am glad to find by the Latter part of Your Letter that Capt. Coram has redressed all Your Grievances nor will it I dare Say be his fault if they are not Entirely prevented for the future.

I am satisfied Mr. Butienter will do every thing on his part to furnish Messrs. Bolzius and Gronau with necessaries to make the rest of their Voyage Comfortable.

Your Zeal for pursuing Your Voyage with the News you mention is very laudable and I hope God has raised You up not only to be as a Father to the Saltzburgers on bord the Purisburg but to many others that may follow them from Germany or Piedmont to Georgia.

What You mention about the 300 Vaudois shall be Communicated to the Trustees for Georgia and the Society for Promoting Christian Knowledge at their first meetings and their resolves signified to Mr. Wolters.

I Pray God to send you a good Voyage and safe arrival in Georgia of which I shall be very glad to be informed from your own Hand by the first Opportunity with an Account of the more remarkable Occurences at Sea and of the Welfare of all Your ships Crew who will be happy in having the Company of a Gentleman so curious [scrupulous] as Mr. De Reck.

It will give me a singular pleasure to transmit to Your uncle [J. von Reck] at Ratisbon any Account of Your welfare while You are abroad & Mr. Oglethorpe the Director of the Georgian affairs will be very glad to see a Gentleman of Your Genius so like his own for grand Enterprizes.

May God almighty give you long Life & Prosperity is the wish of
Sir   Your most obedient humble Servant   H. Newman

Bartlet's Buildings 13 December 1733.   To the Reverend Mr. Butienter at Dover.

Reverend Sir: I received Yours of the 11th Current and congratulate You on Mr. [Ph.] De Reck's safe arrival with the Saltzburgers at Dover.

I hope all the hardships they have suffered in their Voyage will be intirely redressed by Your and Capt. Coram's Care.

I have just now been with Mr. Symmonds the owner of the Purisburg who shewed me the articles between him and the Trustees for Georgia by which the Daily Allowance to the Passengers was 4 Days in the week Beef and Pudding 2 Days Pork and Peas I Day Fish and Butter with 14 Ounces of Bread and 3 or 4 Qts. of Water or Beer to each Person daily which is a larger Allowance than is given in our Men of War where the Sailors never Complain. I told him that as they are great Lovers of Bread more than English People it would be reasonable to indulge them to have as much as they can eat and more than the Value would be saved in other Articles that he took a Memorandum to order the Captain to do it.

As to Rum Brandy Sugar Spices and Oatmeal Candles etc. he says they are plentifully provided by the order of the Trustees over and above the Ships allowance that the Capt. might not if he were inclined hinder them of reasonable refreshment.

Sir John Philipps and both his Sons being consulted about the plate are of Opinion it should be sent forward under the Care of Messrs. Bolzius and Gronau notwithstanding the Supply they have brought from Germany. they may hereafter be in two Congregations of Saltzburgers who may want the present Supply.

My humble Service to Messrs. Bolzius and Gronau wishing them & their little Flock all manner of Prosperity and to hear by all Opportunities from them.

I remain   Reverend Sir   Your most Obedient humble Servant
Henry Newman

P.S.: My humble Services to Captain Coram.

Bartlet's Buildings London December 15, 1733.   To James Oglethorpe Esq. in Georgia.

Honoured Sir: I have deferred troubling you with a Letter till I could send you some Account of the Saltzburgers designed for

Georgia and partly from an Expectation of the happiness of seeing you in England though they that wish well to the Colony of Georgia would not be sorry if you could spare a little more of your time and prudent advice to that infant Establishment before you left it.

The Trustees for Georgia have undoubtly fully acquainted you from time to time of the Steps taken by them in conjunction with the Society, my Masters for procuring 300 Saltzburgers for Georgia, but their Labours hitherto have not prevailed with more than 40 of those People to venture to meet the Terrors of the Ocean. If this small Transport prove successful under your Patronage and the Blessing of divine Providence, there is reason to expect that more of their Countrymen & other persecuted Protestants will follow them as Ships can be prepared to carry them to America.

The Society were not a little concerned how to obtain a proper Director to carry them to Holland, but Heaven seems to have pointed out a Gentleman born for that Service [Ph. von Reck] so well qualified that when you have read the enclosed Letter from him, wrote partly on the Voyage to Holland, and finished on the Day of his Arrival at Dover, I believe I need say no more to make you in love with a Man that devotes himself with so much Zeal to the Service of the English Nation, with no View that appears, but the Glory of following your and other great Examples for the general Welfare of Mankind. He is the Nephew of Monsieur [J.] de Reck, the King's Hanover Minister at Ratisbourne, under whose care he has had a liberal Education,[24] which joyned to his great Genius, if it please God to spare his life, may in time form him for some signal Service.

Every Body that wishes well to old England are pleased to read in our News papers now & then of the Prosperity of Georgia, under Mr. Oglethorpe's administration. And Governor Belcher in his last Letter to me laments that he is not like to see you in New England, though the People of that Province are all prepared to contend who shall shew you most respect, if you vouchsafe to honour with your Presence in your return home.

May Almighty God prosper and defend you in all your Enterprizes, for his Glory and the good of Mankind is the sincere Wish of
                    Sir   your most obedient Servant   Henry Newman

Bartlet's Buildings December 18, 1733. To Mr. Von Reck at Dover.

Dear Mr. Commissary: I received your kind Letter of 27th

Current N.S. with your Letter of the same Date to Monsieur Symonds which I immediately forwarded to him wherein you seem to have dexterously balanced Accounts for all past differences relating to the Conduct of the Captain of the Purisbourg [Fry] and I hope he will let you know his Entire satisfaction for the paternal part you have acted in your charge hitherto which is abundantly confirmed by the Reverend Mr. Butienter who is safely returned and desires his compliments may be acceptable to you and Your Fellow Travellers particularly to Messieurs Bolzius and Gronau by whom also I beg mine may be accepted with thanks for their Letter of 18 Current which I hope to answer with some Books I am endeavouring to procure for their service on the Voyage to enable them to acquire the English Language. If the Ship should be detained by contrary winds till Saturday next they will go nigh to reach their hands.

I shewed all Your Letters to me to Mr. Vernon and others of Your friends here who are all very much pleased with Your Zeal and Discretion in discharge of Your Trust and hope the many good Qualities You are possessed of Presage Your being an Instrument in the hands of Providence of doing much good wherever Your Lot may be cast.

You cannot more regret Your not coming to London than your Friends here do their not having the Pleasure to see a Man that so easily naturalizes himself to all Countries. For fear Mr. Vernon should not be able to write to You by this Post he commissions me to send his humble Service to You at the same time that I assure You of being

Dear Sir   Your most Obedient Humble Servant
Henry Newman

P.S. My humble Service to Capt. Coram[.] I herewith return Mr. Symmonds's Letter to You. Once more I wish You a Prosperous Voyage.

Bartlet's Buildings London 18 December 1733.   To the Reverend Mr. Urlsperger at Augsburg.

Reverend & Dear Sir: My last was of the 4th Current since which I have received Yours of the 7th Current N.S. and am glad to acquaint You that the Court here have reported their Orders to the British Minister at Vienna to renew his Sollicitations to his Imperial Majesty in behalf of such Emigrants as may come Your way to be admitted into the City of Augsburg.

The Purysburg after the Embarkation of the Saltzburgers at Roterdam under the Care of the Indefatigable Mr. [Ph.] Von Reck was detained about 14 Days by Contrary winds and thro' the Carelessness of the Dutch Pilot ran on a Bank of Sand in the Maese [Meuse, Maas] where she lay 2 days without any other Damage than being hindered proceeding on her Voyage and frightening the new mariners.

But soon after their putting out to Sea they happily arrived at Dover the 11th Current where all their Sorrows were turned into joy and nothing was wanting to give them a proper welcome to the British Shore, they were all invited the next Day to Dine in a large house lent by the Magistrates of the Town for that purpose where they were regaled with roasted Sirloins of Beef and Plumb pudding having marched thro' the Town in great Order to the place of Entertainment singing of Hymns with Mr. Commissary [J. von Reck] at their head and 2 Clergymen closing the Precession[,] abundance of the Neighbouring Gentry resorting to partake of the Pleasure of this new Sight.

I cannot Express to You the Praises Mr. Von Recks conduct has brought upon him for his Paternal Care of these Emigrants both in Holland and on their Voyage hither, so that the Trustees for Georgia and the Society are highly pleased with his prudent behaviour which Surpasses their Expectation from a Gentleman of his Years and they hope pressages if it pleases God to spare his Life, his being a usefull Instrument in the hands of Providence of much good to these and all other Colonists that may hereafter resort to Georgia.

His Friends here would have been glad to have seen him at London if that would have Consisted with his proceeding in this Ship which now waits only for a fair Wind. In the mean time the Saltzburgers all lie a shore refreshing themselves till they can depart and Mr. Butjenter who went to see them on the part of the Society has by their Order made proper Presents to Mr. Commissary & Messrs. Bolzius and Gronau and to every Man & Woman above 20 Years of age a present of 20 s to those under 20 & above 12 a present of 10 s & to all under 12 Years 5 s to lay out as they please I am

Reverend & Dear Sir   Your most humble Servant
Henry Newman

My humble Service to Mr. Vat.

*Bartlet's Buildings 22 December 1733. To the Reverend Messrs.
Bolzius & Gronau.*

Reverend Sirs: I received Your Obliging Letter of the 13th
Current from Dover, the Contents of which were very acceptable
to the Society as they gave them assurances of Your Constancy
and Steadiness to pursue the Voyage You had undertaken to
Georgia notwithstanding the discouragements that happened at
Your first Embarkation.

This was a confirmation to them of the good Character Mr.
Professor Franck had before given of Your Zeal and abilities for
the Service You had so laudibly devoted Your Selves to, wherein
they doubt not but the God who has inclined You thus to con-
secrate Your Labours to his Glory will continue to guide and
Support You and make You Blessings not only to the Flock You
have accepted the Charge of, but to the Country where Your Lot
maybe Cast.

Mr. Ziegenhagen having Signified to the Society Your desire of
Acquiring the English Language in Your Voyage, I was ordered
to procure and send to You as a present from them Mr. Lodowig's
Dictionary and Grammar printed some Years since at Leipswick
[Leipzig] I am glad to acquaint You that tho' they are very scarce
I have met with them and sent them Yesterday morning by the
Dover Coach directed to You at Mr. Minet's[,] Carriage paid 2/6 d.
My humble Service to Mr. [Ph.] Von Reck and assure him of my
best Wishes attending You and him And all Your Fellow Travel-
lers that it may please God to Grant You and them health and a
prosperous Voyage to Georgia and the Multiplication of his
Blessings after Your arrival, of which I shall be glad to be in-
formed whenever You have opportunity to give that pleasure to
Reverend Sirs   Your most Obedient humble Servant
Henry Newman

My humble Service to Capt. Coram.

*London 21 December 1733. To the Reverend Mr. Prof. [G. A.]
Franck. at the Orphan House at Hall in Saxony.*

Reverend Sir: I received in due time Your most acceptable
Letter of the 19 November N.S. which being laid before the Society was read and heard with that attention which is always given
upon reading a Letter from Professor Franck.

In answer to the first part of it, the Society think themselves
very much obliged to You and have ordered me to send their
hearty Thanks for the Care You have taken to recommend to the

Trustees for Georgia and to the Society, two Gentlemen so well qualified for Minister and Catechist to the Saltzburgers in Georgia as Messrs. Bolzius and Gronau are represented to be.

The Society have agreed to allow the Minister 50 £ Sterling per Annum and the Catechist 30 £ to commence from the time of their departure from Hall and prevailed with the Reverend Mr. Butjenter one of their Members to go down to the Port of Dover where the Transport ship now lies to see that they are provided with all necessaries particularly to advance to them half a Year of their Salaries and to present them with such Plate and other Furniture for the Holy Communion as could be got ready at so short a Warning and also 20 s to each Person above 20 Years of age, 10 s to every Saltzburger above 12 and under 20 Years old, and 5 s to all under 12 Years of Age. One of the Trustees for Georgia [Capt. Coram] was likewise sent to Dover on the part of their Corporation to welcome them to England to see them provided with all necessary accommodations on board the ship Purysburg.

In answer to the latter part of Your Letter about contributing towards building an Hospital in Prussia for old decrepid Saltzburgers the Society entertain so high an Esteem for every Proposal that comes from Professor Franck that they heartily wish it were in their power to comply with this, but in the present Instance, it is with great regret that they direct me to inform You that the Principal Benefactors towards relieving the Saltzburgers have confined their Bounty to such of them as shall accept of a Settlement in His Majesty's Dominions in America, agreeably to which the Society have already engaged to settle more Saltzburgers in Georgia than the Sums hitherto collected will enable them to do, notwithstanding the liberal Assistance given by the Trustees for that New Colony out of the Parliamentary Funds towards defraying the charge of their Transportation.

You will easily be convinced of this when I acquaint You that the Society have out of the Casual Benefactions for this Service at several times since March 1732 remitted to the Reverend Mr. Urlsperger at Augsburg 1482 £ 4 s 6 d Sterling part of which has been applied towards relieving those Emigrants who went to Prussia. beside this, the Society have thought it incumbent on them to secure an effectual Provision for supporting a Minister, a Catechist and Schoolmaster to those who have left their Native Country and Friends for the sake of Religion by appropriating 2000 £ for that Service and they have accordingly lodged that Sum in the publick Funds. So that considering the Charge of the present Transport of 40 what remains will not suffice to defray one Third or perhaps One Quarter of the Expence of the number to compleat 300 which they have commissioned Mr. Urlsperger to collect and send to Roterdam, where the Trustees for Georgia receive them at their Expence.

But as in all Cases of this nature it cannot be foreseen what may be the Extent of the charge accrewing to it the Society rely on the same good Disposition of their Contributors who have enabled them to go so far as they have, to raise such further supplies as shall be necessary to accomplish the Deign of making a good Settlement for these Emigrants in Georgia.

The Society congratulate You on the gracious reception his Majesty of Prussia has given you in your late Journey to Berlin. God be praised for all the Effects of his Royal Bounty to Your University and Orphan house, and may they continue to make his Virtues illustrious to all Posterity.

I am also to thank You in the name of the Society for all the Charitable and liberal Remittances You have annually made in favour of the Protestant Mission to the East Indies, particularly to the Establishment at Madras which owes its Origin to that at Tranquebar and both of them their beginning to Your most worthy Father [A. H. Francke] whose Memory is dear to all good Christians.

I doubt not but Mr. Ziegenhagen has transmitted to You the many agreeable accounts which came to my hands from the Missionaries of both Establishments last Summer which seem to promise a glorious Harvest in the age to come, and by the order of the Society, I am now preparing to send to them every thing they have desired, on a presumption of having leave so to do by the East India Company here.

The Reverend Mr. Rodde at Narva has earnestly recommended to the Society the promoting an Impression at Hall of *Arndtius de vero Christianismo* in the Russian Language by Russian Types in your Printing House, presented many Years ago by good Mr. Ludolph Secretary to Prince George of Denmark.[25] The Society would be glad to have Your Opinion of that Proposal and whereabouts may be the charge of an Edition of 1000 Copies and the difference of that and 2000 Copies and whether You will not think a Book of less Bulk and Expence more proper.

   I am Reverend and Dear Sir Your most Obliged humble
           Servant Henry Newman

P.S. My humble Service to Your Fellow Labourer Mr. Freyling-hausen upon whose recovery from a fit of Apoplexy I heartily congratulate him and you.

Bartlet's Buildings 22 December 1733.   To Capt. Coram at Mr. Minet's at Dover.

Sir: If the Contrary Winds have still detained the Purysburg I hope the packet of Books sent by last Thursdays Coach will get to Messrs. Bolzius and Gronau before she sails.

I have just now a Letter from Mr. Urlsperger at Augsburg of the 17 Current N.S. by which Mr. Schumacker who attended the Saltzburgers as their Chaplain to Roterdam was just then returned and had left 2 Ships at Nimuegen laden with Tirnbergers in their way to Franckfort from the Reverend[26] of Cassant [Cadzand] that many of them were sick dead or dying.

That they Expect a large Embarkation of 1000 more Saltz-burgers as soon as the Snow falls upon the mountains of that Country So many being minuted down for leave to depart in the several Courts of the Archbishoprick.

Mr. Urlsperger desires me to acquaint Mr. [Ph.] Von Reck if he had not left Dover that he hopes before he leaves England he will transmit to him an Account of the laying out of the Guilders he received from him at Augsburg as well as of what he took up on Credit at Franckfort and in Holland which please to acquaint him with.

Pray give my humble Service to him and to Messrs. Bolzius and Gronau and assure them of my hearty Wishes for a prosperous Voyage to them and all the Colonists.

Capt. Bonner is just now arrived from Boston by whom advice is come that Governor Belcher has gained another point from the Assembly viz. an Order to rebuild the Fort at Pemaguid according to Orders from hence but then as I'm Informed they intend that at Casco Bay should be demolished and transplant the Garrison to Pemaguid as a Barrier between us and new France of which You may have a more particular Account in the Prints as also of their having voted 3000 £ for Governor Belcher Children and 76 Thousand pounds to Supply the Treasury to pay two Years Arrears to Officers etc.

I wish You a safe return and all the pleasures Spiritual and Temporal of the approaching Festival Being
Sir   Your most humble Servant   Henry Newman

Mr. Butjenter got well to Town, not a little pleased with his Journey.

Bartlet's Buildings London 27 December 1733.   To Mr. [Ph.]
Von Reck at Dover.

Dear Sir: Your detention at Dover by contrary wind gives me once more the Opportunity of Saluting You and Your Fellow Travellers.

Since my last to You I received a Letter from Mr. Urlsperger the 17th Current N.S. wherein he desires me to tell You if You had not left England that he wishes You could transmit to him if You had not Already done it An Account of Your Disbursements on the Saltzburgers from Augsburg to Roterdam.

And upon reading that part of his Letter to the Society relating to such an Account I was ordered to desire You would favour them with a Copy of it in that Language which is most familiar to You; Mr. Urlsperger writes that there are among the Saltzburgers some Proselytes from Popery[.] the Society hope there are not many such and desire You and Messrs. Bolzius and Gronau would keep a Watchfull Eye over them that they do not by any misbehaviour disorder those who have sincerely left their Country for the Sake of Religion particularly to have a care of the Wirtsberg Cook [Schweikert] whom the Society are informed would have imposed upon a Shopkeeper at Dover if You had not prevented him since the misconduct of such man may do more mischief than the good behaviour of ten honest men may bring credit to the reputation of the Emigrants.

I am glad to understand by Messrs. Bolzius and Gronau's Letter of the 24 Current that they have received Ludowig's Grammar and Dictionary, and doubt not their making the best use they can of them. Our German Missionaries to East India commonly get the English Language in their Voyage thither, but I hope Your Voyage will not be long enough for that purpose, beside the Conversation will be mixed which on bord an English Ship is for the most part entirely English.

The Society hope by Captain Coram's and your prudence all difficulties with the Captain of the Purysburg will be wholly removed.

I once more wish you and Your Company health and a Prosperous Voyage and remain

Dear Sir   Your most Obedient humble Servant   Henry Newman

Mr. Urlsperger sends his Complements to You and the Ministers & to Mr. Zweibler [Zwiffler] the Apothecary and please to give mine to Mr. Ortmann the Schoolmaster.

Bartlet's Buildings 27 December 1733. To Capt. Coram at Mr. Minet's at Dover.

Sir: I received Your favour of the 22nd Current and this day Communicated it to the Society, who have ordered me to thank You for it.

And as there were some particulars in it proper to be known to the Trustees for Georgia, Mr. Vernon borrowed it to lay before them at their Meeting next Saturday.

Every Body is pleased at Your Resolution to Stay till a fair Wind carrys the Purysburg out of Dover for fear of some accident which Your Experience and Prudence might soon set right.

Pray let me know in Your next whether Mr. [Ph] Von Reck understands English enough to read my Letters without an Interpreter. You will See by the Enclosed Print that the Town is not uninformed of what passes at Dover relating to the Saltzburgers which I find Yours to Mr. Vernon Confirms.

Your observation of a Persecution rising in Saxony was new to the Society, and they wish it may not prove true, but the Tyranny of Antichrist will not cease without making some extraordinary Convulsion in Christendom, out of which Evil God Almighty will undoubtedly bring good to the Protestant Interest.

I wish You a safe return to your Family and Friends and remain
Sir   Your most humble Servant   Henry Newman

Bartlet's Buildings London 28 December 1733. To the Reverend Mr. Urlspreger at Augsburg.

Reverend Sir: My last was of the 18th Current since which I received Yours of the 17th Current & Mr. Vat's of the same date both which relating to one Subject were this Day laid before the Society and deliberately considered.

The Society have never been informed of any misunderstanding between the Commissary Mr. [Ph.] Von Reck and Messrs. Bolzius and Gronau the Minister and Catechist but the contrary has appeared to the Society in Mr. Von Reck's desiring that Orders might be given that the Captain of the Purysburg should afford them as good accommodations in the Great Cabbin as the Commissary himself had, which the Trustees for Georgia have taken care of.

Upon receit of Your last Letter I have desired Mr. Von Reck if still detained at Dover to send You an Account of his Disbursements on the Road to Rotterdam and of the application of the Money he took up on account of the Saltzburgers at Augsburg, Franckfort and Rotterdam.

The Society are not at present in a Capacity to undertake for receiving more than 300 Saltzburgers including those already in their way to Georgia but in order to encourage further Contributions to so good a Service they are about to publish an Account how the whole that has been already collected has been applied[,] and to render such an Account more perfect I am ordered to desire You would with Your first Convenience transmit to the Society a partial Account of all the Sums you have disbursed for the Saltzburgers including the Transport to Rotterdam as also what further Engagements You are under on Account of them and the Tirnbergers and that You will likewise let the Society know what Sum remains in your hands.

I have herewith Sent a particular Account of the Remittances made to You amounting to 1482 £ 4 s 6 d Sterling wherein if there be any mistake You will be so good as to rectify it.

I don't find but our Family of Saltzburgers at Dover are all very easie under their confinement by a Contrary wind where I doubt not but Care is taken to Supply all their wants.

I have by order of the Society sent Ludowig's Grammar and Dictionary in 2 Volumes each to Messrs. Bolzius and Gronau to enable them in their Voyage to acquire some insight into our Language, so as to be able to converse in it when they come to Georgia.

I shall take Care to remit to You the Value of the 5 Medals by the first Opportunity; please to let Mr. Vat know that his friend Colonel Valogne is well just returned from Bath where he left his Friend Governor Shute well and that we both send our humble Service to him. I remain

Reverend & Dear Sir   Your most humble Servant
Henry Newman

NB. Enclosing Copy of my Letter of the 6 November 1733 which Mr. Urlsperger's of the 17 December N.S. signifys had miscarried.

Bartlet's Buildings London 8th January 1733/34.   To Mr. [Ph.] Von Reck at Dover.

Sir: I received your favour of the 11th Current N.S. and had this day the opportunity of laying that and your Account before the Society who were not a little pleased and have ordered me to give you their Thanks for them and for your great care and Diligence in conducting the Emigrants and to desire that you would pay the Ballance in your hands of 184 Guilders and 36 Cruitzers or whatever it be to Mr. Isaac Minet at the Exchange[.] he and you shall Agree upon taking 2 Receits for the same viz. one for your self and the other to be transmitted to William Tillard Esq. in Spital Square London without Bishopsgate the Receit to Express that Mr. Minet will be accountable to Mr. Tillard or his order for the same in behalf of the Society. I am also to desire if you have leisure for to send a short Account of the Families of the Saltzburgers in the present Transport their Names circumstances and what they have left behind in the Arch Bishoprick of Saltzburg.

You signify that your Experience in the first Transport has informed you in the means of conducting 200 Persons at very near the same Expence as this has Cost, the Society would be glad to know in what articles and by what measures you would propose to save expence in a future transport to Roterdam.

They are glad to observe your chearfull Resignation to God Almighty's pleasure in detaining your Ship by contrary Winds; We see but a little way into the Designs of Providence and as there are come advices of several Ships falling of late into great distress by the bad weather they have met with perhaps your detention at Dover has prevented the like disasters to the Purysburg.

You will always have the best wishes of the Society to attend you in your Voyage to Georgia wherein joyns most heartily.

Sir   Your most Obedient humble Servant   Henry Newman P.S.: I have forwarded both your Packets to Germany and shall send your Account to Mr. Urlsperger[.] my humble Service to Messrs. Bolzius and Gronau.

Bartlet's Buildings London 8 January 1733/34.   To Mr. Isaac Minet Merchant at Dover.

Sir: Mr. [Ph.] Von Reck having signified to the Society for Promoting Christian Knowledge that he had a Ballance of Money in his hands of 104 German Guilders and 36 Crutizers [Kreutzers] which he is ready to pay to their order I have wrote to him by this Post to desire him to pay it to you at the Exchange you and he shall agree upon which you are desired to receive and to give him two Receits viz: one for himself and the other for Mr. Tillard Esq. in Spital Square London to be accounted for to him in behalf of the Society.

Your goodness in favouring the Saltzburgers on Several occasions induces the Society to hope you will excuse the trouble of this application from

Sir Your most humble Servant Henry Newman

P.S. You may please to Signify by a line to Mr. Tillard or me when You have settled this affair.

London 8th January 1733/34. To the Reverend Mr. Richard Lowther at Roterdam.

Reverend Sir: I read your favour of the 4th of Last Month N.S. and sent the Society's thank for it the 4 of the same month O.S. The Purisburg arrived at Dover 11th of December where they have been ever since detained by Westerley winds but that detention has given them opportunity to furnish themselves with every necessary that the Emigrants could be thought to want in their Voyage as the Enclosed Print will inform you which I have sent to You by order of Severall Worthy Members of the Society that You may Republish what may be thought proper in the Holland News Papers to inform other Emigrants that may hereafter be inclined to go to Georgia what usage they may Expect in their way thither.

The Clergymen are now received into the great Cabin and the Captain of the Purisburg treats them and Mr. [Ph.] Von Reck with great Respect[.] While the Emigrants are at Dover they are kept on Shore at a house taken for that purpose where they Seem much pleased with their quarters.

The Prince of Orange set out this day sennight for the Bath by easy journeys and Expresses come almost every day bringing an Account of his Journey's agreing very well with him so that 'tis Supposed he arrived there Yesterday if not before.

I am Reverend Sir Your most humble Servant
Henry Newman

Bartlet's Buildings London 15th January 1733/34. To the Reverend Mr. Urlsperger at Augsburg in Suabia.

Reverend Sir: I received Yours and Mr. Vat's Letters of 17 of last month & answered them the 28 of December O.S. only one p[art] of Mr. Vat's I could not answer till I had the Direction of the Trustees for Georgia who have Signified their answer thereto in the words hereafter written which You will please to Communicate to Mr. Vat.

I have received none from Mr. Vat or You since the Letters of the 17th last month but this Day Mr. Ziegenhagen laid before the Society a translation of Your Letter to him of the 7th Current N.S. relating to several particulars of importance which I am ordered to lay before the Trustees for Georgia to direct an Answer to those Articles wherein they are chiefly concerned and the same will be done at their next meeting.

The Society were very much concerned at the Account you gave of Mr. Vat's indisposition and would be glad to hear by your next of his recovery[.] in the mean time I am ordered to desire that You would defray all the Expences of his Journey from Switzerland to Augsburg and during his residence there till further Orders, there being a Prospect of another Transport sometime next Spring in which he will be serviceable if it please God to recover him. The Society are very sensible of your great civilities to him and thank you for receiving him into your house where they hope he will want for nothing that may be necessary for his recovery, the charge of which they desire You will put to their Account as mentioned above.

My Last informed you of the arrival of the Saltzburgers under Mr. Von Reck's care at Dover the 11th December[.] I am glad to acquaint you that after they had been detained 4 Weeks at Dover by Contrary winds and plentifully refreshed themselves to their entire Satisfaction they Sailed the 8th Current in the Morn with a fair wind which has continued by Gods mercy ever since, so that there is reason to hope they are now far on their Voyage out of the Channel.

I must observe to You that while they were detained at Dover by Contrary Winds there happened several Storms in which many Ships were lost on the Coasts of England and Ireland but they Escaped them by being in a good Harbour which tis hoped is a presage of their safely proceeding under Gods Protection the rest of their Voyage.

Of their reception in England and their Satisfaction while they Staid here You will judge somewhat from the Authentick Account given in the inclosed News Papers some extract of which You may perhaps think proper to publish in Germany as I believe has been already done in Holland.

Mr. [Ph.] Von Reck sent a particular Account of his Draughts on You and his Disbursements on the Emigrants, open to me, to be forwarded to You which consists of several Sheets of paper and Shall be sent to You by the first Convenient Opportunity after I have examined a Copy of it with the Translator.

I shall follow your directions in Superscribing my future Letters to You and desire you will direct all Your Letters to me here, but You need not take Notice of my being Secretary to the Society for fear of giving umbrage to some people that may malign our Correspondence.

I heartily wish to hear of Mr. Vat's recovery and pray God to give you & him an happy new year and many more to follow this with his choicest Blessings on you and Your family, and remain
                Reverend & Dear Sir   Your most Obedient humble Servant
                                                          Henry Newman
The Society are now engaged in dispatching a Cargo of money Books etc. to the Missionaries at Fort St. George[27] and Tranquebar which in a manner Engages all my time.

The Resolution of the Trustees for Georgia in answer to that part of Mr. Vat's last Letter from Augsburg of the 17th Current N.S. which relates to the Quantity of Land given to the Emigrants for Georgia in the following words viz. "The Trustees for Establishing the Colony of Georgia in America have made no distinction between the Saltzburgers and those of the English who went in the first Embarkation and have even granted them better Terms than the English in later Embarkations.

"The Trustees gave no more than 50 Acres for every family to those English who went at first, and grant no more than 40 to those who go at present. If they give the same number of Acres to a Single man it is with expectation he will marry and have a family. They will give 50 Acres to every person among the Saltzburgers for the same reason. As there is no distinction in the property, there will be none likewise in their Rights whether Civil or Religious."

To which the Trustee who made the Report adds, That it is likewise to be observed that the Grant of 50 Acres to every Person who Settles in Carolina or other of our Colonies is to such who transport themselves at their own charge whereas the Saltzburgers are defrayed at great Expence from Augsburg to Georgia and are maintained at the Expence of the Trustees in Georgia till such time as they can get in a Crop from the Land given them for their Maintenance.

Bartlet's Buildings 22 January 1733/4.   To the Reverend Mr. Urlsperger at Augsburg.

Reverend Sir: My last was of the 15th Current in answer to Yours of the 17th of December since which I have had none from You till this day I received your favour of the 18 Current N.S. which I was glad to see in Mr. Vat's hand and that he is recovered so much from his late illness as to be able to assist.

According to Your promise of my last I have sent as above a Copy of the minute of the Trustees for Georgia relating to Mr. Goebell's kind offer in behalf of the Emigrants gone to Georgia[.][28] I am sorry it is not in the power of the Trustees for Georgia nor of the Society to do that Service for the absent Emigrants, but the Trustees will be moved at their next meeting by one of their own Members to send forthwith to Georgia to Enquire what Effects the Emigrants lately gone thither have left behind them in Saltzbourg and to have a proper Authority from them to demand those Effects.

I am sorry to find You have not received mine of the 18th of the month of December which gave You an Account of the arrival of the Saltzburgers at Dover, their reception there and of the praises Mr. [Ph.] Von Reck's conduct brought upon him both from English and Germanes and also the good Behaviour of Messrs. Bolzius and Gronau together with the presents made to the Saltzburgers by order of the Society of all which if Your next does not inform me that You have received it, I may give You the trouble of a Copy tho' I believe You have had the Substance of it in other Letters.

I hope to Send Mr. Von Reck's Account of Disbursements by the next Post in the King's Packet under cover to his Uncle [J. von Reck] at Ratisbone. The Purisburg sailed the 8th of this month from Dover with a fair wind and since we have heard nothing of her, but that the wind continued fair for a week at least so that 'tis hoped she e're this got into a warm Climate. My humble Service to Mr. Vat and please to believe that I am with great respect

Reverend Sir   Your most humble Servant   Henry Newman
I have no reason to believe that any of Your letters directed to me here ever miscarried but they always came a day before hand by being so directed and at half the Charge.

London Bartlet's Buildings 23 January 1733/34.   A Monsieur Monsieur [J.] Von Reck Ministre de Sa Majeste Britan. à Ratisbone.

Honoured Sir: Your worthy Nephew [Ph. von Reck] who has accepted the Office of a Conductor to the Protestant Emigrants from Saltzburg gone to Georgia sent to me before he Sailed from Dover the Account herewith sent of his Disbursements on that Service from Augsburg to Roterdam with a desire that I would transmit it to the Reverend Mr. Urlsperger at Augsburg. You will excuse the Liberty I have taken to cover it to You in his Majesty's Packet that it might come at less charge to Mr. Urlsperger.

The Society here who have been Sollicitous for the welfare of the Emigrants think themselves obliged to You for the many good Offices You have done them for which they have been informed by Mr. Urlsperger and they are not a little pleased that the Gentleman who has conducted them to England and is gone with them to Georgia has the honour to be Your relation, since he has acquired the esteem of every one and has given perfect Satisfaction to those under his Direction. He has I doubt not informed You of what has occurred here till the day of their departure from Dover which was the 8th Current O.S. under a fair wind which continued several Days, so that 'tis hoped we may hear early in the Spring of their safe arrival in Georgia of which I shall with Pleasure acquaint you as soon as I shall be able being

Sir   Your most obedient humble Servant   Henry Newman

Bartlet's Buildings 22 February 1733/34.   To the Reverend Mr. Urlsperger at Augsburg.

Reverend and Dear Sir: I am ashamed to see my last was of the 22nd January but Mr. Vat can inform you how buisy I am in buying Things for the Mission in E. India Shipping and Writing Letters at this time of the Year[.] Since that I have your favours of the 8th & 15 February and this Day yours of the 18th Current N.S.[.] the two former have been laid before the Society and read[,] in answer to which the above Copy of mine of the 18th December which You signify never came to your hands is repeated and the Extract from Mr. Robinson at Vienna to my Lord Harrington one of His Majesty's Principal Secretaries of State, added.

Your Last Letter is now translating and will be Communicated to the Society and Trustees for Georgia at their next meeting from whom you may Expect to hear as soon as I receive their Commands. In the mean time I am glad to hear Mr. Vat is so much recovered as to be Capable of Business, and hope it may be our Lot to see some of the Tirnbergers here, concerning which Your next Letters I expect will more particularly inform

Reverend & Dear Sir   Your most humble Servant
Henry Newman

I sent Mr. [Ph.] Von Reck's Account to his uncle [J. von Reck] at Ratisbone the 23 of last month in the King's Packet to be forwarded to You. My humble Service to Mr. Vat. When you have an Opportunity pray favour me with a Copy of your Picture to be put up in our Society's Chamber such as Mr. Von Reck brought over but presented when I could only get a Sight of it.

I have desired Mr. Mayer to request his friend Mr. Munich to pay you 25 Florins for the 5 medals, which be pleased to Signify when it is done.

Bartlet's Buildings London 1 March 1733/34. To the Reverend Mr. Urlsperger at Augsburg.

Reverend & Dear Sir: Your Letters of the 18 and 22nd of February N.S. have been laid before the Trustees for Georgia and the Society for Promoting Christian Knowledge & both of them observe that the answer of the Tirnbergers to the proposal sent to them by You is not yet known and therefore till they have received from You the said answer and a Compleat List of such persons as shall Signify their desire to Settle in Georgia they cannot Send their further Resolution and Instructions thereupon.

I am ordered to observe to you that the Expence of a Transporttation of Persons from Germany to Georgia amounts to a much greater Sum than was at first imagined and that a large Sum being placed out in one of our publick Funds viz. 2000 £ to answer the Salaries of the two Ministers that are lately gone with the Saltzburgers, makes both Societies apprehensive that they are hardly at present in a Condition to transport and Settle so large a number as 270 persons but they are using their utmost Endeavours to raise Contributions for the fulfilling their Engagements of making up the Number to 300.

In the mean while it is recommended to You not to Suffer any to be put into the List that are not Emigrants from Saltzburg on Account of their adherence to the Protestant Religion[.] the collection made by this Society must be applied Solely for their Use.

My humble Service to Mr. Vat of whose perfect recovery I shall be glad to hear and please to believe that I am

Reverend Sir   Your most obedient humble Servant
Henry Newman

P.S. I hope you have received the Account of Mr. [Ph.] Von Reck's Disbursements to Dover which I sent to you under Cover to his good Uncle [J. von Reck] at Ratisbonne.

Bartlet's Buildings 16 March 1733/34. To the Reverend Dr. Pardo Principal of Jesus College Oxon [Oxford].

Reverend Sir: According to your desire last month I have this day directed a Roll of 26 Memorials to the Reverend Dr. Conybeare in favour of the Emigrants from Saltzburg being one for each College and Hall addressed to the Vice Chancellor & Heads of Houses with a Letter to the Dean of Christ Church Dr. Smyth D. Cockman Dr. Pardo Dr. Foulkes and Dr. Felton requesting Your good Offices to recommend it to the Vice Chancellor for his approbation and recommendation of it to all the Heads of Houses in your University. The packet goes by one of the Coaches hence next monday morning Carriage paid here.

I did not till since sending it away know that Dr. Conybeare is to be in London next week to preach at the King's Chappel or I would have directed it to you in his Absence, but if the Dean [Conybeare] has left Oxford before it comes to hand pray let Your Servant enquire after it and please to open it as if it had been directed to you.

While the Society are meditating to wait on the Vice Chancellor in London they were advised that he was returned to Oxford to attend the Prince of Orange and now it is said he is coming to London again but I hope before he leaves Oxford you will with the other Gentlemen mentioned be able to wait on him with the Memorial to recommend the Affair in the most respectfull manner.

I am,   Reverend and Dear Sir   Your most humble Servant
Henry Newman

Bartlet's Buildings 21 March 1733/34. To the Reverend Dr Cockman near Bartholomew Close.

Reverend Sir: I took the Liberty of troubling you the beginning of the week with a Letter directed to the Dean of Christ Church [Conybeare] your Self and Several others Members of the Society at Oxford requesting that you would be pleased to recommend a Memorial from the Society therewith sent to the Vice Chancellor and other Heads of Houses that they would vouchsafe to encourage in their respective Colleges and Halls a charitable Collection in favour of the Protestant Emigrants from Saltzburg.

But while this Letter and 26 Copies of the Memorial were in the Stage Coach on the Road to Oxford, it seems the Dean and many of those worthy Gentlemen on whose good Offices the Society depended were set out with the Vice Chancellor for London which will necessarily delay for some time the Design of the Memorial, but if while You are in Town the Society may be favoured with your good Company they will be glad to see you particularly at their meeting here next Tuesday morning by 11 or 12 o'Clock to have your advice therein. I am

Reverend Sir   Your most humble Servant   Henry Newman

Bartlet's Buildings 21 March 1733/34. To the Reverend Dr. Smyth Provost of Queen's College at Oxford at his House in the Square at Paddington.

Reverend Sir: I acquainted the Society how kindly You had prepared the Vice Chancellor of Oxford to expect an Application from them in favour of the Protestant Emigrants from Saltzburg which they are very Thankfull for, and if it can consist with your Conveniency to Favour them with your company here next Tuesday Morning by Eleven or 12 o'Clock they will be glad of your advice thereon. I am

Reverend Sir   Your most humble Servant   Henry Newman

Bartlet's Buildings 26 March 1734. A Monsieur Monsieur de Münch Banquier à Augsburg en Suabia. To the Reverend Mr. Urlsperger.

Reverend Sir: My last was of the first Current since which I have your Favour of the 18th Current N.S. enclosing one from the Magistracy of Franckfort and Communicated it to the Society who have directed me to acquaint You that they readily concur with those worthy Magistrates of Franckfort in their Opinion that those Families of the Tirnbergers in which the Men are dead and none left but Women and Young Children are unfit for undertaking such a Voyage or for the managing and Cultivating the Lands to be granted to them in Georgia and therefore are not to be desired by us. but as to the objection of the unhealthiness of the Place from the Neighbourhood of the Sea they answer that the Trustees have given orders that the Transport already gone are to be Settled above 20 Miles from the Sea; and that care will be taken to fix the poor Tirnbergers as near as possible to their Countrymen. that the Air is good and the Land high and Dry,[29]

which was not the Case in Cassant [Cadzand] and therefore the
Society are Still willing to hope that such of the Tirnberg Families
where the men are living and fit for Labour and the undertaking
of such a Voyage will accept of the Offers made them and that
whenever such a Number of them shall come to a Resolution, You
will immediately transmit to the Society such a List of Persons and
Qualifications as you sent me before: And when they are ready to
set out on their Journey You will dispatch Mr. Vat to Franckfort
in order to conduct them and give me notice of every thing relating to the Transport that Shipping may be provided at Rotterdam
to meet them. I am
    Reverend Sir   Your most humble Servant   Henry Newman

*Bartlet's Buildings London 2 April 1734. A Monsieur Monsieur
de Münch Banquier à Augsburg en Suabia. To the Reverend
Mr. Urlsperger.*

Reverend Sir: I received by last post your Letter of the 22nd
March N.S. and this day acquainted the Society with the Contents
of it, who finding the Tirnbergers are unwilling to go to Georgia
have ordered me to desire you would let them know whether you
have good reason to believe any more Protestant Exiles will come
from Saltzburg and if they should whether it's likely that they will
be better disposed than the Tirnbergers to go to Georgia, or if
they should shew an Inclination to come whether they be permitted at this time to leave Germany.

They request Your answer to this in as short a Time as Possible
because if there be no immediate Prospect of Saltzburgers willing
to go to Georgia, they shall look upon themselves disengaged from
their former Proposals and at Liberty to receive other Persecuted
Protestants who have been driven from their Country for their
Religion and have actually made application to be sent to Georgia.
    I am   Reverend Sir   Your most humble Servant
                                Henry Newman
My last Letter was of the 26 March and pray give my humble
Service to Mr. Vat.

*Bartlet's Buildings 9th April 1734. To the Reverend Dr. Pardo
Principal of Jesus College Oxon [Oxford].*

Reverend Sir: Upon reading Your Letter of the 7th Current
to Sir John Philipps, the Society have ordered me to acquaint You
that they exactly followed Your directions in a former Letter by

writing to your Self and the rest of the heads of houses who are members of this Society desiring you and them to recommend the miserable Case of the Distressed Saltzburgers to the Vice Chancellor [Dr. Holmes] and to deliver to him the Memorials enclosed for his approbation and consent to make and receive collections in your University.

The Society thought this the most proper way of applying to him especially since Dr. Smith and Mr. Wilson last year laid the Case fully before him and at his desire it was put off till this time. I endeavoured twice to wait on the Vice Chancellor at his Lodgings when he was last in Town, but he was constantly engaged that I had not the honour of Seeing him.

The Society therefore desire you will be so good as to take an Opportunity as soon as possible to set this matter in a proper light to the Vice Chancellor and they doubt not but upon recollection You will find that they have done every thing in their power to make this application in as respectfull a manner as possible.

They desire also that you will return their Thanks to the Vice Chancellor for his kind readiness in sending and recommending the Memorial relating to the Saltzburgers to every College in Oxford and present their Service to you[,] at the same time I assure you that

I am   Reverend and Dear Sir   Your most obedient humble
Servant   Henry Newman

London 9th April 1734. A Monsieur Monsieur Christopher Charles Louis de P. feil [Pfeil] Secretaire de l'Embassy de son Altesse le Duc de Wirtemberg à Ratisbonne en Allemagne.

Sir: I had the honour of your Letter of the 21st last Month N.S. addressed to me as Secretary of the Trustees for establishing the new Colony of Georgia with a Packet for the Trustees but tho' I have not the honour to Serve them in that Capacity I took Care immediately to convey the Packet to them and you may be assured that it will be Submitted to their Consideration at their next meeting from whom you may expect an Answer by their Secretary Mr. Martyn, in the meantime You will please to be assured that I am with great respect

Honoured Sir   Your most obedient humble Servant
Henry Newman

To the Reverend Dr. Cockman Master of University College Oxford.

Dear Sir: I have been with the Society for Promoting Christian Knowledge this Day, who are in hopes to hear soon from You how you have Succeeded in your application for the poor Saltzburgers. It is thought proper to acquaint you that this Society and the Trustees for Georgia have endeavoured to procure a further No. of those Exiles to settle in Georgia but they have hitherto refused that Offer and the Archbishop of Saltzburg uneasy at the Loss of so many of his Subjects and the War now breaking out he has thought fit to use milder methods for which reasons we do not expect at present any more Emigrants from thence and being Sollicited by several of our poor persecuted Brethren expelled for the Profession of the Gospel from the Valleys of Piedmont and who are yet unsettled.

We hope you will not think that it will be at all wrong in us to apply some part of the money Collected for the use of the Saltzburgers towards Settling the other Exiles in Georgia[.] some of the most considerable of the Contributors in this Town are of this Opinion and we should be glad of your Thoughts in this Affair. It may perhaps be proper to Communicate this to your Brethren and Heads of houses if you think it will not put any Stop to the present Collection.

Our Design in these Contributions is to assist poor persecuted Protestants[.] when we first Set about it, those from Saltzburg were the most distressed[.] whenever they are disposed to accept of the Offers made to Settle them in Georgia we hope to have money for that Use. Till then we think we may receive other Distressed Protestants who have Suffered great Calamities and are disposed to Settle in Georgia.

I shall be glad to have a Line from you to Lay before the Society next Tuesday and if You have any Doubts, I shall take their Directions to Obviate them but pray let nothing that I have said be mentioned to anyone who shall take Occasion to put a Stop to your Charitable Design of assisting the Persecuted & Distressed Protestants. I am

Your most Obedient and most humble Servant   Tho. Wilson
Bartlet's Buildings
16 April 1734.

Bartlet's Buildings 23 April 1734.   To the Reverend Dr. Cockman Master of University College Oxford.

Dear Sir: I laid before the Society Your Letter and they order me to acquaint you that they have received a Letter from Mr.

Urlsperger this day who tells them that there is a prospect of more Emigrants from Saltzburg who perhaps will be prevailed upon to Settle in Georgia. This has occasioned the Society to think Seriously of this affair and continue in their former Resolutions of making Collections for the persecuted Saltzburgers only, till they are assured no more of these poor people will be drove out of their Country and accept of the Offer made them of Settling in Georgia.

The Society therefore beg You would encourage the Collection upon the footing of the Memorial now lying before You [.] and they will lay it out for the use of the poor Sufferers.

You will therefore be so kind as to acquaint the Dean with the reason of the Society's still keeping to their former Engagements of Settling a further No. of Exiles of Saltzburg in Georgia as soon as ever they shall be desirous of going thither.

The Society thank You heartily for your kind Intentions to wards the poor Sufferers and doubt not but they will soon hear of the Success of your endeavours for their Relief.

I am Your most Affectionate and most Obliged Humble Servant
Tho. Wilson

To the Worshipful Dr. Holmes Vice-Chancellor, and the rest of the Heads of Houses of the University of Oxford.

The Memorial of the Society
for Promoting Christian Knowledge

That they have (by Permission obtained from his most Gracious Majesty) undertaken a Collection for the Relief of the Distressed Subjects of the Archbishoprick of Saltzburg harrassed by many Cruelties and at last driven out of their native Country for refusing and rejecting the Errors of the Church of Rome and for their Stedfast adherence to the Protestant Faith.

They have laid before the Publick two Several [separate] Accounts of the Sufferings of these Persecuted People which by the Blessing of God has moved the Hearts of many pious persons to contribute towards their Support. Several of the Right Reverend the Bishops recommended this Charity earnestly in their Dioceses and considerable Sums have been by their Sollicitations returned to Us. Many of the Nobility Bishops Judges and Gentry and all Ranks of People have chearfully contributed to this good Work, as has also the University of Cambridge who have distinguished themselves by a Liberal Collection.

That notwithstanding the money thus gathered (of the application of which some Account has already been given and a further is preparing to be published) They find themselves unable effectually to carry on the good Work they have so happily begun[,] their Stock being near Exhausted by the Settlement procured for a number of those Emigrants in his Majesty's New Colony of Georgia[,] By the provision they have made for the Support of Two Clergymen gone over with these poor People[,] and by several large Summs remitted to Augsburg which have been laid out for the immediate Relief of such of the said poor Sufferers at their first arrival there in the Depth of Winter Stript and destitute of all things.

That they are now called upon afresh to Exert their Christian Charity in behalf of a considerable number whom they are Engaged to bring to the Trustees of Georgia to be Settled in that Colony.

They can't better hope to Execute this Charitable Design than by applying in this Emergency to the University of Oxford, whose Learning & Zeal in the most Dangerous Times have been the Great Bulwark and Support of the Protestant Cause in this Kingdom and who have upon all proper Occasions Enlarged their Compassion and Charity for the good of the Protestant Churches abroad whenever persecuted and Oppressed.

That Nothing can more Strongly express the rage of Popish Persecution than the present cruel and Barbarous Treatment of the protestant Subjects of Saltzburg where their Prince out of hatred to their Religion would sooner see his Country depopulated than Suffer a Protestant to live there.

All therefore who have a Sincere Love for Truth and a Value for their Holy Religion will exert themselves for the Relief of these distressed Emigrants and You Gentlemen who under a Popish Prince [James II][30] felt the Effects of their cruel Superstition will have a more Sensible Compassion for those who now Suffer what You were providentially freed from by the late happy Revolution.

With Confidence therefore, we apply to the Vice Chancellor and Heads of houses desiring they would recommend the miserable Condition of these poor Exiles to their Respective Societies, who we assure our Selves will upon this Occasion give the World as Shining a proof of their Benevolence as they have upon all other Occasions done of their Learning Zeal and Charity.

Henry Newman Secretary

Bartlet's Buildings London 11 June 1734. A Monsieur Monsieur de Münch Banquier à Augsburg en Suabia. To the Reverend Mr. Urlsperger.

Reverend Sir: I received your favours of the 5th and 22nd of April and now of the 27th of May N.S. in answer to mine of the 2nd April O.S. which have been all laid before the Society, and read, who are glad to observe by your last that there are still some Protestant Emigrants expected from Saltzburg, particularly from the District of Hallein, some of whom you doubt not may be disposed to go to Georgia but you are desired to remember, and to let such persons know that the Trustees for Georgia have signified to the Society that there can be no embarkation till August next, that being the best month for Strangers leaving Europe to arrive in that Climate, when the Rains which Usually fall there in our Autumn will be over and the most healthfull Season to Strangers ensue.

The Society were in hopes 'ere this to have sent you some Account of the safe arrival of the Saltzburgers gone to Georgia in the Purysburg, but hitherto they have only heard by a ship who met them at Sea near the Coast of Carolina that they were all well making in under a fair wind for the Land and that they desired the Master of the Ship who brings this intelligence to let their Friends in England know as much, So that we every day expect to hear of their happy arrival, by Ships that were expected to Sail soon after this Ship that met them.

In answer to your former Letters concerning the Salaries of the Minister and Catechist the Society knew of nothing of the Promises You had made to them of 60£ . . . . . . and 50£ when they advanced ½ a Year's Salaries to them at Dover in their way to Georgia supposing that you had agreed with the Minister for 50 £ per Annum according to my Letter of the 18 of May 1733 in the following words viz.

"The Society also desire You would recommend a person Qualified to accompany these Protestants as their Minister to whom the Society are willing to allow a Salary of 50 £ per Annum till the Settlement may be in a Condition to support him."

After which the Society upon your importunity to send over a Minister and Catechist with the first Transport, left it to your discretion the 3rd July last to do as you should think fit presuming you would find a person qualified for 50 £ per Annum as mentioned above, which is the utmost allowed to all the Mission that the Incorporated Society for Propagating the Gospel in Foreign parts have in the Province of South Carolina particularly the English Missionary on the Georgian Establishment, and to Catechists and Schoolmasters they allow only from 10 to 20 £ per Annum but the Society have agreed to allow the Catechist 30 £ per Annum and joined to him a Schoolmaster at 10 £ per Annum and the Society could allow no more without giving Offence to the Benefactors and the Missionary there Settled on the English Establishment, and therefore pray let those worthy persons know that the Society hope they will not think it want of respect to them that they do not exceed the Allowances they proposed.

I am   Reverend Sir   Your Most Obedient humble Servant
Henry Newman

I thank you for your kind present of 4 Pictures one of which is put up in the Society's Room. My humble Service to Mr. Vat.

Postscript

Reverend Sir: Since writing the above there is advice come from South Carolina to Mr. Symond's Merchant owner of the Purysburg that the Saltzburgers arrived off of Charles Town just as Mr. Oglethorpe was embarking there for England but upon their arrival he put off embarkation and having put Pilots on the Purysburg to carry her up the River Savannah he returned to Georgia to give directions for their better Settlement of which tis hoped the next Ships may bring more particular advices.

I am once more   Reverend and Dear Sir   Your most humble
Servant   Henry Newman

Bartlet's Buildings London 18 June 1734. A Monsieur Monsieur de Münch Banquier à Augsburg en Suabia. For the Reverend Mr. Urlsperger.

Reverend Sir: My last was of the 11th Current in answer to yours of the 5 and 22 of April last and 27 May, since which I received Your favour of the 17th of this month N.S., and this day the same was read to the Society expressing Your Surprize at hearing nothing of the arrival of the First Transport of Saltzburgers, but before the Society broke up they received the Pleasure of having the Account herewith Sent Communicated to them by Sir John Philipps from a Letter he had that moment received from Mr. Oglethorpe dated at St. Hellen's Road near the Isle of Wight at which they were not a little rejoiced, and Ordered me to acquaint You with the contents of it by the first Post.

I had the honour of a Short Letter at the same time from Mr. Oglethorpe wherein he tells me he has several Letters from Mr. [Ph.] Van Reck and the Congregation of Saltzburgers whom he left well pleased and in good Health at Eben-ezer that he thought it not proper to send them by the Post since he was coming to London and would deliver 'em with his own hand. You may be sure I shall with pleasure forward them which are directed to Germany, and if they are large and numerous I shall make so free with the King's private Secretary as to get them forwarded in the King's Packet to Baron [J.] Von Reck at Ratisbone from whom you may expect to receive those directed to Augsburg.

My humble Service to Mr. Vat, I hope by the present appearance of things it may not be long before he may be employed according to his own desires. Governour Shute and Colonel Valogne at St. James's very Lately Enquired after his health which I was glad to inform them of, and may you always enjoy your health for the publick Good is the wish of

Reverend and Dear Sir

Your most Obedient humble Servant   Henry Newman

Bartlet's Buildings London 25 June 1734.   To the Reverend Mr. Urlsperger at Augsburg.

[Marginal Note:] Enclosing Mr. [Ph.] Von Recks and Bolzius' Letters from Savannah in Georgia 22 March 1734. see their Letters Entered at Large in No. 2 Saltzb. Letters received folio 42, 43. Approved by Sir J. Philipps, Mr. Vernon & Dr. Vernon at Sir John Philipps's after the Committee was up, and forwarded by Mr. Reiche the King's Private Secretary under cover to Mr. [J.] Von Reck at Ratisbonne.

Reverend & Dear Sir: My last of the 18 Current acquainted You with Mr. Oglethorpe's arrival from Georgia and a Copy of his Letter to Sir John Philipps, since which I have received and laid before the Society the Letters from Mr. Von Reck and Mr. Bolzius of the 22nd of March last and by Order of the Society I send You Copies of them, but the principal Letters accompanying the Journal of Mr. Von Reck and Mr. Bolzius are not yet come to hand, which I have sent to Mr. Oglethorpe for, but he not being at home, I can only tell you that I suppose they are in some part of his Baggage not yet arrived but you may expect to hear further of them. In the mean time I received with these a packet for Baron

Von Reck at Ratisbonne which I have forwarded by this Post with the same Apology as I have made above for not sending the Journal etc. I received also a large Packet in High Dutch for Mr. Ziegenhagen the Contents of which have not yet been communicated to the Society.

I cannot but congratulate you on the safe arrival of the Saltzburgers in Georgia and doubt not but the kind reception they have met with will dispose some of those who have refused the invitation formerly given them to wish themselves with them, and had they gone when they were invited and peoples Zeal for assisting them was warm there is reason to believe there would not have been wanting such assistance from the Charity's of many good persons in this Kingdom as would have enabled the Society to have defrayed the Expence of their Voyage to Georgia.

The Society make no doubt but that the same Spirit of persecution Subsists in the Arch Bishop and his Clergy but as the Instances of Persons that have lately left their Country on that account are grown rare the Society have not been furnished with Materials to set out a 3rd Narrative of the Sufferings of the Saltzburgers in order to excite a new warmth of Charity for them which has been much abated by the refusal of the Tirnbergers to go and by not hearing of any number of Protestants disposed to come out. And the Collections made by the Society are already so much exhausted in transporting the small number already sent to Georgia together with the allowances made to the Minister Catechist and Schoolmaster that they dare not undertake for sending anymore.

This is not only the Case of our Society but the Trustees for Georgia on whose assistance the Society very much relied having been so much drained by the necessary Expence for supporting their Colony during the first Year of their Settlement that the Society are informed the like assistance as was given in the Transport of Saltzburgers already gone is not to be expected from them for the future. It is therefore recommended to you to be as frugal as you can in managing the remainder of the Saltzburg Cash you have in your hands and that good Mr. Vat may be duly considered for his pains and Expence in making a Journey to Augsburg, that you would advise the Society what Gratification may be proper to be made to him beyond what he has already received.

Pray give my humble Service to him and I beg you to be assured that I am Reverend Sir Your most humble Servant H.N.

Bartlet's Buildings London 25 June 1734. A Son Excellence Monsieur [J.] de Reck Conseiller de Sa Majeste Britanique et Son Envoy à la diete de l'Empire à Ratisbonne.

Honoured Sir: Yesterday I had the honour of receiving from your Nephew [Ph. von Reck] the Letters of which a Copy is on the other side but the Principal Letters accompanying the Journal of Mr. Von Reck and Messrs. Bolzius and Gronau are not yet come to hand which I have sent to Mr. Oglethorpe for, but he not being at home I can only tell you that I suppose they are in some part of his Baggage not yet come to hand but you may expect to hear further of them.

I have herewith forwarded the Letter I received for you according to his desire and hope the advices of it may be much to your Satisfaction till you can be advised of his safe return to Europe[.] in the mean time I remain with great respect Sir Your most Obedient humble Servant Henry Newman

Bartlet's Buildings London 2 July 1734. A Monsieur Monsieur de Münch Banquier a Augsburg en Suabia. To the Reverend Mr. Urlsperger.

Reverend Sir: My Last was of the 25 of June, since which I have the favour of yours of the 1st Current N.S. and this day laid the same before the Society upon reading which they were led to consider the Contents of their last Letter; particularly the Paragraph containing these words viz. "that the Society are already so much exhausted in transporting the small number of Saltzburgers already sent to Georgia together with the allowances made to the Minister Catechist and Schoolmaster that they dare not undertake for sending anymore." I am now ordered to desire you would understand the Society to mean that their assistance can't be given to any more *at present* but if those few Saltzburgers who were disabled from attending the last Transport to Georgia remain still at Augsburg the Society don't at all mean to exclude them the benefit of following their Friends thither if they are so inclined.

And as to Mr. Vat if there had been Emigrants enough to form a new Transport for Georgia and things in other respects had answered expectations as the Society are very sensible of his Capacity and integrity they would have been very glad he might have been the Conductor of them to Rotterdam, but now with regret they find themselves under a necessity of discharging him from any such Views and to gratify him as you shall advise according to

what I mentioned in my Last for his trouble and expence in coming to Augsburg attending there to be employed in the Service
mentioned and his return to Swisserland.

Dear Sir: You will perceive that to save time I have been
obliged to make use of my Clerk's hand to signify the pleasure of
the Society by this Post. My Letter of the 25th of last month was
sent under Cover to Mr. [J.] Von Reck at Ratisbonne to have the
benefit of going in the King's Packet and with it went one to you
from Mr. Bolzius in Georgia.

Mr. [Ph.] Von Recks Journal is not yet come to hand with some
other Letters daily expected.

My humble Service to Mr. Vat and believe that I am
Reverend Sir   Your most obedient humble Servant
Henry Newman

*Bartlet's Buildings London 26 July 1734. A Son Excellence
Monsieur [J.] de Reck Conseiller de Sa Majeste Britanique et Son
Envoy a la Diete de l'Empire a Ratisbonne.*

Sir: I had the honour of your Letter of the 22nd Current N.S.
by last Post and tho' I have not the honour to be Secretary to the
Trustees for Georgia I immediately acquainted them with the
Contents of it so far as concerns your Worthy nephew and shall
deliver Your Letter to him upon his arrival in England which I
hope maybe 'ere long.

I have been very much concerned for the delay of sending the
Original of your Nephew's Journal, but it was detained sometime
in Mr. Oglethorpe's Baggage on bord the man of War so that he
himself could not come at it; then I was ordered by the Society
(my Masters) for Promoting Christian Knowledge to make an
exact Copy of it which was finished but yesterday, in doing which
the Copies had the Misfortune to have some Ink spilled upon it
by the Copier, for which he and I ask your Pardon but to rectify
the same as much as in his power the Copier has transcribed so
much of it as was blemished by the Ink which I wish may be sufficient atonement.

I shall be glad to do any Service I am able to your Nephew
when it shall please God he arrives in England to convince him
and his good Uncle how much I am both his and
Sir   Your Excellency's most obedient humble Servant
Henry Newman
Mr. Urlsperger at Augsburg will be very glad to be acquainted
with the contents of your Nephew's Journal.

*Bartlet's Buildings 30 July 1734. To the Reverend Mr. Url-
sperger at Augsburg.*

Reverend Sir: I received Your Letters of the 19th & 22nd July
Current N.S. and upon reading them this day before the Society
they are come to the following resolution of sending a Transport
of between 40 & 50 Saltzburgers to be conducted by Mr. Vat with
all convenient Speed to Rotterdam and from thence to come over
with them to Gravesend in the River of Thames in the passage
Sloop where there will be a Ship ready to receive them to Georgia.

This Transport the Society would have consist First of those
Five persons left in Germany which you mention in your last.
Secondly those whose Friends and Relations are already gone and
who are desirous to follow them and if there should not be
enough of both these then the Transport is to be filled up with
such other Saltzburgers who are willing and ready to accept of this
Offer. The Society desire that this Transport be made ready as
soon as possible and they hope they will be in England by the latter end of September at farthest.

Mr. Vat is desired to be very frugal in their Expences upon the
Road etc. lest the Sum of money in their Hands should not hold
out, for we formerly told you that the Trustees of Georgia have
been so much drained by the necessary Expence for supporting
their numerous Colony during the first year of their Settlement
That the same assistance is not to be expected from them in this
as in the former Transport. Mr. Vat will have in a very short time
full Instructions how to deport himself upon his Journey and Voyage hither, and when he comes here it will be Considered what
acknowledgment to make him for his care and pains in this affair.

You will be so kind as to acquaint the Society with the Steps
You take in this affair from Time to Time and wishing You Suc-
cess in your pious endeavours for these poor persecuted Protestants
I remain   Reverend and Dear Sir   Your most humble Servant
                                                    Henry Newman
My humble Service to Mr. Vat. Reverend Sir, Last Post the 25
July I sent to the Baron [J.] von Reck at Ratisbonne his Nephew's
Journal on the voyage to and at Georgia till Mr. Oglethorpe came
away and have desired him to acquaint you with the contents of
it. this day the 30th July I saw a Letter from Mr. [Ph.] Von Reck
dated at Ebenezer in Georgia the 5th of May to Mr. Oglethorpe
by which he and all the Saltzburg Colonists were well going on

chearfully diligently & successfully in their new Settlement where I hope God will abundantly bless them.

Your single Letter to me of the 19th cost 3 s postage because it came under cover to Mr. [John Baptista] Mayer.

Whereas your last of the 22nd cost but 1 s because it came directly to me tho' both were marked fr. à Cologne.[31]

Bartlet's Buildings London 15 August 1734. A Son Excellence Monsieur [J.] Von Reck Conseiller de Sa Majeste Britanique et Son Envoy à la diete de l'Empire à Ratisbonne.

Sir: I had the honour of your Letter of the 22nd of July last N.S. and am glad by this to have the pleasure of congratulating your Excellency on the safe return of your Nephew [Ph. von Reck] from the Province of Georgia to whom I delivered the Letter you was pleased to recommend to my care. During his short stay in London, he was very agreeably entertained his Friends with an Account of his Travels particularly the Gentlemen of the Society for Promoting Christian Knowledge who have a true Sense of the favour done them when you recommended your Nephew to the good work of conducting the Saltzburg Emigrants to Georgia, the Society having found by experience how well he was qualified for such an undertaking, which has been executed by him with great discretion and Zeal for God's Glory as will fully appear to any one who shall attentively read his Journals to and from Georgia Shining with a Spirit of true Piety. The Society are exceedingly pleased that out of his unwearied Zeal for doing Good he is willing to conduct a Second Transport of Saltzburg Emigrants to Georgia.

In his return to England he brought several Letters from America, among which I had the honour of one from His Excellency Governour Belcher of New England referring me to his Son a Barrister at Law in the Middle Temple who had another Letter both which make Honourable mention of your Nephew[.] I have herewith sent Copies of them that you may have the pleasure of knowing in how amiable a light he appeared abroad.

Mr. Oglethorpe who had been an Eye Witness of his Prudence and Zeal in Georgia took Care to get him presented to His Majesty, and if his impatience to pay his Duty to your Excellency and his other Friends in Germany had not prevailed above all other inclinations he would have had any other respect shewed to him that his Friends here could have expressed[;] but when I mentioned your Advice he told me he was only going to Germany in order to return to England to obey your commands more effectually. Your nephew Embarked Yesterday in the Roterdam Sloop for Holland.

I hope God Almighty will bless him wherever he goes and remain Sir Your Excellency's most obedient humble Servant Henry Newman

Bartlet's Buildings 20 August 1734. To the Reverend Mr. Urlsperger at Augsburg.

Reverend and Dear Sir: My last was of the 30th July since which I have none from you, but according to the advice of your last the Society have procured 2 Royal Passes from His Majesty Viz. one for Mr. John Vat to act as Commissary and the other for Mr. [Ph.] De Reck to act in the same capacity for conducting any number of Saltzburg Protestants who may be disposed to come to Rotterdam in order to proceed to Georgia.

These Papers will be sent under cover to the Baron [J.] Von Reck to be used as there may be occasion which the Society have this day ordered me to acquaint you with, at the same time I assure you of being

Reverend Sir   Your most humble Servant   Henry Newman

My humble Service to Mr. Vat. Mr. [Ph.] De Reck happily arrived here from Georgia the beginning of this month and is gone hence last Wednesday to wait on you and his other Friends in Germany.

London 20 August 1734. A Son Excellence Monsieur [J.] Von Reck Conseiller de Sa Majeste Britanique et Son Envoy à la Diete d l'Empire à Ratisbonne.

Sir: My last was of the 15 Current O.S. since which I have the honour of one from your Excellency of the 19 Current N.S. enclosing a Second Letter to your nephew [Ph. von Reck] which I have herewith returned as the most ready way to reach his Hands at Ratisbonne whither he set out hence last Wednesday via Rotterdam.

Since he went hence His Majesty has granted Two passes Viz. One for Mr. [Ph.] De Reck and the other for Mr. John Vat to act as Commissaries in conducting any number of Saltzburgers that may be disposed to come from Germany to Roterdam in order to proceed to Georgia[.] this I am directed by the Society humbly to acquaint you with, and that the Ships of Georgia will be ready to proceed thither the beginning of October next.

Sir: In answer to your Enquiry the Secretary to the Trustees for establishing the Colony of Georgia by a Charter from His Majesty is Benjamin Martyn Esq. but I serve as Secretary to a Voluntary Society known by the name of the *Society for Promoting Christian Knowledge* who concern themselves on this occasion only as Sollicitors of Charity on behalf of their Persecuted Brethren the Protestants of Saltzburg and to do what they can to secure the Religious Interests of Those Confessors, on which account you will perceive that it is not in their Province to address the Court here for any Civil or Military Employment in favour of your Nephew tho' they were never so much inclined to it, beside his not being a native of England or naturalized disqualifies him for it, but I am well assured that the Trustees for Georgia who act in another capacity not Subject to these limitations have made an offer to your Nephew of any Encouragement they can give him, that he please to accept of in their new Colony of Georgia, concerning which Mr. De Reck will be able more fully to explain, to whom I beg Leave to refer and remain with all possible respect
Sir   Your Excellency's most obedient humble Servant
Henry Newman
The Passes abovementioned will be sent to your Excellency to be used as there may be occasion.

Georgius R.

Georgius Secundus, Dei Gratia Magnae Britanniae, Franciae et Hiberniae Rex, Fidei Defensor Dux Brunsvicensis et Luneburgensis Sacri Romani Imperii Archi-Thesaurarius et Princeps Elector etc. Omnibus et Singulis ad quos presentes hae Literae pervenerint, Salutem. Cum nobis Significatum sit, Dominum Philippum Georgium Fridericum de Reck quinquaginta aut plures Archi-Episcopatûs Salisburgensis olim Incolas, Viros, Faeminas, Infantesque, ab augustâ Vindelicorum in Angliam brevi deducturum, Cumque a Nobis humillimè petitum fuerit ut dicto Philippo Georgio Friderico de Reck cum comitatu Supramemorato, Literas Nostras Salvi Conductus concedere velimus, idcirco nos omnes et Singulos Reges ac Principes cujus cunque Dignitatis aut Ordinis, Status Ordinesque Amicos Nostros et Foederatos, Classium Exercituum que Ductores, Urbium et Arcium Praefectos reliquosque in universum Officiales et Ministros Eorum quoscunque (id quod Subditis Nostris quorum ullo modo intererit, firmiter injungimus) rogandos duximus ut praefato Philippo Georgio Friderico de Reck cum Comitatu Supradicto, et Sarcinis Eorum quibusunque non Solum liberam et Securam eundi transeundi; commorandique potestatem faciant, verum et omnibus Humanitatis Officiis eosdem excipiant adjuventque et eisdem, si opus fuerit, novas in super Salvi Conductus Literas concedant id quod nos, pari vel alio Officiorum genere Servata cujuscunque Status et Dignitatis ratione grate agnoscemus, Dabantur in Palatio Nostro apud Kensington 16° die mensis Augusti Anno Domini 1734. Regnique Nostri Octavo

Ad Mandatum Serenissimi Domini Regis
Harrington

Pass for Mr. De Reck and a number of Saltzburgers to come to England.

Memorandum: A Pass was made for Mr. John De Vat in the words abovementioned Mutatis Mutandis Dated the 12 of August 1734.

Minute of the Common Council November 14, 1733.

Resolved: "That Mr. Vernon and Dr. Bundy be desired to propose to the Society for Propagating Christian Knowledge That on paying over Three Thousand pounds to the Trustees they will engage under their Seal to pay Three Several [separate] Salaries of Fifty, Thirty and Ten pounds per Annum to the Minister Catechist and Schoolmaster of the Saltzburgers in Georgia.

Extract of a Letter sent to Mr. Oglethorpe dated November the 24, 1733.

"Among the First Rank is Mr. Philip George Frederick de Reck their Conductor the Reverend Mr. Boltzius[32] their Minister Mr. Gronau their Catechist, and Mr. Ortmann their Sent as a Schoolmaster to the Children, the Society in Bartlett's Buildings have engaged to pay to the Minister a Salary of 50 £ a year the Catechist 30 £ a year and the Schoolmaster 10 £ a year; And as the Schoolmaster's Salary is but small the Trustees think it right to allow him Twenty acres of Land for a House, Garden and pasturage."

This is all that appears in the Books of the Trustees for Establishing the Colony of Georgia relating to Mr. Christopher Ortmann.

Benj. Martyn Secretary

London 29th August 1734.   A Son Excellence Monsieur [J.] De Reck Conseiller de Sa Majeste Britanique et Son Envoy a la Diete de l'Empire à Ratisbonne.

Sir: I acquainted your Excellency the 20th Current that His Majesty had granted 2 Royal Passes for impowering Messrs. De Reck and Vat to conduct a Number of Saltzburg Emigrants from Augsburg to England. They are now by order of the Society for Promoting Christian Knowledge herewith sent to your Excellency to be conveyed to the Reverend Mr. Urlsperger at Augsburg as occasion may require.

I hope your Nephew [Ph. von Reck] is safely arrived 'ere this at Ratisbonne of which the Society will be glad to be informed. Mr. Vat's Pass was bespoke at My Lord Harrington's Office before it was known that Mr. de Reck was arrived in England, which is the reason of its being antedated. I have by this Post sent a Letter for your Nephew under your Excellency's cover, which came to my hands this week, but whether from America, or Germany I know not. May I presume to send my humble service to Mr. De Reck at the same time I assure you of being

Sir   Your Excellency's Most Obedient humble Servant
H. N. Secretary

Bartlet's Buildings London 1 August 1734.   To Messrs. Bolzius & Gronau Missionaries at Ebenezer in Georgia.

Reverend Sirs: I received your favour of the 22nd of March last and acquainted the Society with the Contents which[33] were very acceptable to them. Mr. Ziegenhagen has since communicated an Extract of your Journal to 22 March by which the Society are very much Concerned to hear of the Barbarity of the Capt. of the Purysburg to the Saltzburgers on their Voyage after all the Caution used at Dover to prevent it. The Trustees for Georgia I hear have resolved never to employ him again for the Sake of his ill usage of the Saltzburgers and if Mr. [Ph.] De Reck at his return with some other person can be found to bear Testimony such as the Law requires[34] to convict him of his Scandalous usage of those honest Confessors who had suffered so much already in their own Country he will I believe be prosecuted here with Severity to be made an Example of to deter others from the like Cruelty[.] at least nothing will be wanting in the Friends of the Saltzburgers here to prosecute him, if he can be Legally convicted. I am

Surprized at your improvement in the English Language, in so short a time under so many disadvantages; may your happy Talent still continue to make You more usefull to the English as well as to the Saltzburgers. In your Journal you mention some civilities shewn to you by the Jews lately gone to Georgia and some of them now and then looking into your Assemblies,[35] pray signify in your next any occurrences of that kind, some Jews in Germany having lately discovered an inclination to read the N. Testament on which Several have embraced Christianity and given hopes that the Time is not far off when it may please God to bring those antient People of God into the Christian World.

My most humble Service to your worthy Colleague Mr. Gronau.

If Mr. De Reck be come for England pray let the enclosed for him be returned directed under Cover to

Reverend Sir   Your most humble Servant   Henry Newman
Mr. Quincy will inform you of the reception of the Indians brought over by Mr. Oglethorpe.

Mr. Ziegenhagen being in the Country for his Health I doubt will not write to you by these Ships. Mr. Butjenter is well & would have been glad of a Line from you, Reverend Sir.

London 3 August 1734.

Reverend Sir: This moment Mr. [Ph.] Von Reck is happily arrived by the way of Boston which has obliged me to break open this packet to take out his Letter. I am glad to tell you he is in good health and if it be possible he will write to you by this Ship. I am

Reverend Sir   Your most humble Servant   Henry Newman
[Newman's letter of 20 Aug. 1734 to J. von Reck, which was previously entered, was inadvertently repeated here in the same hand.]

Bartlet's Buildings London 20th August 1734. A Monsieur Monsieur de Münch Banquier à Augsburg en Suabia To the Reverend Mr. Urlsperger.

Reverend Sir: My last was of the 30th of July since which I have none from you, but according to the advice of your last the Society have procured two Royal Passes from His Majesty, viz. one for Mr. John Vat to act as Commissary, and the other for Mr. [Ph.] De Reck to act in the same capacity for conducting any number of Saltzburger Protestants who may be disposed to come to Rotterdam in order to proceed to Georgia.

These passes will be sent under cover to the Baron [J.] Von Reck to be used as there may be occasion which the Society have this day ordered me to acquaint you with at the same time I assure you of being

Reverend Sir   Your most humble Servant   Henry Newman

P.S.: My humble Service to Mr. Vat.

Mons. [Ph.] De Reck happily arrived here from Georgia the beginning of this month and is gone hence last Wednesday to wait on you and his other Friends in Germany.

[Newman's letter of 29 Aug. 1734 to J. von Reck, which had been entered in his own hand above, was inadvertently repeated here by his clerk.]

Bartlet's Buildings 5th September 1734.   To Mr. Philip George Frederick de Reck at Ratisbonne.

Dear Sir: I hope this will find you Safely arrived at Ratisbonne, whither I have sent a large packet under cover to your good Uncle [J. von Reck] directed to you the 29th of last month, together with a Royal Pass from His Majesty impowering you to conduct a number of Saltzburgers from Augsburg to England. This accompanys a Mapp of Georgia which Mr. Oglethorpe desires me to convey to you in the best manner. Your Friends here will be glad to hear of your Welfare and what prospect you have of returning to England at the head of a good Transport. My Humble Service to His Excellency the Baron and please to believe that I am

Dear Sir   Your most obedient humble Servant
Henry Newman

Bartlet's Buildings London 20 September 1734   A Monsieur Monsieur De Münch Banquier à Augsburg en Suabia.   To the Reverend Mr. Urlsperger.

Reverend Sir: I have laid your Letter of the 16th Current N.S. before the Society who immediately applied to the Trustees of Georgia and said that a Ship [Prince of Wales] is already prepared to convey over the Indian Chief [Tomochichi] and his attendants under the direction of Capt. Dunbar a Gentleman remarkable for his humanity who has great concerns himself in Georgia and will therefore be further induced by his own Interest to show particular regard to those who shall be sent thither. The Society hopes according to your Advise that the Transport will have set out before this comes to hand[,] if not I am to desire you will consider whether it will be possible for them to reach Rotterdam by the 15th of October O.S. that being the utmost time they can have the benefit of this Ship. If they are already set out you will be so good as to Send a Messenger if necessary to hasten them, for tho they may be received so late as the 15th the Ship will be ready for them by the 8 and it will be some inconvenience to wait longer. I have wrote by the Society's order to the Reverend Mr. Lowther and Mr. Wolters at Rotterdam who took care of the former Transport to receive these and provide for their conveyance to Gravesend.

The Society hope your next will inform them of the number and quality of the Saltzburgers to be Transported and will be glad to hear of Mr. [Ph.] De Reck's safe arrival at Augsburg to whom I wrote the 5th Current but directed my Letter with a Map of Georgia to him at Ratisbonne for the convenience of going in the King's packet but I sent the Royal passes for him and Mr. Vat to conduct the Saltzburgers in packet directed to the Baron [J.] de Reck 29 August O.S.

My humble Service to both these Gentlemen if they are Still with you and please to believe that I am.

Reverend Sir   Your most obedient humble Servant

Henry Newman

Bartlet's Buildings London 20 September 1734.   To Mr. D. Wolters Agent for His Britannick Majesty at Rotterdam.

Sirs: The Society for Promoting Christian Knowledge having received advice from the Reverend Mr. Urlsperger at Augsburg that a Transport of Saltzburgers consisting of above 20 persons mostly young single people would be ready to Set out thence by the 15th Current O.S. for Rotterdam in order to embark for Georgia and having thereupon desired a Conference with the Trustees for Georgia concerning the best means of conveying them, I am ordered by the Society to desire you would in concert with the Reverend Mr. Lowther if he be at Rotterdam assist the person who shall conduct them whether Mr. Vat or Mr. De Reck to bespeak accommodations for them till they can embark and for engaging one or more Rotterdam Sloops to give them passage to Gravesend where a Ship [Prince of Wales] commanded by Capt. Dunbar bound to Georgia will be provided by Mr. Simonds to receive them. Capt. Pigeon and Capt. Cranwell are named to the Society for honest Masters of Sloops but if you know of any other that you can recommend which may be sooner ready, the Society desire you would make the best agreement you can for the comfortable passage of these people to Gravesend and your or Mr. Lowthers or the Conductors Bills on William Tillard Esq. in London for payment of their passage will be punctually complied with I am

Sir Your most obedient humble Servant Henry Newman

The Society are informed that the usual price for passage money to Gravesend is 5 s per head, including their Diet.

I shall be obliged to you for a Line as soon as you have any intelligence of the arrival of these People in your Parts.

To the Reverend Mr. Richard Lowther Minister of the English Church at Rotterdam.

Reverend Sir: The Society for Promoting Christian Knowledge etc. etc. as above to Mr. Wolters Mutatis Mutandis.

Bartlet's Buildings 24 September 1734. To the Reverend Mr. Lowther at Rotterdam.

Reverend Sir: My last was of the 20 Current Since which the Society have advice of the 23 Current N.S. from Augsburg that 54 Saltzburgers Set out that day for Rotterdam under the Conduct of Mr. Vat and tis very likely Mr. Von Reck may follow them.

By this Post Mr. Simonds has sent Credit for what there may be occasion to assist Mr. Vat in bringing them to Gravesend on Messrs. Curtonne and his Son and de Normandie at Rotterdam who will be ready to give what assistance they can to Supply their wants etc. while in Holland and in their way to England, wherein the Society request and doubt not of your good offices together with those of Mr. Wolters to favour their Transport. I have covered a Letter for Mr. Vat to Mr. Wolters Supposing he will of course apply to him on his arrival. I am

Reverend Sir Your most obedient humble Servant
Henry Newman

wrote to the Effect above to Mr. Wolters at Rotterdam Mutatis mutandis.

[Marginal Note:] If Mr. Von Reck comes with or follows this

Transport the Society desire you will Shew him all the respect due to his great Merit for his past Services.

Bartlet's Buildings London 24 September 1734.   To Mr. John Vat under Cover to Mr. Wolters at Rotterdam.

Dear Sir: The Society were this day very agreeably Surprized to hear you was set out from Augsburg the 23 Current with 54 Saltzburg Emigrants to Rotterdam in order to proceed to Georgia and immediately gave me orders to bespeak of Mr. Simonds what Credit you may have occasion for to Supply them with necessaries in Rotterdam and in their way to Gravesend in one or more Sloops where Capt. Dunbar of Georgia in a Ship belonging to Mr. Simonds will have instructions to receive them. By this Post Mr. Simonds has promised to write to Messrs. Courtonne and Son and de Normandie at Rotterdam to Supply you accordingly and to take your Bills on William Tillard Esq. in London for what you shall draw on him for this Service but the Passage money is not to be drawn for till they arrive in England where Mr. Simonds tells me the Capt. of the Sloops will be content to take your Bills. The Reverend Mr. Lowther Minister of the English Church at Rotterdam and Mr. Wolters His Majesty's Agent there are both wrote to by this Post to give you all the advice and assistance they can which from the experience the Society have had of their goodness when the former Transport arrived they have no reason to doubt of.

I hope it may not be long 'eer I may congratulate you in person on your Safe return to England and if Mr. Von Reck has overtaken you in your Journey from Augsburg, the Society doubt not of your having treated him with all possible respect in regard to his great merit and the obligations they are under to him for his past Service and that they will be glad to see him again in England.

I don't write to him because it is uncertain whether he attends this Transport but if he be with you pray shew him this Letter and assure him of my most humble Service. I received his Letter from Rotterdam Sometime since and another from Ratisbonne of the 20th Current N.S. but till I know where to direct to him I forbear writing. I have a Letter by me of old Date from Germany which I believe is from his uncle at Ratisbonne and keep it till I have the honour of Seeing him. Pray let me hear of your motion and of the welfare of the Emigrants under your care and the names and Qualities of those 7 not included in the List of 47 you sent from Augsburg to compleat the number of 54.

I pray God to prosper your voyage and send you safe to your Friends here remaining

Dear Sir   Your most obedient humble Servant
Henry Newman

Bartlet's Buildings 25 September 1734.   To Mr. Verelst at the Georgian Office in old Palace yard Westminster.

Sir: The Copies of the Letters from Mr. Urlsperger herewith sent of the 20th & 23 Current N.S. with the List of the Saltzburgers Set out last Thursday was Sennight from Augsburg were yesterday laid before the Society here who have ordered me to desire you would communicate them to the Trustees for Georgia at their meeting this Evening and that you would procure an Estimate of the whole Charge of Transporting the 47 Emigrants from Gravesend to Georgia mentioned in the List to which may be added the other 7 to compleat the number of 54 as soon as they are known, I have wrote for them by last Post. The Committee of the Society here have adjourned themselves to Friday morning next in hopes of receiving then by you the favour of the advice of the Trustees on these Letters with the Estimate. I am Sir Your most humble Servant

Henry Newman

London 1 October 1734.   A Monsieur Monsieur Von Reck à Augsburg.

Sir: I have this moment received your Letter from Augsburg of the 16 of September and have communicated the same to the Gentlemen of the Society who approve of the reasons alledged in your Letter and that of the Reverend Mr. Urlsperger for your not accompanying the Transport now Sett out for Georgia. We make no doubt but your residence this winter in Germany will be of singular Service to the Colony of Georgia and we are endeavouring to put our Selves into a Condition to receive another Transport of Saltzburg Emigrants in case any such offer, which at present we are not able to do but we are in great hopes by the Blessing of God to be ready by next Spring to receive a new Transport and that it will Suit your convenience to put your Self at the head of them. The Experience of what is past encouraging us to hope for Success where you are concerned.

All your Friends here Salute you with great affection and none more than

Sir   Your most humble and obedient Servant   Ja. Vernon

Bartlet's Buildings London 1 October 1734.   To the Reverend Mr. Lowther Minister of the English Church at Rotterdam.

Reverend Sir: I received your Letter of your 8th Current N.S. this day and immediately read it to the Society who were then Sitting and they direct me to acquaint you that they are not at Liberty to consent to the Sending the Phillipsburg Family to Georgia on their account, their Collections being made for the Relief of the persecuted Protestants drove out from the Arch Bishoprick of Saltzburg and I believe I can inform you with Some Certainty that the Trustees for Georgia are not in Cash to entertain any Families from abroad on their account.

Sir John Philipps, Mr. Vernon one of the Trustees for Georgia and Mr. Copping being present when I received these orders, desired me particularly to Send their humble Service to you at the same time I Subscribe my Self

Reverend Sir   Your most obedient humble Servant

Henry Newman

The Society thank you for your kind readiness to assist in embarking the Saltzburgers now on their way to Rotterdam[.] they were at Franckfort the 4 Current N.S. and expected to arrive at Rotterdam in ten days under the conduct of Mr. Vat. Mr. [Ph.] Von Reck does not go now.

London 1 October 1734.   To Mr. [Ph.] Von Reck at Augsburg. In a P.S. to Mr. Vernon's Letter entered above.

Dear Sir: I take leave to Subjoin a line to congratulate you on your happy meeting with the Reverend Mr. Urlsperger and to acknowledge the receipt of your Letter in a Postscript to Mr. Urlsperger's of the 27 of last month. I pray God to direct you in all your aims for his Glory in your future conduct and shall be glad to hear from you as occasion offers. The Transport under the care of Mr. Vat was at Franckfort the 4th Current all in good health, of which I Suppose he informs you.

I had the honour of a Letter from Baron [J.] Von Reck of the 27 of September N.S. acquainting me of your health and Journey to Augsburg. My humble Service to His Excellency when you write to him. I find you have not yet received the Map of Georgia Sent by the way of Hanover by Mr. Reich in a Gentleman's Baggage it being too large to be sent by the Post.

Tomochichi the Indian King etc. are all in good health being to go over with the Saltzburgers in the Prince of Wales commanded by Captain Dunbar whom you know. I am with all possible respect

Dear Sir   Your most obedient humble Servant
Henry Newman

Bartlet's Buildings London 1st October 1734.   To Mr. Vat at Rotterdam.

Dear Sir: I received your favour of the 4th Current N.S. from Franckfort this day while the Society were Sitting and they were glad to understand by it that you and the Saltzburgers were got so far well and hope you will arrive at Rotterdam in 8 or 10 days as you expect[.] Mr. Deymar [Degmair] the Minister who is So kind as to attend the Transport to Rotterdam will be welcome to come over with them to Gravesend if he is So inclined.

The Prince Frederick[36] commanded by Capt. Dunbar is appointed to be at Gravesend by the 12 Current to receive the Saltzburgers but if you arrive there before She can get to that Station, care is taken to bespeak Lodgings on Shore at Gravesend for the Emigrants and that notice shall be given to you either by the postmaster or some of the Kings officers attending there, of the house they are to be Lodged in as soon as the Sloop or Sloops you come over in shall anchor before the Town.

The Society are glad to understand by Mr. Urlsperger that you are inclined to go to Georgia with this Transport and have taken Care to bespeak proper accommodations for you in the Prince Frederick only I must acquaint you that the Indian King Tomochichi and his Queen and Nephew are promised the great Cabin, but Mr. Simonds has promised that the State Room or some other convennience Shall be made up for your Reception, if not part of the great Cabin. My Letter of the 24 of last month will inform you that Mr. Simonds has taken Care to furnish you with Credit on Messrs. Courtonne and Company at Rotterdam for your Bills on William Tillard Esq. in London to defray all expences of your Transport to Gravesend and when it pleases God you arrive there and have disposed your family on Shore as mentioned above the

Society will be glad to see You here to give you further instructions. In the meantime they have desired two of their members to visit the Emigrants at Gravesend to See that nothing be wanting for their comfortable Support in the Prince Frederick, and therefore I am to desire that you would write to me both from Rotterdam and in your way thence by every post, that I may be able to advertise them of your motion.

I find Mr. [Ph.] Von Reck cannot be ready to return this year as he proposed. I pray God to direct you to conduct these honest People to his Glory and your own Comfort assuring you that I am
Sir   Your most humble Servant   Henry Newman
You are recommended to the Reverend Mr. Lowther and Mr. Wolters at Rotterdam for their good offices and I have this day a very kind Letter from Mr. Lowther to assure the Society he will give you all the assistance he can.

Bartlet's Buildings 1st October 1734.   To the Reverend Mr. Urlsperger at Augsburg.

Reverend & Dear Sir: I this day received your acceptable Letter of the 27th of last month while the Society were Sitting and the Reverend Dr. Guerdes being present interpreted it to them and Mr. Vernon being also present wrote a Letter to Mr. [Ph.] Von Reck which is Sent herewith open that you may know the Contents and forward it to him wherever he may be, he having forgot to Signify where Letters may be directed to him.

I need not repeat any thing said in Mr. Vernon's Letter or my Postscript to Mr. Von Reck but assure you that according to your advice all possible care has been taken by the Society for the commodious transportation of the Emigrants now on their Journey to Rotterdam and also for Mr. Vat's accommodation in the great Cabin or State Room; Effectual Credit has been sent as you desired by Mr. Simonds to meet Mr. Vat at Rotterdam, and Mr. Deymar [Degmair] who is so kind as to accompany the Saltzburgers to Rotterdam is invited to come with them to England if he pleases.

I am
Reverend & Dear Sir   Your most obedient humble Servant
Henry Newman
I received your Letters of the 20 & 23rd of September in due time and my last to you was of the 20 September O.S.

Bartlet's Buildings 9 October 1734. To Mr. Deacon at the Tower an Officer of the Customhouse.

Dear Sir: The Saltzburgers designed for Georgia were Yesterday was Sennight[37] at Dusseldorp in their way to Rotterdam in order to embark for Gravesend, therefore as the Prince of Wales Commanded by Capt. Dunbar cannot be there till the 12th Current nor then if the wind don't favour him I am ordered to request you would desire Mr. Carkass to Signify to the proper Officers of the Customs there to give notice upon the arrival of the Rotterdam Sloop at Gravesend to Mr. Vat the Conductor of the Saltzburgers that Mr. Lott Merchant at Gravesend is desired to take Care for their accommodations on Shore, till the Ship falls down or further directions can be given therein, by which you will oblige the Society as well as

Sir   Your most humble Servant   Henry Newman

Bartlet's Buildings London 9 October 1734. To the Master of the Post office at Gravesend in Kent.

Sir: The Society for Promoting Christian Knowledge who are Sollicitous for the welfare of the persecuted Saltzburgers having notice that a Transport of them designed for Georgia and daily expected at Rotterdam in their way to Gravesend in order to embark on the Prince of Wales commanded by Capt. Dunbar who has promised (Wind and Weather permitting) to be at Gravesend the 12 Current to receive them out of the Rotterdam Sloop or Sloops that may bring them over, but for fear any accident should prevent the Ship falling down at the time intended, I am ordered to desire you would Signify to the Mayor of the Town and to other Civil or Parish Officers that the Society intend to defray all the Expence that may be occasioned by the Saltzburgers Landing at Gravesend to refresh themselves till they can embark on the Prince of Wales. Mr. Lott is wrote to, to provide accommodations for them and it is not doubted but their inoffensive behaviour will recommend them to the Protection and good Offices (if occasion require) of Mr. Mayor and his Brethren. there will be 57 of them besides their Conductor Mr. Vat who has His Majesty's Royal pass for this Service. I shall be glad to be favoured with a Line from you when they arrive directed to

Sir   Your most humble Servant   Henry Newman

Bartlet's Buildings 11 October 1734.   To Mr. John Vat at Mr. Wolters His Majesty's Agent at Rotterdam.

Dear Sir: Your Letter from Franckfort and that of the 12th Current N.S. from Dusseldorp are come Safe to hand and the Society are glad to understand that you are come so far with the Transport of Saltzburgers. I have little to add to what I send [said?] in my Two last which you will find at Mr. Wolters but if you don't find the Prince Frederick now called the Prince of Wales at Gravesend You are to apply your Self to William Haffenden Esq. Mayor of the Town for directions where to meet with the person Sent by the Society to prepare accommodations for your reception with the Saltzburgers on Shore, Mr. Lott whom I mentioned to you being too much engaged on his own affairs to be able to attend such a Service.

When you come to London you will be welcome to accept of a Bed with me during your Stay here to give me the pleasure of your Conversation and to convince You that I am
Dear Sir   Your most humble Servant   Henry Newman

Bartlet's Buildings London 11 October 1734.   To William Haffenden Esq. Mayor at Gravesend.

Sir: I received your favour of yesterday and according to your advice the Society will Send a person down to Gravesend to pay for the Subsistence of the Saltzburgers when they arrive and the Society will esteem it as a favour if you give me Leave to recommend that person to your advice for any assistance you may be pleased to give him by your protection as the chief Magistrate of the Town.

Mr. John Vat is the Conductor of the Saltzburgers and has His Majesty's Royal Pass for that Service. You will go nigh to hear of their coming into the River before I shall. The Prince of Wales commanded by Capt. Dunbar who is to carry them to Georgia falls down to Gravesend next week. Mr. Lott it seems has declined taking care of these people[.] I shall be glad to know how much may be a reasonable allowance by the day for Subsisting and Lodging about 58 persons who are accustomed to live as plain Day Labourers that Credit may be furnished accordingly. I am
Sir   Your most humble Servant   Henry Newman

Bartlet's Buildings 12 October 1734. To William Haffenden Esq. Mayor of the Corporation at Gravesend.

Sir: I have just now Letters from Rotterdam that the Saltzburgers arrived there all in good health last Sunday Sennight and were to embark on the 2nd [sic] Brothers Capt. William Thompson for Gravesend as last Saturday but the wind not favouring I Suppose they are detained there. Mr. Wiseger[38] had my Letter to you last Saturday to request your advice and assistance in bespeaking Quarters for them but by one means or other he has been obliged to defer Setting out till tomorrow morning Tide. In the meantime your good offices if they arrive will be thankfully acknowledged by the Society. I am

Sir   Your most humble Servant   Henry Newman

Bartlet's Buildings 14 October 1734. To the Reverend Messrs. Wilson and Butjenter in Separate Letters. Mutatis mutandis.

Reverend Sir: Just now I have received a Letter from Mr. Vat at Rotterdam of the 22nd Current N.S. by which the Saltzburgers with him to the number of 56 arrived all in good health there last Sunday was Sennight and by agreement if the wind had permitted they were to Sail from Rotterdam to England as last Friday in the Two Brothers commanded by Capt. William Thompson So that by the first favourable Wind that is between N.E. and S.E. they may be expected at Gravesend.

I doubt they will arrive there before it will be possible for Mr. Weiseger to get thither, for instead of Setting out this afternoon Tide, he does not go till tomorrow morning. I am

Reverend Sir   Your most humble Servant   Henry Newman

Bartlet's Buildings 15 October 1734. To Mr. John Vat or in his absence to Monsieur Wolters Agent for his Britanique Majesty at Rotterdam.

Dear Sir: I received your Letters of the 19th and 22nd Current N.S. and this day read them to the Society who were very glad to hear of your and the Saltzburgers Safe arrival at Rotterdam[.] I scarcely think that you will Stay there long enough to receive this Letter but I Send it to take its chance to inform you that William Haffenden Esq. Mayor of Gravesend will inform you of the Place or places appointed for receiving the Saltzburgers while they refresh themselves on Shore and that Mr. Weisinger a German from Pensylvania who is going to Georgia is going down tomorrow to bespeak Quarters for them thou I doubt by your last whether it will be possible for him to be there before you will, in which Case if he is not pray inform your Self from Mr. Mayor and the Reverend Mr. Harris Minister of Gravesend where you may be best accommodated. My humble Service to the Reverend Mr. Lowther and Mr. Wolters to whom the Society are much obliged for their Civilities to you and your Transport. Wishing you Safe to England I remain

            Sir   your most humble Servant   Henry Newman
To William Haffenden Esq. Mayor of the Corporation at Gravesend.

Bartlet's Buildings 15 October 1734.

    Sir: Something has happened that prevents Mr. Wisiger [Weiseger] from Setting out to Gravesend till he has attended the Trustees for Georgia which cannot be till tomorrow Evening, which obliges me to request that in case the Saltzburgers should arrive at Gravesend before he does, that you would favour Mr. Vat their Conductor with your Advice for providing them with refreshments ashore till the Prince of Wales fall down to receive them which will be upon the first notice of their arrival at Gravesend. I should have been glad of a Line from you upon what Terms they could be Lodged and dieted in a plain manner per head for a few days, the Charge of which will be defrayed by the Reverend Mr. Wilson and Mr. Butjenter on the part of the Society who will Set out hence as Soon as he hears of the Saltzburgers being arrived. I am

            Sir   Your most humble Servant   Henry Newman

Bartlet's Buildings London 15 October 1734.   To the Reverend Mr. Thomas Harris Rector or in his absence to his Curate at Gravesend in Kent.

    Reverend Sir: The Society for Promoting Christian Knowledge being daily expectation of the arrival of a Transport of Saltzburgers from Rotterdam have ordered me to recommend them and their conductor Mr. Vat to the good offices of Mr. Mayor and your Self to advise Mr. Vat about quartering them ashore in Some good house where they may be refreshed in a plain & frugal

manner without being liable to the impositions too often practised on Strangers in Seaport Towns. The Reverend Mr. Wilson
is So kind as to promise to go down to Gravesend as soon as he
hears of their arrival, with full power to defray all Expences that
may be reasonably occasioned by their short residence on Shore
till the Prince of Wales commanded by Capt. Dunbar can fall
down the River to receive them. I wrote to you last Saturday to
this Effect by Mr. Weisinger a German going over to Georgia, but
Something has prevented his going to Gravesend as soon as was
expected which is the reason of my troubling you with this from
  Reverend Sir   Your most humble Servant   Henry Newman

*Bartlet's Buildings 17 October 1734.   To Mr. John Vat Con-
ductor of the Saltzburgers in the Two Brothers at Mr. Haffenden's
Mayor of Gravesend.*

Dear Sir: As I expect you may be at Gravesend by that time,
this can be there. I have only to acquaint you that as the Prince
of Wales can't be ready to fall down till the beginning of next
week Mr. Symonds by Mr. Oglethorpe's advice proposed to me
this Evening that much charge might be saved if instead of landing at Gravesend the Two Brothers came up to the Red house
above Greenwich[.] the Saltzburgers might be embarked immediately in the Prince of Wales but as I have not the directions of
the Society herein I can only say that if you don't like the accommodations that may be offered to you at Gravesend, it must be left
to your discretion to take the utmost frugal measures for their
Comfort.
  The Commissioners of the Customs have very readily Sent an
order to their officers at Gravesend to permit the Landing of all
the Baggage of the Saltzburgers and to shew them all the Civilities
they can, but then they [the baggage!] are to be locked up in Some
Warehouse at the Charge of the Society 'till they can be reshipped,
whereas if you come up the River there will be no occasion of
landing the Baggage, because it may be immediately conveyed
into the Prince of Wales and the Saltzburgers themselves may take
possession of their Several apartments in the ship[.] as I said this
must be left to your discretion after consulting Mr. Mayor of
Gravesend and the Reverend Mr. Harris to whom you and your
Transport have been recommended it being impossible for me to
have the orders of the Society therein till their next meeting.

Pray let me hear from you as Soon as possible after your arrival in the River that measures may be taken accordingly. Mr. Weisinger a German from Pensylvania was to have gone down last monday to bespeak Quarters for you at Gravesend, but Something has prevented his going till this day and perhaps he is not yet gone which is another disappointment that the Society could not foresee and necessarily leaves the affair to your Discretion. I am
Sir   your most Humble Servant   Henry Newman
P.S. Mr. Symonds at the same time tells me he has spoke to Mr. Chaille the owner of the 2 Brothers and that he is content that Capt. Thompson bring the Saltzburgers up the River to embark in the Prince of Wales for the Convenience of the Saltzburgers being a well Wisher to them.

Pray give my humble Service to Mr. Mayor and excuse the trouble given to him on this occasion.

Bartlet's Buildings 18th October 1734.   To Mr. Symonds in Nicholas Lane London.

Sir: I shall take the first Opportunity to acquaint the Society with what you propose and believe for the reasons you give that it will be complied with, if they are not gone too far in preparations at Gravesend to Countermand the Saltzburgers landing there. I am
Sir   Your most humble Servant   Henry Newman

To Mr. Vat.

Dear Sir: Upon the Reputation of Mr. Symonds the Society are of opinion that it will be more for the Comfort of the Saltzburgers to come up to the Red House instead of Stopping at Gravesend where they may embark on the prince of Wales in half the trouble and expence they will at Gravesend and give the pleasure to many thousands of Seeing them that can't See them So far from London as Gravesend.

Pray give my humble Service to Mr. Mayor of Gravesend and the Reverend Mr. Harris, and defray any Expence they have been at in bespeaking Quarters for the Saltzburgers or those they have troubled on that Account.
I am   Sir   Your most humble Servant   Henry Newman
If you want money draw on William Tillard Esq. in London and your Bills will be duly paid.

Bartlet's Buildings 18th October 1734. To William Haffenden Esq. Mayor of the Corporation at Gravesend.

Sir: By a Letter of last Tuesday from Mr. Vat at Rotterdam the Saltzburgers were then all on bord the two brothers[.] Captain Thompson who commands the Ship promised to Sail next morning wind permitting, so that I believe it will not be long before they arrive at Gravesend. I am now by order to acquaint you that upon a Representation from Mr. Oglethorpe and other Gentlemen of the Trustees for Georgia that it will be more convenient for the Saltzburgers to embark with their baggage on the prince of Wales directly at the Red House above Greenwich without landing them, [than?] to Set 'em ashore at Gravesend where their Baggage is to be housed by the Custom House officers, and then reshipped with some trouble to the Indian King and his attendants as well as to themselves, and the Ship's Crew, beside Expence. The Gentlemen of the Society have therefore agreed that the two brothers bring them up to the Red house to embark there, and for this purpose I have wrote this day to Mr. Vat by the Gravesend Tilt-Boat Samuel Gladwell, desiring him Mr. Vat to defray any expence you or the Reverend Mr. Harris have been at for their reception at Gravesend, and to excuse the trouble that has been given you on this occasion, especially Since it will give many Charitable Gentlemen who have Contributed to their relief the opportunity of Seeing them which they could not have done at Gravesend. Your favour in dispatching them up the River as soon as they arrive will oblige the Society as well as the Trustees for Georgia[.] Please to Communicate this Letter with my humble Service to Mr. Harris and believe that I am

Sir   Your most Humble Servant   Henry Newman

London 25 October 1734.   To the Reverend Mr. Lowther & Mr. Wolters the King's Agent at Rotterdam.

Gentlemen: Upon the representations you have made of the kind readiness of the Honourable Magistrates of Rotterdam to admit the Saltzburgers to lie before their town in their way hither, in order to proceed to Georgia, I am ordered to desire you would give the thanks of the Society for promoting Christian Knowledge in the most respectfull manner to their Honours for that instance of their Charity to our persecuted Brethren.

They arrived last Tuesday in the Thames in good health with Mr. Vat their Conductor, who has made likewise a report of your great Civilities to them for which I am ordered to desire also you would accept their thanks of the Society.

They are to embark to morrow in the prince of Wales and by a Special Warrant from My Lord Mayor permitted to land next Sunday to walk in procession to the German Church in Trinity Lane to return God thanks for their Safe Arrival in England, and after divine Service is over they are to be entertained at Dinner, the City Marshall and his officers being ordered to attend them to prevent any disorders.

I am Gentlemen   Your most obedient Humble Servant
Henry Newman

Bartlet's Buildings London 29 October 1734.   To the Reverend Messrs. Bolzius and Gronau at Ebenezer in Georgia.

Reverend Sirs: My Letter of the 1st and 3 of August last acknowledged the receipt of your Letter of the 22nd March and acquainted you with the Contents of Mr. [Ph.] Von Reck since which the Society have had the Satisfaction of perusing Extracts of your Journals communicated to the Society by Mr. Ziegenhagen the Continuation of which accounts will be always acceptable to your friends here.

The advices which you and Mr. Von Reck have conveyed to your friends in Germany have prevailed with another Transport of Saltzburgers Emigrants consisting of 56 men women and Children to follow their Brethren already in Georgia and if it please God to grant them a Safe passage as they are mostly young people and all well disposed persons their arrival at Ebenezer will be a great Comfort and addition to the Strength of the 1st Planters.

The Society have taken such effectual Care in the Choice of a Ship furnished with all proper necessaries & accommodations commanded by Capt. Dunbar a Gentleman recommended for his good Humanity and Experience by Mr. Oglethorpe that they hope there will be no reason given for any of the Complaints which were but too justly made against the conduct of the Captain of the Purysburg who as I mentioned in my last is discharged from ever Serving the Trustees for Georgia or the Society in the like Capacity again.

The opportunities of making remittances to Georgia occurring but Seldom the Society have ordered a years Salary from the

1st May 1734 to 1st May 1735 to be sent over in the Prince of Wales Commanded by Capt. Dunbar viz.

To the Reverend Mr. Bolzius in Sterling money	50.—
To Mr. Gronau	30.—
To Mr. Ortmann Schoolmaster	10.—
	90.—

This Sum they have been advised to send over partly in Bills of Exchange and the rest in pieces of 8 and Copper halfpence in the following proportions (till they can hear from you which is most for your advantage viz.

	£
In a Bill of Exchange drawn by Mr. Simons on Mr. Montaguet [Montagut] Merchant at Purysburg one of which Bills is here inclosed for	50.—.—
In pieces of Eight consisting of 140 pieces and 145 pieces in one Bag Weight 134 oz. 14 dwt. 5 s 2-3/8 d per Ounce	35.—.—
In halfpence and farthings 5 £ 1 s 6 d cost	5.—.—
	90.—.—

You will be pleased to accept in good part these allowances which the Society have made and wish they were in a Condition to do better, but they are the same that are made to the Ministers of the English Church in the Town of Savannah and there is no doubt you will have all the encouragement from the honourable the Trustees for Georgia that is in their power by Setting out of Glebe Lands and otherwise; And by this Ship they have sent to you Mr. Bolzius a young Lad [Henry Bishop] who is bound to serve you for Seven Years and who having been brought up in a Charity School here under Mr. Akers bred at Hall we hope will be usefull to you many ways and especially in writing for you in the English Tongue and if you find him towardly you will have it in your Power to fit him for preaching hereafter the Gospel among the Indians and they hope in a short time to Send such another to Mr. Gronau.

The Society have ordered a large parcel of English Books to be sent to your fellow Labourer the Reverend Mr. Quincy at Savannah with a desire that you may be furnished out of them with such a number as you may have occasion for.

In the Ship Prince of Wales are likewise sent a Box of German Books for the use of your Congregation of which you will be more particularly advised by Mr. Ziegenhagen.

The Medicines and other Things desired by Mr. Zwiffler are sent as per Invoice in No. 3, 4, 5, 6, 7, 8, 9 and to this is added as a gratuity and encouragement to him from the Society the Value of 10 £ in pieces of 8 in a Bagg put up in No. 1 containing 44 Spanish Dollars and 1 Ryal Weighing 38 oz. 10 dwt at 5/2 3/8 per ounce in regard to his Service as as Surgeon and apothecary to the Saltzburgers at Ebenezer.

The particulars desired by you of 3 pair of Tongs 3 Shovels Some Sugar Soap white Starch and travelling Beds having been signified to the Society by Mr. Ziegenhagen one of the Trustees for Georgia Mr. Vernon being then present promised to recommend their being allowed out of their Stores at Savannah or sent in the prince of Wales.

Thus having acquainted you with the orders of the Society I take leave to add my own hearty wishes that it may please God to bless you long with health and Success in all your Labours for the Spiritual and Temporal benefit of the Flock under your Care and at the same time to Congratulate you on your and their happiness in being attended with the good Success of Mr. John Vat the conductor of the Present Transport to Georgia with which wishes I remain

Reverend Sirs   Your most obedient humble Servant
Henry Newman

Mr. [Ph.] Von Reck is well at Augsburg or Ratisbonne and seems designed God willing to See you next Summer.

[Invoice enclosed in this letter follows.]

London 21 October 1734

Invoice of Goods Shipped by order of the Society for Promoting Christian Knowledge at London on the Prince of Wales Capt. Dunbar Commander and consigned to Messieurs Bolzius and Gronau at Ebenezer in Georgia. marked and numbered as follows viz.

B.G.   No. 1. A Box Containing 173 oz. 4 dwt in Spanish
at Ebenezer.   pieces of 8 and the Value of 5 £ in Copper
                    half pence and Farthings.

H.P.B. B.G.   2. A Box of Books Weight 1 cwt. from Germany.

B.G.   3. A Worm Tub with Some parts of a Small Still
                  for Mr. Zwiffler the Apothecary at Ebenezer
                  in Georgia.

              4. A Hogshead containing other parts of the Still.
                  4 Sieves

2 Small Square pieces of marble and
1 Small marble mortar to prepare medicines.

5. A Trunk of medicines as per List herewith sent.
6. A Box containing

Iron Bars and Grate for the Still in which
there is a piece of Canvas that came over as
a Cover to the Box of Books from Germany.

7. An Herb Press.

8.⎫
   ⎬ Two Hampers containing
9.⎭

Chymical Glasses with Two tin plates to line
the Cheeks of the Press
1 Shagreen Box with appurtenances for Cup-
ping
1 Doz. Glister pipes and
1 Doz. Bladders

To Mr. Quincy at Savannah.

10. A Box of Books weight 1 cwt of which Mr.
Quincy is desired to furnish Messrs. Bolzius
and Gronau with what they may have occasion for.

NB. 1. Spring Fleam (or Snapper) with
its appurtenances was delivered to Mr.
Simonds directed to Mr. Zwiffler to be
sent by Capt. Dunbar in his Baggage
after packing up the 10 parcels abovementioned.

The 173 oz. 4 dwt of Silver above-
mentioned includes for Messrs.
Bolzius Gronau and Ortmann
                   134 oz. 14 dwt &
For Mr. Zwiffler   38 oz. 10 dwt
                   173 oz. 4 dwt

By order of the Society   Henry Newman Secretary

Estimate for Fifty heads of freight whereof 35 men able to bear
Arms and for all necessaries with them and for their Maintenance
12 months in Georgia.

	£ s d
For Iron ware containing working Tools and necessarys already bought	118.14—
For 35 Musquets and Bayonets Tower Proof at 9 s 3 d ea.	16. 3.9
For 5000 Flints 1 £ 5 s 9 yards brass wire to fit the Touch holes 2 s 6 d Two Steel Turn Vices 1 s Two Scowering Rods 2 s 1-1/2 Cases for the arms 9 s	1.19.7-1/2
For 7 cwt of Dropshott at 1 4/6 and 7 cwt Bullets at 16 s in 28 Barrels at 4 d each with 20 s for Custom and Charges	12. 6.4
For Seven Barrels of Gunpowder at 3 £ each	21. —. —
For 35 powder Flasks at 8 d ea. and 35 pouches at 12 d each	2.18.4
For a ream of Cartridge paper	—.10.—
For 26 Sutes of large Bedding at 23 s and 35 Watch Coats at 8.6 each	44.15.6
For 300 Yards of Sail Cloth or old Duck at 6 d a yard	7.10.—
For Turners Ware	3.19.6
For Potters Ware	1. 3.6
For Haberdashery Ware	2. 2.6
For Tin ware 10/5 and 3 Doz. Tin Lamps at 8/6 a Doz.	1.15.11
For 120 Gallons rape oyl for the Lamps	9.19.—
For 50 pounds of Spun Cotton for Ditto	4.2
For 36 pair of mens Shoes at 30 s a doz. 12 pair of womens at 25 s a doz. 12 pair of Boys at 22 s and 12 pair of Girls at 16 s half of each price	7.13—
For Fishing hooks and Lines	2. —.—
For 38-1/2 cwt of Beef at 20 s per cwt	38.10.—
For 18 Bushels of Salt to Salt it at s 4/2 a Bushel 3 £ 15 s – d for Salting 6 Oxen 9 s	4. 4.—
For 40 cwt of Biskitt in ten hogsheads at 10/6 per cwt	21. —.—
For 6 Bushels of Greets at 5/6 a Bushel	1.13.—
For 10 Casks for the Beef at 12/6 ea. 10 for the Bread at 7 s ea. & 2 for the Greets at 12 s 6 d ea.	10. —.—

330. 2.1-1/2

Brought Over_____£ 330. 2.1-1/2

For 28 or 30 Cheshire Cheeses containing 10 cwt 28 s per cwt Casking 1 £ .... 15. —.—

For 10 Firkins of Butter at 24 s ea. binding and pickle 10 s .... 12.10.—

For Store on board 3 cwt Molasses at 9 s per cwt 1 cwt Sugar 40 s .... 3. 7.—

For 7 pounds of Cinnamon at 8 s a pound & 7 pounds of black Pepper at 1 s 6 d pound 3. 6.6

For Ten Gallons Molasses Spirits at 2 s a Gallon in ten Stone Bottles at 12 d each .... 1.10.—

For Store on board Garden Stuff for Refreshments...... 3.13.—

For 4 half Hogsheads of Vinegar .... 4.4.—

For 8 Ditto for Georgia .... 8.8.—   12.12.—

For 2 Barrels and 1/2 of Crown Soap at 4.16.— a Barrel of which drawback to be deducted 9. —.—

for medicines .... 7.17.—

For 8 Doz. of Candles and package .... 2. 7.—

For Charges of Shipping about .... 3. 3.—

For 8 doz. of Knives at 2.16 a doz. .... 1. —.—

For Some Grain and Garden Seeds .... 6. —.—

For 5 Tons of Strong beer at 6 £ a Ton in 20 hogsheads at 10 s each .... 40. —.—

For Six horses and 6 Oxen in Georgia .... 42. —.—

For 35 Cows with Calf two Bulls, 35 Sows and two Boars in Georgia .... 103.12.—

For Poultry 40 Turkeys and 40 Fowls 3. 6.8

For freight and provision for 50 Heads from Rotterdam to Gravesend at 7/6 per head .... 18.15.—

present money   619. 1.3-1/2

For freight of 50 heads from Gravesend to Georgia at 5 £ each to be paid when Charter party is fulfilled .... 250. —.—

For nine months provision in Georgia more than carried over, for 50 heads of 4 £ 10 s ea. to be paid when fully Supplied .... 225. —.—

Total........ 1094. 1.3-1/2

Bartlet's Buildings London 29 October 1734. To the Reverend Mr. Quincy at Savannah in Georgia.

Reverend Sir: My Last was of the 1st of August Since which I have none from you but had the pleasure of Seeing one you wrote to Mr. Oglethorpe, by which I was glad to find you had recovered your health in England and was returning to Georgia where I hope this will find you in good health.

You will See by the Invoice enclosed what the Society have sent to your Self and the missionaries at Ebenezer by the Prince of Wales and when any Supplies of the like kind are wanting which are in the power of the Society to transmit to you or them you will please to Signify it.

Mr. John Vat the worthy Conductor of the present Transport of Saltzburgers is the Bearer of this and will I hope find Georgia as agreeable as Mr. [Ph.] Von Reck did or rather more agreeable considering the improvements which are daily accrewing to it by the Labour and Industry of the Inhabitants. I hope it may prove a happy Asylum to So many distressed people of our own and other nations by the many friends Providence continually raises up for the encouragement and Support of it. You will See by the enclosed printed Accounts how the Society have been obliged to exert themselves in favour of the Transport of Saltzburgers now Sent, and the blessing of God has Signally accompanied their Endeavours So that if the persecutions in Saltzburg are continued and more are permitted to follow these You may go nigh to See another Transport next Summer under the Conduct of Mr. Von Reck who is now with his friends in Germany and does not want [lack] inclinations to return to Georgia.

The Letter herewith Sent was left with me by Mr. Copping who has promised to Write to you by these Ships. The book entitled *plain reasons for being a Christian* is not a Society book but a present to you from Mr. Edwin Belke an honest plain man who has been a great Benefactor to the Society and hearing that you lived among Heathens in America he imagined the Subject of it might recommend it to your perusal, I don't know who is the author but what I have read of it, he Seems a hearty Friend to the Cause he recommends.

May you long enjoy your health and continue to animate the Foundation of a new British province with all those Christian Virtues which may render it in time the Example and Glory of our Plantations is the wish of

Reverend and Dear Sir   Your most obedient humble Servant
H. Newman

*Bartlet's Buildings 30 October 1734. To the Reverend Dr.
Guerdes in New Broad Street London.*

Reverend Sir: The Society for Promoting Christian Knowledge
being informed of the great Zeal your Self and the Consistory of
the German Church in Trinity Lane have shewn for exciting the
charitable Benevolence of the Congregation there to the persecuted Saltzburgers when they attended divine Service in that
Church last Sunday, and of the kind entertainment given to them
at Dinner the Same day besides distributing a large Bounty to
every Saltzburger that attended, I am ordered to desire you would
accept the Thanks of the Society and that You would make the
Same acceptable to the Consistory of that Church for your and
their great Charity on this occasion, and remain
   Reverend Sir Your most humble Servant Henry Newman

*A Son excellence Mons. [J.] De Reck Conseiller de Sa Majeste
Britannique et Son Envoy a la Diete de L'Empire a Ratisbonne.*

Sir: Having by last Post received a Letter from your Nephew
which required an immediate answer, I take the Liberty of covering the answer to your Excellency that it may be sure of a conveyance as Soon as possible to his hands and under a flying Seal
that you may be acquainted with the Contents desiring you will
please to Seal it when you give it a proper address from
   Sir Your Excellency's most obedient humble Servant
            Henry Newman

London 5 November 1734

*London 1 November 1734. To the Reverend Mr. Urlsperger at
Augsburg.*

Reverend Sir: I received your favours of the 4th of [*sic*] 11, 18, &
21st of October N.S. upon reading which having read no particular commands from the Society, I delayed writing till I could acquaint You with the Safe arrival of the Saltzburgers and Mr. Vat
their Conductor at London[.] Mr. Degmair whose Letter I now
cover informs me that he has from time to time acquainted You
with the progress of this Transport to the time of their leaving
London so that I need only refer You to his Letters for information of those particulars which you might desire to know, and the

Same Apology Mr. Vat when he left London the 29th of October in a great hurry to follow the Ship to Gravesend desired me to make to you for his Silence Since he left Rotterdam praying that you would accept Mr. Degmaires account as if written by Mr. Vat himself. Give me leave only in General to Say that all the Members of the Society who had the Pleasure of Seeing the Saltzburgers of this Transport either on Shore or on bord the Ship were exceedingly pleased with their Modest Looks and Behaviour as You will find in Some Measure expressed in the Accounts just now out of the Press herewith Sent.

Mr. Vat in his hurry at departing had only time to desire me to Send his humble Service and hearty Thanks to Mr. Senior Urlsperger Burgomaster Morel, and to the Reverend Clergy of Augsburg to Mr. Von Münch of Augsburg and Messrs. Munch of Franckfort for their great Civilities to the Saltzburgers and himself.

I am likewise desired by him to Convey to you a Box marked S.U. containing 6 Pewter Dishes and 12 Plates weighing 38½ pounds as a present to Mr. Urlsperger, these are to be Conveyed to you by Mr. John Herman Zur-hurst Merchant in this City as soon as I can meet with him. Mr. Ziegenhagen went the day before Yesterday to Gravesend and returned last night exceedingly pleased with the Behaviour of the Saltzburgers not having had an opportunity to See them before by reason of his obligations to attend at Court or at Kensington. One of the Saltzburg Women being brought to bed of a Daughter in her way from London to Gravesend,[39] Mr. Ziegenhagen Christened it. I have a packet of Books for you which the Society desire Your Acceptance of being ordered Sometime before upon a Letter from Mr. Vat but the Sending them away has been delayed partly for a New Edition of Dr. Hales's Vegetable Staticks which is now out, and partly for a Map of Georgia, which Mr. Oglethorpe has promised to Send to me as Soon as compleated but is not yet come to my hands.

The Society received by Mr. Vat your kind Present of the History of the Saltzburg Emigrants in High Dutch neatly bound for which I am ordered to Send their Thanks to you and beg at the Same time You & Your Lady would accept of my most thankfull acknowledgements for all your Civilities to Mr. Vat during his residence at Augsburg. If Mr. [Ph.] De Reck be with you pray give my humble Service to him and be assured that I am

Reverend and Dear Sir   Your etc.   H.N.

A Son Excellence Monsieur [J.] De Reck Conseiller de sa Majeste Britanique et Son Envoy a la Diete de l'Empire a Ratisbonne. London 5th November 1734.

Sir: Having by last Post received a letter from your Nephew [Ph. von Reck] which required an immediate answer I take the Liberty of Coverning [covering] the answer to your Excellency that it may be sure of a Conveyance as soon as possible to his hands And I send it under a flying Seal that you may be acquainted with the Contents desiring you will be pleased to seal it when you give it a proper address from   Sir   Your Excellency's most obedient
humble Servant   H.N.

Bartlet's Building 5 November 1734.   A Monsieur Monsieur [Ph.] de Reck

Dear Sir: Mr. Vernon this day laid before the Society your letter of the 7th of October V.S. in answer to which I am ordered to acquaint you that what the Society have collected for relieving the Protestant Saltzburgers is appropriated only to them and cannot be applied to the relief of any other whatsoever and the Charges of the late Transport under the Conduct of Mr. Vat requires more money than they have at present in hand, and therefore the Society are in great hopes this letter may come time enough to prevent going on with the Project mentioned in your letter to Mr. Vernon, and desire that you will never engage them in any matter that you don't previously acquaint them with.

The Saltzburgers to the number of 56 under the Conduct of Mr. Vat came up to London and by their Behaviour have gained the Esteem of all who could have the happiness of seeing them; They sailed for Georgia last Friday the 1st Current from Gravesend and tis hoped are in good way of getting clear of the Coast; Mr. Vat is gone with them.

I received your Letter of the 7th October V.S. by last post and was surprized not to find Mr. Vernons letter enclosed to which you refer me, but that is since explained by his having received his letter. I pray God to direct you in all your undertakings, for his Glory and remain
Dear Sir   Your most obedient humble Servant   H. N.
P.S. The Original of this Letter is sent under Cover to His Excellency your good Uncle at Ratisbonne and this Duplicate under Cover to Mr. Wolters for fear the Original by some accident should be delayed going to your hands.

London 5 November 1734.   To Mr. Wolters Agent for his Britanick Majesty at Rotterdam.

Sir: The inclosed letter to Mr. [Ph.] Von Reck is to prevent his bringing over hither a Troop of people from Bohemia and Moravia and therefore you are desired to forward this Letter as soon as you know how to send it or keep it till you do know with certainty to convey it to his hands.

Mr. Vat with the Saltzburgers under his Care sailed last Friday the 1st Current in the Prince of Wales from Gravesend and tis hoped are in a good [way] of getting clear of the Coast.

He particularly desired me at his departure to give his humble Service and Thanks to you and the Reverend Mr. Lowther and to Messrs. Courtonne for all your and their great Civilities to the Saltzburgers and himself while at Roterdam which you will please to acquaint them with assuring them of my humble Respects from
Sir   Your most humble Servant   H. N.
P.S. Your letter of 5th Current N.S. to Mr. Vat came in good time to prevent my writing to you about paying for the Beer delivered to the Saltzburgers on board the 2 Brothers.

Georgia Office Westminster the 6th November 1734.   A Monsieur Monsieur Philip George de Reck a Ratisbonne.

Sir: Mr. Vernon having this day laid before the Board Contents of your letter dated from the Fronteirs of Bohemia the 7th of October, being the first meeting after receiving it, the Trustees were very much surprized at the Contents of it having had no previous Notice of your Intentions of bringing any Persons from Bohemia, and as the Trustees are at present in no Condition to contribute any thing to the sending over either them or any other Persons to Georgia they desire you will immediately put an absolute stop to your Proceedings.
I am Sir   Your most humble Servant   Harman Verelst,
Accountant
P.S. Our friends the Saltzburgers sailed last friday from Gravesend with the Indians for Georgia.

Bartlet's Buildings London 26 November 1734.   To the Reverend Mr. Urlsperger at Augsburg in Suabia.

Reverend and Dear Sir: My last was of the 1st Current acknowledging the Receipt of your favours of the 4th, 11th, 18th and 21st of October N.S. Since which I have received your kind favours of the 8th Current and another without date enclosing an account of the Persecution in Carinthia etc. dated the 7th September which have been translated and laid before the Society who heartily wish it were in their power to remove the Oppressions so justly complained of there and in Bohemia.

Mr. [Ph.] Von Reck signifys that he expects to conduct another Transport of Saltzburgers in the Spring to Georgia, if that be found practicable, the Society will use all their Endeavours to accomodate them as the last were, with every thing in their power.

The Transport under Mr. Vats Conduct sailed from the Downs the 8th Current O.S. with a fair Wind which lasted at least ten days so that I hope by this time they are well advanced in their Voyage.

I this Week received a letter for Mr. Degmair which by the Writing I suppose was from you, I immediately forwarded it to him, and am glad to acquaint you that he is well and daily improves in acquiring the English Tongue, as indeed I find all Germans do with a facility beyond any other nation that comes among us.

Having begun my Letter on the wrong side of the Paper before I discovered my mistake, you will excuse me if to save the time of transcribing it, I for once desire you to read my letter in the Hebraic manner. The Cargo for our Missionaries in E. India is now the Subject of the Society's Consideration, which will for some time engage the Attention of

Reverend & Dear Sir   Your etc.   H. N.

P.S.: I had a letter lately from Mr. [Ph.] Von Reck at Windhausen[40] which I should acknowledge the receipt of but he does not inform me how to direct to him; Mr. Vernon had another.

Bartlet's Buildings London 3 December 1734.   To Mr. [Ph.] Von Reck at Windhausen.[40]

Sir: Your letters of the 1st of November O.S. dated at Windhausen to Mr. Vernon and me came safe to hand and the Contents of your letter to Mr. Vernon having been laid before the Trustees of Georgia and the Society here, they are glad to find you had received the Letters wrote by order of the Trustees and Society to prevent your bringing hither this Winter the People you mentioned from Moravia and Bohemia, an Undertaking which might have brought upon the aforesaid Gentlemen insuperable Difficulties it being impossible in their present Circumstances to have known what to do with them much less to have sent them forward to Georgia.

The Society had no expectation of your return this winter but of your continuing in Germany in order to conduct a third Transport of Saltzburgers in case any of them had offered themselves by next Spring, by which time the Society are not without hopes of being in a Condition to send a few more of those people (Saltzburgers) for whom only they have made their Collection; but tis with great Concern the Society observe that you are going upon an enterprize, all the motives of which they are not apprized of, and of the Event [outcome] of which they are very dubious.

Your telling oppressed People that God hath made choice of the English Nation to deliver them, and that the day of their deliverance will 'ere long appear, is to raise Expectations of which your Friends here see no foundation but which may be attended with Consequences very dangerous to your Self, and not less fatal to the people whom you are endeavouring to serve.

The Society suppose that you build your expectations on the Willingness of the Trustees to give Lands in Georgia to persecuted Protestants but then you seem to forget the Consideration of the great Charge of carrying them thither, and of maintaining them till such time as they may be in a Condition of maintaining themselves.

Your proposal of bringing over to Georgia Manufacturers in Glass and Earthen Ware has been laid before the Trustees who do not think it proper to encourage Manufactures in Georgia which may interfere with those of Great Britain.

Ortmann has a liberty of continuing in Georgia if he has a mind to it.

Mr. Vat with the late Transport of Saltzburgers in the Prince of Wales commanded by Capt. Dunbar sailed from the Downs the Eighth of November O.S. with a fair wind which lasted about ten days so that it is hoped by this time they are above halfway over the Sea. Tomo Chachi etc. went with them.

I sent some account of the good Behaviour of the Saltzburgers to Mr. Urlsperger which I doubt not he acquainted you with.

The Society have sent to the Press an Extract of your and Mr. Bolzius's Journal[41] to and at Georgia for the Satisfaction of those who have been or may hereafter be Benefactors to the Saltzburgers settled there, a Copy of which will be sent you as soon as published.

I sent some time since a Map of Georgia drawn out by order of

Mr. Oglethorpe for you, it was recommended to the Care of Mr. Reiche, who sent it by the way of Hanover, but you don't take any notice of its being ever come to your hands.

Dear Sir: I remember with pleasure your agreeable Conversation when in England and when Providence shall favour me with a return of that pleasure in a way consistent with your own Comfort and the Satisfaction of our Friends as above signified it will be a happiness that will be very esteemed by

Sir   Your most obedient humble Servant   H. N.

Bartlet's Buildings London 13 November 1734.   To the Reverend Messrs. Bolzius and Gronau at Ebenezer.

P.S. To a Letter dated 29 October 1734.

Reverend Sirs: Above is a Copy of my last by Mr. Vat who sailed with the second Transport of Saltzburgers in the Prince of Wales Capt. Dunbar Commander the 8th Current from the Downs, and as they have had a fair wind ever Since tis hoped they are before this time clear of the Channel. I sent the 1st Bill of Exchange drawn by Mr. Simond as above for 50 £ in my former letter, now I enclose the 2nd for the Same Sum hoping they will both go safe to your hands and meet with do honour of which please to advise by the first opportunity

I am   Your most & etc.   H. N.

Postscript to Mr. Quincy's Letter of the 29th October 1734.   London 13 November 1734.

Reverend & Dear Sir: Mr. Vat sailed from the Downs the 8th Current with a fair wind which has continued ever since and I hope he will be safe with you e're this reaches your hand, of which I shall be glad to hear, and that you enjoy your health. I am

Reverend Sir   Your most humble Servant   H. N.

Mr. Wilson wrote a long letter to you some time ago which he should be glad to know whether it came to your hands.

Bartlet's Buildings 3 December 1734.   To the Reverend Mr. Urlsperger at Augsburg.

Reverend & Dear Sir: My last was of the 26 November acknowledging the Receipt of all your letters unanswered which had come to my hands. I am now by order of the Society to send open to you their answer to Mr. [Ph.] Von Reck's letter of the 1st of last month to Mr. Vernon, because they think it proper that you should be acquainted with their Sentiments in an Affair of so much importance and hope they will be seconded to Mr. Von Reck by a Person of your piety, Experience & Great Prudence. The Society are sensible of the commendable Piety and Zeal of Mr. Von Reck, but apprehend those laudable qualities may carry him too far, for want of Experience in the World.

How far it may be proper to communicate this to the Baron [J.] Von Reck, the Society leave to your Judgment.

Be pleased to signify in your next the Opinion you have of the people our friend Mr. Von Reck is so sollicitous for, what they are and whence they came. I am

Reverend Sir   Your most humble Servant   H. N.

P.S.: I have herewith sent a Copy of Mr. Von Recks Letter to Mr. Vernon that you may the better judge of the answer which is sent to it under a flying Seal that you may seal & direct to him when you have perused it.

Bartlet's Buildings 2 January 1734/5.   To the Reverend Mr. Urlsperger at Augsburg. Under Cover to Baron [J.] Von Reck.

Reverend Sir: My last of the 3rd December acknowledged the Receipt of your former letters. I am now favoured with your letters of 13 and 20th of last month N.S. Translations of both which have been laid before the Society, and they are glad to understand by your former that Mr. [Ph.] Von Reck was timely prevented by your advice from engaging him in a Project that must have been attended with great distress to himself and his friends if it had been prosecuted.

The Narrative you published in Germany of the Saltzburgers here the Society hope will be attended with good Effects on those who peruse it both Saltzburgers and others[.] The Society have published here Extracts of the Journals of Messrs. Von Reck and Bolzius to and at Georgia a Copy of which I herewith send you, because you may perhaps have occasion to make some extracts of it known in Germany. I shall send to you by Mr. Zurhorst via Hamborough [Hamburg] a packet of them to be distributed to His Excellency Baron [J.] Von Reck at Ratisbonne, his worthy Nephew [Ph. Von Reck] to whom we are obliged for great part of the Contents of it, and to such other of the friends to the Persecuted Saltzburgers as you may have opportunity to present them to.

The Society are glad to hear that the 2 Bohemian Brethren have so far succeeded at Ratisbonne as to have audience from all the Protestant Embassadors there and that the Compassion of their friends have prevailed on them to prepare Memorials to the Evangelick Body, and to the Courts of England and Holland in their favour a Copy of which Memorial to the Evangelick Body presuming it will be there first presented, the Society would be glad to see as soon as you can procure it, with the names of those who subscribe it.

The Insolence of the Roman Clergy here is surprizing, by the number of their private Mass Houses which we hear of daily, and if this be owing to too great an indulgence in the Conduct of our Magistrates in not putting the Laws in Execution, I hope the Parliament who are soon to meet will take Cognizance of them and animate them at least to put the laws in execution, which of themselves are very indulgent, compared with the rigorous Persecutions the Romanists every where use against the Protestants.

In the mean time My Lords, the Bishops, like good Pastors have exhorted their Clergy to use their diligence in guarding their Flock against the deceitfull Insinuations of the Roman Emissaries, and the Bishop of Londons excellent Letter on that occasion, herewith sent, the notice of which in the manner you shall use it may be a Consolation to our friends in your parts.

Pray give my humble Service to Mr. Commissary [Ph.] Von Reck when you see or write to him and let him know the Contents of this, and that I have received his favour of 18 December N.S. from [omitted] which not requiring an immediate answer I defer acknowledging till I can send him or you the Mapps of Georgia you have so often desired.

Mr. Degmair tells me he had fully answered all your Letters before that to me wherein you complained of his Silence was come to hand.

I cannot close my letter without desiring you and Mr. [Ph.] Von Reck to accept my hearty wishes that God may make the ensuing year and many more of them happy and successful to you Both for advancing his Glory and give me leave to assure you these are the Wishes of the Society as well as of

Reverend Sir   Yours etc.   H. N.

P.S.: Mr. Vernon & Mr. Oglethorpe are well and always pertake of the advices you send. The Box of Pewter from Mr. Vat formerly mentioned as a present to your Lady is markt S.U. No. 1 and a packet of Books as a present from our Society markt S.U. No. 2

directed to Monsieur De Münch at Augsburg are delivered to Mr. Zurhorst to be forward by way of Hamborough [Hamburg].

London 30me de Janvier 1734/5
A Monsieur Monsieur Philip Geo. Fred. De Reck á Augsbourg.

Monsieur: Je suis bien aise d'apprendre par votre Lettre de 13 de ce mois que vous etes arrivè en bonne Santè á Augsburg où votre presence Sera de grande Utilite, en cas que nous ayons une nouvelle Emigration hors de l'Archevechè de Saltzbourg comme vous paroissez vous y attendre. Mr. Newman ecrit par cet ordinaire a Monsieur Urlsperger Sur ce Sujet par l'ordre de la Sociètè Qui Souhaite que vous n'entrainer aucun Entreprize sur leur Conte, qu'apres avoir Consulti le dit Reverend Pasteur qui est leur ancien Correspondent & Membre de leur Sociètè, & à qui de tout tems ils ont Confiè la direction de leur Affaires en Allemagne.

L'Affaire des Moraviens a pris un autre plis. Le Sieur Spangenberg etant passé au frais du Conte [Zinzendorf] en [est?] partie, pour preparer un Etablissement pour les bons Gens dans la Georgie, Le Sejour de Hernhutt, devenant trop etroit pour le nombre qui arrive de Boheme & de Moravie pour jouir de libertè de leur Conscience.

Nous avons reçu une Relation envoyè par le Sieur Urlsperger des Souffrances dès Bohemiens, mais Comme il ecrit en Allemand, et dans un Caractere que nos Gens ne sont point accoutumè de Lire,[42] il faut le traduire et Cela cause du Delay, mais on y fera les reflections necessaires qui Seront Communiquè au Sr. Urlsperger la Semaine qui vient.

J'espere que vous aurez vu la Carte de la Georgie Avis nous etant Venu quelle etoit arrivèe a Ratisbonne.

J'ai Communiquè votre Lettre á Messrs. les Trustees qui Seront bien aise de vous revoir ici, à la tête d'un nouveau Transport de Saltzburgeois, à qui Ils Seront pret à donner des Terres aupres de leurs Confreres qui sont dejà dans la Georgie. Je Suis Monsieur
       Votre tres humble & tres Obeissant Serviteur   Ja. Vernon
je vous prie de Saluer Monsieur Urlsperger bien Affecteusement de ma part.

London, 30 January. 1734-5. To Mr. Philip George Frederick de Reck at Augsburg [above letter translated by the editor].

Sir:

I am very glad to learn by your letter of the 13th of this month that you have arrived in good health at Augsburg, where your presence will be of great usefulness in case we have a new emigration from the Archbishopric of Salzburg as you seem to expect. Mr. Newman is writing by this post to Mr. Urlsperger on this subject by the order of the Society, who wish you not to begin any undertaking on their account without having consulted the said Reverend Pastor, who has long been their Correspondent and a member of their Society to whom they have always entrusted the direction of their affairs in Germany.

The affair of the Moravians has taken another turn. Mr. Spangenberg has departed, at the expense of Count Zinzendorf, to prepare an establishment for the good people in Georgia, the abode at Herrnhut having become too small for the number that are arriving from Bohemia and Moravia to enjoy their freedom of conscience.

We have received an account sent by Mr. Urlsperger of the sufferings of the Bohemians; but, since it is written in German in a script that our people are not accustomed to read, it must be translated and that causes delay, but we will give it the necessary consideration which will be communicated to Mr. Urlsperger next week.

I hope that you will have seen the map of Georgia, word having come that it has arrived at Ratisbonne.

I have communicated your letter to the Lord Trustees, who will be glad to see you here again at the head of a new transport of Salzburgers, to whom they will be ready to give lands near their confreres who are already in Georgia. I am, Sir

Your very humble and very Obedient servant   Ja. Vernon
Please give my most cordial greetings to Mr. Urlsperger.

Bartlett's Buildings London 21 January 1734/5.   To the Reverend Messrs. Bolzius and Granou [Gronau] at Ebenezer in Georgia.

Reverend Sirs: The Society and Trustees of Georgia hope by this time that you have had the happiness of seeing the Transport of your Countrymen under the Conduct of Mr. Vat which no doubt will be a great Comfort to you. This comes to you by the hands of Mr. Spangenberg who with 9 others have a settlement given them near You in Georgia.

The Society is glad to find that this Gentlemen is well known to you and they doubt not but you will receive him as a fellow Labourer in the same work of confirming and converting many to righteousness and the Knowledge of our Saviour Jesus Christ. His design is the same with yours, and you will be able to forward it, for by this time tis not doubted but you have made some progress in the Indian Language and the Society heartily recommend this Study to you, that in time by God's Blessing a Door may be opened to those that sit in darkness, that so you may be the happy Instruments of bringing many to the Knowledge of a Saviour and Redeemer.

The Society have published an Extract of your Journals, and hope you will send from time to time an exact account of your State and Condition, which will be a Satisfaction to your Benefactors here and be a means of sending more of your Countrymen to you. The Society wish you all manner of Success in your pious designs and I am in their name

Reverend Sirs   Your most obedient humble Servant   H. N.

Bartlet's Buildings London 21 January 1734/5.   To the Reverend Mr. Urlsperger at Augsburg.

Reverend Sir: The Society thankfully acknowledge the favour of your last Letter of the 30 of December and are pleased to hear that you joined in giving Mr. [Ph.] Von Reck those Precautions which we Sent under Cover to you and will always be highly necessary in undertakings of so nice [delicate] a nature. Besides the Trustees of Georgia as well as the Society must always have proper notice given them of an intended embarkation that they may consider the State of their cash.

If the Society shall hear from you that no more Emigrants from Saltzburg are expected They may perhaps with the consent of His Majesty and of their Benefactors extend their Care and Charity to the Relief of those pious Confessors, whom you Say in your last have been lately driven out of Carinthia, but till then they don't think themselves at Liberty. Mr. Spangenberg with nine other persons are arrived here and the Trustees have made them a Grant of Lands. Mr. Spangenberg has given the Trustees and Society the most Solemn assurances that Count Zinzendorff and his Herenhutischen Brethren agreed in Doctrinal Points with the Evangelick Body, So we hope upon further inquiry you will find that these People differ with the Lutherans only in point of Discipline, in which the *Fratres Bohemi* as we understand always did differ from the Lutheran Church.[43]

In relation to these peoples preaching the Gospel to the Indians Mr. Spangenberg has been told, and is Satisfied with it, that no

Body will be allowed to go among the Indians, without a Lycense obtained from the Trustees for Georgia who will enquire carefully into the Characters and Qualifications of those to whom they shall grant any, a precaution highly necessary for preventing Disturbances in the Province; And the Gentlemen of the Society & Trustees have talked very freely upon this head to Mr. Spangenberg and have also wrote to Messrs. Bolzius and Gronau to cultivate a good Correspondence with those people for promoting and enlarging the Kingdom of our Common Lord and Master. They will be Settled upon the River *Ogichie* [Ogeechee] a days Journey from the Saltzburgers.

The Society have not seen the larger account of your Correspondence in relation to the Count which you mention as Sent to Mr. Ziegenhagen So that they cannot be more particular at present. They Shall always be glad to be informed by you as to this or any further affair that relates to the persecution of our Protestant Brethren and will Study all means in their power to assist them under their Afflictions. I am in the Society's name

Reverend and Dear Sir   Your most humble Servant

Henry Newman

P.S. Reverend Sir: Since the above was written yours of the 13 Instant N.S. has been received and will be laid before the Trustees and Society at their first meeting and their respective answers to your Queries will be transmitted to you by the first opportunity. In the meantime they hope to receive a fuller Account of those Saltzburgers who intend to Emigrate as Soon as you have received it from Mr. [Ph.] Von Reck himself to whom I beg my humble Service may be made acceptable from

Reverend Sir   Yours as before   H. Newman

Bartlet's Buildings London 31 January 1734/5.   To the Reverend Mr. Urlsperger at Augsburg.

Reverend Sir: I have received both your Letters of the 13th and 24 of January Current N.S. and in answer to the Queries in the former the Society are of opinion that no Overtures of any kind be made to induce the Saltzburgers to leave their Country; but should they be driven out by Persecution purely on account of their Religion then you may offer them (when they are actually come out of the Territories of Saltzburg) the Same Terms as have been made with those Settled in Georgia not exceeding the number of 200.

The Society are at present of opinion that the allowance to be made to Mr. [Ph.] Von Reck for Alimentation money Should begin when any number of Saltzburg Emigrants not exceeding 200 shew themselves disposed to go to Georgia under his conduct and the Society desire in your next you would inform them what this Alimentation money should be per diem or per week, or if he should not go what may be proper to allow him for his travelling Charges from Ratisbonne to Augsburg and back again.

Your Letter of the 24 of January relating to the Bohemians etc. is not yet translated, but as Soon as it is the Contents of it will be laid before the Society and you may expect their Answer. In the mean time the Letters enclosed to Mr. Vernon and Mr. Ziegenhagen were delivered as directed, and Mr. Vernon will write by this Post to Mr. Von Reck desiring him not to enterprize any thing relating to the Society or an Emigration from Saltzburg without consulting you and I beg my humble Service may be acceptable to him.

I hope you have received ere this my Letter of the 21st Current Sent as this is in the Kings packet, and shall be glad to hear when the parcels Sent by Mr. Zurhorst by way of Hamborough [Hamburg] have reached your Hands. I am

Reverend Sir   Your most obedient & humble Servant
Henry Newman

Bartlet's Buildings 31 January 1734/5.   A Monsieur Monsieur Philip George Frederick de Reck at Augsburg.

Dear Sir: I cannot cover the enclosed from Mr. Vernon without sending you my hearty Salutes and covering at the same time a letter I received for you some months ago, but while there were any hopes of your speedy return to England I kept it, and wish the delay of it going to your hands may have been no prejudice to your Affairs. The Seal of the Cover of it was so fastned to it that to prevent breaking it open I was obliged to cut off the Cover which is the reason of your finding it with a double Seal. My humble Service to His Excellency the Baron [J.] Von Reck and please to be assured that I am

Sir   Your most obedient humble Servant   H. N.
P.S.: There is no new yet from Georgia of the Arrival of the last Transport.

Bartlet's Buildings February 7th 1734/5. To the Reverend Mr. Urlsperger at Augsburg.

Reverend and Dear Sir: My last was of the 31st of January according to which a Translation of your Letter of the 24th January and the melancholy narratives accompanying it have been laid before the Society who have ordered me to write to you by this post as mentioned in the Extract of their minutes herewith sent to which I beg leave to refer you. There's not advice yet of the Arrival of the last Transport of Saltzburgers but we hope every day to receive agreable news concerning them of which you may expect to partake as soon as it comes to the hearing of

Reverend Sir   Your most humble Servant   H. N.

P.S.: My humble Service attends our good friend Mr. [Ph.] Von Reck.

Bartlet's Buildings February 19th 1734/5. To the Reverend Messrs. Bolzius and Gronau at Ebenezer in Georgia.

Reverend Sirs: I received your favour of the 12th December in answer to mine of the 1st of August and communicated it to the Society who were very glad to hear of your health and of that of the Survivors of your Flock. I hope long e're this you will have received mine of the 29th October last by Mr. Vat, and that the Transport with him are safely arrived in Georgia to the Joy of their Brethren at Ebenezer.

The Society have ordered a large Packet of the Printed Extracts of Journals herewith sent to be forwarded to the Reverend Mr. Quincy at Savannah with several hundred of Books against Popery to be distributed in Georgia as you and he shall think proper to prevent any Perversion to the Romish Communion which may probably happen even in your retirement from the Countries where they reign and persecute as they please, and if any people popishly affected should come into your neighbourhood the Society desire you would make your Observations of their behaviour and acquaint them with them and also any other remarks which relate to your Settlement that may be of use to the Society to be informed of

Pray give my humble Service to Mr. Vat and let him know that his friends here are very impatient to hear of his safe arrival with the Saltzburgers under his Conduct in the Prince of Wales.

There is little Prospect yet of another Transport in the Spring but Mr. [Ph.] Von Reck is labouring to collect one in Germany, and the Society will not be wanting to give them all the Assistance they can in Case Mr. Urlsperger and Mr. Von Reck should recommend another to their good Offices.

I hope your Salaries for one year remitted partly in a Bill of Exchange and partly in pieces of 8 by the Prince of Wales are got safe to your hands and shall be always glad to hear of your Success in not only edifying the Saltzburgers to whom you are so much devoted but of your instructing the poor natives of Georgia in the Knowledge of true Christianity which will be exceeding acceptable to the Society and to the Trustees for Georgia[.] May you never want the benign Influence of Heaven to support you in the Wilderness where your lot is cast to make every thing tend to the Glory of God and agreeable to your own desires is the Wish of

Reverend Sirs   Your most humble & obedient Servant   H.N.

P.S.: Pray give my humble Service to Mr. Quincy if I should not happen to write to him by this Ship as I fully intend when I hear of the Parcel of Books being delivered to Mr. Simmonds, and pray let him know that Mr. Copping is well tho' not above a fortnight recovered of a dangerous Fever.

Bartlet's Buildings 22 February 1734/5. To the Reverend Mr. Quincy at Savannah in Georgia.

Reverend Sir: On the Dolphin Scooner Capt. Luske Master there is a Box of Books sent directed to you by order of the Society for promoting Christian Knowledge containing 500 Copies of the extracts of the Journals of Messrs. [Ph.] Von Reck and Bolzius with several small Tracts against Popery of which the Society desire your acceptance to be distributed as you may have occasion; and that you will present Messrs. Bolzius and Granau [Gronau] with 100 Copies of the Extracts of Journals, and as many of the other Books as they may at present or hereafter want. The Society will be glad to hear from you that they are misinformed of the resort of Papists to Georgia and that the Tranquility of that new Colony is not like to be disturbed by their Correspondence from hence with some Popish Clergy.

The Box of Books is Sent recommended to the Care of Mr. Causton to be delivered to you the Freight and other Charges for it being paid here to Mr. Simmonds. The Society will be always glad to hear of your Welfare and of the Missionaries at Ebenezer and of the safe Arrival and prosperity of the last Transport of Saltzburgers under the Conduct of Mr. Vat (in the Prince of Wales) to whom I beg my humble Service may be made acceptable from

Reverend Sir   Your most humble Servant   H. N.

Bartlet's Buildings 15 March 1734/5.   To the Reverend Mr. Urlsperger at Augsburg.[44]

Reverend Sir: The Society have received your Letters of the 15th February, 24th March, 3 & 11 M. with the enclosed papers And in particular the Memorial of the Persecuted Protestants in Bohemia which has moved great Compassion in all that have read it, and have stirred up the Endeavours of the Society to do them all the Service they can from hence, and we have now the Pleasure of informing you that by last night's post His Majesty's orders were sent both to his British & German Ministers at the Court of Vienne, to joyn with those of Denmark Prussia and Holland in their Sollicitations with the Emperor in behalf of these oppressed People and to make their Sufferings a *Causam Communem* with those other Protestant Powers. I[n] relation to the printing of the Memorial and the Account of the Sufferings of the Bohemians, the Society look upon it to be a matter of a very nice [delicate] Nature and ought to be thoroughly weighed before they can come to any resolution thereupon, especially while it is not known what answer the Emperor will give to the powerfull and united Application now making to him.

The Society have the pleasure to be able to answer your Query about the Arrival of the Saltzburgers at Georgia[.] They sailed from hence as you know the 8th November O.S. and were upon the Coast of Carolina near Charles Town the 16th of December and soon after got into the River Savannah, and they all landed at the Town of Savannah in good health upon the 28th of the same month.

The Society in answer to the Queries in your Letters of the 24th order me to acquaint you that they are in no doubt but that the Carinthians are as great Objects of Compassion as the Saltzburgers, but the Difficulty lies here that the Charitable Collections were made for the Saltzburgers only. If you had informed us that no more Saltzburgers were to be expected the Society would have immediately set about an Application for leave to apply the Collections for the relief of such Carinthians and Austrians as are actually drove out by Popish persecution but you not being positive in this the Society cannot give any opinion about it, but there will be time for a further Communication between the Society and you, because the best time for the Transport to set out from Augsburg will be the beginning of July in order to be ready here for an Embarkation by the beginning of August.

In relation to the Number of which the Transport should consist the Society are of Opinion that there will be no Objection against its being under 100 because there will not be room for more than 100 in one Ship.

As to the people of Berchtolsgaden, the Society have always looked upon them in the same light as Saltzburgers.

You will observe by the above that the Advertizements you intend for the publick must be altered especially with regard to the Account of the arrival of the 2nd Transport, and the Time when it will be proper for a 3rd Transport to embark from hence for Georgia. The Society hope, if Mr. [Ph.] Von Reck is pleased to conduct the next Transport that he would be very carefull not to pick up any Stragglers upon the Road, and that it consists of those Persons only who have past your Examination and Approbation at Augsburg, I am

Reverend Sir   Your most obedient humble Servant   H. N. P.S. dated 18th March 1734/5.

Dear Sir: Since what has been writ above, the Parliament by a Vote has put the Trustees of Georgia into a Condition to take some of the Austrians and Carinthians on their Account which will take off the Difficulty that lay upon the Society And therefore you may now safely engage what persecuted Carinthians or Austrians offer to come along with the Saltzburgers and Berchtolsgaders, but their Charge must be kept distinct from that of the Saltzburgers. It will be time enough for the Transport to set out in July.

I remain   Reverend Sir Your &   H. N.

Bartlet's Buildings 29 April 1735.   To the Reverend Mr. Urlsperger at Augsburg.

Reverend Sir: Your Letter of the 7th of March came to my Hand this week by the Penny post, being I suppose Conveyed to England by Some private Hand, and by the last foreign Mail I received Your favour of the 28 of April N.S. both which were this day Communicated to the Society, who in answer to the first have desired Mr. Ziegenhagen to write to you to Send over 25 of the Screw Medals wherein the Protestant Mission to East India is described and that you would Send over those of the best Sort of them with your first Convenience and charge the Society with them.

In answer to your Last of the 28th Current the Society have particularly desired me to Let you know that they are very impatient to hear from you whether they can depend upon any Number of Saltzburgers coming out of the Arch Bishoprick Saltzburg during this Summer in order to embark for Georgia. This is so earnestly desired that they would be very glad to receive your Answer to this enquiry as Soon as possible.

I have herewith Sent the Society's Printed Circular Letter to their Members in Town and Country that you may know the Contents of it and especially those Facts therein related which you Your Self have had so great a Share in. Be pleased to communicate it to our Common Friend the worthy Mr. [Ph.] Von Reck with my humble Service and let him know that I hope to See him 'ere long at the head of another Transport of Emigrants.

Mr. Manitius Sets out next week Passenger to Bremen in his Way to Hall, and Mr. Degmair will follow him in a Short time, they are both very well & have made a wonderfull Progress in acquiring the English Tongue for the time they have been in England. I am

Reverend Sir   Your most humble Servant   Henry Newman

Bartlet's Buildings London 6 May 1735.   To the Reverend Mr. Urlsperger at Augsburg.

Reverend Sir: In answer to your Favour of the 28th of April I am ordered to acquaint You that application has been made as you desire for His Majesty's Gracious Intercession at Vienna for Liberty for the Wives and Children of the Carinthian Colonists now at Ratisbonne to follow their Husbands in their Voyage to the Colony of Georgia.

Your Letter of the 2nd of May has been also received and put into the hands of James Vernon Esquire who will do his best to procure a Satisfactory answer from the Trustees to Herm. [Herrn Ph.] Von Reck's Demand.

I must put you in mind that the Colonists bound for Georgia Should be at London before the end of July Old Stile. I wish you

and Mr. Von Reck all manner of Prosperity remaining his and
Reverend Sir   Your most obedient humble Servant
Henry Newman

*Bartlet's Buildings London 13 May 1735.   To Mr. John Vat at
Ebenezer in Georgia.*

Dear Sir: I received Your Letter of the 10th of February by
the Ship James Capt. Yoakley Master who carrys this Answer.

The Society were much concerned to hear of your indisposition
after the fatigues of your Voyage and will be glad to hear by your
next that you are happily recovered.

It was a great Pleasure to the Society to hear that the Transport
of Saltzburgers under your Care performed their Voyage in so
Short a time, and that they all lived to arrive in Georgia though
two of them who were but in an ill State of health when they Left
England died it Seems soon after their arrival.

The Accounts you gave of the ill choice the first Transport of
Saltzburgers had made for a Settlement at Ebenezer affected the
Society So much that I was ordered to Send a Copy of your Letters
to the Trustees for Georgia which was immediately done, and
Mr. Vernon has acquainted the Society that the Trustees will do
every thing in their Power to make you and the Saltzburgers that
you Conducted easie, and they doubt not but full instructions are
Sent over in this Ship for that purpose to the Gentlemen their
Agents at Savannah.

The place you wished they might be Settled in it Seems belongs
to the Indian Natives, but the Society hope some expedient may
be fallen upon to accommodate them entirely to your and their
Satisfaction.

The Society being informed of Capt. Dunbar's civility to the
Saltzburgers, You are desired to give their Thanks to him for it,
wishing You all manner of Prosperity I remain
Dear Sir   Your most obedient humble Servant
Henry Newman

*Bartlet's Buildings London 13 May 1735.   To Messrs. Bolzius &
Gronau at Ebenezer in Georgia.*

Reverend and Dear Sirs: Your Letter of the 6th of February last
came safe to Hand and was So acceptable to the Society that they
gave leave to print it in one of our News Papers that the publick
might have a Share in the Pleasure they had, and especially that

those who had been Benefactors to the Saltzburgers might participate of the joy with which the last Transport under the care of
Mr. Vat was received by their Brethren at Ebenezer.

The Society are not a little concerned to hear that the choice
of a Settlement by the first Transport is not acceptable to those
of the last Transport but whatever difficulties may arise on that
Score, they doubt not but the Trustees for the Colony of Georgia
will do all they can to remove them by permitting them to pitch
their Habitations in Some other place more eligible on account of
the Soil and distance from the Sea, or neighbouring Settlements
and that full instructions will be Sent over in this Ship to the
agents of the Trustees to make all the Saltzburgers as easie as the
Nature of a New Settlement will admit.

I have by order of the Society shipped in the James Capt.
Yoakley Master 2 Casks of Half pence and farthings to the Value
of 85 £ of which No. 1 Marked B. G. contains the Value of 45 £
for 1/2 a year of your Salaries ending the 1st of November 1735.
Viz. for

Mr. Bolzius	25 . — . —
Mr. Gronau	15 . — . —
Mr. Ortmann	5 . — . —
	45 . — . —

And the Copper money in B.G. No. 2 amounting to 40 . — . —
is to be applied as Mr. Ziegenhagen shall direct it being bought &
Shipped by his order.

The Society are So Sensible of the Fatigues you daily undergo
that they have ordered me to desire your acceptance of Twelve
Doz. Bottles of Vidonia Madera which I have packed up in Two
Hogsheads marked B.G. No. 3, 4 and hope they will go Safe to
Your hands to refresh you and your Friends under the Toils of
your new Settlement.

Mr. Urlsperger gives no certain advice when another Transport
may be expected, but in general that he hopes Mr. [Ph.] Von Reck
maybe enabled Sometime this Summer to conduct another Transport which is very Sollicitous to accomplish So as to be here in
August next, but of their number no Judgment can as yet be
made.

The Society are very much pleased to hear of the Patience,
Resignation and Industry of the Saltzburgers under your care
and they hope by the Blessing of God on your and Mr. Vat's
endeavours they will in fine be made as easie as the circumstances

of Settling a new Colony will admit. In the meantime they pray God to continue the Blessing of Health to you and your Flock and desire you would advise them of your proceedings by all opportunities and wherein they can further contribute to your being happily established. What hopes you have of gaining the Savage Natives to the Knowledge and Love of the Saviour of Mankind? whether they have any relish for Divine Truths, and the pleasures of a civil Life, especially those who have been in England and had the happiness to return to their native Country with the last Transport?

Whether they have improved so much by their Voyage to England as to be able to impress on their Countrymen a Love of Temperance, civility, and a Sense of their dependance on God for every Comfort they enjoy.

The Society hope your incessant Labours and pious Examples joined with those of the Saltzburgers may produce Some advances to So good an Effect, though little has been the fruits of Christian Instruction upon the Savages in other parts of America. May God Almighty Support and direct you in the prosperous discharge of the Trust reposed in you is the wish of the Society as well as
Reverend Sir   Your most humble Servant   H. Newman

London 12 May 1735.

Invoice of Goods shipped on the James Capt. Yoakley Master for Georgia by order of the Society for Promoting Christian Knowledge for the use of the Reverend Messrs. Bolzius and Gronau at Ebenezer, viz.

Cask
B.G.

No. 1 Copper half Pence   £ s d
    3 cwt 2 dwt_____at 10:16:4 per cwt_____37:17:2
    Farthings—2 [sic] cwt___at 10:12:4 per cwt_____ 5: 6:2
        Half pence to Make up_____ 45 £_____ 1:16:8

45: –:–

No. 2 Copper half Pence   £ s d
    3 cwt_____at 10:16:5 per cwt_____32.9.–
    Farthings___2 [sic] cwt___at 10:12:4 per cwt_____ 5.6.2
        Half pence to Make up_____ 40 £_____ 2.4.10

40: –:–

85: –:–

Hogshead

B.G. No. 3.4 Containing 12 Doz. Bottles of Vidonia Madera as a present from the Society to Messrs. Bolzius & Gronau.

Cash and Cartage Charges at the Custom house etc. Gratis Shipped by order of the Society

Henry Newman Secretary

Note No. 2 is Shipped by order of Ziegenhagen and all the above-said parcels were Sent to Mr. Simond's who is desired to take the Bill of Lading for them.

Bartlet's Buildings London 20 June 1735. To the Reverend Mr. Urlsperger at Augsburg.

Reverend Sir: My Last were of the 29 April and 6 May since which I have received your favour of the 2nd, 9th & 19th May and of 22nd of June N.S. to me and of the 6th of June to Mr. Ziegenhagen N.S. Translations of all which have been communicated to the Society and Copies of them Sent to the Honourable Trustees for Georgia for their consideration of those parts which related to them, and as they have not Sent me instructions to answer the Principal Enquiries you made I hope you have their answers by their own Secretary only I understand by what I have heard in conversation that they are come to Some resolutions in favour of Mr. [Ph.] Von Reck which I hope may be agreable to him, but I dare not take upon me to acquaint you or him with the particulars without authority from that Board, nor can I doubt but that you have or will soon receive them from their own Secretary together with Instructions about placing the Meridians and Parallels to the Map of Georgia or rather that a Copy as engraved here will be Sent to you.

The Engraver[45] you have employed a Copying the Print made here of Tomochachi and his Nephew has shewn his Skill So well that it is thought here he has in the Portrait of the Old Man exceeded the original Print made here.

Mr. Robinson the Kings Minister at Vienna is directed to Sollicit in his Behalf Name for Liberty for the Wives and children of such Carinthians who being at Ratisbonne, have intitled themselves Colonists for Georgia to accompany their husbands and fathers and he signifying that a List of the names is Sent by this post to Vienna by the way of hanover. I inclose a Copy of the said List that if it be Defective you may rectify it, by Sending a more perfect List to Mr. Robinson who will have notice to receive it from you, or if you think more proper from Baron [J.] Von Reck.

As to Such Saltzburgers or Berchtolsgaders as shall be disposed to accompany the persecuted Protestants, which the Trustees for Georgia intend to take into their care, they (the Saltzburgers or Berchtolsgaders) must lose no time in Signifying their names ages and professions by you or Mr. Von Reck to the Society that care may be taken for their Safe conduct hither in their way to Georgia.

Capt. Dunbar, who carried over the last Transport is just now returned from Charles Town in South Carolina but having left Georgia Several weeks before he does not bring any Later advices thence than we had by the Ship which arrived about a month ago[.] I wish you all manner of Prosperity and remain

Reverend and Dear Sir   Your most Obedient humble Servant

Henry Newman

P.S.   The young Man *Jastrum*[46] whom You give Such an Extraordinary Character of, if he comes to England I shall be glad to give him any assistance in my Power to recommend him to Business.

A List of Carinthians now at Ratisbonne who have resolved to go to Georgia.

Bartholomeus Globish his Wife and 7 Children	9
Christian . . . and his Wife	2
John Unterwald his wife and 5 Children	7
Clement Leitner his wife & 2 Children	4
Gregorius Koffler his wife & 1 child	3
Matthew Egenter his wife & 3 Children	5
Nicholaus Neidheart his wife and 3 Children	5
Francis Santer his wife and 3 children	5
Simon Moser his wife and 4 children	6
Matthias Auer	1
Lorentz Minchler	1
Rupert Glotz and Balthazar Glotz	2
Jacob Bingel, Geo. Krawol, Geo. Stegobereg	3
And. Geuiter, Caspar Walter, Richard Mittereg	3
George Scheibler, Thomas Scheibler & 1 child	3
Matthias Edeg, Nicholaus Moser, Helena Hussen	3
Martin Ausserulammer his wife & 1 child	3
Vincent Reiger	1
	66

Bartlet's Buildings London 8 July 1735.   To the Reverend Mr. Urlsperger at Augsburg.

Reverend Sir: My Last was of the 20 of June O.S. since which

I have received your favour of the 30 of June N.S. with an Extract of Mr. [Ph.] De Reck's Letter to you but you forgot to Signify the date as also an Extract of a Receit given to Mr. Von Munch for 500 Florins taken up on account of the Society, which I am Ordered to assure You will be reimbursed or the Value in Sterling Money here in what Manner you please.

In Answer to Mr. [Ph.] De Reck's Letter the Same is to be communicated to Morrow to The Trustees for Georgia and in the mean time Orders are given for procuring a Royal Pass as desired for Mr. De Reck to Conduct a number of German Persecuted Protestants not exceeding 100 Men, Women and Children with their Baggage from Ratisbonne to London in their way to Georgia and I hope by that time you have received my Last, you will hear of Some good Effects of the further Application made to the Imperial Court by Mr. Robinson in behalf of the Wives and Children of the Carinthians that they may be permitted to accompany their Husbands to Georgia.

I took Care to Convey your Letter to Mr. Degmair who was at Kensington with Mr. Ziegenhagen, and I believe you will Soon hear of his leaving England upon the News of his father's Death.

I am Reverend Sir   Your most obedient humble Servant

Henry Newman

I shall be very glad and So will the Society to hear of your finding Benefit in the use of the Waters at Lindau.

London 11 July 1735.   To His Excellency the Baron [J.] de Reck at Ratisbonne.

Sir: At the desire of the Reverend Mr. Urlsperger at Augsburg Her Majesty's Royal Pass is bespoke for Mr. [Ph.] De Reck Your Excellency's Nephew to conduct 100 Men, Women & children of German Protestants with their Baggage from Ratisbonne to England in order to proceed to Georgia. I was promised to have the pass Sent to me this Evening from His Grace the Duke of Newcastle's office to be forwarded to Your Excellency but as it is not come I thought it my Duty to acquaint you that it is ordered & will be Sent as Soon as it is in my Power, and request you would be pleased to Let your nephew know that he may dispose himself for the Journey to Holland as if he was actually possessed of Her Majesty's Pass which he may expect next Post under your Excellencys Cover. I am

Sir   Your Excellencys most obedient humble Servant

Henry Newman

I beg my humble Service may be acceptable to your nephew.

Bartlet's Buildings London 15 July 1735.   To the Reverend Mr.
Urlsperger at Augsburg.

Reverend Sir: My Last was of the 8 Current Since which I have
none from you, but the Duplicate of the Receit you have given
Mr. Von Munch for 500 Florins being Transmitted to his Correspondents here Messrs. Meyer and Jansen having been tendered
as a Bill of Exchange on the Society at 8 5/6 florins to a pound
Sterling making 56 £ 12 s — d the Same has been paid and I am
ordered to acquaint you with it.

Mr. Degmair has desired me to cover his Answer to the Letter
you Sent me concerning his Fathers death, which I doubt not but
he considers with a becoming Resignation to the wise Providence
of God, and in answer to your Queries concerning his behaviour
here, I can assure you that to the best of my knowledge it has
given great Satisfaction to all that were acquainted with him[,]
and his Industry has been so remarkable in acquiring the English
Tongue that he has been employed by the Society in translating
our German Papers into English and as we believe him to be a
Sincere Lover of Christ and his Religion we doubt not of his
being very usefull at his return in the Service of the Church and
particularly for Carrying on a Correspondence with England so
necessary for the welfare of the Protestant Cause.

I shall be glad to hear of your return in good health from
Lindau, and am
            Reverend Sir   Your most humble Servant   Henry Newman

Bartlet's Buildings London 22nd July 1735.   To the Reverend
Mr. Urlsperger at Augsburg.

Reverend and Dear Sir: My Last was of the 15 Current O.S.
Since which I received your Letter of the 12 Current N.S. and in
answer to your enquiries concerning the Success of our Applications on behalf of the Carinthians that their wives and children
may have Liberty to follow them to Georgia.

I am to inform you that repeated Orders have been Sent to His
Britannick Majesty's Minister at Vienna for the Liberty of the
Wives and Children to follow their husbands and fathers to Georgia and that we are in hopes that these Sollicitations will meet
with the desired Success, but in Case that fails and that the

Carinthians do not Set out for Georgia, the Society in that case have no objection against filling up the number of Emigrants with Bohemians, and even in Case Some Carinthians should come, they are willing as many Bohemians may join them as will make up the number one Hundred which must not be exceeded, regard being had to give the Preference to Such families as are grown up to Man's Estate.

I received at the Same time the Extract of Mr. [Ph.] Von Reck's Letter of the 5 of July from Ratisbonne and am glad to find So near a Prospect of his return to England where he will be welcome to all his friends particularly to

Reverend Sir   Your most obedient humble Servant
Henry Newman

Bartlet's Buildings London 23 July 1735.   To the Reverend Mr. Urlsperger at Augsburg.

Reverend Sir: Having wrote to you the 15th Current by the Post my thoughts concerning the good Behaviour of Mr. Degmair during his Stay in England, I have nothing to add but that I hope it will please God to return him Safe to his friends at Augsburg and to make him instrumental of much Glory to his Great name wherever his Lot may be cast.

The Society desire your Acceptance of a packet of such books as have been lately printed by them or imported into their Store as present from their Friends, which goes in Mr. Degmair's Baggage. I wrote to you by yesterdays post, in answer to Your Letter of the 12 Current N.S. which was Sent in the Kings Packet to Ratisbonne, to which I beg leave to refer you for the rest and remain

Reverend Sir   Your most obedient humble Servant
Henry Newman

Bartlet's Buildings 2 August 1735.   To Mr. John Vat.

Sir: Above is a Copy of my Last, to which I have now to add that Mr. [Ph.] Von Reck is expected daily with a Transport of Carinthians and Austrians about 50 or 60 to be sent to Georgia on Account of the Trustees but the Society bear the Charges of their Transport to Holland till they can embark.

Mr. Oglethorpe did me the honor to call here this day and tells me he hopes to be ready to embark in a months time for Georgia;

I am Sorry to understand by him that You continue to want Your health and that you have been obliged to reside at Purysburg for the recovery of it.

I hope your next will bring me a more favourable Account of your health and proceedings to be laid before the Society who will be much pleased to hear of your doing well by

Sir   Your most humble Servant   Henry Newman

All your Friends here are well and Mr. Hales our old friend particularly Salutes You with his best Wishes.

I am glad to hear by Capt. Dunbar tho he brought me no Letter from you that you and the Saltzburgers begin to be better reconciled to the Soil and Climate of Georgia.

London 2 August 1735.   To the Reverend Messrs. Bolzius and Gronau at Ebenezer in Georgia.

Reverend Sir: On the other Side is a Copy of my Last, I now cover a Bill of Loading which I could not get time enough to Send in my last, which I hope you have received with the Goods mentioned in the Invoice Inclosed. Mr. [Ph.] Von Reck is now Setting out from Ratisbonne with 50 or 60 Carinthians and Austrians persecuted for the Sake of Religion and daily Expected to be sent to Georgia whither Mr. Oglethorpe is preparing to return in hopes to Remove some difficulties complained of in your Last. He did me the honour to call on me this day and hopes to be ready to embark in about a months time. Mr. Urlsperger's last Letter to me was dated at Lindau in the Lake of Constance where he was drinking the mineral Waters for his health. Mr. Ziegenhagen is well at Kensington but drinking also the mineral Waters for establishing his health. I wish you, Mr. Zwiffler, Mr. Ortmann and all your Flock, all manner of Prosperity Assuring You that I am, Reverend & Dear Sir

Your most humble Servant   Henry Newman

Bartlet's Buildings London 19 August 1735.   To the Reverend Mr. Urlsperger at Augsburg.

Reverend and Dear Sir: I this day received your favour of the 11th Current N.S. and immediately communicated it to the Society it being the day of the Society's Meeting, and in answer to it, I am particularly ordered first to acquaint You that Mr. [Ph.] Von Reck must come away without delay with such Emigrants as can be ready to come.

Secondly, That the Society and the Trustees do not think it fit that any Body should be Sollicited in their Names to go to Georgia. Only that they declare they are willing to receive Such as desire of themselves to go.

That is to Say.

1. So far as the Fund the Society have in hand will go.
2. On Condition that the people that Offer themselves are fit for the Purposes of a New Colony according to former Directions.

Thirdly. That Mr. Oglethorpe as I Signified in my Last of the 15 Current cannot Stay Longer here than the latter end of September and that if the Transport cannot be here by that time it must not come this Year at all.

The Society are glad to hear of your return in Safety and good health to Augsburg and that you have had the Pleasure of Such an agreable Relaxation from the multiplicity of your Affairs as to visit Lindau. I am

Reverend and Dear Sir   Your most obedient humble Servant
Henry Newman

Bartlet's Buildings 12 August 1735.   To Mr. [Ph.] De Reck Junior at Ratisbonne.

Dear Sir: I received Your Letter of the 8th Current N.S. and this day communicated it to the Society who have ordered me to inform You that they are assured from the Trustees of Georgia that if You do not Set out immediately upon receit of this Letter with what number of Emigrants You have now ready, You must not think of going this Year & will lose the Benefit and advantage of Mr. Oglethorpes Company who is determined to go over to Georgia next Month.

As for the difficulty You lie under in relation to their Wives and Children detained behind[,] You may propose this Expedient to 'em viz. to leave 3 or 4 grave and elderly persons to wait their release which is now pressed for at the Imperial Court by the British Minister there, and the Trustees for Georgia will at any time receive and transport those Women and Children after their Husbands and Parents; the Society therefore earnestly recommend it to You that upon receipt of this Letter You will come away with those that are ready and willing to come with You.

I am Honoured Sir   Your most obedient humble Servant
Henry Newman

P.S.: My humble Service to His Excellency Your good Uncle [J. von Reck] let him know that I had the honour of his Letter under Your Cover and hope to acknowledge it by next Post.

London 14 August 1735. A Son Excellence Monsieur [J.] de Reck Conseiller de Sa Majesté Britanique et Son Envoyè a la Diete de l'Empire à Ratisbonne.

Sir: I had the honour of your Letter of the 8th Current N.S. and thank You for the Notice You therein favoured Me With of your nephews [Ph. von Reck's] disposition to return to Georgia. I wrote by the last Post to him of the absolute necessity there is of his using all possible dispatch in conducting Without loss of time the Protestants that design to embrace the opportunity of being transported thither under his care and the Society for Promoting Christian Knowledge have ordered me to acquaint Your Excellency with the contents of my Letter to him, that You may be pleased to urge his departure from Germany as Soon as the Nature of his undertaking will permit. I have the honour to be

Sir   Your Excellency's most obedient humble Servant

Henry Newman

Bartlet's Buildings London 15 August 1735.   To the Reverend Mr. Urlsperger at Augsburg.

Reverend and Dear Sir: My Last were of the 22nd & 23rd July in answer to Your favour of the 12 of the Same Month N.S. at Lindau Since which I have received none from You but hope this will find you returned in good health to Augsburg.

I had a Letter of the 8th Current from Mr. [Ph.] De Reck concerning the Slow progress of the Transport of Emigrants coming under his conduct for Georgia.

I have wrote to him in the Terms enclosed and by order of the Society Send a Copy that you may see of what Consequence it is to the Success of his Voyage that all possible dispatch be given to him and that the Society promise themselves nothing will be wanting on your part to hasten them.

I expect every hour to hear from You which I hope will advise of their departure and am

Reverend Sir   Your most obedient humble Servant

Henry Newman

Bartlet's Buildings London 26 August 1735. To the Reverend Mr. Urlsperger at Augsburg.

Reverend Sir: My last to You was of the 15 Current O.S. which I hope you have received with a Copy of Mine to Mr. [Ph.] De Reck who I hope has also received the Original advertising him of the necessity of departing forthwith with the Carinthians etc. of persecuted Protestants that are disposed to go to Georgia this Season or they will be too late to go with Mr. Oglethorpe.

I have this morning received the inclosed from Monsieur Bolzius by Capt. Tompson from Georgia which I wish May bring You agreeable News. I received at the Same time one from Mr. Vat of the 30 of May at Ebenezer by which he was well, but he complains of the Choice the Saltzburgers have made of a Situation at Ebenezer for Soil etc. of which I Suppose they write to You particularly. I am

Reverend Sir   Your most humble Servant   Henry Newman

My humble Service to Mr. De Reck if with you. I have covered by this Post a Letter to him from Mr. Vernon.

Bartlet's Buildings 5 September 1735. To the Reverend Mr. Richard Lowther Minister of the English Church at Rotterdam. Sent on a C. L.

Reverend Sir: Mr. [Ph.] Von Reck according to our last accounts from Augsburg was to Set out with a 3rd Transport last week consisting of Saltzburgers and Austrians etc. Emigrants for the Sake of Religion in their way to England in order to proceed to Georgia. The Trustees for that Colony undertook to recommend them to the good offices of Mr. D'Ayrolle's for obtaining a free pass from the States[47] for them through their Territories, but as they must pass by You any favour You are Pleased to Shew them in advising Mr. De Reck how to address the Magistrates of Rotterdam, for their Countenance will be gratefully esteemed by the Society, I am

Reverend Sir   Your most humble Servant   Henry Newman

London 9 September 1735. To the Reverend Mr. Richard Lowther Minister of the English Church at Rotterdam.

Reverend Sir: Since my Last of the 5th Current the Society having received certain advice of a Transport of 40 Saltzburgers and other Protestants being Set out from Augsburg the 6th Current N.S. in their way through holland to England in order to proceed to Georgia, I am ordered to recommend them to your good offices for obtaining any favour from the Magistrates etc. at Rotterdam to facilitate their Embarkation for England, the charge of which will be defrayed by Mr. [Ph.] Von Reck their Commissary who has the Queens pass. I have wrote to the Same Effect to Mr. Wolters His Majesty's Agent, and doubt not of your favour to advise and assist the Commissary and his Small Transport by which You will oblige the Society. I am

Reverend Sir   Your most humble Servant   Henry Newman.

London 9 September 1735.   To Mr. Wolters Agent for his Britannick Majesty at Rotterdam.

Sir: The Society for Promoting Christian Knowledge having certain advice of a 3rd Transport of Saltzburgers and other Protestants to the number of 40 being set out from Augsburg the 6 Current N.S. under the Conduct of Mr. [Ph.] Von Reck in their way through Holland to England in order to proceed to Georgia, I am ordered to recommend them to your good offices for facilitating their Passage through the States Territories conformable to the Royal Pass Her Majesty has granted to Mr. Von Reck and to advise him in the best means for embarking for England he defraying all Charges that may accrew thereon, and Whatever favour You shew the Commissary and his Transport on this occasion will be thankfully acknowledged by the Society. I am

Sir   Your most humble Servant   Henry Newman

Bartlet's Buildings London 9 September 1735.   To Mr. Commissary [Ph.] Von Reck under cover to Mr. Wolters.

Dear Sir: I am glad to understand by Letters of the 5th and 8th Current from Mr. Urlsperger at Augsburg that you are returning to England with a Transport of Protestant Sufferers for the Sake of Religion. The Trustees for Georgia will give directions for your Embarkation in Holland, and where You are to Land in England in order to Embark for Georgia which I hope May be the same Ship with Mr. Oglethorpe Since you are like to arrive here in time. In hopes 'ere long to kiss your hands.[48] I remain

Dear Sir   Your most obedient humble Servant   Henry Newman

Bartlet's Buildings London 7 October 1735. To the Reverend Mr. Urlsperger at Augsburg.

Reverend Sir: My last was of the 26 of August Since which I have been favoured with Yours of the 1st, 5, 8th 15 and 22nd of September N.S. relating to Mr. [Ph.] Von Reck's last Transport of Saltzburgers and Austrians now happily arrived here in their way to Georgia, for Your dispatch of whom The Society thank You, and nothing has been wanting in the power of the Society or the Trustees for Georgia to accommodate them with necessaries during their Stay here and in their way to Georgia.

They are designed to embark to Morrow in the London Merchant Capt. Thomas Master the Same Ship that carried Mr. Oglethorpe to Georgia for his first Voyage, and they Sail in Company with the Simonds another Ship which carrys Mr. Oglethorpe over With a large Transport of English and Some Bohemians that came hither before Mr. Von Reck's Transport arrived. The Lords of the Admiralty have appointed the Hawk Sloop man of War to convoy them through the Voyage.

According to Your desire Signified in Your Letter of the 15 of September the Society have ordered their Treasurer Mr. Tillard to pay Mr. Ziegenhagen 269 Florins, and I believe he acquaints You by this Post that the Same is accordingly done.

I was in pain to Send You this acknowledgment of the receipt of your Letters abovementioned, but a continual hurry has prevented me till now, which I hope You will excuse.

The Society also thank you for the 3 Copies of the Account of the Emigrants dedicated to them in High dutch and English which they received by the hands of Mr. [Ph.] Von Reck.

May You long live, and rejoice in the Success of your Labour for these pious Confessors is the wish of the Society as well as of
Reverend and Dear Sir   Your most obedient humble Servant
Henry Newman

Bartlet's Buildings 8 October 1735. To Mr. Harman Verelst at the Georgia Office near old palace yard.

Sir: The Society here having under consideration the giving Instructions for Writing to Mr. Vat at Ebenezer it will be a pleasure to them to know whether the Things desired in the enclosed List are Sent by the Ships now going or any part of them, and if they are not, whether they can Still be Sent on Board the London Merchant.

I Sent a Copy of Mr. Vat's Letter to Mr. Oglethorpe who I presume acquainted the Board with Such parts of it as related to the wants of the Saltzburgers, unless Some more other Letters to him or them had desired the Same things.

Your Answer by the Bearer will oblige Sir   Your most humble
Servant   Henry Newman

Bartlet's Buildings 9 October 1735.   To Mr. Harman Verelst at the Georgia Office.

Sir: I thank you for your Short Letter of this day and now only take Leave to add by way of Memorandum Some other Things Mr. Vat desires that were not mentioned in the list I Sent you Yesterday viz.

That the People at Ebenezer are in Great want of Linen and Shoes.

Earthen Ware and Utensils for the Kitchen.

1. Handsaw to each Freeholder, instead of 8 for all.

Some large Coppers for Boiling of Beer the River water being very bad in Summer.

That the Grinding Corn by Such hand mills is very tedious and therefore they desire Some Mill Stones of a midling Size. And That Some fishing Tackle may be Sent.

If you please to let these things be added to the List I sent, it will represent their wants more compleatly to the Honourable Trustees. I am

Sir   Your most humble Servant   Henry Newman

Bartlet's Buildings London 10 October 1735.   To Mr. John Vat at Ebenezer in Georgia.

Dear Sir: I received your Letters of the 30 February and 30th of May Last by Capt. Thompson which were both read at the Society and as all or most of the Articles therein mentioned were properly cognisable by the Trustees for Georgia to give Such Directions thereon as Should be thought proper was I ordered to Send Copies of them to that Board which was immediately done And Mr. Verelst now Sends me word that Several of the things you desire are Sent over in the Ships now going and what is deficient will be Supplied hereafter.

This Paragraph by Special direction of the Society:

I acquainted the Society with your expectations of their giving Instructions for your Support in regard to your Stay in Georgia till next Spring, in answer to which I am directed to acquaint You that Mr. Oglethorpe will on his Arrival take care that you have a proper Allowance as well as to obviate all difficulties that you complain of, in respect to the Situation of the Saltzburgers.
[Marginal note:] This paragr. by Special direction of the Society.

Mr. Von Reck returns by the *London Merchant* with a Transport of 58 Persons, Men, Women and children near half of which are Saltzburgers and the rest Austrians that have been persecuted for the Sake of their Profession of the Protestant Religion, which will make a considerable addition to the strength of those already at Ebenezer, when it shall please God to grant them a Safe arrival.

Mr. Oglethorpe returns in the *Simonds* a Larger Ship than the *London Merchant* freighted with 120 odd passengers most of which are English, and the rest Foreigners, and Mr. Von Reck tells me there are 120 Passengers in the *London Merchant* including the Saltzburgers etc.

Capt. Dunbar is now with his Ship in North Britain receiving Passengers on Board, and is designed to meet Mr. Oglethorpe at Maderas, and thence to proceed to Savannah under Convoy of the Hawk Sloop Man of War Commanded by Capt. Gascoigne.

I am very glad Mr. Oglethorpe has condescended to revisit Georgia and hope his Presence and Prudence will dispell all the difficulties that have arisen on the Settlement of that new Colony. I Look upon it as one of the greatest pieces of Self Denial this Age has afforded that a Gentleman of his Fortune possessed of a Large and Valuable Acquaintance a Seat in Parliament, with a Genius to make a Figure in any Senate in the World, Should renounce all these Pleasures to cross a perilous Ocean for the Sake of establishing a few distressed families undone by Idleness, Intemperance, Sickness, with other ill habits and all Oppressed with Poverty,[49] to Found a Colony in a Wilderness wholly uncultivated, abounding with Pine Barrens Crocodiles, Bears and Wolves with other Animals, of no apparent use to the Creation but to punish the Posterity of fallen Adam.[50] This is not to be Accounted for by narrow Souls because it does not yield any immediate visible reward but the distant expectation of an uncertain Glory like that Julius Caesar enjoyed when he Landed in Britain about 1700 Years ago with a more promising People than are now the first Planters of Georgia, except we take in the expectation after this Life which are as unbounded as they are Questionable. But whatever be the reward you will justly be intitled to a Share in the Glory whether present or future, by the Share you have in the inter-prize, and while it pleases God to give you health I hope your courage will not fail you to persist in what you have Laudably begun, not for Your own Sake So much as for the Sake of others. Let me hear from you by all Opportunities and believe that I am

Dear Sir   Your most humble Servant   Henry Newman

Bartlet's Buildings London 10 October 1735.   To the Reverend Messrs. Bolzius and Gronau at Ebenezer in Georgia.

Reverend Sirs: In August Last I received your Favours of the 2nd of April 5 & 19th of May by Capt. Tomson all which were communicated to the Society, and gave them not a little pleasure to observe with what Alacrity and resignation to the Divine Will you bear the fatigues of Settling the new Colony of Georgia; and whatever discouragements have Attended the first Attempts on your Plantation the Society hope Mr. Oglethorpe will make You and the People easie, by indulging them in another Choice of a Situation, the Fertility of which may make you and your People ample Amends for your disappointments hitherto.

The Society have Sent in the London Merchant another half years Salary in Copper Money viz. from the 1st of November next to the 1st of May 1736. According to the Invoice inclosed by which you will find that it is to lessen your Trouble, I have as well as I could distinguished every parcel According to the Service they are designed for. viz.

No. 1 is entirely for Mr. Bolzius's Salary 1/2 a year_____ 25. — . —
2 is 15 £ for Mr. Gronau and £ 5 for Mr. Ortmann 20. — . —
3. is entirely for Mr. Zwiffler a Gratuity from the Society for his care of the Saltzburgers etc. in regard that he has no Lot yet assigned him_____ 20. — . —
4 is from the Reverend Mr. Ziegenhagen to be ap-plied to Such uses as he Shall direct_____ 30. — . —

£ 95. — . —

The Society desire your Acceptance of a Small packet of Books recommended to the Care of Mr. [Ph.] Von Reck, consisting of as follows viz.

2 Setts of Arndts true Christianity in 3 Vols. each Set.

2 Dr. Pearce's Sermon at St. Sepulchres 1735.

2 Mr. Drew's . . . . Reformation Sermon 1735.

2 Dr. Comber's Plausible arguments of a Romish priest Answered.

2 Mr. Parsons's Sermon at the Funeral of the Earl of Rochester.

2 Bp. of Chester's View of the Articles of the Protestant and Popish Faith.

2 Dr. Berriman's Family Religion recommended 1735.

2 Bp. of Dromore's Sermon preached at Dublin, 23 Oct. 1733.

"Beside these a large packet of the Society's Books have been presented to the Reverend Mr. Wesley consisting of the Books mentioned in the List enclosed out of which he will Spare you what he can at your desire.

The Society flatter themselves that Mr. Spangenberg and his Companions will demean themselves So as to cultivate a good Correspondence with all the Protestants in Germany.

They are glad to hear that the behaviour of Mr. Ortmann and his Wife recommend them So much to your esteem as your Letter Signifies and hope they will continue to do so.

The Society recommend you to the good Providence of God and his gracious direction not doubting of Mr. Oglethorpe's good Offices on all occasions to advise and assist you to the utmost of his power to which his own inclinations very Strongly lead him he having Said many good things both of you and your people Since his return to England.

May the bountifull Giver of all Good prosper all your Labours to his Glory and the Welfare of your Flock are the Wishes of the Society as well as

Reverend Sir   Your most obedient humble Servant   H. Newman

Bartlet's Buildings London 9th October 1735.   To the Reverend Mr. S. Quincy at Savannah in Georgia.
[Marginal Note:] Per Mr. [Ph.] Von Reck in the London Merchant Thomas Master.

Reverend Sir: I received your kind letter of the 4th of July last by Capt. Tompson, which I communicated to the Society, who thank you for the Account of it and what you mention of some Bills of Carolina Currency which Messrs. Bolzius and Gronau received of Mr. Montagute in part of payment of the Bill on him, the matter has been recommended to the Trustees for Georgia, and they verily believe they have given proper directions about it, and that Mr. Oglethorpe will take Care to see Justice done therein.

The Society desire you Acceptance of a packett of such Books, as have been lately imported into their Store, sent by Mr. Von

Reck, to which I herewith add their Circular Letter, by which you will be informed of what they have directed me to signify to all their Members upon every Branch of their designs.

The method you have taken to form a Religious Society, they hope will be attended with good Effects, especially if you can spare time to inspect and conduct them in their Conferences to their Edification.

[Marginal Note:] NB. The names of the Volunteer Missionaries are Mr. John Wesley Fellow of Lincoln College Oxford Mr. Charles Wesley Student of Christ Church College. Mr. Ingham or Ingram of Queen College Oxford Who sailed in the Simonds with Mr. Oglethorpe.

The Reverend Mr. Wesley a Corresponding Member of the Society with other well disposed Gentlemen go over Volunteers with Mr. Oglethorpe and will undoubtedly be always ready to assist you in your Labours to cultivate a Sense of Religion among the Europeans in your Settlement and if possible among the Natives who for many Ages have lived in the utmost darkness.

This is a Work that requires great pains and prudence to prosecute, and till they that profess Christianity can be persuaded to lead better lives, becoming their holy Profession, the difficulty of gaining the Heathen to embrace a Religion which they see makes but little Impression on the Morals of those who profess it will without a miracle be insuperable.

This hath been the constant Complaint of our Missionaries in the East Indies as the grand Obstacle to the Success of their Labours, but if it please God to interpose the powerfull Influence of his holy Spirit to turn the hearts of those who profess Christianity to shine in their lives, such Examples would bear down all Opposition, and recommend the Truths of the Gospel, so as to give Joy to those who have the honour of propagating them, in which that you may always succeed is the Wish of the Society as well as

Reverend Sir   Your etc.   H. Newman

Bartlet's Buildings London 13th October 1735.   To the Reverend Mr. Wesley Going to Georgia.

Reverend Sir: I hope you received the packet of Books in due time by the Society's Messenger a list of which is enclosed by which you will see the Society desire you will spare what you can to supply the present wants of Messrs. Bolzius and Gronau, and in return I am sure you will be welcome to receive any out of their or Mr. Quincy's Store which you may happen to want.

I hope you long since received the Society's Circular Letter for this Year, but my Clerke having omitted to enter it in the Index for that purpose I have herewith covered a Copy of it, for fear it has been forgot.

I heartily wish you and your fellow Travellers with Mr. Oglethorpe a prosperous Voyage, and that it may please God to bless you with health and Success in the high Errand you have undertaken for his Glory, of which it will be a great pleasure to the Society to be as freqently informed as Opportunitys offer by Reverend Sir   Your most obedient humble Servant   H. Newman

Bartlet's Buildings 15th October 1735.   To the Reverend Mr. Gronau at Ebenezer in Georgia.

Reverend Sir: After I had sent the 15 £ of Copper on board the London Merchant for your Salary 1/2 Year from the 1st of November next on Account and by order of the Society for promoting Christian Knowledge; the Society considering your great Zeal and Faithfullness in assisting the Reverend Mr. Bolzius and discharging the Office of Catechist among the Youth of his and your Flock, have unanimously agreed to augment your Salary from 30 pounds to 40 pounds per annum and accordingly desired Wm. Tillard Esq. one of their Treasurers to honour your Bill or Bills on him for the 5 pounds deficiency of the ensuing half year which will end the 1st day of May next. If you please therefore to draw a Bill or Bills on him for that Sum according to the Exchange that governs for that value in Georgia, and give him or me advice thereof Your draught will be duely honoured. Be pleased to signify in your next whether your receiving you Salary in Copper money for the future or drawing Bills for it on our Treasurer may be most for your Benefit.

I refer you to my letter of the 10th Current by Mr. [Ph.] Von Reck for the rest and remain
          Reverend Sir   Your most humble Servant   H. Newman

Bartlet's Buildings London 31 October 1735.   To the Honourable Mr. [Ph.] Von Reck in the London Merchant at the Downs.

Dear Sir: I received your favour of the 27th Current and immediately forwarded in the Kings Packet the enclosed Address to your Excellent Uncle at Ratisbonne. By your long Silence your friends here imagined that Mr. Oglethorpe and you were gone through the Downs. I made your Complements to Mr. Vernon etc. here who return theirs. Yesterday I received a small Box of Medicines from Germany by the hands of Mr. Ziegenhagen for Messrs. Bolzius and Gronau but as you expect to depart the first fair wind I shall reserve it to be sent with the next Ship which Mr. Simond's Clerke tells me will be Capt. Tompson. Pray let me hear from you by all Opportunities particularly if you are detained by contrary winds on the Coast of England. Please to write from Madera. Once more wishing you a good Voyage I recommend you to the Divine Protection and remain

Dear Sir   Your most obedient humble Servant   H. Newman

Bartlet's Buildings London 14 November 1735.   To the Reverend Messrs. Bolzius & Gronau at Ebenezer. [Marginal Note:] per Capt. Tompson

Reverend Sirs: Having wrote largely to you the 10th of last Month by Mr. [Ph.] Von Reck in the London Merchant who is now at Cowes in the Isle of Wight with Mr. Oglethorpe waiting only for a wind, I have by this Ship to add that Capt. Tompson signed 3 Bills of Lading the 20th of October at Gravesend for the parcells sent to you in his Ship, one of which Bills Mr. Verelst delivered to Mr. Von Reck, another to the Captain and the 3rd is left with me. The parcells mentioned in the Invoice sent by Mr. Von Reck are marked

B.G.S. No. 1[,]   B.G.O. 2[,]   B.G.Z. 3[,]   B.G.F.Z. 4[.]
Since which I have recommended to the Care of Mr. Simonds a Small Box of Medicines from Germany sent to me by Mr. Ziegenhagen to be forwarded by Capt. Tompson's Ship which I wish safe to your hands and remain

Reverend Sir   Your most humble Servant   H. Newman

Bartlet's Buildings 14 November 1735.   To Mr. John Vat at Ebenezer in Georgia By Capt. Tompson.

Dear Sir: I have wrote largely to you the 10th of last month by Mr. [Ph.] Von Reck in the London Merchant now at the Isle of Wight where they have waited some time for the Hawk Sloop Man of War, and now she has joined them I hear they wait at Cowes only for a Wind to sail with Mr. Oglethorpe in the Simmonds for Madera and thence to Georgia whither I heartily wish them all a good Voyage.

I hope all things may answer your Expectation in the Service
you have generously undertaken, by the measures Mr. Oglethorpe
will take on his arrival.

As to publick Affairs here the Suspension of Arms between the
Emperor and France 'tis hoped will pave the way to a general
Peace by a Congress in some place not yet agreed upon, but sup-
posed will be at Aix la Chappelle. I beg leave to refer you to the
prints herewith sent for the rest praying God to direct you in the
Station his providence has placed you for his own Glory and your
Comfort and remain

    Dear Sir Your most humble Servant H. Newman
P.S.: Enclosed are 7 London Journals from 4th of October to 15
Current Inclusive.

*Bartlet's Buildings 26 November 1735. To the Reverend Mr.
Urlsperger at Augsburg.*

Reverend Sir: My last was of the 7th of October since which I
have received none from you[.] Mr. Oglethorpe and Mr. [Ph.] Von
Reck have been wind bound at the Isle of Wight for several Weeks
but by the last advice are gone off all in good health and by this
time 'tis hoped they are clear of our Coast. Upon their departure
out of the Thames last Month the Society promised in our news
papers an Account of their Expences on the Emigrants from Germany to Georgia, and in order to make it complete they would be
favoured with an Account of the Application of the undermentioned Summs which stand in their Books as in your hands viz.

	£ s d
Ballance of Mr. Urlspergers money as by the printed Account 4 October 1734 which I formerly sent to you ___	100.11.6
1734 Dec. 18 Paid Mr. Wilson 23 £ 5 s 6 d which he paid Mr. Vat for the Saltzburgers, being for 198 florins the Evangelical Body remitted to them as by Mr. Urlspergers letter of the 21st October 1734	23. 5.6
Jan. 14 Paid Mr. Ziegenhagen the amount of 302 Florins part of 500 fl. ordered by the Society the 15th Oct. last for the use of the Protestant Mission to East India as by Mr. Urlsperger's letter 21st Oct. N.S. 1734 ___	35. 4.3

		£ s d
1735 July 14 Paid Messrs. Meyer & Jansen for Mr. Urlspergers draught on the Society for 500 florins he received of Mr. Münch for the use of the Saltzburgers which 500 fl. at 8-5/6 of a florin per pound Sterling is _____		56.12.–
Oct. 7 Paid Mr. Ziegenhagen for 269 florins desired by Mr. Urlsperger in his Letter the 15th Sept. 1735 _____		30. 9.6

£ 246. 3.2

The Society will be glad to be favoured with your answer to this with your first Convencience and to know whether the Spirit of Persecution be so abated in Saltzburg as to leave no room for expecting more Emigrants from thence, and whether the prospect of a general peace does not alleviate the Sufferings of the Protestants in other parts of Germany. I heartily wish you all manner of prosperity and remain

Reverend Sir   Your most obedient etc.   H. Newman

P.S.: I have no late news from Mr. Vat or the Missionaries at Ebenezer.

Bartlet's Buildings 26 November 1735.   To the Reverend Mr. Ziegenhagen at Kensington Square.

Reverend Sir: The Committee of Auditors could not go through their Examination yesterday because some Accounts wanted to be fairly transcribed for their signing in order to make a general Audit from the Beginning of the Receipts and Payments on Account of the Saltzburgers, and therefore they have adjourned themselves to next monday morning to finish it, against which time I beg to have the Articles of your Account which are entered on our Minutes transcribed in the Book herewith sent with an Account of the Application of the Ballance of 5 Guineas in what terms you please that it may be closed and Mr. Philipps's Servant will bring it to me if he hath it by Saturday night.

Mr. Dean Maddox has promised to be here next Tuesday to go on with the Instructions to the Missionaries and therefore I beg you will be here that nothing may delay their being prepared to go by the first Ship to Fort St. George.

I am   Reverend Sir   Your most humble Servant   H. Newman

P.S.: Please to return the Copy enclosed with the Transcript because that is already mark by the Auditors.

Bartlet's Buildings 28 November 1735. A Son Excellence Mons. [J.] De Reck Conseiller de sa Majeste Britanique et Son Envoyè à la Diete de l'Empire à Ratisbonne.

Sir: I am glad by this to acquaint your Excellency that your Nephew Mr. [Ph.] Von Reck after having been detained by contrary winds at the Isle of Wight for seven [several?] Weeks is sailed in Company with Mr. Oglethorpe for Georgia[.] They design to stop by the way at Madera to refresh themselves which will prolong the Voyage 14 days or three Weeks so that if it please God to grant them a safe Passage I hope some time in April next you have a letter from him in Georgia.

You will excuse the liberty I take in covering the enclosed to your Excellency for surer Conveyance and believe that I am

Sir Your Excellency's most obedient humble Servant

H. Newman

Bartlet's Buildings December 9 1735. To Messrs. Bolzius and Gronau at Ebenezer in Georgia.

Reverend Sirs: Yesterday your very acceptable Letter of the 1st of September was laid before the Society who are extremely pleased with the agreable Account you give of your ministry and more particularly rejoyce in the happy Success of your Endeavours among the Children who progress in the Knowledge of their holy Faith, gives them good hopes that the same Zeal which hath animated their Parents will be transmitted to the latest Posterity.
[Marginal Note:] NB. This letter being returned from Deal Capt. Tompson having sailed through, was sent by Capt. Peircy via Charlestown 27 December 1735.

You have their hearty Approbation in what you are doing and their most earnest prayers that you may continue to see the same good Fruit of your labours.

The Gentlemen of the Society take an Interest in whatever administers to your Comfort; I am therfore commanded in their name to congratulate you and your Fellow Labourer on your respective marriages,[51] which must be attended with the most advantageous Consequences both to your Selves and the Congregation[.] You have chosen Companions that will by God's blessing make the Journey of Life easy and comfortable, and contribute to Your spiritual as well as temporal Interest. They will if possible still farther endear you to the people under your Care as you have given so kind a pledge of your Attachment to them, and will by this means be most disposed to receive your Christian Exhortation and advice.

I am glad the money and wine sent by Capt. Yoakley came safe to hand. They have sent by Mr. [Ph.] Von Reck in the London Merchant another half years Salary and some Books which they believe will be acceptable to You.

Give me leave to add my own Congratulations upon these happy Events and my Prayers that Almighty God will be pleased to continue his Protection that you may see the blessed Effect of your pious and Christian Zeal in present Satisfaction, and enjoy the reward hereafter promised to those who turn many to Righteousness.

I am   Reverend Sirs   Yours etc.   H. Newman

P.S.: My humble Service to Mr. Vat & let him know that I hope e're long to receive a letter from him in answer to mine by Mr. Von Reck & Capt. Tompson Commander of the 2 Brothers.

Bartlets Buildings 2 January 1735/6.   To the Reverend Mr. Urlsperger at Augsburg.

Reverend and Dear Sir: My last was of 26th November since which I have received none from You, but have taken the liberty to repeat mine as above for fear of any Miscarriage. The Society having engaged to publish to the world an account of the distribution of the Summs of money received and distributed by them on Account of the Saltzburg Emigrants would be glad you could transmit to them the best Account you are at present able to give, consistent with your other various and important Affairs.

The beginning of last month I received a letter from Mr. Bolzius at Ebenezer of the 1st of September by which he and his Colleague were both happily married to the 2 Daughters of Mrs. Kroherin who went over in the first Transport with Mr. [Ph.] Von Reck. All their Congregation are well and rejoice in an happy Tranquility not doubting but on Mr. Oglethorpe's Arrival every difficulty will be removed that could give them any uneasily. Mr. Vat is well but I have no letter from him. May you enjoy an happy new year and that it may please God to bless you with many of them for the Advancement of his Glory and the publick Good is the sincere wish of

Reverend Sir   Yours etc.   H. Newman

P.S.: I fear I have forgot to thank you as I now do for your kind present of 3 Copies of the History of Emigrants which the Society received from Mr. Von Reck and for your care in transmitting the Mission Screw Medals which came safe to hand.

Bartlet's Buildings 27th January 1735/6.   To Mr. [Ph.] De Reck at Ebenezer in Georgia.

Dear Sir: Just now I received the enclosed from your good Uncle to be forwarded to you by the first Opportunity, which Mr. Simonds tells me will offer in a day or two by the way of Charles Town.

I hope this will find you and Mr. Oglethorpe happily arrived at Savannah with all the honest People you carried with you, and that you found Mr. Vat, and the Reverend Messrs. Bolzius and Gronau in good health with the *Wives of the* later [latter]. Mr. Vernon, Mr. Ziegenhagen, Mr. Butjenter, Sir John Philipps, and all your friends of the Society are well, but I have not had one line from Mr. Urlsperger since you went, the reason of which silence he has signified to Mr. Ziegenhagen to be occasioned by the death of his Colleague in the Rectorship of St. Anne, and the ordaining of another to succeed him, but I don't hear of any more Emigrants. I wish the general peace which Providence seems to open a door for to Europe may produce any relaxation to the tyrannical Oppressions of the Romish Clergy to those Excellent Persons who dare to suffer Persecution for the sake of Conscience: You have tasted of the Tryal your Self and therfore know what it is, at least in some measure. I pray God to give me Grace to follow the Example of those Confessors you have led to the Wilderness of America if it should ever be my lot to be called to it or to renounce the Dictates of my own heart against him that made and redeemed me.

Pray give my most humble Service to Mr. Oglethorpe and to the Reverend Mr. [John] Wesley and his fellow Labourers that attended him. The Reverend Mr. Quincy having desired leave to come home the Society for Propagating the Gospel have readily granted it and that Mr. Westley [John Wesley] or his Brother [Charles] should succeed him as a Missionary from that Society, concerning which I suppose he will have a proper Signification from the Reverend Dr. Humphreys Secretary of that Body.

I beg my humble Service may also be acceptable to Mr. Vat and to Messrs. Bolzius and Gronau from

Dear Sir   Your most obedient humble Servant   H. Newman
P.S.: My humble Service to Your Brother.

Bartlet's Buildings February 12th 1735/6.   To Reverend Mr. Urlsperger Senior Rector of St. Anne's at Augsburg.

Reverend Sir: The Society are concerned to understand by Mr. Ziegenhagen that the reason of you long Silence hath been occasioned by the death of your Colleague, and the filling up of his place, which they will be glad to hear is to the Satisfaction of your Self and your flock.

I am now by order of the Society to acquaint you in answer to your Enquiries by Mr. Ziegenhagen, that they are mindfull of their former Engagements of transporting 300 Emigrants to Georgia, of which upwards of one hundred and fifty are already gone in 3 Transport, and if the remaining number should come down not less than 50 or more than 60 at once and offer themselves to go to Georgia, upon your advising the Society thereof, and that they can be here before the Month of August, care will be taken for their Embarkation.

I am likewise ordered to acquaint you that the Society are not in a Condition to allow any thing to maintain them in Germany till such time as they are all ready to set out together from thence in their way to Holland.

Dear Sir, The Society are very sensible of your fatigues and trouble in assisting these pious Confessors, and hope God will reward your labours. I have no advice from Georgia since my last of the 2nd January but hope in April we may hear of the safe Arrival of Mr. Oglethorpe and Mr. [Ph.] Von Reck with their Respective Transports of which I shall be glad to acquaint you.

May it please God to continue your health for the ensuing year and many more for the Advancement of his Glory is the wish of the Society as well as

Reverend & Dear Sir   Your etc.   H. Newman

Bartlet's Buildings 13 February 1735/6.   A Son Excellence Mons. [J.] De Reck Conseiller de sa Majeste Britanique et Son Envoyè à la Diete de l'Empire a Ratisbonne.

Sir: I have had the honour of your letter of the 23rd January enclosing a packet for your Nephew, which came in good time to be forwarded to him by a Ship going to Charlestown in South Carolina about 100 Miles northward of Ebenezer, so that he will have it within a few days as soon as if it had been sent directly to Georgia. The last letter I had thence was dated 11th September when the new Colony of Saltzburgers were all well.

I shall be glad to send your Excellency an Account of their Welfare, especially while your Nephew is at the head of them to assure you that I am

Sir   Your Excellency's most obedient humble Servant
Henry Newman

Bartlet's Buildings London 19th March 1735/6. A Son Excellence Mons. [J.] De Reck Conseiller de sa Majeste Britanique et Son Envoyè a la Diette de l'Empire a Ratisbonne.

Sir: By a Ship just now arrived from Carolina advice is come that Capt. Rous who commanded her met with Mr. Oglethorpe and Mr. [Ph.] Von Reck in their respective Ships the 1st of February last about 30 leagues on this side Georgia[.] The Captain went on board of Mr. Oglethorpe's Ship and received a letter from him signifying that all the Passengers on board both Ships were in good health, not one person dead but several Children born in the voyage. That a Ship from Scotland and another from Ireland laden with Passengers were happily arrived in Georgia before Capt. Rous left Carolina. More particular Advices are every day expected in the mean time it is with pleasure I congratulate you on your Nephews having so far accomplished their Voyage and hope it may not be long before you will hear of their safe Arrival from

Sir   Your Excellency's most obedient humble Servant
H. Newman

London Bartlet's Buildings 13 April 1736. To the Reverend Mr. Urlsperger Senior Rector of St. Anne's Church at Augsburg.

Reverend Sir: I received your favour of the 16th February last in due time and acquainted the Society with the Contents of it as soon as I could get it translated, in which they observe with pleasure the Continuance of your unwearied Application for the Service of the persecuted Emigrants[.] I am glad now to acquaint you that Mr. [Ph.] Von Reck with the 3rd Transport to Georgia under his care happily arrived at Savannah River the 4th of February, all the Emigrants being in good health, the Ship Simmonds that Mr. Oglethorpe was in, arriving also at the same time with his Crew all in good health. That both Ships were preparing to sail to Alatamaha River about 10 Leagues southward of the Savannah where they were advised of finding a good Settlement. This advice comes from Charlestown in South Carolina, and

therefore brings no letters from Mr. Von Reck but a Ship is daily expected from Savannah by whom you may expect to hear more particularly. By the same advices the Society are informed that all the Saltzburgers at Ebenezer are well, and Mr. Oglethorpe intended to visit them before he set out for Alatamaha to do every thing in his power to make them easy in their Settlement there. The News Papers here advise that the Court of Vienna is disposed to make some Concessions in favour of the Protestants in the Treaty of Pacification now under Consideration[.] Pray what Grounds are there for that advice? Your Correspondents at Ratisbonne can very well inform You, and you will be so good as to enable me to acquaint the Society with them in your next letter to
Reverend & Dear Sir   Your etc.   H. Newman

London Bartlet's Buildings 16 April 1736.   A Son Excellence Mons. [J.] De Reck Conseiller de sa Majeste Britanique et Son Envoye a la Diete de l'Empire a Ratisbonne.

Sir: The 19th of last month I had the honour to acquaint your Excellency that Mr. [Ph.] Von Reck and his Brother were met in good health about 30 Leagues on this side Georgia there are now letters come advising that Mr. Oglethorpe and They safely arrived at Savannah the 4th of February and were preparing to sail with both their Ships to Alatamaha River 10 Leagues southward of Savannah to settle the People they carried over. A Ship is daily expected from thence with more particular Advices, by whom I hope to receive letters for Your Excellency. The News papers here signify a disposition in the Court of Vienna to make some Concessions in favour of the Protestants in the Treaty now under Consideration. I shall be glad to hear such a piece of News confirmed by your Excellency and desire you may be assured that I am with the greatest respect
Sir   Your Excellency's most obedient humble Servant
H. Newman

Bartlet's Buildings 11 May 1736.   To the Reverend Messrs. Bolzius and Gronau at Ebenezer in Georgia.
[Marginal Note:] By Anna Maria Capt. Scott for So. Carolina.

Reverend Sirs: I received your letter of the 8th of January by Capt. Snelling via Charlestown but being daily in Expectation of hearing from you more directly in answer to what I wrote by Mr.

[Ph.] De Reck, I have now only to acquaint You by order of the Society that they are endeavouring to get Credit for 50 £ to be sent to you for your Salaries 1/2 Year from the 1st Current to the 1st of November next which I doubt not but Mr. Oglethorpe will readily furnish you with, if you should happen to want it before a regular Credit come to your hands from

Reverend Sir   Your most humble Servant   H. Newman

P.S.: Please to let me know whether a Standing Credit on which you may draw Bills or the Remittance of money in Specie, be most acceptable. Mr. Ziegenhagen wants also to make You a Remittance.

Bartlet's Buildings 11 May 1736.   To Mr. Harman Verelst at the Georgia Office.

Sir: By order of the Society here I have sent 1/4 of a hundred of the Accounts they have lately published of their Receipts and Disbursements on Account of the Saltzburg Emigrants which you are desired to make Acceptable to the Honourable Trustees for Georgia, and if any more Copies may be wanted please to command them.

There being no Ship going to Georgia Directly for some time the Society would be glad to know whether they may be favoured with a Credit for 50 £ Sterling to pay the Missionaries and School Master their Salaries in Georgia at the Current Exchange upon taking their Bills on Wm. Tillard Esq. for that Sum towards their Salaries from the 1st of this Month; the Bills to be drawn by Messrs. Bolzius and Gronau.

I am   Your most humble Servant   H. Newman

London Bartlet's Buildings 31 May 1736.   To His Excellency Baron [J.] de Reck Minister from His Majesty K. George at Ratisbonne.

Honoured Sir: The Bearer Mr. Michael Jastram hath the Character of a worthy Gentleman from every body that knows him, and being to take a Journey to Venice I take leave to recommend him to Your Excellency's Protection & favour at Ratisbonne in his way thither if occasion require.

Your Nephew's [Ph. von Reck's] last letter to me is dated 11 March at Ebenezer expressing some uneasiness at the Situation of the Saltzburgers at that time, but I hope by the representation now made to the Trustees for Georgia in the name of the Society for promoting Christian Knowledge every difficulty may be removed that can give them uneasiness.

I hope to receive particular advices from him by a Ship daily expected of which your Excellency will be informed God willing by

Sir   Your Excellency's most obedient humble Servant

H. Newman

Bartlet's Buildings 31 May 1736.   To the Reverend Mr. Urlsperger at Augsburg.

Reverend Sir: I received your letter of the 14th Current N.S. with the melancholy news of the sudden death of your Son when he was in his way to pay his duty to his Father from Hall. The ways of Providence are not to be penetrated into, but you have the Satisfaction of reflecting that you have spared no pains or Cost to bestow a pious Education on him to fit him for the Service of his Maker if it had pleased God to have lengthened his Life. I am ordered to send you as I now do the Condolences of the Society upon so afflicting an Accident, and to assure you of their wishes that it may please God to bless You and the remains of your family with more durable Comforts wherein joyns most heartily

Reverend & Dear Sir   Your etc.   H. Newman

Bartlet's Buildings 3 June 1736.   To Messrs. Bolzius and Gronau at Ebenezer in Georgia.

Reverend Sirs: I received both your letters of the 10th and 16th of March last. Mr. Gronau's letter of advice about the 5 £ Bill payable to Mr. Purry was punctually complied with to compleat his Salary for 1/2 Year the 1st of May last. As for the Current years Salary's to your Selves and Mr. Ortman the Society have applied to the Trustees for Georgia to furnish you with a Credit for 50 £ which is accordingly done to the 1st of November next.

I must now acquaint you that all your Complaints of every kind from Mr. [Ph.] Von Reck and your Selves have been particularly represented to the Trustees in the name of the Society by one of their own Body who is also one of the Trustees, and they have reason to believe that such effectual directions will be sent by the Ship that carries this as will remove all Difficulties from the Saltzburgers in the power of the Trustees; The particulars of those Instructions being an Affair that goes from the Trustees Office I must desire leave to refer you to the Advices you will receive from their Secretary.

I made your Complements to the Society who return theirs with wishes for your Success in all your Labours at the same time I assure you of being

Reverend Sirs   Your most obedient humble Servant
H. Newman

P.S.: By the Invoice inclosed You will see what I have sent by the 2 Brothers on the Account of the Society and on Account of The Reverend Mr. Ziegenhagen, All the things therein contained being recommended to the Care of Mr. Verelst of the Georgia Office.

Drugs sent Mr. Zwiffler at Ebenezer in Georgia for the use of the Saltzburgers according to his Letter of the 13 March 1735/6 by order of the Society for Promoting Christian Knowledge. [Marginal Note:] 1736 4 June No. 7.

Antimon. Crud	—	—	—	—	—	lb. 6
Sal. Nit	—	—	—	—	—	7
Cinna C. Natu	—	—	—	—	—	1
Rad. Jallap	—	—	—	—	—	1
Ciner Clavel	—	—	—	—	—	4
Crem. Tart.	—	—	—	—	—	½
Gum. Assa Faedit	—	—	—	—	—	?
Sem. Bardan	—	—	—	—	—	½
Rad. Liquorit	—	—	—	—	—	6
Succ. Liquorit	—	—	—	—	—	½
Writing Paper	—	—	—	—	2 Quires	
Brown Do.	—	—	—	—	4 Do	
Linnen, Rasors, and Scissars	—	—	—			
A Box.	—	—	—	—	—	

London 28 May 1736.

Invoice of Goods shipped in the Two Brothers Capt. Thomson Commander, on Account and by order of the Reverend Mr. Ziegenhagen and Consigned to the Reverend Messrs. Bolzius and Gronau at Ebenezer in Georgia Directed and marked as in the Margent Viz.

To Messrs. Bolzius & Gronau:

No. 1.   A Chest containing 4 cwt weight of Books.

2.   A Box of Books.

3.   A Box of wearing Apparel.

4 A Ten Gall. Wine Cask ⎫ 5. ⎱ ⎬ 2 Six Gal. Wine Casks ⎨ 6. ⎰	containing 6 cwt weight of Copper farthings and 137 Ounces in pieces of 8 con- sisting of 158 New & old pieces of 8 pack up in No. 4.

N.B.   In the aforesaid Box of Druggs there is a small Box directed to Messrs. Bolzius and Gronau containing the Ingredients for Red Ink, with a few of the Society's Books to fill up the Box.

Bartlets Buildings 3 June 1736.   To Philip Geo. Fred. De Reck Esq. at Ebenezer in Georgia.

Honoured & Dear Sir: I received the Journal of your Voyage to the 5th of February and your short but melancholly letter of the 11th of March referring me to your longer Letter to Mr. Vernon, all which were laid before the Society, and referred to a special Committee who have taken the pains to represent all your Grievances to the Trustees in the most effectual manner, and Mr. Vernon has assured the Society that nothing in the power of the Trustees will be wanting to remove whatever may occasion un-easiness to the Saltzburgers, but for the particulars of their Reso-lution, I am obliged to refer you to the Letters You will receive by order of the Trustees and shall be glad to hear that they are agreable to your Wishes and those of the Saltzburgers.

Your good Uncle the Baron [J. von Reck] at Ratisbon was well not long since, and I have signified to him an Account of your health by a letter I had lately the honour of writing to him.

I hope the Countrey proves agreable to your Brother and wish every thing may conduce to both your Satisfactions being with great respect

       Sir   Your most obedient humble Servant   H. Newman
P.S.: My humble Service to your Brother.

London Bartlet's Buildings 3 June 1736.   To Mr. John Vat at Savannah in Georgia.   By the 2 Brothers Capt. Thompson.

Dear Sir: Not having had one line from you by Capt. Thom-son nor the Simonds nor any of the late Ships I concluded you were coming home, and still believe I have a chance to see you by the next Ship because of your silence. and that I doubt the Climate of Georgia is not so agreable as you expected it would be, tho' Mr. Oglethorpe's presence there must very much alter things for the better as he is the Life and Soul of the Colony.

If this find you in Georgia pray let me know by a line or two now and then if not by every Ship that you are alive, whether you want any thing and how the Countrey agrees with you, what prospect there is of its answering the Expectations of those who wish well to it, in respect to the products of Rice, Raw Silk, Naval Stores and any other things for the Benefit of Great Britain where all our hopes and wishes in Trade must center at last. I pray God to direct you for the best, whether you come home or stay at Georgia, and if you will let me know wherein I can serve you I shall with pleasure receive and obey Your orders being

Sir   Your most humble Servant   H. Newman

P.S.: My humble Service to Col. Purry & his Son[.] Mr. Johnson of Lombard St. is well and wonders at your Silence to him as well as me but we both infer from it that you are coming home.

London Bartlet's Buildings June 4th 1736.   To the Honourable James Oglethorpe Esq. at Savannah in Georgia.

Sir: I cannot let this Ship return to Georgia without paying my Duty to the Person who hath given Birth to the Cultivation of the Countrey, and to whose wise Maxims for establishing it, the future happiness and Prosperity will under God depend.

America is much beholden to Gentlemen of your Genius for without such, the Original Natives, and their Children might have slept on to the End of the world, and never heard of any other way of living than in Wigwams or one remove from the wild Beast of the Forest. I hope the Examples which the Europeans under Your discipline will set to the Natives will give virtue an Amicable Lusture, and let them see that the Christian Religion is not only in name but in reality an Improvement of the Divine and Social Virtues that adorn the humane Species.

I own it is easier to speculate upon things of this kind than to put them in practice, but I hope you are the *White Man* that will according to the Tradition of the Indians, deliver them from the darkness that hath for many Ages overspread them.[52]

Some Representations to the Trustees in behalf of the Saltzburgers have been put into Mr. Vernon's hand to be recommended in the name of the Society here, and the Trustees will I doubt not recommend to you what shall be thought proper upon every head and your own love of Equity and Justice will incline You to what is best.

Mr. [Ph.] Von Reck and Mr. Vat will I hope always merit your

Protection; they have listed themselves under your Banner, and I hope will never repent the Choice they have made.

If you will honour the North parts of the Continent of America with your presence, at least as far as Boston, before your return to great Britain, I am sure they will be proud of seeing you every where, and if the King himself were to go thither I believe it would hardly be possible to shew him more respect than they will do Mr. Oglethorpe but I don't say this to discourage You, for you know the World too well to be instructed how to avoid Complement when it becomes a burthen.

I heartily wish you health and a safe return to Great Britain in God's good time, that I may have the honour personally to assure you that I am with all possible Respect

        Sir   Your most obedient humble Servant   H. Newman

P.S.: I have herewith sent and pray your Acceptance of the last Account of Money's received and distributed by the Society on Account of the Saltzburgers.

Bartlet's Buildings 8 June 1736.   To the Reverend Mr. [John] Wesley at Savannah in Georgia.

Reverend Sir: It was a pleasure to all your friends here to be informed of your safe Arrival in Georgia. I hope the Country in all respects answers your Expectations. Mr. Broughton was so kind as to promise me a sight of your Journal, which I hope e're long I may see, tho' I suppose it chiefly related to the Occurrences of your Voyage thither, but any remarks from a Gentlemen of your Taste must be acceptable. I wish you find the Indians as tractable to Religious Instruction as you expected, but the method of conveying that Instruction is so labourious that it seems insuperable without a Miracle (considering the Brevity of humane Life) for you must either learn their Language or they yours before you can instill the first Rudiments of Christianity into their minds. To do the former there is neither Dictionary nor Gramar to lead you, and you must endure the Mortification to live Savage as they do at least a Year to make any Proficiency in it; but where those difficulties have been surmounted as in New England it served only for a small district not so big as Yorkshire[.] beside the barrenness of their Language, would puzzle a learned Man that is Master of it to express divine Truths in the Clearness they are made to appear in a Language that hath been for several Ages polishing. For which reason the people of

N. England seem now convinced after 100 Years experience that the shortest way to instruct the Indians is to teach them English & good manners in order to instruct them in the Christian Religion. The Adult People will hardly be capable of this but the young of both Sexes may by a method of Instruction be led into a way, that may gradually diffuse the Language and common Civility among themselves till as in some parts of Ireland the Old wild Language may fall into Oblivion[.] But of all these difficulties you are by this time fully apprized, and I can suggest nothing but what you better know already. I can only wish that your Patience and Courage may never fail you, for you will find great Occasion for both to make any considerable Progress in an undertaking of this nature, and after all, 'tis the Blessing of God only that must crown your Labours with any thing like Success[.] Our Zeal in N. England went so far as to build at a great Expence a Brick College at Cambridge to accomodate the Indians, and a large printing House, where a quarto Bible was printed in the Indian Language under the Conduct of the famous Mr. John Eliot voluntary Missionary among them, but the Indians could never be prevailed with to accept of an Education there gratis, to be obliged to quit their old wild liberty one only excepted Caleb Cheschanneuch by name who lived to be a Batchelor of Arts and then killed himself with drinking tho' all possible Care was taken to restrain him from it, so hard it is to refrain inveterate habits or Correct a Byass which Nature hath interwoven in their very Constitutions; and as for the Impression of the Bible the Language was so much altered in 70 or 80 Miles distance that a Chinese Bible would have been as edifying to the Natives as Mr. Eliot's Impression. I shall be glad to hear you find it otherwise in Georgia, That you enjoy your health and all other Blessings necessary to enable you to pursue your laudable designs, and desire you to be assured that I am

      Reverend Sir   Your most humble Servant   H. Newman
P.S. I have herewith sent a Letter from Mr. Thorold and the Account I have sent to our Members by order of the Society relating to the Saltzburgers. I hear Mr. Quincy is gone to N. England for which reason I don't write to him, but if I am misinformed pray let him know the reason of my Silence & give my humble Service to him.

London Bartlet's Buildings 8 June 1736.   To Mr. John Andrew Zwiffler Apothecary at Ebenezer in Georgia.

Sir: I take the same liberty to write to you in English as you do to the Society in High Dutch; you did not name the quantities of the Druggs you desired, but as the Society were pleased to leave it to me I have sent something of every thing you desired, and when you want more of either sort, dont let your Modesty prevent your naming the quantity necessary for so good a use as preserving under God the health of the Saltzburgers at Ebenezer. The quantities I have sent were by the advice of a Physician and a Chymist and thought at present sufficient but if I had sent a greater value I find the Society would not have disapproved of it. I am

Sir   Your most humble Servant   H. Newman

Bartlet's Buildings 16 June 1736.   To the Reverend Mr. John Wesley at Savannah in Georgia.

Reverend Sir: I wrote you the beginning of this Month by the 2 Brothers Capt. Thomson to whom I delivered ten large and small packets for Georgia, but Yesterday the Society had the pleasure of seeing a letter from You of the 23rd of March to Sir John Philipps signifying your want of Bibles and Common prayer Books, whereupon I was ordered to provide imediately a Box of 40 Bibles Minion[,] 40 New Testaments Long Primer etc.[,] 40 Common pray. Books minion[.] These are all packed up in a Box directed to your Self by the 2 Brothers which I hope may go safe, if I can be so happy as to ship them before the Ship falls down[.] Writing is very troublesom to Sir John Philipps and therfore hopes you will excuse his not answering this and a former letter which he received and desires you will accept his humble Service and thanks for. With his and the good wishes of the Society that the Blessing of God may always attend your Labours for his Glory, wherein joynes

Reverend Sir   Your most humble Servant   H. Newman

London 13th July 1736.   To the Reverend Mr. Urlsperger at Augsburg.

Reverend and Dear Sir: The Society not having heard from You since the 14th of May N.S. which I answered the 31st of the same month they infer from it that you have no prospect of a Transport of Emigrants this Year and have thereupon ordered me to acquaint you that they think the Season too far advanced for sending any Transport of Emigrants to Georgia this Year[.] I have received no advices from Georgia but letters are daily expected only I can assure You that the Rumours of Hostilities between the Spaniards & English in that part of the World are by the last Advices from thence without foundation. However the Rose man of War from So. Carolina has joyned Mr. Oglethorpe at Alatamaha to give his new Settlement there the better Countenance as there may be occasion.

I hear you intend to drink the Waters for establishing your health which the Society will be glad they may do, and hope your next will inform me of the Success of them.

I have committed to the Care of our Friend Dr. Guerdes a large Packet for you from Governor Belcher of New England which will be conveyed by some friend of his from
Reverend Sir   Your most obedient Servant   H. N.

London 30 July 1736.   To Mr. Dolon Deners at Eton Bucks.

Dear Sir: I received your favour of the 3rd Current with the Bill for 4 Guineas on Mr. Malmazet which was duely paid the delay of receiving it was owing to my being out of town and a report of it being made to the Society I am ordered to desire as I now do that you would make their thanks acceptable to the Charitable Benefactors and if your letter has been misunderstood as to the Application of those Benefactions, whether to the Society's designs in general or to the Protestant Emigrants from Saltzburg in particular You will please to rectify it. In the mean time I have paid them to Wm. Tillard Esq. one of our Treasurers for the Saltzburgers and herewith covered his Receipts Mr. Ziegenhagen having signified that he had sent his parcel to you and could not easily convey them franked I have therfore done it.

Col. Waller is dead in Ireland and so is the Gentlemen employed by the Society for recovering your debt which I fear is desperate, however another Gentleman has been lately empowered to see if any thing can be done.

We daily expect letters from the E. Indies and Georgia the advices of which you may expect to partake of either by Mr. Ziegenhagen or my hands.

The anniversary Sermon at St. Sepulchres with the Account of the Society annexed is not yet published for want of the Copy from the Author Mr. A. D. Denne who is gone in to the Country and has had too many Engagements hitherto to favour us with a Transcript of it but gives hopes of soon conveying it. When it comes out how may I convey some Copies to our Friends at Lausanne with such other things as have been emitted since my last Packet to them? I made your Complements to Sir. John Philipps and his Sons who return theirs by

Dear Sir   Your most obedient humble Servant   H. N.

London 22nd September 1736. To The Reverend Mr. John Wesley at Savannah in Georgia.

Reverend Sir: The Reverend Mr. Arch Deacon Rye of Islip having lately sent a Benefaction of three Guineas to the Society to be laid out in Bibles and sent to you and your Brother to be disposed of in Georgia that money has been laid out in purchasing 20 Bibles Minion which are packed up in a Box directed to you at Savannah and sent to the Georgia Office to be forwarded by the first Opportunity. In the same Box are packed up a few of the Society's Books lately imported into their Stores of which they desire your Acceptance by

Reverend Sir   Your most humble Servant   H.N.

London 1st February 1736. A Monsieur Monsieur Urlsperger Senieur Ministre de l'Evangile dans le Eglise de St. Anne à Augsburg en Suabia.

[Marginal Note:] Approved by the Society 1 Feb. and ordered to be sent forward which was done this day in the King's Packet under Cover to Baron [J.] de Reck at Ratisbonne.

Reverend and Dear Sir: I received in due time your favour of the 15th of last month N.S. and as soon as I could get it translated acquainted the Society with the Contents of it and by their Order sent a Copy of it to the Honourable Trustees for Georgia to answer such part of your letter as related to them and at their last meeting they came to the resolutions signified in Mr. Martin the Secretary's Letter of the 27th of last Month of which the Society have ordered me to send you the enclosed Copy.

As to what concerns the Society they have ordered me to acquaint you that their fund for relieving Saltzburgers is at present so low that they are not in a Condition to engage for any new Transport but if you be under any Engagements for assisting or supporting any persecuted Saltzburgers now at Augsburg or in your Neighbourhood which you have made in expectation of Assistance from the Society either for Old Persons Invalids or Apprentices the Society will once for all answer your Bill on

Wm. Tillard Esq. for any Sum now [not?] exceeding 40 pounds Sterling to make the 10 pounds you signify to be remaining in your hands up to fifty pounds as a final discharge of the Society from any further expence for Emigrants from Germany and they would be very well pleased if some part of this money might be given to the poor Woman mentioned in your letter who had made her escape from Saltzburg with 7 Children.

In answer to your Enquiry about the Tenure of Lands in Georgia in regard to the Succession of the female Sex[53] you see by Mr. Martin's letter that the Society have no Authority from the Trustees to give you answer to them.

Wishing you all manner of prosperity in your Zeal for relieving poor persecuted protestants I remain

Reverend & Dear Sir   Your most humble Servant   H. N.

London 1 February 1736/7.   A Son Excellence Monsieur [J.] De Reck Conseiller de sa Majesté Britanique at Son Envoyè a la Diete de l'Empire à Ratisbonne.

Sir: I congratulate your Excellency on the safe return of both your Nephews Messrs. De Reck from Georgia. The[y] arrived here the Beginning of December last and in 3 Weeks set out hence for Hanover intending from thence to pay their Duty to your Excellency at Ratisbonne[.] I shall be glad to hear of their arrival there and beg leave to assure you that I am

Sir   Your Excellency's Most obedient humble Servant   H. N.

P.S.: His Majesty continuing indisposed with a Cold the Lord High Chancellour this day opened the Session of Parliament with a Speech in his Majestys Name which I doubt not will be sent to Your Excellency.

Bartlet's Buildings 15 February 1736/7.   To Mr. Martin Secretary to The Trustees for Georgia.

Sir: The Reverend Mr. Guerdes having made a Report to the Society for promoting Christian Knowledge concerning Mr. [Ph.] Von Reck's unexpected Expences in terms inclosed I am ordered to desire you would acquaint the Honourable Trustees for Georgia with it and that you would signify their Pleasure thereon to

Sir   Your most humble Servant   H. N.

Bartlet's Buildings 24 February 1736/7.   To Mr. Verelst Accomptant at the Georgia Office in Palace Yard.

Sir: I hope you have received a hogshead of wine in bottles containing a Gross marked B.G. No. 3. This accompanys two small Casks marked B.G. No. 1, 2 containing 3 cwt of Copper Half pence and No. 2 has 373 oz. 9 dwt in foreign Silver added to the half pence for the Missionaries at Ebenezer to which there is now added another small parcell marked B.G. No. 4 containing Books and packets left here for the Missionaries.

If I can get them time enough I shall send half a Tun more of half pence but they can't be delivered till next Wednesday. If the Charge of Including them in your permit be inconsiderable I should be obliged to you to get them inserted and if the Ship stays till this day se'enight hope to get them on board in the way that you shall advise

Sir   Your most humble Servant   H. N.

Bartlet's Buildings 24 February 1736/7.   To the Reverend Mr. Ziegenhagen at Kensington.

Reverend Sir: I just now received the enclosed from Mr. Obuch and am glad to find by his Letter to me that all their Cabbins are built in the Steerage and that the Capt. is so good as to let them stay in the Nottingham though the Wind is contrary.

They complain of being crowded with people and Chests but those Inconveniences will lessen every day when they get out to Sea.

I have sent to the Georgia Office to go in the Peter & James Capt. Dymond the following parcells for Georgia viz.

	£ s d
B.G. No. 1. 2. Two small Casks containing 3 hundredweight of Copper half pence at 10 £ 10 s 6 d per cwt _____	31.11.6
In No. 2 is a small Bagg of pieces of 8 marked B.G. 69 oz. 18 dwt which at 5 s 3¼ d per Oz. cost _____	18. 8.6
	50. –.–
In the same Cask No. 2 is a large Bagg marked B.G.Z. cont. 303 oz. 11 dwt at 5/3¼ consisting of Rials and double Rials Spanish _____	80. –.–
	130. –.–

B.G.   No. 3. A hogshead containing a Gross of Vidonia Madera Wine in Bottles

No. 4. A Packet containing 2 Parcells from Germany which you left here and the Society's Packets to the Missionaries.

And I have desired Mr. Verelst to get a permit for adding 1000 weight of Copper half pence to these parcels but I can't receive them at the Tower till next Wednesday or Thursday so that if the ship sails before they must be left to go by the next[.] however what is now sent pays their Salaries to the 1st of May next which I have chose to forward without waiting for the other that they may be sure of something[.] And when all is on board I shall get them Insured[.] I am

Reverend Sir   Your most humble Servant   H. N.

P. S.: The sooner I have Your letters for Georgia the better.

Bartlet's Buildings London 25th February 1736/7.   To Mr. [Ph.] Von Reck at Ratisbon.

Dear Sir: I had the honour of your two letters from Hanover of 17/28 January and Zellerseld 2/13 February with those enclosed which were immediately forwarded according to their respective Directions. Viz. to Mr. Vernon Mr. Oglethorpe Dr. Guerdes Capt. Coram and I think to Mr. Ziegenhagen; I hope the inclosed may be an agreeable Answer to one of them and wish I could have congratulated You on the Success of a kind Motion Dr. Guerdes made in Your favour to our Society but it seems other demands on account of the Saltzburgers in Georgia have disabled them from complying with the said Motion and for the same reason you will find they have by their last letter to Mr. Urlsperger proposed closing their Account with him entirely. The Society were very much concerned to understand that your Journey through Germany has been attended with great fatigue and want of health to your Self and Brother. I have received the 2 Cocay [?] Nut Shells[54] which you was obliged to leave behind because the Artificer who had them was not to be found and I shall obey your orders for transmitting them to You. I beg my most humble Service may be acceptable to his Excellency the Baron Von Reck and that you will believe that I am

Dear Sir   Your most obedient humble Servant   H. N.

P.S. my humble Service to Your Brother.

Invoice of Goods and Money shipped on Board the Peter and James Capt. Dymond Master for Savannah in Georgia and consigned to the Reverend Messrs. Bolzius and Gronau at Ebenezer in Georgia by order of the Society for Promoting Christian Knowledge Viz.

B.G. No. 1⎫ Two small Casks containing 3 hundred ⎫ £ s d
       2⎬ wt. of Copper half Pence at 10 £ 10 s ⎬ 31.11.6
        ⎭ 6 d per hundred _____ ⎭

In Cask No. 2 is a small Bagg of pieces of ⎫
8 marked B.G. cont. 69 oz. 18 dwt ⎬ 18. 8.6
which at 5/3¼ per oz. cost _____ ⎭
                                    ─────────
                                      50. –.–

In the same Cask No. 2 is a large Bagg ⎫
marked B.G.Z. containing 303 oz. 11 ⎪
dwt at 5 s/3¼ d per oz. consisting of ⎬ 80. –.–
Spanish Rials and double Rials to be ⎪
applied as Mr. Ziegenhagen shall direct ⎭
                                    ─────────
                                     130. –.–

3. A Hogshead cont. 12 Doz. of Vidonia
   Madera Wine in Bottles.

4. A Packet containing 2 parcells from
   Germany and three packets for Messrs.
   Bolzius Gronau & Wesley by order of
   the Society at London.

5⎫ 5 Casks containing 8 hundredweight and ⎫
6⎪ 1 quarter weight of Copper half pence ⎪
7⎬ which at 10 £ 10 s 6 d per hundred- ⎬86.16.7½
8⎪ weight came to _____ ⎪
9⎭   ⎭

To Messrs. Bolzius and Gronau
their Salaries for half a Year
from 1st May 1737 to 1 November following _____ 45 £ 0 s 0 d
Mr. Ortman's Salary for said time 5   –   –
For Ditto a present from the Society in regard to his ill State
of health _____ 5   –   –
                                     ─────────
                                      55   –   –

By order of Mr. Ziegenhagen to
be applied as he shall direct_____ 31.16.7½
                                  ─────────
                          £ 86.16.7½

London 2nd March 1736/7.   By order of the Society
                                Henry Newman Secretary.
Received March 2nd 1736/7 on Board the Peter and James George
Dymond Commander five firkins marked B.G. 5 to 9.
                                        per Robt. Davies

*Bartlet's Buildings London 2 March 1736/7. To the Reverend
Messrs. Bolzius and Gronau Missionaries at Ebenezer in Georgia.*

Reverend Sirs: I have waited an Opportunity of sending the 4
first Articles mentioned in the enclosed Invoice ever since November last which has brought us so near to the 1st of May 1737 that
the Society have ordered me to send your Salaries for another
half year ending the 1st November 1737 which is accordingly
done in the 5 Cask of Half pence now added to the Invoice[.] I
have received both your letters of the 8th October and 4th December and according to your desire in the former have sent the
latter remittances to you in Copper halfpence and shall continue
to do so till you desire it may be otherwise. I acquainted the
Society with your desire of Physician or Surgeon but it will be
by great chance that any can be prevailed with to go to Georgia
on the terms Mr. Zwiffler has done and therefore it is wished he
could have found Encouragment to stay there by some Grant from
the Trustees in his favour.[55] The Society were glad to understand
by your letter of the 4th December that it hath pleased God to
restore you and your flock to so good a Measure of health as you
and they now enjoy[.] May it long continue and may all the
Difficulties you mention to Mr. Ziegenhagen be in good time
removed. I hope you have long since received the 1/2 years
Salary to 1st November last which the Society paid to the order
of the Trustees in June last but it seems the Credit they sent to
Mr. Oglethorpe did not arrive in Georgia till after he was come
away and it was forgot to impower any other person to execute
their orders but I hope that is effectually done 'ere now[.] Mr. [P.]
Von Reck and his Brother are gone to Germany but Mr. Vat
continues here in good health[.] Mr. Oglethorpe enjoys his health
and seems designing to return to Georgia which I wish may be
for the Advantage of the people. God Almighty continue to support you under all your trials and give you patience till you see
the end of them is the wish of

Reverend Sirs   Your most obedient humble Servant   H. N.
P.S.: The Gross of wine No. 3 now sent is for refreshing your
Selves and such of your Congregation as you shall think most want
it. And the Packet in No. 4 You will please to accept and to convey that for the Reverend Mr. Wesley. Pray what is your real
Opinion concerning the Inclinations of the Indians to embrace
Christianity or Instruction in the first Principles of common
Morality. I received Mr. Zwifflers letter of the 12 October last

but don't answer it because he seems to be on his return to England.[56]

London 2 March 1736/7.   To the Reverend Mr. John Wesley at Savannah in Georgia.

Reverend Sir: I received your letter of the 7th September last by Mr. [Ph.] Von Reck who with his Brother arrived here in December and after a short Indisposition having recovered their health returned to their friends in Germany in January. The Society desire your Acceptance of a small packet sent in the Missionaries Parcel B.G. No. 4 by this Ship. I shall be glad to hear of any Success you and Mr. Ingham have in your Conferences with the Indians and to understand that they are more tractable to Instruction in Christianity than they used to be in New England. Mr. Oglethorpe is well and will I hope be able to see you again this Summer to perfect the Establishment he has taken so much pains to encourage. I have not yet seen your Brother [Charles] since his return tho' I should be glad of it which I attribute to the Death of your and his good friend Sir John Philipps the Beginning of January last and presume he has never been this way since. Mr. Quincy is in the Countrey somewhere in Suffolk and I hear enjoys his health much better than he did in Georgia.

May you never want it for the good use you make of it is the Wish of the Society as well as

Reverend Sir   Your etc.   H.N.

London 3rd March 1736/7.   To Mr. Verelst at the Georgia Office.

Sir: I have your letter of this day and shall attend at the Custom house to morrow as you desire to get the Cocquet [cochet, seal] for the Silver tho' I thought the Swearing to it and the Copper half pence would be dispensed with as it came from the Georgia Office.

I have added to 3 cwt which I sent to your Office 8 cwt 1 quarterweight of Half pence in five Casks marked B.G. No. 5, 6, 7, 8, 9 and have shipped them from the Tower on the Peter and James according to the Receit enclosed from the Captain's Mate and shall be obliged to you to get me 2 Bills of Lading from Capt. Dymond including the 5 last parcells and the 4 sent to your Office.

I shall acquaint Mr. Ziegenhagen as you desire and am   Sir

Your most humble Servant   H.N.

NB. a Copy of the Receit enclosed is entered under the Invoice folio 94.

Bartlet's Buildings London 22 March 1736/7. To the Reverend Messrs. Bolzius & Gronau Missionaries at Ebenezer in Georgia.

Reverend Sirs: The Society being desirous that you should never be disappointed of your Salaries as they become due or even before they are due recommend it to your consideration whether it may not be more for the Interest to receive your Salaries in Georgia upon the Exchange that governs there by your Bills drawn on William Tillard, Esq. one of the Treasurers for the Saltzburgers, who will duly honour them, & by such method You can receive Payment in such Specie as you best like, & I know in some parts of America the Merchants greedily catch at good Bills payable in London.

My Letter of the 2nd Current by this Ship will inform you of the remittances made now in Copper money & Spanish Silver to pay your Salaries to the 1st November next, before which time you may have Several Opportunities to let me know your own inclinations on this head; And as to the Charge of 8 d. in the pound to those that take your Bills if that is insisted on and you can't have Money nor Credit for Goods without it, the Society would rather be at that charge themselves than you should be straitened for want of your Subsistence or if that might be saved by Paying the money here beforehand and sending you Bills hence they would readily do it, but then you will be under the necessity of taking your Payment in something you may chance to lose by, as you did formerly. Mr. Ziegenhagen just now gave me the Letters enclosed to be forwarded if possible by this Ship.

I wish you all manner of prosperity & remain
Reverend Sir   Your most Humble Servant   H. Newman

London 24 May 1737. To Mr. Zwiffler Apothecary at Ebenezer in Georgia.

Sir: This day I received the enclosed Letter from the Reverend Mr. Ziegenhagen and though I have reason to doubt whether you are still in Georgia[56] I choose to Send it forward, for fear it should be some Commission which the Writer of it desired might be executed there before your Leaving the Province. My humble Service to the Reverend Messrs. Bolzius & Gronau, and believe that I am

Sir   Your most humble Servant   Henry Newman

I have no Letter from Messrs. Bolzius and Gronau but what I have answered.

London 19 July 1737.   To the Reverend Mr. Urlsperger at Augsburg in Germany.

Reverend Sir: I received your favour of the 6th May and according to your desire the Society have directed the Payment of Forty Pounds out of the Saltzburg Fund to the Reverend Mr. Ziegenhagen pursuant to their Resolution Signified in their Letter of the 1st February last.

There were 4 Paragraphs of your Letter which related to the Trustees for Georgia, and a Copy of them having been laid before that Board, they have Signified to the Society by one of their worthy Members that the Succession of the Females is the same among the Saltzburgers as it is among the English, calculated for the Service of the Colony in its Infancy, and cannot be broke into without injury to their Original Constitution, and all the Grants already founded upon it, but as the Colony grows in Strength this may receive another Consideration.

They have granted the request of the 2nd Paragraph that the 3rd Transport of Saltzburgers have the same assistance in Cattle with other Priviledges as the former had, and that Orders are given accordingly.

In answer to the 3rd Paragraph the Trustees will always continue their particular regards to the Saltzburgers as Confessors for Religion, worthy of the compassion of every good Christian who Sympathizes in all the hardships that befall his Suffering Brethren.

To the 4th Paragraph. It is true that the Parliament have granted 20,000 £ to the Colony of Georgia, but they have at the same time appropriated it to such Uses as the Trustees have not power to deviate from, but as those uses are all directed for the defence of the whole, the Saltzburgers will not fail to be partakers of that benefit with the rest of the Colony.

I am glad to add that Mr. Oglethorpe is preparing to return to that Province to complete whatever may be thought necessary for the happy Settlement of it, in order to which, His Majesty has lately honoured him with the General Command of all his Forces in South Carolina and Georgia to quel any attempt from their Enemies to disturb their Tranquility.

The Society thank you for the advices Subjoined to your letter concerning the motions of the Protestants in Carinthia, Bohemia, Transylvania etc. and wish the present tendency to a Rupture with the Turks may contribute to Lessen the Spirit of Persecution in the Romanists, and to encourage the growth of true Piety among all Professors of Christianity.

May you long live in health to Succour those that are persecuted for the Sake of Truth, is the wish of the Society and of

Reverend Sir   Your most obedient humble Servant

Mr. Vernon returns his humble Service to you, I have just now received from Mr. Ziegenhagen 3 Copies of a new Edition of your Pourtraiture for Governor Belcher Dr. Colman and my Self, for the last of which I thank you & shall forward the 2 former by the 1st Conveyance.

Capt. Coram Sends his humble Service and Thanks to you for your Picture which is also come safe to his hands.

London 22 July 1737.   To Mr. Philip George Frederick De Reck à Ratisbonne.

Dear Sir: I thank you for your kind remembrance of me in your Letters to Capt. Coram from whom I have now enclosed a Letter. I am heartily sorry and condole with you for the Loss your Friends have Sustained by Fire at Zellerseld. I hope the same Providence that preserved their Persons will make up their Losses by some unseen prosperity.

Mr. Martin the Secretary of the Trustees for Georgia sent the enclosed to be forwarded to you; Mr. Oglethorpe is well, preparing to return to Georgia at the head of a Regiment with a Commission to command all the Forces of South Carolina, as well as Georgia. I had a Letter this week from Messrs. Bolzius and Gronau by which they & the Saltzburgers were all well; Mr. Zwiffler for want of Passage directly home is gone to Philadelphia in order as Supposed to take passage for London.

Mr. Vat is returning to Swisserland, having finished his Expedition to Georgia and sends his humble Service to you.

I shall be glad to hear of your and your Brother's Welfare, My humble Service to him from

Dear Sir   Your most humble Servant   Henry Newman

I hope your good Uncle the Baron [J. von Reck] excuses the liberty I take of covering your Letters to him.

London 19 July 1737.   To the Reverend Mr. John Wesley at Savannah in Georgia.

Dear Sir: I cannot forward the enclosed from Sir Erasmus Philipps without thanking you for your Letter to the Society of the 30th March last, and letting you know their concern for the difficulties you met with. Mr. Oglethorpe being appointed General of His Majesty's forces in South Carolina and Georgia, is preparing to return thither by whom I expect orders of the Society to trouble you with a fuller answer; this takes its chance by the way of Charles Town, with desire that you may be assured that I am

Reverend Sir   Your most humble Servant   Henry Newman

London 9th August 1737.   To the Reverend Messrs. Bolzius and Gronau Ministres en Georgia.

Reverend Sirs: Mr. Ziegenhagen is well & just now puts the enclosed Letters into my hand to be forwarded to you by this opportunity which being round about; We take leave only to salute you by it, in hopes of Writing more largely & directly by Mr. Oglethorpe sometime next month and in the mean time desire you may be assured that all your friends here are well and that I am

Reverend Sirs   Your most humble Servant   Henry Newman

Georgia Office Westminster.   To the Reverend Mr. Ziegenhagen his Majesty's German Chaplain at Kensington.

Reverend Sir: The Trustees out of a Regard to the Recommendation of Mr. Thielow who is going to the Saltzburgers in Georgia to be Assistive to them in his Profession, have agreed to defray the Charge of his Passage and Bedding for him and to Subsist him for three years in Georgia in the following manner viz.

Six pounds of Beef a week.
Two pounds of Rice. ⎫
Two pounds of Peas. ⎬   a Week.
Two quarts of Flower ⎭
A pint of Strong Beer a day.
A quart of Molasses a week.
Four pounds of Cheese. ⎫
Two pounds of Butter. ⎪
Two ounces of Spice. ⎪
Two pounds of Sugar. ⎪
A Gallon of Vinegar. ⎬   a Quarter.
Six pounds of Salt. ⎪
Three quarts of Lamp Oil. ⎪
Three pounds of Soap. ⎭
And a pound of Spun Cotton a year.

In Consideration whereof the Trustees hope that he will on his part be assistive to all other Settlers in the neighbourhood of Ebenezer that may want his help.

Please to communicate this to him & let him know I will wait on him next Thursday morning at Mr. Newmans about Eleven to acquaint him when & where he goes on board. I am

Sir Your most Obedient Servant Harman Verelst
3 October 1737.

London 6 October 1737. To the Reverend Messrs. Bolzius and Gronau at Ebenezer in Georgia. By the Georgia Pink Capt. Daubas.

Reverend Sirs: I received your joint Letter of the 31st of March last and communicated the Contents of it to the Society who were not a littled pleased to find you and your Congregation so chearfully resigned to the Will of Heaven under the difficulties you and they conflict with in your new Settlement. I hope you received mine of the 2nd of march last.

The Current half year for your and Mr. Ortmann's Salaries expiring the 1st November I am ordered to Send your Salaries for the half year ending the 1st of May 1738 which is accordingly done in the Georgia Pink in the Quantities mentioned in the Invoice enclosed, to which there is added a present of 5 £ to Mr. Ortmann in regard of the smallness of his Salary.

Mr. Professor [G. A.] Frank and Mr. Ziegenhagen who are always Sollicitous for your Comfort, in order to Supply your Loss of Mr. Zwiffler, have prevailed on Dr. Thilo a Learned Gentleman of great acquirements in the Knowledge and Practice of Physick to relinquish all advantages he could propose to himself by staying in Germany, to embark on a Voyage to Ebenezer to assist You and your Congregation in the Quality of Physician Surgeon and Apothecary, wherein the Society have endeavoured according to their present Abilities to give him all the encouragement they could, either from their own narrow Fund or by recommendation of him to the favour of the honourable Trustees for Georgia concerning which Mr. Ziegenhagen who has with great Zeal always appeared on his and Your behalf, will more fully inform you.

You will See by the Invoice that a Present to him of 20 £ in Copper half pence is included in the remittance now made to you, which he chose rather to receive in Georgia than here. To which Mr. Ziegenhagen has added the Value of 85 £ to be applied as he himself shall direct. To these are added 3 Iron Corn Mills with a Ring fitted to all of them to facilitate the grinding with one hand, and a Box containing all the Medicines you desired in your last Letter.

Professor Frank has very strongly recommended Dr. Thilo to the Society and therefore they hope what he has denied himself of worldly considerations on this occasion may be abundantly made up to him in Blessings of every kind, not doubting of your friendly Treatment of him so far as may be in your power to reconcile him to the difficulties he must expect to encounter in his first Settling an uncultivated Wilderness.

The Society and all your friends here are very much concerned at the menaces of an Invasion from the Spaniards on your Settlement but the Government here have taken measures to enable Mr. Oglethorpe to return with so powerfull a Support of Regular Troops as they hope will effectually prevent the Spaniards giving you or the Province of Carolina any molestation.

May you and your flock never want the Divine Protection is the wish of the Society as well as

Reverend Sirs   Your most humble Servant   Henry Newman My humble Service attends Mr. Ortmann.

Mr. Vat is still here but purposes soon to return to Swisserland. I expect a Bill of Lading every Hour from Mr. Verelst at the Georgia Office, if it don't come time enough to be covered to you, I presume he will Send it under cover to Mr. Causton at Savannah.

London 4 October 1737.

Invoice of Money and Goods Shipped by order of the Society for Promoting Christian Knowledge on the Georgia Pink Capt. Henry Daubas Commander to Georgia & consigned to the Reverend Messrs. Bolzius & Gronau at Ebenezer for the cases hereafter mentioned. That is to say

B.G. No. 1 to 8 8 Casks with Copper half pence marked
　　　　　　　　and numbered as in the Margin No. 1 to
　　　　　　　　7 containing 2 hundred weight each and
　　　　　　　　No. 8 one cwt which in all 15 hundred
　　　　　　　　weight at 10 £ 14 s per hundred_____ 160.10.–
　　　　　　　　To be applied as follows Viz.
　　　　　　　　Mr. Bolzius's Salary 1/2 a year
　　　　　　　　　ending 1st May 1738 _____ 25. –.–
　　　　　　　　Mr. Gronau's Salary for said time 20. –.–
　　　　　　　　Mr. Ortmann's Salary said time_ 5. –.–
　　　　　　　　　　　　　　　　　　　　　　　　 50. –.–

A Present to Mr. Ortmann
from the Society —————— } 5. –.–
A Present to Dr. Thilo from
the Society —————— } 20. –.–
                                                    ———
                                                    75. –.–
By order of the Reverend Mr.
Ziegenhagen to be applied as } 85.10.–
he shall direct ——————
                                                    ———
                                                    160.10.–

No. 9   A Chest containing 2 Iron Corn mills with 2 handles to each & 1 Iron malt Mill with 1 handle. Note to this last Mill a square hopper of Slit Deal is to be made at Ebenezer that will hold a Bushel or more of Grain. This Mill is not only to grind Malt, but Indian Corn & prepare it to pass the other Mills to make it fine to what degree is desired.

No. 10   A large Iron Ring covered with Matting to be applied to the Malt Mill, or either of the other Mills occasionally for grinding with Expedition & to enable one person to grind as much as two.

No. 11   A Box of Medicines containing
3 lb. of Venice treacle.
2 lb. of Oxy croce[?]
1 pint of Camphirated Spirits
3 lb. of Nerve Ointments
By order of the Society   Henry Newman   Secretary

Bartlet's Buildings 8 October 1737. To the Reverend Mr. Ziegenhagen at Kensington.

Reverend Sir: I have herewith sent a Copy of the Invoice of what goes to the Missionaries in Georgia in the Georgia Pink for which Mr. Verelst promises I shall have the Bill of Lading by Tuesday next. I sent a Bible as you desired to Dr. Thilo by Mr. Behren and have added to the Invoice the Box of medicines consisting of all desired in their last Letter. I have bespoke the Value of 200 £ to be insured by Mr. Verelst, and he has in compliance with your desire promised to pay for Dr. Thilo's Bed to the Upholsterer on account of the Trustees.

The Ships for Fort St. George were dispatched last night and the Letters you gave me are Sent by the Godolphin Capt. Steward Commander. I am

Reverend Sir   Your most obedient humble Servant
Henry Newman

London 10 October 1737.   To the Reverend Mr. John Wesley at Savannah in Georgia.

Reverend Sir: I received your favour of the 11th of June last by which the Society were glad to hear of your welfare notwithstanding the discouragements you meet with on the part of the Heathen in your Neighbourhood and the Province you are engaged in.

I shall acquaint Mr. Archdeacon Rye of your thankfull acceptance of his Benefaction.

As to the Circumstances of good Sir John Philipps's death which you enquire after, He had dined the 5th January last with some of his dear friends who were always welcome to his Table, and they had left him about half an hour when his Sons being gone out his Servants below Stairs hearing a noise in the dining-Room they immediately run up to See what was the matter and found him fallen from his Chair with his head against one of the Legs of the Table, upon which he seems to have expired in a moment without fetching one Breath or giving one Groan, for tho' a Surgeon and Apothecary were immediately sent for, they could not get one drop of Blood from several Veins which I saw opened in his Arm, Temples, and among the Jugular Veins. This was just Such an Exit as I believe Sir John himself would have chose, who had the happiness of an habitual Preparation for Death; To leave the World without the Anxieties of a Lingering Sickness, and the Agonies that usually attend the Separation of Soul and Body, is a Privilege every Good Man is not to expect.

May it please God to fit us for our departure from this frail State, though it should happen to be in a manner of Sudden (resembling a translation) as the Instance I have given, and that you may be Instrumental in bringing many to Salvation by Jesus Christ is the sincere Wish of

Reverend Sir   Your most obedient humble Servant
Henry Newman

Sir Erasmus Philipps is gone to Montpellier for his health. All your friends of the Society are well. Your Brother will inform you of Mr. Oglethorpe's preparations and when he intends to set out for Georgia with a Regiment which the King has given him.

Bartlet's Buildings 13 April 1738.   To the Honourable General Oglethorpe at the Georgia Office.

Sir: The Society for Promoting Christian Knowledge have ordered the Box of Arabick books herewith Sent to be Conveyed to you earnestly wishing they may answer the generous purpose you have desired them for, among the Mahometary Africans in Carolina, but if by any means your Religious Intentions should be frustrated, the Society will be thankfull for the return of any part of them to be distributed among the Greek Christians in Palestine & for whom they were orginally designed.

They Likewise desire you would accept their Thanks for your Benefaction of Two Guineas by the hands of Mr. Peter Bohler towards the Charge of binding these Books, to whom a Copy of the new Testament in Arabick was delivered for your Self of which also they desired your Acceptance.

The many Instances of your Goodness to the Saltzburgers at Ebenezer leave no room to doubt of the Continuance of your favour to them, but in obedience to the orders of the Society I am humbly to recommend them & Their Ministers Messrs. Bolzius & Gronau with Mr. Ortmann the Schoolmaster to your favour & protection as occasion may require.

That you may reap Satisfaction adequate to all your Toil for Serving the first Planters of Georgia & Succeed in all your enterprizes for the Glory of God the Honour of His Majesty, & the true Interest of Great Britain, as well as your own Satisfaction, are I am sure the wishes of the Society but of none more than
Honoured Sir   Your most obedient humble Servant
Henry Newman

Bartlet's Buildings London 20 May 1738.   To the Reverend Mr. George Whitefield at Savannah in Georgia. [Marginal Note:] Sent in the Blandford Man of War with Mr. Oglethorpe.

Reverend Sir: I thank you for the Pleasure you gave to the Society and many of your friends here beside my Self in your Letters of the 2nd of March to the Reverend Dr. Hales and of 6 March to me the Accounts of which were very acceptable concerning the disposition of the Officers and Garrison in General at Gibraltar to encourage a Sense of Religion and Virtue there.

According to your advice the Society have sent to Dr. Cunningham the Chaplain there, a large Packet of Christian Monitors and Several Tracts against Popery to which I believe Dr. Hales has made a Considerable addition.

Mr. Oglethorpe's safe arrival in your parts will I hope give new Vigour to the Inhabitants of Georgia and compleat their Settlement on a happy foundation notwithstanding all the menaces from the Spaniards to disturb them.

I wish the inconveniences that attend an infant Settlement may not be too hard for Your Constitution, though Mr. Wesley & Mr. Ingham have surmounted them. Messrs. Bolzius and Gronau at Ebenezer will be glad of your acquaintance and the Society will be glad to hear from you of the Situation of the Saltzburgers, and that they have at last obtained a happy establishment with a prospect of Subsistence in recompence of all their fatigues to acquire it.

I received the enclosed from Mr. Thorold who is much your friend, and will with the Society be always glad to hear of your Welfare wherever the Providence of God directs you to whose Protection I take leave to commend you and remain

Reverend Sir   Yours etc.   H. N.

Bartlet's Buildings London 20 May 1738.   To the Reverend Messrs. Bolzius and Gronau at Ebenezer in Georgia. [Marginal Note:] By the Blandford Man of War.

Reverend Sirs: The last Letters I received from you are dated 28 July 1737 nor have I had any advice of your receiving the last from the Society dated 6 October 1737 Accompanying Remittances in Copper money of your Salaries etc. to the 1st Current. I have now by order of the Society packed up at the Mint in the Tower in 5 Small Casks 9 cwt 1 q. 7 lb which at 10 £ s per cwt including your Salaries to the 1st of November next* as per Invoice which will be sent with it in Capt. Tompson Master of the 2 Brothers the first Ship that will Sail from London to Georgia.

I received and forwarded your Letters of July last to Mr. Ziegenhagen, Mr. Butjenter, Dr. Guerdes and Mr. Verelst and I hope they embrace the opportunity of writing though it be but short, as I am compelled to against my inclinations for fear of not Writing at all. I write this in haste to go into the Trustees packet which is immediately to be Sent to Portsmouth from hence. Wishing you all manner of Prosperity I remain

Reverend & Dear Sirs   Your most humble Servant
Henry Newman

	£ s d	cwt q lb		£ s d
*Salaries to 1st No-vember next	50. –.–	NB. 5.1.–	is	55. 2.6
a Present to Mr. Ortmann	5. –.–	4.–.7		42.13.1½
remitted by Mr. Ziegenhagen	42.12.–			
	97.12.–	9.1.7 at 10 £ 10 s		97.15-7½

P.S.: I hope the Reverend Mr. Whitfield may be Safely arrived in Georgia from Gibraltar long before this which goes in the Ship with General Oglethorpe.

Memorandum Sent enclosed in Mr. Bolzius's Letter—a Letter to Henry Bishop Apprentice to him.

Bartlet's Buildings 20 May 1738. To Mr. Verelst at the Georgia Office in old Palace yard.

Sir: I am in too great a hurry to write my Complements to General Oglethorpe, but shall be obliged to you if you give my most humble Service acceptable to him. I leave it to you to dispose of the Letters herewith sent as you please; only please to observe that one is for Charles Town which you will convey accordingly it being just now come to hand from a particular friend in Ireland.

I heartily wish you a good Journey & safe return to London being

Sir Your most humble Servant Henry Newman

London 8 August 1738. To the Reverend Messrs. Bolzius & Gronau at Ebenezer in Georgia.

Reverend Sirs: My Last was of the 20th May acknowledging the Receit of your Letter of 20 July 1737 since which I have received yours of the 30 January and 19 February last by which the Society were glad to understand Mr. Thilo was Safely arrived and enjoyed his health, and that you had received the remittances by the Georgia Pink Henry Daubas Commander. This accompanys the remittances by the Two Brothers Capt. Thomson according to the Invoice and Bill of Lading herewith Sent.

The uncertainty of Opportunities directly to Georgia and the Trouble and Charge attending the Remittance of Copper money makes it wished that there had been 2 half years Salaries sent instead of One by this Ship, but now the Money is not to be had at the Mint, the last Coinage being all disposed of and this was taken up 2 or 3 Months before Opportunity presented of sending it. If there should be a War, those Opportunities will be still more rare and the risque greater; You will be pleased therefore to Consider whether you can't have your Salaries as duly paid to you in Georgia by Draughts on William Tillard Esq. our Treasurer as formerly mentioned, the Society allowing the Poundage of 6 d for each pound you draw, to the Merchant or Person that supplies you with money for your Bills, the Society being at a much greater expence in the way your Salaries are now Sent.

I hope Mr. Oglethorpe's arrival in Georgia has established tranquility to the Colony and that they will enjoy a Completion of their wishes in this World under his Administration. My humble Service to Mr. Thilo and Mr. Ortmann wishing you & them all manner of Prosperity I desire you may be assured that I am

Reverend Sirs   Your most humble Servant   Henry Newman
I have Enclosed the Captain's Bill of Lading.

London 8 August 1738.   To the Reverend Mr. Whitefield at Savannah in Georgia.

Reverend Sir: I received your favours of the 19th of April and 20th May last which were acceptable to the Society as they informed them of your health & safe arrival in Georgia.

The Society have formerly attempted what you recommend of getting the Prayers in the whole Duty of Man printed Separately for the use of Devout Persons but the Proprietor of the Copy declined consenting to it for fear of discouraging the Sale of the Book they belonged to; They have therefore ordered me to desire your acceptance of a quarter of an hundred of *the Bishop of London's Family Devotion* which will be Sent you I hope in this Ship the Two Brothers but if not, in the next by Capt. Daubas, in the mean time you have the good wishes of the Society and all your friends here that it may please God to give you health and Success to your Labours.

The Society will be glad to hear from You by all opportunities and of the Welfare of the Saltzburgers at Ebenezer.

I wrote to you the 20th of May last by the Blandford Man of War in which Mr. Oglethorpe went, who I hope is safely arrived 'ere this and that the Colony of Georgia may be always prosperous under his Administration.

I am   Reverend Sir   Your most humble Servant
Henry Newman

Please to accept of the enclosed new Edition of the Account of the Society in which you will See a List of such Books as are dispersed by them.

London 29 July 1738.

Invoice of Money shipped by order of the Society for Promoting Christian Knowledge on the Two Brothers Wm. Thomson Commander to Georgia & consigned to Messrs. Bolzius & Gronau at Ebenezer for the uses hereafter mentioned, that is to say.

5 Casks with Copper half pence marked & numbered as in the Margin.

B.G.

No. 1   containing __ 2.1.–
   2   containing __ 2.–.–
   3   containing __ 1.–.7
                5.1.7
   4⎱ containing __ 4.–.–
   5⎰

$£$ s d
9.1.7 which at 10.10.– per cwt cost 97.15.7½
    cwt q lb $£$ s d
    5.1.– is _____ 55. 2.6
    4.–.7 is _____ 42.13.1½   97.15.7½
    9.1.7

On the undermentioned Accounts viz.

Mr. Bolzius's Salary a year ending the ⎱ 25. –.–
1st of November next _____ ⎰

Mr. Gronau's Salary Ditto time _____ 20. –.–

Mr. Ortmann's Salary Ditto time _____   5. –.–
                          50. –.–

Mr. Ortmann a Present _____   5. –.–

By order of the Reverend Mr. Ziegen- ⎱ 42.12.–
hagen to be applied as he shall direct ⎰
                                 97.12.–

By order of Ditto a Small bagg of Span- ⎤
ish Rials containing 75 oz. 3 penny- ⎬   20. –.–
weight which at 5 s 3⅜ d per Ounce ⎦

                            $£$ 117.12.–

NB. This Bagg is packed up in Cask No. 1 & sealed with the Seal of this Letter.

By order of the Society.   Henry Newman

London 19 December 1738.   To the Reverend Messrs. Bolzius and Gronau at Ebenezer in Georgia.

Reverend Sirs: Herewith I send a Copy of my last of the 8th of August, and now acknowledge the Receit of yours of 26 August by the Reverend Mr. Whitefield, who I am glad to tell you is happily arrived by the way of Ireland and this day delivered your Packet to Mr. Ziegenhagen which I hope he may acknowledge the receit of by this Ship. The difficulty of returning your Salaries from hence in Copper money has induced the Society to order me to repeat my last Letter, and if no opportunity of Sending them till next Spring offers you may then expect a year's remittance at once unless you can fall into the method proposed in the Letter herewith Sent.

The Society were glad to understand by your Letter and Mr. Whitefield's Report that your Flock enjoys such a measure of health as God is pleased to bless them with, may you and they long partake of his Goodness which is extended to the uttermost parts of the Earth is the wish of all your friends here particularly of

Reverend Sirs   Your most humble Servant   Henry Newman
Mr. Whitefield seems resolved to return to Georgia in the Spring.

London 23 December 1738.   To the Reverend Messrs. Bolzius and Gronau at Ebenezer in Georgia.

Reverend Sirs: Since my Writing the 19th Current by this Ship I have conferred with Mr. Simonds an eminent Merchant here who trades much to South Carolina & Georgia, & he tells me his Brother in Law Mr. Montague[57] at Purysburgh or Mr. [Charles] Pury at Savannah will readily take your Bills on Mr. Tillard for the Current Exchange that shall then govern at Savannah without expecting 6 d in the pound for Commissions as is taken at Charles town and that he has wrote to them by this Ship accordingly. This goes undercover to Mr. Oglethorpe for surer Conveyance to you from   Reverend Sirs   Your most humble Servant
Henry Newman

London 22 February 1738/9.   To the Reverend Messrs. Bolzius and Gronau at Ebenezer in Georgia. By Capt. Shubrick.

Reverend Sirs: I received your Letters of the 4th of October from Ebenezer, and 20th of October from Savannah, by which the

Society were glad to understand that you and your Flock are in health, that your Salaries in Copper money were got safe to your hands by Capt. Thomson. Since which the Society are glad to find you are in the way of receiving your Salaries with less risque and charge than heretofore, as well as with more certainty as you want it by drawing Bills of Exchange on Mr. Tillard who has accepted and paid to Mr. Simond your Bill of the 7th of November last for 50 £ being ½ a year's Salary for your Selves and Mr. Ortmann due the 1st May 1739. I had bespoke ½ a year's Salary in Copper money, but now the Society have countermanded that Order and directed me to desire you to continue your draughts for your Salaries on Mr. Tillard as opportunities offers, & that you would add to your next draught the Sum of 10 £ remitted by Messrs. Schalkhauser, Flugel & Jastram Merchants at Venice to the Society for the benefit of the Saltzburgers in Georgia, which the Society desire may be applied to the Service of the Orphan house at Ebenezer.

I am also ordered to desire you would with your Convenience transmit to the Society a Copy of your Journal or an Abstract of it so far as relates to the progress of Religion, and the Education of Youth at Ebenezer.

Mr. Whitefield is safely arrived in England and designs to return to Georgia the approaching Summer in the mean time he has been a painful Sollicitor for Benefactions to your Church and Orphan house of which you may expect an Account from him, he being now in the West of England for that purpose and to visit his Relations in those parts.

Bartlet's Buildings 2 April 1739. To the Reverend Messrs. Bolzius and Gronau at Ebenezer in Georgia. By the Carolina Capt. Somerset.

Reverend Sirs: My last was of the 22nd of February by Capt. Shubrick of which I have herewith sent a Copy and refer you to it because I have little to add but that all your friends of the Society here are well, and hope to hear of your happy progress by all fair Opportunities. The Saltzburger [Sanftleben] who returned to Germany last year is now returning to you with an Addition of half a Doz. more to your Congregation.[58] I wish them a good Voyage and desire you may be assured that I am

Reverend Sirs   Your most humble Servant   Henry Newman
The Parliament have lately given 20 thousand pounds towards encouraging the Settlement in Georgia.

Bartlet's Buildings 14 July 1739.   To the Reverend Messrs. Bolzius and Gronau at Ebenezer in Georgia.   By the 2 Brothers. Captain Thomson.

Reverend Sirs; My last was 2nd April by the Carolina Capt. Somerset since which I have yours of the 11th & 12th December 1738 and 15 March last but have not been able to send you any Letter by reason of the embargo first on all outward bound Ships and now of a Declaration of War with Spain[59] published the 10th Current a Copy of which is enclosed and the Society will be glad it may not affect you as it is partly the pretentions of Spain to Georgia that has drawn the Government into a War. This will very much interrupt our Correspondence for the future and therefore I doubt not but you will embrace those Opportunities which present of letting the Society know of your welfare. I just now received the enclosed from Mr. Ziegenhagen who is well with the rest of your friends here, and desire you may be assured that I am always   Reverend Sirs   Yours   H. N.

P.S. Your Letters abovementioned desiring another Transport of Saltzburgers or other Protestants from Germany have been for sometime under the consideration of the Trustees for Georgia, but I have no directions yet for encouraging you to expect another Transport as Things are now Circumstanced.

Mr. Tillard paid your Bill for 50 £ payable to Messrs. Montagut & Purry for your half year's Salarys due at the 1st of May Last.

London 14 July 1739.   To His Excellency General Oglethorp at Savannah in Georgia.

Sir: Capt. Thomson having obtained a Protection by The Interest of the Trustees to proceed to Georgia notwithstanding the Embargo, I embrace the opportunity to Salute your Excellency with my best wishes for the Success of your administration in this Critical Juncture, when our Court have been obliged as you will find by the Prints sent in this Ship to declare War against Spain the 10th Current which has occasioned for the present no small Ferment in this City, but Messrs. Bolzius & Gronau with the honest Saltzburgers under your Government being under no apprehensions of danger while you protect them, have desired another Transport of their Brethren or other German Protestants may be sent over to Ebenezer, in which the Society have taken no Step nor can do as things are circumstanced especially till they are informed of the inclination of the Trustees to favour them with their Passage and other encouragements formerly allowed to their Brethren.

I doubt not of Seeing the Confines of Georgia settled on a better Foundation for the future by your wise conduct, wherein all your friends here wish happy Success, but none more than

Sir   Your Excellency's most Obedient humble Servant

Henry Newman

 PART II.

Inward Correspondence

Extract of a Letter from the Reverend Mr. Urlsperger at Aug-bourg 14 April 1732, N.S. To the Reverend Mr. Ziegenhagen at London.

The Remittance of 125 £ Sterling for the Benefit of the Exiles of Saltzburg has raised an Hearty Joy amongst us. The Lord be praised, and may He Bless the Benefactors for this their Charity, in Time and to Eternity.

The Distribution of this Sum has been made with the Approbation of Our Protestant Magistracy in the following Manner.

To Memmingen were sent ‗‗‗‗‗‗‗‗ 100 Florins.
To Kauffbeyren [Kauffbeuren] ‗‗‗‗‗‗ 250
To Nordlingen ‗‗‗‗‗‗‗‗‗‗‗‗ 200
(These three Imperial Cities have joyntly with the City of Augsbourg made the Affairs of these Exiles their Common Cause.)
Into our Publick Fund raised for the Benefit of
 Exiles, have been delivered ‗‗‗‗‗‗ 350
For the Maintenance of above 30 Ancient and for
 the most Part Sick People ‗‗‗‗‗‗ 179

Amounting in the whole to ‗‗‗‗1079[1]

And here I must observe that our Hospital having no Stock at all, but Subsisting only by the Free Gifts and casual Benefactions makes me hope it will not be taken amiss in England that Part of their Charity is employed this Way.

We have still but a dismal Prospect of Saltzburg for We are informed that the Number of those that have been ordered to depart amounts already to 40,000 wherefore we thought it necessary to mention particularly the deplorable Case of these poor People in the Publick Prayers of our 6 Protestant Churches, and that in the following Manner.

"The Prayers of this Congregation are particularly desired for the Poor Saltzburgers who for the sake of the Protestant Religion are driven out of their Country; the Lord would be pleased with his Powerful Arm to preserve, Protect and Guide both them that left their Country and them that must leave it hereafter. That the Lord would be pleased abundantly to satisfy their Hunger and Thirst after his Word and Sacrament; and furthermore to raise far and near Hearts Charitably inclined to provide for their Bodily Sustenance; but above all things to preserve'em Graciously from Back-sliding and Hypocrisy and from every Sin and Offence whatever, and to bring them at last with all true Believers to Eternal Salvation through Jesus Christ Our Lord."

In Answer to Your Question about the Trees: what sort of Bread the First Exiles found in the District of Rastadt?[2] My Colleague the Reverend Mr. Hildebrand examined amongst others one Fellockner, who eat of it himself and declared that it was not real Bread, but somewhat resembling Bread Viz. An Excrescence which grows on certain Trees like a Bark as thick as a Finger, white in Colour and Sweet in Tast like Sugar, but soft as New Rolls of Bread. Every one of their Number had got as much of it as was sufficient to satisfy his Stomach for a whole Day.

I have added an Account of 13 of these Saltzburgers in what Manner and by what Means they came to the Knowledge of True Christianity which I hope will not be unacceptable to You.

St. James's Place.[3] 23 May 1732.   To Mr. Newman.

Sir: I am sorry my health as well as other Business hinders me from attending the Committee to Day, and taking my leave of them, for in all probability the King will set out next Monday or Tuesday.

If the Account of the Saltzburgers is out of the Press, I should be very glad to have three or four Copies of it this Week.

May GOD abundantly bless the Designs of the Society in General, and more particularly their Christian Compassion, Zeal and Labour with and for the New born but hard persecuted Members of Christ in Saltzbourg: may he put his Name and Seal to the Account that is or will be published by the Society in behalf of

this distressed Flock of Christ, and by making it more than ordinary Successfull & strengthen the Heart and Zeal, both of the
afflicted and them that are willing to reach out their Hands to
the Afflicted. And, I think there is somewhat particular in this
Affair, which gives the Society the more Ground to expect such
a Divine Approbation and Concurrence in this their Undertaking.
The Promise, that is given to two or three, who are gathered together in the Name of Christ Viz to request a Blessing for themselves and others,[4] cannot but be justly and strongly applied to
two or three that are gathered together not only in the Name but
also for the Name of Christ, Viz: to promote and Support the
Knowledge and Service of his blessed Name.

If during my Absence the Society should be enabled and resolved to send some Money to Augsbourg for the Relief of the
Saltzburgers, I would humbly recommend to them two proper
Persons, by whose Hands the Money might be remitted. The One
is Mr. John Baptista Mayer, A German Merchant in the City, living in New Court, Broad Street, behind the Royal Exchange,
whose Correspondent at Augsbourg Mr. Münich is a very Considerable Banker, neither of them hath taken any agio on the
Money I sent to Augsbourg. The other Person is Mr. Schick, Clerk
of His Majesty's Privy Council for His German Dominions; whose
Correspondent is Mr. Gullman, the King's Resident at Frankfort,
who is very desirous to serve the poor Saltzburgers in any manner
he can. but I submit this entirely to the Consideration of the Society remaining with my humble Respects to the Gentlemen present, especially to Sir John Philipps and Mr. Vernon,[5]

Sir   Your very humble Servant   Ziegenhagen

Extract of a Letter from Mr. Münch at Augsbourg 7 July 1732.
To Mr. Mayer at London.

I have paid immediately to Senior Urlsperger the 50 £ Sterling
You was pleased to draw upon me for the Relief of the Poor and
Necessitous Emigrants of Saltzburg, this Payment I made Charges
free with all the Pleasure and Good Will, the same shall be done
on the Benefactions that may hereafter be remitted, they are well
applied for such Good and Poor People, who are wanting of
Succours suitable because though this City is not backward in
providing for them; yet their great Number requires also help
from abroad; some Days ago 500 more of these poor Exiles set out,
and to morrow or next Day 800 of them are expected, and so it

goes on continually. The Lord be pleased to excite more and more Charitable Hearts, to enable us to assist these Good People with sufficient Relief.

Please to Command freely[.] I shall Joyfully serve these poor Exiles for nothing, who for Christ's and Conscience sake endure great Misery, quit their Country, Estates and all they have in the World, Parents & Relations, undertake a long Voyage with Wives and Children, amongst whom are Women big with Child and Suckling Children, yet they never murmur against their Sovereign, but always speak well of him, bearing with Patience to be driven out of their Country for Religion's sake; There are already more than 20,000 Souls gone away out of the small Country of Saltzburg, and we hope to see a great many more, perhaps also out of other Provinces, Therefore the Benefactions that arrive from abroad, are a great help to such a Multitude, may the Lord reward the Benefactors.

Extract of a Letter from Mr. Münch at Augsbourg 14th July 1732 N.S. To Mr. Mayer at London.

I have paid immediately to the Reverend Senior Urlsperger the 40 £ Sterl. You have drawn on me free of all Charges and Commissions, for which I debit You in Your Account.

Yesterday, arrived 347 more Emigrants from Saltzburg, amongst whom are many Children, Women in Child-Bed, and others big with Child; and to Day 500 more are expected, being arrived all ready in the Neighbourhood, these People have suffered much because of the Rain and bad Weather we have had for these 3 Weeks past, and are worthy of Commiseration; only it is to be wondered at, that they shew so much Patience and Resignation, may the Lord more and more comfort and protect them.

There are actually [presently] two more Provinces that quit the Romish, and will embrace the Evangelick Religion, if this happens they will be forced likewise to depart, it would be good for this Distressed People to make General Collections, could not such be obtained likewise in London?

Translation of a Letter in High-Dutch.[6] To Mr. Newman.[7]

Honoured Sir: With my most humble Respects to Sir John Philipps, Baronet and Sir Richard Ellis, I return first of all sincere Thanks to them in the Name of our Dear Saltzburgers for the Charity of 50 £ and another of 40 £ Sterling, which according to our Money makes 437 Florins 30 Creutzers 345[8] they have sent as a Relief to these dear Confessors of the Gospel; They are worthy of it.

I had presently a fair Opportunity to put out these Talents committed to my Charge to Use, partly among 900 Exiles, who 5 Days ago went some Miles from hence towards Memmingen, partly among 900 Exiles more, who arrived here Yesterday and to Day, and are to rest till after to morrow; Those that came Yesterday, I welcomed with a Sermon in the upper Burying Ground of the Protestants, and taking Occasion both of the Gospel for the 5th Sunday after Trinity, Luke 5th and of the Hill I stood upon, The Saltzburgers and a great Multitude of other People standing on a Lower Ground. I discoursed on those Words: *Launch out into the Deep and let down Your Nets for a Draught.* The Good People had not Rested for 8 Days and yet were so desirous to hear the Gospel that they could never hear enough of it. Among them was a Woman brought to Bed but 5 Days before. After I had finished my Discourse, I asked them in the Presence of several Thousands among whom were many Papists, divers of which heard the Word of GOD with Tears, Whether they were not sorry for having left their Native Country? Whether they believe sincerely in GOD the Father, Son and Holy Ghost, in whom they were Baptized? Whether they were resolved to continue stedfast in the true Doctrine they now acknowledged, to hear the Voice of Christ only in his Word, to follow him, and to hope for Salvation only through his Merits? To these Questions They Answered Yes! often Clapping their Hands together over their Heads, to express their Earnestness.[9] Afterwards they were conducted to their Quarters, singing all the Way Spiritual Hymns. All these abovementioned came out of the Jurisdiction of Grossart [Grossarl], belonging to the Arch-Bishoprick of Saltzburg, from whence some of those that went before'em were, most of them, very Poor but of good Cheer.

This Morning being Monday, My Wife and other faithful Persons are to go and enquire carefully in all their Quarters (most of which I my self visit also) who are the most indigent amongst them, in Order to provide Shoes, Stockings, Shirts, Neckcloths etc.

In the Afternoon the Ministers of the Gospel preached the Word of GOD to the Exiles in their Severel Quarters without the City, I had also an Opportunity to do the same in Shooter's Field,[10] in the Presence of many Thousands of other People a great Number of whom were Papists, both Rich and Poor, and several

Regulars of Divers Denominations, on these Words: Psalm 23:5. Thou preparest a Table before me, in the presence of mine Enemies. Which Psalm I have Chosen for Exiles, and has furnished me with many Discourses to these poor People. Though the Number of Hearers was great yet was there so deep a Silence that I was not in the least disturbed. Severel Papists were again observed to shed Tears. I meddled not with Controversey, only set Christ before them as the Good Sheppard of his Sheep.

Before I concluded with a Benediction, I proposed again several Questions to the Exiles, which they Ansered chearfully and exactly, though in few Words. Give me leave to mention, that as 600 of them, who were left behind the Day before, arrived here towards Noon, I went to meet them in my Chaise. And no sooner had they seen me and observed that I was a Protestant Minister, but they came running towards me, offered their Hands. and cried out Oh! Preach to us, Oh! Preach to us. I spoke to several who manifested a great Power of Godliness. One said upon my asking him Whether he had left any thing behind? replied YES I have, but what are all things belonging to the Body, in Comparison with Eternal Goods. I am only sorry for those who staid behind for the sake of the Belly, and would not come away along with us, though they knew better things, but they will come still I hope.

Dear Sir: When I began on Monday Morning to write to You and thank you in the Name of the Exiles for transmitting the Bills of Exchange, a certain Gentleman sent a Piece of Gold Value 10 Ducats for them, and added these Words: "I desire You not to let my Name be known, for having no Mind to sound a Trumpet with my Mite, I am satisfied with the Assurance that Our heavenly Father who sees all things in Secret will reward us openly for Alms given with a faithful Heart. Matthew VI."

Two Hours after another Benefactor sent 25 Florins with the same Request, and as I went towards Evening to Shooter's Field to preach there, an Eminent Protestant Banker met and told me he had Orders to pay me next Morning a Bill of 2000 Florins, which were collected by two Foreign Protestant Congregations, and left to the Disposal of the Protestant Ministry here, and another Bill of 40 £ Sterling more which came from England. You may suppose Sir, my Heart was excited to praise the Lord, especially since the very next Day 100 Florins more were sent to my House in three several [separate] Parcels. This was indeed the fulfilling of what I preached from the Words: *Thou preparest a Table before me in the Presence of mine Enemies.*

On Tuesday I had another Discourse on the 23 Psalm in the Celebrated Garden belonging to Mr. Schaur[11] before 100 Exiles, to make the Deeper Impression upon them, of the most happy Condition of those who have a saving Knowledge of their Good Shepherd Jesus Christ. In the Afternoon a very Edifying Sermon was preached by my Colleague Mr. Hildebrand in St. Ann's Church, upon Revelation III V. 14 22 before the Exiles, who could not be looked upon without the Greatest Commotion of Heart.

Wednesday Morning they were brought again to this Church, when I discoursed upon the following Words: Psalm 23:6 *Surely Goodness and Mercy shall follow me all the Days of my Life and I will dwell in the House of the Lord forever.* After Sermon they Sung, by themselves, two Spiritual Hymns, One beginning with these Words: *O My faithful GOD and Father* etc. and the other *Come and hear the Sacred Story etc.* which caused a Flood of Tears in the Congregation. In the Afternoon most of Our Ministers preached to'em, when 600 Florins part of the abovementioned 2000 were distributed among them, I was with them in the Gardens of Mr. Benz and Mr. Wolf, in Company with our Protestant Syndic.[12] who is the Chief of the Protestants here.

This Morning being Thursday they parted from Us in good Order, dismissed with many Tears, Prayers, and Benedictions. They retain their former Excellent Character [reputation] of being Pious, Quiet, Meek, and Patient. Our Protestant Congregation has abounded in their Charity towards them and during their stay here, were continually sending Benefactions to all their Quarters about the City.

To show the Contentedness of these People, I beg leave to mention one Instance: It happened that a Shoe-Maker took Measure of an Old Man to make him a Pair of Shoes; several came and said they also wanted Shoes, and being told they should have them Gratis, they then said no more about it, giving this Reason upon Enquiry, because there might be some amongst them, who stood more in need than themselves.

I come now to Answer some Particulars of Your Letters of the 13th and 23rd of June Viz.

1st. I have Received all the Sums of Money which Mr. Ziegenhagen sent by Bills of Exchange, for which I gave him Receipts.

2nd. The Honourable Society and every Body else may be assured that the Benefactions are managed with all possible Care and Faithfulness; and that I don't know how any One at this time can bestow his Temporal Goods better than towards the Support of these true Confessors. Therefore be not weary to send Daily and we will not be weary to distribute faithfully Your Charity and Contributions as hitherto our Mites. The Harvest is great and We Expect Daily [any day] 3000 more in one Company.

3rd. It is computed that there are actually [by this time] gone out of the Country of Saltzburg about 15000 Persons.

4th. Those who left Saltzburg since March last are all going to Prussia and most of them also who came in the Winter and were here and there received, are desirous to go along with their Countrymen to Prussia. I hear just now that the King of Prussia[13] hath given Orders to receive 10,000 more, besides them he hath already Received.

5th. They all go as we are informed to the Prussian Lithuania.[14]

6th. My Correspondent[15] from Ratisbon informed me three Months ago that the Whole Number of these Exiles amounted to 40,000 but the last Exiles say that about 15,000 more left behind in the Country, however no Body can tell the exact Number because they increase Daily. It is but a Week ago that We had a well attested Account that about 1000 Persons belonging to the Salt Mines have openly Professed the Protestant Religion and desired Assistance from Ratisbon. This is a great Blow to the Arch Bishop.

Concerning the Course of Exchange, I receive here for 1 £ Sterling 8 Florins 45 Creutzers and accordingly for 50 £ Sterl. 437 Florins 30 Creutzers. This Account I likewise send this Day to Mr. Ziegenhagen at Pyrmont. I acknowledge with all Thankfulness that I have Received a Packet of Books from the Honourable Society by Mr. Machtolff, who continued with me 6 Weeks.

I admonished our Dear Saltzburgers to pray for their Benefactors in England, which they most readily promised to do. They are indeed very fervent in their Devotions, One of them said this Morning at taking leave: "Sir, when we were in the greatest Perplexity, we betook our selves unto Prayer and then GOD was our help." I beg likewise the favour of Your Prayers that GOD may vouchsafe Grace and Strength unto me to labour furthermore in his Service for the Benefit of these poor Exiles.

Yesterday I had Letters from Tranquebar[16] dated 18 Oct. 1731 with advice that the Sum of 1100 Dollars[17] collected in this Place

Anno 1730 for the Benefit of the Mission came safe to their Hands; The inclosed Copy concerning the great Commotion in Bohemia, I recommend in Secret to the Consideration of the Honourable Society I have sent the same to Pyrmont. May GOD Bless and prosper the Society and You in particular! I recommend myself to their devout intercessions with GOD and remain under a Multiplicity of Business.

Most Honoured Sir   Your very Obedient Servant   S. Urlsperger Augsbourg 17, July 1732.

N.B. A Copy of a Letter from [name omitted] was enclosed in this Letter and is entered in this Book Folio 21. [See p.   ]

Extract of a Letter from Mr. Munch at Augsbourg 24 July 1732. To Mr. Mayer at London.

I took up presently Your Draughts of 37 £ 17 s and 102 £ 2 s Sterling and paid to Senior Urlsperger the Value of it without any Provisions or Charges, and I wish suchlike Providings more and more may be made, being very glad to serve the Poor Emigrants. The Lord will also reward You for Your Trouble, and all the Benefactors. We expect in a few Days some hundred more of them.

Translation of a Letter in High-Dutch [from Urlsperger].   To Mr. Newman. Augsbourg, 1st Aug. 1732 N.S.

Sir: I hope You have Received my last of the 17th of July, containing a particular Account of what passed here with the Exiles of Saltzbourg the 13th, 14th, 15th, 16th & 17th Current. Now I acquaint You that Yours of the 30th of June came safe to my Hands, and that the two Bills of Exchange have been readily paid without deducting any L'Agio by Mr. Munch (Who was Knighted a few Days since by His Imperial Majesty) this Gentleman's Fore-Fathers have been likewise Exiles and upon that Account he hath a very great Compassion towards the Exiles of Saltzburg.[18]

In my last Letters, You have seen some Particulars what Profit was made with the Charities from England, and a more particular Account shall follow as soon as possible. Some Days ago 1800 Exiles passed by here both on the Left and the Right Side, but none of them coming hither, I sent some of the Charity Money in my Hands unto faithful Persons, in order to distribute it among the needy, which hath been done accordingly, as I am informed this Minute by a Letter from Memmingen, for which these Dear

Exiles have often Blessed the Benefactors. To Morrow in the Evening or the next Morning we expect about 700-800 Exiles. I have ordered my Domesticks to get ready a Good Number of Shirts etc. to distribute them to those who are in want. Though I am Drinking the Mineral Waters, Yet it shall not hinder me to visit these new Guests as soon as they come and to take Care What Land the Seed of Charity sent from London can best be sowed in. I assure You Sir, You cannot see a more pleasant Sight than to be present at the Arrival of such a Company. The Rich among us are glad to be able to serve these Strangers.

If the Report which is spread almost throughout Germany shall come to Your Ears in England that the Elector of Bavaria had sent for me to preach before him I desire You'll not believe it, as being without Foundation. However there are a great motions up and down in Bavaria, a great many Persons, and even whole Families being come to us, and having embraced the Protestant Religion.

In Carinthia the Fire of Professing the Gospel is breaking out too, as I have been credibly informed; what happeneth in Hungary You will see with Grief by the enclosed Copy of a Letter I received but last Night from Good Hands.[19] Let us Sincerely and incessantly pray for these our Afflicted Brethren. Popery is very much enraged again by the Motions in Saltzburg. However the Lord is with us. Therefore fear not, thou little Flock.

Concerning the Evangelick Body at Ratisbon, it hath made all possible Remonstrances to his Imperial Majesty who also several times sent dehortatory Letters to the Arch Bishop of Saltzburg, but they have done no Good hitherto in the main Business, and though the Ambassador of Saltzburg [Zillerberg] seemed to make the Sincerest Promises at the Diet, Yet all things were found otherwise. Only One thing is done, Viz. the 70 and odd Prisoners, every one of whom continued stedfast, notwithstanding the Greatest Torments they endured, have been set at Liberty and obtained leave from the Arch Bishop to take as much of their Goods along with them as they can carry, but in this very Point many unjust things are Committed. Nor do I see any other way for these good People to get any part of what they left behind'em in Houses, Land, Cattle etc. unless the Emperor appoints a local-Commission, consisting of Members of the Empire of both Religions. I hope to be able very soon to Communicate to You an exact List of all the Exiles.

Being much obliged to the Honourable Society for their Confidence in me, I commit them, all the Benefactors in General, and the Trustees in particular (to whom our Dear Saltzburgers return many Thousand Thanks, and promise faithfully to pray for them) together with You Dear Sir, and Mr. Martini to the Grace of GOD and remain with all Sincerity

Dear and Honoured Sir   Your very Obedient Servant

Samuel Urlsperger

Translation of a Letter in High Dutch [from Urlsperger]. To Mr. Newman. Augsbourg. 12 August 1732.

Dear and Honoured Sir: This Day arrived here 800 Exiles, and at the same time upon the Right and the Left two Companies of each 900, one towards Memmingen, the other towards Rein a little place in Bavaria, I have therefore little time to write.

1st. I acquaint You that I have Received all Your Bills of Exchange, as mentioned in my Letters of the 17th, and 31st of July,

2nd. The last return of 208 £ 16 s (for which GOD be ever praised) I have Received at that Instant when there were 900 here who came upon the 2nd, 3rd, 4th, 5th and 6th of this Month; as a great many in this Company were very poor and miserable who wanted [needed] particular Care to be taken of them, I obtained leave for them to stay a Day longer, in which time I took Care to supply the most needy with Shoes, Stockings, Shirts, Neck-Cloths etc. I distributed likewise some Books amongst those who could Read, and Ready Money, upon which they frequently broke out into this Language, *Good GOD, do they think of us at such a Distance? pray GOD reward them.* The Contentedness, Patience, and unspeakable desire of these People to hear the Word of GOD is not to be expressed by any Pen.

3rd. 13,000 Exiles are actually gone to Prussia upon the 3rd of August. 5000 were lodged in the County of Oettingen to be examined before the Prussian Commissary,[20] previous to their March, and since that time within Six Days 5000 more went away, including the Company that arrived here this Day. So that in the whole there are 23,000 gone from Saltzbourg, we shall hear to Day how many Thousand remain behind.

4th. Inclosed is the Copy of Minutes taken by the Mayor when the last Company of Saltzburgers were here upon which You may depend.

5th. In Bavaria a great Fire from the Gospel begins to glow, of which We have many Accounts, however we must be Silent as Yet.

6th. Eight Days ago a Person from the Valley of Tefferegger [Deferegen] in Saltzburg came hither, who professed the Protestant Religion here a Year and half ago; he went three Weeks since into the Country, and when he told the People in the Valley of Gross-bentzer that the Exiles were so well Received by the Protestants, They wept and the next Day 230 professed openly the Protestant Religion which none had ever done in that Valley before.

7th. A Minister of the Gospel from Ratisbon[21] lodges at present in my house, whom GOD has employed these 5 Years past in the Business of Saltzbourg; he came hither incognito to confer with Me in the Name of some Protestant Ambassadors upon some weighty Affairs. But I have not time to say more.

My most humble and devoted Respects, and many Thousands Thanks in the Name of the Saltzburgers to the Society, and in particular to Sir John Philipps, and the Rest of the Gentlemen who collect.

I commit You to the Grace of GOD   Sam. Urlsperger

The Copy which I lately sent concerning the Motion in Bohemia, I desire may be kept Secret.

Extract of a Letter from Mr. Munch at Augsbourg 7 August 1732. To Mr. Mayer at London.

Sir: I have Read the Favour of Yours of the 14 July past O.S. and according to Your desire I acquaint You that I have paid Mr. Urlsperger the Pounds Sterling every time as high as the Course of Exchange could bear, he having received for the   Flor.

		Flor.
125 £ St.	Flor.	1079.15
50	at 8.45	437.30
40	at 8.45	350.—
102.2 s } 37.17 {	at 8.43	1219.54

So that upon Calculating it Your Self. You'll find that I paid rather too much than too little, which I hope will please You as well as the Collectors, I have no design to get any thing by these poor Emigrants who are worthy of Compassion, much less to lessen the Christian Benefactions that are bestowed upon them.

The 208 £ 16 S Sterling which You have drawn again upon me have been paid already to Senior Urlsperger in Florins 1820.2 after the Rate of 8.43 per pound Sterling and have been very acceptable, when just 1000 such Emigrants were here, it is observable that amongst these there are several who had a Spight against the first Emigrants for their Professing the Evangelical Faith, railed at that Religion and wronged very much such as quitted the Country on that Score, but were convinced afterwards of the Gross Errors of Popery, and were brought by heavenly Grace to the Light of the Gospel, so that they Chearfully professed it, willingly quitted their house and home, chusing rather to undertake the fatiguing Emigration.

It can be therefore well said that GOD does now great Wonders and imparts his Grace to so many calling them with Power and Might to save them from eternal Damnation; the Patience and Resignation of those People is very particular and almost incomprehensible that they Submit to their Fate as well in their Country as out of it, without the least Murmurings or Signs of Impatience and never have lost the Respect and Obedience due to their Sovereign or his Subalterns.

There appears likewise a Divine and Gracious Direction in that these poor Exiles are received every where with Good Will and Charity, and that Money is sent for their relief from remote parts where they do not pass; which Love and Charity GOD will not leave unrecompenced.

It is easily to be imagined our Ministers by Such frequent Passages are very much occupied, The Spiritual Consolations giving much more Comfort to these People than Temporal Benefactions, and time would not permit Senior Urlsperger to write so often as he could wish to do, but he tells me that he has written to Secretary Newman the 17th and 31st of July last, under cover to Resident Gullman in Franckfort,[22] giving him a particular Account as much as was possible.

Extract of a Letter from Mr. Münch at Augsbourg 28 August 1732. To Mr. Mayer at London.

I send You here enclosed a Letter from Mr. Urlsperger for Mr. Newman which please to deliver. The Confession of the Gospel is now broke out in the Bishoprick of Bercktolsgaden,[23] and there have been 1200 already Registered as protestants of the Augustan [Lutheran] Confession.

Translation of a Letter in High Dutch [from Urlsperger]. To Mr. Newman. Augsbourg 18, 25 August 1732 N. S.

Dear and Honoured Sir: My Letters of the 17th and 31st of July as also of the 12 August I hope You have received with the

Papers enclosed in'em; the two first I sent by the favour of Mr. Resident Gullman at Franckfort, and the third under cover of Monsieur de Munch to be by him forwarded to Mr. Mayer at London, in which Letters I have acknowledged the Receit of all your Bills of Exchange and given a full Account of what then related to the Exiles, and I showed likewise how the Money Received hath been disposed of by me.

At present I have nothing further to acquaint You with but only this, that in Berchtolsgaden from whence no Emigration hath happened hitherto, 1200 have professed the Protestant Religion and these are of those Artists whose famous Turnery Ware is sent through the World. But how it is with these People and likewise with the Tirnbergers,[24] You will see by the Enclosed which contains an Extract of a Letter from my Correspondent at Ratisbon to me, and of another Letter from the Tirnbergers to my Correspondent who as I have mentioned in my last was the chiefest Instrument in the Affair of Saltzburg. And all these are People who did not stir a Year ago. [To the right of the above paragraph is the following note:]

The Extracts of Letters here referred to are entered at length in this Book Fol. 23 & 24 [See pp. 253-54].

Since the last Company of Exiles, who were here the 11th August and consisted of 900 Persons, we had no other but 'tis said that another Company is to come in a short time and that the Remainder who have Professed the Protestant Religion are obliged to stay the whole Winter in the Country: of which however we have no Certainty but this I may affirm with Truth, that in the Neighbouring Popish Countries a great stir is observed and that many by the Example of the Saltzburgers are become doubtful of their Religion. I could produce many Instances but I dare not venture yet to commit them to Writing. This I may say, that within a few Weeks many Proselites are come to me, particularly a certain Monk, who secretly Corresponded with me before. arrived here 4 Days ago and having changed his Habit is conveyed into a Place of safety. Here a Question arises whether it may be done with the consent of the Honourable Society that in such sudden cases as with the aforesaid Monk, I dispose of some of the Money they have intrusted me with for You may be sure GOD will do great things and Babylon will receive a blow of which it is not aware.

The Letters of the 25th and 28th of July, one of which being written by You my Dear and Honoured Sir, and the other by

Messrs. Philipps, Vernon & Tillard, I have received with the enclosed Bill of Exchange of 121 £ 8 S Sterling which Monsieur de Münch very readily paid without any Deduction. It would not be amiss, if You Honoured Sir did insert in your next Letter to me a Thanksgiving to be made in Your Name to Monsieur de Münch.

Pray excuse me to the other Gentlemen who Honoured me with their Letter that I have not sent'em an Answer immediately, for the Multitude of other Affairs hinders me as yet, but I'll do it as soon as possible.

'Tis very agreable to me that the Society hath been pleased to Print an Extract of my Letter of 17th of July may GOD prosper it with many Good Fruits.

The 2000 Florins came from the Protestant Congregations at Vienna, at the Protestant Ambassadors Viz: the Swedish and Danish Ambassadors there, to which the Protestant Merchants settled in that City likewise resort, but they desired for many Reasons to forbear making it Publick. In all there hath been gathered in these two Congregations 600 Florins for the Saltzburgers.

Tis well known now that the Number of these who are actually gone and are still to go out of Saltzburg (the People of Tirnberg, Berchtolsgaden, Pischerwisen, Gehrn, and of Grossbentzer Valley being not yet comprehended in this Number) will amount to 28 or 30,000 Persons, the full number of them cannot yet exactly be known. hitherto they go almost all, as I am informed, into the Prussian Lithuania, a Matter of 2000 of them may be are still here and in other Protestant Imperial Cities and Countries hereabouts. The Commissioners of Zeeland (one of whom being the High-German Protestant Minister at Middleburg hath Preached Yesterday in my Church) have not been so successful as to get 400 Saltzburgers which number they were sent for, but must be satisfied with 45 Persons.[24] The Reasons are because these People love to stay together, and because in Zeeland they can have no Land of their own to cultivate.

If GOD should bring out more of these People, as it appears very plain it will be, it will require mature Deliberation where to place them. However GOD will show ways which no human understanding knoweth yet. If the motions about Religion continue in Bohemia, as I am very well assured of: there is no other Way of taking their Parts, but by Intercession, because the Emperor's hereditary Dominions are to be considered quite otherwise than according to the Peace of Westphalia;[25] but the Lord, who is more powerful than all the Powers upon Earth, and in these our Days particularly manifested his Glory to the Comfort of afflicted Consciences, will give Counsel in this Affair also, for with him is Counsel and Understanding.

The Evangelick Body at Ratisbon doth still continue strenuously to espouse the Cause of the Exiles of Saltzburg. but the Arch Bishop having now made Application to the Aulick [Catholic] Counsel of the Empire, the Affair is brought under new Difficulties. Nevertheless in the mean time many Advantages have been obtained.

My next Letter will show how far the Confidence the Society puts in me concerning the discreet and Prudent Management in the distribution of all such Charities as already have been, and (thro' the Goodness of GOD) may further be paid into my hands, hath been fulfilled; at least I do my Endeavours for that purpose, and call upon GOD at the same time for the Spirit of Wisdom. May GOD reward the Society, the Trustees, and all the Benefactors for the Good they bestow upon these our persecuted Brethren. I wish the Society could see such a Company here and converse with them. A Multitude of other Business calling me off from this, I commit You to the Grace of GOD, and remain with my very humble Respects to the Honourable Society.

Your most humble Servant   S. Urlsperger

P.S. If any of these Extracts were to be printed concerning the People in Tirnberg & Berchtolsgaden, the Names of Ratisbon and Augsburg must not be mentioned as far as they have any thing to do with the Correspondence, lest the Enemies get any Advantage of it. Not long since a Romish Priest preached: *Ah! call upon all Saints that the defection may not come amongst our Clergy.* In the last Company of Exiles were many persons whose Children were taken away by force from the Mother's Breast, at their departure.

N.B. You may depend upon my Accounts because I have them either from the first hand at Ratisbon, or know by my own Experience.

Translation of a Letter in High-Dutch [from Urlsperger]. To Mr. Newman. Augsbourg. 1 September 1732 N.S. Hebrews 6:10

Dear and Honoured Sir: Being informed by Your two last Letters that You have received mine of the 17th of July and 1st of August N.S. I hope You have by this time got likewise my Letters of the 12th and 25th past, Your last of the 8th of August O.S. came to my hands the 29th Ditto N.S. with three Bills of Exchange which were immediately paid by Monsieur de Munch with his usual Readiness and disinterestedness Viz. 1991 Florins 34 Creutzers at 8 Fl. 45 Creutzers per Pound Sterling. In a Multiplicity of Affairs, which both my Ministerial Functions and other extraordinary Occurrences daily increase, I add only as much as the present Circumstances permit.

First You will always receive in my Letters particular Receits for the Sums of Money You transmit, as I enclose now three Receits.

Secondly. The use I make of this Money as I hinted several times, is as followeth:

1st. I buy good Books and get them bound for the Exiles.

2nd. I cause Linnen to be bought which is made up into Shirts and Shifts of several Sizes, that are distributed afterwards amongst them that are in need.

3rd. I buy Shoes, Stockings, Neckcloths, and other necessary Clothing.

4th. I assist Women in Child-Bed, Babes, Sick Persons etc.

5th. I have put out to Nurse several little Children which their poor Mothers were not able to carry along with them, especially the Children of those who did not know where to find their Husbands. Such a little Boy Thomas Trickel by Name lives in my House, who is very pretty, and hath very good Parts, and may perhaps be brought up to the Ministry for his Countrymen at the Charge of the Society.

6th. I distribute Ready Money among the New-Comers according to their several Circumstances, and sent likewise certain Sums to Memmingen, Obermergen and Oettingen. I have given something to our Protestant Hospital for Poor People, because a good Number of very Pious Exiles 6 of whom are 80 and above, and a Woman 96 Years old, have been taken in there. And I have likewise given some Money to our Infirmary, because several very infirm Persons of the Exiles have been taken into this House also.

7th. I have always contributed to the Common Cash collected here, to provide the Exiles with Victuals and Drink, during their stay here, and to send them further to Oettingen.

8th. Several Young People of the Exiles have been put out Apprentices with this Money.

9th. Some Saltzburgers that were forced to leave their Country
last Winter, and are gone this Summer back again to fetch
their Children, have received some Money towards the Expences of their Journey.

10th. If any Proselytes come out of Bavaria; who have been convinced by the Saltzburg Exiles to turn Protestants, they receive
some Money to get soon to a safe Place, because they would
be in Danger here.

Thirdly. No Company of Exiles having been here since the
14th August N.S. nor any at Memmingen, nor Donawert [Donau-
wörth], my Cash for the Exiles is in very good Circumstances
because I have still in my hands the greatest part of Your three
last large Remittances, Viz.   £   s d
    14 July _____208 16 –
    28   _____121   8 –
     8 August _____227 12.6
I am now like a Mother, who having both her Breasts full of Milk,
can hardly stay 'till her Child comes to suck. For no Company
of Exiles having been here these 16 Days I am thinking it is not
right, and wait as it were with Pain to make them Partakers of
the English Charity but GOD knows best, how to regulate everything; He fills in the mean time the Magazines, that nothing may
be wanting afterwards when new Companies arrive.

Fourthly. There are spread several Reports why no more Exiles
have been this Way all the while, but We cannot depend upon
these Reports. The discourse runs now almost entirely upon the
Affair of Tirnberg and Berchtolsgaden of which in my last of the
25 of August.

Fifthly. Yesterday I sent a certain Saltzburger to Nuremberg
who was forced to fly the Country for having assisted two Merchants [Zwilling and Moedlhammer] at Saltzburg, who have a
secret Inclination to the Protestant Religion, two Chests of Protestant Books (marked Chests of Cheese) were sent to them from
Nuremberg who forwarded them to the Mountainous Countries[.]
both the Merchants having been betrayed are condemned to Publick hard Labour on the Fortifications of the City of Saltzburg and
being Rich one of them was forced besides this to pay a fine of
1100 Florins and the other 1300 Fl.

Sixthly. Part of the Company of Exiles that were here in the
beginning of August went from hence towards Nuremberg and
coming through a Popish Place belonging to the Teutonick Order
called Ilingen were miserably pelted with Dirt and Stones by the

People so that the Bailiff of the place could hardly make them leave off. When the last Company was coming hither it happened that at a Place belonging to the Bishop called Kirchlein, very famous for frequent Pilgrimage, which lyeth 5 Hours Journey from hence, an Exile Woman very big with Child went to the Mistress of a Publick House into the Kitchen and begged heartily to let her have some warm Broth for Money because She had had no warm Victuals for several Days, but the Woman of the House flew in a Passion against her and bid her be gone out of the Kitchen immediately into the Stable her Lodging for She could not endure her Sight. The Poor Woman went away with Sadness, and hardly entring the Stable fell into Labour. This was a great Trouble, at last they suffered her to come out of the Stable which stank prodigiously into a Barn to lie down upon the Straw. She would not be delivered by a Popish Midwife, and in the mean time the Child was born upon the Straw. A Protestant of this City, who went to meet this Company on Horse back hearing of it, hastened back to Augsbourg and fetched a Protestant Midwife with Bedding and Medicines in a Chaise. The Midwife found the Child under the Straw but the Woman extraordinary Weak; however by the Mercy of GOD She hath been brought hither with the Child and recovered so well that She cannot praise GOD enough for herself and her Child. Some Days ago She and her Husband, and Maid Servant desired the Sacrament from me and having previously examined'em and found them Answer mine Examination excellently well, I distributed it unto them, during which holy Performance Tears of Joy were continually running down their Cheeks. I can assure You Sir of this Family again that I did not hear them complain of any Hardship, but they are continually full of Praises and Thanksgivings. This Family hath been liberally refreshed by the English Charity and I shall do so still. many more particulars might be added, but I want [lack] Time and hands.

Pray Sir give my humble Service to Mr. Mayer and let him Know some of this News who is so ready to send over Your Bills of Exchange. next Thursday I design to write again adding only now that two Days ago I had certain Intelligence from Leipzig that 30 Bohemians went through that City towards Berlin to implore the King of Prussia's Protection in matters of Religion for themselves and other Bohemians but they will meet with the greatest Difficulties because the Peace of Westphalia does them no Good and because most of the Bohemians are Slaves.[26] My most Obedient Respects to the Honourable Society and particularly to

Sir John Philipps, the honourable James Vernon Esq. and William Tillard Esq.

The Lord is GOD! The Lord is GOD. In whom I am

Yours entirely   S. Urlsperger

P.S. For the last Bills of Exchange I return Thousands of Thanks in the Name of the Saltzburg Exiles. May GOD bestow Graciously all the Blessings he hath promised in his Word. Amen.

Translation of a Letter in High Dutch [from Urlsperger].   To Mr. Newman.   Augsburg 8 September 1732.

Dear and Honoured Sir: I hope You have Received my last of the 1st of September N.S. which contained several particulars: though it's now 8 Days since I had any Letter from You, I would not entirely neglect the Post, but give You such further Account of the Exiles for the Information of the Honourable Society, as my Affairs would permit.

'Tis probable that We shall have no more Transports this Year out of those Countries from whence any Exiles have already come. Possibly some may come from Tirnberg and Berchtolsgaden, unless they should move towards Ratisbon. Should they not come this Way, I should take care to convey some of the English Charity to my Correspondent at Ratisbon [J. von Reck], *who as I formerly told You has been the Chief Instrument in this Affair,* that they may have the same Relief as though they were here. The Circumstances of these Saltzburgers and particularly the People of Tirnberg and Berchtolsgaden, You will learn by my Abstract from Ratisbon since the 2nd of September to which You may give entire Credit.

I must explain the latter part of the Abstract, which mentions how ill the Protestants left behind are treated by severe Oaths imposed upon them. They are to Swear.

Ist. That the Protestant Faith is new, Heretical and Damnable, on the Contrary that the Roman Catholick is the only right and true Faith without which no Man can be saved.

IInd. That the Popish Mass is a Sacrifice for the Sins of Men both Living and Dead.

III. That without the Intercession of the Virgin Mary and other Saints no one can be saved.

IV. That there is certainly a Purgatory in which we can make atonement for our Sins, and thereby obtain Mercy.

V. That we cannot be saved by Faith alone but by Good Works also.

VI. That the Popish Sacrament of the Lord's Supper, under one Species [bread] is more efficacious than the Protestant Sacrament under both kinds [bread and wine].

VII. That all these and other such Articles must be believed in order to Salvation.

VIII. Some of the last Companies affirmed that All who are any Ways suspected must take this Oath and must at the same time Damn such as were already gone out of the Country.

I have added an Abstract of a Sermon preached by one P. Fyffer[27] a Popish Priest in the Cathedral upon the 11th Sunday after Trinity, to which he invited the Protestants 8 Days before hand. It was heard by some intelligent Protestants who drew the inclosed Abstract. You will read the wretched Stuff with Amazement, which contains nothing but Falsities, Ridicule, and Blasphemy the usual Weapons of the Papists when They cannot prevail against the Truth by Force or Fraud. Their Impudence is surprizing that they dare affirm that the Exiles said Christ despaired upon the Cross, though the contrary has been assured Them in Print and the Exiles have answered the same Question here in Publick before many Papists. This part however of Mr. P. Fyffer's Sermon is true that some Young Persons are come hither, who have had Children by each other without ever being Married by a Priest; but we must observe in Answer to it.

First, That of so many Thousand Persons we never pretended to say They were all Perfect and that none amongst them had been formerly Wicked.

Secondly, When they observed that these Young People were enclined to the Protestant Religion, the Catholicks would not marry Them though they had their Parents consent. but insisted they should embrace the Popish Religion and

Thirdly. We have had many Instances of such as repented their former Sins with thousands of Tears, earnestly desiring forgiveness of them by the Blood of Jesus Christ, so that we may apply to them what St. Paul says of the Corinthians I. Chap. 6:11 such were some of You, but Ye are washed etc.

I hope in a few Days to receive an Account from the Ministers belonging to the Danish [Möllenhof] and Swedish Ambassadors at Vienna, how they fare at Schemnitz and the Places there abouts which I will Communicate to You as soon as I can, but with this

Caution, that the Names from whence these Accounts come may not be published.

The Popish Clergy do what they can to suppress the Motions which the Emigration has caused, particularly in the Bavarian Mines, and to prevent the like for the future; to which End the Government of Bavaria sent a Rescript some Weeks since to the United Magistracy in this Place, forbidding our Protestant Booksellers to carry any Books relating to the Exiles to sell at Munich, upon pain of having their Books and all their other Goods Confiscated, and the Transgressors be further Punished.

They observed this Order here, but the Treaty of Westphalia being an universal Law of the Empire, cannot be forbid to be made Publick, and being so little known amongst the People, it has been Printed here and a great Number of them sold at Munich. As a Protestant Bookseller's Shop at Munich was searched by the Examiners, there happened to be a Popish Priest amongst them who finding the Treaty said 'Ay! this is the very Devil upon which the Saltzburgers and other Protestants depend." The Bookseller Answered. No, most Reverend Sir this is no Devil, but the Treaty of Westphalia upon which the other said no more.

We had lately another Rescript from the Elector of Bavaria, requiring the Magistracy at Augsburg to examine a Protestant Baker's Wife, who is charged with having said "The Elector of Bavaria is turned Protestant, and all Bavaria will turn likewise," which we have no Reason to doubt because the Protestant Ministers have mentioned it in the Pulpit, but the Story has been found to be altogether without Foundation; it shews however how vigilant the Popish Clergy are; but we sing this Hymn, let'em lay their Designs ever so wisely, GOD orders them a contrary Way, All is in his Hand.

The Tirnbergers in a Letter to My Correspondent [J. von Reck] at Ratisbon have this Expression: "We have at Present but little Comfort (they mean because every One is against them upon Account of their declaring themselves Protestants.) and have nothing but hardship to expect, nevertheless we rejoice that we have publickly professed the Word of GOD." And this indeed is the only true Matter of Joy! may GOD bring many others with Comfort to the Gates of Zion, for as much as the Sound still continues. *Yet there is Room.*

I shall send You a very agreable Account of a Circumstance which happened to one Joseph Scheitberger[28] who has been banished from Saltzburg ever since the Year 1686 has all along by his Writings been a Spiritual Father to his Countrymen and is yet living. (By the By) he has likewise been assisted by the Charity from England.

I beg you will assure the Honourable Society with my humble Respects & Prayers that their Charity brings Praise to Almighty God, and is a great Refreshment to many miserable Creatures and that the Money is distributed with all Faithfulness, and that You will be so kind to Communicate part of this Account to Mr. Meyer [Mayer] the Banker.

I write in hast, and send the Letter immediately away having no Copy for want of Time and Assistance. I commit You to the Divine Grace; and with my humble Service to all the Benefactors and Acquaintance, remain

Yours entirely   Samuel Urlsperger

Where is Mr. Ziegenhagen? I have wrote 3 times to him in Germany, but don't know whether my Letters be come to Hand.

P.S. The Moment I was concluding my Letter I Received Yours of the 18 of August O.S. with a Bill of Exchange for 100 £ and will Answer as much as the Time will permit.

1st. Monsieur de Munch is still willing [still wishes] to do as He has hitherto done. He has just now wrote to me in these Terms. "Whereas the Honourable Society and Mr. Newman are not weary in bestowing their Charity upon the Confessors of Christ, and send them so much Money, which I hope the Lord will reward. I shall likewise not be weary in the Affair, to give my Assistance with all my Heart, and desire You'll give my humble Respects to the Honourable Society and to Mr. Newman."

2nd. I return Thanks in the Name of the Salzburgers and other Protestant Converts who are now come to me for Refuge.

3rd. I can as yet send no more exact Accounts than what I have already given which I received from the Prussian Commissarys.[29]

4th. By my last You will see that we cannot certainly tell how many are yet behind in the several Districts that openly profess the Protestant Religion. I am afraid some may be deterred from it, but it will be of little Service to them. We have Examples of this kind who yet afterwards became more Zealous.

5th. I hear the New Confessors in Tirnberg and Berchtolsgaden amount to above 2000.

6th. The King of Prussia has indeed given Orders to receive all that are already come except Miners, which all these Tirnbergers are.

7th. The Rest are dispersed in the Imperial, and other Protestant Principalities and Countries, but several afterward went from thence into Prussia.

8th. They go into the Prussian Lithuania, which lies 20 Miles[30] from Koningsberg [Königsberg].

9. Most of the Exiles hitherto have been Farmers, with their Men, and Maid Servants, and few Trades-Men, they were such Farmers as lived in their own, some of them very fine and large Farms.

10th. The Number of Children was very great, more than that of the Adults.

I have thus answered all your Letter, and shall always continue these Accounts as I am able, but Sir I have not been able so much as to read my Letter over again.

Abstract of a Letter[31] *dated Ratisbon 2 September 1732. (enclosed in the foregoing Letter to Mr. Newman)*

Just now arrived here the third Deputation from the People of Tirnberg and Berchtolsgaden, who brought again several Letters. The Young Man is come back who I told You went from hence with a Letter into Saltzburg from me and laid it before the Chancellor of the Court, who read it Publickly to the Commissaries, who often looked upon one another. The Chancellor asked the Young Man several Questions concerning the Substance of the Letter, and said at last this was both Preaching the Protestant Religion, and giving useful advice, to all which the Young Man Answered with Courage from a lively Faith and hath now brought a Pass-Port along with him; wherein he is called a Protestant. The Case of the Tirnbergers is very Good, of which I give this short Abstract.

First, they have sent 4 Deputies with their Memorial to the Arch Bishop, in which they made mention of about 800 Persons upon which the Commissaries came to Tirnberg, August the 8th about those Miners in the Salt Mines, called them together and proposed several Points concerning Religion, but they could not be moved, though they examined them one by one.

Secondly, Whereupon their Names were set down as Protestants, being told at the same time how to behave themselves for the future.

I. They should endeavour to sell their Goods as well as they could.

II. To forbear Meeting upon a severe Corporal Punishment.

III. They should do their Work in the Mines diligently and leave no Fraud or Damage behind them (of which they are very much afraid at Saltzburg because one single Miner could do them unspeakable Damage by Water) but they answered modestly: *it should be far from them to do so etc.*

IV. They should give Notice of their Departure before hand. The Preachers of Repentance[32] and Jesuits, as Foreigners will now try what they can do with them. But it will be in vain.

The People of Berchtolsgaden are gone almost a whole Year very little to their Churches, and not at all to the Mass but they meet together, read, Sing and Pray, and no Body forbids it. Their Dean would not venture to receive their Memorial and therefore they could have no Answer, now they will go to the Prince himself with their Memorial. And then I hope the Lord will open one Door after another, and bring his Flocks out of other Folds. Blessed be the Lord who alone doth Wonders, and blessed to his Glorious Name; all Lands must be at last full of His Glory. Also,

I have been informed that the two Merchants at Saltzburg Messrs. Zwilling and Moedthamber [Moedlhammer] have been severely punished and must Work in the Fortifications for bringing Protestant Books into the Country. God Grant that the English Magazine [funds] with You be filled more and more with his Blessings.

If the Tirnbergers should be permitted to come hither to us, perhaps I shall make Application to You, though it should be told from whence the Fountain flows. All is quiet now about more Companies coming, it seems the last that was to go, remained in the Country. 'Tis said, they are very hard upon them, especially with taking of Oaths.

Abstract of a Sermon Preached by a Popish Priest at the Cathedral Church at Augsbourg 11th Sunday after Trinity 1732. (Enclosed in the foregoing Letter from Mr. Urlsperger to Mr. Newman).

His Proposition was:
The Glory of the Lutheran Church, which she hath from the Exiles of Saltzburg.

In our Catholick Church our Glory are the 12 Apostles with their Faith and Miracles etc. What is the Glory of the Lutherans? What Miracles have they? the greatest Miracle of our Time, than which none hath been greater yet, Viz. That so many Thousand

Saltzburgers turn to their Church. This is such a Wonder that the whole World must Wonder at it. We will consider therefore

First, The Wonder in it self.

Secondly, As it is an holy Wonder.

1st. The Lutherans say: is it not a great Wonder that above 2000 Men have been enlightened at once? I must confess it is a great wonder; but I read in their Printed Papers, that before Luther's time Faith had been already in Saltzburg. which sheweth that they have not been enlightened suddenly, and thus You see that they contradict themselves. Pray consider whether it be a Wonder that Tares are found in a Field, which the People see and cut off but leave the Root in the Ground, therefore it grows again and brings Seed, which falls down upon the Ground in several Places, and spreads it self far and near, and so it happened in this Case.

Further observe, that these People were stubborn from their Childhood, who would not hear when others were willing to instruct them in the Catholick Faith, but resisted, remaining stiff in their stubborn free will, and their scandalous Lives, and chosing rather a Religion that is agreable to their Nature and Affections where they may live as they please. and are not tied down to any Injunctions of Fasting, of making Auricular Confessions, of mortifying their Body with Works of Penance etc. but where they say it is impossible to keep the Commandments of God, and therefore we may do what we please.

But what do I say? surely they keep bravely in an unmarried Life, the Commandment, increase and multiply;[33] this was evidently seen by several Hundreds of Children, which they brought along, as so many Witnesses, in Waggons. Pray see this is the great Wonder of the Lutherans, of which they speak in their Form of Prayers for the Festival of Peace, of which they have made Pictures and added Explications to'em, exclaiming with David: it is a Wonder before our Eyes Hallelujah. it is of no great Reputation if People Praise their Goods too much, as we see by the Mountebanks, who make a great Noise of their Medicines to make the People buy them, though it is very poor Stuff; and thus do the Lutherans, they proclaim and make a Noise about this matter in single Sheets, in Prints, in so many Books etc. There is likewise published a Print and called a Wonder by Lutherans that a Child of two Years old jumped down three Stories high. Mind the Wonder! the Child of two Years old did know what Faith is and that the Lutheran Faith is the true Faith; the Child of two Years hath Committed himself to the Providence and Protection of God, and jumped down three Stories high; mind the Wonder, it was a little Country House of a Farmer, three Stories high, such as they have in Saltzburg. My Christian Believers, what must our Wonder be at this and that Altar? before this and that Image of Grace? they are nothing, our Wonders are all Lies, but with them all is true, and all together a great Wonder.

II. We consider it as an holy Wonder. Mind how the Lutherans praise this Wonder. They say: these People behave themselves very devoutly, O how fervently do they Pray, how humble and contented they are, their Piety & Zeal cannot be looked without Tears. Behold these are the holy People! But Mind, the Lutherans say nothing at all of the Behaviour of these People in their Country. There they sat together round a Table, with a Salt-Cellar upon it, they dipt their Fingers into the Salt, and bound themselves with an Oath,[34] not to leave one another 'till they obtained their Will, they gathered together before the Magistrate's House, as Rebells, attacked him, with several threatning Words, saying there were 800 Men sent from Ratisbon, and 1800 from other Places to their Assistance and were already on the Road; the Popish Heads who resisted them should soon tumble about on the Ground, particularly mentioning their Prince, and other Great Men.

Is not this Rebellion? moreover they spoke Blasphemously of Christ, that he despaired on the Cross, because he cried out: *My God, why hast thou forsaken me?* likewise of the Blessed Virgin Mary, that She was in the lower most part of Hell. Saint Nepomuck they have called a Hangman, they have disfigured the Image of St. Augustin and other Saints. And yet these are Pious People.

Concerning the Doctrine they have received, Pray, who hath instructed them? I tell You a learned Man, Viz: a Smith. But they have their Faith from the Scripture, You say, but how can that be, when they cannot read? But some could read, and read the Scripture unto others that could not. Well, but who hath explained the Scripture and proved that this is the True Faith, for the Lutheran Ministers themselves say that the Scripture ought to be explained? I Answer, the abovementioned Smith and some others amongst them have explained the Scripture. Ha, Ha, this and his Companions are the Learned Men, who can explain the Scripture: 'tis likely they have made a true Explication thereof. What are the most learned Men of our Church and all the Fathers of the Church? They are nothing in Comparison to these People. but who can be so void of Sense and Understanding and believe that these People can explain the Scripture and prove from the same the Articles of Faith. I have often told the Lutherans, and repeat it again to Day, that they shall prove from the Scripture by one Single Sentence only, but without Explication, that my Faith is a false Faith, I say without Explication, I will turn a Lutheran this very Day.

But it is certain these People have the Lutheran Faith, Why? because their Examination is printed here, there You may Read it: But I find only Nine have Answered, and therefore are all the rest of the same Faith, for all that many may be Anabaptists and I don't know what. And then why must We believe they have Answered so? two Ministers have Subscribed their Names, therefore it must be true, and yet these are themselves interested in the Case. but so it is, what they say, all is true, what We say, is nothing but Lies. Let it be so, but I desire the Lutherans to consider what they do in seducing so many People. May God convert them that they may return to our true Church!

Copy of a Letter from Mr. [name omitted] dated 10 May 1732. (Sent by Mr. Urlsperger to Mr. Newman enclosed in a Letter dated 17 July 1732).

There is an extraordinary Motion throughout the whole Kingdom of Bohemia, whose increase can't be related in few Words: This Week the Papists have intercepted again 18 of such as were going to leave their Country to which they were encouraged by a Man of Limburg, who having visited our Country [Saxony] at his return home gave his Friends an Account of what he had seen and heard here. This made such Impression upon their Minds that they expressed a great desire to be Members of such a Congregation as their Friend had described. In short they left all at once and made the best of their Way towards us with two Waggons but before they could leave the Borders of Bohemia, they were pursued by their Enemies, and all of 'em brought back again in Bonds and Fetters, except one, who narrowly escaped out of their Hands in the following manner:

After they had been taken they were forced to take up their Lodging in a Gentleman's Stable, and being all Manacled and fast tied together two by two, it happened that a Son's Foot was but slightly tied to that of his Father's, and by endeavouring found means to free himself from his Shackels. This done he went towards the Door, but stumbling over one of their keeper's Feet, and being forced to stride over the Body of another that lay before the Door, he would have paid Dear for it, had not been both fast asleep: then opening the Door as soft as he could went out and bolted it after him. When he came into the Yard he had another difficulty to surmount, which was a pretty high Wall, he must climb over it if he would save himself, and finding nothing but an old Broken Ladder of one Pole and but a few Spikes, he made shift to get to the Top of the Wall, and jumping down the other side sprained his Foot. However God was pleased to strengthen him so much that he reached Zittau by 12 at Noon.

The next Morning as the rest were carried back again, they went on Chearfully in their Journey, singing along the Road David's Psalms and other Divine Hymns, so that many that heard'em were moved at it. They stretched forth their hands to those that Guarded'em, expressing their thanks for giving them an Opportunity of confessing their Lord Jesus openly.

It was remarkable that the very same Person who made his escape, in his Prayers with the Rest of his Brethren upon the Road, after they were taken, made this most earnest Request to God that but one might escape to give an Account of their Circumstances to our Congregation to excite us to remember their Afflictions before the Throne of Grace.

The Officer who Guarded'em asked him why he Prayed with the Rest that were no Lutherans yet, and at the same time gave him such a stroke with his Fist as made the Blood gush out of his Mouth: However the Lord was pleased to hear and to fulfill his Prayers afterwards and although all-wise Providence has suffered these Young and unconfirmed Babes in Christianity to meet with such hard usage, yet that doth not deter others, but We hear every now and then that here 50 and there 100 are ready to leave their Country.

Let this Relation engage You to consider amongst Your selves, whether it would not be advisable and of Great use,

1st. To Print a short Account of the Present Persecutions of the Protestants in Bohemia, in Order to excite all true Believers to pray for these Poor People, especially since they meet with a far more Rigorous Treatment than the Poor Saltzburgers do.

2nd. To acquaint other Protestant Princes of the Empire with the Bohemian Persecutions by whose Intercession the Emperor might perhaps be Prevailed upon to grant these Poor People either a free Exercise of Religion, or Liberty to leave the Country.

3rd. To compose and print a Letter for the Comfort of these distressed People as well as for an Encouragement of their Constancy and Patient waiting the time of their Deliverance. but whether this might not be looked on as an Incentive to Sedition or Rebellion and whether it would not be of more Weight if a Foreign Divine were the Author of such a plain and edifying Composition.

A Speedy Answer upon these Questions is most earnestly desired.

P.S. Pray Assist our dear Bohemians with Your Advice, many are ready to depart and would be glad to live in a Place where they might serve God according to their Consciences without disturbance. In short all wait for it. Do as much as lies in Your Power to procure'em a Place of Retreat. The Father of Mercies will reward Your Love.

Extract of a Letter from Vienna dated 26 July 1732. (Enclosed in a Letter from Mr. Urlsperger to Mr. Newman dated 1 August 1732.)

The great Pains You have taken in favour of the Poor Schemnitzer (Schemnitz is a City of the High Lands in upper Hungaria[35]) I acknowledge with all due Thanks. Whilst I was expecting an Answer, I heard the Melancholy News that not only the Question is disputed about in General, whether the Hungarian should be permitted to send for Ministers from foreign Countries, but the Schemnitzer in particular have been actually forbid to supply the Vacancy of their Primariate with another Minister. nay, it looks as if they had a Mind to deprive'em entirely of the free Exercise of their Religion and suffer so many Thousand Souls to Famish for want of Spiritual Food. This has put those People into the utmost Consternation, who for the present are obliged to abstain from calling any Ministers from abroad.

As much Reason we have to rejoice in these our Days at the Providential Retreat of the Saltzburgers and their Deliverance from their Oppressors, so much ought we to be concerned for the Distress of the Numerous Hungarian Church and pray to GOD for their Relief.

In the District of Eisenburg they continue to lock up all the Churches and Schools of the Protestants, but as these strenuously insist upon their Priviledges, some of'em resist by force in many Places, so that the very Women tear off the Seals of the Church

Doors, and oblige their suspended Ministers to Preach to them. Last Week the Struggle about a Reformed Church ran so high that one Person was killed on the Spot and Six Wounded, all which may make the Case of the Protestants but worse. The Lord have Mercy upon the Oppressed and endow'em with the Wisdom and Patience of the Saints which is the only Means of holding out and overcoming the Hour of their Affliction.

Extract of a Letter from Ratisbon dated 21 August 1732. (Enclosed in a Letter from Mr. Urlsperger to Mr. Newman dated 18th and 25th August 1732.)

But above all things to come to the main Point, my Joy at my happy return to this Place increased when I was informed that among others a Messenger from Tirnberg waited for me with a Letter which I here enclose for You. God be praised a Good and Joyful Harvest appears! The People at Berchtolsgaden are increased already above 1200 who are ready to go away when the time cometh. O how did the Messenger Rejoice when I Informed him of the Love of the People at Augsburg towards them, and what Good Tidings will it be to them at Berchtolsgaden when they hear that they can live among the Protestants as a Free People, whereas they have lived hitherto as Slaves, but especially that the Bonds are Broken which have tied up their Conscience.

Those at Tirnberg have been called every one by himself before the Commissioners,[36] and examined about their Religion, and at the same time Notice was given them to make their Departure known 8 Weeks before, and in the mean time to sell their Goods etc. These People as I understand by the Messenger have a great Faith, and Fear no Man. But Reverend Sir hear how Strangely I am dealt with in this Affair. A Young Man of Tirnberg, who hath been here last, having received a Letter from me (to his Countrymen) delivered the same unto the Chancellor of the Court [Christiani] and all the Commissioners, and confessed openly without any Fear, that he had been at Ratisbon and desired a Passport from the Commissioners to go thither again, but they gave him no Answer but that it could not be done 'till the Commissioners met together and refused likewise to return the Letter. Yet I hope the Lord by his particular Mercy will turn all this to our Benefit. If it goes well I expect that Young Man very soon again with more Good Tidings.

Abstract of a Letter from Tirnberg Received the 20 August 1732.
(Sent to Mr. Newman together with the foregoing Extract.)

Whereas we heard of [from] Tobias Wurndle who hath been lately at Ratisbon that You were very desirous to hear very soon of us and our Profession of Faith, and whether we belong to the Augustan Confession, therefore we acquaint You by this Letter that we are come to this Resolution and have sent accordingly Four of us to Saltzburg with a Short Memorial to the Gracious Lord Chancellor of the Court [Christiani], who hath delivered it to the Arch-Bishop. Whereupon Commissioners being sent have called the People before them, and first of all the Officers, who were Stedfast, but not every one of them etc. but the Miners some few only excepted, have openly professed the Evangelical Religion.

The Commissioners are still here and knowing very well that the Gracious Ambassadors and the Reverend Mr. Hard expect an Answer from Us, we are resolved to send a Messenger as soon as the Commissioners are gone, and acquaint You with more Circumstances. For in Berchtolsgaden are not only our Fellow-Miners, but also other Tradesmen and People, especially in Pischerwisen and Gehm [Gehrn], who will [wish to] live and die by the Augustan Confession. Pray serve us, God will reward You for it in Eternity.

To Sir John Philipps. Baronet.   Isle of Man 24 August 1732.

My most Worthy Friend   Had it not been for the Letter to the Secretary of our Society, which by a Good Providence came into my hands, I had not [would not have] had this Opportunity (which I should have been much concerned to have lost) of contributing my Mite towards the Relief of the Persecuted Saltzburgers.

I know You will be some way or other concerned for these distressed People, and therefore I beg You will order somebody to call for the Value of the Underwritten Bill, and apply it to their use.

One cannot Wonder at the Evil Treatment these People have met with, since they, making the Gospel of Christ the Rule of their Faith and Manners, must of Necessity upbraid continually their Persecutors with perverting its Design by their Antichristianism, of which this is a flagrant Instance.

I pray God Support and Comfort these Good People under this severe Persecution; and open the Hearts of their Brethren of all the Reformed Churches to contribute to their Relief.

This Persecution though Grievous to them, will end in the Glory of God. and the Good of Others, as well as of themselves.

The World will see that the Number of those that Oppose the Corruption of the Church of Rome, is much greater than was imagined.

It will show that Church in her true Colours, since Her Corruptions are so many and great as can be perceived and abhored by a People who have been deprived of all the Advantages of Learning and Correspondence with other Reformed Churches.

By their Patient Suffering for Righteousness Sake, many of their very Adversaries will no doubt of it be converted, and such of their Brethren who have stayed behind against the Sense of their own Conscience, will be awakened and follow their Example.

And One would hope that many amongst our Selves, who are too Prone to Infidelity, will, through the Grace of God, be convinced of the Truth and Power of that Faith, which has enabled this People to Suffer the loss of all for Christ.

But however that be, this People's present great Distress will give all good Christians a fair Occasion of exercising a Practical Relief of one of the Fundamental Articles of the Christian Faith; *The Communion of Saints:* by contributing to the Necessities of their Suffering Brethren.

I am, My most Honoured Friend Your Son's and Your Most Affectionate & Obliged humble Servant Tho. Soder and Man[37]

Translation of a Letter in High Dutch. Augsbourg 25 September 1732. To Mr. Newman.

Sir: In hope You have Received my Letters of the 1st & 8th of September N.S. I continue according to my Promise, the Account concerning the Emigration. When the Tirnbergers who are Miners with their Officers, to the Number of 730, had obtained leave upon certain Conditions to go out of the Country, they sent Deputies to the Evangelick Body at Ratisbon with a Memorial to implore their further Assistance, and in particular their Advice, how they should regulate their Journey.

Whereupon the *Gentleman I often mentioned already,*[38] *hath sent them* back again to their Brethren with comfortable Instructions. The People of Berchtolsgaden have delivered their Memorial unto the Abbot, their Prince,[39] but have not yet received an Answer.

The Arch Bishop of Saltzburg hath published a New Patent in his Dominions, promising therein a *Salvum Conductum* to those of the Exiles, who will return and look after their Possessions and Children; but forbidding at the same time in the severest manner, to keep Meetings, to turn the Simple Papists from their Religion; or to have any Dangerous Correspondence: but the Master of a House should be permitted to Read, Sing, Pray & Exercise Devotion with his Family. However the Patent being expressed in very ambiguous Words, will give but little Comfort to the Poor People.

In the several Districts of Saltzburg they tell those that are still in the Country, very dismal Stories about the Exiles, as how unmercifully the Protestants dealt with them, by selling the Young People like Cattle, or sending them to the Gallies; and by starving the old People: how the Poles had cut to Pieces 1500 of the Exiles, that came into the Prussian Lithuania etc. but the Deputies of Tirnberg at their return into Saltzburg have been furnished with Sufficient Accounts by Word of Mouth, in Writing, and in Print, whereby they can give truer Information to their Countrymen.

A Fortnight ago a Popish Deacon came from Saltzburg to Ratisbon, in order to embrace the Protestant Religion, likewise an Officer among the Miners out of another Popish Country, who told my Correspndent that if they did but know in his Country of the kind Reception the Exiles met with among the Protestants, we should hear of wonderfull things next Year.

Some Weeks ago an Exile of Saltzburg, Thomas Keeswurm, went back to Saltzburg to fetch his three Children, who had been detained there by force, but he could get neither of them, and was forced to go away with being called a Lutheran Dog. And yet this very Man was provided with Letters of Recommendation from the Evangelick Body at Ratisbonne.

Just now I received Letters from our Protestant Merchants, who went 10 Days ago from hence to the Fair at Saltzburg, and were desired by me to enquire as well as they could into several Particulars. The Extract of those Letters You will find in the Postscript.

In case the Tirnbergers should come out of their Country this Year, and take their Tour towards Ratisbon, I will send some Relief to them out of my Magazine [funds]. It being known that God hath made use of me in assisting the Exiles of Saltzburg, I receive many Petitions from several Places, to contribute something towards the Support of the distressed Exiles.

I have not yet Received any further Account concerning Hungaria from my Correspondent. A Protestant Minister on the Borders of Bohemia, Mr. Liberda. a very good Man, who Preacheth in the Bohemian Tongue, hath Petitioned the King of Prussia in behalf of 2000 Bohemians, who within a few Years went out of their Country, to grant them some Land in Prussia, and hath obtained his Request. My Brother in Law spoke 4 Weeks ago in Saxony with this Gentleman, and gave me the Information.

With my most obedient Respects to the Honourable Society, and particularly to the Trustees in behalf of the Saltzburgers etc. I commit You to the Grace of God and remain

Dear and Honoured Sir   Your most humble Servant

S. Urlsperger

P.S. There are still some Proselytes coming out of Bavaria. I have not yet finished the Account of Joseph Scheitberger.

Extracts of several Letters from Saltzburg dated 22 September 1732. (Sent enclosed in the Foregoing Letter from Mr. Urlsperger to Mr. Newman.)

It is the Common Report here that the Farms of the Exiles are almost all let out to Tenants, but to Strange and mean Persons, who will hardly be able to pay the Rents; and that some of them are gone away already, after having cut down and Sold the Corn; and let nothing behind them. We asked the Landlord, whether there were many more in the Country, who desired to leave it, and he Answered; certainly, many more, and that no Man would live 'till they were all gone: and though they took the Oath, nevertheless their Hearts were otherways inclined.

Here in this City they talk little or nothing of this Affair; especially at the Time of the Fair. Some Weeks ago the Representatives of the several Counties were called together to consent to an extraordinary Tax of 3 per cent as they say but they came to no Resolution, and are to meet very soon again. The Tirnbergers make themselves ready to depart at Michaelmas, if they can have leave to go so soon; they wait only for an Order it being agreed to, that they may depart. 'Tis certain that about 2000 Persons in Berchtolsgaden are desirous to depart out of the Country, but they don't know here the true State of the Matter, because they have their own Abbot, who is a Prince.

They say positively that about 200 People of *Hallein,* a City, three Hours Journey distant from Saltzburg, are desirous to go out of the Country.

Translation of a Letter in High Dutch [from Urlsperger]. Augsburg 6 October 1732 N.S.   To Mr. Newman.

Dear and Honoured Sir, To acquaint the Honourable Society with the Present State of the Affair concerning the Emigration within and without the Dominions of Saltzburg I have sent the Enclosed Abstracts of Letters, the Exiles who are actually arrived in Prussia are quartered near Welau, and thereabouts in a District of 12 [German] Miles, the King being at great Expences to accomodate these People.

I have Received Your last of the 8th September, and being informed thereby to my Satisfaction, that my Letters, except the two last, are all come to Your Hands, I hope by this time those also are delivered to You.

The Prussian Commissary, Mr. Gobel,[40] hath Communicated to me the Minutes of the Depositions the Prisoners have me [made?] containing about 12 or 15 Sheets of Paper in Folio[.] the Question is whether I should send them to You by the Post, knowing the Postage of Letters is very Dear in England? I desire Your Answers upon this Point. 'tis astonishing what Particulars are related therein, which makes me think that if they were soon Published in England, they would have great Influence upon the People. the Rest of Your Letter I'll Answer another Time; Just now there Goeth a Report that we shall have another Transport in a few Days, but it wants Confirmation; not knowing whether You have received the Memorials which the Tirnbergers and the People of Berchtolsgaden delivered to the Evangelick Body at Ratisbon I have enclosed Copies of them.

May God prosper the New Designs in Georgia, but our Saltzburgers will hardly be persuaded to go that Way because they have no inclination to go to Sea.

I recommend You to the Grace of God, and remain with my most Obedient Respects to the Honourable Society.

Dear and Honoured Sir   Your most humble Servant
Samuel Urlsperger

Extracts of several Letters dated Saltzburg 20th, 22nd and 30th September 1732.   (Sent enclosed to Mr. Newman in the foregoing Letter.)

Being arrived here the 15th Instant, God be praised in Good Health, I endeavoured after the dispatch of my own Affairs, to learn the Present Circumstances of things relating to the Emigration. But I could hear very little of it in this City, only I was informed that the whole Number of those who have left this Country on Account of their Religion amounted to 21446. A certain Friend whom I may believe told me likewise that about 15 Families in this City were good Protestants, which were by Order of the Arch-Bishop watched very narrowly and that no doubt this small Number would in time increase more and more, because the dissolute Life and Behaviour of the Priests both within and without the City was very well known and a great Scandal to honest People. I have seen myself here in the City, not only in the Publick House where I lodge, but in others also, where I had been about my Affairs, such dissolute Priests, especially of the Petriner, who were not only in Company with Drunken People, who had Dancings and made a great Noise, but some were Drunk themselves, and others behaved themselves most scandalously Lewd with Women.

And here I must acquaint You with a Relation [an account] a very grave Understanding Papist (who seems in his Heart to value the Protestant Religion more than that which he Professes) gave me with his own Mouth; viz: A certain Regular in Saltzburg put this Question in a Letter to one of his Order in a celebrated City in Suabia: what he thought the Cause of the great Defection from the Catholick Religion in the Bishoprick of Saltzburg: but added presently: he supposed that his Friend would expect that he, who lived in Saltzburg and could know all Circumstances better, should resolve him this Question; and therefore to confess the Truth ingenously, he must say the Chiefest Cause thereof were they (the Clergy) partly because of their Laziness and Negligence, and partly because of their Scandalous Lives. Upon which he goeth on in his Letter and acquaints his Friend that a few Weeks ago the Visitator going from Saltzburg in Order to make a Visitation in the Country (for now the Superiors begin to be more careful) came towards Evening to a Ministers House, and understanding by his Maid Servant that he was at a Wedding in the Publick Inn, he sent for him twice, however not telling who he was, but he did not come, but sent Word to the Stranger in his House that he should stay; at last after Eleven o'Clock at Night, whilst the Visitator waited for him at his House, the Minister advanced towards his House with a Fiddle in his Hand, and a Company of his People after him, Singing and Dancing and carrying their Liquor along with them. The Visitator seeing all this reprimanded the Priest severely, when he came into the Room to him; but he made

Answer; he was forced to please the Country People in this Manner, else he could not subsist for his living was too small to maintain him. The Heads of the Clergy in Saltzburg, the Deans and Prebendaries and such like were more to blame than He and his Brethren, for they received the Income of the Livings, and appointed Curates in their Places, whom they would rather suffer to Starve than to give them any thing to live upon etc.

But I return to relate what I have observed myself furthermore. As there was but little News to be heard of the People on the Mountains in Saltzburg, I went with a Friend, a very Grave and understanding Papist, and no Enemy to the Protestants, without whom I could not have obtained the Chancellor's Passport, which is necessary for Passing through two Forts, into the Country, through Hallein and Tirnberg to Berchtolsgaden, where a Servant in the Inn told Us a secret that his Master the Innkeeper was at that time in the Castle, where a Council was held by the Jesuits and other Clergy, about the Present Circumstances of the City and Country, because there were in this little Tract of Land some Thousand Protestants already, some indeed lived in secret still, but others made Publick Profession of it, and the Number of them increased Daily, who desired either a free and regular Exercise of their Religion, or leave to go out of the Country before the End of the Year; and because a great many Artists, especially in Fine Work, were amongst them, it was very hard for the Prince to let them go out of the Country.

This Servant used a very particular Expression to me, not knowing I was a Protestant, viz: He went hastily to the Window which lieth towards Werffen [Werfen] and opening it, pointed with his Hand and said, that Way all are Protestants, and then running to another Window which looketh towards Austria, opening likewise he said; and that Way all are Protestants too. He said farther that he had been but two Days before in the District of Saalfeld [Saalfelden], and heard that many who last Year had been deterred from making Publick Profession by severe Threatening had now Publickly given in their Names as Protestants, which was a great Grief to the Magistrate of that District, and that the Mountainous Country in General was still full of those People, which was the Reason that very few Rich Men ventured to buy their Possessions, for fear the King of Prussia would at last make Pretensions [claims], and cause many Alterations; 'tis certain that all over the Country they are very much afraid of the Brandenburgers as they call Them.

On my Journey into Saltzburg I met many Jesuists who coming from Munick, Ingoldstadt etc. are willing [desire] to try their skill in the Dominions of Saltzburg, but hitherto their Labour hath been without any great Effect.

I met likewise a Weaver, who being a Subject of the Bishop of Augsburg, had been in the Country of Saltzburg to buy a small Farm which the Exiles had forsaken, but the Magistrate in *Goldeg* [Goldeck], upon his saying among other things that he heard of more People who Professed the Protestant Religion, was so enraged at him and threatened him with Imprisonment, that he had no further desire to buy any Farm, and because the Weaver thought I was a Papist, he said Sir, Indeed I don't like to live in this Country, for wherever You come You are sure to find People who will not salute You with the Salutation: Blessed be Jesus Christ; who have no *Scapulare* make no Cross, neither take off their Hat before the Images etc. in short Sir, believe me the whole Country is Good for nothing.

My Fellow Traveller to Hallein, to the Mountainous Countries and Berchtolsgaden, enquired of me, how the Exiles had been received, and whether I had spoke with any of them about the Purgatory, the Mass etc. and on my giving him a full Account of all things he was silent a while, and then Smiling desired me, by an Opportunity he named himself, to send him all the Accounts of what passed with the Exiles, and to send him Letters for the People in the Mountains, which he would forward to them with all possible Diligence; certainly this Gentleman hath some secret Design; he was very Desirous to know the true Circumstances of the Commotion in France, and knew more of it than I.

The Situation of the Mountains. especially in Summer Time is very Pleasant and the Cattle looks better than in Suabia, which convinces me that these Good People had no want of Temporal Things.

An Account from a Protestant Merchant who is at the Fair at Saltzburg dated 30 Sept. 1732. (Sent by Mr. Urlsperger enclosed in the foregoing Letter to Mr. Newman.)

Yesterday I had a visit from N. N. of Berchtolsgaden, who told me that the Names of all the Protestants in the Dominion of Berchtolsgaden were set down, that they were allowed to read their Books, none of which had been taken from them, in their Houses, but they were forbid upon severe Penalties to hold any Meetings as they had done hitherto, which made others that could not read Sigh and complain that they were now deprived of the Opportunity of hearing the Word of God. They Promise themselves no Good from the Present Prince, he having sent Letters to Rome to desire Instructions how to behave himself towards these People, and likewise to Vienna with Proposals, whether he might not send his Trades People beyond Sea that the Manufactures might not be set up in Prussia, that the Misery of these People increased because the Papists are forbid to employ the Protestants by whom they have hitherto got their Livings, the worst is that they are all Poor People. That a Number of 200 or 300 Papists of Berchtolsgaden were sent down to work in the Mines in the Room of the Protestant Tirnbergers.

The Postmen who come from the Mountains, have Orders here to deliver all the Letters they bring from thence to the Curators and Deans, they are searched in the City Gates, and all the Letters from thence are taken from them, so that their Friends here can have no News from them, of which they are very desirous. This is a fine Behaviour according to the Treaty of Westphalia. Just now I got the Memorial which the Protestants of Berchtolsgaden sent to the Evangelick Body at Ratisbon, in which only 150 Persons are mentioned who Profess the Protestant Religion, but there is no doubt that if the secret Disciples see the good Success of these, they will likewise appear only [openly?].

Translation of a Letter in High Dutch.   To Mr. Newman.

Dear and Honoured Sir, I do not doubt You will have received in due time my Letters of the 25th past, and 6th Current. So as I have received Yours of the 8th of September[.] it seems that at a certain neighbouring Country there are great Commotions, on Account of Religion, yet I will not give any Account thereof, 'till I have received more and more certain Advices. The Advices which I herewith send You are given to me from Faithful Men whom I had instructed what Enquiries to make. I am expecting [awaiting] what is to be done with the Protocolls and assuring all the Gentlemen of my humble Respects.

I remain   Yours entirely   S. Urlsperger

N.B.: The Original of the above written Letter was read Wednesday the 1st November 1732 O.S. and therefore may be supposed to have been wrote about the 29 October 1732 N.S. a Fortnight before.[41]

The Following Account came to hand enclosed in the foregoing Letter to Mr. Newman.

Faithful Accounts from Saltzburg about some Affairs in that Country.

The Affair concerning the Emigration is managed at Saltzburg by Eight Deputies. Viz:
2 Are of the Arch Bishop's Treasury Viz: Baron of Rehling, commonly called the Suabian Rehling, who is said to be of a Severe Temper against the Protestants; and Baron of Aver, who they say, does not concern himself much about them.
2 of the Cathedral Chaples Viz: The Dean, Count of Thurn, who is a very great Enemy of the Protestants, and Count Truchses of Scheer, who minds the Temporal part of the World, more than the Spiritual.
2 Of the Consistory. Viz: Mr. Freylander and Mr. Hueber who is also Censor of Books.
2 Of the Court Viz. The Chancellor, Mr. Christiani,[42] who is Praeses of his Deputation, and Mr. Meichelbeck, Secretary of the Aulick Counsel.[43]
These Eight Deputies meet in the Palace three times every Week, about this Affair, namely, Mondays, Wednesdays and Saturdays from 9 o'Clock, 'till 12 in the Forenoon, but in extraordinary Cases they meet oftner.
The Imperial Troops who have been above 4 Months in this Country, and consisted of two Battalions of the Regiment of Warmbrand, of the whole Regiment of Stahremberg, and two Battalions of Prince Eugene's[44] Dragoons in all 3600 Men, received daily 8662 Florins 17 Creutzers, for which Expences the Arch Bishop now demands a Sum of 1100,000 Florins from the States of the Country, but they will not consent to it.
From the District of Rastadt [Radstadt] 3962 Persons are gone away, and they say the Deputies have received Advice that 442 Persons who had their Names registered before as Protestants, have declared themselves again for the Popish Religion, and begged Pardon.
There are now on the Mountains and other Districts, Preachers of Repentance of the Society of Jesus, with a design to convert the People, and they going from house to house meet with many Protestants against whom they shew their inveteracy on all occasions. For Instance

N: N: of the Ziller Valley (who sells Brandy at Saltzburg in the Fair-time) told me with Tears in his Eyes, that in his way through Gastein he had seen a Peasant who died a Protestant, and was not allowed to be buried in the Church-Yard but was cast into the open Fields, and there left for his Friends to put him into the Ground at Night. This happened a Week before the last Fair at Saltzburg.

The same N.N. informed me that the Peasants in the Ziller Valley, who belong partly to Saltzburg and partly to Tirol, disregarded the Popish Mass and Invocation of Saints, being instructed by their Parents to pray to God alone for if he forsook them, the Saints could not help them.

He said further: that a Year ago, in the beginning of the Motions in Saltzburg, their High Bailiff would [wished to] force them to make an Irruption from their Valley into the Mountains, kill the Lutheran Dogs, the Peasants there, and take their Goods, but the Peasants of the Ziller Valley answered that as long as the Protestant Peasants did them no harm, they would do them none neither, but in Case they should come and attack them, they were ready to defend themselves. That their high Bailiff was very severe to them, but their Gracious Lord who is a Baron and one of the Council at Inspruck [Innsbruck] took their parts, and assisted them in many things.

He added That formerly he had been very often on the Mountains, but he thought the Grief of his heart would not let him go thither again, much less to make any stay there because the best People and almost all his particular Friends were gone away.

Mr. Seiler, Clerk of the Cathedral at Saltzburg, saith, that he missed the People very much in gathering the Tythes, he was sure the People that went away were the best of People, having paid their Tythes and other Taxes very exactly and most of them if not all even before the appointed time, whereas the Popish Peasants were commonly, three, four or more Quarters behind hand.

The People who come from other parts in the Room of the Exiles, are not fit for the Business, and most of them go away again, several repassing through the City of Saltzburg. I have my self in one day Viz: the 8th of October spoke with three of those People, namely, a Smith from Passau, with his Wife and one Child, a Tailor from Wallerstein with his Wife and three Children, and a Peasant from Ottobeuren near Memmingen, who unanimously affirmed that no Body could live in this Country upon Account of the hard work there was to do, the Peasant from Ottobeuren added, it was not only the poor Livelihood that one could get but the very bad Christianity of the Catholicks, also, which offended him, and made him leave the Country again, especially for the Sake of his Children, who were brought up as Good Catholicks, but the People of this Country were nothing less than Catholicks, for at the Celebration of the Mass all sat still, none of them standing up, just as if there was no Mass that no Body did stay 'till the Priest gave the Blessing and Sprinkled the holy Water, and therefore both the Blessing and the holy Water were given by the Priests rather to the Pews of the Church than to the People. That none at all made the Cross; That he had worked very hard 8 Weeks in Hattau [Hüttau], beyond Werffen [Werfen], and could get but 6 Creutzers a day.

How very watchful they are in some Districts of Saltzburg, to hinder any Protestant Books from being brought into the Country, may appear from the following Instance: Michael Neumayer a Pedlar, who carried about Books, and Goods of Gemund in Suabia, and sold them on the Mountains of Berchtolsgaden and Saltzburg these 13 Years, and was never found Guilty or even Suspected of dealing in forbidden Books or other Goods, came the 1st of October into Abtenau where the High Bailiff presently stopped him, and took his and his Companions Pack from them, telling him at the same time, that all his Goods were forfeited and confiscated, because he used constantly to bring Books into the Country and that now he would not permit any more Books to be carried into the Country. The Pedlar answered him as well as he could that he never brought any forbidden Book into the Country, producing at the same time both a Certificate from the Dean of Hallein, who having examined his Goods already, attested thereby that he had no forbidden Goods, and the Notes of some Ministers and Clergymen, who lived in Abtenau and some other Places up farther on the Mountains, and bespoke several Books of him. he desired the High Bailiff to examine his Goods, and take away what he found forbidden, yea he begged of him at last to return him his Packs again, and promised he would go no farther, but would carry'em back: all was to no purpose, he was forced to leave his Packs, and go away empty handed with his Companions. the 4th October they arrived with great Lamentations at Saltzburg, but there being no Sitting of the Aulick Council 'till the 7th of the same Month, they could not deliver up their Memorial 'till then, and are still in fear, not knowing what Sentence the Aulick Council will give in this Matter.

28 September. An Edict of the Arch Bishop was Publickly read in the District of Werffen, Signifying that the Children of those Exiles, who were killed in the Prussian Dominions, being on the Road to return home, it was ordered, that the Tenants of the Farms belonging to these Children should bring them up, and whosoever was not willing to bring them up, should pay so much money as was Sufficient to educate them at Saltzburg in the Work-houses and Hospitals.

The Publication of such an Edict is attested also by the above named Peasant of Ottobeuren, a strict Papist, who was present at the Reading of it; and saith further, that the new Tenants had been ordered already to pay 1 Florin, 1 Florin 30 Creutzers or 2 Florins each, Entrance Money, according to the value of the Respective Farms.

A certain Counsellor of the Consistory at Saltzburg hath bought 11 Copies of the Treaty of Westphalia, of a certain Protestant Bookseller at the Fair of Saltzburg, and advised him to bring no more of them into the Country, because it was not a Book for common People, and though a Peasant did read that such and such a Religion must be tolerated in the Roman Empire, yet he could not understand the following Articles, the true meaning of which was even disputed at Ratisbon, and therefore it was better not to bring it at all among the Common People. It seems the Good Gentleman smells something.

John Walch, a Shopkeeper of Tirnberg informed me that October the 1st the Protestants were called together at Tirnberg, and told that they should continue their Work during this Month, and then they might depart.

Some Questions answered from Saltzburg concerning several particulars relating to the Commotions in that Country.

Q: How many Thousands are Actually departed?
    A: I cannot get any exact Information thereof, but only that the Number amounted to 18 or 21,000.

Q: Who takes care of the Farms of the Exiled Persons?
    A: Most of them are taken by Suavians, Tyrolians, Austrians, and Stirians, but many have left them again.

Q: Whether there was no talk in Saltzburg of the Religious Motions in Bohemia, Moravia, and other Countries?
    A: Very little and almost none at all.

Q: Whether many Jesuists and Preachers of Repentance travel about in the Country?

A: Yea.

Q: Whether the Jesuists are about to establish themselves in the Country?

A: Though they desired leave to set up a Seminary, and their Petition was Backed by the Intercession and Recommendations of the Bishops of Brixon [Brixen], of Trient and the Cardinal Sintzendorf, yet they could not obtain it.

Q: Whether those of Tirnberg and Berchtolsgaden will not depart from thence this Year?

A: The Tirnbergers are ready to leave the Country, but it is uncertain when those of Berchtolsgaden will leave it.

Q: What sort of a Man is the new Prince of Berchtolsgaden?

A: He is a Baron of Nothasst [Notthaffel] a Bavarian, and very Severe, which gives but little hopes of any Good to be done.

Q: Whether there are no Citizens at Saltzburg who declared themselves for the Protestant Religion, or are willing to declare?

A: There are a few Protestants in the City, but in secret, who desired a friend of theirs to send to them two Catechisms used at Augsbourg and one used at Franckfort.

Q: Whether the Arch Bishop himself is so severe or only some of his Ministers?

A: Chiefly the Arch Bishop himself, who will not rest 'till all Suspected Persons are turned out of his Dominions. However some of the Ministers are not behind hand in their Zeal, who are pretty well known.

Q: How many may there still be in the Country who profess the Protestant Religion?

A: Tis not to be known, but if it should please God to order things so that people might Freely and Publickly declare themselves, I believe very few would continue Papists.

Q: Whether a great many of those, who did not depart with the other Exiles returned to Popery?

A: They talk of a great many, and a certain Secretary shewed a Protestant the Minutes of the Deputation, in which 442 Persons only of the District of Rastadt are set down as new Converts.

Q: What is the Contents of the Oath which is to be taken by those who stay in the Country?

A: The order for taking the Oath hath not been put in Execution, nor is it taken by any Body hitherto, but those who declare themselves willing [who wish] to stay must make their Confession of Faith in the Church.

Q: Whether the Merchants Zwilling and Moedlhammer must still work on the Fortifications, and whether they were forced to pay one 1100 Fl. and the other 1300 Florins, as a fine for clandestinely bringing Protestant Books into the Country, from Nuremberg?

A: Mr. Zwilling hath been confined three Days in the Castle, but was discharged upon paying a certain Sum of money and building a new Altar: but Mr. Moedlhammer hath never been confined there, but cleared himself presently with a Sum of Money, as hath done likewise Mr. Wimber the Leather-Dresser for the Court.

Q: Whether they don't say the Arch Bishop repented of his Proceedings?

A: Not at all, but he will begin again by having his People registered.

Q: Whether the People in the Dominion of Saltzburg have received any Notice, how well the Exiles were received?

A: Very little, for such things are kept secret by all means, on the contrary, reports are spread that they meet with the worst of Treatment.

Q: Whether the Books published about the Exiles kind Reception by the other Protestants are brought into the Country?

A: Very few, and as far as I know 6 Accounts only published at Augsbourg are brought into this City, and 6 only sent into the Country.

Baron Imsland, who has been High Bailiff at Steuffeneg near Hoeglworth is said to be gone off for Male-Administration, and they believe at Saltzburg that he will go likewise to the Protestants. His Lady a Baroness of Hueffstein lives with her Mother at Saltzburg.

Translation of a Letter out of High-Dutch [from Urlsperger]. **To** *Mr. Newman.* **Augsbourg 3 November 1732.**

Dear and Honoured Sir: From my Correspondent at Ratisbon I received the following Account: last Week arrived here two Pious Families from Tirnberg, where all things are ready for a departure. These People work still in the Mines till the End of October; and then I fear, when 750 of them depart, the want of such understanding Men will cause a Deal of Trouble in the Territories of Saltzburg; there has been other Miners from Tyrol, who made a Tryal of the work, but came off with broken heads, and seeing the great danger in these Mines, they shew but a little desire to make another Tryal. From Carinthia is gone out a Man, and his Wife with Seven Children, a very Pious Family; he Saith that a great Number of them stayed behind, expecting [awaiting] to hear good news from him; that the Minister and Magistrate of his Parish, had formerly Caned him Soundly upon the Account of his Profession of the Gospel. Now he is full of Comfort and Confidence.

The following Paragraph with a Copy of Verses made no doubt by a Papist hath been inserted here in the Publick News Paper, published by a Papist. Viz.

"Augsbourg 22nd October 1732. According to our advice from France, they say that Philip 5th King of Spain, designs to make another descent in Africa toward Promoting the Christian Faith, and then was fully resolved to lay down his Crown, and retire to the Hermitage at Ildephonso upon which a certain Pen, alluding to Acts the 8 v. 27-40 (where it is said that Philip the Apostle, as soon as he Baptisted the Aethiopian, was carred away by the Spirit of God, and was seen no more) Set down the following Thoughts."

Quos dolet amissos Tibi, Petre, Iuvavia natos
    Africa mox quaestu, cum locuplete dabit
Suis ab Occasu, nigricans ubi vespera regnat,
    Res nova! Meridiem lux radiata subit.
Quod veluit reliquis ars et natura Phillippus
    Catholicus, Mauros, Aethiopesque lavat.
Has Simul abluerit: rapiet, procul inde Philippum
    Spiritus insolitio constituetque lore
Quintus esinstauras Carole Vestigia Quinti,
    Qui Monachus, dominans ante Monarchus erat;

Ponere, stat fixum fragilis Diadema Coronae.
Sic Age, Sic Finis Rite coronat opus.

From Vienna I received this certain Information Viz.: "In the Church of Schemnitz, a well known Place in Hungary, the Vacant Place is not yet filled up, and there are still two great Obstacles to be removed, namely whether they shall have for the future any Minister at all, and whether a Foreigner may be put into the place. The Churches which have been taken away from the Protestants stand still unemployed, and many Children die without being baptized, for want even of Popish Ministers, because ours that are ejected are forbid upon pain of the severest Punishment to exercise any Ministerial Function. There is some hopes given of Restitution, but with what Foundation time will soon shew. According to the Common behaviour of our Enemies we have always more to Fear than to hope from them and yet we may say; Fear not thou little Flock etc. They say here that the Arch Bishop of Saltzburg himself hath declared to those of his Subjects that remained in the Country that he would grant the private Exercise of their Religion, but I fear if the poor People accept of his Declaration they will put themselves in a worse Condition than before."

The following is an Extract of a Letter from Elbingen[45] to me about the Saltzburgers.

The dear Saltzburgers have been also with us, a true Salt. both Protestants and Papists have manifested their Love to them in a Particular manner, especially a Popish Tavern-keeper sent 150 Bottles of Wine two miles after them. Other Papists indeed are envious, because they have lost a rich Bleecher or Whitstler amongst them thereby, for when the first Saltzburgers passed through Elbingen, he was excited to turn from the Popish to the Protestant Religion, and when they pressed hard upon him he procured a Soldiers Warrant from the King of Prussia, and then they let him alone. When the last Company came through this Place, a Popish Maid Servant quitted her Service and went along with the Saltzburgers, being now in Service at Frederick's College at the University of Köningsberg [Königsberg].

What happenth in Bohemia, the following Account will inform You.

A Fortnight ago about 500 some say 9000 Persons were stirred up, to meet together openly in private houses. and as it happened, with the greatest Zeal. three (Popish) Priests came amongst them, who used at first hard Words but no violence. One of the Congregation answered, saying to one of'em, Father, hitherto we served the Devil and the World, and ye said nothing to us having been entangled perhaps in the same Abominations with us; but now when we desire to serve the Lord Jesus, You speak against us; why don't You rather go into the Publick houses, and drive out the Drunkards, Dancers, and those that Fight one with another. You may know that no body shall hinder us from this our undertaking. A Jesuit hereupon saying Ye are Lutherans, was answered by one of the Congregation: We don't know Luther, we will serve the Lord Jesus Christ. two of the Priests having staid about 6 or 7 hours amongst them, went away, but the Priest of the Parish continued with them to the End, and at the breaking up of the Congregation, he gave them the Blessing, and wished them increase and Constancy in good things. In the mean time the Enemies took several Resolutions to mischief these Persons, for which Purpose a few days after, three Companies of Soldiers entered this district, which belongs to the Government of Opiezin, and seized the People, and cast them into several Prisons, in a very Barbarous Manner, which their Barbarity they executed more particularly on the three Chiefest of them; For having Pulled off their Cloaths to the Shirts, and tied their hands and Feet fast to a Bench, they beat them with Willow sticks, laid before hand in warm Water, so severely that the Blood gushed out through their Shirts. and when it begun to thicken they Poured warm Water upon them, those that beat them, doing it by turns, and when they were quite Senseless they cast them into the worst of Prisons[.] many of'em escaped and came to us; May God Assist them altogether.

P.S. Augsbourg 10 November 1732.

Whereas the foregoing Letter from November the 3rd was not sent away that Day, I received since both Yours of the 15 September, and the 13th October with the Bill of Exchange for 50 £ Sterling, and I hope that my two last of the 6th and 16th of October with the inclosed are come safe to Your Hands. I add now,

1st. That October 28th I sent Franco 'till Amsterdam a Packet made in a Roll, addressed to You, containing Minutes, two Medals for Sir John Philipps Baronet and You, the Life of Joseph Scheitberger [Schaitberger] in form of a Letter etc.

2nd. That I send You the latest Account I had about the Affairs in Saltzburg, as also in Hungaria and Bohemia, of which last Kingdom, daily more informations come in, which I'll communicate another Time.

3rd. That I'll send very soon a Receit for the 50 £ Sterling because the Money is not yet Paid; may God Graciously reward dear Mr. Hollis for it.

4. Concerning the main Point of Your last, I cannot answer yet, but will endeavour to inform myself, in the most Private manner, and send the best Grounded Information I can get to You. The Multitude of these extraordinary Affairs in Religious matters obliges me to take a proper Student into my House, for my Assistance therein. If Opportunity serves, pray Sir favour me with the Accounts You have Printed in the Affair of the Saltzburgers, I rejoice heartily at Mr. Ziegenhagen's Health, and committing You to the Grace of God, remain

Yours   S. Urlsperger

P.S. You can hardly believe what great Joy the Charity of the English Nation towards the Persecuted Protestants hath Raised amongst us.

Abstract of a Letter from Ratisbon dated 3 November 1732. (enclosed in the foregoing Letter to Mr. Newman)

How can we assist the Poor Bohemians? It seems to me these dear People want [lack] proper advice. It is not advisable that they hold their Meeting in so Public a manner as yet. But How? why don't they address themselves to the Protestant Body and use their good Advice! is there no way to acquaint them with good Advice that they may begin the Cause of God and the Work of the Lord in a quiet manner.

The Tirnbergers have obtained leave to depart[.] three of Them provided with a fine Passport (Wherein they are called Protestant Subjects) are come hither, and gave us Notice thereof. This day two of them are going back again in hast into the Country where the High-Bailiffs etc. are really afraid of these People, having desired them not to accuse them before the Illustrious Evangelick Body. They (Viz. the High Bailiffs etc.) in open Court have made a Memorial to the Evangelick Body here, with all their Titles, that by their Requisitory Letters to the Bishop of Passau, and to the Elector of Bavaria, they might obtain Permission for these People to come by Water to Ratisbonne! the States [General] of the united Provinces [of the Netherlands] have now taken them under their Care and Protection. their Ambassador having made already an Agreement upon favourable Conditions with them, Viz. To bear all the Charges of their Journey from Tirnberg to Cassanot [Cadzand] near Middleburg, to maintain them some Years, provide them immediately with Houses etc. which Conditions I can another Time communicate to You if You desire them. the abovesaid Ambassador has already written to the Elector of Bavaria and the Bishop of Passau in their behalf, and if he gets a Favourable Answer, then they can at the End of November depart in the Name of the Lord. A Saltzburger who is Settled here and was lately in Saltzburg [says?] the Number of those, who Profess the Protestant Religion, increaseth even in Places they did not think, and that they likewise are desirous to depart; and I hope God will also provide a Time for them. He says that in Austria[46] there was a very great Number of those People; but they enjoy upon the Emperor's Order Perfect Rest, and use their Books openly. I hope the Flock of Christ will now gather together more and more.

One of the Inhabitants of Berchtolsgaden returning from Ratisbon with a Letter from me into the Country was immediately made a Prisoner, but as soon as they had read the Letter, they gave it him again, as also his Liberty. 'Tis likely if those of Berchtolsgaden can obtain leave from their Sovereign, they will come along with the Tirnbergers.

Those that die there in the Protestant Faith are not allowed a Place in the Common Burying Ground, but must be buried in a Forest, or sunk into a muddy Lake. and as I spoke with one who was grieved upon this Account, I told him, wheresoever a true Christian is buried, the Ground is consecrated already. May the Lord carry on his Work more Gloriously from Day to Day that we may be able to thank his holy Name more and more with Chearful Hearts.

Abstract of a Letter dated Dresden October 12 (Enclosed in the foregoing Letter from Mr. Urlsperger to Mr. Newman.)

The King of Poland and Elector of Saxony [Augustus II] hath appointed Commissaries to receive into his Dominions the Exiles of Bohemia, who come in great numbers. they have many Obstacles, and 'tis thought they will Petition the Evangelick Body at Ratisbonne to make Intercession.

Translation of a Letter in High-Dutch.   To Mr. Newman.   Augsbourg.   17th November 1732.

Dear and Honoured Sir: I thought it necessary to send immediately the inclosed to You, hoping to be able next Thursday, being the 20th Current, to acquaint You with more. Just now I received Your Letter of the 24th October last.

Adieu   T. S. Urlsperger

Copy of a Memorial delivered by the Dutch Minister [Gallieris] residing at Ratisbonne to the Minister of the Elector of Bavaria, and of the Bishop of Passau, about the Protestant Tirnbergers. (Enclosed to Mr. Newman in the foregoing Letter from Mr. Urlsperger.)

Whereas, Their High-Mightinesses, the States General of the United Provinces have been Pleased most Graciously to resolve to receive into their Dominions 300 Families of the Saltzburg Exiles, who shall depart from Tirnberg[47] or Berchtolsgaden the 29th of next Month (being November 1732) or sooner, if possible, and Whereas the said Exiles, being destitute of necessary Carriages, are not able to carry along with them upon Waggons their few Goods, small Children and Sick People, His most Serene Highness the Arch Bishop of Saltzburg hath been most Graciously Pleased to grant them leave to depart from Saltzburg Water, upon the Rivers Saltza and Inn, and afterwards upon the Danube to Ratisbonne. But those Exiles being obliged likewise to pass through his most Serene Highness's Dominions the Elector of Bavaria's, which 'tis wished may be done without any Offence or Molestation, the Illustrious Minister from the Elector of Bavaria (at [omitted][48]) is desired by the Minister of the States General with all due Respect that he would be pleased favourably to use his Intercession and Good Offices with his Serene Highness his Master (with whom the States General live in Good Harmony) in order to obtain for those Exiles a most Gracious Permission, for them freely to Pass upon the Danube (as far as the Territory of his most Serene Highness reaches) to Ratisbon, and afterwards by Land to Nuremberg, and that necessary Orders be given every where to this Purpose, that these poor People may not only be able to travel with Speed and Safety, but find also all proper relief and Assistance to reach the Places of their future abode before the rough Winter Season breaks in upon them.

The same Memorial Mutatis Mutandis, was presented to the Bishop of Passau.

Conditions upon which their High-Mightinesses, the States General of the United Provinces have offered to receive 300 Families of the Saltzburg Exiles into their dominions.

1st.   The Saltzburg Exiles shall be looked upon as Brethren by their High Mightinesses out of Christian Charity and Commiseration, they shall be received with open Arms of Love, and fully enjoy all the Rights and Privileges, which the Natives actually now have and enjoy.

2nd.   Their High Mightinesses will not only protect the Exiles against every Body in their Evangelick Religion, but give them also able Ministers and School Masters in order to instruct them furthermore in their own Language; and provide them with a Competent Salary out of the Publick Cash.

3rd.   The Exiles are to live in a Country where diligent and able Persons at any work whatsoever may get a Sufficient Livelihood and Maintenance, for which Purpose every one shall be assisted according to his Ability & Skill with necessary ways and means.

4th.   The Exiles may be assured that their High Mightinesses will not only provide them in the said Country with necessary Dwelling Places, but also with Sufficient Food, necessary Cloaths and Furniture, untill they shall be able fully to maintain themselves with their own Labour though it should last 4 or 5 Months.

5th.   The Exiles shall be exempt many Years from Paying any Civil Taxes, or they shall receive in lieu of this Immunity Yearly a certain Sum of Money, in order to hinder other Persons from defrauding the Publick Revenue under their Names.

6th.   All and every one, Men, Women and Children, shall from the time they are received (by the States Commissary) be assisted with Money, as agreed upon, for their daily Sustinence. Nor,

7th.   Need they take any more care of their future and entire Transport, Viz. 'till they come to the place of their future abode, but they shall with all the Furniture and Goods they carry along with them be transported thither in the most convenient manner, without being put to any trouble or Charge.

8.   The Children, Old, Sick and disabled Persons shall be provided for and maintained in the best manner like the Natives.

9.   The Exiles shall not be liable to any Servitude, but all together and every one in particular shall enjoy an entire Freedom.

10. If the Exiles shall behave themselves as faithful and Obedient Subjects, they may not only be assured of all necessary Comfort, Help and Assistance from their Magistrates, but will find them moreover so gracious and ready to assist them, as they neither should nor could have expected and found in other Places.

That these are the true and Genuine Conditions, which their High Mightinesses the States General of the United Provinces, my high and mighty Masters have sent to me, their Counsellor and Minister here at the Diet of Ratisbonne, I attest hereby in the best Form, by Subscribing my Name and putting my own Seal to it. Ratisbon 4 Nov. 1732.

LS   Gallieris

Translation of a Letter in High Dutch [from Urlsperger]. To Mr. Newman. Augsbourg. 27th November 1732.

Dear and Honoured Sir: The Offers made by the Honourable Society as well as by the Trustees for Georgia to the Saltzburg Exiles as great and favourable as they are, so little prospect have I as yet that they will be accepted by them. For,

1st. They are very much afraid of the Sea.

2nd. The States General have actually engaged all the Tirnbergers.

3rd. The King of Prussia has lately declared by his Commissary, that He is willing to receive all the Saltzburgers, who should be obliged to leave their Country.

4th. They are very desirous to keep together, and in a Body.

5th. Those of Berchtolsgaden being for the most part Trades-People and Artificers in Turney Ware have already the King of Prussia's Promise of being received and Settled in His Residence at Berlin: besides it would be highly necessary in case a Number of Saltzburgers should come out next Spring (as there is great Probability they will) thoroughly to acquaint them with the State and Condition of Georgia before they leave their own Country. But I own it will be very difficult to give them such a full Information, The Passages in Saltzburg and Berchtolsgaden being so narrowly guarded that it is no easy matter for the Exiles to send a Letter safe to their Friends left behind in the Country: however means might be found to get over this difficulty. If the Reports that go about should prove true, that the greatest part in the

Emperor's Hereditary Dominions are resolved to declare themselves Protestants,[49] no doubt there will be many who will readily accept the abovementioned Offer.

I should be glad if You Mr. Secretary could inform me,

1st. in what manner the Money sent hither has been collected.

2nd. Whether there has been a General Collection in the Kingdom besides the Money sent already hither by the Society.

3rd. What Sum the General Collection did amount to.

4th. Whether the Money of the General Collection is still in the Kingdom, or where it is sent to.

5th. Whether the Presbyterians did shew themselves well disposed in this Affair.

Before I conclude, I must acquaint You that I have allmost every day Opportunity to improve far and near to good purpose the English Money, I am entrusted with amongst our dear Saltzburgers, which will certainly be a great Blessing to the Society as well as to the Benefactors, and this moment comes a Man of good credit to me telling me that designedly the Tirnbergers had been detained so long that their Voyage which in a proper Season was no more but 4 or 5 days would now require three Weeks, and not without great danger of their lives, my most humble Respects to the Gentlemen of the Society.

Tuus Totus   Samuel Urlsperger

Abstract of a Letter from Ratisbonne 25 November 1732. (Enclosed to Mr. Newman in the foregoing Letter)

The Tirnbergers upon Intercession of the Dutch Minister have got leave to make their Voyage hither passing through the Dominions of the Elector of Bavaria, and are said to break up at the End of this Week. the Dutch Minister acquainting our Magistrate with it desired some Preparation might be made for their Reception, which was agreed to.

The People of Berchtolsgaden who have desired leave to go out of the Country meet still with many difficulties, particularly upon Account of their pretended [alledged] Servitude. However their Adversaries continue to injure and vex them, especially by depriving them of any opportunity of getting any work to do, and consequently obliging them to spend their little Substance that in case they can't be prevailed upon to alter their Minds, they may go out empty handed.

*Translation of a Letter in High Dutch [from Urlsperger]. To
Mr. Newman. Augsbourg. 11 December 1732.*

Dear and Honoured Sir: By what I communicate unto You at present, the Gentlemen of the Honourable Society will see

1st. How the Case of the Tirnbergers and their leaving the Country stands.

2nd. That they are declared already to be Subjects of Holland.

3rd. In what manner the Ambassador of the States General at Ratisbon hath acted in this Affair at the Court of the Elector of Bavaria, and the Bishop of Passau, and what Answer he received from them.

4th. When they are expected at Ratisbonne.

5th. What Preparations have been made for their Reception.

6th. What Money partly myself out of the English Cash, partly other rich Families of this Town have sent to Ratisbon for their Relief.

7. How the Affairs of those of Berchtolsgaden stands at present, who 'tis said have engaged themselves to settle in the Dominions of the Republic of Holland, though the King of Prussia would have them go to Berlin, from whence a certain Person some days ago privately passed through this City in his way to Berchtolsgaden, on purpose to persuade those who have declared themselves Protestants to go to and Settle at Berlin, the King intending to build them a Street of their own, and allow them Ten Years free of Taxes.

8. That His Majesty the King of Great Britain as Elector of Hanover is desirous to have a Thousand of the Exiles to settle in the Dukedom of Saxen Lawenburg [Sachsen Lauenburg].

As for the Question whether some of the Exiles might be prevailed upon to go to Georgia, I have told You my humble Sentiments of it in my last of the 17th & 27th Current [*sic*] N.S. and am now in Expectation of Your Answer thereupon, besides which I think it is necessary.

1st. To consider Whether we shall have new Transports of the Exiles from Saltzburg next Spring of which there is great Probability.

2nd. That the Gentlemen of both Societies[50] would be pleased to draw up some Articles (after the manner of the States General) backed with the Royal Authority, and send them to me with full power to publish them (all which must be done without loss of time)

3rd. That the Society would employ their Endeavours to get leave that the King's Minister at Ratisbonne may Correspond and act in Conjunction with me in this Affair.

As much as I wish England may enjoy the Pleasure of getting a Number of the Saltzburgers for Georgia (who knows what a Fire God may light by these People in the West Indies[57]) so difficult I am afraid will the Circumstances mentioned in my last of the 27th make it. However God having hitherto in this whole Affair done above our thoughts and Expectations, I still hope he will graciously be pleased to do the same in this Particular Branch of it, at least I shall not be wanting to do anything in my Power.

At the End of February next there will be a great Fair in the City of Saltzburg, whither some of our Merchants intend to go: whom I shall charge to enquire as much as possible, what may be done in this Affair.

Just now I receive the favour of Yours of November 22nd therefore I add briefly,

1st. That I am glad You have informed me more fully of the State of Georgia, and what hath hitherto been done for Establishing a new Colony there.

2nd. What I have mentioned above of the People of Tirnberg and Berchtolsgaden will convince You, that it is no more res integra.

3rd. In Case the Bohemians should continue to declare themselves Protestants, a good Number of them might perhaps be got for Georgia. But the Event of this is as yet uncertain, especially when as I have been Yesterday informed from Hall [Halle], the Reverend Mr. Liberda is arrested at Dresden, and it is to be feared he will be delivered up to the Emperor on Account that a great many of the Bohemians have desired his Instruction and Advice, with a Number of whom he went last Summer to Berlin, to Petition the King of Prussia for their Reception into his Dominions. this Mr. Liberda, a Man full of Faith and the Holy Ghost, hath preached the Word of God in the Bohemian Language to those who for Religion's-Sake retired Ten Years ago into Lusatia: and his preaching has been attended hitherto with great Blessings.

4th. It is true that the Accounts from other Places are not of equal Credit. However the Society is at Liberty to Publish for the Edification of others what comes well attested to them, and they think proper.

5th. I must do Justice to my Correspondent at Ratisbon [J. von Reck] and freely own that he is a very Useful and Serviceable Man and shall acquaint him with Your Respect for him. My most humble Respects to the Honourable Society and particularly to Sir John Philipps, whom I value very much.

Tuus Totus   Samuel Urlsperger

(Ist) Exract of a Letter dated Ratisbon 27th November 1732. (Enclosed in the foregoing Letter from Mr. Urlsperger to Mr. Newman). [Probably from J. von Reck]

I have taken occasion to talk with a Minister at the Diet about the desire of the Society in England whereof You mentioned some thing in Your last, and my humble Opinion is that since nothing can be done in this Affair at present it is better to keep it secret, and even not to mention any thing in General of it in Publick Prints 'till God favours us with a more hopefull Prospect. But God be thanked for the Resolution of the Society, because it prevents our anxious thoughts, where to Settle the Exiles in case a considerable number of them should come next Spring.

The Berchtolsgaders are not to leave their Country this Winter, being not much pressed upon for their departure: but all those who have a Mind to go out are ordered to pay 5 Florins every one of them on Account of the Pretended [alleged] Personal Servitude which they complain much of as not being able.

There's a Report here as if they are actually Printing at Constantinople, a Translation in the Turkish Language of Arnd's true Christianity,[52] and two other Books of Devotion wrote by a Divine called Millar.

You have prevented [anticipated] my desires and Expectations in sending me so liberal an Assistance for the Poor Tirnbergers both in Money and Books (among which is particularly Arnd's Paradise Garden,[53] most of the Exiles being very desirous of it.) the Dutch Ambassador has taken particular Notice of this Charitable Contribution, and set down both the Sums of Money and Number of Books: the Tirnbergers are to stay here about three days during which time care will be taken to instruct them both by Preaching and Catechizing.

(II) Extract Dated Ratisbon 9th December 1732.

I can not Sufficiently express the Joy and Satisfaction which Your enabling me to give some more Relief to the Poor Tirnbergers hath caused unto me. I have received the Letters of Exchange for 350 Florins, and was agreably surprized at seeing a large Chest and two others with Cloaths and Books brought into my house: upon which occasion I could not help thinking on the Ark of the Covenant being brought into the House of Ebed Edom. By this means You have taken care to provide against the want of necessaries and Cloaths for the Poor Exiles, and for the Preservation of their Health too having sent also some [omission] for having mentioned this the Gentleman in very tender and expressive Words wishes all manner of Blessings to the Benefactors, which takes up most part of his Letter.

The Tirnbergers are actually arrived in the Territories of the Bishop of Passau in their way hither, where some Priests made an attempt to bring over again some of them to the Roman Religion, but seeing it was to no purpose they desisted.

The Papists very much grieve for the loss of a certain Nobleman Mr. de May, who turned Protestant some time ago, and above two hundred of the Jesuists Scholars assembled before our Church last Sunday, on purpose to see him or rather to get him again.

(III) Extract dated 9 December 1732.

The Elector of Bavaria has granted the Tirnbergers free Passage through his Dominions through the Intercession of the Ambassador of the States General, and prohibited to extort any Taxes from them on condition that they shall not stir from their Ships or go separately [secretly] into Houses.

They actually set out the 30th November notwithstanding the rough Weather: at their Arrival in the Arch-Bishoprick of Passau, they were received very civily, and all good will shown to them, Phisicians also being sent to take care of their Sick. whether they can continue their Voyage hither by Water, or be obliged to go by Land, on Account of the Ice that comes up the [omission][54] cannot be known before the return of a Messenger who is soon expected back.

The Ambassador of the Elector of Brunswick[55] hath orders to treat with a Number of 1000 Exiles to go to and settle in the Dukedom of Lawenburg [Lauenburg], if there are any more to be got.

Extract of the Dutch Ambassador's Letter to the Tirnbergers dated 17th November 1732. Sent to them by one of their Countrymen who with three others had negotiated in their behalf.

The Dutch Ambassador acquaints the Tirnbergers that they being actually [now] Subjects of the States of Holland, the Elector of Bavaria and the Bishop of Passau had at his request granted them free Passage through their respective Dominions, and that they had already given the necessary orders to their Magistrates and Officers to facilitate their Passage as much as possible. wherefore he desires them to make all the hast they can; and not to accept of any offers that might be made to them by some other Powers.

Signed   Marines Gallieris

Translation out of High-Dutch of a Paper enclosed in a Letter from Mr. Urlsperger.   To Mr. Newman dated 25 December[56] 1732.

Some Grievances of the distressed
Protestants in Hungary.

1st. Five and Forty Churches of the County of Eisenbach [Eisenburg?[57]] have been taken from'em, (Since November 1732) some of which are of the so called reformed Religion.

2nd. The Reverend Mr. Vogt Pastor of Chemnitz [Schemmitz] dying the 24 May last, they are not permitted to call another in his Place though they have sent two particular deputies to Vienna for this purpose; but they can't be heard.

3rd. The two Protestant Superintendants. the Reverend Mr. Fahnsmith at Leutschall [Leutschau, Löcse] and the Reverend Mr. Antoni at N— are forbid to form a Consistory or to confer Ordination.

4th. At Cashau [Kaschau] a privileged Place in upper Hungary the Church wants repairing but the Papists have forbid it.

5th. They are forbid to teach either Philosophy or Divinity in their Schools or any thing else besides Grammar.

6th. At the Churching of Women the Priest won't give'em his Benediction except they turn Papists.

7th. They are strictly forbid to send any of their Scholars to the Universities abroad though some go clandestinely under the Pretence of improving their Trades.

8th. Formerly when it happened that People of different Persuasions married together, they brought up their Children, the Man agreably to his and the woman to her Opinion: but now if the Father be a Papist he is obliged to bring up all his Children his own way.

9. On their great Festival particularly that of *Corpus Christi* they force the Protestant Tradesmen to go in procession with them.

10. When there is a Marriage agreed on among the Country People, and the Bridegroom only happens to be a Papist; the Priest won't marry'em except the Bride turns Papist too.

11. In those small Towns where both Protestant and Popish Ministers are allowed, the Protestant dare not visit a dying Person of his Persuasion, 'till the Popish Priest has been with him to persuade him to embrace his; in order to which he uses all Endeavours to make him Pray with him, but in case the sick man refuses it, then the Protestant Minister is permitted to give him the Sacrament.

Memorandum: What is entered above came inclosed to Mr. Newman in a Letter dated 25th December 1732 but the rest of the Letter not being of much Consequence is not translated from the Original High-Dutch and therefore is not entered her.

Extract of the Reverend Mr. Urlsperger's Letter to the Reverend Mr. Ziegenhagen dated at Augsburg 19th January 1733.

Whilst I am in expectation of an Answer from Mr. Newman upon my Letter of December the 11th N.S. I chose in the mean time to acquaint You with some things which I think most material at present.

By one of Mr. Newmans Letters it seems the Society is somewhat inclined to think that I dont approve of their Design of sending some of the Emigrants to Georgia; but far from it, For I like the Design very well, but I must needs mention the Difficulties it is attended with which however may not be insurmountable.

In Order to Forward the Design it will be requesite.

1st. To have a Short and plain Description of the Country they are going to, tho that may be Sufficient what has been sent already by Mr. Newman.

2nd. A Specification of those Conditions upon which they are to be sent and what they are to expect.

3rd. The Societys Leave or orders for printing here those Conditions as well as there.

4th. For Corresponding with a publick Minister at Ratisbon about concerting the proper Measures in this affair.

5th. To improve [utilize] me to provide a Proper person to

direct their Course and be their Guide from Place to Place or a Commissary for their Journey.

6th. above all it will be necessary to make a publick Declaration in a Conference of the Evangelick Body at Ratisbon either by the Kings Minister or by the well known Mr. [J.] Van Reck Minister of Zell, that His Britannick Majesty was graciously pleased to receive [omission] Families of the Protestant Emigrants and to settle'em in Georgia.

In Writing this it comes just now into my Mind wether it would not be more expedient to go to Ratisbon for some days my self (where I can be in 24 Hours with the Flying Post) to forward Matters the better. I am concious I can spare very little of my time, but when the good Success of any thing depends very much upon the first settingout about it, and a great deal of Time is Spent in Corresponding by Letters I submit my self to the Orders of the Society.

The Resolution the Society has been pleased to take about the Money left still in my hands, viz to save it for the Benefit of those that are to go to Georgia, this Resolution seems to be somewhat different from their former Sentiments. It is owing to my good Husbandry that there is any money left[.] Had I laid out all the Remittances for the Relief of the present Exiles I am sure I had [would have] acted agreeable to my Instructions.

If this Principle should prevail to deny any farther Charity from the very time the Exiles are engaged with any Protestant Power their Case would certainly be very deplorable. We in this City have almost none of those Saltzburgers, yet above 12000 Florins have been given by the Protestant Inhabitants and they are still willing to do more, let them be engaged with what Protestant Prince soever. The People at Ratisbon have freely maintained the 800 Tirnbergers for four Weeks together and been very liberal to at their Departure, notwithstanding they knew very well of their Engagement with the States of Holland some Months before.

As for the Questions I put in my Letter to Mr. Newman I think they are very innocent in themselves and were Suggested to me by several Persons and among the rest by a Royal Person. For in the French News we had it positively said that there had been a general Collection made throughout the Kingdom. The occasion of that particular Query about the Presbyterians was owing to a Letter written from London and communicated to me not to mention the cold and indifferent Treatment the Exiles met with from some of the Reformed Religion called Calvinists in a famous City, in Germany.[59]

You may assure the Society of my most earnest desire to serve them in the best manner I can.

The most material contents of the Reverend Mr. Urlsperger's Letter to Mr. Newman dated at Augsbourg January the 22nd 1733.

1st. Some further Particulars relating to one of the Difficulties which Mr. Urlsperger is of opinion do attend the Society's Design viz. The Saltzburgers fear of the Sea.

By the by he mentions some Reasons which had facilitated the Tirnbergers Engagement with the Dutch.

An Account of some things which he thinks still requsite [*sic*] in order to Forward the aforementioned Design as

1st. That a Short and Accurate Description of Georgia might be drawn up in two or three Sheets by a fit person in English & afterwards translated and sent to him, in order to be printed & dispersed among the Saltzburgers, Containing a Description of the Climate, Fruitfullness and other necessary Circumstances & Qualifications of the Land and its Soil, Particularly what the yearly produce of an Acre of Land is.

2nd. Where the Colonists could live in the mean time that Houses were building for'em.

3rd. That the Trustees would provide'em not only with a Minister but a Catachet and Schoolmaster too; And that in Case these should come to die they would take Care to get others in their Room [place].

4th. That the Trustees might be pleased to explain themselves Specifically how much they had in Mind to allow for a man a Woman and a Child, during their Voyage, and after their Arrival in Georgia and put the same in the conditions of Agreement.

Mr. Urlsperger humbly desires for some Reasons which he Specifies that the Society might be pleased to Keep to their former Resolutions viz. to let other Exiles that are not resolved to settle in Georgia also have a share, after their Engagement with other Powers.

Translation of a Letter in High Dutch.   To Mr. Newman.   Augsbourg, 12 February 1733.   No. 12069.

Dear Sir: Though I wrote but the 19th Instant a Letter to you, yet I would not be wanting [would not wish to fail] to communicate unto you the edifying example of a new exile out of Carinthia, hoping that it will not be disagreable at all to the Honourable Society to read something thereof. besides this I have been credibly informed that His Prussian Majesty hearing that I had sent 333 Rix dollars out of the private Cash for the exiles to Prussia, for the reasons which I have mentioned in my former he expressed his satisfaction thereat, and spoke in a most Gracious manner and in very Gracious terms of the Honourable Societies zeal, Compassion and Good nature towards the oppressed Protestant Exiles to others that were present etc.

I am   Dear Sir   Yours etc.   S. Urlsperger

Extract of a Letter from a private Correspondent dated Ratisbon 21st January 1732/3.

These are only to let you know that yesterday the Courier returned from Berchtolsgaden being accompanied by six of the Inhabitants of that Country, who having a great desire to come hither, secretly made off with him. some Accounts which partly the Tirnbergers conveyed to them as also some Catechisms etc. which I lately sent them, have contributed very much to strengthen all the Protestants in those parts. their number is said to amount at present to about one Thousand. we heartily wish that God may bless us again with some good and pious Christians, and possibly those may come hither in time according to their desire. they have not liberty to sell any of their Estates or Goods, and when they were registered, the Magistrates valued'em at half the price only of what they are really worth. I believe the Event will correspond with my opinion and thoughts of the Berchtolsgaders; viz. that they would be parted and not all go to one and the same place. most of'em are resolved that if could but come to Ratisbonne themselves, they would then declare whither they were inclined to go. a Considerable party of'em will probably go to settle in Holland, some to Prussia, particularly the Artificers, others into the Hanoverian Territories, where they will want nothing in regard of their Souls as well as their Bodies but both will be well provided for. A small Transport of about 50 persons is going soon to Prussia. I have provided these exiles likewise with some Books and assisted'em in what manner I could, and shall continue so to do, especially if I am to expect and can rely upon some farther supplies from others.

Extract of a Letter dated Ratisbonne 3 of February 1732/3.

After my last some other Deputies from the Miners in Berchtols-gaden are arrived here, who will apply to the Evangelick Body for their Assistance and Intercession, that Liberty to leave the Country may be granted them the sooner, for at present they have not leave to sell any of their Goods, or to depart the Country though before the Commission they publickly declared themselves Protestants in due form and their Goods and Possessions have been Appraised. In the mean time they are permitted to read Protestant Books at home, yet with this restriction that there shall be no more present at their devotions than what belong to the Family, and it is reported that one of the Saltzburg Exiles had lately the misfortune to be arrested there for no other crime but because he came thither, and prayed and sung with some of the Inhabitants, but it is not yet known what will be done with him, neither what crime they will charge him with. the Saltzburg Exiles which arrived here a few days ago, unanimously report that the Inhabitants of of the whole Bailiwick of Gastein have declared themselves Prot-estants and petitioned for leave to depart the Country. they add that there was many more still who would do the same.

The Melancholy Accounts that come in out of Hungaria con-tinue to be the same. after the Papists have taken away all the Protestant Churches except two in the County of Eisenberg [Eisenburg?], they go so far as to force the People by the Soldiers which are quartered in those parts to go to Mass; it was upon such an occasion that it happened that one of the Soldiers kicked a poor Countryman on the Belly in so violent a manner that he immediately drop't down on the Ground and lay Languishing 'till the Evening, when he died of the hurt he had received[.] as they were going to bury this man, the Popish-priest would not permit it. moreover they would [wished to] punish the Widow, for not having fetched the Priest before her Husband died, and given him an opportunity to try whether he could not convert him after their manner and way. It is said that the Papists intend to deprive also the other Protestants in other Counties of their Churches in the same manner as they have those in the County of Eisenberg, and that they meditate besides to shut up all the Schools throughout the whole Kingdom, thereby to oblige the Inhabitants to send their youth to the Jesuists to School, all repre-sentations that have been made against these their Proceedings, and all intercessions for the poor oppressed people have been fruitless hitherto.

Extract of a Letter dated Ratisbon 5 February 1732/3.

A certain tradesman of this City having been at Saltzburg has brought me the following piece of news, Viz. That he had been credibly informed that in the Bailiwick of Radstadt 700 Persons had declared themselves Protestants, and would leave the Country sooner than was expected.

Translation of a Letter in High-Dutch. To Mr. Newman. Augsburg 19 February 1733.

Dear Sir: The following particulars I have only to communicate to you at present; as for the rest I refer you to my former Letters, and desiring you to give my most humble Respects to the Honourable Society.

I am   Yours   S. Urlsperger

An Account of a Man and his Family that left Carinthia for Religion's Sake as it was given by them at Augsburg 15 January 1732/3.

John Gruber, late of Simitz (which is thirty Leagues distance from Saltzburg) in the Bailiwick of Allwecken in the Principality of Carinthia, Husbandman, 39 years of Age with his wife Magdalen Pucklerin,[60] likewise 39 years of age, and seven Children, Viz. Mary 18, Apollonia 15, Ulrick 12, Matthew 10, George 7 & 1/2, Margaret 3 & 1/2 and Helena 2 Years old, appeared the day of the month abovementioned and produced their Letters of Admission, as also a Pass-port from a Bailiff in Carinthia called Lorentz Sholler, bearing date 27 September 1732. He left the Country on account of the Protestant Religion which he and his Family profess and related.

That because he publickly declared himself a Protestant, he underwent a hard persecution, and had to struggle with great difficulties which were laid in his way. The Bailiff would [wished to] keep back (from him) his wife and Children, and all that he had but was at last prevailed on by his continued prayers and repeated instances to grant him his request since he could be persuaded by no means to return to the Roman Catholick Religion. but before he set out on his journey, the parish priest named de Esoh [Esch?], who is also the Bishop's Confessionary together with the said Bailiff Sholler came to his house and renewing their remonstrances to him, to continue a Roman-Catholick assured him, whether he persisted in his resolution to adopt a Blind Heretick Faith and forsake the true Roman Catholick one, by which only Salvation was to be had and obtained. whereupon the Deponent answered that he was surely convinced in his conscience that in the Protestant Religion where the word of God and Christ was taught pure and uncorrupted he should be eternally saved, and that he would live and die therein. Seeing therefore that nothing was to be done with him but that all their labour would be lost whatever way they went to work, they cursed him, saying that now he was sure to go to the Devil, and would be eternally damned. The Constable and Archer[61] whom they had brought along with them were ordered upon this, to search every corner in his house and having discovered the Protestant Books which he had concealed under the Bed, being Spangenberger's and Luther's Postil or Collection of Sermons to be read and used in Families;[62] Aviller's Art to die well and happy, a Book of Hymns printed at Zittau, Religious Dialogues (or a Book called a Conference upon matters of Religion, the Holy Bible, and some others, the Constable put'em all together into one Bag. But the Priest looked again under the Bed himself to see whether there was any more Books concealed under it, and insisted that he must have Scheitberger's Heretick Book also, and would have the Deponent deliver it to him. and though the said Book, together with Martin Lodinger's Writings (of comfort) and Letters, and Martin Luther's large Catechism lay under the Bed as well as the others beforementioned, yet these Inquisitors did not see'em, hereupon the Priest and the Bailiff grew so angry, that they turned their Canes upon him, and placing themselves on each side they begun to beat him most severely in such a manner that the Silver Ring at the end of the priest's Cane flew off, which being not found again, the priest insisted that the Deponent should pay him two Florins for it. two days after they found it laying in a Drawer which was left open that day when they searched the House, after the Deponent and his Family had taken great pains to themselves to look for it every where else. they went afterwards with the Books to the Bailiffs' House, where they carried the Deponent along with them, to whom they intimated that the Books which they had taken from him should all be burned, for which purpose necessary preparations were accordingly made in an adjacent Garden, whither they went together, and the Archer lighted the fire. the Bishop all the while looked out at Window of the Cloyster to be himself a Spectator of the Burning of the Books. the Priest and Bailiff pressed very hard upon the Deponent and would [wished to] force him to throw the Heretick Books into the fire himself, which he refused to do saying though you burn my Books, yet all that is contained therein will remain and be left impressed in my Soul, and be preserved in my mind and I know most part of the Prayers and Hymns which stand in these Books by heart, whereof I could say above an hundred to you which you will never be able to rob me of. The Archer then threw all the Books into the fire, one after another and they did not part 'till all were burnt to ashes: when they came to the Bible which was left last of all, the Priest turning to the Deponent said that this he thought he could throw into the Fire with a safe Conscience, but he shewing little inclination to join with him in that; answered That he should be guilty of the greatest sin if he had the wickedness to burn that Book which contained what Christ had preached and taught with his holy mouth, as it is related in the Holy Gospel which also says that these his words shall be preserved for ever, though even Heaven and Earth should pass away and be no more, therefore they themselves might burn it, if they could take it upon their Conscience the Priest took the Bible along with him and did not deliver it up to the flames. the Deponent added that before the burning of the Books the Priest asked him, How many Sacraments he thought there was, and he answering he believed in two Viz. the holy Baptism, and the Sacrament of the Lord's Supper. the Priest replied that is even downright Lutheran Heretick Doctrine. the Bailiff put a question to him also, to which he desired an answer. Viz. Which of the two he believed to be true, That Luther is in Heaven or in Hell? the Deponent's answer was Judge not and you shall not be judged, condemn not and you shall not be condemned. He certainly believed that the beloved and dear man of God Martin Luther is in Heaven, since all his writings are founded on the word of God, and because he seeing that the Traditions of men had got the uppermost in the Church of Rome, and that on the contrary the Gospel of Christ was laid aside, he endeavoured to root out the Tares that were sprung up in the said Church. upon the Deponent's expressing himself in this manner, both the Priest and Bailff fell into a rage, damning his Soul and Body. The Deponent having in hast sold what goods he could amounting to the value of Seventy Florins ready money and taken whatever came first to his hands, retired out of the Country, and came hither with his Family. He is a very sensible man, and extraordinarily devout, the like I scarce have met with in any other, he can say whole prayers by heart word for word; upon his being asked whether they who profess the protestant religion in his Country are not threatened with being put into prison, and used in the same manner with those in the Arch-Bishoprick of Saltzburg, he answered: that as yet they had not proceeded so far against many of them, and had not much made use of such measures: as for him he had not been at all afraid of being imprisoned and the like; he was ready for Christ's sake to suffer imprisonment and any torments whatsoever adding what follows in German Verse.

> Don't be afraid of, nor tremble at the Thorn, Band of Monks,
> Commit thy Cause to God,
> Though they put thee into prison
> Yet God looks down upon thee from Heaven on high,
> And says, O ye wicked Tyrants,
> You touch my Eye-Apple.
> But alas, you will pay dear enough in Hell
> For your rage and Fury to injure others.

His possessions he would willingly part with for the sake of the word of God even though they amounted to some Thousands, and not look back for them as Lot's Wife did, but with unshaken Constancy presevere in the profession of it, amidst the hardships and dangers it should be attended with. when he passed through Saltzburg, he was ordered to produce his Pass-ports. The Papists there seeing his Children heartily pitied them that they should go to the Devil, and pressed upon the Deponent to sell'em all or any of them to'em and leave them there but his answer was: that in God's name he would take them all along with him and sell none of'em. Lastly he related: that there's a great many protestants in Carinthia. the Inhabitants of about 13 Bailiwicks having declared themselves Protestants, and the number of'em is about 12 or 13000 Men. that some of'em had been imprisoned already, and these good people heartily wished that some other Protestant power might interpose in their behalf and procure'em liberty to leave The Country. they are all substantial men and the most part Husbandmen. my Correspondent at Ratisbon inserts the following particulars concerning the same John Gruber in his Letter dated Ratisbon 10 February 1733. Have given to John Gruber from Carinthia what you [Urlsperger] sent him out of your private Cash for the protestant Exiles, who with Tears in his eyes and a pious heart received this Benefaction, acknowledging the providence of God in this particular, who thus takes care of him and his Family. he promised not to forget in his prayers the Benefactors in England, but to beseech God that he may bless'em for ever and ever. the said Gruber has particularly gained the favour of all the protestant Ambassadors by his pious behaviour and one must wonder to hear how many fine Spiritual prayers his seven Children known by heart. he has had the honour with his family to dine at the Saxon Ambassador's Table, who also gave handsome Benefactions to them.

Extract of a Letter dated Ratisbon 10 February 1732/3.[63]

The Deputies from Berchtolsgaden wait still for a Resolution upon their Petition which they have presented to the Evangelick Body which I shall send you by the next Mail in its regular form, the most material things contained in it are,

1st. They Petition the Evangelick Body that the same would be pleased to interpose with their Prince, not to hinder any longer their leaving the Country, but to grant'em liberty so to Do.

2nd. At the same time they complain of their being refused and denied decently to bury their dead; as also of some particular Taxes which they are obliged to pay who go out of the Country being already exacted and levied upon some of'em though not a penny of all that belongs to them had been delivered to'em, and several other hardships.

Part of the said Berchtolsgaders, namely the miners with their Families, amounting to 550 Persons are said to go into the Hanoverian Countries, but where the Artificers are to settle is not yet known. Mr. Urlsperger adds, That at that moment he had received the news that eighteen of the most substantial Families, had engaged with Nurembergers.

Extract of a Letter in High-Dutch from the Reverend Mr. Urlsperger. To the Reverend Mr. Zeigenhagen. 5 March 1733. No. 2. No. 12088.

I hope you have received my Letter dated January the 19th N.S. as I have Mr. Newman's answer dated February the second. I shall write to him again as soon as I have received from him some further advice concerning some particulars in his Letter. in the mean time I praise God that he hath blessed with such good success the well meant Letter which I wrote to the Honourable Society concerning the distribution of the money sent for the relief of the poor exiles, and I long to know what resolution the Parliament has taken upon the Honourable Society's Petition. one Mr. Grinn a Protestant Minister in Hungaria, who is banished that Country has been here for some Weeks, and a friend at Vienna writes to me that we may soon expect more of the like Visitors.

Extract of a Letter from Ratisbon 15th February 1733.

The 3 Instant I sent one *Simon Lerchner* upon an Errand into Berchtolsgaden, who returning on the 14th at shutting of the City Gates related that being informed as he was going to pass through Saltzburg on his way to Berchtolsgaden, that they rigorously examined and searched every Passenger at Saltzburg, he took the precaution to tie with a string the Letters which he was entrusted with about his Body upon the bare skin, hoping to save'em in this manner, tho' upon his being asked who he was and he answering that he was a Journeyman Brick maker, and shewing the Pass-port he had from the Magistrates at Ratisbon, they let him go without searching him at all. upon his departing the 11th from Berchtolsgaden he met with two Fellow-Travellers who intended likewise to return to Ratisbon having been in the Arch bishoprick of Saltzburg on account of some Business. these latter when they passed through Saltzburg were examined and even their Cloaths searched very strictly, but nothing being found upon them they were sent their ways. All the Avenues in the Arch Bishoprick of Saltzburg are closely guarded by Soldiers, so that there is even no passing into it by means of taking some By-ways, and in case they deprehend any body travelling out of the common Road, he is outlawed, whereof the Inhabitants give warning to the Travellers. for these reasons every body that travels in those parts, must take care that he carry no Letters about him, in case he will not [does not wish to] be arrested, etc. and have severe Punishment inflicted on him. He likewise related that three Penitential Preachers[64] go about the Country and visit the Inhabitants in their Houses one after another to strengthen'em in the Roman Catholick Religion. But that the more pains they take the more of those who have any knowledge of Religion embrace the Protestant because it is impossible to hear these Penitential Preachers teach such abominable and Blasphemous Doctrines as they do without being surprized at it, they endeavouring by any means to make the ignorant People believe amongst other things that the blessed Virgin Mary ought only to be worshipped and adored because God hath made her Queen Regent and yielded the Government of Heaven and Earth unto her. God the Father is an elderly man, decayed with age, and hath lost his Hearing. God the Son cannot see any more, and God the Holy Ghost knows nothing of what passes, therefore the Mother of God only Lords it now over all.

Extract of a Letter from Ratisbon 26 February 1733.

You mention in your Letter that you can hardly believe the truth of the Deposition concerning the Blasphemy beforementioned, which is registered in the Protocol I have sent you, but if you please to look into the Protocol, where the Depositions of some Durenbergers [Tirnbergers] are registered by two notaries according to the desire of the envoy of the States General, you will find there an account of other Blasphemies to the same purpose and Intent.

Having most humbly requested His [Prussian] Majesty to let me know what his Will and Pleasure is, in case any more of the Saltzburgers should leave their Country hereafter, I have received orders to engage all that may come from those parts,[65] and because it has not been possible to build so many Houses for the reception of such a multitude as are necessary: they are in the mean time to be settled in other of His Majesty's Dominions as in Marchia [Mark Brandenburg] and Pomerania[.] by the way I must inform you that 4000 Rixdollars have lately been sent to Professor Franck[66] at Hall from Denmark of the relief of those exiles that are already gone into Prussia.

Augsburg 3rd March 1732/3.

This Evening one Christian Schartner, a Journeyman Mason, late from the Bailiwick of Radstadt in the Arch Bishoprick of Saltzburg, about Twenty Seven years of age, came to my House and having shewed me his Pass-port he related at the same time

1st. That some of those who gave in a year ago their names as Protestants before the Magistrates, but afterwards were enticed to remain in the Country when others left the same, had already taken a certain oath to persevere in the Roman Catholick Religion, which the Priests read unto them on a Sunday sitting on a Chair before the Altar, and they who are to swear kneeling down before him, and touching with their fingers the holy Gospel which lies open[.] of this the deponent hath been an Eye witness himself.

2nd. That though they had continued for three Sundays following to swear people in the Bailiwick of Radstadt, nevertheless all are not sworn yet.

3rd. That though many take this oath, yet they retain their former sentiments in religion, and are very desirous of getting Protestant Books.

4th. That a certain Peasant, who assisted at the divine service in the Church, hearing those who had formerly professed themselves Protestants, take the oath to remain Roman Catholicks for Temporal ends, he publickly and with a loud voice declared that woe would be unto them but that he had been put into prison for so saying.

5th. That those who comply to take the oath, are nevertheless punished and fined for having formerly declared themselves Protestants.

6th. That many others of'em are determined to leave the Country rather than to take the oath and that they cannot be diverted from this their resolution whatever endeavours the Capuchins who go about the Country use to alter their minds.

7th. That those who have occupied part of the Lands of the Exiles are like to be sent away again next Spring, because they have the same sentiments about religion with the Exiles, for which reason it is related that one of the Bailiffs said; we have driven out one Devil, but ten others are come in his place.

8th. That the Priests in the Arch Bishoprick of Saltzburg have wrote to those in the Country from whence these new Colonists came, advising'em to be well upon their guard, for that they had many of these very People amongst'em, which they had been pestered with so much of late; and partly had still living amongst them.

9th. That all those who have a mind to stay in the Country are obliged to enlist themselves in some Fraternity or other and wear a certain Mark of distinction.

After I had set down the forementioned Account, which the said Shartner gave, I read it over to him again and asked him whether this his disposition was true in fact, admonishing him at the same time to tell nothing but the truth but he affirmed that so he had done. One of the Exiles that lived in this City whom I am acquainted with told me that he knew this Shartner and his Relations who last Summer went to Prussia very well and that they are honest People.

No. 12101. Extract of a Letter in High-Dutch from the Reverend Mr. Urlsperger at Augsburg. To the Reverend Mr. Zeigenhagen. Augsburg 16th March 1733.

I hope my last Letter dated March the 5th N.S. is come safe to yours hands, as is yours of February 9th O.S. As to the project of the Honourable Society, I have not yet received any further Instructions by the hands of Mr. Newman, which in their last Letter were promised, thereby to enable me to take proper measures in case an occasion should offer to get a Transport for Georgia. in the mean time I have wrote a Letter to the Prussian Commissary [Göbel] who at present is at Ratisbon, to inform my self about the following particulars; Viz. whether in regard of the repeated orders which he had received from His prussian Majesty to contract with all the Exiles that may leave their Country and are inclined to settle in Prussia, some difficulties might not perhaps be laid in the way towards getting some of them for Georgia. in answer to this he expressed himself in the following terms Viz. That for his part he would not in the least hinder me, in making my best endeavours to dispose some to go to that Country, but that on the contrary he should be glad if any would take such a resolution, and that I could firmly assure the Honourable Society of this being his intentions, he added that in the mean time he should be glad to Know what number of exiles they were inclined to engage with. I suppose the Commissary had already acquainted the King His Master with the thing, and therefore wrote this by His consent otherwise he would not have expressed himself in this manner. besides this I have desired a friend of mine who went to Ratisbon four days ago to confer before hand with the Envoy of Brunswick-Zell, (Baron [J.] de Reck) about this affair in case things should fall out so that we should get some of the Exiles for Georgia, who at first looked upon the thing as quite impracticable but afterwards changed his opinion. For my friend informs me in his Letter of the 10th Instant that the Baron de Reck had invited him that same day to dinner and told him as soon as he came there that what he told him the day before had quite prevailed with him and that now he was entirely of his opinion; desiring him at the same time to procure him a more particular Account of Georgia etc. my friend added he thought a Letter to his Gentleman would meet with a Favourable reception, for which reason I design to write to him Thursday next in my own name. But after all it is highly necessary that the Honourable Society lose no time in writing me a full answer to several particulars which I have mentioned in my Letters partly to Mr. Newman, partly to you, that in case some new transports set out from their Country (for the Berchtolsgaders have been engaged already before their leaving the Country) all things may be in readiness. it will be requisite also to give full power to some Envoy at Ratisbon or to a Commissary, and thereby authorize him to write to the respective States, the Dominions of which the Exiles at their first setting out as well as afterwards in the Progress of their Journey are to pass through, requiring them to grant'em a free passage. I have talked about this affair some time ago with two Exiles, who are men of good Sense and much esteemed by their Countrymen, to whom the Voyage to Georgia appeared not to be attended with so many difficulties as they at first imagined, and who have promised me to serve as instruments in disposing some of their Countrymen for such a resolution, and to meet for that purpose the first that shall leave the Country. but at the same time they earnestly desired me to let 'em have in time something in print concerning the condition of the Land[.] I have also desired the Banker Mr. De Münch to think with me upon a Person that is fit to be employed as a Commissary. But as for a Minister and a Catechet it will be somewhat difficult to find proper Persons for these Offices tho' I hope God will shew us ways and means even in this particular. I have wrote upon this Account to Hale [Halle] and to Tubenger [Tübingen], there being likewise in the latter Place a Number of young Students in Divinity of very good Parts and a sincere piety. But what if we could not get an unmarried Man, who is of a sincere Piety, good Learning, and at the same time, knows a little English, have we leave to choose One who is married, and hath no Children, or not above one? Would he be allowed a Salary that he and his Family can subsist by? We must even think of such particular Circumstances in due time. Upon the whole, I must mention this once more, that because often matters admit of no delays, it will be necessary that the Honourable Society be pleased to grant me full power to do in this Affair what is just and right before God. But as for those things which don't press so much for a quick dispatch, I shall not be wanting to acquaint them therewith in due time, and take not one step before I know what is agreable to the Honourable Society.

From Ratisbon I have received the following News by a letter dated March the 12th viz. That the Saltzburg Exiles are not permitted to take their Rout thro' Bavaria to Ratisbon. Further, that the Land-Grafe [Landgrave] of Hesse-Cassel is also desirous to have a number of the Exiles. That the King of Prussia got no more than Seventy Persons of the Berchtolsgader that have left the Country, they being parted, and some of them going to Holland, some into the Hannoverian Countries, others to Nurenburg.

That the Passages into the Archibishoprick of Saltzburg are shut up and so closely guarded as ever, for which reason no letters can be conveyed a little way into the Countrey without running great hazard, the Papists there talking immediately of sending to the Gallies, those upon whom any Letters shall be found. & Lastly that we may soon see some families of Exiles out of Austria.

By Letters from Elbingen [Elbag] I am informed that after the King of Poland's Death[67] the Dissenters in that Country undergo again a hard Treatment from the Roman-Catholick Clergy. The same melancholly Account is given me by a Friend, who is Secretary to an Officer in the Emperour's Service at Cassovia,[68] concerning the Protestants in Hungary, who continue to lie under great Hardships and to be severely persecuted, but I hope the Lord himself will send us help and relief.

I have wrote this Letter to you only at present because I am in Expectation of some fresh Letters from Mr. Newman desiring you in the mean time, to give my humble respects to the Gentlemen of the Honourable Society.

P.S. One thing still I have to add, that it seems to be necessary for me to go my Self to Ratisbon for some days after I have received proper Instructions from England, which as yet I wait for as I have mentioned before in my Letters.

Translation of a Letter from the Reverend Mr. Urlsperger to Henry Newman dated Augsperg April 27 N.S. 12151 Read May 1, 1733.

Honoured and Dear Sir: In answer to your Letter dated March the 27 O.S. the time will not permit me to acquaint you at present with any thing else except the following particulars.

1. That I am fully convinced of the Honourable Society's good disposition and intentions towards the Protestants Exiles.
2. That I shall patiently wait the Resolutions the Parliament may come to as also for the Description of Georgia.

3. That I shall not be wanting [fail] to send you from time to time an Account of the State of this Affair.

4. That Mr. de Reck is Envoy of the King of Great Britain for the Dutchies Zell, Calenborg [Calenberg], and Saxe Lauenberg [Sachsen-Lauenburg] at Ratisbon.

5. That I am heartily glad to hear that so good News is come from Mr. Oglethorpe.

6. That yesterday I sent some relief to Ratisbon for the 800 Berchtolsgaders who are to settle in the Hannoverian Countries, in money, Linnen, Books etc. and likewise a certain quantity for 100 others, who are going to the King of Prussia's Dominions. These People deserve to be relieved.

P.S. The Elector of Bavaria is not so willing to grant a free Passage thro' his Dominions to the Exiles as he was last year but he will have them go by water as the Tirnbergers and Berchtolsgaders have been obliged to do. It is not known yet what will become of some thousands of Saltzburgers, who have been registered as professing the protestant Religion. The Medals shall be sent according to desire.

The Contents of an Abstract of a Letter to the Reverend Mr. Urlsperger dated Ratisbon April the 7th 1733.

Mr. Urlsperger's Correspondent mentions (1) That the 800 Berchtolsgaders who will settle in the Hannoverian Countries are to set out next Week as they have been promised and have obtained leave at last from the Archbishop of Saltzburg to embark at Hallein. But that those who are to go into the Prussian Dominons are resolved to go thither by land. (2) That the Passages into the Archbishoprick of Saltzburg are shut up still, and all Travellers examined very strictly whether they carry any Protestant Books or Letters about them. (3) That they expected some more penitential Preachers in the Archbishoprick of Saltzburg.

The Contents of an Abstract of a Letter to the Reverend Mr. Urlsperger dat. Ratisbon April the 14th.

The Commissary which the Envoy of the Elector of Hannover [J. von Reck] sent to Berchtolsgaden set out on Wednesday last for that place, and hath taken along with him some thousand Florins to assist therewith those that cannot pay the money which the Majestracy at Berchtolsgaden pretends [claims] to have an [on]

Account of the Allegience[.] Mr. Gobel the Prussian Commissary is likewise gone to Berchtolsgaden.

Here follows an Account of one Simon Hoser who had been at Saltzburg April 11th to fetch the Child he had left behind him and to dispose of some of his Goods.

He related chiefly what happened to himself upon his Journey and could not give any particular Account of the State of the Affairs in the Archbishoprick, because he was not permitted to go about the Countrey, but some of his Relations were sent for and called to the Bailiff's House at Saalselden [Saalfelden].

The Contents of an Abstract of a Letter from Mr. E— [Esterlin?] to the Reverend Mr. Urlsperger dated at Ratisbon April 16th 1733.

Mr. Urlsperger's Correspondent informs him of the ill treatment he had met with from some Roman-Catholicks, they having got Information that he was very much engaged in the Cause of the Exiles.

He informs him farther that 12 Exiles out of Austria were arrived there and that 20 more of them were upon the Road, the rest impatiently waiting the Success of these their forerunners. No body it seems minds yet these dispersed Sheep; but God will take the more Care of them.

He adds briefly in relation to those that in Carinthia have been imprisoned on account of the Protestant Religion. That the chief Man amongst them having been pressed upon very hard by the Roman-Catholicks in his prison to confess whether he knew of any more such People being in that Countrey, He said at last if he should speak out he must tell them that most part of the Inhabitants were Protestants which declaration struck them with so much terrour that they began to treat him with more lenity.

The Contents of an Abstract of a Letter to the Reverend Mr. Urlsperger dated Ratisbon April the 21 [Probably from J. von Reck].

By and by there arrive some Exiles out of Austrian Countries and we also hope some Provision will be made for them. The Berchtolsgaders are expected the 24th of this Instant at Passau whither five Ships with necessary Provisions sailed three days ago.

The said ships carry English Banners, which is a thing very seldom seen upon the Danube.

To Mr. Henry Newman.

Dear Sir: I had the Honour of Writing to You the 30th of March last, and having since received a letter from the Reverend Mr. Hales dated at London the 1st Instant, acquainting me that the Society for Promoting Christian Knowledge have lately written to Mr. Urlsperger at Augsburg, to send three hundred Emigrant Saltzburgers, or other persecuted Protestants, to Rotterdam, in order to their being transported thence to settle in Georgia, and also that the Society, upon Mr. Hale's Recommendations have thought fit to honour me with the Care of conducting them from Augsburgh to Rotterdam, in Case Mr. Urlsperger should not be otherwise provided of another Person. I cannot but return my hearty thanks to the Society for the great trust they are pleased to confer on me. In pursuance of that trust, and that no time be lost, I have this day acquainted Mr. Urlsperger by letter that I am in readiness of receiving his Commands according to the directions of the Society for that purpose, and of my going hence upon his notice to meet with Chearfullness such worthy Friends in those parts of Germany.

It is with great Satisfaction to hear that both Colonies of Mr. Oglethorpe and Mr. Purry are going on with good success And that the Parliament of Great Britain have given ten thousand Pounds to the Society for establishing the Colony of Georgia, to which plentifull Contributions from private hands are daily added. I beg the favour of You Sir to make my respects acceptable to the Gentlemen of the Society and to be assured that I am with great truth.

Dear Sir   Your most obedient & most humble Servant   John Vat
Swisserland Biel June the 26th 1733.

To Mr. Henry Newman from Mr. Urlsperger covering Copies of the several Letters following.   A Copy of a Letter from Mr. Senior Urlsperger of Augsburg dated at Poltzig near Zeitz the 15th of June 1733.   No. 12227.   Read 3rd July 1733.

To Monsieur [J.] Van Reck at Ratisbon.

After I had sent my last to Your Excellency which was some days ago I received the day following another Packet of Letters from England, whereof here is the translation which plainly shews how much His Britannick Majesty, the Parliament the Honourable Trustees for Georgia, and the Society are in earnest to promote farther the Interest on the Emigrants and send a number of them to Georgia. I am sorry I was not in Augspurg in order to go my Self to Ratisbon to confer with your Excellency upon this importent Affair; in the mean time I beg the favour of your Opinion about the following particulars as soon as possible; My humble Opinion is Viz.

1st. That the Accounts I have of Georgia after having added some material Circumstances relating to the Emigrants shall with all proper Precaution be published in print Viz. at Ratisbon Augsburg etc.

2. That before the said Additions are ready, these Accounts shall not be published.

3. That some of the Emigrants of Saltzburg, Carinthia and Austria being either in Augsburg or Ratisbon may beforehand be acquainted with the Contents of the Accounts of Georgia.

4. That Your Excellency would take the trouble upon You to enquire of the Dutch Envoy [Gallieris] what the Transportation of a certain Number of such persons from Ratisbon to Rotterdam including the Salary of a Leading Commissary might amount to.

5. That in the mean while I shall look for leading-Commissary as well as for a Minister.

6. That in Case the Number of 150 Persons could not be had all at once, single Persons may be engaged and maintained till the number is completed.

I shall stay here about 17 or 18 days longer and dare hope your Excellency will farther honour me with Your letters before my departure, and if with the help of God I get safe home it will be of an absolute necessity for me to make a Tour for some days to Ratisbon. May the Lord promote his Works every where and for ever bless Your Excellency and all Your Family.

A Copy of a Letter from Monsieur [J.] Van Reck at Ratisbon dated the 18th of June 1733 directed to Mr. Senior Urlsperger.

I had the honour to receive the day before yesterday Your Letter from Poltzig of the 12th Instant, and return you my humble thanks for the Contents thereof. The same day our Embassy received a Royal Rescript dated the 16 Instant by which His Majesty informs us that the Society in England de Propaganda Christi Cognitione is come to an Agreement with the Trustees of Georgia and unanimously resolved to engage 300 Exiles of Saltzburg Berchtstolgaders etc. as Colonists for Georgia and to provide for their Transportation as well as future Maintenance, and that we should likewise keep a Correspondence with You Reverend Sir on this Subject to promote so good an Intention[.] we shall on our Parts not be wanting [not fail] to conform ourselves entirely to His Majesty's order desiring you would be pleased to communicate to us your Thoughts thereupon viz. How the matter in agitation might be best effected.

A Copy of a Letter from Monsieur [J.] Van Reck at Ratisbon dated the 22nd of June 1733 directed to Mr. Senior Urlsperger.

I answered the 18th Instant yours dated at Poltzig the 12th Ditto by the way of Augsburg, and hope it is safe come to your hands. Friday the 19th Ditto I received another of the 15th Ditto together with the Translation of the Writings of Mr. Newman by which I see what further Resolutions the Parliament hath taken about the Society. The French News make mention of ten thousand pounds Sterling allowed for that purpose[.] I acquainted you in my last with the Royal orders we received here. Whereupon we have this day made appear the necessity and usefullness of soon publishing a full Account of Georgia which may perhaps have a good Influence to strengthen the Resolution of those Exiles that are still behind. Hitherto their exact Number is not known to us nevertheless some of them arrive daily & just now are come 10 of them from Saltzburg. You are Reverend Sir sensible that I should be much pleased to see the Publication of the Account of Georgia supported Regiâ Authoritate. The Dutch Minister *M.*[69] *Gallieris* is in Holland and will not return hither before the End of next month when I shall take the Opportunity to enquire about some necessary matters. The Charges for a Leading Commissary will not be very great and I shall propose in due time a proper Person for it. Indeed the last 800 Persons gone into the Hannoverian Dominions stood us in above 10 Thousand Florins. This here I send directly by way of Zeitz according to the direction you gave me wishing it a Safe arrival and your happy return. I remain

Reverend Sir etc.

To Mr. Newman.

Dear and Honoured Sir: I received in due time your letters of the 4th, 11th, 18th and 29th of last month here in Poltzig near

Zeitz seven German Miles from Hall at the House of Count Henchler a Lord of great merit and intimate Friend of the late Professor [A. H.] Frank where I have began to drink the Waters and from whence I constantly keep Correspondence with Monsieur [J.] Van Reck Envoy at Ratisbon as may be seen by the Copies I send you herewith; at present I should be glad to be informed.

1st. If the number of an hundred and fifty persons for a Transportation could not be got at once, whether it would be agreable to the Society to engage those who daily arrive and maintain them till the whole Number of 150 shall be complete.

2nd. As one Minister of the Gospel being hardly sufficient for the whole number of 300 and falling sick or dying would leave his Flock quite destitute whether it would not be necessary to look out for two besides a good Schoolmaster, which is the main point intended for these poor Emigrants persecuted for the sake of the Gospel, they would be at the Loss of being instructed in a Plantation where no other German Teachers are to be had.

3rd. Whether such Ministers for the Exiles should not be ordained at Augsburg. I wish you all manner of divine Blessings and remain with Esteem. Poltzig the 29th June 1733.

P.S.

I am of Opinion that it would be of great Service to the Society to make intercession with the King's Minister in order that the Publication of the Account of Georgia (according to the Proposal and Advice of the Deputies at Ratisbon) might be published Nomine Regis, and then I should take Care that nothing that could give Offence be inserted. I believe I shall return to Augsburg by way of Ratisbon which will be but 18 [German] Miles about [out of the way], to have the Opportunity to talke in a more ample manner with the Honorable Hannoverian Minister [J. von Reck] about ordering the matter in Agitation. Farther I could wish Viz.

1st. That the Account of Georgia had been more full, and they had spared no Cost to send us a complete Relation thereof[.]

2nd. That altho' I have made mention of the necessity of having Two Ministers thither I should be glad if I was so lucky as to have One of them a Man of Piety and Learning that may be relyed on.

3rd. That I shall do my best to induce the Evangelick Body at Ratisbon to grant the same Allowance for the Colonists of Georgia as was granted to them that went into Prussia, Hannover, and Holland.

4th. That I still have in my Cash above 2000 Guilders.

5th. That I shall herewith shortly declare my Opinion concerning Monsieur John Vat, being now by reason of my Indisposition not able to do it, in short I must own that the Issue of these Affairs has overjoyed me and move me to praise God for it. As soon as I shall be able to write my Self, I shall inform You farther about the matter. My most humble respects to the Honourable Society. I shall if it please God in a few days be at Hall[e] where I shall perhaps be able to find out a good Man for Georgia. I remain

Dear & Honoured Sir   Your humble Servant   Samuel Urlsperger

Poltzig 29th of June 1733.

The Answer to this will I hope meet me again in Augsburg.

Copy of the Translation of a Letter out of High Dutch from the Reverend Mr. Urlsperger to Mr. Newman. Ausburg the 10th of August 1733. No. 12280. Read 14th August.

Honoured & dear Sir: Being some Days ago returned home from a Journey into Saxony to my beloved City of Hall[70] for the sake of my Health, which thank God, is mighty well recovered; I shall inform You of several Points beside what I have mentioned in my last of June the 10th N.S. Viz.

1. That I have received several of Your Letters partly in Saxony and partly after my Return home viz. of May the 29th of the 3rd, 6 & 10th of July, together with the Bill of Exchange of £ 300 Sterling, exactly paid by Mr. Münch.

2. That I've thought fit to set apart the said 300 pounds, solely for the Benefit of such Emigrants as shall resolve to go over to Georgia.

3. That there is as yet but a small Number of those Emigrants, who are willing to embark for Georgia, the Reason hereof is that no whole Transport is lately arrived from the Saltzburg Territories; but only some single Persons now and then, some following their Parents others their Children, gone before'em into Prussia.

4. It seems that besides former Difficulties attending the Transport into Georgia which we have in some Measure found means to remove; Two new ones have been Started. The first of which is that in all the News Papers here and at other Places,

there has been Published a large Letter from Pennsylvania, by which every-Body is warned not to imbark for the said Colony. The reason alledged is that New Comers meet with but very Course Reception there: the 2nd Difficulty is caused by the returning of 50 Tirnberger Emigrants from Zeeland to Ratisbone which happened last Week, The said Emigrants making loud Complaints of their being ill used by the Dutch, who keep none of the Conditions Stipulated with 'em: insomuch that, if they had found ways and Means, all the rest would certainly have come back again. The Evangelick Body is highly concerned about it, and it may prove of very bad Consequences, not only in regard to the Emigrants, but to the Papists also: be the Complaints ill or well grounded.

Now to come to an Answer of Your Letters (leaving what remains to the 13th Instant.).

1. I shall, if Possible, make use of Mr. John Vat, who has been proposed for Marching Commissary; & whose Letter to that purpose I received Yesterday, tho' Mr. Rect [J. von Reck], the Envoy here residing, has proposed another Person, that is here.

2. The reason why the Publishing of the Description of Georgia has hitherto been delayed is that the Envoy thought it (as it really is) necessary it should be done by Royal Authority. Nevertheless, as you've been pleased, Sir, to observe to me in your last of July the 10th that the Trustees were Authorized by King & Parliament, also as to this Point, I shall this very week take the Opportunity to propose this affair to the Envoy; that so the said Description may forthwith be published under Royal Authority; without waiting for any further special Order from Court.

3. We shall strive to follow the Intention of the Trustees as to the Number, they've marked down of Emigrants for the Transport of one Ship.

4. I shall also imploy all my Skill in the Choice of Persons fit for the Ministry and School teaching: and I have already found out Two very learned and pious Students in Divinity,[71] who, out of pure Love to Promote the Honour and Interest of the Kingdom of Christ, are resolved to serve this new Colony, the one in Preaching and the other in Catechising or teaching in the School.

5. When I shall have a Sufficient Number of Emigrants for the Transport of One Ship, I shall forthwith transmit an exact

Specification of the same according to the Direction given by the Honourable Commissioners. This Sir is what I've thought fit to Communicate to You in great Hast. The next Post the rest shall follow. Wherewith I remain,

Honoured and dear Sir, Your most humble Servant
Samuel Urlsperger

To Mr. Henry Newman. enclosing Copies of the several Letters following from Mr. Urlsperger Translated out of high Dutch. Augsbourg Aug. 17th 1733. No. 12298. Read 28 Aug. 1733. Ordered as on the Minutes of this Day.

Honoured and Dear Sir: In Pursuance of the Promise I made in my last of the 10th of August (which I hope you've received) I communicate the following Pieces (which in part contain many Arcanas) from whence the Illustrious Society as well as the Honourable Trustees for Georgia may understand,

1. That the Description of Georgia has hitherto been delayed by reason of the Envoy's waiting for Orders, to know under what name & Authority it is to be published but your last Letter informing us that the Society is pleased to leave it to our Discretion I shall,

2. By the first Opportunity send the said Description to Ratisbone, according to the Discretion of the Hannoverian Envoy; to have it printed where it most shall suit our Conveniency &

3. Seeing that as well the Imperial Court as all the Catholicks in general look upon this Emigration with jealous Eyes, we that live in a Country and City where both Parties are equal, must use all imaginable prudence, not to give the least occasion to publish Popish Clergy of new Troubles; the late instance of the Reverend Mr. Esterlin at Ratisbone making us Cautious, & shewing what may happen in other Places in the like Case.[72] It would be therefore not only Convenient but even necessary that by some short Order written in Latin or French I should be authorized by his Britannick Majesty as Serving in this Commission the Trustees for Georgia, by the Request of the Society, whose Correspondent Member I am, to forward a Transport of Protestant Emigrants for Georgia, the said Writing being made to entreat & require all persons to assist me herein and so forth; But in such Writing no use should be made of the Title of Commissary in Respect to me. You can't imagine what a World of Troubles & obstacles we

find in our way, having seriously set about this Work, and with all this we have hitherto but Eight Persons resolved to go to Georgia; The number of new Emigrants being very thin. What happened to the Zirn-bergers [Tirnbergers] at Cassant [Cadzand] in Zeeland is to be looked upon as an Obstruction of the Whole work of Emigration; But, we trust, in God in spight of the malicious Enemy's will forward and protect his own Work; The Examples of Moses and Aaron must encourage us, who had no smaller difficultys to struggle with under their Charge of Commissaries in the Israelites Emigration out of Egypt. I recommend You to the Divine Protection, & next my humble Submission to the Illustrious Society & Honourable Trustees, I remain

Honoured and dear Sir,   Your most humble Servant
Samuel Urlsperger

P.P. i.e..... A Paper promised.

The Tirn-bergian [Tirnberger] Emigrants, lately returned from Holland are already so far provided for that by their Industry they may gain their livelyhood well enough here with us; the difficulty would be how to provide for more, as 'tis likely they will not be the last of that Swarm, seeing they are kept so meanly where they are. Their Conduct, no doubt, had [would have] been more prudent, if instead of coming away, they had by some Deputies represented their Case. But considering their Circumstances tis hardly possible that the rest abide there, by the Infraction of all that was stipulated with'em & drawn up in Writing before their Departure from hence. We shall see the Issue of all this at the Arrival of the Dutch Minister here; who is upon his Journey hither.
Ratisbone August the 11th 1733.   Weismeyer

A short Account of the Tirnberg-Saltzburgian Emigrants returned from Holland with their Grievances.

Last Friday Evening the 31 July we were surprized to see Sixty Saltzburgers Emigrants returning hither from Holland: every Body being desirous to know the reason of their unexpected return they alledged those as follows.

1.   That after their safe Arrival to the Isle of Cassand [Cadzand], which was the Place assigned them, they were obliged to pass 5 nights aboard the Ships in a raw cold Season at Sea, nobody being allowed to Stir from thence, till

2.   They had assigned'em for Lodging Hog-sties Bake Ovens, & other such uncommon places.

3.   Having hardly Time allowed'em to rest & recover themselves after so long and tiresome a Journey.

4.   They were obliged to do the hardest and toilsomest work for the Inhabitants and after their Labour, instead of their deserved Hire they received Scoldings and ill Usage, so that

5.   Besides their working for nought, they were obliged to Subsist by their own Means; by which they found'emselves necessitated.

6.   To complain of such hard usage to the Commissary who

7.   Instead of hearing their just Complaints, with Threats of Bastonado, sent 'em back again

8.   That all Provision was put at so high a Rate for them that they were obliged to pay for a pound of Beef 10 Crusers for a pound of Butter 18 Crusers for a Pot of Beer 6 Crusers etc. insomuch that

9.   They found themselves obliged to think on a timely retreat for to escape of the last Misery that threatned them; whilst they yet had something of their own means left.

10.   They also found way to escape by the assistance of two German Soldiers who were Lutherans; all the Ship masters being under severe punishments forbidden to transport any of them, so they returned hither by way of Brussels.

11.   Lastly as to the Exercise of Religion, or Divine Worship, they had 3 Leagues to go to Church; so that during all their Stay there, few of'em had been above 3 Times, many but Twice, & some not at all been able to attend Divine Service.

These Grievances seem outrageous, & not to be believed by the Reverend Mr. Urlsperger, as he writes in the Margent.

Charles Albertus, Elector of Bavaria.[73]

Dearly beloved! As we are credibly informed that Your Ministers of the Church & some of your Citizens take upon them to draw away by promises and allurements several of our Subjects from the Roman Catholick Religion to the Confession of Augsburg; a late instance whereof is a certain Goldsmith's Daughter, named Dendacher, who being first set upon by one of your Citizens a Shop keeper whose name is John George Rothier, & afterwards by two of your ministers, Esterlin & Roser was prevailed with, thro' Promises and Presents, to go to Confess in the Lutheran way, was upon that against her Will almost forced to embrace the Lutheran Religion, insomuch that the said Shop keeper promised this Young Woman that if she would turn Lutheran he not only would marry her but also provide for all her Family, tho' he himself was unable to do it, only by idle Promises to ensnare this poor Creature. Besides this we are informed that your Citizens married to Roman Catholick wives & vica versa, endeavour to withdraw and detain their Consorts from their accustomed Worship. Therefore it being well Known to you, that this is a grievous infraction of the Laws fundamental to the Roman Empire, by which every one must enjoy the free and undisturbed Exercise of his Religion, and above all this, being forewarned by an imperial Rescript to you, to the same Effect. I do Expect from your prudent Behaviour that you will put a Stop to these unlawfull proceedings, and by your Connivance not oblige me to look out for other more effectual means to hinder it.

Given at München the 24 June 1733 by special Commission, To the Magistrate of Ratisbone.

Copy of a Letter from Mr. Esterlin at Ratisbone dated August the 11th 1733.   To Mr. Urlsperger.

The Lord be praised for your safe Return to Augsburg again, my Joy hereat is exceeding great as it renders me the occasion to continue our Correspondence, & thereby to communicate my Hearts Thoughts to You. In order to answer shortly to the Points of your last Letter, I must tell You

1.   That all the Envoys of the Protestant Princes here are very much astonished at the unexpected Return of 50 Tirnberg Emigrants from Cassant [Cadzand] who prompted by their Impatience did not well consider what they did in leaving their new Settlement, & come back again here to Ratisbone, where their Inconstancy is exposed to the Censure of the whole World. One may easily imagine what a secret Pleasure this inconsiderate Flight affords the Adversaries of our holy Faith, and what a damp it will cast upon the remaining Work of Emigration, to see these fickle Persons lusting after the Flesh Pots of Egypt. Nevertheless our Trust is in the Lord, who can best promote the work he has begun. These People returned hither so well cloathed that nobody could taken'em to be Peasants; which makes me think that they fared not so ill as they would fain make us believe they did. I perceive also that they are grown more stubborn than they formerly

were, which I have taken the Freedom to tell'em when I, instead of
their Coming to see me, have been obliged to hurt [hunt?] after
them at the same time shewing them that these were not the Master of a humble and meek Spirit, & that the Trial God had put
them upon, ought to make them more Supple. The Envoys give
little Attention to their Complaints and it is remarkable that none
of'em dare come to the Envoys, particularly to the Saxon Envoy,
to tell them their Case; & I my self, with much ado, persuaded
Two of the last returned to go & shew themselves to the said Envoys. I really pity those poor dispersed Sheep and think they are
to be treated with some Care and Tenderness. I foresee already
they won't have a farthing more out of the Emigrants Purse; there
has been given to them a great deal of Charity, and they have
imagined that they should always fare as at Ratisbone, where they
daily found the Table spread for them; and lived without any
Care, which has perhaps spoiled them. The first Emigrants did not
find it so, & were yet Contented. I shewed them some days ago the
Example of the Israelites Emigrants, whom God by a Strong Arm,
brought out of Egypt, & how they murmered and complained in
the Wilderness against the Lord about their Dyet remembring
the Fish they did eat in Egypt freely Numbers: XI, but the Lord
was highly displeased there at, & his Anger was kindled. This I
think may find place here Intra muros peccatur et extra. That
there lyes some fault on both sides. Yet I think these Emigrants
should have had more patience & stayed for to see the Issue of the
Affairs; for we know that a Work of this nature cannot be done all
at once. All that now can be done is to put up our Fervent Prayers
to God that this Tryal may have a happy Issue. These poor
Wretches must learn here to put their Hands to work; and I am
afraid some of them will find it a hard matter to subsist. God in
his Wisdom turn all to a good End! As to the Second Point; to
your demand, if there is any Appearance of any Transport to be
Expected from Gastein, I can give no positive answer, Seeing that
200 persons have along while ago a time set for their Departure in
the Court of Judicature[74] of Gastain and for all that are still detained in order to make them spend what they have before they
leave the Country. Mr. Lerchner from Saltzburg, who has been
with You in Augsburg, & whom I often have bid take more care of
himself; lies now in Chains in the Goal at Lintz; he will find it a
hard matter to get loose again. God be his Comfort! The News
from Austria tell Us that several hundreds have made an open
profession of their Faith. They write from Passaw [Passau] that a

great number of Protestants Subjects have discovered [revealed] themselves. In the mean while, we are here attacked on all Sides from Vienna as well as from Bavaria by divers Rescripts, concerning me personally, who am plainly named therein. I send you herewith a Copy of the last Rescript with my Answer to it.

Copy of a Letter from his Excellency Mr. [J.] Reck at Ratisbone dated Aug. the 4th 1733.   To Mr. Urlsperger.

Our last news from England concerning a Transport are very favourable, I could wish that the desired Description with the Conditions entered upon with the Colonists might forthwith be printed; but have as yet no order for it. For the present I know but 3 or 4 Persons inclined to go for Georgia; unless there may be found some among the Tirnberger - Emigrants, that are returned hither from Holland, being 60 in all, for which there is but little appearance. I fear this Return will prove a great advantage to the Roman Catholicks, who will not fail to make use of these Clamours of the discontented Emigrants to deter the rest from following their Steps. I can hardly believe all that they complain of to be well grounded, chiefly what they say of their not having received any of the Collected money. In Case any further Orders should come to me from England about the aforementioned Affairs, I shall immediately go to work about it.

Copy of a Letter dated Aug. the 6, 1733. [Apparently from J. von Reck at Ratisbon].

To Day being a Post-Day, I shall have no opportunity to consult the Envoy von Hugo concerning the Contents of your Letter of the 3rd of August; to morrow I intend to see him. In the mean while I referr to what I have told you in my last viz. that I see [?] have received no Orders as yet, for the Description of Georgia so that we hardly shall come f[?] a Resolution[75] from Saltzburg cannot so soon be expected; as it was thought. We are here at great Charges with those lately returned from Holland, whose Clamours leave a very noisome Eccho after them. I hope next week to know the Opinions of the whole Body of the Protestant Envoys concerning the Demands You've proposed.

Copy of the 13th of August 1733   [apparently from J. von Reck at Ratisbon]

I send You the inclosed which is the Resolves of his Excellency the Envoy Hugo, Minister from the Court of Brunswick Luneburg, upon Your Demands; to which Resolves I also consent referring You to what I have said in my last. We expect the Dutch Minister Gallieries here in 12 days at farthest, when we shall hear what Reply he will make to the Complaints of the returned Emigrants.

Resolves of his Excellency von Hugo upon the Demands of Mr. Senior Urlsperger.

1. I hold it to be very usefull to have the Description of Georgia, as you have proposed it in your Letter of the 3rd August, published in print.
2. As also therein to declare His Majesties Intention of settling a new Colony there; together with the Stipulated Terms, upon which the Colonists are to be received and,
3. To make known therein the Intention of the Society for promoting Christian Knowledge; but I am of Opinion that
4. In this matter, as to the Intention of the said Society, you can not be careful enough how to propose it, not to give a Handle to our Adversaries to complain of any the least Infraction of the *Instrumentum Pacis*,[76] in drawing away any foreign Princes Subjects. Wherefore
5. It will be a necessary Precaution always to confine and limit the said Intention to such persons as are not obnoxious to the Breach of the said Law or *Instrumentum Pacis*. I beg leave to your own best Judgment to decide whether it would not be to the Purpose to insert, at the Beginning of the said Description, the Commission You're charged with to provide a sufficient Number of Emigrants for the said Transport, that so they that are inclined to go thither, may know where to address themselves. In my Opinion, it would also not be amiss; for to present [prevent] unnecessary Troubles, to have the said Description printed not at Augsburg or Ratisbone but somewhere else.

Copy of a Letter from Mr. Urlsperger at Augsburg the 20th Aug. 1733. To Mr. Henry Newman.

Honoured & dear Sir: The foregoing is a Letter which the Reverend Mr. Fisher [J. G. Fischer] of the Colony of Emigrants at Cassant [Cadzand] in Zeeland wrote to my Correspondent at Ratisbone (I desire you Sir not to make use of any Body of those underlined Names) Out of which the true State of the Case of those Tirn-berger Emigrants is dispersed. I trust in God that if we may have a sufficient Number for the Transport into Georgia, no occasion will be given for the like Complaints either against the Commissaries, or the usage they shall meet with there. I must add this to what I have said in my last of August the 17th that in Case an Order from the King should not be attainable an Order only underwritten & sealed by the Trustees and the Society will suffice. This is in great haste. Mr. Zeigenhagen's Letter shall be answered with the next Post. Putting my self under the Patronage of these two Illustrious Bodys & entreating God almighty to bless their great & noble Endeavours, I remain

Sir   your most humble Servant   Samuel Urlsperger

P.S.: Without waiting for Orders from Court for the publishing of the Description of Georgia under the C [ourt] Authority, I shall after tomorrow send a Draught, which I have made of the same to the Envoy at Ratisbone that so the same may be printed in intertiolow[?] according to the advice of the Hannoverian Envoy, being first accompanied by the Envoy at Ratisbone, of which I do not in the least Doubt.

Copy of a Letter from the Minister [Fischer] of the new Colony of Tirnberger settled on the Isle of Cassand [Cadzand] in Zeeland. Groede the 20 June 1733.   To Mr. Urlsperger.

Reverend Sir: I've taken the Opportunity of the first Post to answer without any Delay to yours Dated June the 4th. As to those Emigrants, who have taken the Flight from hence, tho' they at their Arrival did encounter many and great Hardships; Yet are they not so excusable as I could wish they were. Seeing they ought to have waited with Patience & Submission for help from above, & Stayed for high Orders of the Sovereign, who will not fail to mend any thing that's amiss. Concerning the Carpenter, & that Youth led astray by him: The first lives in Strife with his wife & continually rambling & Sauntering about the Country. The Second is also an idle Young Lad, who instead of following the Business he is brought up to, viz. Agriculture, likes better to take a Gun and run about like a fine Sportsman, which he used to practice here till it pleased the Magistrates to forbid him from the like Sport; which also made him Fly. The number of such Emigrants unsettled Emigrants whose Faith is but temporary (I'm afraid) is too great; chiefly among those of the first Swarm, whose chief Design has not been to change their Religion but the Air for a better, in which they may happen to find themselves deceived: &

these dangerous Persons (which God forbid) might prove the ruin
of many others. Among these deluded Souls, some were found last
week attempting to escape, but being timely discovered, & closely
pursued, they were overtaken by the Soldiers & brought into the
Prisons of Sluys. You may easily imagine, good Sir, what Grief
this Causes me, besides my daily Troubles, in tending my poor
Flock Scattered up and down throughout this Island. I cannot dis-
own my well grounded Fear that our Commissary's uncouth Be-
meanour [sic!] towards the Emigrants is the Chief, if not the only
Source of all this Disorder, treating those poor Wretches with
nothing better than Bastanadoes, bitter Invectives, & Curses,
whereas it would be much more becoming a Man in that Station,
like a Father to take part in their Concerns, & as their Interpreter
to speak a good word in their Behalf and by these means to gain
their Affection. It has pleased their High Mightiness to appoint
several Switzers to serve as Interpreters betwixt the Inhabitants
and the Emigrants and to assist them upon any other Emergency
but our Commissary makes use of them gratis contrary Purpose.
Our Emigrants are for the most part unacquainted with the Dutch
Language; many of them too weak and too much worn out for
hard Labour[.] the Peasants of the Island are not always endowed
with Patience enough for to bear & forbear with the poor Strangers. Many of them have been obliged to take up with a Barn or a
Bake-hutt for their Lodging; where they do not find Shelter from
the Injuries of Wind and Rain; others there are that must pay for
their Lodgings[.] the Dearth of Provisions is that which most puz-
zles our Emigrants, especially those that have numerous Familys to
provide for, they are quartered here and there in Farmer's Houses,
at a great distance one from the other; the knitting of Stockings
which is the Daily Employment of most of them, is very unprofit-
able in a Country, like this, where Provision fetches so high a
rate[.] But, notwithstanding all this, I hope in God that with Time
and Patience every thing will become easier to us. In all the abovementioned disorders nothing can be laid to the Charge of their
High Mightiness, who at the desire and Instances of their Subjects
have, not without vast Expences brought these poor refugees into
their Dominions: but the Fault lies Chiefly in the Mistake of the
Dutch Subjects who thought in this to serve themselves and their
Friends too, imagining that these people were all Young and robust persons whom they might imploy in their hard works as Servants, and who thereby might earn wherewith honestly to subsist,

the Servants Wages in these Parts running very high but, contrary
to the Inhabitants Expectation, it happens that a great number of
these Refugees consists of old and sickly Persons, & little Children!
'Tis a lamentable Case that the most of them are forced to leave
all they had in this world behind them. The sad and mournfull
Condition of our Emigrants has hitherto obliged me to preach
upon unusual Texts as particularly last Sunday upon that of St.
John's Ch. VI. v. 67, 68, 69 which was not performed without
shedding of Tears, considering how my sheep leave me one after
another, and with visible Danger both of Body and Soul wander
thro' Roman Catholick Provinces[.] But all I can say or do herein
is to throw my self down before the Lord, imitating the great Examples of Moses and Aaron, who intreated the Lord to take Care
of his wandring People.

We cannot enough praise the Lord for the Liberty we here enjoy of Performing Divine Service in all the neighbouring Reformed
Churches: but whether we shall be so happy as once to have a
Church of our own, we must wait to know when these Commotions are over. I really consider what a hard Struggle these poor
people must experience, which none can imagine but they that are
in the like Case. May it please the Divine Goodness to Strengthen
them that remain with us, and thro' Sufferings teach them patience and all other necessary Virtues. They are certainly worthy
our Pity and Compassion and as new Proselytes claim our Indulgence and Forbearance. As to the Huntsman's Complaints viz.
that for 20 Days successively their Diet-Money had been withheld,
is in my Opinion not so, but some few days were deducted by orders of our Superiors in regard of the free Entertainment and
Presents the Colony received in Sundry Places: further that the
Collections were not distributed amongst them, is only part true
because all the Charitys gathered upon the Road, were parted
among them off hand: bating those we received at Cologn, [omis-
sion] & Nimweg where we were advised to Set apart this Money to
be kept in Store for to Assist the poorer Sort in their want; Seeing
there were some that needed it not, & others that perhaps would
not make the best Use of it. The Magistrates of Dordrack [Dor-
drecht] and Rotterdam keep the Collections there gathered in
their own hands but tis reported that they will forthwith send
Deputies hither to provide Cattle and other necessary things for
the Colony with the said Means. The former Collections abovementioned have been at our Arrival Lodged in the Hands of the

Magistrates of Sluys, our immediate Superiors. God have Mercy upon Us & crown this his work with a happy Issue, according to his infinite Wisdom and Goodness. Amen!

Copy of Mr. Urlsperger's Letter Augsburg 3rd Sept. 1733. No. 12318. Read 18th Sept. 1733. To Mr. Henry Newman.

Dear & Honoured Sir: Our first Georgians arrived here Yesterday, & were met by many People in their Coaches and on Foot. They are 25 in number and this Day we expect 10 more of them. The Roman Catholicks here not having consented Yet to Lodge these People in the Town, a rich Gentleman named Schauer (who is the Author of the famous Balsam called after his name) has received these 25 Persons for a few Days gratis in his House joining to his Garden without the City, where I saw them Yesterday, and having joyned in Prayers with them I endeavoured to Strengthen them with the Word of God, which I repeated likewise this Morning taking for my Text the 50 Chapter of Joshua and I have explained them at large the Charitable Intentions of the Society and the Trustees. I find them all firmly resolved. 50 more would have come along with them had it not been for an old Man amongst them, who told'em at Memingen that if they were bent to go to that Country not 15 would arrive alive. But God will Send no more of them. Several of our Evangelical Inhabitants here especially divers illustrious Familys have made them poor People already Presents of Cloaths Shirts etc.

I send You herewith Copies of Letters which I received from Messrs. Hugo & Recht [J. von Reck], & what those Ministers have wrote to Our Magistrates, as likewise what Mr. Schorer, who together with one of our Evangelical Burgher Masters Mr. Morrell bestirrs himself exceedingly in this affair, has wrote to me. You'll also find here inclosed a Copy of a Letter from the Prussian Commissary Mr. Gobel and an Abstract of my Memorial to our Privy Council here. Now we must stay till we can get 70 of these Emigrants together, and then I can send only for the Minister, whom I am almost sure of, & take Care of his Ordination here.

As to the Harbinger or Commissary for the Journey whereof Mr. Recht [J. von Reck] makes mention in his Letter, I shall write to the latter notwithstanding, that I am authorized by the Society to engage one and to defray his Charges. But what Method must I take when after having got together a Transport of 70 Persons & sent them away, & the Season advancing, & Winter coming on, shall I take more of them and send them away? And shall I send the Minister away immediately with the first Transport together with the School Master? I am for the affirmative by reason that if one could find means to send away this Year another Transport of 100 more, then one could send with them the other Minister. This amongst other Questions must be resolved.

I must moreover acquaint You that it seems as if the Roman Catholick Members of the Magistracy here are not willing that our Georgians should lodge in the City since they have taken time till to Morrow to deliberate upon this Subject. My humble Respects to the Trustees & Society & I am etc.

Extract of Mr. Schorer's Letter from Mindelheim to Mr. Urlsperger of the 29th August 1733.

Being with the Divine Assistance safe arrived here with my poor Saltzburgers, I received an hour after the Favour of Your Letter with the inclosed papers, whereupon and after having Lodged these People, I summoned their Chiefs, in order to be apprized of the answer they Promised me Yesterday. In the beginning it seemed as if they all inclined to go and Settle somewhere else contrary to the hopes I had conceived from the Discourses I had with some of them in our Journey; But having read to them several Articles contained in the Subject sent from Georgia and represented to them Briefly the rest, especially the Conditions upon which they are to be received in that Colony, they all fell into an Admiration of the Charity that was shown them, & of the Temporal as well as Spiritual advantages that would be granted to them, assuring that they could never sufficiently praise the Divine Providence for those Favours. That however they must put off to the next Day giving a final Declaration, in order to converse once more upon the Subject with the rest of their Brethren. Whereupon I dismissed them after having exhorted them to pray to Almighty God with Fervour to please to direct their Thoughts where to go. In these Circumstances I can't write to You any positive answer by this Post, since it will be proper for me to go with them tomorrow to Memingen, & to cause their Depositions to be taken before the Magistrates of the City.

Extract of Mr. Schorer's Letter from Memingen to Mr. Urlsperger of the 1st Sept. 1733.

This is to acquaint You that Twenty odd of the last Saltzburge Transports have resolved to go and Settle in Georgia and those I shall Conduct to day to Mindelheim and proceed with them to-morrow please God to Augsburg. I had conceived great hopes to get more of them but several Circumstances have prevented it. Nevertheless several more have declared that if they could not obtain any Settlements in the Territory of Ulm, they would likewise repair to Augsburg, and thence Transport themselves with their Country men to Georgia.

A List of the 25 Persons that have resolved to go to [omission] in Georgia, but have little or nothing.

1. Family. Thomas Schwandel [Geschwandel] a Miner from Gastein 38 Years old his wife Margaret Holferin[77] 21 Years old and one Child 1 Year old.

2. Family. Lorenz Haber [Hueber] from Gastein, Labourer 51 Years old, Mary Mandelleithem his wife 45 Years. John his Son 10 Years Magdalen his Daughter 13 years, Mary, 7 Years, and Margaretha 5 Years old.

3. Family. Paul Schweckhofer [Schwaighofer] a Weaver from Mittiersel [Mittersil] or Binfgan Pintzgau 48 Years old. Margaret Prin Wingerin [Prindlinger] his wife 40 Years old Mary his Daughter 6 Years. Thomas his Son 4 Years Ursula his Daughter 1 Year.

4. Family. Barthel Risser [Rieser] a Farmer from Gastein 46 Years old, his Wife 40 Years with 3 Children.

Single Persons

Anna Hoferin 25 Years old.

Christian Steiner from Gastein a Miner 29 Years old.

Balthasar Fleiss from Gastein a Miner 27 Years old.

Tobias Lackner from Gastein a Coalburner 40 Years old.

Mathias Millersteiner [Mittersteiner] from St. John [St. Johannes] a Weaver 41 Years old.

Extract of a Memorial presented by Mr. Urlsperger to the Privy Council at Augsburg which is composed of Evangelical and Roman Catholick Members of the 2nd Sept. 1733.

A Transport being arrived unawares from the Bishoprick of Saltzburg, at Memingen by the way of Lansperg [Landsberg] & Mindelheim, whereof 25 persons have resolved to transport themselves to Georgia, and who as I am informed are to arrive here this Day, my request to the Honourable Privy Council is that these 25 Persons may have leave to Enter this City and to dwell here at their own Expence like other Travellers untill a Transport of 70 Persons of them can be formed. His Majesty and King of Great Britain having taken them under his Royal Protection, I have so much the less reason to doubt of the Complying with this my Request since we are ready in case it be required to give to the Honourable Magistrate or to their Deputy appointed for that purpose an exact List of these Persons at their Entry into the City, whether any are received into the Evangelical Almhouse or lodged in the Houses of the Evangelical Burghers, and that the same shall be repeated at their leaving this Place.

Extract of a Letter from the Envoy Mr. Recht [J. von Reck] to Mr. Urlsperger of the 1st Sept. 1733.

As for what relates to the desired Requisitorial Letter to the City of Augsburg the Ministers here are [sic] at the Diet are willing to grant it. As for proposing a Commissary for the Journey, I have indeed for my own private account offered me and there is such a Person here [Ph. von Reck], whose fidelity and Capacity may be depended on. But he must be engaged by the Society and have his expences defrayed. The Cash established here for the Emigrants will on this occasion be as charitable as they have been before to those people.

P.S. I could have wished some Member of the Society have drawn up the Paper to be published, for we are not so well acquainted here with all particulars & Circumstances. The piece hereby enclosed, called the *Directorium,* will shew the remarks sent me by the Envoy v. Hugo, upon the Project transmitted to us. You are however at Liberty most Reverend Sir to range the materials as You think proper. I only am of opinion the printing it should be hastened. It was not possible for us here to write sooner nor to send the things desired for the Requisitorial Letter in form we received but to day.

Copy of the Requisitorial Letter from the Electoral Ministers at Ratisbon to the whole Magistracy of this City of 1st September 1733.

You are without Doubt already informed that the English Society for promoting Christian Knowledge, having first made an Agreement with the Company of Georgia at London, have resolved to receive in Georgia 300 persons from among the Emigrants of Saltzburg Berchtolsgaden & other Places, and to take care of their Transport as also of the means of subsistance to be assigned for them; and that accordingly they have thereupon corresponded with the Senior of the Clergy at Augsburg, the Reverend Mr. Urlsperger, and furnished him with proper Instruction for engaging such Persons. And whereas His Majesty the King of Great Britain, our most Gracious King and Sovereign, has been pleased to approve of this undertaking and has ordered us to assist the said Senior in all possible manner in the executing of what the Society have committed to his Care; And Whereas he will find it a difficult matter at once to gather so many of those People as will be required for a Transport nor can they be conveniently transported till the Society shall have received advice of their being all together and made the necessary disposition for their Reception at Rotterdam and for their further conveyance, during which time those People, who are to have their subsistance from the said Senior, cannot better stay and continue together any where else than in Your City. We therefore most humbly desire of You to allow them to Stay there till a Transport may conveniently be ordered for them and after having first assured You that your Compliance with this Request will be particularly acceptable to His Britannick Majesty, we offer on our part all agreeable Services in return, remaining always etc.

von Hugo. [J.] von Reck.

Extract of a Letter from the Prussian Commissary Mr. Göbel at Ratisbon to the Senior Mr. Urlsperger of 1st September 1733.

We have all received advice here of the Emigrants who are on the Road to the Number of 250 Souls; I should be glad You could obtain that some of them might be sent over to Georgia; The Danish Envoy here will take care of the rest, and go to meet them at Nurenberg, though he is expected [expecting?] first nearer Orders from his King by every Post. I hear another Transport of 600 persons is to set out speedily from the Country of Saltzburg, when I am to meet them as far as Augsburg; for I have the Kings orders to take another certain number of them under my Care, who are to be distributed in the Electoral Marches, [Mark Brandenburg], the Old, the Middle, and the New and other Provinces, there being a little more room left in Prussia. May it please Almighty God to move the Hearts of the other Powers to shew the like Christian Charity and Compassion towards the future Emigrants and to take care of their Subsistance. The poor People of Dürenberg [Tirnberg], who are returned from Holland and as they unanimously say have offered [suffered?] great misery are now straying about like Sheep lost from their Flock; The Burghers and Inhabitants of this City, who had given them quarters before, have now received them again; Nobody gives himself any Trouble about them, time will shew after the return of the Dutch Envoy hither, how they may be provided again. May the Almighty be their Protector.

Copy of a Letter from Mr. Urlsperger Augsburg 7th September 1733. No. 12325. Read 11 September 1733. To Mr. Henry Newman.

Dear Sir: My last of the 3rd of September as well as the foregoing I hope You have received. You See here a Copy of my Letter of the 6th September to the Envoy Von Recht [J. von Reck] out of which You may understand how hard and unfriendly the Roman Catholick part of the Magistracy here behave towards the Emigrants and now particularly towards the British Colonists of Georgia. Notwithstanding my Memorial and the Letter of the Brunswick Envoy in their behalf I send herewith the Declaration of the Roman Catholick part of the Magistrates at Augsburg together with an Account of the Emigrants Newly arrived made by two of the Senates of both Religions and Lastly the Pasports of these People.

How the Declaration of the Roman Catholick part of the Magistrates may be answered (which the Evangelical part of the Magistracy have done already and likewise intend to do by a new memorial to the whole Secret Council wherein I shall request again the Lodging of the said People within the Walls of the City). You will see out of a Copy thereof That which seems most necessary is that the Honourable Trustees for Georgia together with your Honourable Society would most humbly Petition His Majesty of Great Britain as soon as possible that He would be pleased to give forthwith orders to his Envoy at Vienna to represent to his Imperial Majesty how the Roman Catholick part of the Magistrates at Augsburg have hitherto refused to Permit that the Emigrants of Saltzburg and new British Colonists of Georgia might come into the City and be Lodged in the Evangelical Hospital and other Houses of Protestants where they could be cheaper and better maintained notwithstanding there are but 300 in all to be received as such and tho' they come in small Numbers from time to time and are to be sent away at 3 different times and that the Magistracy have been required [requested] to Grant it by the Brunswick Embassy Yet they have refused it[.] that therefore his Imperial Majesty might be desired in the name of the King of Great Britain to order by way of Rescript the Magistracy of Augsburg that these and all other following British Colonists may always be received in the City and that the Roman Catholick part of the Magistracy may not any further oppose it but conform therein with the Evangelical part of the Magistracy, it being a Civil Case and agreeable to the common Law of nature as well as Christianity And this under pain of incurring the displeasure of his Imperial Majesty. These are in my humble Opinion my Thoughts according to the Love I bear to the Georgian Colonists.

Now I must tell You that besides the 25 Georgians this very day 16 more have engaged to go thither out of the Emigrants lately arrived[.] amongst them is a Family of 11 Persons, The Father George Buchman [Buecher] out of the District of Leightenberg [Lichtenberg] is an excellent prudent and brave man and as I believe has brought along with him near 200 £ Sterling in Effects[.] he is a Farmer and a Salt-Peter boiler about 35 Years of Age, his wife 33 Years, has Six Children the 1st about 13 Years the 2nd 9, the 3rd 7, the 4th 5, the 5th 2 Years and 1/2 the 6th 10 weeks old besides a Servant of 33 Years and another who is a Miller of 28 Years and a maid Servant of 17 Years. If God should bring this man to You to London (as I doubt not but that the Ship is to bring them from Rotterdam first to London that the nation may see a transport of Saltzburgers) You will wonder at him. This very moment when this was wrote I received the favour of Yours of the 14th of August but as the Post is just going away and I must still to [go?] to Ratisbon about some things necessary according to the Contents of Your Letter, I can now add no more than that I shall Use my Utmost Endeavours to do according to your Orders and I shall Write to You for that very reason every Post day if possible.

All depends upon this Whether the People of Tirnberg will accept of the invitation I intend to give them or no, if they accept of it then the Transport will come quickly, I having now 40 People together: But if they do not accept of the invitation then it may last 3, 4, 5, 6, 7 and more weeks till a new Transport comes out of Saltzburg and we can Compleat the number required. In short one can fix no time. Perhaps some of the last Transports may change their minds and come to Us tho' they have refused at first. I remain in great hast,

   Honoured and Dear Sir,   Your most Obedient humble Servant

                S. Urlsperger

For want of Room I do not send the Accounts and Pasports abovementioned but only a Copy of the Declaration of the Roman Catholick part of the Magistracy and the Answer of the Evangelical part thereupon.

A Copy of the Declaration of the Roman Catholick Part of the Magistracy at Augsburg.

The Catholick Part of the Magistracy cannot consent to Senior Urlsperger's Memorial *communicated* Yesterday to the Catholick Privy Council, by the Intendant of the City and Privy Councillors of the Augustan Confession nor to the Letter, delivered at the same Time, from the Brunswick Embassy concerning *300 Emigrants* to Engage them to go to Georgia, and to have their *Rendesvouz* here any further than if they were provided with Authentick Passes from their Magistrates to give them the Liberty to pass through this Jurisdiction, and if required a quick abode *without the Walls* of this City for a *short Time;* likewise to allow them to go to Church, and Visit the Ministers of the Augustan Confession; but to take these People into the City cannot be Allowed because they would immediately, especially if they stayed long, draw *continually more of such* Emigrants hither and come *privately [secretly] into the City,* and then they could not *answer* for their *good Behaviour,* and it was to be feared that easily such Trouble and Reproof might happen to them, as lately did to the City of Ratisbon from the Emperor and the Elector of Bavaria.[78]

   ANSWERS upon this Declaration.

1. *communicated* Answ: How can the Catholick Magistracy say that the Elector of Brunswick's Requisitorial Letter, as well as Senior Urlsperger's Memorial hath been communicated only to them when the former plainly shews that it hath been directed to the whole Privy Council. The Catholicks will by this Word accuse the Protestants, as if they again had done something upon their own accord among themselves.

2. *300 Emigrants.* This no Body desires of the Catholick Part of the Magistracy.

3. *Rendesvouz.* Immediately they call it a Rendesvouz or a Place of abode. And if it could be called a Rendesvouz, It is my Opinion that they ought to have so much Regard for so great a King and besides an Elector of the Empire, as for the Venetians, whom they formerly allowed a Rendesvouz for their Recruits of all sorts of People, which they received in this City without any Scruple.

4. *If they.* Why do the Catholicks say *if?* when already two days they have had in Hands as well the Relation of the Deputies as the Passes, whereby they can see plain enough that all is in due Order.

5. *without the Walls.* I hope that not only the Protestant part [of the Magistracy] here but also the Elector of Brunswick's Embassy will insist upon taking the People into the City because they can live at half the Expence in the City as they do now and can be better Instructed.

6. *Short time.* Why this? so long as every time there is a Transport together they will be sent away and yet they would set the Time according as they please, and then they must go whenever it comes into their Hands.

7. *continually more.* No more than what will go to Georgia and no others than what are gone out according to the Treaty of Peace.[79] Here they will not stay.

8. *privately into the City.* Here one may see how they instruct the Protestants, notwithstanding We have offered when such Persons arrive from time to time, and go away again, to deliver every time a Specification [list] of the People.

9. When did ever his Imperial Majesty reprove the City of Ratisbon for having harboured Emigrants even many 100 for several Weeks within the Walls. About 900 Tirnbergers and 900 Berchtolsgadners have been above a Fortnight in Ratisbon and the Emperor hath not objected, nor could he say any Thing against it. But the Imperial Rescripts to Ratisbon did concern quite other Things, of which some Citizens were accused.

A Copy of Senior Urlsperger's Letter to the Envoy the Lord von Recht [J. von Reck] dated 6 September 1733.

Having received Your Excellency's Letter of the 3rd September this day after the Sermon and a great Communion, I hope You have likewise received mine dated Yesterday, together with the adjoined Pieces. Herewith I send Your Excellency, for weighty reasons, by an Express, 1. the Declaration of the Catholick privy Council to our Protestant Privy Council, to which I have added in the margin as answers for an Explication; 2. the Deputies of both Religions for the Emigration Affair, and [omission] and *Protocol* concerning the Emigrants that arrive and 3. the Passes they have brought with them. After the Passes stands a NB [nota bene] which Mr. Morrell hath added to this Copy. But in the Original delivered to the Privy Council are no such Lines. Further I let you know that the Protestant Privy Council keep themselves *Protestants* and that tomorrow they will write about this Affair to the Brunswick Embassy and send all these Pieces, which I send here privately, along with it. Likewise will the Protestant Magistrate, who had Yesterday Evening an Extraordinary Meeting, acquaint the Emperor with the unfriendly Behaviour of the Catholick Magistracy, and desire a Resolution upon their Letter sent a Year ago about the like Affairs. I for my part will write tomorrow by the Post to the Society and the Lords Commissioners for Georgia and acquaint them with all this, & if the Brunswick Embassy will do the same, I hope His Majesty the King of Great Britain by his Envoy at Vienna will dispose the Emperor to Command the Catholick part of the Magistrates to take his Majesty's Subjects into the City. I do not know whether Your Excellency and his Excellency the Lord von Hugo will represent the Case here once more. For this is certain that the People will cost me as much again without the City as in the City. Not to mention the great Trouble it will occasion to me and others if they should be without the City. I have sent an Express with the Consent of some *Procerum Evangelicorum* [Illustrious Protestants] here, however not under my name, nor publickly for reasons I will tell You another Time. This Day the Georgians have been in my Church and I have spoke a word with them from my Heart. My humble Service to his Excellency the Baron von Hugo and to Your Excellency. I remain etc. S. Urlsperger

Copy of a Letter from Mr. Urlsperger Augsburg the 14th September 1733. No. 12288. To Mr. Henry Newman.

Honoured & Dear Sir: I hope my last Dated the 7th instant is now come to Your Hands. Notwithstanding I have delivered another Memorial according to the Copy adjoyned, concerning our 40 Georgians that they might be taken into the City, the Catholick Magistracy continues still to oppose it contrary

to the Laws of Nature and Nations, which Causes great Trouble and Charges. If according to the Envoy's the Lord von Recht's [J. von Reck's] Letters, the Emigrants either from Tirnberg or Berchtolsgaden or from both places, Should resolve to go to Georgia, the first Transport would soon be completed. The Lord von Recht [J. von Reck] hath presented a very honest Man [Ph. von Reck] to be a Commissary whom I have taken. However his Salary doth not begin before a Transport is ready. He will if required, go with them to Georgia. An Experienced Apothecary [Zwiffler] who understands these People's Manners, Language and Nature, is likewise resolved to go with them to Georgia and to stay there, if not only the Diet Money is allowed him like the Emigrants upon the Journey and some Sustenance in Georgia till he hath established himself. For the Rest, our Emigrants behave themselves here like Christians, and some of them are very well grounded. Yesterday they hear [sic] again to hear my Sermons, in which I propounded these 3 Sentences Matthew VI. 33 Seek Ye first the Kingdom of God etc. Philippians IV. 11, 12, 13. I have learned etc. and 1 Thessalonians V. 24. Faithfull is he that calleth You, who also will do it etc. My humble Service to the Trustees of Georgia and to the Society, I remain honoured and dear Sir
Your very humble Servant   Samuel Urlsperger
P.S.: All my Colleagues in the Ministry do likewise daily preach the word of God to these People.

Copy of Senior Urlsperger's Memorial to the whole Privy Council in Augsburg 8th September 1733.

Although I expected a Resolution, agreable to my Petition and to the Brunswick Embassy's Letter to the Magistrate, upon my Memorial to the whole Privy Council, grounded in the Common as well natural as Christian Law, as also upon my Motives delivered afterward concerning the Colonists engaged for Georgia, to be taken in a moderate and circumspect Manner into the City, Yet instead of a gracious and agreeable Answer from the whole Privy Council, I have seen by a Writing called Promemoria from the Catholick part of the Privy Council delivered to the Privy Council the 4 September this instant Year, and by them communicated to me, that the Catholick part of the Privy Council hath been pleased without any Reflection upon my Petition and Reasons agreeing with Christian Religion in general and the Party of the State here to permit the Great Britannick Colonists only to live without the

Walls of the City and but for a short time, and not to let them come into the City under the pretence 1. That these Emigrants would continually draw more after them hither, 2. That they would come privately [secretly] into the City and 3. that it was to be feared that this City would have the same Reproof and Trouble from the Emperour and the Elector of Bavaria as lately the City of Ratisbon had. But as to the first reason, the requisitorial Letter as well as my Memorial shew plainly that no more than 300 engaged for Georgia shall be taken in, whereof hardly 100 come together at once, besides it is known that so many of the Emigrants at first received here are gone from time to time with those lately arrived, that of 3. we hardly count 1/3 more; as to the 2nd reason, we have engaged our selves, and do still, to give an accurate Specification of the People when they come and when they go away; 3. what hath happened with the City of Ratisbon hath no Relation at all with this Case, the former concerning particular [private] Accusations, and here is the Question about the Reception of the great Britannick Colonists, according to the Constitution of the Empire, into this City in order to maintain them the cheaper, to provide the necessaries the better for their long Journey, and to converse the more with them[.][80] his Majesty the Emperour will not be displeased[81] but rather be Surprized that what is permitted to all People that live Honestly quiet and without troubling the publick Should be refused to a few Colonists of one of his Imperial Majesty's powerful and faithfull Alliances and an Elector;[82] Especially whereas the Catholick part of the Privy Council well know that at the End of last Year, and this Year, as well the Tirnbergers as Berchtolsgadeners every time about 900 Heads have lived for some weeks in Ratisbon before the Face of the most Excellent Embassy of the whole wholly [Holy] Roman Empire without the least Resentment of the Emperor because all was agreeable to the Constitution of the Empire. I therefore most humbly beseech the Whole Privy Council of the City once more, to give me a gracious Resolution and to take these People into the City who have legitimated themselves on Account of their Passes last Thursday, and so much the rather because we are ready, if required, to produce the Original Letters written in the English Tongue, some of which I received Yesterday, whereby appears that in England they desire no other Colonists but such as are gone out with Permission of their Magistrates, and that they insist upon that, as soon as a Transport of 70 or 80 Persons are together, they might be sent by a Commissary to Rotterdam. Recommending myself to your Grace and wishing You a happy Government I remain etc.

Extract of a Letter from the Envoy the Lord von Recht [J. von Reck] to Senior Urlsperger dated at Ratisbon 8th September 1733.

The Behaviour of the Catholick Magistracy is very unfriendly and deserves a good Reprimand, which also will follow when the Letter from the Evangelick Magistracy here to the Evangelical Corps[83] here is come which is not yet come. In the mean time Yesterday hath been wrote to England and Hanover about the Consequence of it. The Holland's Minister Mr. Gallieris hath communicated to me a Resolution concerning their Stubborn Emigrants which is now to be copied, and next time I shall communicate it to You etc.

P.S.: Concerning the Commissary [Ph. von Reck] I have proposed, he is a Young Man of 21 Years of age and hath been above a Year in my house here, whose nature and Conduct I know, & can say for Truth that he leads a vertuous & Christian Life before God, and Men, he is of a fine Stature and great understanding, educated at the University at Helmstedt where he hath studied the Laws, he understands the French Language and something of the Italian. His Christian Name which at present I will but mention is George;[84] This Young Man after he had heard of the Society in England and of the Design of a Colony in Georgia, he shewed a great Love for this work and would willingly contribute to it as much as in his Power, if it was but to conduct the Colonists, as to a Commissary. And to shew his disinterested Mind, he desires no more than to be exempt from Charges in Travelling, and his free Diet, which may come to one Guilder per Day. If the Society should approve of this, the Question would be when he is to appear? Then he may come immediately & receive his Instructions to conduct the Emigrants if required to Georgia.

Extract of a Letter from the Envoy the Lord von Recht [J. von Reck] to Senior Urlsperger Ratisbon the 10th September 1733.

I hope You have received my Letter dated the Day before Yesterday, and Yesterday I received Yours dated the 7th instant whereby I see that the Transport of 70 or 80 Persons is to be sent soon and that in England they Expect the Tirnbergers that are returned. As to these people they can Stay no longer here because the City cannot have them. The Prussian Commissioner Goebel seems as if he did not care for these People, and Yet he will not explain himself and perhaps under Hand makes them engage themselves, which will appear within a few Days. Those 8 Families Consisting of 50 Persons that are returned from Brunswick are upon the Road yet and not here, neither dare they stay here long. If the Account of Georgia together with the Conditions of Reception in print were now ready, they would come in right Time, in order to distribute some Copies of them among the Emigrants. This Day the Proposal hath been made to some of the Tirnbergers, who have desired 2 or 3 Days time to consider. as soon as I have their Resolution I will let You know it, if need by an Express. When the Berchtolsgadener Remigrants[85] appear here, the Question shall be asked them whether they will go to Georgia, and the Success thereof we will leave to Providence.

Copy of a Letter from Mr. Urlsperger. Augsburg the 17th September 1733. No. 12344. Read 25 September 1733. To Mr. Henry Newman.

Dear & honoured Sir: This Day I have received Yours of the 28th of last month whereby I see that You have received mine of the 17 and 20 August, and hope that my Letters of the 3, 7th & 14th are likewise come to Your hands. The small Number of 41 Heads is not yet increased. Mr. Von Recht [J. von Reck] writes to me that the Tirnbergers in Ratisbon have no mind to go to Georgia, yet perhaps they might resolve still. In the mean time I have sent the Draught to Nurnberg, Franckfurt, Ulm, Ratisbon Memmingen, Kauffbünen [Kaufbeuren] etc. in order to make it known. The Catholick Part of the Magistracy continues in their unfriendliness and will not permit these People to lodge in the City. A Difficulty happens about the Persons that should go with them as Ministers and Catechists.[86] But God will help Us to overcome this too. The Ministers of his Majesty of Great Britain correspond with me twice a Week and do their utmost in this affair. I shall acquaint them with what You have wrote to me. Time will not admit to write more at present. Here is inclosed the Draught according to your desire. My humble Service to the Society, in great Hast I remain

Yours   S. Urlsperger

Dear and beloved Mr. Ziegenhagen must have Patience with me yet.

Copy of a Letter from Mr. Urlsperger. Augsburg the 21st September 1733. To Mr. Henry Newman.

Dear Sir: Here is inclosed a Letter which I was obliged to publish because several even of our Protestant People reckon it as a great Inhumanity of me to occasion these People to go to America. The Condition of our Georgians is as I have mentioned in my Letter of the 17th instant. The chief Matter will be whether another Transport will come from Saltzburg. If so be, then there is Hopes of getting a Transport of 100 Persons at least: But if not, it will be very hard to bring a Transport together; And then I must ask again: in case no Transport can be expected what to do with these 42 Heads? Whether to send such a small Number away with a Minister and a Catechist in hope that in the Spring more will follow them? or because it is to be feared that these few People will not go, whether they shall be dismissed? The unfriendliness and Obstinacy of the Catholick part of the Magistracy, who hath hitherto[87] will not permit these People to Lodge in the City, hath been a great Hindrance and will be yet further as much as lies in their Power. Further I must mention that these People cost daily Six Guilders to maintain them and would not Cost half so much if they did lodge in the City. This is all at present. However I must acquaint you that I expect to Day the person who is to marry my Eldest Daughter this day Sennight. He hath been Minister in the German house at Venice 5 Years and a half, and is now Minister and *Adjunctus* to the Superintendent in Grafentha [Gräfentha] in the Dutchy of Saxony Salfeld. He was 2 years *Amanuensis* to Professor [A. H.] Franck deceased:[88] He understands the English Language very well and could do me very good Service if he was here. This I mention on Account of Mr. Ziegenhagen. His name is Wenndrich, and it is he that sent me the first Collection for the Saltzburgers from Venice.

Yours   S. Urlsperger

Copy of a Letter from Mr. Urlsperger Augsburg the 31st August 1733. No. 12321. Read 11 September 1733. To Mr. Henry Newman.

Dear and Honoured Sir: I herewith acquaint You that the 27th instant the Bavarian Sheriff in Landsperg, Baron von Mondel, hath writ to our Evangelick Councellor Mr. Morell, that on the

28th Ditto a new Transport of Saltzburgers Emigrants 247 Heads were arrived, who could go the next Day to Mindelheim and be on Sunday the 30 in Memmingen. As soon as I heard this, I desired [requested] an Officer in the Evangelick Service, a prudent man, to come to me, whom I gave Commission to go on the 28th to Landsperg 5 Miles distant from hence in order to acquaint the Transport with his Majesty's Commissioners and the Society's Designs; The same Day I also gave Notice of it to the Brunswick Embassy at Regenspurg. This Officer Mr. Shorer [Schorer] set out, and the 29th in the Morning I had a Letter from him containing as the Copy here adjoyned shews. I sent his Messenger back with my Draught of Georgia, of the Conditions, of the Provisions made already, and of the dissolved Objections against this Affair, with Instruction to force no Body but to stay for a voluntary Resolution from the Emigrants. And Nicolaus Forstreiter (one of the two that were Examined at Berlin in September 1732 by Rau and Reinbeck or Reinbeck and Roloff) a very prudent Saltzburg Emigrant, arriving here the Same Day from Regenspurg; I sent him together with another well experienced Saltzburger, whom I know to Memmingen, in order to talk with their Countrymen according to their Conscience about the Georgian Affair; On the 29th I wrote again to His Majesty's ministers at Regenspurg as the Copy page 4 shews from whom I expect impatiently their Approbation of any Draught and their Orders. This day the 31st instant about 10'o Clock, I received another Letter from Mr. S[c]horer dated the 29th according to the Copy adjoyned, and not knowing whether I shall receive his third Letter before the Post departs tho' he makes me Hopes of receiving it this Forenoon; I must, in case these People keep to their Mind to go to Georgia, make some Questions and Admonitions that I can give a right Information and Answer from the Honourable Commissioners and the Society to these People and others that may present themselves hereafter. For they will [wish to] know for certain:

1. Whether they and their Children and Posterity shall be maintained in the Doctrine of the Evangelick Church according to the Augustan Confession of Faith and their Symbolical Books, and constantly be provided with Preachers and Catechists?

2. Whether they shall not be parted too far from each other but left together as near as possible to the End they may be able to assist one another the better and be at Divine Service and have their Children instructed?

3. Whether People will have Patience with their Manners and Language and not jeer them as the Inhabitants of Zeeland have done, as 'tis said, those from Tirnberg; and

4. Whether it is certain that they shall not be made Servants of but have Goods and Land for themselves, except those that may have a mind to go to Service. Although now I shall not Scruple, but, according to the instrument and Conditions sent to me from England, answer these Good People as they would it, Yes. Yet it will be better to send me once more a positive answer upon these 4 Questions, in order to convince them the more. Further I must make this Question and Admonition, because it is so late in the Year, whether these 300 Emigrants (in case 300 are complete) could not be divided into two Companies, so that 150 men were imbarked in each ship: or if for the future but 200 Persons should come, always 100 in a Ship. If there comes 300 Emigrants I shall, according to your Permission, send two Parsons and one Schoolmaster with them and if but 200 one Parson and one Catechist. However, because nothing certain can be said yet of the Emigrants, other Things are likewise uncertain: and this causes great Difficulty, Trouble and Labour, e.g., the vocation, Ordination etc. of a Minister only, all which cannot be done, unless there be a Transport of 70 Persons at least compleat who ought to have a Minister, and yet one dare not Stay so long. But if God but gives Grace that his Kingdom may be spread out, all will be overcome. God be with You my Service to the honourable Commissioners and Society, I remain

Yours   S. Urlsperger

Extract of a Letter dated at Ratisbon 25 August 1733.   To Senior Urlsperger.

Among the Saltzburg Emigrants that are returned from Holland, is only one provided with a Pass from the Magistracy there all other having left that Country privately [secretly] because they were not permitted to depart. The reason of his Departure, as is expressed in the Pass, is that he did not understand nor could learn to work as the Country People there and that his Wife could not be used to the Victuals of that Country. However, according to his own Saying, this was not the only reason, but rather these Circumstances mentioned in my former, which have Caused him and his former Country People [compatriots] to return and would occasion many more to follow them, if they had liberty to go. Now and then there comes likewise Emigrants here from Austria, two of which had been imprisoned there for several Weeks and Condemned for some Time to publick work because they had brought their Young Children privately [secretly] out of the Country and seduced others, as they imputed them, from the Catholick Faith, Yet after they had suffered this Punishment, both were dismissed and provided by their Magistracy with a Pass, wherein their aforementioned pretended Crime is mentioned but also attested that besides they had behaved themselves honestly and, because they were resolved to go into Lutheran Countries, they had ben discharged of their Oath of Allegiance by the Commission ordered for the Reformation of Religion in their Country. But neither have they got any of the Money for their houses which are sold Judicially, nor was it permitted them to take the least of their Goods with them, much less were their Children permitted to go with them but all kept back without distinction, grown or not grown.

Copy of a Letter from Mr. Shorer Intendant of the Prebends, dated at Landsperg 28 August 1733.

The Saltzburg Emigrants arrived here on Friday the 28 August N.S. They consist in 247 Heads with 12 Waggons of their own 23 Horses and 2 great Baggage Waggons which they had hired as the adjoined Piece numbred A. further mentions. Among these are, according to the Specification of their Magistracy 73 Persons, some of which have taken nothing from their Country and some not so much as 10 Guilders. And having already been 14 days upon the road and spent what they had, a great many of them are very poor & want Charity. Therefore they beg Your [omission] to assist them upon the road to Memmingen to wit in Mindelheim. For they are not able to pay 5 Waggons hired here, each 5 Guilders to Mindelheim. They are provided with Passes & Certificates, as the Copy marked B & C tells, and in Condition like the former Transport being very desirous to come into Protestant Countries to hear Gods holy word. I have also called before me the wisest among them and told them that the honourable Commissioners of Georgia and the Society for Promoting Christian Knowledge under the Authority of His Majesty the King of Great Britain out of Christian Love were resolved to send 300 Protestant Emigrants to Georgia and at the same time gave them a Description of the Condition and situation of that Country and laid before them the Condition upon which they were to accept of this Offer, recommending to them after their earnest Prayers to God to consider of it and to give me their Answer. Immediately one of them said that as there were many unmarried Persons among them he believed they would resolve to go thither and the next Day I would have their further Resolution. Those from Niederie [Niedersill] who have neither Relations nor Acquaintance yet abroad seemed particularly Enclined to accept of the Proposal. The General von Walpergen's former Coachman's Father, Mother and Brother are also in this Transport, which latter signified that it was indifferent to him whether he went to Prussia or Georgia, provided that he might but have the Word of God, for which reason he had left his Country. I was pleased with this Declaration but am afraid they will easily change their Mind when upon the Road they come to People who perhaps may make a great Difficulty of this Voyage to Georgia. But in case some of them should resolve to go to Georgia. I desire You to send me a further Instruction about it and especially the Description of Georgia. The adjoyned Piece marked A is the Specification of the Persons arrived viz. from Niedersil 27 among which are 11 Children.

St. Johannes	2
Werffen [Werfen]	4
Radstall [Radstadt]	1
Raueiss	1
Wagrain [Wagrein]	3
Leigktenberg [Leichtenberg] or Saalfeld[en]	22 among which are 12 Children.
Gastein	213 among these are 44 Children
	273

A Copy of a Letter from Senior Urlsperger to the Envoy the Lord von Recht [J. von Reck] dated 29 August 1733.

I hope Your Excellency hath received my last dated 27 instant, and thereby seen that unexpected new Transport of Saltzburgers Emigrants are arrived at Landsperg, and to go from thence to Memmingen. It was impossible to bring this Transport hither 1. because the Sheriff in Landsperg would not have suffered it, it being contrary to the Orders about their March. 2. The March Rout being not made for Augsburg, the Roman Catholicks here would have been against it. I have therefore Yesterday sent the

Protestant Intendant of the Prebends [Schorer] to Landsperg, who hath writ and sent to me what Your Excellency finds in a Copy here inclosed. Now your Excellency may easily judge, how I have longed Yesterday for an Answer upon my Questions, but have received none. At present I humbly desire your Excellency to send me your Orders, if it is necessary, by an Express. According to my Opinion 2 Things are necessary, especially if more or less should resolve to go to Georgia, 1. A letter to the whole Magistracy here that the Colony may stay here till they can depart; and 2. requisitorial Letters in general to the States were [where] they must pass through from Memmingen to here.

Extract of a Letter from the Envoy the Lord von Recht [J. von Reck] to Senior Urlsperger dated 25 August 1733.

Thursday last I received Yours of the 15 instant, but could not answer upon it at that Time, because the Envoy the Lord von Hugo was in the Country. Yesterday I received a Packet by the Post, which as is likely, hath been here since Sunday, In which I found without any other writing from you Your Draught that is to be published either in Writing or Print, about Georgia, perhaps only to have the Opinion of our Embassy about it. All this I have communicated this Morning to the Envoy the Lord von Recht [von Hugo?], but it being a little tedious and Time not allowing us to speak about it before this Evening, I must desire You to have patience till next Post, especially because it will be necessary to speak to the Envoy from Holland [Gallieris] likewise, who arrived here again yesterday, and to enquire about the Emigrants that are returned from Holland. Then I shall have the Honour to give You a full Account etc.

Copy of a Letter from Mr. Urlsperger Augsburg the 1st October 1733. No. 12357. Read 9th October 1733. To Mr. Henry Newman.

Honoured & Dear Sir: Yours of the Seventh of September I have received, the 28 Ditto and thereby seen that likewise mine of the 31 of August and 3 September are come to Your hands. Our Colonists are increased to the Number of 57 Persons small and great ones. I have read to them the Contents of your last and the Resolution upon my Questions, which they were glad of, and they are all resolved to go to Georgia and are well contented that 60 Persons should go first. I for my part give many Thanks for the plain and Circumstantial [detailed] Answer which was very necessary for the Satisfaction of these People. A new & unexpected Difficulty hath happened about a Minister and a Catechist so that at present I am sure of none:[89] however I do not doubt but I shall find some; only what makes the Case difficult is that I can call none formally together a Transport is together.[90] But let us take Heart, for he that gives the Sheep will also send a Shepherd. *Dominus providebit.* The Roman Catholick Part of the Magistracy continues in the old way nevertheless they have allowed that every time when I desire 3 Persons may come to me into the City through a certain Gate. This Catholick part hath sent an Excusing Letter to his Majesty's Embassy at Ratisbon but falsely Grounded. The Embassador his Excellency [J.] von Reck writ to me the Day before Yesterday that the Embassy will make a Remonstrance once more. It is said that another Transport will come from Saltzburg; which will soon appear. If it be so, then there is hopes of getting the Transport soon compleat. As to the money I have in Cash, it is as I have mentioned some weeks ago, except what the Maintaining of the Colonists these 4 weeks hath cost. In case I should want, I hope M. von Münch will supply me with some. This moment Mr. von Münch sends me word that I shall always have as much money out of his Cash as I want. Here is inclosed a Recept for 300 pounds Sterling. At present I have nothing more to write than to commit You to God's Grace and to recommend me to the Trustees and the Society, remaining

Your most humble Servant   S. Urlsperger

Copy of a Letter from Mr. Urlsperger Augsburg the 12th October 1733. No. 12370. Read 16 October 1733. Ordered as on the minutes. To Mr. Henry Newman.

Honoured & Dear Sir: I herewith acquaint You that our Georgians to the Number of 51 Persons, great and little ones, are still in the same Condition. some Days ago they were removed from Schaurer's Garden to the Place where the Citizens shoot at the mark [Schiessgraben], where I must pay 4 Kreutzers for each Person daily, more than before. The Envoy Mr. [J.] Von Reck writes to me that another great Transport is expected from Saltzburg whither they were gone, some of them to fetch their Children, some their wives, and some their Certificates of Baptism who unanimously declare that all these that are resolved to say [stay] in that Country are forced to take a Solemn Oath for the Popish

Religion, and those which afterwards repent and declare themselves to be Protestants are looked upon as such that have broke the Laws of the Empire and sent away. Some of these Persons mentioned that went to Saltzburg have been successfull in their Business and some not. One of them who hath been there twice could not get his Son; the same happened to another. We heard for certain that some thousands in the Austrian Countries, as Steyria [Styria] etc., have declared themselves Protestants and desired the Emperor to maintain them in the exercise of their Religion in the Country. But they have got no Resolution yet. In the mean time these People are treated tolerably well[.] I long for Letters from London, and for a Resolution what to do with them 51 Persons; wherefore I refer to my former Letters and remain

Yours   S. Urlsperger

Copy of a Letter from Mr. Urlsperger Augsburg the 15 October 1733. No. 12382. Read 23 October 1733. To Mr. Henry Newman.

Dear Sir: I have received Yours of the 7th & 18th September O.S. and thereby seen that You have received my Letters of the 7th and 14 September.[91] I expect the full Power and Instruction from the Lords Commissioners of Georgia by the way of his Majesty's Embassy at Ratisbone, likewise the good Effect of the Representation made to his Imperial Majesty by the Minister of Great Britain at Vienna. Viz. that our Magistrate by his Imperial Majesty's Order shall take the English Colonists into our City. According to the order of the Society I have writ to Mr. Vat at Bienne and told him that he should keep himself ready to come hither upon my first Letter, which I send after this. But I am afraid that the Envoy Mr. [J.] Von Reck, who upon my Desire hath proposed a Commissary, before the Society mentioned Mr. Vat and afterwards recommended him in his Letter, for his sufficient Qualities, will take it ill that another Commissary shall go with him. However I shall represent the Case so that the Envoy Mr. Von Reck shall also in this Point be convinced of the Society's good Intention. Further I must Confess You to my Great Sorrow, that that Countreyman [Buecher], who hath a Family of 11 Persons and is pretty well Stocked and hath promised me several times that he would go to Georgia because he looked upon it as a Calling of God hath before Yesterday and today declared to me and to the Burgomaster Morell that he and his Family never will go to Georgia, in which Declaration he persists, notwithstanding all Remonstration made to him, in which I have spent many Hours. After this manner the Transport of 51 Heads is decreased to 40. If the Embassy at Ratisbone were of my mind I would send this Transport away the sooner the better, because I hope that when once a Transport is gone and more Emigrants should come they will follow the others the sooner and if these People should stay here long I am afraid ignorant People will disuade them from it. How heavy now their Commission is upon me, Yet God will help. Nothing can be more by Force because it would be contrary to the intention of the Society, and the Catholick Magistrate would be glad if any Hindrance should happen to this Transport. If, with the Approbation of the Embassy, I should send this Transport of 40 or 41 I shall let You know the Day of their Departure as also in how many Days they may come to Rotterdam. I shall also send the Apothecary with them in Case he will be contented with the Treatment as an Emigrant[.] My Respects to the Society, I remain

Your humble Servant   S. Urlsperger

P.S. This moment the Apothecary Zweibler [Zwiffler] came to me and is resolved to go with them of which more in my next having no time now.

Copy of a Letter from Mr. Urlsperger.   Augsburg the 22nd October 1733.   To Mr. Henry Newman.

Dear Sir: I have received Yours of the 28th September, and thereby seen that You have received mine of the 17th and 21st September and hope that by this Time You have received my Letter of the 15th instant. By this I acquaint You as follows: The full Power from the Trustees for Georgia was sent me Yesterday from Ratisbonne. I wish to be more able to serve God in this Case and to satisfy those that have Love for it. Now it is resolved that the Transport shall Set out from hence the 30th instant if Possible. It is grown again to the number of 50 Heads. I likewise ordered Mr. Vat to be here as soon as possible. How soon the Colonists may arrive at Rotterdam I cannot tell Yet; but think it will require about 4 Weeks time. The Peasant [Buecher] who does not go with them as I have mentioned in my former, hath disuaded none, but all the rest are constant and very glad to depart. I have invited some prudent Politicians and Merchants to a Conference to-morrow in order to make the March Rout in the best and Cheapest manner. The Electoral Brunswick Embassy have perfectly agreed with me that the Transport should go away the sooner the better. My Endeavours about a Minister and a Catechist being disconcerted, it causes me Grief and Sorrow till it is redressed. I am glad to hear that a Schoolmaster [Ortmann] and his wife will go to Georgia. But as to a Catechist it cannot be altered. For the Society hath expressly told me by You my dear Sir, that they did agree with my Proposal, that two ministers and one Catechist should be sent, in Case one Minister should die that another might be in the room of him. Afterwards You wrote to me that because the Number of the Colonists or Emigrants being but small in the Beginning the Society thought it proper to take a Catechist that had studied, and in Case of Death could minister, and therefore should be ordained. And in this latter Case I was likewise pleased with the Society's Opinion, and I expect every Post Day Letters from Saxony [Halle] that a good Minister and Catechist hath been found who if they cannot come hither they can find their Sheep in Rotterdam, since I have writ to all Places where it is necessary and appointed the Time and Day. If now a Minister and Catechist come from Germany the Society may Consider whether they will send the Schoolmaster proposed which I would be glad of on Account of the English Tongue. In the Calling of these two Persons I look chiefly upon the Georgians but not only upon them but also for the Service they may do to the Americans [Indians], and therefore the Society should not grudge if they do something more for the spreading out of the Empire of Christ. The Lord God will reward what the Trustees and the Society does in this Case. I recommend myself to You, and give many Thanks for the Continual Confidence You repose in me, and I remain

Your humble Servant   S. Urlsperger

Copy of a Letter from Mr. Urlsperger Augsburg the 29th October 1733. No. 12389. Read 30 October 1733. To Mr. Henry Newman.

Dear Sir: The Commissary for our Georgian Colonists, Mr. [Ph.] Von Reck, Cousin to the Envoy Mr. [J.] von Reck is now arrived from Ratisbone. He is a Young active Man, and shews a great deal [zeal] to serve these People and the Lord Commissioners for Georgia and the Society. His Uncle the Envoy assures me that he will give all Satisfaction; and the Dutch Envoy hath given him very good Instructions how to behave himself on shore and at Sea during the Voyage. Upon my Desire to the Envoy of Saxony each Person Man and Woman above 14 Years of Age is to have out of the Emigrants Cash at Ratisbone, 4 Guilders and each Person of both Sexes under 14 Guilders 2 Guilders And about 160 Guilders which private Persons have given them, have been distributed among them this Day; of other Things You will be informed more in my next, being at present occupied with writing many Letters and Instructions. The Transport of 40 Persons or Heads will certainly set out the 31st instant if God pleases. They must have about 4 weeks time to Rotterdam, according to which, measures may be taken in sending the Ship from London. The Diet Money is regulated so that a Man shall have 15 Kreutzers, a Woman, 12 and a Child 8 per Day. The Commissary Mr. Von Reck desires no more than his Diet and Travelling Charges. He is to go with them to London, and if required [requested] to Georgia. I wish that by this Opportunity he may one time or other be Employed either in Civil or Military Service. He appears to be a fine Person, and gives goods hopes to discharge his Office duly. I have writ twice to Mr. Vat at Bienne but got no answer Yet. I likewise must acquaint You that I take no further Care in Providing the People with Money than till they are arrived at Rotterdam. Afterwards I leave them to the Lords Commissioners [Trustees] and to the Society. Mr. Von Münch who at present is very sick assists with good Counsel, Money and Recommendation to Franckfurth, Mentz [Mainz] and Cologn and Burgomaster Morell hath likewise done much Good and is willing and able to Serve upon the least notice. 26 of the people have Yesterday received the Sacrament in my Church, and to Day I married one Couple of them[92] of which more in my next. I have told the Commissary to write often to You during the Voyage to the End that in London they may know how they proceed in their March. I recommend my Georgian Children to your further Love and to the Prayers of all Good Christians in England and remain

Dear Sir   Your humble Servant   Samuel Urlsperger

P.S. It is reported for certain that another Transport is coming from Saltzburg. Our Transport of 40 Heads will perhaps be increased by the Tirnbergers of which 100 are gone anew from Holland.

Copy of a Letter from Mr. Vat at Bienne Switzerland 30 October 1733. No. 12397. Read 6 November 1733. To Mr. Henry Newman.

Dear Sir: I had the honour of giving You an Account on the 25 September last of Three hundred and Three Score and Fourteen Waldenses[93] gone down the Rivers Thiele Aar and Rhine in order to Settle in some Province in Holland. And on Monday last I received Your most Acceptable Letter of the 18 September as also at the same time Two more from Mr. Urlsperger Dated the 15 and 21 Instant. Now in Pursuance of the Commands both of the Society and the Reverend Mr. Urlsperger, I purpose to proceed to Morrow by the way of Zurich St. Gall and Lindau to Augsburg in order to Conduct in Conjunction with the Gentleman recommended by Baron Von Recht [J. von Reck] and approved of by the Society such Emigrants from Saltzbourg to Rotterdam as shall be Ready for the First Embarkation for Georgia. And I am highly obliged to the Society for the Honour they are pleased to bestow on me by my being joined in Commission to a Gentleman of Fine Parts, of Learning and ingenuity, I being of Opinion that his Knowledge will be of greater advantage to the undertaking than by my being Solely intrusted with so weighty a Commission.

I am exceedingly pleased with the Compliment of the university of Oxford, of the Degree of Doctor of Divinity to Dr. Hales[94] in Regard tho' absent to his Great merit and his Service both to Religion and Learning; And I beg the Favour of You Sir to present him my best Respects.

I shall lay hold on all Opportunities of Writing to You and of letting You know where You may direct Your Commands to me, which I desire may be henceforth to Augsburg till I shall thence change my Situation.

I beg also to present my best Respects to Colonel Valogne to whom I gave on the 5th of this month an Account of some Disturbances lately come to pass in the Lordship of Erguel the Inhabitants whereof in Military Affairs are Subject to this City, which since are increased and are like to be worse and worse[.][95] And I remain

Dear Sir   Your most Obedient and most humble Servant
John Vat

Copy of a Letter from Mr. Urlsperger Augsburg 2nd November 1733. No. 12398. Read 13 November 1733. To Mr. Henry Newman.

Sir: The Trustees and the Society will perceive by the inclosed Report how far we are got with the Assistance of the Divine Providence & after many and incredible Difficulties in the sending off those Emigrants of Saltzburg that are to Settle in the Colony of New Georgia & who departed from hence the Day before Yesterday.

I hope to be able to inform You in my next of the Costs expended for their Entertainment [maintenance] during their Stay here as well as for their Equippment and Transportation to Wertheim.

A Draught of a Project has been given to these Colonists, in what manner they could according to our Opinion enter London when they arrive there.

There is Still a Report of 6 or 800 more of them had left the Country of Saltzburg which however I cannot give Credit to.

I desire You Sir to make my humble Service acceptable to the Trustees and the Society and am

Sir Your etc. S. Urlsperger

P.S. I hope that with the Divine Assistance our Colonists will be at Rotterdam in a month's Time from the 31st October. This Minute when the Post is upon its Departure I receive Your Letter of the 8 & 9th October.

Copy of a Letter from Mr. Urlsperger Augsburg 5 November 1733. To Mr. Henry Newman.

Sir: I have received your Favours of the 9th October and hope that my Letters of the 22nd & 29 past as well as that of the 2nd Instant are Come to Your hands. Our Saltzburgers have left this Place with much chearfulness I send You herewith divers Copies viz. one of an intimation read from the Pulpit concerning Our Colonists. One of the Certificates given them by our Clergy and one of the full power and Instruction, which I have given to our Commissary, who Conducts these People. Since the Departure of the Latter I have received 2 Letters from his Cousin the Envoy [J.] Reck at Ratisbone who is very glad that his Kinsman has an Opportunity of Recommending himself to the Trustees and the Society, and assures me that he will give Satisfaction.

I send You Moreover a Copy of a Deposition made by the most ingenious of these Emigrants to Whom our Colonists belonged before the Magistrates at Memmingen, As also an Extract of a Letter from Franckfurt which I received this Minute from Mr. de Munich [Münch], in pursuance whereof I have wrote to Day to Mr. Gullman the English Resident at Franckfurt, and to the Two Brothers Von Munich, as also to our Commissary [Ph.] Van

Reck in what Manner they are to Act with relation to these People of Tirnberg newly arrived there. Should more or less of them resolve to go to Georgia with our Colonists, the Said Commissary Reck will acquaint You therewith Sir from Francfurt.

NB   It will be necessary that some body or other of the Trustees or the Society would take the trouble to write to Rotterdam to desire some Friend or other to Assist your Commissary when our Colonists embark there. I am ever etc.   S. Urlsperger

Extract of a Letter from Franckfurt of the 2nd November 1733.

Most of our last Emigrants of Saltzburg or Tirnberg who were sent to Zeeland are received there and the rest returned hither. 40 or 50 of them arrived here the Day before Yesterday, but in a miserable and deplorable Condition, being Sick of Fevers and Consumptions and the Children looking like shadows whereas they appeared in their march thither to be all hail and healthfull People which causes a great Astonishment. Thus these Poor Emigrants are become greater Objects of Charity and in such a light our Senior Dr. Munden represented them Yesterday to his Congregation and recommend them to their Compassion. They are lodged here in the Alms house or Hospital where they are well taken Care of. We should be glad if some of them would go with Your Colonists to Georgia And if Your English Commissary with whom we intend to Speak on our subject, would do the Charity to take them with him for we are persuaded they would be dealt with there with more reality than they have been by those interested People of Zeeland.

Depositions made by some of those Emigrants from Saltzburg who were Settled at Garstein [Gastein] and who left their Country in the Month of August 1733 to the Number of 250 Heads.

The first of their Country Men of the Diocese of Garstein who professed publickly the Evangelical Religion, having been last Spring expelled their Habitations by the Soldiers and [omission] & obliged to retire out of their Country, and some of them having accompanied them voluntarily They, Deponents, who did not publickly own themselves to be of that Religion were in hopes that they would have been suffered to remain quietly in their Habitations. But it happened this Spring that 2 Jesuits were sent to them into the Mountains who make a chief Enquiry after all those who were any ways suspected to be Protestants, and asked these Deponents what Religion they were of? And whoever professed himself to be Evangelical was marked down But who owned to be a Catholick was obliged to Corroborate his Confession with an Oath and especially the Articles of the Purgatory and the Adoration of Saints, and that these Articles were true and necessary to Salvation. That they the Deponents did not directly declare themselves to be Lutherans, but Evangelical Catholicks in hopes they should be suffered to remain in their Country But this would not content them, and they were asked what it was they believed? whereupon they and others who had more Talents and Grace replied They believed all our Saviour taught, the Evangelists described, and the Apostles Witnessed. Upon which they were forbid to assemble any more and whoever did not Obey was obliged to pay a Fine of 10, 20, 30 to 100 Florins under pretence of having acted against the Treaty of Westphalia. Moreover they were Required to deliver up their Books of Devotion under Promise that whoever did deliver them up voluntarily should not be fined. And tho' Several of them have parted with them, they were nevertheless Obliged to Pay a Fine; Several of them however had kept their Books. But those that had been given up were carried to Saltzburg and not returned them[.] Except this they had met with no other bad usage nor ever been obliged to go to Mass, on the Contrary they were permitted to Assemble in their Houses where they read and Offered up their Prayers. About Easter it was signified to them for the first time to prepare for their Journey, & the 2nd intimation for that Purpose was given them about a Fortnight before Midsummer and they left to choose whether they would go together thro' Bavaria, or Separately and by single Families to Regensberg [Ratisbon] by Water. In the first Case their Passage thro' Bavaria would cost them 200 Florins and they must go apply to the Bishop of Saltzburg for it by way of Petition.

And being resolved to depart all together, they applyed accordingly to his Grace the Bishop and Obtained A favourable Resolution with leave to depart the Country in 6 Weeks, during which time their Propertys were marked down and the Day of their Departure fixed to the 16 August when it took Place accordingly. On the road thro' Saltzburg they were joyn'd by other Protestants from the Neighbouring Villages and Countrys and a Guard of Soldiers were given them for their Security to escort them. They were not molested at all during their Journey[.] on the Contrary all the Assistance imaginable was given them both in the County of Saltzburg and Bavaria[.] But only their Expence were Considerable when they arrived near the Capital of Saltzburg and where they kept a resting Day they were obliged to pay a Florin per head which amounted to about 230 Florins whereof 200 Florins were kept to defray their Passage and the Maintenance of such of them as were poor. But 30 Florins were Continued to their Elders, who laid out that money for Horsehire to draw their Baggage along to which those who were able always Contributed their quota parts in proportion. They all Say that their Emigration has been very greivous to them by reason they had left them behind them very considerable Effects and that on their Journey they had been frightened with dismal Reports, Viz. that some of those who left the Country before them had been murdered by the Poles, others had been hurried or beheaded others drowned etc. And these Stories having made Impression upon several of them Some thereof were returned so that of 300 heads that had been together but 225 remained constant in their Resolution.

That they who made their Confession and Depositions had never been daunted, but always reposed their Confidence in Almighty God who they were sure would never abandon but bless them, if not in this world tho' in the next.

Copy of the Translation of the Certificate given by the Evangelical Clergy of Augsburg to these Emigrants of Saltzburg that went from thence the 31 October to Settle in Georgia

We the Seniors Ministers and Deacons of the Evangelical Lutheran Churches of Augsburg certify herewith that the Bearers hereof who for the greatest part have been obliged to retire out of the Bishoprick of Saltzburg on Account of their Religion which they professed as true Protestants of the Augsburg Confession[96] and now are going to Settle in Georgia under the protection of Great Britain have been 8 weeks in this City, during which time we have Preached them the Gospel both in their Lodgings and in the Publick Congregations and always found them very desirous of hearing the Word of God and to lead a good Life in Conformity to the Same. There was a perfect friendship and Harmony amongst them and they shewed themselves very grateful towards their Benefactors. They were brought so far in the Knowledge of God and themselves that 26 of them of both Sexes after a Sincere and hearty Confession of their Sins, and having received the absolution of them were admitted on the Day Simon and Jude to the holy Communion which was administered to them conformably to the Institution of our Saviour in our Great Church of St. Anne in a publick Congregation and whereof we hope they were worthy Partakers. And as their [they are] now upon the Point of Learning [leaving] Us, in order to repair under the divine Protection into the Countrey to which they are invited, we thought incumbent upon Us and conformable to Christian Charity to wish them not only all imaginable Blessing and to pray to God to replenish them with his Grace in the new Colony where they are going to etc. but to give them likewise this Certificate concerning the Purity of their Faith and their Lives, humbly and Charitably requesting all those of whatsoever Quality and Condition they are whom these Presents may reach to Shew them all imaginable Benevolence and Assistance in their Journey and to rest assured that Almighty God will not leave unrewarded such Acts of Charity etc. Augsburg the 31st October 1733.

Seniores Pastores, & Diaconi of the Evangelical Church at Augsburg

Copy Translation of an Intimation given from the Pulpitts of the Church of St. Anne concerning the Emigrants that are going to Georgia.

Whereas those Emigrants of Saltzburg that are to settle in New Georgia and that have been here sometime will Set out under the Divine Protection on Friday next and being desirous to receive before their Departure the Holy Communion according to the Institution of our Saviour, Be it known to this Congregation that a due Permission being obtained for that Purpose from the Magistrates these Emigrants after having confessed themselves on Tuesday next are to receive the holy Communion on Wednesday following being the Festival of Simon and Jude in this public Congregation. Wherefore this Congregation is desired to Offer up their Prayers to the Throne of the Almighty to the end that these Emigrants may not only worthily partake of the holy Communion, but that they may also may be blessed by Providence in their Journey as well as in the new Colony where they may Settle thro' the Merits of our Saviour Jesus Christ Amen.

Copy of Translation of the Instruction given by Senior Urlsperger to the Commissary [Ph.] Reck etc.

Instruction to the Commissarys who are to conduct these Emigrants of Saltzburg and others who have been expelled their Country for their Religion, consisting of ‗‗‗‗ Heads and who by Vertue of a Commission granted to the underwritten by the Lords Trustees for the Colony of New Georgia and the Society de propaganda Christi Cognitione, have been received in order to be sent as subjects of Great Britain to New Georgia by the way of England.

1. The Commissary will please to Conduct these Colonists conformably to the Rout given them and to take Care as much as Possible that they may take up their Quarters every night in Evangelical Places.

2. They will take all imaginable care of them on the Journey that they may all keep together and that their Baggage may be always carried along with them.

3. In Case by reason of the badness of the Roads these Colonists when marching on Foot must be Separted from their Baggage One of the Commissarys is always to remain with the Baggage and the other with the People for the Security of both.

4. These Gentlemen are to take Care that the Colonists begin their March early in the Morning in order to reach in good Time the Place when they are to dine and to lay and that they may not be obliged too much in the night time, And to the End that the Quarter for them may be duly prepared. They are

5. To take due Measures that every Day, unless it be a Post Day a Person be sent on purpose before hand to the Place where they are to be Quartered, in order to give due Notice of their Arrival and of the Legitimation of the said Commissarys and to make the Necessary applications for their Reception, As also to bespeak the Quarters and to wait for the Transport at the Gate in order to conduct them to their Quarters, and Since

6. A proper Allowance per Diem has been granted to every one of these Emigrants a Commissarys are to take Care that this allowance may not be exceeded but that their Dinner and Quarters may be regulated with as much Frugality as possible to the End that their money may be saved as much as is feasable.

7. The Commissaries will take Care that these Colonists may not prejudice their Health with Drinking unnecessary Brandy in the Morning, but to Provide Supps for them before they Break up in

the Morning and to the Children a Sort of a P[omission] when they have occasion for it. They will likewise avoid giving them at their Dinner too much Pork meat, but rather Beef and Pannades [panades] to which they have been most used in their own Country.

8. In Case one of them should fall sick on the Roads they are to take advice of Mr. Zwibber [Zwiffler], who goes along with them; But should there be a necessity the Patients sickness increasing to consult the Physician of the Place they are to do it. And in Case the Distemper should come to such a Pass that the Patient should not be able to Walk them, they are to take Care that he may be provided at a reasonable Rate with good Accommodations, or left in an Hospital to the End that the whole Transport may not be stopped in their March.

9. The Commissaries are to keep an exact Diary and to send Us the same from Time to Time that we may see how they Advance in their Journey and what has passed in the divers Places thro' which they passed.

10. Whenever the Colony travels by Water, things are to be disposed, if Possible, in such a manner that the People may be landed every night in order to Sleep on shore and to be provided with warm Quarters and to have in the meanwhile the Ships and Baggage well Guarded.

11. The Emigrants are to perform Mornings and Evenings their Prayers to obtain of God almighty Prosperity in their Journey and to thank Providence for the Protection they have enjoyed hitherto.

12. The Commissaries are to keep a true and exact Account of all the Expenses Ordinary and Extraordinary in order to be able to justify themselves both at London and here. when the Transport arrive at Rotterdam, the Companies [commissaries?] are to take Care that the Baggage be safely transported out of the Ship when [which] they arrive in, into that which is to Carry them to London, and that for more Security some lodge [?] be left on board of the former[.] some body else is to accompany the Baggage to the new Ship where a 3rd Person is to remain to receive the Baggage and to take Care that nothing may be lost, and for the Security of the whole it will be in our Opinion proper that all the Pieces of the said Baggage may be told [counted] at the Derembarkation [disembarcation] as well as the Reimbarkation, where it may appear if any thing is lost. And the Commissary's are to admonish the Colonists to take every one Care of their little

Bundles that they may not be lost and that every One may know where to find it.

13. The Commissary's are to require the Colonists that when they travel by Water they may not expose themselves to any Danger.

14. The Commissaries are to take Particular Care of the Directions that are given them for the several Bankers, where they are to receive their money & to use their endeavours that they may not lose by the several Species of Money.

15. Should anything happen during their March which could not been foreseen, nor Instructions given upon it Accordingly, The Commissaries are to Use their Judgement and in such Case We must[97] to their good Conduct and Fidelity not doubting but they will do always the best for the advantage of the Colonists; But in particular the Commissaries are herewith impowered in Case they meet any Disturbance [?] on the Road coming from Holland, who have an Inclination to go along with them to London and Georgia to receive upon the same Foot and same Conditions as the rest of the Colony and conformably to the printed Scheme which to that End I have given them Copies of.

Done at Augsburg the 31st October 1733.

Samuel Urlsperger

P.S. I have likewise agreed with the said Commissary [Ph. von] Reck that the Expences of his Journey from Regensburg [Ratisbon] hither and those from thence to the Place where he is to conduct the Colonists by order of the Trustees as well as the Expences for his return to Regensburg shall be defrayed and that moreover for his Trouble and Pains he is to enjoy 1 Reichesthaler [rix dollar] or 1 Florin 30 Kreutzers in lieu of board Wages per Diem.

Copy of Translation of the full Power given by Mr. Urlsperger to the Commissary Mr. [Ph.] Reck.

Be it Known, to whom it may Concern that the Lords Trustees Established by his Britannick Majesty for Settling the Colony in New Georgia and the Illustrious Society de propaganda Christi Cognitione have authorized me by vertue of a Commission dated the 12 September 1733 whereof the Words are following (inserantur Literæ Patentæ) to receive 300 of these Emigrants of Saltzburg or others who have been expelled their Country for the Sake of the Protestant Religion, as Subjects of Great Britain to be

Settled in Georgia and to take Care of the Spiritual as well as Temporal Concerns of these New Colonists and having in Conformity to their Commissions received already _____ Heads of these Emigrants who have offered voluntarily to go and Settle in New Georgia and to undertake the Journey & appointed for their further Transportation to that Place the Bearer hereof Mr. Philip George Frederick von Reck who not only has shewed a Charitable Desire to serve these Colonists, but has also been recommended to us from Persons of great Distinction as a very Prudent and dextrous [omission] every where fit for this Commission For which Reason we have appointed him to be Commissary, to the End that he may Conduct and Transport these New Colonists together with their Baggage and Effects from hence to Rotterdam and thence to London, and to take the best Care imaginable of them and to [omission] their Interest as much as possible as well in their March as in their Quarters. Be it therefore known that the said Mr. Von Reck is to be looked upon as a lawfull Commissary to Transport the said Emigrants and that he is to be received as such in the same Manner as If I myself were present to transact their Commission And to that End we humbly intreat all Persons of what Condition and Quality soever to receive the said Mr. Von Reck as a Commissary to Conduct these Colonists and to grant them every where a free Passage with his People to assist them as much as possible in the execution of his Commission, to suffer him to take up Quarters for these Emigrants & to furnish him with all necessaries for which he'll pay at a reasonable Rate ready Money, and in short to give him all the aid and Assistance And to rest assured that the above named Lords Trustees and the illustrious Society de propagandæ Fide will acknowledge such a Favour on all Occasions and that on my [part?] I shall always do the same.

In Witness whereof I have set hereunto my hand and Seal done at Augsburg the 31st October 1733.
Samuel Urlsperger Senior of the Evangelical Ministry and Pastor of the Church of St. Anne at Augsburg, as being commissioned by the British Trustees and the Society de propagande to take care of the Spiritual as well as Temporal Interests of the within mentioned Colonists.
P.S. This Minute I receive a Writing from Regenspurg [Ratisbon] which bears the following Title.

A short Deposition made under Oath of the Reasons why a great many of the Emigrants of Dernberg [Tirnberg] who were

Settled in Holland are returned probably to move several great or lesser Powers who were very much irritated against them to a Compassion for them.

The Post being upon its Departure I am not able to read their Writing. But I rejoyce in my Soul at the hopes that our Georgian Colonists will have a better Lot, and that their Affair will Succeed provided we continue to implore the Divine Providence for its Blessing.

Copy of a Letter from Mr. Urlsperger at Augsburg 9th November 1733. No. 12399. Read 13 November 1733. To Mr. Henry Newman.

Honoured & Dear Sir: I have received Letters from our Commissary Mr. Von Recht [Ph. von Reck], dated at Rothenburg on the River Tauber the 4th November, where he was happyly arrived with his Transport and the Magistrate there had given each Person 30 Kreutzers and lodged them free in Sundry Inns. The 5th I reckon they are arrived at Marck Steft [Marksteft] near Wertheim and imbarked the 6 in order to proceed down the River Mayen [Main] to Franckfurth and so further. Letters from Franckfurth Say that the Tirnbergers arrived there are in a miserable Condition and a great Part of them sick. Therefore I have this Day by the Post wrote to the Commissary not to take any sick Persons in his Transport First that they may not hinder the others in their Journey and Secondly that if some of them should die Suddenly they might cause no ill Report. This was my Advice because I heard that there were so many Tirnbergers in Franckfurth that the Franckfurters might keep them till they were recovered, and that in next Spring one could send a Transport of 60 or 70 Persons, if they are so many and have a mind to go. At last I have received a Letter from Mr. Vat dated the 30 October wherein he tells me that he hath received my Letters of the 15th & 21st in which I told him that the Transport would set out the 30th October with out fail But notwithstanding he writes to me that he would Set out the 31st October[.] I have not seen him Yet, As soon as he comes, I shall accordingly to the Order of the Society send him after that he may be in Rotterdam when they embark. Mr. [Ph.] von Reck hath hitherto administered his Office well. While I am writing this I receive Yours of the 16 October upon which I answer that the Case with Respect to a Minister and a Catechist could not be altered, because I just now have received

Letters from Professor Franck that according to my Desire he had got 2 good faithfull and well qualified Persons[98] who are resolved to go to Georgia one of them as Minister and the other as Catechist and that both were Yesterday gone to Werningerode in order to be Ordained there according to my Instruction sent them, and afterwards to go to Rotterdam. That in this Case I have acted according to the Order of the Society and the Promise grounded upon this order made publickly to the Colonists in the Name of the Society, I have wrote in my former. For their chief objection being this: In Case our Minister should die on the Voyage or in Georgia where shall we get another Catechist? This Objection could not be removed but by answering them thus: The Minister shall have with him a Catechist who hath studied Theologie and shall be ordained likewise to the End that if the Minister should die they presently might have another in his place Which hath been the Chief Motive that induced the People to go thither because Spiritual Things are chiefly what they want. Dear Secretary! You know very well that when the Society hath wrote me their Resolution and I have begun to act accordingly and given my word, I can not go from it. Besides all was so difficult with this Transport that I was obliged to write every Post Day to Ratisbone Saxony, and London. The Charges for this first Transport are very great, But the People having staid 8 weeks here, and being obliged to do many things in order to recommend the Design, which most People would not apprehend, it is no wonder. God will help that all may be provided. For the rest you may be assured that this Transport of 40 Persons hath caused me more Trouble than many 1000 of Emigrants that passed through here last Summer. This I say not to complain in the least but only to move the Society to have more patience with my Infirmity. And am obliged to send these Lines away without taking a Copy thereof. I remain

      Honoured & Dear Sir   Your humble Servant   S. Urlsperger
P.S. Many Writings are published in which the Hollanders are reproached for having treated the Tirnbergers very ill, and not kept their Promise. This makes a great Noise among People and made me act more cautiously with the Colonists to Shew the people that they should [not] be used so in Georgia.

Copy of a Letter from Mr. Urlsperger Augsburg 12 November 1733. To Mr. Henry Newman.

    Honoured & Dear Sir: The Day before Yesterday Mr. Vat arrived here from Bienne and because you have described him several times to me to be a very honest Man and will Stay here for the Orders of the Society, I have taken him into my house for Saving him the Charges he would be at if he should lodge in an Inn. Because the Transport is but small and the Commissary Mr. Von Reck hath hitherto acted according to the Instruction and is provided with good Recommendations and Credit, he did not think it necessary to follow the Transport. It is still reported that a Transport of 800 Emigrants is coming; if this continues I could well Employ him for I find him to be a Person that is very honest and willing to serve poor and miserable People. I am assured that the King of Prussia will take no more Emigrants unless the Nobility in every Circle [district] will pay the Charges of the Transport and take them upon their Estates. If these 800 Emigrants should not come this Winter we shall See how it will be with the Tirnberger Emigrants, as well with those that are already in Franckfurth, as those perhaps which will come thither, for I believe they will all go successively from Holland. Yesterday I received a letter from our Georgian Emigrants that they were imbarked the 6 instant at Marksteft and reckoned to be in Franckfurth the 11th. Whence they may, according to Mr. Vat's Reckoning, be at Rotterdam within 14 days. Hitherto all things hath gone well. The Commissary and the Transport agree very well together. The Candidate of the Ministry [Schumacher] preaches and prays diligently with them. At a Popish place they would [wished to] hinder them from passing through there, But the Commissary behaved himself bravely and got the victory of them. At some Places great Love hath been shewn to the Transport; partly in defraying them and partly in making them presents. At one Place they have been received with the Sounding of Trumpets from the Church Steeple. In other Places the Ministers and School Boys have received them and so forth. It will go well further. Now I long to hear by the next Post how many Tirnbergers will go with the Transport from Franckfurth. The following two Letters wherein the Qualities of the Minister and the Catechist are described, I have Copied for Mr. Ziegenhagen to report them to the Society. I Praise God for having given us so good Men. And the Society will never repent the Charges bestowed upon it. I have copied the Letters whole having no time to make an Extract of them. Mr. Ziegenhagen will Extract the necessary part of them. I have heard nothing yet; whether the English Minister at Vienna hath done any Thing in the Affair known. This Day we had the Sorrowfull News that that

Man and his wife who refused to go to Georgia after they had Promised, have declared themselves Roman Catholicks before the Magistrate to day. But the Man seems to be inconstant in Matters of Religion and the Truths he heard of [from] Us will trouble him enough. So goes all through Examination. I commit You to God's Protection and remain

Dear Sir   Your most humble Servant   S. Urlsperger

Copy of a Letter from Mr. Rodde.   Narva,[99] 6th September 1733. To Mr. Ziegenhagen.

Reverend Sir: Since Your last dated the 15, 26 April 1729 in which You told me that the Society for Propagating Christian Knowledge had chosen me a Member of Correspondence, I have had none of Yours, tho' I have wrote twice to You viz. the 3rd July 1729 inclosed in a Letter to Mr. Ruperti and in the year 1730 with Mr. Lichtenstein Student in Theology and a Letter inclosed to the Society. However I believe You have not quite forgot me because You have been so good as to send me 2 Circular Letters from the Society dated the 30 December 1730 together with a Packet of English Treatises when You was in Hall by the Inspector Mr. Grischow,[100] which I have received and give You many Thanks for the sending of them. And because I know very well that You are full of Business as well of Your own as in the Care you take in propagating the Knowledge of Jesus Christ so I own that I have very Little time to employ in corresponding with Your Friends which I can judge by my small Circumstances and therefore you are excused by me. Yet I must desire an Answer from you to let me know 1. Whether the Society hath received my Letters dated 1729 and 22nd October 1730 by Mr. Lichtenstein the 21st June 1731 by William Astell Esquire and the 29 October 1732 by a Merchant from here, and how the Society hath taken them. 2. Whether nothing hath been amiss or wrong in the Title to the End I may take better Care for the future. Especially I desire you to recommend that Business to the Society where of I have made mention to you in my Letter 21 June 1731 concerning the Printing of John Arnd's true Christianity[101] in our Country Language which would be a great Blessing in this Nation, & at present is the best Opportunity to perform the same. For a Studiosus Theologie a native Russian who hath been some Years in Halle a Person that fears God [omission] and hath imployed his Time there and translated the said Book whole into the Russian Language and because he stays there till next Easter he is willing to undertake the Correction and only the Charges of Printing are wanted. For in this Country nothing can be gathered neither by a Collection nor Subscription. *Sat Sapienti* If now the Society would be pleased to be appoint a certain Sum for the Printing of the said Book, which I hope they will not refuse. For Great Britain hath Yearly great Commerce with this Empire, therefore it is reasonable that they also take Care of their Souls and Contribute something for the Propagation of the Knowledge of Jesus Christ. There is a certain Number of Russian Letters in Halle in the Orphan House, which are come from England, For the late Secretary Ludolph[102] hath given them to the Orphan House 30 Years ago, in view that something could be printed for the Benefit of this Country. I have made a Trial with them by printing there Professor Franks: *Beginning of Christian Doctrine* which is to be found in his Book called: *Witness of the Word Work and Service of God* to which are annexed some of our Hymns translated into the Religion [Russian?] Language and to be Sung in the same Tune as we Sing them, and at the end of it is a prayer out of John Arnd's Garden of Paradise to wit the 9th of the Psalms of Praise of the Love and Blessings of our Lord Jesus Christ. This Small Treatise hath already caused great Blessings among this Nation, and what could we not expect of the true Christianity?[103] But because this Nation will not bestow much upon Books, and the Poor are not able to do it they must be given them; and this Book containing no Controversies, it will have no Contradiction as I have tried it already with a few Sheets in writing which were very well received. Therefore I hope the Society will Shew their Bounty in this Point as they have done many others for the Spreading out of the Kingdom of Christ. I hope also that Professor [G. A.] Franck and Mr. Freylinghausen[104] will write about it to the Society and you, and that You will Contribute to this Affair all what is possible. And God Almighty will order all for the Honour of his holy Name. It seems as if God makes preparations for the Spreading out his Name in the most remote Parts of this Empire. For Mr. Wise hath this year been called to Catharinenburgh in Siberia by a Congregation which mostly consists in German Miners, and another Lutheran Minister namely Millies is gone with 500 Persons, most Protestants, of all sorts of Handicraft to *Kemtschatke* [Kamchatka] a new found Country beyond Siberia, which is properly the Northern part of the Kingdom *Japan* and the Country *Jedso* or *Jesso* in order to Establish there a Colony. This Mr. Millies was called by the General Venediger to go to the conquered Places in Persia, but this was hindered by the Death of the said General and the good hopes are lost viz. that they might have brought the Arabian New Testament and Psalms among the Inhabitants in Persia and given us the best Account of their Reception. Of this we hope to have an Account next Year of the German Officers when they return hither from Persia. It is certain that the Copies have been distributed among the Inhabitants, some especially their Clergy have contradicted them, others have taken them and read them. It is enough that we have given them the word of God in their Hands and now we will let him take Care whose Word it is and he will not Let his word be without Fruit but in a proper Time give Us a rich Harvest of this Seed. I commit You to God's Protection[.] Mr. Lichtenstein gives his Service to You, he is in the School in Crohnstadt[105] and is married. I remain

Reverend Sir   Your most humble Servant   Caspar Mattias Rodde
P.S. This Opportunity being in Hast I have no Time to write to Mr. Newman, therefore I beg him excuse me and to communicate to him out of this, what You think proper.

Copy of a Letter from Mr. Vat Augsburg 12 November 1733. No. 12403. Read 23 November 1733. To Mr. Henry Newman.

Dear Sir: The last Letter I had the Honour of Writing to You was dated at Bienne the 30th October And the next Day having Set out thence and gone through Solothurn, Olten, Arau [Aarau], Melingen, Zurich Winterthur St. Gall, Roschach, Lindau Memmingen and Mindelheim, I came safely and in good health, God be praised, to this Town, on Tuesday last the 10th instant in the Morning and immediately waited on the Reverend Mr. Urlsperger of [from] whom I learned that Mr. von Recht [Ph. von Reck] with about 40 Emigrants Saltzburgers went hence the 31st of October in order to go to Holland and England. As it was not in my Power to get hither before their Setting out Since Mr. Urlsperger's Notice of the 21st of October did not come to my Hands till the 26th I hope my being here now will be of no Detriment to the undertaking but that another Transport may be made by the Time I may have the Societies Commands upon this Letter and that of the Reverend Mr. Urlsperger to which I wholly referr. Mr. Urlsperger hath been so kind as to give me an Apartment in his own house and to admit me to his Table and This in order to be nearer at Hand upon any Occasion and likewise to save Expences, I am,

Dear Sir   Your most Obedient & humble Servant   John Vat

Copy of a Letter from Mr. Lowther at Rotterdam 13 November N.S. 1733. No. 12404. Read 13 November 1733. To Mr. Henry Newman.

Sir: Your Favour of the 26th ultimate O. S. I received last tuesday and after Considering the Contents, I waited upon the Burger Masters to acquaint them therewith to bring such a Number of People into the Town without their Consent, was what I could not venture to do, when the Saltzburg Emigrants arrive in our River they will have orders to lay a Day or two between Dort and this Place or at least so long, till our Magistrates be Satisfyed that they are in health and then if the ship be not arrived they will give their orders for their proper reception. I shewed them your Letter to assure them they should not be any Expence to the City that Mr. Vat, who had the Direction of them was Supplied with Credit to Defray all their Expences here, they told me they saw that very well, but the health of their City must be taken Care of.

If I am rightly informed, Your ship will be here long before them if they set out at the time You mention, in all Probability it will be a month before they can be here, if so, I hope they will not have any occasion for Quarters here[.] whenever they come I shall See Mr. Vat and If I can be any way Serviceable to them I shall with Pleasure do it. I shall ever be glad of any Opportunity of Serving the Society and shall be proud whenever they have any Commands on this side the Water to receive them[.] Pray Sir present my humble Service to all the Worthy Gentlemen of the Society and assure them of my prayers for a Blessing upon all their good and pious undertakings. I am with great Respect

Sir   Your most humble Servant   Richard Lowther

P.S. Sir I desire you will pay my best Respects to Sir John and Mr. Phillips with my hearty Thanks for all favours when in London   Your   H. S.   R. L.

Copy of Translation of a Letter from Mr. [Ph.] De Reck. Franckfurt 15 November 1733. No. 12415. Read 20 November 1733. To Mr. Henry Newman.

Sir: In Pursuance of the Orders I have received from Mr. Urlsperger authorized by the Society for Promoting Christian Knowl-

edge to receive and Transport 300 Saltzburg Emigrants to Georgia a
Province in Carolina, I sat out last day of last month for London
with 37 Saltzburgers in the Quality of an English Commissary.
Having reached Franckfurt the 13th Current I have the honour
of acquainting You with it. Our Transport altho' as yet small I
have reason to hope will be encreased in Holland, where the
Cassan Duremburgers [Cadzand Tirnbergers] unsatisfied by the
Usage they receive from the Hollanders watch all Opportunities
to Escape and return the same way they came. There are here
some returned already 80 but as they are most of them afflicted
with some Contagious Distemper they might perhaps infect our
sound People I doubt not but that our Little Transport will be
followed by another more considerable For the same Day that
I left Ratisbone to go to Augsburg, I saw at my Uncle's [J. von
Reck's][106] a request of 800 Saltzburgers addressed to the Evangelick
Body for their intecession for leave to depart their Country, Besides these there are near 20000 men in the lower Austria desiring
the same Grant. But notwithstanding the intercession of the
Evangelick Body's Sollicitations for these Poor People and the remonstrances of almost all the Evangelick Princes, The Emperor
would not so much as vouchsafe to answer to 'em, so that the time
the Crisis of present affairs and the Catholicks themselves will give
perhaps the opportunity of saving so many souls who else would
perish in Popish Prisons. They have respected the very name of
the English and I have gone through many Catholick Provinces
without *Litt. Requis* [*litterae requisitoriae*] which otherwise would
have been most necessary on these occasions.

A Candidate in Theology [Schumacher] a man of Probity by the
Direction of Mr. Urlsperger accompanies me to Rotterdam where
he will be relieved by another Preacher and Catechist who will go
quite to Georgia[.] I Embarkt at Mark Steft [Marksteft] a place
situate in the Markgravat [Margravate] of Anspack Baraith [Ans-
bach Bayreuth]. I have prepared a Vessel there which will by the
Grace of God Carry us over the Main the Rhine and the Moselle[107]
to Rotterdam. I shall be glad to find there an English Vessel and
Captain to Transport Us. The Hollanders being so little favourable to the Saltzburgers, and more especially to a Transport which
in all Likelihood will occasion some Deserters from the Durem-
bergers [Tirnbergers] Mr. Urlsperger gave me at Augsburg 500
Florins but as I have at present not above 150 left, I have taken
up of Mr. Münch Banker and a Relation of mine 200 and have
received Letters of Credit to Serve me in Case forced. I Conclude

& wait your orders at Rotterdam where I shall be please God the 23 Current and have the Honour of being

Sir   Your most Obedient humble Servant
Ph. Geo. Fred. De Reck

P.S. Of the Reception in divers places and of our Voyage I will inform You by word of mouth.

Copy of a Letter from Mr. Ortmann at Rotterdam the 11 November O.S. 1733. No. 12416. Read 20 November 1733. To Mr. H. Newman.

This Day at 3 o'Clock in the afternoon we are arrived safe at Rotterdam thank God; I have been informed by Mr. Lowther that the Saltzburgers are Still upon their Voyage & are expected in a few Days, whatever happens I shall do my Duty. I hope God will bless my Zeal in seeking all Opportunity to Serve you as I desire. I shall think my self a great gainer which You will understand further by my Actions which will testify more than by words. I return your Honoured for Your Goodness most humble Thanks and remain

Your most Dutifull obedient humble Servant
Christopher Ortmann

P.S. My humble Service to the Reverend Dr. Guerdes and Mr. Ziegenhagen.

Copy of the Translation of a Relation about the first Transport of Saltzburgers to Georgia. Enclosed in Mr. Urlsperger's Letter of the 12th November 1733. N.S.

Relation

To the Lords Commissioners or Trustees for the New Colony in Georgia in South Carolina in America and to the Society for Promoting Christian Knowledge concerning the Departure of the first Transport of Saltzburgers.

The Georgian Colonists and Subjects of Great Britain having Been maintained upon the Trustees's and the Society's Charges from the 3rd September to the 31st October first 9 and 10 Kreutzers and afterwards 13 Kreutzers for each person. They at length, according to a Special Order that, be there never so few, they should be sent away, set out the 31st October[.] And that both the Trustees and the Society may know how they have been kept here and at Last dismissed, I shall besides them [those] many Particulars mentioned already in my former Letters, give here an Account of what has further happened.

The Chief Matters which were to be observed in respect to these People were Spiritual Things and in order to Obtain these All Ways and Means have been provided for them. I gave to all those that could read, Bibles, Hymn Books, Catechisms and John Arnd's true Christianity, which I had still in my Emigrant Magazine [stores], besides the Small Treatises. They come very diligently to Church on Sundays, Holydays and work Days to hear the Sermons preached in the afternoon on the Catechism, in which they always were Spoke to and teached in particular[.] When they did not come to Church one of the Ministry went to them and discoursed to them a Text out of the Bible of which I made the beginning upon Joshua 1. Every Day one Hour they were taught the principles of the Protestant Religion by an ordained Minister, and a Schoolmaster was ordered to teach their Children the Alphabet etc. the same did those Parents that could read. About 10 or 12 Days before 26 persons of them received the Sacrament[.] I desired [requested] 3 Candidates of the Ministry that each of them should spend one hour every Day to prepare these People for the worthy receiving of the Lord's Supper which they willingly did. The same did my Colleague to St. Ann's parish and 2 Days before the Communion he held a general Examination with them and, I likewise ordered my Discourse with them to that Purpose and catechised them and prayed with them when I visited them or as it often happened they came to me in my House[.] The Day before the Communion I ordered them to come to our Church where one part of them confessed to me and the other part to my Deacon [Hildebrand] Simply heartily and devout their Penitence and Desire for God's mercy and earnestly promised that they would keep the Evangelick Doctrine and never go from it but lead a good Christian Life upon which we announced God's Mercy in Jesus Christ and the remission of Sins. The same Evening my Deacon went to them and prayed with them and Explained to them our Agency at the Communion to the End that when the next Day it was read to them they might understand it the better. On the Day of their Communion which was on a Friday viz. the Day of St. Simon and Jude on which we have no publick Communion in our Churches here, but our Colonists only had Permission of our Protestant Mission to have a Communion[.] they were Conducted into the Church and placed in the Pews about the Altar. I preached the Sermon upon the Gospel on that Day out of St. John XV v. 17. *These Things I command You* that Ye *love one another.* Comming into the Pulpit I used the words out of Isaiah LXV v. 8 *Destroy it not for there is a Blessing in it* and made the application upon our small Georgian Flock[.] in the Introduction I shewed how the Disciples and the Believers in the time of Christ were Considered and that in regard to them one might also Say *destroy it not for there is a Blessing in it* and that Christ in his last discourse had chiefly recommended that this Blessing should not be destroyed. The Proposition was: The exercising of Love among Believers as one of the chief means not to be destroyed with the Blessing received. Hereby I shewed

1. That those Persons to whom Christ had Spoke and speaks still these words are Believers, that is Blessed, and in what their Blessing consists.

2. That Christ is [in?] the Words of the Text, speaks of the Brother of Love[108] as a Duty proper to Believers and Blessed in the Lord.

3. In what the Brother Love consists, or how it is exercised.

4. Which is the Ground of exercising of Brother Love.

5. How necessary it is and

6. That it is a way not to be destroyed with his Blessing. To each of these Parts was joyned an Examination for the Colonists; In the application I chiefly aimed at them, and the Conclusion was an Exhortation for the worthy receiving the Sacrament. During the Communion was Sung: *Lord God we praise thee* and the Organs and other Instruments played very devout & edifying the whole Congregation almost being in the Church[.] But no Body except We two Ministers of St. Ann's parish received the Sacrament. after the Divine Service they went home and during their Dinner time they had a Candidate of the Ministry with them, who entertained them with good discourses[.] after dinner they came to Church again to hear a Sermon which was likewise for their Edification. After this Sermon another Candidate went to them, who Sung and prayed with them and held a discourse of the Death of Christ and the Blessings thereof.

The next Day a Couple of the Georgian Colonists[109] was married in our St. Ann's Church in the presence of all their Country People[110] from Saltzburgh that are going to Georgia and the Burgomaster Morell and two others. They had a sober wedding Day with an edifying Discourse which was Concluded with the Hymn: *My Soul exalt the Lord thy God,* and with prayer.

They have likewise been provided with all manner of necessaries for Life. They have had no want for Sufficient and wholesome

Victuals and Drink. When any of them was sick, I sent a Physician to him to have him cured. Some of them being not well and some very bad cloathed, I have provided them with necessary Clothing to Cover them with against the Cold as well in the Day as in the Night time.

As soon as the Day for their Departure viz. the 31st October was fixed, I sent for the Commissary *George*[111] *Philip Frederick* von Reck from Ratisbonne hither. After his arrival I made him acquaint with the Burgomaster Morell, who is constantly Deputy of the Protestant Magistrate for the Emigrants, and afterwards I prescribed him to the Colonists, who rejoyced for him. He brought a Passport with him for his Colonists from the Electoral Brunswick Embassy; however I thought it necessary to desire also a passport of the Privy Council of the Augsburg Confession for these people that had been here 8 weeks which they likewise have obtained. And I as Attorney of the Trustees and the Society, delivered to the Commissary the Letter of Attorney, the Instruction and the other Orders under my name and Seal likewise I delivered to him 500 Guilders ready money, and procured him by a Banker Mr. von Münch Direction and Credit at *Franckfort Cologn* and *Nimwegue*. As to the Minister [Bolzius] and Catechist [Gronau] they both or at least one of them will go from Saxony to meet the Transport at Rotterdam[.] I have sent them the Vocation and Instruction, but expect [am waiting] as Yet to know their names. However I have given the Commission of this Affair to such good Theologians and Pastors in Halle and so well known to me that there is no Doubt of getting one or two good persons. From hence to Rotterdam, I have sent with them a very good Candidate of the Ministry at Ulm Mr. *Shomaker* [Schumacher] whom I know, and have presented him to the Colonists and delivered to him the Attestation and recommendation given by the whole Evangelick Ministry in a solemn form to the Colonists, because 1. it was promised them 2. that they should not fear to be left without a Minister as some would falsely persuade them.

When the 31st October approached the Burgomaster Morell, who hath been indefatigable to serve these Colonists early and late, and I together with the Commissary [Ph.] Von Reck went to the place where the Citizens shoot at the mark to get every thing in good order. Their Baggage was loaded upon a Waggon and bargained for by the hundred weight to be carried to the Ship lying at Mark Steft [Marksteft] not far from Wertheim where they came upon the River Mayne [Main]. The People went in two Waggons covered over that they and especially the small Children might be secure from the great fog, Wind and Rain, and not fall Sick on their great Journey. For the Commissary and the Candidate we hired a Coach to Mark Steft, to which place the Commissary sent a Letter before hand, according to the direction given him by the Holland's Envoy to a Banquier there that a good Ship might be ordered for 41 Persons to be in readiness[.] before they departed I held a short discourse of Exhortation and gave the whole transport the blessing which the Candidate Mr. Shomaker did likewise in a short Prayer. In the Presence of two Members of the Magistracy of both Religions the Colonists were counted Head by Head that the Catholick part might See that none of them was kept here as they mistrusted. The Burgo-Master Morell hath kept a very accurate Protocol in every point of which I have four Copies. One I have given to the Commissary, the Second is for the Trustees and the Society, the third remains here, and the fourth I will send to the Electoral Brunswick Embassy. Lastly I must mention that the Colonists have received much Good here. The Ministry gave each person one Guilder of that money which they have had in their Hands since a year ago. Some rich People gave each person half a Guilder, others gave them Linnen and other Cloaths. At last was distributed among them the Money locked up in a Box, given from time to time by the Protestant Augsburgers that visited them, 173 Guilders, likewise out of the Emigrant Cash at Ratisbone 142 Guilders. It is to know further that a Family of 5 Persons [Rieser] is remained here because a Boy hath broke his Leg[.] as soon as he is cured they will come after. I have several other Things to mention but the Time will not admit of it. The Lord be with them by Water and by Land.

Copy of a Letter from Mr. Urlsperger Augsburg the 16 November 1733. No. 12417. Read 20 November 1733. To Mr. H. Newman.

Dear Sir: Yours of the 23 of October I have received this Moment when the Post went away. Hereby I send the whole protocol. Mr. Vat gives me great Contentment. Next Thursday more[.] I remain Yours in great hast.

Yours etc.   S. Urlsperger
Augsburg 4 September 1733.

Whereas the Reverend Mr. Urlsperger, Senior of the Evangelical Ministry and Pastor of St. Anns here in the Imperial City of

Augsburg and Corresponding Member of the Society for Promoting Christian Knowledge as having full Power of his Majesty's of Great Britain Trustees or Lords Commissioners for the new Colony in Georgia and of the Society of Propagating Christian Knowledge to take & receive 300 Persons of the Saltzburgers and other Emigrants that are driven out of their Country for the Sake of the Gospel, hath given Notice to the Underwritten Deputy of the Magistrate of the Augsburg Confession for the Saltzburger Emigrants that of those three Hundred Saltzburger Emigrants, which came the first of this month through Landsperg in Bavaria to the Imperial City Memmingen about 30 Persons were resolved upon the Proposals made them to go to Georgia and were come hither Yesterday and for Reasons well known lodged in Mr. John Caspar Schaur's Garden, a Citizen and Distiller here. Wherefore He the Reverend Senior desiring me to hear the said Saltzburg Emigrants about their Resolution taken on this account and to set down the names of those that would be Constant in their Resolution to the End that some provisions might be made for the Necessaries of their Souls and Bodies for their further Transportation as also to be sure of them that the Charges bestowed upon them might be not be in vain; I according to this reasonable desire went this Day in the said Schauer's Garden and proposed to them all and to everyone of them the Christian good and Charitable Intention of his Majesty's of Great Britain Lords Commissioners for the new Colony in Georgia and of the Society for Propagating Christian Knowledge towards them the Emigrants, in sending them to Georgia and the Conditions offered them with all their Circumstances and gave them likewise a Sufficient Account of the Nature of Georgia; and though they seemed to be very well Satisfied with all, Yet they might consider of it the better time was given them till next Day.

The 5th September 1733

According to the Yesterdays Resolution upon the Reverend Senior Urlsperger's desire above mentioned the Saltzburger Emigrants, lodged in Schauer's Garden, were this day again examined separately after they had had time to consider whether they were now still of that Resolution they had declared Yesterday that upon the Conditions offered them they would go to Georgia as Colonists and Subjects of his Majesty the King of Great Britain upon which they all and every one of them gave Thanks for the Offer and for the Love shown them & declared free & willingly that after having well considered of it they were resolved upon the Conditions offered them to be Stiled Colonists and Subjects of H. M. K. of great Britain and in the name of God to go to Georgia desiring at the same time that in Case their Friends and Relations that were Still in the Country of Saltzburgh should in Time like them be driven out from thence for the Sake of the Gospel they might be sent thither likewise. Which was promised them with the Condition if they would Willingly resolve to follow them.

The Saltzburg Emigrants that are Resolved to go to Georgia are these following.

1. Hans Grueber born in Gastein in Hoff[112] a Single Man about 45 Years of Age, a Servant [laborer].
2. Paulus Schwaighofer, a Weaver about 43 Years of Age born in Mietosel [Mittersill] on the Mill Brook [Mühlbach] in the Valley of Pantzgau [Pintzgau] & his wife.
3. Margaretta Prindlinger with her 3 Children.
4. Maria Seven years.
5. Thomas 4 Years And
6. Ursula 1 year old.
7. Lorentz Hueber born in Gastein in Hoff 54 Years of Age & his wife.
8. Maria Mändeliter with her 4 Children.
9. Magdalena in her 13th Year.
10. Johannes 10 Years.
11. Maria 8 Years and
12. Margaretta 5 Years of Age.
13. Christian Steiner a Miner, unmarried born, in Hoff in Gastein 30 Years of Age.
14. Balthazar Fleiss a Miner unmarried born in Godauner [Gadaunern] in Gastein 27 Years of Age.
15. Jerg Schwaigger, unmarried a Servant 19 Years of Age born in Gastein near Prembstall[113]
16. Thomas Gehwandt [Geschwandel] a Miner 38 Years of Age born in Hoff in Gastein and his wife.
17. 1 Margaretta Hofer 23 Years of Age with one Child.
18. Margaretta 1 Year old.
19. Anna Hofer Sister to the aforesaid Margaretta Hofer unmarried born in Hoff in Gastein 25 Years of Age.
20. Tobias Lackner upper Servant[114] in Coal Mines unmarried born in Hoff in Gastein 40 Years old.

21. Johannes Mashemer [Mosshamer], Servant unmarried born in the County Zeller in Lainthal[115] 34 Years of Age.

22. Barbara Rohrmoser wife to Peter Kröers[116] who is still in the Country of Saltzburgh with small Children, born in Oberkelback[117] in the County of Saalfeld[en] 36 Years of age.

23. Gertraut Kröer, unmarried daughter to the aforesaid Rohrmoser born in Stockham upon the Heeth[118] in the County of Saalfeld[en] 14 Years of Age.

24. Catharina Kröer, unmarried likewise Daughter to the abovementioned Rohrmoser born in Stockham upon heeth in the County of Saalfeld[en] 17 Years of Age.

25. Maria Kröer unmarried Step Daughter to the said Rohrmoser born in Stockham upon Heeth aged 28 Years. N.B. now married to Johannes Mashemer [Mosshamer].[119]

26. George Beucher [Buecher] Digger in the Saltpeter Mines and Peasant born in the Jurisdiction of Lichtenberg in the County of Saalfeld[en] aged 33 years and his Wife.[120]

27. Ursula Madreiter, 37 Years of Age with Six Children viz.

28. Regina 14 Years.

29. Barbara 11 Years.

30. Simon 9 Years.

31. Matthias 5 Years.

32. Johannes 3 years.

33. Catharine 1/4 of a year.

34. Margaret [Martin] Hertzog Miller[121] unmarried, born in Pintzgau in Schriecking [Schrecking] 35 Years of age.

35. Maria Reiter unmarried Servant maid to the abovementioned Buecher born in Tyrol at St. Ulrich,[122] in her younger Days come to Saltzburg 21 Years old.

36. Christian Leinberger, unmarried born in Loigam [Leogang] in the County of Lichtenberg Servant to the abovementioned Buecher 23 years of Age.

37. Mattheus Mietersteiner, Servant unmarried born in Marckt in Goldegg [Goldeck], 41 Years of Age.

Since now these Emigrants as abovementioned have desired to send their Friends and Relations after them in case they should likewise become Emigrants and hearing that they already for that purpose have been Listed and set down in the Arch Bishoprick of Saltzburg and these Colonists having left several Goods and Things behind them as we have thought proper to Specify it as follows:

Hans Grueber No. 1 above hath left his Father and Mother, namely Wilhelm Grueber and Anna Liechner in Laickau [Luggau] in Gastein likewise 4 Sisters and 1 Brother namely Margaretta, Magdalena Maria Anna and Michael Grueber who perhaps will follow him. N.B. especially he desires that we may endeavour and take Care that his 2 Children namely Johannes aged 10 Years and Peter Grueber 13 Years may be let free and sent after him. He hath left behind him of his own 220 Guilders with Michael Grueber at Dorff[123] in Gastein, whose Estate is called on the Staller 100 Guilders and with his Father 100 Guilders and with Balthazar Schafflinger a Peasant in Unterlarasteig[124] 20 Guilders.

Paulus Schaghofer [Schwaighofer] No. 2 above expects his Brother in Law Viet Prindlinger, likewise his Friend Frantz Roffmayes [Rossmeyer?] with 3 Children. His wife expects her Father Martin Prindlinger and her Sister Mary Prindlinger who is married to Gregori Graffmiller. He hath left 100 Guilders with Rupert Jackberger Röstermaster[125] upon the mühlback[126] in the Jurisdiction Mietersiller [Mittersill], to whom he hath sold his house and owes him these 100 Guilders remaining.

Lorentz Hueber No. 7 . . . . . expects his Brother's Paul Huebers (who is gone with the great Transport to Memmingen) Son and Daughters namely Paul Hueber, Barbara, Magdalena Margaretta and Christina Hueber. He hath left nothing behind him except his Household Goods.

Christian Steiner No. 13 hopes his Mother Ursula Steiner, his Brother Rupert and his Sisters Gertrude Anna and Sibella Steiner will follow him[.] he hath left no Goods behind him.

Balthazar Fleiss No. 14 hath his Father alive namely Christopher Fleiss a Miner in Gastein but cannot expect him because of his Great age but he expects three Sisters Namely Ursula, Gertude and Christian. He hath left behind him 20 Guilders which he hath lent to the Baker in Hoff in Gastein.

Jerg Schwaigger No. 15, whose Mother remained in the Great Transport in Memmingen[.] he expects no Body, hath left behind him 50 Guilders with his Trustee namely Jerg Schwaigger, Baker in Hoff in Gastein.

Thomas Gehwandt [Geschwandel][127] No. 16 expects his Cousin Hans Schock the younger 24 Years old and his wife expects her Sister Maria Trigler, 30 years old unmarried. But doubts whether they will let her have her Bastard with her. NB. This Gehwandt desires if possible that his Bastard Child,[128] Thomas Gehwandt, gotten by [omission] Heiglin and Baptized in Hoff in Gastein, which because they would not let him take it with him he hath left with Michael Wallner in Hinterstorff in Gastein and is 11 years old, may be sent after him. The said Gehwandt hath left behind him 400 Guilders which he hath to demand according to a Writing dated the 14 August 1733 of his Estate called Nieder-berg in Bodauner in Hoff now in the Possession of Hans Kohler in Bodauner.[129] he hath also left to Hans Kohler all Moveables, among which are 1 Cow, 12 Sheep, 1 Hog, 1 year old [horse?] together with several Foals which he desires likewise to be Sold according to the Custom of the Country and the Money Sent him. Likewise he hath to demand of his Brother Blasius Gehwandt living upon an Estate called Muhlback near Pabrugg[130] as an In-heritance after his Father and Mother according to a Note dated the 15 August 1733 100 Guilders.

Tobias Lackner No. 20 expects his Brother George Lackner living near Faschenberg[131] married, likewise his Sister Christina Lackner married to Hans Riste[132] Miner and her 2 Sons Namely Michael and Johannes.

He hath left in his House and Garden which Mattheus Frey-berger hath in Possession about 10 Guilders and with his afore-mentioned Brother George 40 Guilders in Money let him as also 20 Guilders in Turner Tools.

Johannes Masshamer [Mosshamer] No. 21 expects no Body un-less his Brother in Law Martin Thum, living in Haram [Harham] in the Jurisdiction of Lichtenberg and his own Brother Matthias Masshamer living in Lainthal in the Jurisdiction of Zeller, would Change their mind and embrace the Protestant Religion. He hath left 60 Guilders with his Brother in Law and 20 Guilders with his own Brother and expects to Inherit his Cousin Rupert Piberger in Linthal [Lainthal] in the Jurisdiction of Zeller worth about 4000 Guilders.

Barbara Rohrmoser No. 22 Expects that her 2 Sisters namely Gertrude married to Andrew Eps a Peasant in Unterholwegen[133] and Eva Rohrmoser married to Peter Hirschbichler at Grünwald in Margehn will follow her. she hath left with her Husband Peter Kröer at Stockham upon the Heeth in the Jurisdiction of Saalfeld, her Portion brought to him, according to a Note which says she hath by her 300 Guilders.

Maria Kröer, No. 25 hath left behind her according to a Note 100 Guilders and 26 1/2 Kreutzers.

George Beucher [Buecher] No. 26 expects his Brother Martin Beucher Peasant at Maisshofen[134] in the Jurisdiction of Zeller and his Sister Maria Beucher a Baker's wife in Loigam. He hath to Demand in his Country according to Writings which he hath with him about 752 Guilders 16 Kreutzers.

NB. This Peasant hath afterwards abjured his Resolution being Persuaded by Ignorant People and because he was not to be brought to other Thoughts we have dismissed him.

Martin Hertzog No. 34 whose Brothers and Sisters are resolved to Stay in the Country of Saltzburgh. He hath to demand George Buecher at Mietergrissen [Mittergriessen] in Loigam [Leogang] according to Balance produced 115 Guilders 26 Kreutzers.

Maria Reiter No. 35 expects no Body, she hath to demand Wages and Cloaths of her Master George Beucher 3 Guilders worth as also hath the following Christian Leimberger.

Christian Leimberger No. 36 expects his Brother Andreas Leimberger, unmarried a Servant in the Jurisdiction of Lichtenberg 26 Years of Age. He hath to demand 1/2 a Year Wages 5 Guilders & 30 Kreutzers, of his Masters Estate which now Hans Stockel is Possessor of.

Mattheus Mietersteiner No. 37 expects no Body[.] he hath to demand two Years Wages of Martin Schwaigger Peasant on the Lehen at Goldegg [Goldeck] who is gone to Prussia but his Sister Maria Litzner who lives upon the Estate hath perhaps taken this Debt upon her to pay.

The 24 September 1733

Yesterday in the Evening 2 Saltzburgh Emigrants coming directly from their Country hither with the Intention to go with the aforementioned to Georgia they have been heard about it this Day, who after the Christian Intention of the Lords Commissioners and the Society of Propagating Christian Knowledge as also the Nature of Georgia, together with the Conditions upon which they were to go thither, was told them by word of Mouth and likewise communicated them by which is now in print declared free and Willingly after they had considered of it to go also in the name of God thither, wherefore they have also been set down in the Protocol, viz.

38. Peter Grueber, a Servant born in Gastein at a Place called Lehn Grube[135] unmarried 36 Years of Age his Father Michael Grueber and his Mother Magdalena Amoser are Still in Saltzburgh. He hath to demand of his Brother Hans Grueber who lives upon the Saltzburgh Fathers Estate 100 Guilders for Wages and of his other Brother Michael Grueber 50 Guilders of

which he hath received already 8 Guilders. He expects his Brother
Hans Grueber likewise his Brother Thomas Grueber.
39. Simon Reiter, a Servant [laborer] born in Gastein unmarried
aged about 26 Years[.] he expects his Father Michael Reiter Wood-
cutter and his Brother John Reiter a Servant and 6 Sisters, he
hath nothing [to] demand in his Country.

The 9th October 1733

40. Stephen Hirdelsperger who presented himself this Day, a
Servant unmarried about 22 Years of Age born in Saalfeld[en] in
the Jurisdiction of Lichtenberg[.] he came hither as an Emigrant
with a great Transport 1 Year and half ago and since that time
lived with Mr. Münch Banker as Servant declaring freely that he
was resolved to go with the Saltzburgers to Georgia desiring to Set
his Name among others, which was granted him. He expects his
Mothers Sister Ursula Höhle, he hath to demand 12 Guilders of
Hans Hammerschmidt who is gone to Prussia as an Emigrant.
41. George Bartholomeus Roth, born in Wirtzburgh [Würtzburg]
and his Wife,[136]
42. Maria Barbara Oswald likewise born in Wirtzburgh, who
both presented themselves this Same Day as Converts with Recommendations from Ratisbonne, declaring they had resolved to go
with the Listed Saltzburgers to Georgia and the Reverend Mr.
Urlsperger recommended them likewise with respect to the said
Recommendation their Request hath been granted. The said
Rothe hath 2 Children in Wurtzburg and hath given himself great
Trouble to get them from there but in vain because he hath
embraced the Protestant Religion wherefore he desires very much
to Petition the Bishop of Wurtzburg to let him have his Children
be sent after him.

The 19th October 1733

The following Emigrants came from Memmingen viz.
43. Bartholomeus Riste [Rieser][137] a Miner, born in Gastein at a
    Place called Ristehauss about 43 Years of age and his wife.
44. Maria Zugseisen and 3 Children namely
45. Michael 12 Years of age.
46. Balthazar 9 Years.
47. George 7 Years, who upon his Journey hath before broke his
    Leg. Both Parents declare that they are resolved to go to
    Georgia with the Colonists, wherefore they have been Listed.
    The said Riste hath bought his Estate according to a Writ-
    ing dated the 13th June 1709 for 168 Guilders.

Likewise a piece on the Lucauer Mountains and an House and Garden and a Small Place on the Lucau [Lugau] on the Top for 150 Guilders and all household Goods at least worth 50 Guilders all which he hath to demand. His Wife expects her Mother Barbara Webmanig and a Sister Margaretta Zugseisen and Christiana Zeigseisen likewise her Sister and Magdalena Zugseisen likewise her 2 Brothers Hans Zugseisen and Nicholaus Keyl her Step Brother.

The same Day

48. Leonard Rouner [Rauner] born in Hirnstein one Hours ride from Ulm, aged 27 Years arriving here Yesterday and shewing a great Desire to go with the Georgian Colonists to Georgia and the Reverend Senior approving of the Listing tham [them] he hath been listed this Day in the Protocol together with the others after a Second Declaration. He hath his Father and Mother alive Namely Hans Rauner and Barbara Schmeltzer and a Brother George Rauner in Hirnstein.

The 30 October 1733

49. Matthew Braumberger from Aichach in the Electoral Bavaria a Millers Servant unmarried aged 30 Years who hitherto hath worked here in Augsburg in the looking Glass Fabrick and formerly hath learned the Miller's Trade upon the Creutz Mill, presenting himself to the Reverend Senior's Urlsperger in order to go with the Present Transport to Georgia having been converted to the Protestant Religion in the year 1720 and is not permitted to live in his own Country and the said Reverend Senior Urlsperger being pleased to have him also listed with the abovementioned Colonists, He is likewise set down in the Protocol.

Likewise was all the Money given to the Colonists by Charitable People during their Stay here distributed among them[.] each of them except George Buecher and his Family hath 4 Guilders 20 Kreutzers.

And the 31 Day of October which is tomorrow being determined that the said Colonists shall Set out it hath been thought necessary to set down how much each of them have taken with him and it is declared.

Hans Gruebe[r] about ___ 40.	Lorentz Hueber &	
Paulus Swaighofer	his Family ___	54.30
[Schwaighofer] and his	Christian Steiner ___	8.
Family ___133.30	Balthazar Fleiss ___	43.15

Jerg Schwaigger	4.10	Gertrude Krœer	4.46
Thomas Geshwandt		Martin Hertzog	7.31
[Geschwandel] and his		Maria Reiter	10.20
Family and his wife's		Christian Leimberger	19.20
Sister Anna Hofer	50.	Mattheus Mietersteiner	6.22
Tobias Lackner	29.40	Peter Gruebe[r]	27.2
Johannes Mosshamer		Simon Reiter	12.23
and Maria Krœer		Bartholoma Riste [Rieser]	
about	90.	and his Family	39.10
Barbara Rhormoser		Leonard Rauner	5.17
[Röhrmoser]	19.9	Stephen Rickelsperger	
Catharine Krœer	6.13	[Riedelsperger]	40.

At last their Baggage was laden and the Chests marked with Letters and Numbers and Specified what Persons had part in the Chests as follows.

No. 1. P.S.   In which have Part Paul Schwaighofer Martin Hertzog and Mattheus Mietersteiner.

No. 2. S. R.   In this have part Stephen Riedelsperger Johannes Mosshamer, Barbara Rohrmoser with her Children and Christian Leimberger.

No. 3. L. H.   Belongs to Lorentz Hueber and his Family.

No. 4. T. L.   In which are Partners Tobias Lackner Christian Steiner, Balthazar Fleiss Peter Gruebe[r], Hans Gruebe[r] Simon Reiter Maria Reiter and George Schwaigger.

No. 5. B. R.   Belongs to Balthazar[138] Riste [Rieser] and his Family.

No. 6.   Is a red Trunk belonging to Thomas Gehwandt [Geschwandel] and his Family. In which also his Sister in Law and Leonard Rauhner [Rauner] have their Effects.

John George Morell Burghemaster.

Copy of a Letter from Mr. Urlsperger at Augsburg 19 November 1733. No. 12418. Read 20 November 1733. Ordered as on the Minutes. To Mr. Henry Newman.

Honoured & Dear Sir: I send You hereto Annexed the Translation of Two Letters, written by a certain Baron, to a Protestant Divine of this Town by which you'l See as well as his own as the Declaration of other People concerning his Religion and their Petition presented to his I. [Imperial] Majesty and his Sufferings thereupon; likewise his request to my Self & to the Society. This is an Affair to be handled with great Secrecy. I have also Sent the Two Letters to his Excellency [J] Von Reck at Ratisbone and desired his advice thereupon. But in the meantime We have endeavoured to Supply the said Baron with our Information by Letters and Comforts and Bills of Exchange of Fifty Guilders. I send You likewise a Journal of our Emigrant Colonists from this Town to Franckfurth. I committed the Translation of them to Mr. Vat to shew You that his being here is not quite destitute of Employment. You mentioned in Your last to have received my Letters of the 15 and 22 of October And I hope those of the 2nd, 5th, 9th & 12th and 16th of this Month of November will before this be in your hands. It was a great Satisfaction to my mind and visible to others to hear that the Society have approved of the Measures taken concerning a Minister and Catechist. No body will repent it that they are very proper for such a Mission as well as that God will provide for them a full Harvest in Georgia and in America (Mr. Vat tells me many Things which confirm me That God intends something in the West Indies; and I praise the most high God for his leading so powerfully the Heart of the English Nation in these Days. May God who hath the Hearts of the chiefs of the Nations in his Hands, in the present Junctures of Affairs whilst the assistance of the Protestant Powers is wanting, grant that they may at the same Time heartily promote the Protestant Interest or at least remove the Religious Grievances or propose Conditions, Sine qua non. The Boor [George Buecher] who in the Beginning shewed himself so well inclined, but afterwards was persuaded by others and by the Seduction of his Heart to neglect going away with the Transport is returned again to the Papists. He hath given great Scandal. But it happened to him as I foretold it him and it is Likewise to be worse with him For he is gone back against his own Knowledge and Conscience Confer. John VI 66 and following Verses. Nothing is to be heard of the Eight Hundred Emigrants, who were to come to out. Yet it is certain that Many are left behind. Should the Tirnbergers at present at Franckforth resolve to go Georgia Mr. Vat could be employed for Conducting of them to Holland. But it seems nothing can be done with them before next Spring. Should You have an Opportunity I beg the Favour of You Sir to send me an English Bible in Octavo, Dr. Hales's Vegetable Staticks, Dr. Wake's Church Catechism One Small Map of Carolina or Georgia and one print of such a One, as is at the Top of the Georgia Commission and You'l highly oblige me. Now follows an Extract of a Letter from Mr. Mollenhoff Danish Minister [at Vienna].

I have since taken Care to be informed of the Motlein [motion] in upper Carnten [Carinthia] but have not been able to learn any Thing since such Affairs are here very much concealed especially to Us Protestants. Indeed, here are People of Karnten who are Masons by Trade. They come every Year in the Spring to Get a Lively hood by their Hand Labour, frequent our Assemblies come to the Holy Communion, purchase as many Evangelick Books as they can come at, towards Winter return to their own Country and carry those Books with them at the Greatest Danger[.] Some of these People I sent for to my House. But as they are of the Lower Karnten they know nothing of what is done in those Higher Lands.

The Dear Baron hath in the mean Time, together with his trusty Brethren not chosen a good Place of Refuge For Persecution in Hungary against our poor Brethren goeth on worse and worse, and if God doth not direct that in the present Junctures the Persecuting Spirit be abated a little, the Protestant Religion in Hungary is like to be wholly lost[.] Vienna October the 30. 1733.

Now Dear Sir I recommend You to the Grace of God remaining

Sir   Your humble Servant   S. Urlsperger

Copy of Translation of a Letter written from Modern in Hungary 30 September 1733.

Wishing You first from the bottom of my Soul the peace of God the father the Grace of our Lord Jesus Christ and the assistance of the Holy Ghost which giveth us Comfort in all our Afflictions.

It may possibly already be publickly Known at Augsburg in what manner by the Providence of the Father and the powerfull Motion of the Holy Ghost Three Valleys in the upper Cärnten [Carinthia] in one Day and in one hour have publickly and unanimously declared themselves for the pure Gospel of Jesus Christ and have desired nothing but free Leave of going out from His Imperial Majesty by the Administrators of Clagenfurth [Klagenfurt] as Sovereign Lord thereof but have obtained an Answer contrary to their request. Now as we according to what we have seen in other neighbouring Countries, were intent to obtain such free

Leave by the Interposition of the Envoys of the Evangelical Body at Ratisbone having even sent some Deputies thither and being at Drawing up a Confession of our Faith according to the unaltered Confession of Augsburg at my House at Croneck in order to send it to the said Deputy about 12 and 1 of the Clock at Night 50 Soldiers came upon Us who were to seize me and such others as were with me; But God did hide us like David, Since being in a room between the Privy and the Stables for Three Days and nights they did not discover Us, but took away all my Papers, Library, Money, and all they could carry off. In these Circumstances I together with 3 Farmers, my Neighbours who had the like fate resolved to instruct our Wives and Children how they should behave themselves and sent them to their Friends, who for fear of the Jews,[139] are not yet publickly known and under the Protection of God to retire where we could. Thus we came on Foot and disguised through Crain and Windishmarck in Hungary as far as Ordenburg, thence further over the Danube to Modern, having nothing with us excepting a Bible, Arnd's Paradise & true Christianism, and Twelve Gold Ducats and Twenty Groschen in Silver. We now wait here till God shews the Way, where and by what means? We are at the House of a Protestant Friend but likewise under a persecuted State. I came Eight Days ago in Boors Cloths to Vienna in order to learn what was done by the Austrian Council concerning our People; But have been told that his Imperial Majesty will not permit any Emigration but is to send Missionary Jesuits to the Vallier [valleys] to search from House to House after the Ring Leaders and to Instruct the others better in the Catholick Faith. To this End 100 Copies of a Book intitled Vogelfrist order stirb[140] had actually been already sent into upper Cärnten like the Forerunner John Baptist, of the Mission. I had obtained a Copy thereof from the Papist at Vienna. It is a Scurrilous Scumy and devilish Language against Luther and the Protestant Clergy etc. as if that Book was not be answered. And I can assure You that this pernicious Book will give great advantage to the Missionaries for perverting such as have declared themselves Protestants in those Valleys, in Case these poor People are not encouraged by a Speedy answer, by shewing its pernicious Spirit. And I prepare such an answer. For which purpose I provided my self at Vienna with Stratemanns Theatrum Historium, Dr. Jäger Fr. de Ecclesia etc. & brought them hither. As one Book after the other is ready I send it to Nürnberg and the printed Copies are thence to be sent by the weekly Messenger of Nürnberg, to Ortenburg; From which

Place I can afterwards send them to our people by the way of Steyr [Styria] and upper Steüer [Styria] into upper Carnten [Carinthia]. And now our Intention only is to know in what manner we may go through Masten and Austria to Ordenburg in order to Conduct thither our respective wives and Children. Therefore we Earnestly desire, for Christ's Sake that You would be pleased to lett us know as soon as possibly may be, Whether it were not possible that a small assistance might be sent for Us from Augsburg to Vienna? That our Friend where we now are might receive it and bring it hither. I likewise entreat You to let us know from the Reverend Gentleman who corresponds with Scheidberger [Schaitberger][141] and in England to what Gentleman of the Society de Propaganda Fide, in England, I could apply to, Since I am resolved to offer my Service to the new University and Academy at Gonttingen [Göttingen] to be Erected by H. M. of Great Britain, as having found out and put in practice a method of Teaching and Learning by which Young Gentlemen and Noblemen can in Three years time make a greater Progress in Polite Learning, than otherwise in 6 Years. I hope God will by means of Charitably inclined Persons provide me with a small loaf of Bread, which I for the Sake of his Holy Gospel have left behind me. I only beg for God's Sake that you would soon write to me that we may know where we shall direct our Course. May God the Rewarder of all Good pour down upon your self and all Christian Souls, his choicest Blessings both spiritual and temporal for their Love and Charity to Us. In the mean Time we recommend You to his Omnipotent Protection and gracious Conduct and ourselves to Your Prayers and Lovers. And I desire to live and die as your etc.

Another Letter

Your Comfortable [comforting] Letter of the 15th of October last came the 24th safe to hand and we return our most hearty Thanks both for Your Comfortable Admonition and Your gracious Remittance of 50 Guilders; and as we never had any other Intent but to Live and Die according to the Truth of the Gospel and Your Doctrine and Admonition so we have been exceedingly rejoiced at that remitted by which we are enabled to go there to Ortenburg in order to our Safety and to be nearer at hand to our Brethren of the Faith, that we may the better and easier assist them and our Families with Teachings Remembrings Admonitions and Comfortings of whom we hear nothing but the utmost Afflictions. In these Circumstances we beg most humbly to give other Orders for our assistance and Comfort and according to the inclosed Directions because we are by the advice of good Friends obliged to change our abode for our Safety's Sake. As to your Corresponding with me You need not be apprehensive of its being divulged since it is in a very safe way. We have no Complaints against H. I. M. [His Imperial Majesty] as Sovereign Lord of the Country, neither have we taken up Arms, but we have only declared publickly that we are protestants according to the Confession of Augsburg, as our Ancestors likewise did in the Year 1638 upon which Ferdinand the Third Graciously Granted them a Toleration of Six Articles. And now we desire nothing further but the renewal of the said Articles of Toleration or Free Liberty of Withdrawing. But as this has been now denied and resolved better to Instruct the Idiots as we are now called by Emissaries and afterwards to send away the Refractories; To which End the Emissaries are already appointed & 500 Copies of the Calumniating Pamphlets Vriss Vogel[142] actually [now] are deposited at Grätz in order to disperse them among the Idiots, there remain no other remedy but a short Exhortation, concerning these Calamities, to be sent to our Brethren in the Faith that they may not be misled by such a wicked and Impudent Pamphlet. And as there is Danger in the Delay, since such Exhortations as mentioned before cannot be printed as quickly as it is necessary, So I have made an Extract thereof, and I have ordered Six Manuscript Copies to be fairly written, which I shall send as soon as I can provide mony by one of my Fellow Labourers by the Way of Steür [Styria] who is to caution and to Strengthen the Brethren previously. Its Conclusion tends to admonish the Brethren in the Faith that they shall Stand fast in God's Holy and Saving word and venture all their worldly Goods, even Life and take Care not let themselves be drawn away from the Known and publickly acknowledged Truth, by any artful representations Errors and persuasion but to remain faithfull to their Sovereign Lord of the Country and likewise insist on the renewing of the Toleration Articles of the Emperor Ferdinand or free Leave of coming away, & thus to save their Souls which have been redeemed with the precious Blood of Jesus Christ to Eternal Life. God hath provided me for this work with the most excellent Books by the assistance of a protestant Gentleman living at about 12 miles E. distant from hence. This Gentleman if he had sufficient means would do any Thing in the world for us by excepting his Store of Books, he hath nothing above Eating and Drinking which is here plentifull and exceeding cheap[.] On the Contrary our

Brethren in the Faith here must be Content with a Small Pitance, for according to the advices come to Vienna the most Eminent of them are in Custody and have only once a Day a little warm Broth and Bread and Water till the arrival of the Emissaries. The others indeed are at home but are obliged to have soldiers quartered on them, who use them with Severe insolencies. May God assist them all with his heavenly Comforts till their Deliverance draws near. Our people would very willingly go to Georgia. Moreover I beg the favour of You to lett me know not only the Conditions, but also the Name of the Secretary or officer with whom the divine of Augsburg [Urlsperger] corresponds, who likewise changed Letters with a Miner of Saltzburg [Schaitberger] who liveth at Nürnberg and who hath pointed several fine small Treaties of his own Stead some of which came likewise to Us in upper Carnten. The Name of this Corresponding Gentleman [Newman] in England is particularly desired as I hope by his good officer [offices] to carry out my new found out method, All Political Sciences in a short time to teach and to learn[143] and to present it to his Britannick Majesty, and thereby to be Employed, either at the Royal Academy in England or in the Electorate of Hanover, at Göttingen or at Lunenberg [Lüneburg] That I and my Family may be set up again. Therefore You'l highly oblige and rejoice me with such Advices. For should as it is mentioned the Emissaries about Christmas be in upper Carnten my Family and those of my Fellow Sufferers will be the first that must come away. In the mean Time we must endeavour with the assistance of God and the advice and aid of good Christians to see where they may find their Settlement.

Journal of the Travelling Saltzburg Emigrants from Augsburg to Franckfort on the Maine. [Apparently by Philip von Reck]

1733

October 31. After the Emigrant Saltzburger[s] who had resolved to go to Georgia together with Mr. [Ph.] Von Reck, their Commissary & Mr. Schumaker their travelling Chaplain, under the Protection of God, left Augsburg the 31st of October last, at Eleven of the Clock in the morning, they went through Donawerth [Donauwörth] and got the next Day November 1st to Ebermorgen,[144] a Protestant Town, and after having assisted at [attended] an occasional [chance] funeral Sermon they went that day to Harburg also a Protestant Town. The following day November 2nd they went through Dunckelspil [Dinkelsbühl], and after they had received the Blessing of the Minister of the Town, and the Liberalities of its Inhabitants came the 4th at 10 in the Morning safely and in good health to Rottenburg [Rothenburg], two and two going together into the Town, Singing of Hymns, and were lodged in several houses and after they had received at the townhouse [city hall] each thirty Creutzers and the blessing of God, they came the 6th through Marckbareith [Marktbreit] to Marck Stefft [Marksteft] where they were to take Shipping. The Catholick Bailiff of Marckbareith had sent a Messenger of the Town to Mr. Commissary, with this express notification that he with his people should go out of the Territories of the Prince of Schartzberg, and not to pass through the Town of Marckbareith; but as Mr. Commissary did not go out of the road he was in, and in the name of God came near the town, he was suffered to go into and through the town without shewing his Passports to the Bailliff; for most of the Inhabitants of that Town being Protestants and hearing the Bailif's Message sent two Trumpeters upon a Steeple, who played all the while we went through the same. after they were on board the Ship at Marck Stefft, there was a Calm the 6th and the Seventh, but the 8th and ninth the Wind was contrary, so that they got to Wertheim later than they expected by one Day. They took Provisions at Wurtzburg sufficient to wertheim and there what was necessary to go to Frankfort. as they received great kindnesses in the Protestant Towns in their way so they were also very kindly received at Wertheim, although the Inhabitants thereof Suffered much by the Inundations last Year. After the reigning Countess at Wertheim had sent for the Saltzburg Children and Shewed them great favours and a Roman Catholick Gentleman which must not be forgotten who had assisted at [attended] our Emigrants Morning and Evening Prayers, and other Devotions being greatly moved took his Leave with Tears of the Travelling Chaplain and thanked him. They went on Board the Ship the 11th and the Wind was good that Day, But the next Day it was very contrary. Therefore being about Two miles English beyond Hanau They were obliged together with the Chaplain to lay that Night on Board the Ship and Mr. Commissary went that whole Night by Water to Franckfurth, where he waited upon the Commiss [?] Burgomaster who was at the Römer or Town House (of whom and several Senators he was kindly received and gave him Notice of the coming of his Colonists. After that he immediately went with a Secretary of the Senate to the Place where most of the Tirnbergers were quartered and declared there, having first obtained

Leave of the Burgomaster Von Schweitzer, that he had Orders to receive as many Persons of the Saltzburg Emigrants as would willingly go to Georgia and having laid before them the Printed Copies, and Thomas G. Schwandel [Geschwandel], One of the Georgian Colonists having given a good Account of the good Usage they had to that time received, Some of them seemed to be inclined of accepting the offered Conditions, but they excused themselves for that time considering their common Sicknesses, which they contracted in Holland and which according to an Eminent Physicians Opinion proceeded chiefly from Hunger and Affliction. Thus he desired nothing of them but that those who after their being cured should resolve to go to Georgia might set down their respective names and deliver them to him in the mean Time[.] the Colonists came on the 13th at one of the Clock to Franckfort and the Magistrate resolved to defray their Expences and to give them a Bounty towards their Voyage And they were Conducted by some Officers and Soldiers Two and Two in Singing of Hymns into the City and received great and many Benefits both Religious & Civil, especially of the Honourable Magistrate, who gave to every Man 90 Kreutzers, 60 to every Woman and Thirty to every child. Besides 2 Guilders of One unknown and of another unknown Person 20 Kreutzers a piece and Several other Benefactions.

On the 14 their Chaplain carried the Colonists to the Church called Barefooted [Franciscan] where the Minister out of Colossians 1.9 directed his whole discourse to the Colonists. On Sunday the 15 The Colonists went again to Church in the Morning. Their Travelling Chaplain speaks accordingly well of the Magistrates Care in particular, and especially of Mr. Gulman the English Resident, and of Messrs. Von Munch's real Love to the whole Transport[.] Likewise he commends Mr. Commissary's good Success in his Employment. The Magistrates of Franckfurth will, out of Charity Subsist the Tirnberger Emigrants returned out of Holland till they are recovered of their Sickness; and Care will be taken in the best manner for their Religious Concern.

Copy of a Letter from Mr. Urlsperger at Augsburg the 23rd November 1733. No. 12430. Read 27 November 1733. Ditto 4 December 1733. To Mr. Henry Newman.

Honoured & Dear Sir: In answer to yours of the 30 October O.S. whereby I learn that You have received mine of the 29th October N.S. The Number of the Colonists is reduced 40 Persons, because the Peasant [Buechner] known together with his Family is remained here, and some at Ratisbone have been disuaded from their Resolution, and a Family of 5 Persons [Rieser], is obliged to Stay here because a Child of theirs hath broke his Leg which is Still under Cure. I am very Sorry for many reasons that the Colonists do not come to London, and wish I had known it sooner to the End that I might have acquainted the whole Transport with it who hitherto have believed the Contrary. I wrote by this Post to Rotterdam that the Reverend Mr. Boltzius and the Commissary Mr. [Ph.] Von Reck, if it is time enough Yet, may go with the Packet Boat from Helvoet Sluys to London in order to speak with the Trustees and the Society about several Things concerning this present Transported as well as the future. If my Letter comes to Rotterdam and Mr. Boltzius and Mr. Von Reck do come to London, I recommend them by You Dear Secretary to the Trustees & to the Society, likewise to the Reverend Mr. Ziegenhagen, desiring him to be their Interpreter. And Mr. Von Reck when he was here having assured me Several times that if he had but free Passage and Dyet to & From Georgia he would go thither, I leave it to the Trustee's Resolution. If he goes thither it will not be without a great Benefit for the future, especially he being a very active and insinuating Young Man, who is very usefull. Mr. Boltzius and Mr. Gronau, expecting to provide themselves with Necessaries in London I have writ to them to do it Rotterdam or Dover especially if Mr. Boltzius should not come to London. I commend You to Gods Protection and remain etc.

   Dear Sir Your most humble Servant Sam. Urlsperger
P.S. There was no time to Translate this Letter. Mr. Vat gives his most humble Service to the Society and your Self.

 In case Mr. Zweibler [Zwiffler] should go to Georgia where as an Apothecary he might be of good Service I doubt not but he may have his Diet on board the Ship as one of the Reverend Divines.

Copy of a Letter from Mr. Urlsperger. Augsburg the 26th of November 1733. No. 12431. Read 27 November 1733. Read 4 December 1733. Answered 4 December 1733. To Mr. Henry Newman.

Honoured and Dear Sir: My Cash of the Colonists is at this time in such a State that I had no need of taking up any thing

from Mr. Von Münch. I repeat what I mentioned in my last of
the 23rd Viz. that I wish the Colonists might have gone to London,
as they were here put in hopes thereof, yet as that might have been
very detrimental in their whole voyage, so I hope they will be
sensible of its advantage. now it seems to be necessary that Mr.
Commissary [Ph.] Von Reck, especially the Reverend Mr. Boltzius should have gone by the Packet Boat to Harwich and thence
to London as I wrote to them thereupon on Monday last. I herewith send the translation of a Letter from Mr. Mollenhoff preaching Chaplain to the Danish Envoy at Vienna, to whom, as knowing him personally to be a trusty man and upright, I applied and
desired him to push that affair by the Assistance of the Danish
with the English Envoy that the Imperial Court might send a
Speedy Resolution to the Roman Catholick Magistrate of this
Town, for permitting the English Colonists to take their quarters
in this City, and this step was so much more necessary, since our
Protestant Magistrate had, by mistake, made use of their Agent at
Vienna to apply to the Envoy of Hanover. The other Extract
cometh from the Envoy V. R. [J. Von Reck] By which it appears
how Zealous he is for the Protestant Interest. I had Yesterday with
me a very ingenious Journeyman Weaver of Steuermark [Styria]
who last Summer was laid in Iron Chains for 3 weeks in upper
Austria being accused of having been Instrumental in the Declaration of so many Thousand Protestants. At last as nothing could
be proved against him he was banished out of the Country. I have
very certain advices from Saxony That H. M. the King of Prussia
hath bespoken Ten Thousand Copies of the New Testament according to the Impression of Canothuis [Canstein][145] together with
Hymns and Soldiers Prayers upon all Occasions, which are to be
distributed among his Troops and Soldiers are now binding. This
is a Sign of our Time! My Dearest Secretary, I make bold to write
to You one word: It is known to You by many instances The many
innumerable Grievances of the Protestants in Germany against
Popery, and how many under the Popish Power groan and Sigh
for Liberty of Conscience. Now is the Time that Great Assistance
will be required of the Protestant Powers; were it not possible
that by the Representation of some Archbishop or Bp. or Minister
of State in London, to be made to his Majesty of Great Britain
jointly with the States of Holland, such an affair might be powerfully negotiated? Certainly if this affair was taken seriously and
earnestly to Heart God would give his Blessing to it, and Jesus
the King of Kings would be a great rew'd [reward] and Armour to

such high Powers and their Subjects! I know indeed as having dwelt at Court that there they don't always Care to hear a great Deal of Religion; Notwithstanding which, something should be attempted.

I must also lett you know that the old Joseph Scheidberger [Schaitberger], who came out of Saltzburg in the Year 1685 died some weeks ago at Nuremberg having some hours before he died among other remarkable Sayings of his, left his Blessings in particular to his English and Augsburg men as he Termed them for their many Benefactions bestowed on himself and his Country People [compatriots]. Dearest Mr. Secretary This is more than if I had sent You a Bill of Exchange of many Thousands of pounds For the fervent prayers of the Just is very powerfull & availeth much! This Scheidberger indeed was a just man! We have here such a Grayheaded Saltzburger of 86 Years of Age in our Protestant poor House, who liveth in an uninterrupted State of Praising God, which his prayers are likewise directed for England! I remain with all my Heart

Honoured & Dear Sir   Your humble Servant   S. Urlsperger P.S. I have received Letters from Mr. [Ph.] Von Reck dated at Cologn where he arrived the 20 instant in the Morning & set out from thence the same Day. I hope he is at this time in Rotterdam. This Mr. V. R. hath hitherto behaved himself so that the Colonists are contented with him & other People likewise approve of his Conduct. The Candidate [Schumacher] who goes with them to Rotterdam is also very diligent, insomuch that by his Prayers upon their Journey the Roman Catholicks have been moved. The Apothecary Mr. Zweibler [Zwiffler] does likewise his Duty. Time will not admit of more.

Extract out of a Letter from Baron [J.] von Reck in answer to Two Letters concerning some protestants in Upper Cärnten [Carinthia], Dated at Ratisbone 19 September 1733.

Reverend Sir: The Contents of the Secret Letters You were pleased to communicate to me on the 17 Instant have taken up all my Attention, & that very deservedly. As to the advice You desire of me, I am indeed inclined and willing to Employ my good Offices at Hanover for the Learned Gentleman. But I think it would be of great Weight if, by the interception of the English Lords Commissioners and the Society for Promoting Christian Knowledge, that affair was laid before His Britannick Majesty, and that this honest Gentleman might be entertained [maintained] in Germany. I doubt not but that every body, in case any Persons should be driven out of the Lands in question, might give them Entertainment in their respective places. This Affair is indeed very nice [delicate], but our Salvation is concerned therein, for which we are obliged to do our utmost. I earnestly desire You would be pleased to Communicate to me whatever farther Advices you may receive concerning this affair and You may be assured of my keeping it Secret and of my readiness to promote it heartily.

Extract out of a Letter from The Reverend Mr. Möllenhoff Chaplain to the Danish Envoy at Vienna Dated 21st November 1733.

I can tell You now That I have Spoken both to Mr. Von Praun[146] told me that he was ordered by the Protestant Magistrate of Augsburg to apply to the Envoy of Hannover and not to the English Envoy. That the former had desired of him the State of the Case in Writing; which he had delivered to him some weeks ago and he doubted not but the Envoy had sent it to Ratisbon, Hannover, and London. I have got that Memorial as a Copy of that which was read before his Imperial Majesty, and I can truly Say that it is drawn up in Solid and Strong Terms. Yet in this manner this affair goes very heavily and the English Envoy is also to use his best Endeavours therein. Now as to the last mentioned, I have heard our Envoy who was actually fully informed of that affair Say That he the English Envoy [Robinson] had orders from his Court to concern himself therein; That he had some weeks before actually given a memorial to the Vice Chancellor, But that as yet no answer thereupon had been given. Yet that he would next week apply again for an answer. Our Envoy hath promised me that he would put it to his Remembrance. And then I may give You perhaps a more full and agreable Account thereof. In the mean time I wish that God may let us See a glorious Issue of this his work.

Extract of a Letter from a Protestant Envoy [J. von Reck] at Ratisbone Dated 24 November 1733.

It is easy to Conclude that your last of this month, together with my Cousins [Ph. von Reck's] Journal to Franckfort was very acceptable to my Self. May God assist him further that I may hear good things from him out of England. The warlike affairs are now more than were before, in such a Crisis that we protestants may obtain every Thing, if we our Selves would have a mind to it. It would be shameful if we do not take our advantage thereof.[147] I dare say no more. I have this Day remitted to You 50 Guilders which are from a good & trusty Hand and which you may employ to the Use and Service of the Persecuted Emigrants and particularly of the B. V. R. [Baron von Reck] for which I expect a Receipt that I may be enabled upon occasion to do Still more. I remain.

Copy of a Letter from Mr. [Ph.] Von Reck Rotterdam 27 November 1733. No. 12419. Read 24 November 1733. To Mr. Henry Newman.

Sir: I have taken the Liberty to inform You that my Colonists are arrived at Franckfort & I believe my Letter of the 15 Current will come to Your hands by the direction of Mr. Gulman English Resident at Franckfort[.] the Preacher who has accompanied us hitherto is about to return the 2 others being arrived by the Post 8 days before our arrival which was to day.

The Magistrate of this City has been so unchristianlike and so little sensible of the Kindnesses that the Tirnbergs have received from the Crown of Great Britain that he has refused us Entry into the Town contrary to the Laws of Nations[.] I'm not uneasy at it because I can live cheaper in the cheaper[!] ship than in an Inn.

We shall embark (God willing) the 3rd December N.S. on board an English ship called the Purysburg which arrived here the 22 Current. I'm just now informed that the Transport is not designed for London only to touch at Dover where I hope to have the Honour of Seeing You. I shall take up as much money here as will supply our occasions till we arrive at Dover, from where (God willing) I shall transport the Colonists to Georgia. The Reverend Mr. Lowther who has used me with a great deal of friendship and treated me handsomely at Dinner, makes his Compliments to you. I am with much Respect.

Sir   Your most humble & most respectful Servant
Philip George Fred. De Reck

P.S. Mr. Ortmann assists me as Interpreter and makes his Compliments to You.

Copy of a Letter from Mr. Lowther at Rotterdam 4 December 1733, N.S. No. 12434. Read 4 December 1733. Answered 4 December 1733. To Mr. Henry Newman.

Sir: Your Favor of the 16 of November N.S. I received and am much obliged to the Gentlemen of the Society for their Thanks. I shall always be glad of any Opportunity of serving them, being satisfied they are desirous to do all the good they can[.] they may freely Command me whenever they think I can be useful.

Captain Fry is just now under Sail, his People are all in good Health, God send them a safe Voyage, he [Fry] might, I think have been a little more Expeditious[.] the Emigrants arrived here last Fryday[.] he would not take them on board before Wednesday, since which time the Wind has been very fair. The Trustees, if ever they should send a ship upon the same errand again I would advise them to send one that draws less Water, I am afraid this will not be able to get down to Hellvoet with this fair Wind, She draws 12 foot & ½ Water an Easterly Wind drives all the Water out of our River. I thought this hint proper. Mr. [Ph.] Von Reck is I think too Young a Gentleman for an Expedition of this Kind, in my Opinion a grave man ought to be employed, the man is a good kind of Man but he seems to me to have too much Gaiety about him for the People under his Care, I gave him Your Letter.

As to the Minister and Catechist, I offered them any Service in my Power, if the Captain of the ship be not wrote to to Dover, I am afraid them Gentlemen will fair but indifferently on board, I mean as to their Lodging[.] the Captain tells me that they must lay among the common People that his Cabbin is his own, and that he has let it to a Family that is to meet him at Dover I imagined the Trustees had hired the whole ship so thought it proper to mention this to You.

I am much obliged to you Sir for the Account You gave me of the prince of Orange. I sincerely wish him a good recovery, he is a Prince of great Value.

I am glad to hear Sir James Lowther is well, pray You present my humble Service to him and to all the worthy Gentlemen of the Society You may assure them of my Prayers and good wishes for the Propagation of Christian Knowledge. I am with great Respect Sir

Your most humble Servant   Richard Lowther

Copy of a Letter from Mr. D. Wolters Rotterdam 4 December 1733 N.S. No. 12435. Read 4 December 1733. To Mr. Henry Newman.

Sir: Mr. [Ph.] Von Reck having no time left to write to you this night desired me to give You notice that he is imbarked this

Day with the Saltzburgers and the Captain designs with God's leave to sail to morrow for Dover.

I have the pleasure of seeing Mr. [Ph.] Van Reck go away very well satisfied with the little Services I have done him, and I shall always be very proud to execute the Commissions of the Honourable Gentlemen. We have had a conference with the Commissary of the Piemontese,[148] upon the discovery I have made of a Man arrived here who would be very glad to embark next Spring hoping to have a Company of Piemontese, which will be able to work with him he is very fit for to bring up silk worms, and working the Silk and for making Spanish Soap of all Sorts, I have no time to send to You all the Papers belonging to this affair. I will have the honour next post to write to You about this. the time for the departure of the post being very nigh I must finish with assurance of the peculiar Consideration of

Sir   Your most humble and obedient Servant   D. Wolters

Copy of a Letter from Mr. Butienter at Dover 1st December 1733 O.S. No. 12436. Read 4 December 1733. Answered 4 December 1733. To Mr. Henry Newman.

Honoured Sir: I safely received Your favour of 29th November which gave me a great deal pleasure because You therein acquaint me that the ship that carries the Saltzburgers had left Rotterdam 23 November. For while we were in doubt whether or no she had weighed Anchor[.] we were very uneasy for fear our long stay here should be to no purpose But now we expect the Ship every Day and feed our Selves with the agreeable hope that we shall see these Emigrants in a few Days. We shall Study and Endeavour to our utmost to have great care taken of every Circumstance and that the Men shall be carefully provided for; As to the afflictions which the Protestants in Hungary and Bohemia labour under we must Commit them to Providence not doubting but that God will touch the hearts of all Protestant Princes to their Relief, Perhaps some of 'em may share the Benefit of a mansion in America where they may be at ease[.] As to the Silver Sacramental Vessels made for the Saltzburgers they may be sent if thought proper to Mr. Merchant Minet[149] who will forward 'em to me and if I should perchance be on my way to London He will keep them till he receives Directions where they are to be sent and who delivered them to You will be pleased to make my Duty acceptable to the worthy Members of the Illustrious Society for promoting Christian Knowledge and be assured that I shall miss no opportunity of shewing with how much Zeal and Pleasure I shall always be

Honoured Sir   Yours most Strictly   H. A. Butienter

Copy of a Letter from Mr. Urlsperger Augsburg December 7th 1733. No. 12446. Read 11 December 1733. Answered 18 December 1733. To Mr. Henry Newman.

Honoured & Dear Sir: I don't doubt but that You have received my last of the 26th of November. [I have advice from Vienna of the 2nd Instant that there has as yet no answer been given to the Memorial Presented by the English Minister to his Imperial Majesty. It might not be amiss for certain Reasons that it should be recommended by the Court of Great Britain to their Minister at Vienna to insist on a favourable Resolution of the Imperial Court, otherwise every thing will be left undone; And I, should a new Transport of Emigrants accidentally come hither, and not be permitted to come into this City, as I apprehend, might be again put to a great Deal of Trouble] I beg You would be pleased to reflect upon it.

The Baron of Upper C. [Carinthia] whose Letter I lately sent You translated is at this time gone out of Hungary and we daily Expect to hear where he and his Fellow Sufferers are gone and in what manner He and such as are in his Case may be advised. Since the 20th of November last I have heard nothing of our Colonists which was from Cologn; So that I am apt to think since the news papers do not mention any Misfortune that a Packet of Letters directed to my self may be mislaid or lost. The present Duke of Wirtenberg [Württemberg] Charles Alexander was still at Vienna the 2nd instant[.] He keeps all his Employments under His Imperial Majesty and hath given a full Declaration to the States of the Dutchy that he will not make the least alteration in Religious affairs.

The King of Prussia besides the 10000 Copies of new Testaments Hymns and Soldier's Prayers lately mentioned to you hath desired of Professor [G. A.] Franck, Nine thousand more of them with this Express Order to bring into the Account a higher Price than usual, For it was his Will and Pleasure that the Orphanstrophy[150] should get something thereby. They go on in Hungary to Persecute the Protestants with very great Severity; The Lord God have Mercy upon them!

Two Days ago a Popish Chaplain of this Town came to me in the Night time in order to embrace the Protestant Religion and the same Night he was sent to Ulm with a Messenger. And some days ago a German Gentleman teaching here the French Language, together with his whole Family, hath publickly at the Town House declared himself a Protestant and they have received the Holy Communion at St. Ann's Church. I remain Sir, with all my Heart

Your very humble Servant　S. Urlsperger

Copy of a Letter from Mr. Butjenter Dover 9th December 1733. No. 12447. Read 11 December 1733. To Mr. Henry Newman.

Honoured Sir: We have hitherto Expected our Ship from Roterdam to no Purpose not a little uneasy least she may have come to any misfortune for the winds have been favourable and there has been no Storm for those four days and nights[.] God knows what may have happened and how he will assert the Greatness of his Glory, if by any mishap the Enemies of Truth shall take upon them to convert it to their Blasphemous interpretations[.] The munificence and Goodness of the Honourable Society I cannot sufficiently admire and extoll, For if these Things proceed from a sincere Love of God and our Brethren in Christ they are not only acceptable to God but they shew the Constancy of those who bear the Character of faithfull Disciples of Christ and amidst many difficulties shine as light in darkness. I shall readily and joyfully obey the generous Commands of the Illustrious Society and according to the directions I have read both in your and the Reverend Mr. Wilsons Letters will provide Messrs. Bolzius and Gronau with all necessaries. They shall be furnished with Bedding Ecclesiastical Habits Linnen and every thing else requisite; I will buy 'em Paper and whatever else their Voyage requires; being Commissioned by the Society in Mr. Wilsons Letter in the words "You are desired by the Society to supply those Gentlemen with all necessaries as Bedding Cloaths refreshment etc. while at Dover and for their Passage." The Sacramental vessels designed for the Saltzburgers I have not Yet received nor can I account for their delay[.] What is proper for me to know I beg to be informed and particularly what may have happened to our Ship[.] Be pleased to make my Services acceptable to the Society and Continue your Favours to

Sir　Your most humble Servant　H. A. Butjenter

Copy of a Letter from Captain Coram Dover 10 December 1733. No. 12450. Read 11 December 1733. Answered 11 December 1733. To Mr. Henry Newman.

Good Sir: I am very Thankful for Your Kindly Remembring me in your Letter to Mr. Butjenter who received it this Day as he did the Trunk with the Communion plate, Books etc.

About half an Hour past the Ship with the Saltzburgers on board from Holland (as I believe it is) came before this Harbour and put the Colours out, but the Wind being now come to the North West and blows very fresh that no Pilot Boat could get to her & the Master doubtless thought it not very safe as the Wind now is to come to an Anchor in the Open Road is put away for the Downs from which we Expect the Master this Even after his ship is at Anchor there.

Mr. Butjenter is gone to hasten the Bedding & other Things for the Ministers, else I am sure he would have presented You with his best Respects if he had writ to You as I believe he has not by this Post[.] I am in Truth and with great Esteem,

Good Sir   Your most obedient Servant   Thomas Coram

P.S. I hope this will come to your Hands before the Associates go from Your House.

Copy of a Letter from Mr. [Ph.] De Reck Dover 22nd December 1733. No. 12454. Read 11 December 1733. Read 18 December 1733. To Mr. Henry Newman.

Sir: I received yours of 16 of last month with extream pleasure at Roterdam and more especially so because it so kindly advises me of the approbation of the Trustees for Georgia concerning my conduct which is the Object of my wishes and the Sum of my Endeavors. The Reverend Mr. Lowther and Mr. Walther [Wolters] his Majesty's Agent have favoured me with their assistance and facilitated as much as in them lay our Embarkation on bord the Purisburg wherein we sat sail the 5th Current N.S. and by the Imprudence of a Dutch Pilot whom we had taken on bord Struck upon a Bank of Sand on the Point of the Mease [Meuse, Maas] where we were Obliged to wait till the 10th and we toiled 15 days on the River before we could put out to Sea which we have traversed in three days Arriving at Dover the 22nd N.S.

I found on bord our Ship Store of Provisions for three months but the Captain's disposal of it seems sometimes too Sparing. The small Children ought to have thick Supping but he allows 'em neither flower nor Butter to prepare it, so that I have been forced out of Compassion to furnish them among other Things with Butter and flour. The Colonists sometimes pass the whole Evening in the dark and have but now and then a bit of Candle allowed them. The Captain forbad the Ministers and Apothecary from visiting the great Cabbin Obliging them to be always Cooped up in their beds or at the bottom of the hole and altho' he has redressed that a little yet are they afraid lest he should fall into it again. And if I had not bought at Rotterdam some small matter of provision of [omission] the Gentlemen must have taken [would have had to take] up with the common fare of the Colonists. Give me leave then to inform my self if the three Gentlemen abovementioned may not hope for a better treatment for the future. To remedy these complaints I should be glad first that the Captain would regularly give me every Day a little butter and flour for the Children, secondly Candles to light them. thirdly sometimes a little Brandy to the ancient People. fourthly if he is to find us in Provision that it may be something better than that of the Colonists 5thly that the two Ministers and Apothecary may be allowed the privilege of Coming into my Cabbin when ever they shall think proper, As for the rest we are all well enough Satisfied with our Captain[.] This moment there was a Complaint of the Scarcity of Bread 5 Persons receiving but 4 Pounds of Bread per Day.

I am resolved to go my self to Georgia for these two reasons— First to Transport Good Laborious and Industrious Men, Secondly to forward Husbandry, the Vineyards and trade. To accomplish the former in Georgia, I go first to see the goodness and fertility of the Country and to make an Exact description of it. Secondly thereby to Encourage a Number of families in Germany to go and settle there for they Expect from me an ocular proof at my return, thirdly to facilitate by this Experience the other Transport of Saltzburgers and if it please God some of Austria who may be perhaps induced to follow, fourthly to persuade a number of 300 of the Waldenses who are now in Switzerland in the Canton Bern to do the same, almost all good Manufacturers and well Skilled in the Vintage, who are to come out of that Place next Summer at the Cost of their Excellencies of the said Canton to Roterdam[.] Mr. Wolter the Kings Agent at Roterdam who has a good Correspondence in Switzerland and Mr. Dumont a French Minister[151] at the same place and who is Member of Your Society whom I trusted with the Business will Carefully execute the orders and measures therein taken. Of these 37 or 40

Saltzburgers whom I conduct I can assure You that they are accustomed to Obedience, to Labour to Sobriety, and Piety, That they are patient Good husbands and perfectly agreeing among themselves.

To accomplish the latter I go first to instruct and teach the Colonists the best method of tilling the Ground having learned by a Course of some years the economy of it and found that the Earth altho' Covered with Shrubs and Thickets can Yield from her Bosom fruits in abundance and multiply her gifts according to the merits of those her Children who by their Labours Deserve her Encrease.[152] Secondly that I flatter my self with better success than another Person as being acquainted with the ways and manners of these People.

After having ordered all these Things and Set the Business agoing I shall return to England as soon as possible and also to Germany to conduct those who may follow.

It would be very necessary to give the Captain strict orders to behave for the future with more moderation towards me the Minister and the Colonists than he has hitherto done For it may be of ill Consequence after the Saltzburgers having been already disposed to make him really sensible that they are not Slaves and that they hardly knew how to put up his harsh Treatment.

I was acquainted by the Amsterdam papers that Mr. Pury is set out from Carolina on bord the Perl Sloop designing for Switzerland whence he hopes to bring 400 Persons so that there is room to hope that he will have the 300 Persons of the Waldenses in the Colony of Bern, who chuse to place themselves under the Protection of the Trustees for Georgia[.] one of the Principal men of the Waldenses is actually at Roterdam by the Name of Poyas who assured me that these People are well Skilled in the propagation of white Mulberies and in the Manufacture of Silks.

I should be glad of the honour of Speaking to you of it by word of mouth but as this cannot at present be brought about You will be pleased to write to Mr. Walters [Wolters] thereupon, Pardon the Liberty I take in writing to You as by Journal and 'tis by the same Liberty that I make bold to inform You that the Captain has much mended within these few days and the rest will be Settled by the remonstrances and good orders of Mr. Coram.

I adore the Divine Providence which has placed us in such good hands and which so fully and abundantly makes us Sensible of its paternal grace, I admire the cares and Bounties which the Trustees for Georgia and Your Society have for these poor people.

They will endeavour to deserve 'em by their Industry in Georgia[.] I have the honour of returning my most humble and grateful Thanks to Your Illustrious Society for the present of 5 £ Sterling which they were pleased to send me by the Reverend Mr. Butjenter, I hope to find opportunities to render my Self worthy of their favour and I pride my Self in being

Sir   Your most humble and Obedient Servant

Philip Geo. Fred. De Reck

P.S. I reproach my self with not acknowledging Your Letter of the 24th N.S. and cannot find words full enough to shew You how sensible I am of so many favours wherewith the Society are to graciously pleased to honour me[.] Captain Coram has redressed the abovementioned complaints by a generous settling and Ordering of every Point, thank God.

Copy of a Letter from Mr. [Ph.] De Reck Dover 27 December N.S. 1733. No. 12455. Read 18 December 1733. Answered 18 December 1733. To Mr. Newman.

Sir: Herewith is a Letter which I received Yesterday of Mr. Symmonds[.] I make bold to Communicate it to You together with my answer to it, which You will be so good as to forward to him and I take it that the first was written at the first Flowing of Passion; but as I am not willing to have A hand in any mans ruin, I thought it my Duty to soften Mr. Symmonds resentment and to assure him that we are well Satisfied with our Captain as soon as I find he's changed for the better; he has been protesting & promising us every Thing we could wish for[.] Nevertheless! I dont cancel what I have before said being able to verify it both by Oath and Witnesses. But I complain no more of the Captain finding by the Letter herewith sent that he has received Orders no other than what was executed, which was to give us nothing but Salt meat.

Our Passage from Roterdam to Dover lasted longer than we expected. The Ideas that the Saltzburgers have of a Sea Voyage together with their own Infirmities has indeed something contributed to it.

In short we hope what is passed will be no more, so that as soon as God sends us a favourable Wind we Shall proceed with our Captain Fry for Georgia[.] Mr. Butienter I hope will arrive in health at London[.] my Compliments attend him together with my Excuse for not having as yet worte to him[.] we are all thank

God very well and are every Day receiving New marks of Favour from Your Illustrious Society; and the Trustees for Georgia.

I had a great mind to have made my Compliments to Mr. Vernon by Writing but it being too late for this Post I must deferr it till another time assuring You that I am with unlimited Veneration.

Sir   Your most humble Servant   P. De Reck

P.S. The Captain tells me that he has just now bought some Barrels to Store our Fresh Water in for it is a very material point at Sea.

Copy of a Translation of a Letter from Mr. Butienter Dover 11 Demember 1733. No. 12456. Read 18th December 1733.   To Mr. Newman.

Honoured Sir: Mr. [Ph.] De Reck is a very accomplished Gentleman and his Prudence far surpasses his Age which is but 24 Years[.] He came to Us Yesterday in the Evening to Deal in company with the Captain of the Ship to acquaint us with their happy Voyage[.] at the same time he informed me that the Saltzburgers were already provided with Silver plate gilt and Exceedingly handsome for the use of the Sacrament.[153] Therefore I am in doubt what to do with that which has been sent me from London and whether or no I shall give them to the Saltzburgers Ministers, I shall Expect Your answer thereto on Thursday, if I receive none I shall deliver them to the Ministers. Mr. [Ph.] De Reck is resolved for Georgia[.] The Ship Purisburg came into Day about 10'oClock and every Body is well tho' there is a Complaint of the little care and Hospitality of the Captain. This I write in haste because the Post is going off. I shall take all Care that every Thing possible shall be done for their Service. Be pleased to make my Compliments to all Friends who shall ever remain

Honoured Sir   Your most obedient humble Servant
H. A. Butienter

P.S. I have paid Mr. De Reck the 5 £ Sterling.

Copy of a Letter from Capt. Coram Dover 13th December 1733. No. 12460. Read 18 December 1733.   To Mr. Newman.

Good Sir: I thank You for Your Letter of the 11th Instant on which Day the Ship came back from the Downs into this Peere or Harbour[.] Mr. [Ph.] De Reck is a Clever Young Gentleman very much like Mr. Oglethorpe tho not altogether so thin in his face,

Yet as handsome to the full, he looks as much like a noble man as any I have seen, and his behaviour is very engaging being of an Excellent good Humour of good understanding and Prudence, I fancy You had a wrong notion of him, the Ministers and Mr. Butjenter pay him a great difference [defference], his Garb, his Person and his Equipage (which is likely man servant [Schweikert] in a handsome livery) seems to belong to him, in short I can see no fault in him Yet.

The ministers and People seemed to be [omission] against the Captain and made very great Complaints of ill usage at their first coming in here, but Mr. De Reck (who is about 22 or 23 Years of Age) said there was not so much Reason of Complaint as the folks conceived they had for that they wanted water and other Things, when the Seamen were all in a hurry and could not attend[.] however I believe there was not that attention had towards these Emigrants as ought to be and You know as well as I do that the Common heard of (King Tarrs) Masters of Ships are not so Complaisant and Compassionate as they ought to be, and as they would if they had come of worthy Parents[.] But that even this mans behaviour may have been[154] I am persuaded it will be well for the future, his owner Mr. Symmonds has Strenuously ordered him to follow my Directions in every thing and I am sure nothing shall be wanting, and the Ministers and people are Satisfied thereon and because every Thing shall be perfectly right I have resolved to stay here (tho' some of my own affairs require my being at home) to see it performed as it ought to be and shall not embrace the Opportunity of Returning with Mr. Butjenter To morrow morning Early.

I gave Yesterday a Dinner at the Poor house in this Town (lent me for that purpose) to all the Saltzburgers and the English together. I have not time to Enlarge being called away, I am with the greatest respect.

Sir   Your most obedient Servant   Tho. Coram

Many Things are wanting which I will see well provided for.

Copy of a Letter from Messrs. Bolzius and Gronau at Dover 13 December 1733. No. 12462. Read 18 December 1733. Answered 20 December 1733. To Mr. Newman.

Honoured Sir: The Difficult Business upon which You Sir are so much bent for the good of Your Country would without much [omission] have hindered us from troubling You with writing, if we had not been informed both by Mr. Urlsperger's Letters and other reports of the favours You have shewed us and the Poor distressed Exiles[.] we have often returned thanks to God the great Giver of all things that he has raised us up such Patrons in the Metropolis of England who take such strong Care, both of our, and our Churches Welfare both temporal and spiritual and among them Sir You must undoubtedly possess a principal place of Esteem.

And as we have constantly by prayers and Aspirations [sighs] made our Supplication to God that he would be the bountifull Rewarder of You and all our Patrons for their many favours already bestowed on us and those which we may hereafter by the Grace of God receive of them so we hope that God will plenteously afford what we begged for him. Be assured also that in this Transport of Saltzburgers there are some men who give every Day visible proofs both in words and actions of a true and sincere piety[.] what more can be expected of them but that they should gratefully acknowledge the many favours which they had so little reason to Expect and implore their God to preserve their worthy Patrons. I wish Scarcity of time and variety of business and the many Letters which I have to write to Hall and Augsburg as well as other matters which must be procured for our Journey did not hinder our expressing to You Sir the many things worthy observation,[155] of the Earnestness of our Congregations desire of hearing the Gospel of Christ and the holiness of his Life; To tell You their wonderful Patience, there Sparingness their diligent exercise of Prayers and other virtues becoming good Christians which would be much to your Satisfaction and the improvement of others. Let it suffice then that You and the rest of our worthy Patrons who have so laudibly provided for the Salvation of these miserable men be assured that we (who are appointed by God and the Nobles of England, Ministers of the Gospel) are well convinced of the probity of these Exiles and that we very much rejoice at this favourable Opportunity of Going to America under the Patronage of such worthy Gentlemen and in Company with Men so praise worthy.

As for the rest we are to return our hearty thanks for the advancement of half our Salary which we received contrary to our Expectation and occasion in the name of the Illustrious Society by the hands of Mr. Butienter[.] we shall endeavour to apply this

money to the glory of God and the Salvation of other men[.] We
recommend You to God, and ourselves to Your Favour who are
Your most humble Servants   John Martin Bolzius
Israel Christian Gronau
P.S. We take this opportunity also to return Thanks for every
favour received as well as those Things sent us necessary for our
Voyage.

Copy of the Translation of a Letter from Professor [G. A.] *Frank
at Hall in Saxony 19 November 1733.   To Mr. Newman.*

Honoured Sir: The Reverend Mr. Samuel Urlsperger President
of the Venerable Consistory of Augsburg and Rector of the Church
of St. Ann having received both from the Honourable Commissioners for the affairs of Georgia and the Illustrious Society for
Propagating Christian Knowledge command and Power to choose
two Young men of the Confession of Augsburg fit to go as Ministers of the Gospel with the Saltzburg Exiles who are to Settle at
Georgia in America and to impower them to discharge that
function in the Name of both Companies, he by reason of the
Friendship that is between Us, applied to me and desired that I
would with great caution choose out of the Seminary of our Hospitall for Orphans, men duly qualified for that Work, and recommend them to him. I willingly undertook it and by the assistance
of God I found in few days time such who joyfully accepted that
Charge which God called them unto and immediately prepared
themselves first to be ordained according to the Rites of the
Evangelick Church and then to depart for Rotterdam, from whence
with this intended Congregation they are to go over into England.
One of them is the Reverend Mr. John Martin Bolzius and the
other the Reverend Mr. Israel Christian Gronau, both of them
highly valued by me, and Men of approved and undoubted Characters. The first has been with us seven whole Years three of which
he was a vigilant Superintendant of our School[.] The Second has
lived among Us 5 Years and has approved himself a diligent
Master of our Orphans not only by his Teaching but by his
Government of them also. I sent him last Year along with the
Saltzburgers, who were going into Prussia, whom he accompanied
as far as Coningsberg [Königsberg], and then returned to Us. Both
of them have given proofs of their Knowledge and experience in
Sacred matters and have been so remarkable for their Sanctity of
Life, their true Piety towards God their fervency in prayer and

constant self Denial proceeding from the Love of God and an entire Submission to Christ that they were not only an exciting example to their Companions, but also convinced me by several instances that among our whole Society none could be found more fit for the purpose than these whom the Divine Providence directed me to make choice of. Wherefore Sir, I cannot but recommend them most earnestly to You and the favour of the Illustrious Society who have Condescended to do me the honour to elect me one of their Members and who in my Opinion cannot think unworthy of their esteem those who for the Sake of Christ have abandoned their Country and all the advantages they might have hoped for therein and in Obedience to the Will of God have like Abraham undertaken a Voyage into so distant a Country, and are as yet ignorant where their Travels will end. You did formerly shew a great deal of Kindness to the Reverend Mr. Sartorius & the Reverend Mr. Geister, whom I sent over and who at present are propagating the Gospel in the East Indies[156] and that You would be pleased to have the same for these, and to Continue it for the future to'em all, is my most humble request for as I do not repent, nor you neither, I hope, that I have sent the first so I make no doubt but these last will so behave themselves in their Ministerial function that their actions will prove the truth of the Testimony I have given of their worthiness. May Almighty God grant for the sake of Jesus Christ our Saviour that they may safely arrive in America, and there faithfully feed the flock commited to their charge. May they by his divine Permission See that Flock augment every day not only by new Comers out of Europe, but also by the Conversion of the Idolatrous Nations; And may the Illustrious Society from such an Encrease of the Kingdom of Christ, see with delight and pleasure the fruit of their pious and holy endeavours. May the Divine Goodness also grant that a time may come as I have often wished in which Affrica receiving the light of the Gospel may join its praises with those of Europe, Asia and America where it has hitherto shone in a more conspicuous manner so that the whole Earth may jointly sing the Glory of Christ. And would to God this may come to pass in our days, the Illustrious Society under God aiding and assisting those barbarous nations also. I am Sensible however that we ought to Submit all things to the Providence of God who brings them to an issue in his due time. Give me leave Sir, to say something more concerning the Saltzburg-Exiles who to the number of above 20000 are now Settled in that part of Lithuania that belongs to Prussia. The men are handsomely provided for by the most Gracious King who has given them Lands Houses Cattle and Furniture, and besides has exempted them from publick taxes and the Younger sort have been put in a way to get a lively hood in Husbandy. But yet there still remain, and those a great many, decrepit old men besides some lame some blind some sick and some Orphans and Several others who neither can maintain themselves nor be maintained by their poor friends and tho' the King who had them brought thither at his own Expence has done and Still does a great deal in their favour, Yet it cannot be Supposed that he can entirely maintain them. Wherefore well meaning People in Prussia and other places have heartily desired that out of the great Sums Collected in other Kingdoms and Provinces, but chiefly in England for the Relief of those Exiles, some may be bestowed on those Settled in Prussia and that if it were possible some Hospital might be built for the maintenance of those miserable wretches. I know that several have objected to this that it was the Kings part to take care of them as being now his own Subjects. But without mentioning that almost a million of Crowns have been besowed on them out of the Royal treasury that they daily receive marks of the Royal Goodness and Lastly that there never was yet a Prince who maintained all his indigent Subjects out of his own Treasury, I think it is enough in case the charities bestowed on them by the King are not Sufficient to consider that those miserable Exiles, banished their Country for the Sake of the Gospel, deserve to be helped by others also, as it can conveniently be done. And certainly it will little Signify that they have in a manner been helped by many just at their own going forth from their Country if afterwards they are obliged to suffer with the more Severity the hardships of exile, chiefly in a Country, 'tis true fertile and good enough in it self were it well cultivated, but equally unpeopled of men by a former plague and exhausted of money, for want of that vast Trade which flourishes in England insomuch that they could hardly be helped by the Inhabitants, if They had ne'er so much mind to do it. These Considerations and my most humble application to the King of Denmark in their favour did prevail upon him to order that 2000 Crowns gathered for the Saltzburgers in some of his Dominions, should be sent to me to be distributed among the most indigent of those that inhabit Lithuania and the management of that Business has with the approbation of our most Gracious King, been committed chiefly to the Care of a Reverend Divine a religious and well disposed Gentleman. All these things being considered, I hope

Sir You will forgive me the boldness I take to beg earnestly of You to represent them at large to the Illustrious Society together with my most humble request that, if there should remain some of the money destined for the Saltzburgers, which should not be bestowed on those that go to America they would be pleased to take those miserable wretches into their Consideration. I will say no more for fear of being importunate, but will leave all to the Providence of God, who turns and inclines the hearts of men according to his own Will. However I faithfully assure You that, if I was trusted with any money, it should be employed in no other manner than what the Illustrious Society whether it were intended for the Building of a house of maintenance, which should be called by the name of your nation, or whether they would have it distributed in charities among the most necessitous. If the Illustrious Society should think otherwise or some considerable Obstacle should hinder such a Liberality, I shall rest satisfied that I have done what I thought my Duty obliged me to do viz. to represent their deplorable condition to men so conpicuous for their extensive charity, and I hope that if my request find no favour with them they will however forgive me the Liberty I took to make it. Pardon me Sir if I have been too long and too tedious upon this Article.

Now, Sir, to speak a little of our own affairs, one thing in particular I am mighty desirous you should know, which is this, I being gone to Berlin in the Beginning of September upon account of some business relating to our Hospital of the Orphans, which, by the grace of God, is not only in a flourishing condition with all the other Settlements, but also augments and encreases; Our most Gracious King was pleased to send for me to his Royal Palace of Wusterhuss [Königs Wusterhaus], where he kept me seven days and did me the honour of his table and of a frequent conversation with him. I represented to his Majesty both by word of mouth and in writing all my desires in favour of our Hospitall and University; he most graciously granted them all and augmented the Privileges of the Hospital; And after having had the honour to preach before his Majesty he graciously dismissed me, and I returned to Berlin, and during my stay there for the writing and Sealing the Privileges granted to me, I received orders from the King to preach in the new Church of St. Peter the building of which cost his Majesty upwards of Two hundred Thousand Crowns, and at last in Company of one of our Colleagues who was formerly educated in our School, and whom at my request the King was pleased to Joyn to the Society of the Divines of Hall, I returned hither without any danger and in good Health after having received many favours from the Almighty, who at my arrival gave me a new occasion to return him my most humble thanks for that he was pleased to restore to his former health Mr. Freylinghus [hausen], who in my absence had been seized with a violent appoplectic Fit.

I heartily beg your Pardon Sir for having till now delayed to answer Your kind Letter of the 30th of May last year and which I did not receive till some months after; and I think there remains nothing now for me to say in answer to it but to return you my most humble Thanks for that singular kindness the Illustrious Society and You Sir were pleased to shew to the Reverend Mr. Geister and Mr. Knolle the physician and the assistance you gave them to forward their Voyage to the Indies.[157] We humbly acknowledge the goodness of God, by which they safely arrived in those parts and we do not doubt but Success will attend each of their Labours. What relates to that part of the Mission, I have already written and will continue to write to the Reverend Mr. Ziegenhagen. Be pleased Sir to present my humble thanks to the Illustrious Society for their accepting of my desires of assisting, as far as lies in my power the Mission of Madrass and to assure them from me that it will be the greatest pleasure imaginable to me if I can by any means be Serviceable to them in their pious endeavour. I humbly beg the continuation of their favour and yours and assure You that I am for ever

Honoured Sir,   Your most humble and most obedient Servant
Gotthilf August Frank

Translation of a letter from Messrs. Bolzius and Gronau at Dover in Kent 24 December 1733. No. 12468. Read 27 December 1733. To Mr. Newman.

Honoured Sir: We hereby return our heartyest thanks for Your so Singular care in procuring us a Lexicon[158] so very necessary to our instruction in the English tongue which we sincerely promise gratefully to acknowledge. We shall strenuously endeavour if it please God by the assistance of this Lexicon and Grammar to make such progress even on our voyage That we may be enabled to acquit our selves of our Ecclesiastical Duties more amply by their means not only among our present Flock, but among others.

And as by the order and great munificence of your Illustrious

Society these Books have been sent us, We beg Sir you would be pleased to return them our most humble thanks for the present. It shall be our special regard to make our supplications to God the Fountain of all grace and Salvation for every Member of her Illustrious Body And as our Congregation in gratitude for favours received join with us in our prayer there is no doubt but that God will for his Son Christ's sake most plenteously reward them. We take this opportunity of wishing you and the rest of our Patrons a happy Issue out of the Present Year and all prosperity in Commencing the new one wherein we Subscribe our selves.

Honoured Sir   Yours most Sincerely   Israel Christian Gronau
John Martin Bolzius

Dover 24 December 1733

P.S. We Beg you will be pleased to deliver the Enclosed to Mr. Vernon.

Copy of a Letter from Mr. Urlsperger at Augsburg 17 December 1733 No. 12470. Read 27th December 1733. Answered 28 December 1733. Vid. W.L.B. Answered further 15 January 1733/4. To Mr. Newman.

Dear and Honoured Sir: I received both Your Letters of the 13 and 20 November last But That of the 6 is not come to hand which hath not happened before[.] Our Colonists as I hear have received very little kindness from the Magistrates of Roterdam. Mr. Vat is still in my house and is waiting till God shall be pleased to send an opportunity of shewing his service to the benefit of the Emigrants of Saltzburg and we Expect to hear further as we have wrote again for that purpose the resolution of such of the Tirnbergers as are at Franckfort. Mr. Schumacher T. S. [Theologiae Studiosus?] who went with our Colonists to Roterdam being returned this moment says that he left 2 ships at Nimmegen [Nimwegen] loaded with Tirnbergers who left Cassant [Cadzand] and were to proceed to Frankfort but were very sick and many of them dying. May God have mercy upon these poor Creatures! What Mr. Vat relates in his Letter concerning a Transport of Emigrants of Saltzburg likewise of the resolution of the King of Prussia of receiving no more of them into his Dominions is even what I told you several times five or Six Weeks agoe. Now I long to hear that the Ship Purisbourg is sailed from Rotterdam and that she is Gott into Some Port in England[.] I have been greatly affected with Mr. Oglethorpe's falling off the Horse. May God have restored him to his health that he may be Still for along time a Blessing to England

and Georgia. Mr. [Ph.] Von Reck besides the 500 Guilders I gave him here and the 150 Guilders he took up at Francfort hath received One thousand Guilders Current in Holland; But as I have not yet received his Account I hope he will send it me from Dover. Each Medal in a Box costs Five Guilders[.] Prince Charles Alexander now Duke of Wirtemberg who, 21 Years agoe conformed to the Roman-Catholick Religion laid last Week one Night in this Town and ordered the Present of this City consisting in Fish Wine and Oats to be given one moitié thereof to the Protestant Ministers of this Town (which is a Precedent without any Instance) and the other Moitié to the Capucins[.] It is hoped he will make no alteration in Religion in His Dominions[.] Now at the Arrival of my Children of Saltzbourg in England together with their Ministers, Mr. Von Reck their Commissary and Mr. Zwibler [Zwiffler] the Apothecary I do recommend them to Your Care[.] there are besides the Saltzburgers some proselytes from Popery who as I understand by Mr. Schumaker require a greater Inspection than the former. Mr. Schumaker hath been very serviceable to the transport, and he cannot sufficiently extoll the kind care of both Messieurs Von Münch of Franckfort in assisting the whole Transport with Money, Victuals, and Recommendations, especially himself, having sent him hither from Franckfort Gratis by an Extraordinary Post. I must also acquaint you that Mr. Schumaker, who presents his duty to the Society, tells one that the Saltzburg Colonists are very brisk and extreamly well pleased with the Provision made for their Well being both in Religious and Civil concerns for which they are praising God and praying constantly for the Prosperity of the Gentlemen Trustees and the Society.

Since Mr. Von Reck is going for Georgia, it seems to be necessary supposing that it is yet time to give Instructions in what manner he and both Ministers shall act in their Subordination to each other in order to avoid any misunderstanding between them.

Within this fortnight two Romish Priests one of them 63 Years of age, and a very sensible man, have embraced in my study the Protestant Religion both of them have been moved thereto by the motions of the Saltzburg Emigration. Sir I recommend You to the protection of God Almighty Remaining

Honoured and Dear Sir   Your Humble Servant   Sam. Urlsperger

Copy of a Letter from Mr. Vat at Augsburg 17th December 1733. To Mr. Newman.

Honoured Sir: It was with great pleasure I received the favour of your Letter of the 13th of November last; by which you acquainted me with the Society's pleasure that I should remain in this Town 'till They or the Reverend Mr. Urlsperger upon occasion can point out the service I am to undertake either Here or at Franckfort in Conducting a further Transport of such Emigrants as may come out of the Archbishoprick of Saltzburg and shall be willing to go to Rotterdam and thence to be embarked for Georgia. At present there is in this Town but one Family of such Emigrants [Rieser] inclined to go to Georgia, whose Tackle is gone with the first Transport but was detained here by reason of the breaking of a Boy's Thigh which being Cured, the whole Family was Yesterday privately brought into the Poor House of this Town, by which one Moiety of the Expence will be Saved. Yet there is a prospect of Seeing a good Number of Emigrants, which in all Probability will be sent away at the first Snow that falls in the mountains of Saltzburg as usual; Since Burgomaster Morell told me the other Day that according to certain advices there were 1000 of them and above minuted down in some Country Courts of Saltzburg. And as the King of Prussia is not inclined to receive any more of them in his Dominions unless his Subjects do provide them with necessary means of Subsistance, it may so happen that such a Number would not Care to be divided, but resolve to go jointly to Georgia[.] In this last Case I should be glad to know the Pleasure of the Society, whether any Greater Number than 300 might be accepted and likewise the Number of Acres of Land that will be allowed to every Head? and also the Conditions annexed thereto, I am humbly of Opinion that by giving no more than 50 Acres to a Family, whilst 50 Acres in Carolina are given to every Head it may be attended with some Discontent, which might be avoided by allowing a proportionable Number of Acres to every Head.

I am very much concerned at Mr. Oglethorpe's misfortune but I hope he is fully recovered of his Contusion, and I am glad to hear by Your Letter of the 20th That Colonel Purry is returned to England, an Account of the progress of his Colony would be here very acceptable.

I pray God His Royal Highness the Prince of Orange be so well that his Nuptials are compleated by this Time, and I beg Leave to Subscribe my Self

Honoured Sir   Your most Obedient & most humble Servant
John Vat

Copy of a Letter from Capt. Coram at Dover 22 December 1733.
No. 12473. Read 27 December 1733. To Mr. Newman.

I am very thankful for Your favour of the 15th. I gave Your Letter for Mr. Oglethorpe to Mr. [Ph.] Von Reck with your respects to him and the minister and Catechist as desired. The Minister is a little thin man about 31 Years of Age, and resembles Mr. Vat at a Distance[.] The Catechist is about 25 Years a Tallish ruddy looking young man, they seem both to be very sober persons[.] the minister preached before them last Sunday and I suppose will do so at the House hired for those Emigrants by Mr. Butjenter (during their lying here wind bound) at 2 S per week, I supply them with Coles for their firing to wash their Cloaths and to Sleep by, they seem to be harmless Sober folk. I want very much to be at home, yet would fain see those people off at Sea with a fair Wind for Georgia, but am morally sure if I go away one unhappiness or other will fall out which can be prevented or cured if I stay, I have taken effectual care they shall not want for any thing[.] they were very apprehensive of Want of fresh water in their Voyage but I have provided for them and have had empty Wine pipes filled with fresh Water many Days to Soak the taste and Strength of the Wine out of them, ready to be emptied and new filled & put on bord the first appearance of alteration of Wind & have bought them Shoes Stockings, linnen Thread etc. for Shirts and Shifts and necessaries for a lying in woman [Schoppacher] on bord, Because no body is more Sensible than I am of the great good Consequence it will be to this Kingdom to have our Plantations filled with Christian Inhabitants (not with Jews)[159] and I hope to live to see the Thing when there will be as many foreign Protestants follow those Emigrants to our Plantations as those here have Hairs on all their heads; perhaps I do not well in Saying, I hope it but I say again I expect it from the Persecutions going forward in Saxony (if the affair of poland does not hinder it)[160] which I have seen mentioned in many of our News papers[.] I do not remember I have Spoken of it to You when at London[.] it is as follows viz. the Elector of Saxony (a good Christian like his father) it seems has declared he will have none but of the three following Religions in his Dominions Roman Catholicks Lutherans and Calvins. now there are Baptists Quakers and other Protestants who are greatly alarmed at this pious Declaration[.] The Baptists in that Country have sent to the Baptist Teacher at Harleim [Haarlem?][161] to learn out and inform them in Saxony of the manner of Settling in the New Colony of Georgia. The Charge of the passage thither of 60 folks with their necessaries and other matters relating thereunto: The Baptist Teacher came in September from Harleim to Rotterdam to be informed thereof, and on his enquiry concerning it among the Merchants he happened to hear of me there at that time[.] I met him at one of their houses and gave him a true and faithfull Account of the whole in the best manner I could for him to transmit to his friends the Baptists in Saxony; If that Persecution goes forward it will help to fill the British plantations with Christian Protestants and when once some are got there they will soon engage others to follow if they are not ill used there[,] good usage has a wonderfull effect; especially if it be applyed at the first[.] it is on that account I do all I can to wipe off the ill impression The Ministers conceived at their first coming into the King of Great Britain's Dominions I mean in a British Ship and upon the Sea[.] as for the Barron [Ph. von Reck] he is of a healing temper[.] the more I see of him the more I like him and all that I have said of him is not half enough.

I do all I can to give them all true Satisfaction and to engraft in them a high Idea of the excellency of the British Government. I visit the poor Emigrants once every day at the house Mr. Butienter hired for them during their Stay here (the Baron and the two Ministers and Physician remain in the Lodgings he hired for them in the Town)[.] I carry the Children a few apples, and sometimes give them a few plumbs, a pound of malaga Raisins which costs 3 d. fills them with above 5 pounds worth of Love for me[.] they shew it by a Dawn of Joy in their faces as soon as they see me coming. I was very much Disturbed a few days ago by Being told that 3 of the Saltzburgers viz. one old and 2 young men begged on Sunday Night at the house of Mr. Paschal a pilot here, I acquainted the Baron & the Minister thereof they could not believe it, but made Enquiry amongst them all but could learn nothing for the Emigrants deny'd it very strongly so the Minister as well as the Baron would have had them all the men brought to the said house to see if Mr. Paschal could know which of the three or any of them that was at his house & I intended it when Mr. Paschal would fix a time for him to be at home but was last night informed it was 3 Dutch Sea men in their way thro' this Town for Deal had begged and Lodged here in their way thither so I was glad to find the Saltzburgers were not Guilty of begging.

I had another Difficulty Yesterday morning, Ortman and the Barony's Cook [Schweikert] who is of Woortsberg [Würtzburg] not of Saltzburg can speak no English, but Ortmann can, they came to me and told me that the Cook had been cheated of a Guinea by a Merchant in the town who had sold him Cloaths for a Suit of Cloaths and was to pay him for making it 15 S. so I desired the Baron to go with me that the poor man might not be Cheated. Upon hearing the matter fairly, it appeared to me that there was a mistake of the cunning Cook who would have Cheated the Shop keeper of a Guinea by endeavouring to prove he had not received it But the Shop keeper who happens to be of the best reputation of any in the Town made affidavit of his paying a Guinea to the Cook and I saw all the reason in the World to believe it. I am glad the Cook is no Saltzburger.

I am with the utmost respect   Good Sir   Your most Obedient
Servant   Thomas Coram
P.S.: I have received Yours of the 20th with one Enclosed to Mr. Bolzius which I gave him[.] he is very thankfull for the kind Care of them as is the Baron [Ph.] von Reck also[.] I expect the Coach will be here to morrow.

Copy of a Letter from Capt. Coram at Dover Kent 27 December 1733. No. 12474. Read 31 December 1733. To the Honourable James Vernon Esq.

Honoured Sir: I am extremely Sorry the Wind continues still Westerly. I was in hopes Yesterday morning the Wind was coming about to the North and North East Point.

I am afraid if I come away before the Ship sails some dissatisfaction or Confusion will happen though all is in peace and Quietness at present but whilst Mr. Butjenter was here he hired the use of a Large room in a house near the Ship for the Saltzburgers and most of them lye therein. Mr. Ortman and his wife are the Linguisters and they say and misconstrue what they please which gives uneasiness. The Baron [Ph. von Reck] is very good and wise, I should have been wonderfully pleased for you to have seen his Behaviour after I had told him that I would [wished to] when he should think it a proper time have all the Saltzburgers together to sign and Seal the Parchment Instrument I brought down with me and shall bring it up[.] he said it would be necessary to prepare them for it and he thought next morning would be much better than that afternoon, accordingly he and the Ministers went before to the house aforesaid to Prayers with them (as they do every morning and Evening)[.] they had prepared the Saltzburgers over night who had their own and their Children's faces Clean Washed and every thing in order; I took with me the Captain of the Ship [Fry][162] and Mr. Sale who goes passenger, and when we Were there come the Baron made a Speech to them in their or his own Language (for he can Speak but little English but hurried it very fast) in a very Comely and Solemn manner and read the Instrument to them and afterward read an Oration to them that he had drawn up and when he had ended some part of it they all with one Voice Cried Yea, and at the End came one after another and gave me their hands (as a Token of their Strict abiding by their undertaking) as a Dep[uty?] from the Trustees[.] the Manner and Behaviour of the Baron on this occasion was so Comely and with so much Gravity that I dare say Your Honour would have thought him very agreeable for that Service, I have desired him to give me a Copy of that Oration which I herewith enclose and should be glad to see a Translation of it hereafter[.] I still hope the Wind will turn soon[.] I am

Honoured Sir   Your honours most Obedient Servant
Thos. Coram

Copy of a Letter from Christopher Ortmann at Dover in Kent 30 December 1733. No. 12486. Read 8th January 1733/34. To Mr. Newman.

Honoured Sir: It behoves me to return you humble Thanks for the Share you were pleased to grant me of your Salutation particularly in your Letter to Monsr. [Ph.] De Reck. I make bold therefore to renew in your remembrance the Assurance of my humble Service and the passionate desire I have to give You proofs of it beseeching You to believe that I shall hold most dear all Opportunities of shewing and manifesting it. Therefore I most humbly beg the Countenance of Your favours that you will be pleased to afford me the Effects of your protection and favour tho' I shall be far off yet shall always acknowledge them with fervent prayers to God daily, especially with a beginning of a New Year that the Divine Goodness will be pleased to grant his Blessings upon You and the Honourable Society and the Trustees for Georgia,

from your most Dutifull & obedient humble Servant
Christopher Ortmann

My humble respects to the Reverend Doctor Guerdes Mr. Butjenter and Mr. Ziegenhagen.

Copy of a Letter from Messrs. Bolzius and Gronau at Dover Kent 4 January 1733/34. No. 12488. Read 15 January 1733/34. To Mr. Newman.

Honoured Sir: Had we not received many proofs of your Favour to Us we should fear incurring your displeasure by our frequent importunities; You who are ingrossed by such a Concourse of Business[.] But the greatness of Your kindnesses obliges us before we leave England to return thus (since there is no other way) our hearty acknowledgment to You and the rest of the worthy Patrons and certainly it were unjust to pass over in Silence such a multitude of Obligations; Altho we cannot find words to express the Gratitude wherewith we are penetrated in a Sense of so many favours.

But to omit the rest we must not be silent in reference to the Honourable Capt. Coram whose Summ of care has been to provide us and our Congregation with all necessaries for our welfare both while here and in our Passage so that we can easily obliterate every thought of the Misbehaviour of one man [Fry]. May God (whom the Sweetness of your kindnesses compells us to invoke) crown his white head with the Blessings of his grace, and that he would be pleased to accompany with all mercy our worthy Patrons (at whose Cost and by whose Command and direction we have felt so many Sweets) is the sincere wish and Prayer of us all.

Nothing is now so much our regard as the Search after an Opportunity of shewing how sensible we are of all Civilities[.] in the meanwhile we shall to the utmost of our poor Capacities endeavour by the assistance of God to acquit our Selves to the Glory of heaven the honour of our Patrons and the Benefit of other men as well as our Congregation.

Adieu and Continue Favourable to U[s?]   Sir Your most obedient humble Servants   J. Martin Bolzius   Isr. Chr. Gronau

Copy of a Translation of a Letter from Mr. [Ph.] De Reck at Dover 4 January 1733/34.[163] *No. 12489. Read 15 January 1733/34. To Mr. Newman.*

Sir: I believe my last to you of the 11 Current 27 December together with my Letters for Ratisbon and Franckfort and my Account of Receipts and Disbursements have before now reached You. The Acquittances Instruments, Contracts and other verifications, I hope at my Return to deliver to You with my own hands.

You may perhaps find in the Account some Articles bought purely for my own Convenience, but as I am ready to reimburse it or have it placed to my Account the Sum I have in my hand is the Greater and the Expence something less. This Transport has cost a deal of money by reason of some Occasional Necessaries which for the future we shall be able to Shift without.

Some hundreds of these People might be very well managed if the Succeeding Transports were upon their immediate Arrival at Rotterdam to Embark on bord the Sloops which come twice a Week for London. That would not come to above half a Crown a head, whereas this Vessel Stands in, for the passage above 4 times the Sum, besides the great expence it is to live at Dover considering the exorbitant price of every thing. Capt. Coram who Sets out to morrow for London will give you a more exact Account of the Situation of our Affairs and assure you by word of mouth as I do by Letter that I am sincerely

Sir   Your most Obedient Humble Servant   P. De Reck

Copy of a Letter from Mr. Isaac Minet at Dover in Kent 10 January 1733/4. No. 12491. Read 15 January 1733/34. To Mr. Newman.

Sir: I am honoured with the favour of Your Letter of the 8th Instant with the Letter directed to Capt. Coram and in his absence to me also, a Letter directed for Mr. [Ph.] De Reck, the first being gone back to London and the Latter Sailed the 8th Instant in the Purysburg, I send your Said Letter back herein, I should otherwise have readily Complied with your order and been glad to Oblige said Mr. Reck, who I am Sensible hath received great favours and Civilities both from the Laudable Trustees for Georgia and praise worthy Society for Promoting Christian Knowledge all which I do believe he deserves by the Zeal Courage and prudent behaviour, and management and Vigilance with which he hath acted during the time he was here. I hope God will bless his undertaking and prosper all the just and Charitable designs of the Trustees for Georgia. I present my Service to them and am Respectfully

Sir   Your most humble Servant   Isaac Minet

I pray God restore the Reverend Mr. Butjenter to perfect health.

Specification of those Saltzburg-Emigrants who set out from Augsburg the 20 of October 1733 under the Conduct of Mr. Von Reck,

Commissary, and arrived at Rotterdam the 16 November following and Embarked on board the Purrysburg Commanded by Capt. Tobias Fry for Georgia the 21st Ditto 1733.

1. Mr. Von Reck, Commissary
2. J. Martin Bolzius ⎱   Minister
                    ⎰   Joined at Rotterdam
3. Is. Christ. Gronau ⎰   Catechist
4. Christopher Ortmann ⎱
5. Juliana his wife . . . ⎰   Schoolmaster
6. John Andr. Zweiffler . . . Apothecary
7. Christian [Schweikert] . . . Footman to Commissary Reck.

                               Years

8. Thomas [Ge]Schwandel . . . 38 . . . A Worker in Silver mines
     Aged _____
9. Margaret his wife _____ 23
10. Margaret their Child _____ 1
11. John Mosshamer _____ 34 born in Lainthal Zeller
                                 Jurisdict.
12. Mary his wife _____ 28

13. Paul Schwaighoffer _____ 45 born in Mitter-⎱
                                 field [Mittersill] ⎰ A Weaver
                                 in the plains of ⎰
                                 pintzgau
14. Margaret his wife _____
15. Mary ⎱ _____ 7
16. Thomas ⎰ their children   4
17. Ursula ⎰ _____ 1

18. Lorentz Hueber _____ 54 born in Gastein . . .
19. Mary his wife _____   Husbandman
20. Magdalena ⎱ _____ 13
21. John   ⎰ _____ 10
           ⎰ their Children
22. Mary   ⎰ _____ 8
23. Margaret ⎰ _____ 5

24. Barbara Rohrmoserin aged 36 Wife of Peter Kraer [Kröer]
                                 born at Kulback in the
                                 Jurisd. of Saalfeld.

25. Katharine ⎫———————— 17
             ⎬ her Children
26. Gertrude ⎭———————— 14

27. Mary Reiterin ———— 21 Born at St. Ulrick in Tirol.
28. Anna Hofferin ———— 25 Born at Hof in Gastein.
29. Hans Grueber ———— 45 Born in Gastein . . .
                                     Husbandman
30. Christian Steiner ———— 30 Born in Ditto . . . A Worker in
                                     Silver Mines.
31. Balthazar Fleiss ———— 27 Born in Ditto . . . Ditto
32. Jerg [Georg] Swaigger —— 19 Born in Ditto . . . A Labourer
33. Tobias Larkner [Lackner]— 40 Born in Ditto . . . An overseer
                                     of Coal works
34. Peter Grueber ———— 36 Born in Ditto . . Husbandman
35. Simon Reiter ———— 26 Born in Ditto
36. Martin Herzog ———— 35 Born at Schinking⎫ A Miller
                                  in Pintzgau. ⎭
37. Christian Leimberger —— 23 Born in Saalfeld . . .
                                   A Husbandman
38. Stephen Reidelsperger —— 22 Born in Lich-⎫
                                  tenberger in⎪
                                  the Juris-⎬Husbandman
                                  dict. of Saal-⎪
                                  feld ⎭
39. Matthew Mittersteiner —— 41 Born in Marck Goldegg . . .
                                     Husbandman
40. Simon Rauschgot ———— 24 Born in Ditto . . . Ostler
41. Leonard Rauner ———— 27 Born near Ulm . . .
                                     A Labourer
42. Matthew Braunberger —— 30 Born in Bavaria . . .
                                     A Husbandman
43. George Bartholomew Roth 45 . . . A Distiller
44. Mary Barbara his wife —— 32
45. Gilbert Beque[164] ———— 26 . . . A Baker

Men ———————— 26
Women ———————— 7
Girls ———————— 10
Boys ———————— 2
                 ————
Total ———————— 45

Copy of a Letter from Mr. [Ph.] Von Reck at Dover in Kent
11 January
31 December 1733.[165] *No. 12487. Read 8th January 1733/34.*
Answered 8 January 1733/34. To Mr. Newman.

Sir: I acknowledge the Receipt of Your three Letters viz. one of the 13th a Second of the 18th and that of so late a Date as 27th of December which you did me the honour to remit by Capt. Coram. The Care which You will take of the Waldenses and the report that will be made of them the first Opportunity has very much rejoiced me. Mr. Pury who is gone to Switzerland to fetch them told he would accept of none but choice ones such as are of most Service to the Colony but I am of Opinion he won't get one without'em all.

My uncle at Ratisbone [J. von Reck] will receive great Comfort in his old Age if You favour him with the Journal which I shall now and then send You.[166] Beside the two Letters at Frankfort and Ratisbone, to which I make bold to refer you have here added the Account I kept of the money I took up both at Augsburg Frankfort and Rotterdam[.] I assure you that by the experience I have picked up in Conducting this first Transport I shall be able to go through the same Voyage with another of 200 persons at very near the same rates I have managed this with. In my Letter to Mr. Urlsperger which I inclosed in a Packet to my Uncle I have made mention of the present State of Accounts. Scarcity of time prevents my sending two Copies but as soon as You perused it You will be pleased to Communicate it to Mr. Urlsperger.

The 8 and 28 we had a good Easterly wind which Seemed to promise us a good Voyage but as soon as we were on bord and ready to Set Sail, The Wind chopped about to S.W. which is the most Contrary Point it could turn, Thus GOD presented Us with the prospect of a Good, he did not think fit we should enjoy.

We Submit our Selves entirely to his will and brake out always in praise of him be it for Good or ill fortune conscious of whatever befalls us is the appointment of the father of Light. I adore him and entreat him that that year we are now commencing may prove auspicious both to you and every member of Your illustrious Society, God preserve them and You for the publick Good and the good of so many Souls that look upon them as the only prop to save them from being Slaves to their Consciences and in short for the Comfort of this Transport the Society's true Servants in the number of which I have the honour of Stiling myself the most humble and the most Zealous.

The Affair about the Cook who is a native of Wertzburg [Würtzburg] with a merchant here (having throughly examined it) is decided by an Oath the merchant took who has thereby gained himself a demi probation; Tis true the Cook is no very good Christian but as he says that he is ignorant both of the money and Language impute it to a mistake.

The Difficulties of our Voyage are thank GOD pretty well removed and Surmounted. We have nothing to do but to pray to GOD earnestly for a fair wind and a renewall of his Grace with an Opportunity, I believe and I hope, recommend my Self to your favour which I beg you to continue to me, this and the ensuing Year, and for ever, to the man who glories himself in being sincerely

Sir   Your most humble and obedient Servant   P. De Reck

Copy of a Letter from Mr. Urlsperger at Augsburg 18 January. 1733/34. No. 12501. Read 22 January 1733/34. To Mr. Newman.

Honoured and Dear Sir: I have received none of Your Letters since that of the 20th of November S.V. neither is your Letter of the 6th of November last as yet come to hand. But you and the Reverend Mr. Ziegenhagen have since that time as I doubt not received several of my Letters the Receipt of which, and Answer thereupon I long to know. Now I have desired Mr. Vat, who, GOD be praised is restored to Health to translate divers Extracts of Letters viz.

1. Extract of a Letter from Mr. Von Herald one of the Privy-Council of his Majesty the King of Prussia Dated at Berlin 19th December 1733.

His Majesty hath appointed one of his Chief ministers who taketh particular Care of the Emigrants out of Berchtolsgaden of which Thirty Families are remaining here So that they are provided both with Spiritual and Temporal necessaries. And there is actually a proper Schoolmaster appointed for the Children of these poor people. His Majesty hath likewise given gratis the Materials for a School House and one hundred Rix dollars towards the Charges of the Workmen employed therein. In the mean time these Berchtolsgadeners settled here are very pious and quiet, and rejoice heartily at the Progress of their Children in Christianity together in Reading and Writing of which last the old people know nothing at all.

2. Extract of a Letter of the Rev. Mr. Möllenhof, Chaplain to the Danish Envoy at Vienna. Dated at Vienna January the 6 1734.

I can now acquaint You relating to the affair known to you That the English Envoy hath actual Orders from his Court and that he will in few Days make a Representation to the Ministry here. If England had declared themselves for the Emperor[167] as 'tis hoped it will be done out of Hand there is no doubt to be made but it would have a good Effect Yet perhaps may it be the sooner hoped for. May GOD direct as it may be best. Affairs in Hungary have a very melancholy Prospect at this Time. One would think there were other Affairs to be taken Care of; Yet the Clergy are still going on very furiously. They are now more intent than ever upon this, That the Apostates (for this they call such as for 20 - 30 Years and before are come over to us, yet the Law against Apostacy was published not above Two years and Half ago) shall be punished! and at Presbourg [Bratislava] They will with Authority reduce the Schools to That, which teacheth Grammar and they have only Fifteen days Respite allowed them. May GOD give his assistance to these poor people and to all such as are in want thereof.

3. Extract of a Letter dated at Ratisbone January the 14th 1734.

We receive advice from time to time from Austria that many good Souls are there moved by the good Spirit; likewise from Cärnton [Carinthia], where many good men in one District of Sessions, have declared themselves freely concerning the Holy Doctrine of Christ and that they would not forsake it whatever Should be done with them; Seven of them were immediately clapt up in Prison, as a pious Emigrant out of that Country, who went thither and brought over one of his Sons in Safety hath assured me thereof which hath likewise been confirmed by a Messenger. Oh! that it might[168] a greater insight to the Great. I have received a very fine Letter out of Austria, which I showed to one single Envoy, who is a Director but I must keep it in my own hands and no man is to see it, yet I frequently wish in my Soul to speak to you Sir thereupon.

4. An Account of the Tirnbergers which I have ordered to be printed.

Whereas several different accounts, concerning such Tirnbergers as are gone to Holland have been given, partly to the prejudice of the Country where they are Settled, partly to that of the Emigrants, upon which no Judgment can justly be made by reason that no Sufficient Evidence could as yet be had either on one Side or the other, so that the different Accounts might be compared; It is hoped it will not be unacceptable to the publick, to read what hath been communicated by a Letter and may be depended on.

Extract of a Letter from N. Dated 14 January 1734 to N. N. in N.[169]

As to the Emigrants in Holland a great deal of Contradicting and untruth is both Spoken and written. Sometimes Mr. Fisher [Fischer] their Minister is reported to be Dead at other times to be upon the Road to his return into the Country, Then it's said, There was not one Single Emigrant left in Cassant [Cadzand] And it is certain that Mr. Fisher is Still alive and praiseth GOD that altho' he hath been left with a Small yet good Flock which was overheaped with many Benefits, as having received one Thousand Guilders and upwards from the City of Amsterdam towards their Support. He adds that an Orphanatrophy hath been erected, wherein he hath put the Saltzburg Children which he himself instructs in Part. That now some of the Tirnbergers themselves are appointed Schoolmasters, who have a good Foundation to whom the Consistory of Amsterdam allow a Gratuity of one Hundred Guilders etc.[170]

I further acquaint You Sir that I received some days ago Letters from Mr. [Ph.] V. Reck dated at Dover 23 December last telling me the arrival of the Transport there and how generously all fatherly Care was taken for himself the Ministers and the Colonists. Now I long to hear what passed there since the 23 December and at what time the Ship Sailed. Just now as I was going to finish this Letter I received your favour of the 28th past. But for want of Time I can only answer thereupon, that I have not got your Letter of the 18 of December and that it will be expedient to direct your Letter as I told it lately to Mr. Ziegenhagen. I should be glad to give my Letter directly to Mr. Newman, to the post here; But when any Thing of Consequence is to be communicated, by reason of our Jesuits here, I dare not trust,[171] for they know too well the Names of the Society of Mr. Newman and mine[.] if in Time to come Books pamphlets or other like things shall be sent to me, They may be directed to my Self and sent to Mr. John Hörman Zur[horst] Merchant in London who will Consign them to Messrs. Peter Laire & Company Merchants at Augsburg, as receiving Yearly several Chests of Wares from England[.] I remain with all my heart

Honoured and Dear Sir   Your very humble Servant

S. Urlsperger

Mr. Vat returns his humble Service to You and his Friends.

Copy of a Letter from Mr. Urlsperger at Augsburg February 8, 1734. No. 12523. Read 12 February 1734. To Mr. Newman. [Marginal Note: The remaining Cash will be mentioned in my next.]

Honoured Sir: Upon the favour of your Letter of the 28 December last S.V. I answer That upon the request of the Society made last Summer, How much Money I had then in Cash? I already answered whilst I was in Saxony, where I remained for some Weeks I had still about Two Thousand Guilders, and as soon after I received a Bill of Exchange of Three Hundred Pounds Sterling being 2587 [Florins] 30 Cruitzers of our Money for the Georgia Colonists I have added thereto One Thousand Florins from my old Cash which belonged to all Emigrant Saltzburgers in General and have made a particular Georgia Cash of 3587 [Florins] 30 Cruitzers[.] of these 3587 [fl] 30 Cruitzers there still remains—X —Xrs.[172] in Cash. The Overplus hath been employed partly in giving some thereof to the Officer; and his Servants who came in August last into Bavaria with a Transport of Emigrants of which our Colonists came having sent to them the English Proposals concerning Georgia and having desired him to bring hither such of them as should resolve to accept them; partly for subsisting the Colonists who remained 8 weeks near this Town with Victuals Lodging Fire and Candles and providing them with Cloaths Stockings Shoes etc. in order to enable them to go hence to Dover, For three Waggons to carry their Women and Children together with their Baggage to Marck Stefft [Marksteft] where they took Shipping; and also for a Chaise to carry thither Mr. Commissary [Ph. von Reck], Mr. Schumacker the Candidate to Holy orders and the Apothecary [Zwiffler]; For the extensive Correspondence at Ratisbonne at Vienna Saxony London and to a Writer for translating into French and Transcribing several

Things; For *Printing the Account of Georgia* and the Conditions thereof For Books as Bibles Catechisms etc. bestowed on them; For Fees to a Physician Surgeon and Apothecary attending some Sick Persons among them Especially a Boy [Georg Rieser] who broke his Thigh. To which is to be added 500 given here to Mr. Commissary, 150 he received at Franckfort and 833 [florins] 20 X^{rs} he took up at Rotterdam. For Mr. Schumakers attending the Emigrants to Rotterdam and his return hither and to Ulm. For the Vasa Sacra sent with the People to Georgia. For Mr. Vat's Dieting since the 10th of November last to this day and for his illness to the Physician and Apothecary. For the Subsistance of the Emigrant Saltzburger Reier [Rieser] his wife and three Children who were to go with the first Transport but were prevented by the breaking of the Boys thigh. Now I have only to provide out of the Georgia Cash Mr. Vat and the Five Persons remaining here by reason of the Boys Misfortune. Should a new Transport come out of Saltzburg next Spring or Summer and resolve to go to Georgia Mr. Von Münch will be at all times ready to advance the Money necessary thereto. As to the remaining One Thousand Florins reserved to the assistance of the Saltzburgers in General I am to pay the pension of some Saltzburg Children to be educated especially Thomas Trickels a Boy of a Lively Spirit. Likewise something towards the Subsistance of some very old and infirm Saltzburgers Lodged in our Protestants poor house Set up thirty years ago and since kept up merely by Voluntary Charities without any capital Stock and which Saltzburgers were taken in in the Year 1732. And then towards the assistance of some Saltzburg Youths put out Apprentices to some Handicrafts Trades etc. as I mentioned it in July 1732 and as it was inserted in the publick Newspapers of London. Moreover it must not be forgot that not only such Saltzburgers have been assisted as went hence to Prussia but also such of the Tirnbergers as went to Cassant [Cadzand] and of the Berchtolsgadeners who went into the Electorate of Hanover and to Berlin, as I gave You an Account thereof, at the Time I sent the assistances to Ratisbon. As to the Bills of Exchange which the Society were pleased to send over to me I immediately upon the Receit of them sent you my Receipts of their Respective Sums; Upon which Receipts I must refer you till such Time as I can obtain the Original Bills of Exchange from Mr. Von Münch, and afterwards compare them with the Account You sent lately to me. I have desired Mr. Von Münch to let me have those Bills of

Exchange, but by reason of the multitude of affairs, he hath not been able to send them as Yet. Therfore I still must wait for them. Moreover I praise God for the many kindnesses as well of the Society as of the Trustees bestowed on our Georgia Colonists. I only wish that in Case God should be pleased to send a new Transport; The Captain and his Crew would shew themselves more kindly to the Colonists as was lately done from Holland to Dover[.] And this for the Sake of the Colonists themselves, as well as for my own Sake as I have upon Mr. Secretary Newman's Letters, faithfully promised them that they should be most humanly well Used on Bord the Ship, and provided with all Necessaries; But which by the Captains neglect and Fault, hath been extreamly wanting between Rotterdam and Dover as I have learned it by the Letters, both of Mr. [Ph.] Von Reck and Messrs. Bolzius and Gronau[.] I will suppress every Complaint thereof, That the main undertaking may not come under an Evil Report.[173]

Concerning the Salary of both Ministers after I had consulted Ingenious and Knowing Men and the Society's Power, dated at London the 3d. July 1733 given me in the following words: "In answer to your Desire for sending more than one Minister, The Society are inclined to Send Two Ministers and one Schoolmaster with the Colony of Protestant Emigrants that shall be disposed to go to Georgia and that those Persons receive their Ordination at Augsburg or some other Protestant Church in Germany before they depart thence. But as the Charge of their Transportation will be considerable, the Society leave it to your Prudence to Settle the Salaries of those Persons as frugally as you can the Society relying on good Providence to enable them to comply with what You shall Stipulate in their names." I have promised Sixty Pounds Sterling to Mr. Bolzius and Fifty Pounds Sterling to Mr. Gronau, and likewise inserted those Salaries in their Letters of Vocation. Now as I understand that only 25 Pounds have been paid in Dover to the First and Fifteen Pounds Sterling to the other as their Half Yearly Salaries, You may Dearest Mr. Secretary easily Conclude that these good men[174] How this Difference may arise. Therefore I hope the Society will extend their Bounty hereafter to such a Sum as I in the name of the Society promised to them.

I received the following advice of the Danish Chaplain Dated at Vienna the 3d. February 1733. "Concerning the Saltzburg Emigrants designed for Georgia I lately asked our Envoy whether he had Learned Nothing how matters were? His answer was; That

Mr. Robinson [the British Ambassador] a few Days before had told him: That his First Attempt had been Fruitless; But that he would give the Affair another Turn, as soon as I shall learn anything more I shall not fail to give You Notice thereof. We know here nothing how it goes in A. and K.[175] neither have I heard anything since from the Baron.[176] But in Hungary all Executions are Suspended and everything is to remain *pro nunc,* or at this Time in the present State. God grant that it might be said *pro semper,* or always! Yet God perhaps designs greater Comforts to the poor Flock."

Mr. [J.] Von Reck, Envoy at Ratisbonne wrote to me some Days ago. "We Evangelicks have applied to the Imperial Commissioner concerning the Clause of Religion in the Treaty of Reswick [Ryswick, 1697] and we shall be upon our Guard thereupon in the Deliberations concerning the impending War, and that in the best manner. We also are in hopes of obtaining our Ends of the said Commission that it be forever *abrogata et abolita,* or abolished and cancelled! I remain in presenting my humble Respects to the Society.

Honoured Sir   Your most obedient and most humble Servant P.S.: Mr. Commissary and both Ministers as also Thomas Gschwandel, extol the Bounty of God for the great kindnesses heaped upon them in everything in the name of the Society and the Trustees by Capt. Coram and Mr. Butjenter and write that Captain Fry behaves himself quite otherwise at Dover.

[Here are copied, in a minute but legible hand, two anonymous letters sent from Bienne to John Vat and forwarded by him to Newman. The first, written on 23 Nov 1733, concerns an emergency diet of the Swiss Cantons convened to take measures for their common defense. The second opens with a eulogy on the Saltzburg Protestants that likens their courage to that of Luther and then describes a disturbance in Erguel, a small Protestant city endeavoring to defend its civil liberties from its overlord, the Bishop of Basel. This confusing account has been annotated and explained in an appendix by John Vat. These letters are deleted here because they do not directly concern the Georgia Saltzburgers. Vat then continues:]

I was taken ill on the 28th of December last with a Pleurisy, on the right side of the body attended with violent pains a Fever and Vomiting, but by God's blessing the kind cares of the Reverend

Mr. Urlsperger his Lady and Dr. Ploss[177] a very judicious Physician, I was relieved the next day of the great pains and after 12 days keeping Bed I was restored to Health, and altho' I do not Stir out of Doors by Reason of the Severity of the Cold Season, I am God be praised able to undertake any Journey. I beg the Favour of You Sir to make my Respects acceptable to all the Gentlemen of the Society, particularly to the Reverend Dr. Hales and to Collonel Valogne, to whom you'l be pleased to communicate this Letter. I am heartily pleased with the safe arrival at Dover of Mr. [Ph.] Von Reck and the Emigrants out of Saltzburg, and I hope they are by this Time not far off from Georgia, observing that the Easterly winds have blown for some time; and that we shall soon have a good Account of their being Settled in Georgia. The Roman Catholics here will not suffer any Thing to be printed in this Town relating to the Saltzburgers or the Plantations, The Printer of the News being under restraint of a Protestant and a Roman Catholick Inspector of them before the Printing of them. And the Jesuit Scholars are Outragious whenever they can have an Occasion to shew their Malice. On the 15th of January last 3 of them attacked at 11 at night One Meyer a Saltzburger watching a house partly pulled down for its being rebuilt, a Neighbour coming to his Assistance, the Scholars were soundly beaten, leaving a Cloak & 2 hats which the next day were carried to the Town House [city hall] by the Saltzburger. Some few days after some other Jesuit Scholars persued a Protestant Schoolmaster[?] into a Shop & beat him and the people of the Shop most cruelly. We can hear nothing which is doing in the Archbishoprick of Saltzburg nor how soon any of them may have leave to come out of that Country. I am Your most Obedient and most humble Servant. John Vatt Augsburg, February 8, 1734.

Copy of a Letter from Mr. Lowther at Rotterdam February 16, 1733 O.S. No. 12526. Read 12th February 1733/34. To Mr. Newman.

Sir: I received Your favour of the 4th December O.S. and am much obliged to the Gentlemen of the Society for their kind acknowledgement of the little Services I am capable of doing, pray You Sir, assure them of my best respects and that it will always be a pleasure to me to have it in my power to render them any Service.

I likewise received Yours of the 8th of January. I have published as much as I thought necessary of the Letters from Dover, relating to the Emigrants in the Leyden Courant, it being a paper that is sent all over Germany, I doubt not but the treatment those that are gone for Georgia found in England will be an encouragement to many to follow them; enclosed You have the Leyden Courant.

Whenever the Members of the Society think I can be useful to them in promoting any of their good works, I beg I may be acquainted with it believe me no one will be more ready to do it than

Sir, Your most humble Servant Rich. Lowther

Copy of a Letter from Mr. Urlsperger at Augsburg February 15, N.S. 1733/34. No. 12530. Read 19 February, 1733/4. Answered 22 February 1733/4. To Mr. Newman.

Honoured Sir: I hope my Letters of the 18th January last and of the 8th Instant are come to Your Hands, as I received Your favours of the 28 December and 15 January last, but your most agreeable of the 18 December last is not yet come to my hands. By the inclosed Sheets, You'l see what I have already printed Three weeks ago, And now I am about Sending new material to the Press. Viz. Mr. [Ph.] Von Reck's Letter of the 8th January directed to my Self; 2. Mr. Vernon's Letter to Mr. Von Reck; 3. Letters from both ministers to my Self; 4. A Letter from Thomas Geschwandel the Saltzburger, dated at Dover to my Self; 5. The Powers of the Trustees for Georgia Sent to Mr. Bolzius; 6. An Extract of Mr. Newman's Letter of the 15 January last to my Self and the Letters from Dover printed in the Whitehall Evening Post; 7. An Explanation of the Word Trustees; Together with a preface to these pieces, in which there will be an Extract of the Ministers Journal And as soon as printed I shall Send to you a Copy thereof. Concerning the Tirnbergers that are at Franckfort, a worthy Minister there to whom a Letter was sent to that purpose hath given the following answer thereupon; "As now you ask me whether such Tirnbergers as are remaining here were inclined to make such a Voyage and that they in such Case should be Conducted to England by a person now residing at Augsburg? I have Communicated that Affair to Mr. Von Ocksenstern Director of the Reverend Consistory who approve the Opportunity for providing for those Emigrants, as being very Good and Advantagious; But he hath desired to postpone the Answer, till such time as those of them as are upon the Road coming from Holland, Shall be arrived at Franckfort, and then it is intended, to enquire of them all, how they are inclined. But before I mention this Answer of the Magistrate here, I thought fit to write this particular Letter to You to let you know the receipt of Your agreeable Letter and at the same time to acquaint you That that Affair is upon the Carpet Yet by reason of the Cold Season of the Year will meet with some Delays. If there should be Still some Copies in your Hands of the Advantagious proposal upon which the last Transport so Joyfully undertook the Voyage I should desire You to send me the Heads thereof upon a quarter of a Sheet of Paper, in order to represent them to the Emigrants. I beg You wou'd please to give notice thereof, together with my most humble Services, to the Reverend M. Urlsperger."

I shall write this Post an answer to the Reverend Divine and send him my Printed Account of Georgia, together with a Copy of the enclosed Sheet; and assure him that I don't believe the Saltzburgers could be any where better provided for than in Georgia. Moreover I shall likewise desire a good Friend to acquaint the Gentleman, Director of the Consistory at Franckfort with what is necessary. Mr. Von Münch being gone upon a Journey hath not yet sent me the original Bills of Exchange mentioned in my last. In the mean time I do let You know, That of the 3587 f 30 creutzers comparing my Georgia Cash there is no more than nine Hundred forty two florins and Twenty one Cruitzers remaining therein. Mr. Vat is, God be praised, restored to Health; but doth not yet go out of Doors. We have hitherto done what we could for his advantage, And shall not fail of doing it further, and We are very glad to See, That the Honourable Society bestows so much kindness for him; He on his part shews the greatest Respect towards the Gentlemen composing that noble Body. The Advices I have from time to time received out of Dover have most sensibly rejoiced me; GOD be forever praised & glorified, for his directing thus the Heart of men; May he also give his Blessing to the Services done for Tranquebar and Madras,[178] as dark as the times at This Day appear to be yet they are as comfortable, considering that the name of Jesus Christ is more and more made manifest upon the Face of the whole Globe. Upon your receiving any Good Advices of the safe Arrival of the Saltzburgers in Georgia, or of any other Circumstances, several of their Countrymen in these countries are like to follow them. I heartily long to know whether Mr. Robinson the English Minister at the Court of Vienna is like to succeed or miscarry.

Since the writings above, I received this Morning your Favour of the 22nd January last and as to what concerns the Commission to be Given to Mr. Göbel, by the Trustees of Georgia. You will See by the Extract of his Letter hereto subjoined, that his Journey hath been put off for some Time. That Letter is dated at Ratisbone 11 February 1734. "My Journey from hence hath met with another Delay by reason that the Envoy of Berchtolsgaden as well as the Envoy of Saltzburg Strongly protest against my Person, But upon his majesty's Resolution with regard to the representation concerning that Affair which I expect it may take another Turn"! I shall take care to direct my Letters as you desire; And I hope You will punctually advise me of the Receipt thereof. But your Letter of the 18th December is not yet come to my Hands, so that I look upon it to be lost. Wishing You dear Sir all manner of Prosperity I remain

Honoured Sir   Your humble Servant   S. Urlsperger

I beg the favour of You honoureds[179] to make my Duty and respect acceptable to the Gentlemen of the Society and to accept the same your Self as being

Your most obedient and most humble Servant.   John Vat

Credible accounts of the Saltzburg Emigrants

As well of those that are gone to Prussia, as of those which are upon their Voyage to Georgia in South Carolina in America, the Great British Colonists; and of the Tirnbergers who are Still in Cassand [Cadzand].

Prussia

Some People having hitherto been very desirous to hear From the Saltzburgers in Prussia imagining they were not well there, as the ill disposed have reported, it is hereby made known that according to His Majesty the King of Prussia's most gracious Order and Command, one of the Ministers which His Majesty keeps for the Saltzburgers, hath held a General Visitation at all Places in Prussia where they live, especially to the End that the old ones as well as the Youth may be well instructed in Christianity, and that for the latter Schools may be ordered, which also hath been effectuated, insomuch that for the Saltzburg Children 14 Schools are to be Erected and proper and Skilfull Men of the

Saltzburgers to be Schoolmasters. Which the Saltzburgers most humbly have acknowledged as a particular Grace of His Majesty and rejoiced at it. As to the temporal Circumstances of the Saltzburgers, those that have built new Houses in Prussia have ordered their matters very well, so that their houses are very convenient and neat, they gain by breeding of Cattle and make so much Butter and [omission] that they can sell part of it to others, nay they have by their Labour and good Housekeeping made themselves famous in that Country, so that every Body praises them for their good Husbandry and profitable economy. Other Inhabitants in that Country especially the Germans have last Summer employed them in the most of their Business. They also shew still a great Zeal and Devotion in Christianity and are desirous to be more instructed, and are in hopes that more of their Country People [Compatriots] will come to them. All which we are assured of by confident and credible Letters.

Berlin the 19th December last Year.

His Majesty most graciously hath Committed the Special Inspection of the Bishopwiesers [Bischofsweilers?] and Berchtolsgaders, whereof 30 Families are remaining here, to one of his prime Ministers in order to take care of their Improvement as well temporal as Spiritual. And by this good Care a Schoolmaster hath been ordered for the poor Children of these people. And his Majesty hath also most graciously given for the Building of a School not only the materials but also 100 Rixdollars as a help for the Workmen. These Bishopwiesers live here quiet and devout, and are heartily glad to see their Children improve in Christianity, reading and writing, in which latter the Parents are quite unskilled.

London 18 December

After the Lords Commissioners of Georgia, and the Society of Propagating Christian Knowledge, had heard the agreeable News that the Georgian Colonists, who for Religion Sake had left their Country, were from Augsburg happily arrived at Rotterdam the 27 November, and were in hopes to go soon from thence, a Small Vessel was immediately sent to Dover with Things designed for them to Georgia, likewise a member of the Trustees [Coram] and a Minister of the Royal German Chappel [Butjenter] went by

Land to Dover, in order to expect [Await] there the Transport of the said Colonists and to order other matters before their Departure to Georgia, for their temporal and Spiritual Welfare, with advice of their two Ministers and their Commissary Mr. [Ph.] Von Reck who is to go with them to Georgia; Not to forget that the Society out of particular Care for these people hath also taken a Schoolmaster [Ortmann] and his wife, who Speak English and read German and English and hath kept a School here for a long time, to the End they may have no Difficulty on Account of the Language, and that their Children from their Infancy may learn both the German and English Tongue.

Extract of a Letter from Dover from the Commissary Mr. [Ph.] Von Reck to N.N. in A. dated 23 December 1733.

The English Ship Purysburg could not go from Rotterdam before the 5th Instant; so that we were forced to Stay there almost 8 Days. The 8th at Night we were by a Storm and Carelessness of a Pilate we had taken on Board, cast upon a Sand Bank near the mouth of the River Maes [Maas, Meuse] where we had but 2 or 3 Foot deep of Water, and laid there 3 Days till by unloading the Goods we had lightened the ship. The 19th we Sail'd from Helvoet Sluys with a good South East wind into Sea and could have been the next Day about noon at Dover, if GOD Almighty had not ordered it otherwise. For about 10 o'Clock in the morning the Wind turned to South-South West insomuch that tho' we were near the English Coast and could see the side of North and South Forland, yet we could not get the point of Dover that Day but were forced to turn back and pass that night in Sea. The 21st at 4 o'Clock in the morning we clapped on all our Sails and came in 3 hours tho' with half wind in the English Channel. But the moment when we thought to get into the Harbour, a Storm arose and forced us to go again into Sea and casted us upon the French Coast. After some Hours we Sailed back, set up all Flags, cruised before the harbour and thereby gave notice that they should come from Dover to help us, at last a Pilate with some Seamen came in the great Storm to us, and brought us the melancholy News that it was impossible for us to come that Day into the harbour, which obliged us to go back and we casted anchor near Deal. The Storm was horrible, and the half part of the ship sometimes, as it were, buried in the Waves high like Steeples.[180] At last I was with great Danger set ashore and went that same Evening on Horse Back to

Dover 7 English miles from Deal, where I had the honour to see Capt. Coram one of the Trustees and the Reverend Mr. Butjenter the Reverend Mr. Ziegenhagen's Colleague who had stayed for us here almost 3 Weeks and were very glad of our Arrival. The 22nd we saw our ship sailing towards the Harbour, where we set up a red Flag for a Sign that the ship might safely go in, which also happened under singing of Hymns of Thanksgiving and Shedding of Tears for Joy; and I can say for Truth that I never was so joyfull, nor found my Heart so touched of Praise to GOD than by seeing our ship in a sure Port. The Honourable Society hath had a fatherly Care for these poor people. The 23rd at 1 o'Clock they were debarqued and brought into a House whither they went in due order 2 and 2 together with their Pilgrims Staff in their Hands, singing their Hymn of Exile[181] and had there a very good Dinner. After that a Charity of the Society was distributed among them, by the Reverend Mr. Butjenter 1 £ Sterling per Head. Likewise hath this illustrious Society out of a particular Favour made them a present of 5 £ Sterling and of another Benefactor from London was given them half a Crown per Head by the said Reverend Mr. Butjenter. The Ministers who shew right true Christianity, have received here the half of their Salary, they are also clothed and with all necessaries very well provided. The Trustees have, besides a provision for 3 months, provided us with a great Quantity of Turnips, Cabbage, Carrots, Pease, Potatoes, Brandy, Sirrup, fresh meat etc. There are 30 Persons more here to be embarked in our Ship, so that the whole Transport will increase to 72 Heads. I hope by GOD's grace and blessings to do much Good in Georgia, where I also am to have a great District of Land[.][182] As soon as the wind will Serve we go from hence. I commit your E. [Excellency] to GOD's Protection and remain

Your   Von Reck

P.S. My dear Saltzburgers and others are God be thanked all in good Health.

Whereas hitherto of the Tirnbergers that went to Holland many Reports have been spread out, whereof we cannot judge neither to the Prejudice of the Country nor the Emigrants, because we have no credible Account of neither Side, whether and how or what hath happened; It will not be disagreeable to the publick to read here what really one may believe out of a very credible Letter.

Extract of a Letter from N. to N. N. in N. of the 14th January 1734

Concerning the Hollandish Emigrants, many Contrary and false Things have been writ and spoke. One Time it is said, their minister Mr. Fisher [Fischer] is dead, another Time, he is upon his return, then they say there is not one Emigrant more in Cassant [Cadzand] etc. I must let the People talk, tho' I have the most surest and most credible Accounts as well of Mr. Fisher as of some Emigrants, directly from Cassant. And it is certain that Mr. Fisher is alive still and praises GOD that he hath left him, tho' a Smaller Yet a good Congregation, which is loaded with Benefits, they having of the City Amsterdam only received more than 1600 Guilders. He further mentions that he hath Erected an Orphan House, in which he hath put the Saltzburger Children, which partly he himself instructs; and that now some of the Tirnbergers that are capable are ordered for Schoolmasters, each of which hath got a Recompence of 100 [guilders?] of the Consistory in Amsterdam etc. GOD help his People everywhere through Jesus Christ, and help his Inheritance in the whole world, for his Honour and Salvation of Mankind.

Copy of a Letter from Ratisbonne dated the 15th February 1734 together with the adjoyned pieces marked G. & H.

Besides what hath been Lately mentioned about the Memorial to wit, that it hath been delivered by Deputies to the Principal Commissary, it is further to be observed, that on this Account first a particular Account hath been held about it by the Imperial ministers that are here, and a Draught made for the answer upon it, which afterwards hath been brought to all Roman Catholicks in a full Conference, who have adjusted what in the adjoined Piece marked G. follows. And this Memorial was last Thursday at Noon delivered upon Request to the first Deputation viz. the Electoral Saxony, and Saxon Gotha, by the hands of the Principal Commissary; after which the Evangelicks together with the Deputies of the Imperial Towns Saturday last held another Conference; And tho' the latter complains that they had been twice pretermitted, and therefore refused to take Part in what the Evangelick States in both high Colleges had concluded, yet at last they Submitted because the Affair Suffers no Delay, and so reserved other Things for the Future; what hath been done in this Conference is to be seen in the adjoined piece marked H. and the Memorial hath been delivered that same Evening about 5 o'Clock to the Imperial Principal Commission. It causes the adverse Party great Ombrage but if they will have the Deliberation Promoted, they will be obliged to consent to what they did not expect, and we are very sure that it will happen so, wherefore the Proposition concerning the War, which was fixed to be made this Day is deferred.

The Collection made for the City Windsheim the 7th instant did amount to about 600 Guilders, which Sum the Magistracy hath supplyed out of the Common Cash etc.

The adjoyned piece marked G.

His Imperial Majesty's Principal Commissaries had with advice of other Imperial Ministers that are here well considered of the Memorial of those of the Augsburg Confession And tho' they well remembered what in the Year 1689 and 1702 in the same Terms as the present Request on Side of those of the Augsburg Confession had been given Notice of and at last by a General Consent of the Assembly of the Empire agreed, And that they did not doubt but His Majesty the Emperor, whom they immediately would most Humbly acquaint with it, will affirm it and all Roman Catholick Electors, Princes, and States of the Empire would never go absolutely from Things which once was agreed to, to which End their Graces the Principal Commissaries intended to recommend the Affair.

But whereas such Declarations and Insertion, which the Empire assents to *Super Declaratione Belli,* on the Side of those of the Augsburg Confession had never been nor never could be desired of the Roman Catholick Party any other Way but when first on the Side of those of the Augsburg Confession it hath been agreed to that the War against France should be begun with the utmost Forces and prosecuted to an happy End, so it was hoped that those of the Augsburg Confession would not withdraw in the present violent Invasion and insulting of the Empire with their General Consent had been done; then they might be assured that the Roman Catholick party or the most part thereof, by all means would be willing to insert the repeated assurance, in the assent of the Empire made for that purpose, that in those Countries by God's assistance recovered and what on occasion of the Wars with France, in ecclesiastical and Political Affairs hath been altered shall be restored most faithfully ad Normam pacis Westphaliae, and other fundamental Laws.

The adjoyned Piece marked. H.

Acted in the Conference of the Evangelick Body, Sunday the 13 February 1734.

The Saxon Minister proposed the Principal Commissaries Resolution upon the Memorial of the Evangelick Body, concerning Clausulum Art: IV. Pacis Ryswicensis.[183] This being not sufficient, the Question was whether they should not insist upon, that a more plain Explanation might be given, it was agreed that by another Memorial, it should be desired that by inserting an Article in the assent of the Empire which was to be drawn up, the Evangelick Body might be made secure after which the Draught of a Memorial was read and adjusted and the delivering thereof was left to the Envoys of the Elector of Saxony and of Saxon Weimar.

2. The Resolved Conclusion, in November last Year, concerning the Divine Service of the Calvinists, in Degenfeld House at Frankfort was read and agreed upon.

3. Whereas the Honourable Magistracy of the City of Frankfort had sent an Information hither that 270 heads of Saltzburg Emigrants that were returned thither from Holland, were there a Burden to the Publick, it was agreed that they should be told that they might resolve to go to Georgia and when they should resolve to go thither, a List should be sent hither to the End that in Respect to Provision and other Necessaries measures might be taken.

4. It was notifyed that from the Countries of Hessen Castle [Cassel] 400 Rixdollars Collection for the Germersheimer Cash had been delivered here.

Copy of a Letter from Mr. Urlsperger at Augsburg 18 February 1734 N.S. No. 12555. Read 26 February 1733/34. To Mr. Newman.

Honoured & Dear Sir: This moment I received the foregoing Pieces from Ratisbone, It being so late that there was no time to translate them this Post. In the meantime the Paragraphs underlined are of such great moment that I thought it necessary to communicate the same to the Society and the Trustees for Georgia and that to the End that the Society or the Trustees may be pleased immediately to write to Mr. Gulman the British Resident at Frankfort and to recommend that affair to him. I write this Evening to the said Mr. Gulman and to Messrs. Von Münch at

Frankfort concerning the Resolution of the Evangelical Body together with the necessary observations thereupon[.] on monday next will be printed, a peice with the following Title. [omission] and as soon as printed I shall Send copies thereof to Franckfort and Ratisbonne. I wrote last monday to Franckfort; to let the Tirnbergers know that I have already with me an Experienced and Humane Commissary [Vat] upon whose fidelity they may depend. This is in great hast[.] I cannot have at present the Title from the printing House by reason of the Post going off, etc. I remain with all my heart

Your very humble Servant   Saml. Urlsperger

Extract of a Letter of Messrs. The Burgomaster and Council of Franckfort Dated the 30 January 1734 to the Protestant Body at Ratisbonne.

We cannot conceal that there are Still here about Two Hundred Three Score and Ten Heads of such Saltzburg Emigrants as returned from Holland who are maintained by Charitable Collections made at Two different Times; But as they are hardly Sufficient and as the People would be in Length of Time too burthensome to this Town should they wholly bear the Charges thereof, which is not to be presumed to be done by them We have been under the Necessity to apply to some Neighbouring States, for taking Care of some of them, at least till such Time as the Protestant Body at Ratisbone shall have taken a Resolution of Providing for them next Spring, in some other part of the Country; Upon which his Highness the Prince of Hesse Darmstadt sent one of his Officers to enquire into the nature of their Circumstances and put them in hopes of being partly admitted into the District of Dornberg. But as since nothing hath been done thereupon, excepting that he would privately propose such an overture, not only to the College of the Protestant Body, but also to the other Protestant Envoys and use his Endeavours for these poor but worthy Confessors of the Gospel, That at the ensuing mild weather they might be transferred and in a Solid manner Settled somewhere, and that this Town might be relieved of bearing wholly the Charges of their Maintenance etc.

Copy of a Letter of B. V. R. [Baron J. von Reck] Dated at Ratisbonne 18 February 1734.   To S. U. [Samuel Urlsperger]

Inclosed is a Copy of an Extract out of a Letter of the Imperial City of Franckfort concerning the Remigrants out of Cassant

[Cadzand] remaining still there, sent hither, and their Request. As the Protestant Body here have desired me to make an Overture to You Reverend Sir and to endeavour, if it is not already done, that these People might be persuaded to go to Georgia; I have accordingly given You Notice thereof, and committed it to your Care, in order if possible to bring it into that Channel. I always remain Etc.

Copy of a Letter from Mr. Urlsperger at Augsburg 22 February 1734 N. S. No. 12556. Read 26 February 1733/34. To Mr. Newman.

Honoured and Dear Sir: My last was of the 18 Instant, and having since received the Extract of the Letter of the Magistracy of Franckfort and a Letter from B. [J.] V. Reck of which above is a Copy, I am Writing this post to the Magistrates of Franckfort, to acquaint them with the Powers I formerly received of the Society for Promoting Christian Knowledge & of the Trustees for Georgia, for Sending Three hundred Emigrant Saltzburgers to England and thence to Georgia and also of the Resolution of the Protestant States at Ratisbonne, and Copies of the Conditions, upon which the first Transport went lately from hence through this Town, in their Way to Holland and Georgia desiring them to propose to the Tirnbergers the List of their Conditions in Case they should be willing to Accept of them, and if they are Inclinable thereto, to let me have a list of their Age and Professions in order to Transmit it to both Societies [Trustee & S.P.C.K.] for their further Resolutions and Instructions, thereupon I remain

Honoured & Dear Sir   Your humble Servant   S. Urlsperger

Copy of a Letter from Baron [J.] Von Reck at Ratisbone dated 1 March N.S. 1734   No. 12568.   Read 12 March 1733/34. [Omitted]

Copy of the Translation of a Letter of the Magistrates of Franckfort.

Reverend Sir: What You sent us, concerning the Saltzburg Emigrants, to the Number of 270 Heads, lately come back from Zeeland, and still remaining here; together with your Letter of the 22nd February last, came in due time to our Hands and we have not been wanting to [have not failed to] communicate to them the Contents of the Inclosed printed proposals and Papers. But as the said people have no Inclination to go to Georgia, but rather have humbly excused themselves by alleging that they could *not bear the Air of the Sea,* which was the main Cause of their Sickly State of health, as to the greatest part of them, and moreover for that Reason, durst not venture to undertake such a Voyage because *there were few men among them yet many Widows and Children, Which Excuses we have admitted to be well-grounded;* And we have thereof acquainted, last past, our Deputy at Ratisbone, Mr. Joh[ann] Ulrick Bössner, and have desired him to communicate every Thing fully to the Evangelical College and to pray their Compassion for those Fellow Christians who are worthy of Commisseration and to take particular Care for their being at last well Settled, for the burthen of their further Subsisting here falls too heavy upon this Town should it be left wholly upon them; And yet it would be to the reproach of the whole Protestant Interest, as it is a Religious concern, should these poor people, *who for the greatest part of them are not able to get a Livelihood;* be left to go about begging or sent away one after another. Now as your praiseworthy concern hitherto for the poor Emigrant Saltzburgers is sufficiently known to us We thought it fit to acquaint you therewith in Answer to your kind Letter and to request you to intercede where it may be proper that they may be once firmly provided for. We recommend you to the Divine providence and in Expectation of having Opportunities of shewing our Readiness to serve you, We remain. Given March the 13, 1734.

Burgermeister and Council of the Town of Franckfort. To the Reverend and very Learned Mr. Samuel Urlsperger, Senior of the Evangelical Ministry and pastor of St. Ann, our particularly beloved Sir, and Friend at Augsburg.

Copy of a Letter from Mr. Urlsperger at Augsburg 18 March, 1734. No. 12586. Read 26 March, 1734. To Mr. Newman:

Honoured and Dear Sir: I received this Day, the favour of your Letter of the 22nd February last, and on Tuesday last a Letter of the Magistrates from Frankfort, of which the Translation is annexed hereto shewing that the Tirnbergers there have no Inclination for Georgia. As alledging first, that the Sea Air of Zealand, being the chief occasion of their Sicknesses among most of them, they cannot venture themselves to the Seas; And that there are few men, yet many widows & Children among them. As I apprehend these poor people have not a right notion of the Climate of Georgia, which is quite different from that of Cassant [Cadzand]

and that their children would be very proper for Georgia, which Air would inure them accordingly. I wrote Yesterday to Baron [J.] Von Reck at Ratisbone That I was inclinable to send hence to Frankfort, an ingenious Saltzburger [Mayer] who gets his Livelihood in this Town, and who was very instrumental in persuading those that went with the first Transport under Mr. [Ph.] Von Reck, in order to talk familiarly to those Tirnbergers and to see whether the most healthy of them could not be prevailed upon to undertake joyfully the Voyage to Georgia? such as are infirm or old might be otherwise provided for in Europe. Upon which I expect his Answer. Should there be a reasonable number of them inclined to undertake the Voyage, Mr. Vat might go hence with such, as are here, tho' few in number, to Franckfort and conduct them to Holland. I remain

Honoured and Dear Sir.   Your most Obedient and most humble
Servant   Samuel Urlsperger

P.S. I communicated the Extract of Mr. Robinson's Letter, to the Evangelical Secret [Privy] Council here, of which they have as yet not heard one Syllable. They are in hopes of being soon acquainted with the good Success thereof.

Extract of a Letter dated at Franckfort the 18 March 1734 from the Reverend Mr. Munden[184] to the Reverend Mr. Urlsperger at Augsburg.

Yours of the 22nd February I received the 24 and being in a bad State of health, I was obliged to defer the Answer till now. However I have in the mean time endeavoured to Second your praiseworthy, and Godly intention and have obtained from our Magistrates, that the poor and miserable Saltzburgher Remigrants from Holland, have got a Student to Instruct them daily some Hours in the Catechism, reading and praying. This Student, who was born in the Dutchy of Weymar [Weimar] and was obliged to leave his Country because of the Levies of Troops hath a great mind to go to Georgia, and therefore hath read the printed Accounts to the Saltzburgers several times, in order to persuade them, to trust to the Protection Love and Care of the English Nation that would not do as the Hollanders, yet the People would not resolve but Excused themselves on the Account of Sickness, Fear of the Sea and their Sorrowfull Experience (as they say) how they have been misled and the like Excuses more. The same Answer they gave to the Recorder when he asked them in the Name of the Magistrate.

Our Magistrate and Citizens have during the Winter shewed great Love to these Saltzburgers, and a double Collection hath been made for their Maintenance, besides what hath been given out of the Treasury. Several of them are dead [have died] some are Sick still, Those that are in health Separate themselves; some of them are employed to Labour and journywork; 17 of them are going to return to Nurenburgh and Ratisbonne, who were with me this Morning and took their Leave of me with many Thanksgivings. To these the Magistrate hath given every man 1 Dollar and every Woman 1 Guilder. I have also given them something out of my little Proselyte Cash which I have made by the help of our Convent and other good Friends. The rest will perhaps follow these in Time. It is the same with those that are maintained in a neighbouring Village whom the Minister there takes great pains to instruct. Concerning the Estate of their Souls, I must confess that much Weed is among them. Wherefore I went to them with the Minister who with the Consent of the Ministry hath taken the Care of their Souls upon him, and instructed them with due Zeal in what was necessary, which also hath had a good Effect. One of them confest freely that some among them for their Faults ought to be rebuked, but at the same time desired us not to be weary of doing good to those that feared GOD with all their heart. I give you many Thanks for the confident Communication of the remarkable Circumstances of the English Society and their praiseworthy Intention, which private Accounts I shall keep Secret etc.

Translation of a Letter from Mr. Gulman His Britannick Majesty's Resident at Frankfort to Mr. [Gulman, his brother] at Augsburg.

As to the Emigrants come from Holland, They are not to be persuaded to go to Georgia by reason of their bad ailing Stomacks, and for other Ailments for which they are not in a Condition to proceed further by Water without endangering of their Lives, and no Body would be well provided with such sickly persons who are of a very indifferent Aspect, and not in the least disposed in their Body. I have made an Enquiry at the Burgomasters [Schweitzer's] concerning them, who upon Examination, hath found them all in such Disposition and the Magistrates will endeavour to dispose of them in some of the other parts of Germany. Among those that came last to this Town there were near Forty small Children, whose Parents and Relations are Dead. And about Twenty of the People died here soon after their Arrival, For their misery is very great. They dread the Water as much as they would the Hangman, so that partly upon Account of Sickness and partly in their Nauseous Apprehensions, they are not to be prevailed upon thereto. The Magistrates here are now about sending off by Degrees such as are in Health, and about Twenty Four of them are actually [presently] going towards Nuremberg. Which You'll be pleased to Communicate to the Reverend Mr. Senior Urlsperger.

Copy of a Letter from Mr. Urlsperger at Augsburg 22nd March 1734. No. 12605. Read 2 April 1734. To Mr. Newman.

Honoured & Dear Sir: You will see by the Letter of Mr. Gulman, the Resident at Franckfort, to his Brother here and by that of Mr. Munden Dr. & Senior of the said Town, to myself, that nothing is to be done with the Tirnbergers concerning Georgia, And therefore I think no more on sending Mayer the Saltzburger to them. Some Days ago Mr. John Sartorius, born in Hungary, came to me from Jena, with Letters of Recommendation from some Eminent and worthy Divines who are well known to me having likewise publick Testimonals, of his Designs desiring me to recommend his intention and his other private Concerns to the Secret Evangelical Council[185] here to the Protestant Body at Regensperg [Regensburg, Ratisbon], & to the Society for promoting Christian Knowledge, of London. This hath been already done here, where he hath received a handsome Gratuity for himself and his Design.[186] I have likewise given him Letters of Recommendation to Ratisbonne whither he went some Days ago. And by these I submit it to the Consideration of the Society whether they shall think it fit to give him some Assistance? His Memorial given here is hereto annexed, together with a publick Recommendation of the former Swedish Chaplain at Vienna, now Superintendant in my Neighbourhood, likewise one Letter on Mr. Sartorius's by another living in Hungary.[187] Now since the Books by him printed are deposited in Jena, I desire the Reverend Dr. and Professor J. Walsh there, who hath likewise wrote to me, that he would be pleased to take the Inspection upon this work For which End I shall Send to him the moneys of this Town & of Ratisbone for greater Security. Upon your Letter of the 1st March, received this Day, I answer that I had not [would not have] taken any other Proselytes from Popery among the Saltzburgers[,] were the following words not inserted in powers of the Trustees for Georgia; Viz. Ult exules quoscumque sive Emigrantes è patria sua Professionis Evangelicæ Causæ, qui se se Coloniæ Supradictæ aggregare voluerint, et in Americam proficisci, tanquam colonos admittat.[188] That

the former Transport for Georgia hath been of Greater Expence
than was thought at first, arises from this That the Colonists staid
there several Weeks, both Societies desired that they should not
be wanting any Thing that the undertaking might not Suffer by
Evil reports; Mr. Von Reck hath likewise wrote to me that another Time now that he had paid his Prentice money He would
come off with less Expences, which as I observe in his Accounts
might be Saved in several Articles in Holland.

   I am Dear Sir Your most obedient humble Servant
                  S. Urlsperger

  Quos expediueras libros sacros, mira Dei providentia ad nos
rite pervenerunt; In uno fasce 200 in altero 150 exemplaria fuere.
Distribuimus inter pauperiores quemadmodum voluisti acceperunt
verbum Die sitientes cum magna aviditate non solum; sed etiam
privatas reque ac publicas pro Tua incolumitate Deo nuncupant
preces, id quod tibi loco Solatii perscriptum volo Esto bono animo,
Deus Jacob et Eliæ pro Sua gloria nunquam Te deseret. Quantum
ad nos attinet indies omnia sunt arctiora. Ducenta, fere trecenta
templa in Spatio duorum annorum nostra in patria per Jesuitas
nostratibus erepta sunt, et hodienum adhuc furit derus Romanus,
nec facile acquiescere vult, donec omniæ nobis adimat, quæ Supersunt. Neque tibi pepercit Tyrannis ipsorum; post quam enim
libros distributos Tuos vertisset apud quosdam ex proviso domum
affinis Tui duo Jesuitæ quoque ingressi sunt atque in absentia
sororis ac affinis sui riscum et cistam et regerunt, tandemque dua
exemplariæ ex tuis libellis integra invenerunt, quæ non moda
perfide cum magna indignatione fecunt [secum?] acceperunt, sed
etiam altero die Hospitem domus citarunt ad curiam, verbis
indigne, tractarunt, custodiæ tradiderunt publicæ, et post sex
septimanas 100[?] thaloris multarunt, et quidem ea lege, ne Tibi
amplius vel Scribat, vel minimum auxilii adhibeat. Tandem ul-
terius progressi Tuam domum, agros prata, vineam et omnia
utensilia publico fisco adjudicarunt. Nostrum nemo his resistere
potuit, omnia tamen hæc privata propriaque fecerunt auctoritate.
Modo spes nobis affulget melior quam olim, nam etiam nuper
mandatum Cæsareum accepimus Clementissimum ut omnia in
statu quo permaneant: Quod super est, valetudinem Tuam, et qui
Supersunt edendi libri sacri, cura diligenter; mandatoque Cæsareo
instructus Tua præsentia, quam multis vobis [votis?] expetimus
nos exhilara. S. Ex Hungaria. de 24 X br. 1733.

  Magno gentium doctori hoc contigit olim ut quidam Macedonis
speciem referens, seipsi conspiciendum daret, qui ut adcives suos

descenderet, eisque suo adesset auxilio, impense rogabat. Unde divinitus se excitatum sentiens apostolus, res suas mox ita composuit, ut devastatæ inter Macedones ecclesiæ pro ex quæ ipsi concessa erat, gratia, Succurreret, Act XVI, 9 Sq. Si hac nostra tempestate in ullis Christianonem [Christianorum] partibus, in Hungaria certe imprimis, ejusmodi pia suppetiarum Spiritualium desideria ad exteros mitti audires, Benevole Lector, si, quod mihi per aliquos annos datum fuit, quædam Tibi cum afflictis illis intercederet conversatio. Ipse regionis situs ostendit, ecclesiam, quæ ibdiem [ibidem] per Evangelium Christo colligitur, ab arctiore cum reliquis fidei sociis seclusam esse commercis, et contra à diris Evangelicæ veritatis hostibus obsessam. Hi dum unice id agunt, ut omnem, si fieri posset Christi memoriam ex animis hominum deleant, eo nunc licentiæ, Deo ita permittente, progressi sunt, ut observaverunt Scholas et templa quam plurima, expulerint etiam et exavitoraverint [exauctoraverunt?] doctores publicos, Sublatis simul et interceptis Sacrorum Bibliorum aliorumque librorum salutarium exemplaribus. Ut igitur miseri isti fervida ad Deum iactant suspiria: ita fratres quoque suos per Sacra communia obtestantur et implorant, ut laboranti imo labanti ecclesiæ suæ subveniant. Hoc autem commodius, nisi librorum, ædificationi inservientium, subsidio, fieri nequit: hi doctores mutivivis doctonem [doctorum] publicorum vocibus tantisper substituendi erunt, dum, prior docendi libertas, si qua speranda est, Substituatur. Quod præclarum christianæ cantatis, immo extremæ necessitatis opus ut præstet civibus suis, hic ipse Vir Clarissimus, Joannes Sartorius, ex Hungaria ortus laudabiliter adgressus est, nec infelici, quod ego testis adseverare possum, successu. Versant quam plurimi libellos ejus, et pabulum inde avidis animabus suis colligunt. Sed impenso tot millium desiderio nondum Satisfactum est, nisi novis auxiliis ipsis prospiciatur. Sic huic negotium suam qualemcunque operam plane destinavit hic Sartorius noster, cui quocumque modo stimulum addere detrectabit nemo qui Servatoris nostri Regnum amplifica, tum cupit. Dab. Neostadii ad Ascham: A. MDCCXXXIV. Johannes Christianus Lerche. Pastor et Superintendens.

[The preceding letter is so badly copied that only an approximate translation is possible. It appears to have been written by John Christian Lercher, pastor and superintendant at Neustadt on the Asch in Upper Hungary to an exile named John Sartorius.]

Behold, the books that you sent us have arrived safely by the Providence of God: in one parcel there were 200 and in the other

150 volumes. As you have wished we have distributed them among the poor, who are not only thirsting for the Word of God but also praying to God both privately and publicly for your health, and I wish you to find a place of solace. Be of good cheer: for the sake of His glory the God of Jacob and Elias will never forsake you. As far as we are concerned, things are getting worse from day to day. In the space of two years in our homeland two or almost three hundred churches have been taken from us by the Jesuits, and still today the Roman beast is raging, nor will he stop until he has taken away all those that remain. Nor does this tyranny spare you; for after your books had been distributed, two Jesuits entered into your home in the absence of your kinsman and his sister and took away the chest and box. At last they found two volumes of your books which they not only took away with them very treacherously and with great indignation but also summoned the host before the court the next day, treated him most shamefully, and turned him over to the jailer; and after six days they fined him 100[?] thalers with the condition that he not write to you or give you any help. Finally they returned to your home and adjudicated your house, fields, meadows, vineyards, and all tools to the public treasury. None of us could resist them, yet they did all this privately and upon their own authority. Now hope shines upon us better than before, for we have recently received an Imperial edict that everything is to remain status quo. Above all, take care of your health and diligently guard the books which you are editing. Informed of the Imperial edict, rejoice! We are awaiting your presence with best wishes. From Hungary, 24 December 1733.

[The next letter, also from John Christian Lercher, date at Neustadt on the Asch, 1734, is so badly copied that any attempt to render it would be more conjecture than translation.

It begins with a reference to Acts. XVI: 9 ff., which mentions that St. Paul was requested to go to Macedonia where the Christian churches were being destroyed. The author likens the conditions then in Macedonia with present conditions in Hungary, where the Protestant churches are being closed and the enemies of religion are trying to eradicate the memory of Christ from the minds of men. All teachers are being expelled and all books confiscated, and all freedom to teach is abolished. A splendid man named John Sartorius has fled the country and is sending back religious books. Sartorius is the author of the following letter.]

Benevolent Reader: The Great and Manifold Oppression for Conscience sake the Protestants in Hungary are labouring under are known to the World! Now in order to Assist such fellow Christians, as are thus greatly oppressed upon Account of Religion, I, about Four Years since out of hearty Compassion, ventured my all; For I resigned not only my Office of master of a School in Hungary, chiefly consisting of 300 Scholars which at first was reduced to teach no farther than the Donat, by the Enemies of Truth, that I may be better able without my Native Country, to Serve my Country men; But also to lay out about 800 Dollars, out of my own small Estate, toward printing some Books in Hungarian Language because most of my Countrymen do not understand any other Tongue, and the old Books are either lost or have been taken from them; That thereby these poor Comfortless Brethren of the Faith, after about Three hundred Churches had been shut up might receive some Comfort & Instruction And after I had printed about 3000 Copies of each of the following Pieces, Viz. Spiritual protestant Hymns in Number 500; in One Volume; Dr. Luther's Catechism, with Scripture Passages; The Psalms of David, Books of Prayers and Communion; and the new Testament; For which Letters with their accents were cast with no Small Expence; A good part of them having been actually already privately sent to Hungary and distributed among the poor. But the Enemies of Truth have taken it very ill, and so straitned all farther Means for my Subsistance, that now I'm not able to Subsist any longer, much less to discharge the remaining Expences amounting to Three hundred Dollars, Still due for Printing them, neither to finish the Homilies on Sundays and Feast Days — Gospels and Epistles, which I designed to add to the former works. In such Circumstances, my humble Request to the Kind Christian Reader is that he would be pleased to promote such holy Intention and good Work, by his compassionate Assistance. May God reward it both here and hereafter!

Copy of a Letter from Mr. Urlsperger at Augsburg 5 April 1734. No. 12620. Read 16 April 1734.

Honoured & Dear Sir: Your Letters of the 22nd February and 1st March last O.S. came in due time to my Hands, and I answered them the 18th and 22nd March last N.S.

We have since received Advice, That the Arch Bishop of Saltzburg held a Sermon on Sunday the 14 March last at Saltzburg wherein he mentioned, that there had been in his Sheepfold many Stubborn Sheep for which he had shed many Tears, and Prayers— to God that they might be led in the right way, but, that having from their tender years never been brought up in the good right Catholick Faith, but always professed the Evangelick Lutheran Religion (as it is called) and persisted in their Stubborn Opinion despised the Saints and shewed themselves Refractaries, he had been obliged to send them away; Wishing nothing so much as to see his Sheepfold pure, and filled up with none but Catholick Sheep, that then he might say with Simeon, Now let thy Servant depart in peace and freely die! As to the present State of Affairs in that Country. The People are now pretty quiet. Some Jesuits are sent into the Country; but they are enjoined to use the People very gently. Some Weeks before Two Persons died in a certain District of Saltzburg — One of them whose Children were sent out of the Country, as he prayed by himself, some Capuchins telling him something of the Purgatory answered them, I hope God will immediately receive me into heaven. It was observed at their Funerals, That this last who was suspected of being a secret Lutheran, was conducted to the Grave by Threescore and Ten persons and above, On the contrary the former, who was always esteemed a Catholick and professed himself as such, had hardly Twenty Persons at his Funeral. Whereby its supposed that there are Still many such protestants in several parts of that Country. Yet we do not hear that any of them are coming out of that Archbishoprick.

We have likewise advice that the Protestants in Upper Kärnten [Carinthia] in Austria who were desired to go into Hungary, if they would leave their Native Country, to the Number of about Three hundred families have declined that Expedient.

The Tirnbergers returned from Holland have had no assistance from the Protestant Body of Ratisbonne, so that they are left to the providence of God, as to their Settling.

Mr. Mölenhoff Danish Chaplain at Vienna writes the 31st March "I have learned from the Envoy of Hanover that after H. H. [His Highness] the Bishop of Bamberg had desired the Roman Catholick Agent of Augsburg to signify to the Roman Catholick Senate That they should no more obstruct the Emigrants out of Saltzburg but to allow a short Stay in Augsburg to such of them as should go to Georgia; The magistrates answer was actually come in purporting That in obedience to H. H. Commands they should not fail to obey in every respect; Which Answer had been already communicated to Mr. Robinson." with which Answer 1 am well pleased. Were the Word *Short* Stay, which perhaps hath been already added by the other Side, left out it would be more acceptable.

We are in hopes of receiving in a Short Time a good Account of the safe Arrival of our first Transport in Georgia; many people here believe that upon such good Advices some of the Emigrants that are not yet Settled, and were some of the Tirnbergers may resolve to go with the next Transport that may happen to be prepared for Georgia.

I remain   Honoured and Dear Sir   Your most humble & obedient Servant   S. Urlsperger

Copy of a Letter from Mr. Urlsperger at Augsburg 22 April 1734 N.S.   No. 12632.   Read 23 April 1734.   To Mr. Newman.

Honoured Sir: The favour of Your Letter of the 26th March last is come safe to Hand, & I observe that the Lands in Georgia are tolerably high,[189] where the Saltzburg Emigrants are to be Settled; The Family of 5 Persons [Rieser] who were to go with the First Transport are still very desirous of going to Georgia; And Yesterday, one of the Saltzburg Emigrants, who went to Middlebourg in Holland some time before the Tirnbergers were sent to Cassant [Cadzand] came hither, and is also desirous of going to Georgia and there are some Emigrants here, who are not Settled, but get their Bread by Day Labouring, that would join a Second Transport, should a good Account of the First come in, as we now Expect to receive every Post Day from Georgia. You'l see by the Letters and Memorial under, in what State the Protestant Religion appears to be in Austria, and that at this Easter Time, many, very Many, of them are like to make publick Profession of their Faith, not only in that Province, but in several other parts of those Countries where the poor People are in Motion. There is a report in this Town that there is also some such Motion at this Time in the Archbishoprick of Saltzburg. As soon as any of them shall come this way, I'll give you notice thereof and how they are inclined. Mr. Vat is highly obliged to such of his old Acquaintance with You; as honour him with their kind remembrance[.] he begs the Favour of you Sir to make his best respects acceptable to them and he is in hopes of having, in a short Time, an Opportunity of being nearer at Hand, and of the pleasure of waiting upon them, as well as on your Self. I am

Honoured & Dear Sir   Your most Obedient & humble Servant
Saml. Urlsperger

P.S. It would be very acceptable here should Mr. Robinson's Letter to my Lord Harrington[190] in answer upon such Answer as was given by the Catholick Magistrate to His Highness the Bishop

of Bamberg concerning the admittance of the Emigrant Saltzburgers into this Town as should resolve to go to Georgia, be sent
to me, or at least an Extract of that Letter in order to make use
thereof in Case any new Colonists should present themselves.

*Copy of a Translation of a Letter written from Goisen, March the
7th 1734 in Austria to an Emigrant of Austria, living at Ratisbone.*

The Grace and Blessing of GOD be to your friendly Salutation
Dear Brother in Christ, Blasius Engleüthir [Engleitner], besides
manifold Greeting, and wishing you chiefly temporal & Eternal
Welfare from God. I do let you know that thy Dear Wife and
Children are in good health and give thee to know that by GOD
Almighty's Direction, thy Letter which you sent us is come safe
to our Hands, & hath brought us good News from you by his high
Hand and Protection of his Holy Angel[.] which Letter, being
well grounded in the Word of GOD, we have read with great
pleasure. Besides other Circumstances therein contained We have
learned That we should be constant in the Faith and in prayer
and not forsake them but stand fast therein That GOD almighty
might direct us according to his Divine Will. We have also
Learned that Mr. Clostermeyer at Ratisbonne would gladly know
what sort of a Memorial We sent to his Excellency The Land
Captain at Lintz[191] differing only in the Titles thereof. Both
Memorials were put under one Cover and directed as followeth
viz. To the High and wellborn—Herren Herren Christopher Wil-
liam; Earl of the Holy Roman Empire, and Lord of Thurheim
Baron of Bibrach Zell, upper & lower Reichenbach, Lord of the
Lordships of Wineberg, Dornach Maraberg, Fischbach, and Stock-
enfeld, His Roman Imperial and Catholick Majesty's actual Privy
Councellor Chamberlain, chief Master of the Faulcons in his
Hereditary Dominions, and Land Captain in Austria upon the
Enns at Lintz; having sent that packet in the last Week of Septem-
ber last, down to him. The Reason why we took notice in the
Memorial of Hans Lerchner[192] is because the Commissioners of
Religion were with us in the Month of August last, and we were
Examined which Examination was chiefly upon the Account of
the said Hans Lerchner who had given his Answer to about One
hundred & Fifty Questions put to him, which were read to us;
Among others, there was one Question or Article That the Gentlemen at Ratisbone would concern themselves in our behalf, if We
should be sent out of the Country. Upon which we have defended
the said Hans Lerchner; both by word of mouth and by Writings

as well as we could, and declared That we professed our Faith & were resolved to go out of our Country before we knew or had seen the said Hans Lerchner. André Neff thanks you a thousand times for the fine Small Tract of our Lord's prayer. I wish to GOD that I could repay it, or return Satisfaction for the same! Dear Brother in Christ, Blasius Engleüthner [Engeleitner], We let thee know that Michael Neff is still detained in Prison at Lintz; But no Body else out of Camer—Gut [Salzkammergut]. However Michael Neff's wife hath sent in December last a Memorial, both to his Excellency, the Land Captain at Lintz and to the Honourable Commission for Reforming of Religion, altho' no Messenger would venture the carrying it to Lintz, yet a Soldier of Lintz who had been here wellknown, carryed it down; upon which she received Weekly one Florin from the Government of Wilbenst[?] . . . for herself and her Children. But as the Woman was very uneasy at the absence of her Husband, she went hereself, the 1st instant to Lintz with one of her Neighbours in order to present a Memorial praying for her Husband's release, without Compelling him to the Roman Catholick Faith, after she had desired leave of going of the Government and afterwards of Count Von Serau, who declined giving it, & did not severely forbid her going, she ventured however to go to her Husband, but we have not yet learned what Success she hath had or may obtain. XAbove Four Score persons Men Women & Children of the Parish of Ichler [Ischl] have this year 1734 entered their Names as Evangelical, among others also Balthasar Greisshofer and his Family.X We Salute your dear Father and desire him, he would join his Prayers to ours for Imploring GOD Almighty's Mercies, That He may give a Happy Issue to this Affair, according to his Divine Will, and to the Glory of his Holy Name. We desire him likewise to intercede for us at the Evangelick Envoys. XAbout Forty Persons of the Parish of Goison, and Ten Persons of the parish of Hallstadt have declared themselves protestants and entered their names.X We Emigrants have already heard from our Government That the whole Evangelical Body at Ratisbone, concern themselves in our behalf. For which we return our hearty Thanks to GOD almighty! my dear Brother in Christ We let thee also know That Leopold Hueber died the 4th of march last. As he was upon his dying Bed, the Chaplain of Goison came to him, in order to Confess, and to feed him, upon which he Contradicted the Chaplain, & said That he never in his Life believed that there was a purgatory or that the Saints that are dead should be Invocated; And thus he declared his

Evangelical Faith to his last Breath. So that he had no absolution from him neither was he fed by him, nor would they bury his Corps in Holy Ground but he was buried in the place where they use to lay the Children dying before Baptism. (NB whilst this man was in Health he was not Entered down as an Evangelical.) Yet Five Florins were exacted and paid, to the Rector of the Parish, and one hundred Creutzers to the Schoolmaster for his Burying. There are Still many Thousand persons as thou well knowest who are Secretly in their Hearts of the Evangelick Faith, and who have not as yet made publick Profession thereof nor entered their Names; but how God shall be pleased to direct them, Time will show us; Yet we hope to God that Still many more will declare themselves to the true Evangelical Faith, *particularly now at the approaching Easter Confession.* We have very great hopes, and we have heard many Say that they would never more go to the Confession, and were resolved to declare themselves of the Evangelick Faith about this Time. We desire thee Dear Brother in Christ to intercede for us at the Evangelick Envoys. We Emigrants intreat them for GOD's sake, That they would intercede for us with the Emperor that this Affair maybe brought to such an Issue as may be wholesome to the Soul, for which we shall in the mean time heartily pray to God. As to the Children of the first Emigrants, I acquaint thee that there are One Hundred & above, who were found too Young at the Examination, and they threaten us with detaining of them from us. Thy Son Jacob sends his Duty to thee, and desires to know whether he shall Sell the Hamlet & House or something else? We desire thee to Salute all our dear Country Folks of our acquaintance, & to put them in mind as they are delivered from their Hypocrisie and Superstitious way of Living that they would be likewise assisting to us; by making Intercession with all Evangelical Gentlemen living at Ratisbone that we also may be delivered out of the Hands of the Superstitious. Herewith thou art, dear Brother in Christ, heartily saluted from all Emigrants and they Likewise entreat you not to be any ways backwards in obtaining something favourable for us, neither to think the Time too Long. Every Thing will be rewarded both here and hereafter; And therefore all is recommended to God's Providence and Protection. Goisen the 7th March 1734. We remain till Death Thy faithfull Wife, Children, & Brethren, and Fellow Pilgrim, The Emigrants in Saltz Cammer-Gut or Estate

NB. Michael Neff's wife called Catharina, together with her Neighbour is got safe to Lintz, and hath been upon her Knees before the Land Captain[.] she had Leave to see and Speak to her Husband as long as she would; But Michael looks already very pale, so that his wife hardly Knew him. She Delivered her Memorial the 4th March, and was answered That it should be Considered of the next Day in Council; and she returned hither this Day very safely. But Michael Neff is not yet come. About 20 Persons declared themselves Evangelical at Ischl, and entered their names very lately, so that their number amounts to One hundred in that Parish.

Memorial sent to the Imperial Commission High and wellborn, very Gracious, highly honoured and Commanding Lords.

We hereto underwritten workmen of the Imperial Saltz-Cammer-Gut, and Emigrants, cannot omit to Entreat most humbly your Imperial Commission for Reforming of Religion. It is no Doubt before now Known to you Sufficiently, in what manner, in June last His Lordship of Serau & Governor of the Salt mines at Haalstatt [Hallstadt] together with 2 Fathers Capucins & all the Officers of Haalstatt as also the Company of Miners at Ischl & Ausse came hither, and seriously proposed, & recommended to all the Miners, that they should be always good Catholicks, and not thus play the Hypocrites in being inwardly Lutherans and outwardly professing the Catholick Religion, as to his Knowledge some of them had sent their Children into Lutheranisme whilst their Parents remained in these Mountains, and also some Children were here, whose Parents were gone to travell into Lutherum[193] whereby it was easily known of what Religion were the Parents, so were likewise their Children. At the same time he Seriously exhorted us not to be Hypocrites, but every one should go to his Magistrate of his respective District & desire him to give a pasport, that We should certainly obtain a Pasport, & not be detained neither the smallest Twig be laid in our Way, but that we should permitted to go out, & where we pleased, our Wives & Children, & likewise with our Small Effects. At this Declaration of His Lordship of Serau &c. we were greatly rejoiced to see that we had so fair an Opportunity, of leaving our Native Country with good reputation & to go to a place where we could publickly profess the Evangelical Faith; and having a short Time before that Declaration of the Count of Serau learned by one Hans Lerchner, who came hither from Ratisbone, & had told us that in Case we should be drove out of our native Country upon Account of our Evangelical Faith, by our Magistrates, or Sovereign Power, he had heard the Gentlemen at Ratisbone say they would concern themselves in our behalf.[194] We have punctually observed the Order of His Lordship of Serau &c Governor of the Salt,[195] and humbly applyed to the Magistrates, and desired them to be so good, as to give us Passports, that we with our Wives and Children, as also with our small Effects might honourably go out of our native Country, and to live in a Place where we might profess freely the Evangelical Faith according to the Confession of Augsburg; and thus we have made use of this good opportunity, publickly to declare our Faith, which we partly nourished secretly in our Hearts for along Time, & some of us all their Life Time; But we could no longer venture with a good Conscience to deny our Faith; for our Lord Jesus Christ, the Living Son of GOD hath Severely Threatned and admonished Mark VIII. 38 whosoever shall be ashamed of me & my words, in this adulterous & Sinful Generation, of him also shall the Son of man be ashamed, when he cometh in the Glory of his Father with the Holy Angels. And further Christ Saith; whosoever shall Confess me before men, him will I confess also before my Father, which is in Heaven. Matthew X. 32. Luke XII. 8. And the Apostle St. Paul Saith Romans X. 10. For with the Heart man believeth unto Righteousness, and with the mouth Confession is made unto Salvation. Therefore no Blame is to be laid upon his Lordship of Serau &c concerning this Affair much less upon Hans Lerchner who is Prisoner at Lintz because we were long before of that Faith, & Resolution of going out, even before His Lordship of Serau &c was born, and before we had known or Seen Hans Lerchner. Perhaps God Almighty had not appointed the Time till now; For the Holy Ghost saith by Solomon Ecclesiastes III. 1. To every thing there is a Season, and a Time to every Purpose under the Heaven. But this we find lying very heavy upon us, that His Lordship of Serau as Governor & other officers had promised to give us pasports for our going out during the warm Season with our Wives and Children and small Effects Yet did not keep their word, but have detained us already a long Time, so that we are apprehensive of coming into the Cold Autumn, or even into the Cold Winter; As it is well known to the Honorable Imperial Commissioners for Reforming of Religion, who in August last were personally at Hallstatt and Ischel, and Examined us; But we could not in good Conscience deny our Faith, altho' we are persecuted for the same; For we know very well That Christ Jesus Saith Matthew X. 22. 23. Ye shall be hated of all Men for my names Sake: but he that Endureth to the End shall be saved. But when they persecute you in this City Flee ye into another. For altho' we should declare our Selves with a false heart to the popish Catholick Church, or Swear thereto, yet Secretly at Heart be Evangelical or Lutheran according to the Confession of Augsburg; we could indeed for some time deceive men, but God Almighty can never be deceived and thus we should deceive our Selves; For it is written in the Book of Job, Chap. VIII. 13. The Hypocrites Hope shall perish and the Royal prophet David, Psalms XII. 3. The Lord shall Cut off all Flattering Lips and the Tongue that speaketh proud things. And Syrach Chap. XI. 14. 15. 16. 17.[196] Therefore we most Humbly beseech Your Excellencies of the Commission for Reforming of Religion that we may remain unmolested in our declared Faith of the Confession of Augsburg, and not use us too hard, but according to Favour and mercy, And God Almighty will show You mercy at Your last Breath, & on the last day of Judgement; to which all men must appear, be he of high or low Condition, Rich or Poor. We pray You to consider that we for a long time by Night and Day have faithfully worked for His Imperial Majesty, as our Gracious Emperor, King and Sovereign of the Country, and upon pressing occasions in dangerous places, we have ventured health and Life, as it is wellknown to GOD Almighty; We have likewise readily paid our Duties and Taxes to our Magistrates, as they doubtless will certify the same; For it is wellknown to us that Christ Saith, Matthew XX. 21. Render unto Caesar the Things that are Caesars; and unto GOD the Things that are GOD's and St. Paul Romans XIII. 2. Whosoever therefore resisteth the Power, resisteth the Ordinance of GOD, we never resisted our Magistrates, but always paid them their due Obedience. GOD forbid that we should not render them due Obedience in Civil matters, But as to what belongs to GOD and the Faith, we must hearken unto GOD more than unto men as it is written Acts IV. 19. and we shall be very willing furthermore faithfully to Serve His Roman Imperial Royal Majesty, as our Gracious Lord and Sovereign provided Liberty be granted unto us of freely and publickly professing our Faith and performing of Divine Service according to the Rites of the Evangelical Churches of the Confession of Augsburg, as it was done above One Hundred Years ago. But if that cannot be done, we pray for GOD's sake that we may have passports as mentioned above, & leave of withdrawing with our Wives, Children and little Effects and to go where we shall think fitting; By reason that his Lordship of Serau &c Governour of the Saltmines had promised it unto us. We shall offer up to Almighty GOD our fervent and constant Prayers in behalf of the Honourable Imperial Commissioners for reforming of Religion, That he may given them Health Blessing Prosperity, & long Life & to all that belong to them as also a happy Government & after this transitory, Eternal Life, in acknowledgement of so great a Favour to be granted to us. We recommend our Selves, by the Love of Jesus Christ, most humbly to your Excellencies, as our most Honoured, gracious and Commanding Lords, as being

Most Honourable Imperial Commissioners for Reforming of Religion, Your most humble Obedient all Emigrants and Workmen of Salt-Cammer-Good-or Estate, in Austria near the River Ennss.

Extract of a Letter from Ratisbone Dated 13th April 1734 [possibly from J. von Reck].

A Little Time since two men coming from Hanover, were here Saying that some, if not most of the Berchtolsgaden people had a mind to come away from Hanover X. But they have been disuaded from so doing. There is an Account of Hans Lerchner, who is a prisoner at Raab, that he is in Irons, and is working all Day long at the Fortifications thereof, and is obliged to be assisting in cleaning Houses of Offices, and altho' he has but water and Bread, yet he is very patient; He is condemned to work here, for Conscience sake, Eight Years. Some Days ago, Two men out of Stiria[197] were with me, all Evangelical Books were taken out of the House of one of them; for which he had been some time in prison. He saith that in Carinthia there are many Eminent, even Persons of the Administration of Justice inclined to Evangelical Truth. In Austria above one hundred Persons of a Parish [Ischl], where it was not Suspected, have entered their names as Protestants. I learn from young Lerchner, who is one of my Servants, that his Father at Raab is tolerably well, excepting that he must work at the Fortifications thereof. I'll endeavour to send him some money by the first Opportunity. I have had him recommended to the Commanding Officer in Gomorra[?] for I have been told that he hath already set at Liberty some poor People of that sort.

Copy of a Letter from Mr. Urlsperger at Augsburg May the 27th 1734 N.S.   To Mr. Newman

Dear and Honoured Sir: That I did not answer before now your Letter of the 2nd of April last O.S. was chiefly by Reason of the Operations of the French Troops having brought us under a great Consternation, after my dear Native Country of Wirtenberg [Württemberg] was laid under their Contribution and that we were here very apprehensive concerning our Neighbours; and likewise as I was in hope of being able to write something certain concerning *the Emigrants of Saltzburg*. Now I answer your dearest that I am of Opinion that still some *Emigrants are to come out of Saltzburg;* For a few Days ago a Merchant of this Town, who lately came out of the Archbishoprick of Saltzburg Precinct of Hallein, reported that there a Trusty Roman Catholick Correspondent of his had told him that very lately Two Hundred persons of that District, who for the greatest part were Stocking knitters and who had worked for himself, had Entered their Names as Protestants But could not determine the Time of their coming away; So that I doubt not but that most of them will be better *disposed for going to Georgia,* than the Tirnbergers who were frightened by the Dutch Usage, as they said. Neither do I doubt of their having Leave of going out of Germany even at this Time For People are not wanting.

Moreover I must tell you that some weeks ago the Privy Council of the Prince of Hesse Darmstatt wrote a Letter to the E. [Evangelical] Privy Council of this Town desiring them to dispose me to the receiving the Tirnbergers (that are now at Franckfort, who were to be distributed among the Neighbouring Protestant States) for sending of them to Georgia, where they would undoubtedly be provided to the best advantage upon which I have myself acquainted the said Privy Council of Hesse Darmstatt, That the Tirnbergers were not disposed that way; yet should the said Privy Council find means of disposing them for Georgia that I was ready to receive such of them as are in good health, and proper thereto Upon which as it's supposed the French have likewise laid them under Consternation, I have not yet received a full Answer.

As I lately mentioned Letters out of Georgia are impatiently Expected; Upon the arrival and Publishing of which we shall know further. Mr. Vat is as perfectly recovered as can be desired; and he longeth soon to perform that for which hath been called hither And he presents his Duty and Respect to the Gentlemen of the Society and to yourself, and I remain

Honoured and Dear Sir   Your most obedient humble Servant
Samuel Urlsperger

P.S. As to War with us it depends wholly upon a Battle; Should it fall out unhappily on the side of H. I. M. [His Imperial Majesty]

and the Empire, unless God doth particularly protect us, We here are in a manner wholly undone. As Two Years ago there were great motions concerning the Emigrants and as one Body after the other passed round about our Walls here, I preached not only once but several times as it was deeply in my Heart. Believe me these motions and marchings will soon be followed by other motions. Christs Symbol remains true. If God be for us who can be against us.

Copy of a Letter from Lord Harrington at Whitehall 5 April 1734. To the Trustees for Georgia. No. 12614. Read 9 April 1734.

Gentlemen: I send you herewith Inclosed an Extract of a Letter I have received from Mr. Robinson, inclosing one from Augsburg by which You will see what that Gentleman has been able to obtain in favour of those Emigrants from Saltzburg whom You propose to Transport to Georgia, which I hope may be to your Satisfaction, I am

Gentlemen   Your most humble Servant   Harrington

Extract of Mr. Robinson's Letter 26 March 1734 N.S.   To Lord Harrington.

The Enclosed Copy of a Letter from a Magistrate of Augsburg to the Catholick Agent of that Town there, which has been given me in Confidence by the Bishop of Bamberg, will show Your Lordship what has been done and what is promised to be done in favour of the Georgian Colonists, I humbly Submit to Your Lordship if more can be well Expected in the Emperor's present Situation joyned with the Tendency of the Elector of Bavaria's Conduct, and in Truth with regard to the Constitution of the Empire in general, however hard it may be for the Trustees to conceive it, especially under the partial and perhaps warm representations of their Correspondents at Augsburg.

Copie de la Lettre d'un des Membres Catholiques de Magistrat de la Ville d'Augsburg à leur Agent à Vienne. à Augsbourg le 15 fevrier N.S. 1734.

Par celle que Vous m'avés êcrite le 30 me du Mois de Janvier passè, j'ay vû en substance ce que Son Altesse de Bamberg et Wurtzbourg vous a donné à entendre de la parte de sa Majesté Imperiall, Savoir, que comme Sa Majesté le Roy d'Angleterre êtoit toujours resolue de faire transporter des Emigrans Saltzburgeois à l Isle de St. George, et que on avoit fait souis [soumis] main des plaintes à Vienne comme si le Magistrat Catholique ne vouloit pas tolerer ces Emigrans dans cette Ville jusques à ce que la Saison permette de les transporter, c'est pourquoy il seroit très agreable à Sa Majesté Imperiall si pour maintenir et fortifier la bonne intelligence avec La Cour d'Angleterre on pouvoit favouriser en quelque Maniere ces Emigrans, et leur accorder un libre accés jusques à ce qu'ils puissent être transportés. Je n'ay pas voulu manquer de conferer amiablement et en confidence avec le President et deux des autres Membres du Concil Catholique, surquoy j'ay ête chargé de vous marquer en reponse; que Vous pouviés assurer S.A. de Bamberg que de la part du Directoire Catholique et du Concil privé l'on fera tout ce qui sera possible pour marquer de toute maniere, la deference la plus soumise aux Intentions de S. M. Imperiall afin que les Emigrans qui viendront icy par la suite, et qui seront destinés pour la Georgie non Seulement recoivent generalement toutes les marques possibles de bonne volenté mais que aussi sur la requisition qui en Sera faite on leur accordera un Asile dans la Ville pour un tens moyennant qu'ils viennent successivement et non pas trop à la fois, afinque par ce Monsen [mesure] on ne donne point lieu a apprehender des troubles parmi la populace dans une Ville composée des differentes Religions comme cellecy, le principal Soin du Magistrat Catholique ayant toujoures ête de la prevenir; Et pour Montrer que l'on na aucune raison de se plaindre de la parte Catholique die [du] Magistrate de cette Ville pour le passé[.] Je vous envoye eyjoint [enjoint] copie de la reponse fait aux Ministres de Bronsvic [Brunswick] à la Diete de Ratisbonne en date du 17 ême Septembre de l'Anneé passeé, laquelle Je vous prie de communiques à Son Altesse de Bamberg à quoy Vous pouvés ajouter, qué les Emigrans Colonistes qui arriverent icy au Mois de Septembre passé au nombre de 36 Personnes furent logés le plus commodement hors des Portes de la Ville, qu'ils visitoient tous les jours et sans empechement les Eglises de leur Religion dans la Ville, et qu'ils ont pû aller venir librement chés leurs bienfaiteurs.

[Translation of the Foregoing letter, by George Fenwick Jones.] Copy of the letter from one of the Members of the Catholic Magistrate of the town of Augsburg to their agent in Vienna. Dated 15 Augsburg February N. S. 1734.

I understand from the gist of the letter you wrote me on the 30th of the month of last January, that His Highness of Bamberg and Würtzburg let you understand, on behalf of His Imperial Majesty, that, as His Majesty the King of England was still determined to have emigrants from Saltzburg transported to St. George's Island [Georgia] and that many complaints had been submitted in Vienna to the effect that the Catholic Magistrate would not tolerate these emigrants in the town until the season would permit their transportation, that it would give great pleasure to His Imperial Majesty if, in order to maintain and fortify the good understanding existing with the Court of England, we could, in some way, favor these emigrants and grant them free access until such time as they may be transported. I did not fail to talk amicably and in private with the President and two other Members of the Catholic Council, about which I have been entrusted to inform you in reply that you may assure His Highness of Bamberg on behalf of the Catholic Directory and the Privy Council that everything in their power will be done to ensure that the slightest wish of His Imperial Majesty will be carried out so that the emigrants who will come here afterwards and who will be destined for St. George will not only receive all possible tokens of good will in general but also, upon request, will be granted sanctuary in the town for a time providing that they do not come in quick succession and not too many at a time, in order to avoid, by such measures, any dissension arising among the population of the town, composed as it is of different religions, the principal aim of the Catholic Magistrate having always been to avoid such incidents and to show that there are no grounds for complaint about the Catholic Magistrate of this town for things past. I herewith enclose a copy of the answer made to the Ministers of Brunswick at the Diet of Ratisbonne dated 17 September of last year, which I beg you to communicate to His Highness of Bamberg to which you may add that the emigrating colonists who arrived here in the month of September past to the number of 36 persons were lodged most comfortably outside of the gates of the city and that every day they visited the churches of their religion in the city without hinderance and that they were able to visit their benefactors freely.

Mr. Butjenter's Journal at Dover. An Account of the several Occurrences during my Stay at Dover and in my Journey thither.

On the 26th of November O.T. having received Commission from the Honorable Society for Promoting christian Knowledge

I entered on my Journey which towards the Evening of the 27 I had finished.

After we had been some Days at Dover, on the 11th of December about 10 in the morning we saw the Ship that carried the Saltzburg Exiles, enter the Port; I with a Friend that accompanied me on some Business (but more especially on account of my Ignorance of the English tongue) Standing on the Shore near the Ship wellcomed the Saltzburgers. The Exiles seemed very melancholly not on Account of any fatigue on their Voyage, but because there were some who told'em they were carried to Georgia in order to be made Slaves of. We had some difficulty to remove those their apprehensions, but at last we effected it. Having thus a little gained upon the fears of the Saltzburgers, I took home with me the Reverend Messrs. Bolzius and Gronau their Ministers to the house which Capt. Coram had prepared to receive Mr. [Ph.] De Reck who had the day before left the Ship at Deal and had been to visit us. Then returning to the Ship and finding that there were some sick persons and some were Infants and that therefore it would be necessary that they should come on shore. I with the Leave of Capt. Coram hired one Room in a publick house wherein they might repose themselves and take some refreshment, and now the Saltzburgers began to be a Little heartned and to give credit to our persuasions when they were not only convinced by my words and those of their Ministers but by actual experience that their good was wholly consulted. The approach of the Evening closed our interview and discourse of that Day.

On the 12th Day of December in the morning I bought for and gave the Saltzburgers what the Ministers instructed me would be necessary for'em, afterwards I examined the Saltzburgers whom I found much more chearfull than before. A Dinner was that Day provided them by Mr. Coram's direction in the Workhouse where the poor are Maintained which was a greatway thro' several Streets[.] Mr. De Reck led'em they following in two Companies with their Children in their Arms and Singing Psalms while I with the other Ministers closed the procession. I could scarce forbear Tears at this Sight. When we came into the House where the Dinner was prepared when they were all sit down at Table they sung Psalms in praise of the Providence of GOD. When Dinner was served up, after Grace said, they received it with all imaginable Thanks to God, acknowledging his Bounty at every mouthfull. what they left they refused to take away saying that they were Satisfied and that they chose that others that were hungry might have it. They drank very sparingly everyone having about a Pint to his Share tho' many of them refused to have ale. We Dined with them in the same House, where after having Sung Psalms and Hymns to God I returned with the Saltzburgers to the said publick House where after having Sung, Standing in the middle of them, I preached to 'em from the 2 Corinthians 6. v. 17, 18 which they attended with the greatest Devotion and with Sighs and Tears some of them Kissing my hands whilst I spoke to them. Having finished my Sermon and Sung Psalms I distributed the money ordered by the Society for Promoting Christian Knowledge, which they received Kissing my hands with singular marks of acknowledgement, Some of them at first declining to take so much money. Having conversed sometime with them I proceeded to their Minister's Lodgings with whom I consulted about settling the form of their publick Worship.

Their publick Worship as near as may be will be conformable to the Church of England to which End I delivered to'em German Books in which are the forms of prayer, Baptismal Rites, The Lord's Supper, Matrimony and Burial etc.[198]

On the 13 of December in the Morning I proceeded to buy Things that were necessary for the Saltzburgers on their Voyage and Examined whether the Clergymen's Lodgings were convenient on Ship bord. After Dinner I again visited the Exiles and discoursed with'em upon Religion and faith and I perceived a greater part of them wanted to be more fully Instructed in the fundamental Articles of the Christian Doctrine and being asked whether they proceeded on their Journey with Cheerfulness they answered that they were willing to go to the uttermost parts of the Earth, That they were Children of God and that God would be with them. I took my Leave of them with Tears in their Eyes. I took also Leave of Mr. De Reck and Messrs. Bolzius and Gronau who do their utmost to chear up the Saltzburgers this they did in my presence[,] They likewise performed divine Service by reading and explaining to them every morning and Evening the Holy Scripture and concluding with Prayers and Hymns.

On the 14th Day of December about 4 in the morning I began my Journey for London by the Stage Coach in company of a Friend of mine Mr. Junner Matthison whose assistance was of great Service both to me and the Saltzburgers, which Journey by the Grace of God I finished the next Day in the Evening very Safely.

Henry Alard Butjenter

Text 2 Corinthians 6 V. 17, 18.

17. Wherefore come out from among them and be ye Separate Saith the Lord and touch not the Unclean thing and I will receive You.

18. And will be a father unto you and ye shall be my Sons and Daughters said the Lord Almighty.

Copy of a Letter from Mr. Urlsperger at Augsburg June the 17, 1734. No. 12712. Read 18 June 1734. Answered 18 June 1734. To Mr. Newman.

Honoured and Dear Sir: You'l see by the following Pieces that the Protestants of Saltzburg are again persecuted and that such as have Protestant Books are drove out of the Country[.] in all Likelihood the Two persons lately come to Ratisbone will be followed by many more, which time will show. We are not a Little Surprized to hear nothing of the Arrival in Georgia of the First Transport of Saltzburgers, And we cannot Account for the reasons of their Silence.

I should have mentioned Long ago that I received a Letter from the Reverend Professor [G. A.] Francke of Halle desiring that the Two Hundred Reichs Dollars [rix dollars] he had laid out at my Request for necessary Books, Equipage, and Travelling Expences from Halle to England for the Two Divines, gone to Georgia [Bolzius & Gronau] might be brought in the Credit of the Missionaries at Tranquebar or the Value thereof delivered to the Reverend Mr. Ziegenhagen for the use of the said Missionaries in East India. I remain

Honoured and Dear Sir   Your most obedient & humble Servant
Samuel Urlsperger

Since your Letter of the 2nd April last I received none in answer to mine of the 5th and 22nd April and 27th May.

Translation of a Letter written to the Protestant Magistrates of Kaufbeuren from Ratisbone, 7 June 1734.

Inclosed is a Copy of a Memorial, or an Examination privately handed about in this Town, concerning Two Emigrants Saltzburgers, born in the Jurisdiction of Goldegg [Goldeck], namely, Michael Gschwandtner, and George Schönegger, being now in this Town viz. (Ratisbone) the first being forever banished the Country, for having possessed some Lutherian Books, leaving an Estate of about Two Thousand Florins, and the Latter for the same

Reasons altho he had no Estate of his own. Time will shew in what manner they may be assisted, I remain with the Greatest Duty and Respect

Your Excellencies &c.   Most obedient & most Humble Servant

G. Wismeyer

Ratisbone 7, June 1734.

An Examination taken at Ratisbone, on Friday the 14th May 1734.

Michael Gschwandtner, born at Goldegg [Goldeck] in the Archbishoprick of Saltzburg, about 40 years of age Saith that he had been several times examined concerning Lutherian Books and that his House had been Searched for the same; Now that, altho' no such Books had been found in his Custody Yet on the 5 of this Month he was provided with the annexed Order of Emigration, together with Sixteen Florins towards his Journey and an Injunction that he should now go out of the Country and never be seen therein neither be in expectation of receiving one Farthing of his Estate which he valueth to be about Two Thousand Florins, because he had said, at one of his Examinations, that he would willingly be of the Lutheran Religion, and live with his Wife and Children, being very Sorry that he did not go out at First with them, as having had Sufficient Opportunities of overcoming his inclination of Possessing worldly Goods; That now he was resolved to enquire after his wife and Children till he had found out the Place of their abode, tho' unknown to him; his Wife being called Maria Kelbin, his Eldest Daughter Christina 10 years of age the other Barbara, about 8 Years and his Son Urban, about 4 Years Praying that their Residence might be enquired after and made known to him.

George Schönegger of Goldegg [Goldeck] aged 48 Years Deponent Saith; That at two different times by order of the Court of Justice Lutherian Books had been Searched for in the House he Lived in, Yet none had been found by reason that he carried them in his Pockets about him which they did not Search; But the Second Time the Clerk and Officer of the Court took away Kollman's Book of Prayer, as containing according to the Clerk's Opinion Superstitions and forbidden Arts and acquainted the Court thereof; Upon which he received Sentence the same Day as Michael Gschwandtner; That he had left nothing behind him, excepting Two Children one of 8 and the other of 6 Years of age, having got his Livelihood by Peasants work; and that they had given him Six Florins towards his Journey. Of which money so received they had paid Thirty Cruitzers to the Officer of Each of the Eight Justiciary Courts[199] through which they were Conducted, being four florins and one Florin for the Waggon upon which they were arrived from the first to the Second Court of Justice. That having paid for their Assistance very little they were in Possession of the Remainder.

Order of Emigration in German   Schub = Schein

By Virtue of an Order of the Honourable Privy Commission of H. H. Dated the 2nd January of this Current Year, Michael Gschwandter former Proprietor of the Estate called Schertzberg, because He contrary to Prohibition hath kept several Lutherian Books is for Ever Banished the Archbishoprick of Saltzburg and is ordered to be Conducted from Court to Court to Strasswallhen [Strasswalchen] on the Frontiers and this Order of Emigration is given him for his going out.

At the Castle at Goldegg, May the 5th 1734. His Highnesses of Saltzburg Government there and the Court of Justice of S. Veit.
L. S.   John Christophle von Eyss   Governor
For his further passing from this Place out of the Country he was sent forward. Strasswallhen May the 9th 1734 His Highnesses of Saltzburg Court of Justice.
L.S.   John Anton Haussen

Copy of a Letter from James Oglethorpe, Esq. at St. Hellens Road near the Isle of Wight June the 16, 1734. No. 12715. Read 18 June 1734. Answered 20 June 1734. To Mr. Newman.

Sir: I have Several Letters for you from Mr. [Ph.] Von Reck and the Congregation of Saltzburgers whom I left well pleased and in good health at Ebenezer. I thought it not proper to send the Letters by the Post since I shall have the pleasure of delivering them to you with my own hands in a few days. I have wrote at Large to Sir John Phillips to which I refer You and am
Sir   your most Obedient and humble Servant   Ja. Oglethorpe

Copy of a Letter from James Oglethorpe, Esq. at St. Hellens Road near the Isle of Wight June the 16, 1734. To Sir John Philipps, Bart.

Sir: Tis with Pleasure I acquaint You I left Mr. [Ph.] Von Reck the Two Ministers and the Congregation of the Saltzburgers in perfect Health, they are a Sober, industrious cheerful and devout people[.] God has been pleased visibly to protect them in their passage as Mr. Von Reck has mentioned at Large in his Letters to Mr. Newman. They are Settled to their Satisfaction upon a River which they have named Ebenezer, where they intend to create a Stone Pillar "in memory of God's having delivered them from persecution and led them through to the utmost parts of the Earth where they may praise his name with freedom and be a light to the Gentiles." These are the Minister's own words as near as I can translate them: and indeed there does seem to be a door opened towards the Conversion of the Indians; for one of the Indian Chiefs called Tomo cha chi the Mecko of Yammacraw a man of excellent Understanding is so desirous of having their Youth instructed in the English Learning and consequently in the Christian Religion that he is come over notwithstanding his great age to Solicit for a means of instructing their people: he is now with me and has brought with him a Youth whom he calls his Nephew and next Heir. This Child hath already learnt the Lord's Prayer both in Indian and English. I shall leave the Indians at my house in the Country whilst I shall go to Town where I hope to have the happiness of waiting upon You and acquainting You more fully with many things relating to this matter at which I believe you will be as much pleased as Surprised. I am

Sir   Your most obedient humble Servant   James Oglethorpe

Copy of Translation of a Letter from Mr. [Ph.] Von Reck at Savannah Town[200] *in Georgia, 22 March 1734. No. 12725. Read 5 June 25 June 1734.   To Mr. Newman.*

Sir: I hope my poor Letter dated at Charles Town with the Journal of our Ministers will be delivered to You; You will see also my little Journal which I delivered to Mr. Oglethorpe. I intreat and humbly beg You will be so good as to send the Journal to my Dear Uncle [J. von Reck] at Ratisbonne together with the Letter herewith sent.

Let him and me have this Joy and Pleasure I beseech You. The suddain departure of Mr. Oglethorpe for England, the Business which engages me here, and a Sickness that took me two days ago having taken up my Time, so that I have hardly strength to assure You and my dear Patron Mr. Vernon of my most humble respects, and the Veneration with which I remain during Life.

Sir   Your most humble and Obedient Servant   Reck

P.S.: After having recovered a little Strength, put in order the Saltzburgers, regulated their Labours and their affairs, built Houses and sowed some Acres[,] I shall return and God willing I hope to Compass it in a month, when I propose to Leave this Place by God's Help.

Copy of a Letter from Mr. Bolzius dated at Savannah Town in Georgia the 22nd March 1734. No. 12726. Read 25 June 1734. To Mr. Newman.

Most Honoured Sir: The Contents of your Letters which we received in Dover from your own hand gave Us Singular Evidences of your favour for us: therefore I hope surely, that you will accept this humble Letter as a Testimony of our gratitude for your beloved kindness and that you may favourably over see the faults, which I committ in writing the first time in your English Tongue. By the Blessings of God we are arrived on the Coast of Charles Town the 7th of March where our Ship rid at Anchor, till a Pilot came to bring us for Georgia. after we were gone with the Captain and a few other persons in a Sloop for Charles Town, we had the great Honour to wait upon James Oglethorpe Esq. and to dine in the House of the Lord Governour [Robert Johnson]. Mr. Oglethorpe received us kindly and sent unto the people in the Ship some fresh water, meat, good Wine and Garden Fruits to refresh them with after the hard and Salt provisions; what Joy this brought is inexpressible. We departed from thence the 9th of march and arrived the 11th Day at Savannah in Georgia. In this place was pitched a Tent and having received here much Benefit by the fatherly care of our Governour Mr. Oglethorpe, we took a Journey further to our own Country designed for us, for which we have Longed a great while. All our fellow Travellers were very glad to see the green and fruitful Tracts of Land on the Right and left sides of the River and especially the Saltzburg people have a great desire to Labour earnestly and hope by the Blessing of God a happy Success. And tho' our Voyage was attended with much Trouble nevertheless this Voyage was of great advantage and we have much reason to magnify God's goodness, that he showed both to our Souls and Bodies. A Few of them were Sick, because they were not used to eat Ship Victuals, and the great motion of the Sea; but their Health was restored in a short time insomuch that they all live and are well at this time. Even as these persons have received from Us at other times a good testimony of their true

Piety and Christian Virtues so I assure You that we can report the same of them at present: for they have much delight both to hear the holy word of God and to exercise with all their heart that which God demands from men in his holy Scriptures[.] taking this good purpose, they will Show themselves not good Labourers only but good Christians also all their Lifetime. and after this way shall it come to pass that they may adorn the doctrine of Christ and Set a good example to dwellers in the same place or in the Neighbourhood. many amongst them are defective in reading because they have wanted [lacked] good instructions in the Land of Saltzburg, but for the Love of God's word they give their utmost application to learn it, and we will help them therein with all our power. Very few are Stocked with Bibles and Psalm Books after which they have a hearty desire. And if our hearers continue (as we surely hope) to walk before God so he shall perform his dear Promises being in his holy word especially Leviticus 26 & Seq. and you and all our Benefactors shall receive much matter to glorify God and take reward for all your Goodness presented to us. I close up this with my heartiest and best wishes and remain with the greatest respects.

Most Honoured Sir   Your most addicted Servant.

John Martin Bolzius

My beloved Colleague Mr. Gronau presents his respects to You.

Copy of a Letter from Mr. Urlsperger at Augsburg 1st July 1734 N.S. No. 12732. Read 2 July 1734. To Mr. Newman.

Honoured and Dear Sir: My Last Letter to Yourself was the 17th June last and it is now near Three Months since I received any of your Favours even since the Second of April last which maketh us apprehend, that you are indisposed; However we are in hopes of receiving agreeable Accounts of yourself and of our Colonists in Georgia, or that the Reverend Mr. Ziegenhagen will take the Trouble thereof in your absence from London. By the annexed Memorial and Extract out of some Letters You will see, that the Spirit of Persecution is still raging on the Confines of the Archbishoprick of Saltzburg; and that the Body Protestant [Corpus Evangelicorum] of Ratisbonne have resolved to interpose their good Offices with His Imperial Majesty on the behalf of such worthy Confessors of the Gospel. Time will Shew what is like to be done with these poor People, and I remain

Honoured and Dear Sir   Your most Obedient humble Servant

Samuel Urlsperger

P.S. I expect [await] Order if [whether] Mr. Vat stays any longer here.

Translation of a Memorial presented to the Protestant Body at Ratisbonne the 18th June 1734 by Forty Heads of Families of the Workmen in the Salt Mines in upper Austria, Sentenced to be Carried to Hungary and Transylvania there to receive their Reward.   T.T.E.

Your Excellencies s.s. have been already several times troubled with an Account of our hard Oppressions concerning our Faith, under which we Labour at the Instigation of the Romish Clergy of this Country and in what manner we are hindred from Serving the Lord our God, according to our Conscience and the Principles of the Protestant Religion and humbly desired that you would be graciously pleased to Intercede for us with his Imperial Majesty, our most Gracious Lord. But as we have not hitherto perceived the least Mitigation, and rather from time to Time have been molested and more severely persecuted by the said Clergy, till we have seen many of our Fellow Citizens, one after the other dragged to close Prison and at last on the 31st May, 1st and 2nd of this Instant June, Forty men Housekeepers [Householders] together with their Wives and Children according to the enclosed List, were ordered to appear before the High Commissioners of Religion, Commanded that all such whose Names were markt with a Seal in the Margin of the List should be ready on the 6 and 7th days of this month, in order to go on Board Shipping and to be transported to Hungary and Transylvania, there to receive our Rewards, intimating that perhaps we should be deprived of our Lives. Upon such a proposal we fell all on our Knees and prayed that we might rather be carried to the Place of Execution and there to die, then [sic] to be sent into Hungary and the Tartars, if it was His Imperial Majesty's Pleasure and Command, which however we could not believe of so gracious a Sovereign; For we said, *if the Emperor is so gracious and can suffer us in Hungary he may be likewise so gracious as to Suffer us to live in our Native Country.* Now we most earnestly entreat your Excellencies, provided it be the Will of GOD, and it be in your Power, that you would be graciously pleased to interpose your good Offices in this Affair in our behalf But if it is not in your Power, *We shall willingly submit our Selves to the Will of God, and rejoice to be worthy to suffer not for any ill Fact, but for the name of Jesus*

Christ and for the Truth of the Gospel. We die with the most profound respect

Your Excellencies   Most obedient Humble Servants.   All the Emigrants and Workmen of the Saltmines in Austria on the River Enns

A LIST of such Protestant Confessors as are Ordered to go out of the Country.

André Nerff [Neff], Widower, and 5 Children. a close Prisoner at Lintz.

Tobias Leimar, his Wife and three Children.

Blassius Eggenrüther, his Wife and five Children.

Paul Lehrbacher his wife and Three Children. himself close Prisoner at Lintz.

Michael Leserer, his wife and 5 Children. He being in close Prison at Lintz.

Paul Kayser and his wife.

Hans Goller, his wife and 5 Children. In all 39 Souls of Halbstadt [Hallstatt].

Blasius Engleither, his wife and 5 Children.

Michael Osterer, his wife and 5 Children.

André Brümbsperger his wife and 2 Children.

Thomas Kütler, his wife and three Children.

Matthias Gütler his wife and three Children.

Hans Karrman his wife and 5 Children.

Jacob Küttler and his Two Sisters.

Matthias Karrman his wife and Two Children.

Hans Höll, his wife and 1 Child.

Michael Gupfleitner his wife and Three Children.

Hans Köhl his Wife and Six Children and 1 Servantmaid.

Matthias Engelkitner his wife and Seven Children.

Andrè Bäer his wife and Six Children.

Hans Greismauer his wife and Seven Children.

Matthias Leinpacher his wife and 5 Children.

Michael Nerff [Neff] his wife and 4 Children. He in prison these 9 months at Lintz.

Thomas Kayser his wife and 2 Children.

Matthias Buchsteiner his wife 1 Child and 1 Servantmaid.

Matthias Fischer, Widower and one Child.

Joseph Güttler, his mother his nephew and 3 Sisters He having formerly been in close prison.

George Stoymer his Wife and Seven Children.
Michael Kerfer his Wife and 4 Children.
Hans Geissmauer, his wife and 2 children.
Hans Bäer, his Wife and 5 Children.
André Neüpacher his wife and Six Children.
George Glöchel Widower, and 2 Children.
Wolf Ursteger his Wife and Six Children.
Wolf Engelleiter, Widower and Two Children. In all 162 Souls
of Gossen [Goisern].
Matthias Hueber his wife and 3 Children.
Stephen Hueber, his wife and Five Children.
Sebastian Hueber and his wife.
Joseph Heüschober his wife and Eight Persons of his Family. He
being now in close prison.
Matthias Heüschober his wife and 8 Children. In all 34 Souls of
Ilschler [Ischl?]. The whole number of the said Three parishes
being 235 Persons.

*Extract out of the Register of the Protestant Conferences held at
Ratisbonne the 12 June 1734.*

It was resolved to represent to his Imperial Majesty the hard
usage against such of the Inhabitants in the archbishoprick of
Saltzburg as kept Protestant Books, and were found among them,
Likewise for obstructing the passages in the said Archbishoprick,
and for refusing Passports to such Inhabitants as are willing to
withdraw out of that Country. As also in regard to such Protestants
in Austria to the number of 300 as desire to obtain Leave of coming out of that Country, and particularly concerning Hans Lerchner, who is condemned to work at the Fortifications at Raab, for
Eight Years.

*Extract out of a Letter, written to the Magistrates of Kauffbeuren;
by Mr. Weysmeyer [Weismeyer] date at Ratisbonne 21 June 1734.*

That by Reason of some Difficulties relating to the Person of
Mr. Goebel, the King of Prussia had appointed Mr. Von Plotho,
Counsellor of Embassy, to assist at the Conferences of the Electoral
College at Ratisbonne, as his Prussian Majesty's Envoy as Elector
of Brandenburg and afterwards to go to Saltzburg to reclaim the
Effects of such Emigrants out of Saltzburg as are Settled in the
Dominions of His Prussian Majesty.

That several Grievances of the Protestants in the 'Paletinate' [Palatinate] in the Bishoprick of Wirtzburg [Würtzburg] and Principality of Schwartzenburg had been laid before the Evangelick Body at Ratisbonne.

Copy of Translation of a Letter from Mr. [Ph.] Von Reck at Savannah in Georgia 7 March 1734.[201] *No 12741. Read 9 July 1734.*

Sir: Being arrived by the Blessing of God altogether safe and in good Health near Charles Town from which we were about Six Leagues distant we there cast anchor, and the Captain went to Charles Town to procure a Pilot who might conduct us into Georgia, and as I was with him I had the happiness to meet with Mr. Oglethorpe at the house of his Excellency the Governour of Carolina [Robert Johnson], the father and delight of the Colony of Georgia[.][202] he designed to have returned into England in 15 days but is now Setting out for Georgia to Settle the Saltzburgers there. The Voyage is not yet finished for tomorrow we shall continue our Progress, my Journey is not yet Compleat, the little time of my Stay here preventing my committing to Writing what I have to say, I propose to send you a full Account of our Passage as soon as we shall come to the end of it.

The first Ship which goes from hence will carry to you my Packets, my Journal, and my Respects to all my Patrons in particular to Mr. Vernon in the meantime the Ministers have compleated their Journal and sent it to the Reverend Mr. Ziegenhagen; Excuse me then I beseech You that My Journal does not accompany my Letter, once more I beg your pardon for it[.] my Journal shall follow in a few weeks and I remain while I live with great Respect Sir   In hast   Your most humble and most obedient Servant   Reck

Copy of a Letter from Mr. Urlsperger at Augsburg 15 July 1734 N.S.   No. 12743.   Read 16 July 1734.   To Mr. Newman.

Dear and Honoured Sir: Both your Letters of the 11th & 18 June last are come to Hand, and I have seen with great Pleasure, not only the safe arrival of the Saltzburgers in Georgia but also the return of James Oglethorpe Esq. into England. The Lord of Hosts be Praised for all and Particularly for his having opened a Door to the poor Gentiles, who at the beginning of this affair lay at the Bottom of my Heart to come at them. Now I long after seeing such Letters, as are come out of Georgia in order to print them out of Hand for the Publishing of them, as I have already done with such accounts as you sent me, by Inserting of them into the Publick news Papers here which have raised an universal Joy among the Protestants here in this Town. Mr. [J.] Von Reck the Envoy at Ratisbonne was no less overjoyed at the accounts I gave him of the contents of your Letter of the 11 of June.

And now is the Question: In case upon the Publishing of the Letters and accounts expected, there should be 30, 40 or 50 Saltzburgers willing to come out of some Neighbouring Imperial Towns, and be inclinable to go to Georgia, Whether I should send them upon Notice thereof given to the Society under the conduct of Mr. Vat to Holland? Likewise whether I should, in the present Juncture of the Warlike Troubles, especially in regard to the French Army along the Rhine, apply to the English Hanoverian Ministers at Ratisbonne for Passports in their Behalf and also whether in case the Emperor upon the Memorial and Intercession of the Body Protestant in behalf of the Protestants of upper Austria of which I lately Sent you Copies, should permit them to come to this Town, I might admit of such of them as should be inclined for Georgia, as being of the Neighbourhood of the Territories of Saltzbourg and being encouraged to come away by these and of the like good Dispositions. We hear nothing more of those of Hallein by Reason, no Doubt that all Correspondence with that Country is cut off. Upon which the Envoys at Ratisbonne very lately as it is just laid before Complaints before the Emperor. God grant that the Imperial Army on the Rhine be not so unhappy as they were in Italy and Neapolis [Naples], otherwise our Circles would be very much embarrassed! What remains to be answered upon your Letter of the 11 of last month Shall be deferred to another Opportunity. I could wish, that I might be for Two Days with your good Company present at the Report of our Dear Mr. Oglethorpe; However I hope to be acquainted with the Substance thereof by your indefatigable Love Letters.[203] May God Almighty give his Blessing to his Work furthermore! I remain

Honoured and Dear Sir   Your most obedient humble Servant
Samuel Urlsperger

In Case the Letters come from Georgia be inclosed in the Kings Packet, it will not be amiss the Recommending to the German private Secretary [Reiche] that a cover may be put there upon in London directed to Mr. [J.] Von Reck H. M. Envoy at

Ratisbonne. Otherwise if they go by the way of Hanover they may be a long while a coming, as it happened formerly with the full powers; Yet it's of Consequence that those Letters might be here in a Short time.

Honoured Sir   I am Highly obliged to the kind Remembrance of Governor Shute and Collonel Valogne; I beg the Favour of you to return them my Hearty Thanks and to make my Respects acceptable to them and to James Oglethorpe Esq. whose return to England giveth me the greatest Pleasure, which would not be diminished could I hear something of his Knowledge relating to the Carolinian Swiss; And what Success Collonel Purry hath in Swisserland[.][204] And I crave Leave to Subscribe myself.

Honoured and Dear Sir   Your most obedient and most Humble
Servant   John Vat

P.S. P.S. Since the writing of the former Postscript, I think his Britannick Majesty's Passport for a Commissioner and as many Colonists as the Society intend to send to Holland, or Leaving the number of them in Blanc would be less liable to Exception than any other.

Copy of a Letter from Mr. Quincy at Charles Town in South Carolina 29 March 1734.   No. 12749.   Read 16 July 1734.   To Mr. Copping.

Dear Sir: I had the Pleasure of yours by Mr. Bolzius Minister to the Saltzburgers and am very Sorry to hear that you have heard nothing from me. I wrote by a Gentleman who went from hence about five Months ago, and desired him to wait upon you that you might have an Opportunity of informing yourself in as particular manner as you desired about the Condition of this new Colony. I have desired the same favour of the Gentleman who brings this. he comes over with Mr. Oglethorpe in order to get Servants, and then returns to Settle here. I hope this will in some Measure attone for the deficiency of my account of this Place. When I came first to Charles Town and should have employed my time in Writing to my Friends, I was so ill as not to be capable of it and am now in a very great hurry to provide for my voyage to New England expecting to sail today. I intend to make my return from New England, as far as Philadelphia at least, which is 400 Miles, by Land, and Mr. Oglethorpe has desired me to make Observations as I pass through the Country and remit them to him which

I intend to do and as I believe they will not be unacceptable will send You a Copy of the same. We [in Savannah] are now considerably increased in numbers being about 900 and have made great Improvement in Building. I believe we have between 50 & 60. Houses up. a large and commodious Store House, but no Church yet[.] we are forced to make use of a very indifferent Place for that purpose, which will not hold 1/4 of the People, so that I hope they will go about that necessary work as soon as conveniently they can. at present they are employed in making a Fortification round the Town; for in case of a War with France or Spain which we have had much rumour of; we expect to be attacked from the French or Spanish Settlements which are our Neighbours, and have a great influence, especially the French over the Indians. If it please God we are not molested this way I think there is no doubt to be made, but that the Colony will succeed very well. We have indeed lost a great many by sickness[.] I have buried since I have been here upwards of an hundred[205] but when we have overcome the Hardships and inconveniences of a first Settlement we shall I hope enjoy better health. I believe the distemper we have been afflicted with, which is the Flux is more owing to mean Diet, and the being exposed to Heat and Cold and Wet, without any covering but Tents or boarded Huts equally defenceless, than to the badness of the Air, so that when these inconveniences are remedied which Time and Industry must do, the bad effect of them will cease. Amongst the rest that have died, I have lost the poor man that came over with me[.] he proved an exceeding good Servant and would have been of great use to me. Mr. Oglethorpe has been so good as to make me a present[206] of another but he proves but very indifferent. I had a Letter at the same time with yours from Mr. Wilson and Mr. Newman. I shall write more to both those Gentlemen from new England, which Letters may probably arrive as soon as this. Mr. Newman has Relations in New England and will I presume be glad to hear of them. My Relations there had heard of my being in this Country and sent me a Letter being a very obliging one. It will be great pleasure to me to see them again from whom next to my Parents I have received the greatest Obligations in the world. my absence will be Supplied part of the time by a very worthy Gentleman Mr. Jones a Missionary at Port Royal, a very good man whose Conversation and Friendship I highly esteem. I hope to be able to return in about three months and then shall not desire to leave the People any

more while I continue in the Country. I had a Letter from Mr. Vernon since Christmas which is dated in May last, and arrived by a Ship in August but by reason the parcel it was enclosed in, was not opened it came not to my hands sooner. I shall do my self the Honour of writing to that Gentleman from New England and return my thanks for so great a favour. The Commissary [Ph. von Reck] and Minister [Bolzius] to the Saltzburgers have the use of my house in my absence which happens very convenient for them. they would otherwise be exposed to equal Hardships with the rest of the new comers. They appear to be very good People, and I hear like the Place very well. I shall do all in my Power to preserve the People in peace and unanimity now Mr. Oglethorpe has left us, and shall not I hope disappoint the expectations of my Friends, in the confidence they repose in me[.] at least nothing within the compass of my ability shall be wanting for promoting the good of the people, the success of our best endeavours must always be left to the wise over Ruler of all things. I shall take care to transmit frequent Accounts of the Circumstances of the Place as you desire. My humble Service to Mrs. Copping, I am sorry to hear she has an ill State of Health. Service if you please to Mr. Vernon, Mr. Wilson and Mr. Newman with other Friends from
Dear Sir   Your most obedient humble Servant   S. Quincy

Copy of a Letter from Mr. Urlsperger at Augsburg the 19 July 1734. No. 12753. Read 23 July 1734. To Mr. H. Newman.

Honoured Sir: Since my last of the 15 Instant, I have received the accounts above, and it is Submitted to your Consideration of the Societies. Whether it would not tend to the Honour of the British Nation should His Britannick Majesty intercede powerfully in behalf of the Protestants of H. I. Majestys Dominions? We are in Expectation of receiving such Letters as were sent from the Colonists in Georgia and I remain
Honoured and Dear Sir   Your most obedient Humble Servant
Samuel Urlsperger

Extract out of a Letter dated at Ratisbonne July the 12, 1734.

The Letter of Intercession of the Body Protestant in behalf of the Emigrants out of upper Austria dated the 19 June last hath been delivered to his Imperial and Catholick Majesty and hath been entered in the Register and even printed and published; But the effect thereof is as yet expected [awaited]. There are here some Deputies from Carinthia who Solicit likewise such Letters of Intercession. But it is evidently reported That at Vienna They insist That in case any Protestants of the Imperial Hereditary Dominions are inclinded to leave those Countries on account of the Protestant Religion, They should go into some other Imperial Hereditary Country and that such as are now in that Case should be sent to Transylvania and that it should be by virtue of a private Priviledge, of which no mention is made in the Treaty of Westphalia, Yet it is very sure and certain that the Arch Dutchy of Austria belongs to the Roman Empire and is to be put again upon the Foot of the Year Annus Decretorius; But it might happen to those people in Transylvania, as it has to the Tirnbergers in Holland, where they must necessarily, among so many Sects, be brought into Confusion.

Extract out of another Letter Dated at Ratisbonne 13 July 1734.

The Protestants of Austria have sent hither a Deputy and a new Memorial wherein among other Circumstances it appears that one of their Parish Priests, called Matthias made use of one possest of the Devil, who upon all Questions put to him was to prove that the Roman Catholick Clergy was the only way of Salvation and thereby to deterr the protestants from adhering to the principles of Luther and his followers.

Now God hath likewise manifested himself in Carinthia by his Light of the Gospel; For there they begin to declare themselves freely and openly; I have been told by their Deputies that at first about 150 of them had publickly declared themselves to be of the Evangelick Faith according to the Confession of Augsburg, which before their coming hither were joined by three Districts of Jurisdiction (called Gerichtes). Another Man, who went thither to fetch away his wife and Children but could only save one child, his wife being caught and sent to prison, relateth that there were already Several Hundreds nay above Thousands gone to these Jurisdictions to be entered as Protestants, and that there would be many more.

N.B.   These People together go into By-Chappels and sing and pray unmolested, and are permitted to read Evangelical Books freely. Some of their Priests at these Events wept bitterly. One of their Deputies hither was a prisoner for 23 weeks. very lately 13 young men who had declared themselves were forced to be Soldiers.

Copy of a Letter from Mr. Urlsperger at Augsburg the 22nd July 1734. No. 12759. Read 30 July 1734. To Mr. H. Newman.

Honoured and Dear Sir: I hope you have received my last Letters of the 15th Instant since which Time yours of the 25th of June and 2nd of July, together with the enclosed are come to Hand; and I am in great Expectation of receiving the Journals of Mr. [Ph.] Von Reck and Mr. Boltzius. The Resolution of the Society, as far as I can see seems to be these: At present to receive none other Transport so that at last it cometh upon the Answer, which you shall be pleased to give me, upon the Question contained in one of my Late Letters Viz. Whether such Persons, as could not go with the First Transport or the Friends of such; as are already gone to Georgia, and should upon Publishing the Accounts come from thence resolve to go to the Number of 20, 30, 40 to 50 would be received upon the same Conditions as the first and whether they should be by Mr. Vat conducted to Rotterdam Or whether the 5 Persons remaining here, since the first Transport went from hence altho they have a great Inclination of following their Country Folks, should be left here? I return my hearty Thanks to the Society for the Copies of Messrs. Bolzius and Von Reck's Letters to them; Their Letters to my self refer themselves to their Journals of which upon their reception You shall have Extracts. Thus much I can see by those Letters that notwithstanding the many Troubles the Ministers work chearfully among their Flock. I have long ago lamented the Loss occasioned by the Conduct of the Tirnbergers; However it seems to me a Transport of 30 to 50 should break no Squares and consequently That such a small number might hope for the assistance of the Trustees in the same manner as the first.

Mr. [J.] Von Reck Envoy at Ratisbonne, writeth the 20th Instant, "I expect [am waiting] to See more particulars of the Georgian Colonists. There are here some Deputies of Austria and Carinthia who assure that the Emigrants out of those Countries which [wish] for nothing more than to be Transported to Georgia, and there is no doubt that after they shall hear how happily the First Transport is got to Georgia they will have a much greater Desire of going thither." I am with all Sincerity

Honoured and Dear Sir   Your most humble Servant
Samuel Urlsperger

P.S.: As to Mr. Vat, He hath had his Diet and Lodging at my House at the Charge of the Society and his own Horse hath been at publick Stable to the latter end of April, and since that Time at

Grass and he designs to make use of it at his return to Swisserland. I paid his Physician and Apothecary when he was Sick; and he hath received Twenty Five Guilders in money. His Journey from Swisserland hither costs him Twenty Florins and his return to Swisserland will cost as much. Now it is to be known what the Society designs that I should give him more. Certainly he is a very pious and understanding Gentleman; a man of great experience in such Things and worthy to be loved and esteemed from all true Christians.

Copy of a Letter from Mr. Urlsperger at Augsburg 2 August 1734 N.S. To Mr. H. Newman.

Honoured Sir: Referring my Self to my Letters of the 1st, 15, 19 & 23 of July last these Lines are only to accompany the Memorial and Extract of a Letter, concerning the unhappy Condition of such Protestants of Austria and Carinthia as have made a publick Confession of their Faith. I was in Expectation of receiving the Journals of Mr. [Ph.] Von Reck and Messrs. Bolzius and Gronau but they are not as yet come to Hand. However I beg Leave to join here to the inclosed Letter which I desire may be conveyed to the Reverend Mr. Bolzius in Georgia, I am

Honoured Sir   Your most obedient Humble Servant
For the Reverend Senior Urlsperger   John Vat

Memorial of the Protestants of Carinthia delivered the 4 of July 1734. To the Protestant States at Ratisbonne.

We the publick Confessors of the Gospel Truth in the Province of Carinthia cannot omit most humbly to represent to your Excellencies, the Envoys of the Protestant States of the Roman Empire at Ratisbonne in what manner we have by the Powerful Assistance of God and by the Free Conviction of our Conscience been roused up to adhere publickly both with Heart and mouth to the pure Doctrine of the Holy Gospel, according to the Confession of Augsburg; For which reason and indeed upon that account only, We have Suffered very heavy Persecutions at our Courts of Justice chiefly by the instigation of our Clergy here; Now we most humbly beg Leave to mention in few words, That we cannot comprehend considering that many Thousands of Protestants who profess the same Faith, as we do, especially in some other Provinces belonging to his Imperial Majesty, our most Gracious Sovereign, especially in Silesia and Hungary where they are protected, Why we should be persecuted with being cast into dark prisons and the like Hard Dealings, because We profess the same Publick Faith as they do, altho' we are no less Faithfull in Civil affairs, than any other Subjects of His Imperial Majesty, as our Sovereign Lord, to whom the Great God King of Kings and Lord of Lords may grant long Life Health victory a happy Reign and pour down his choicest blessings and Favours to a great Age, thinking our Selves bound at all Times to shew our most Dutifull Submission, by venturing our Estates Body and Life, in all Faithfulness to his Service. Yet it is not unknown That among others Simon Seigel, Locksmith, our Brother and Confessor of the Faith hath been detained in Prison at Vienna for Eighteen months, and before that sometime at Klagenfurth and as far as is come to our Knowledge remains still a close Prisoner upon no other account as That of the Evangelical Lutheran Religion. On the 25th March 1733 Christopher Lägler went to the Parish Priest in order to make his Confession and to receive absolution. The Priest asked him whether he did well understand the Five Commandments of the Church? upon which Lagler answered They could be of no Benefit for Salvation. Then the Priest called him a heathen and dismissed him without his absolution pretending That by Order of the arch Priest and the chief Officer or Governour he durst not absolve him, But he gave a Bible to the said Lägler to carry home to his own House telling him That therein he would find every thing concerning the Rosary and the Purgatory; But after Lägler had Signified to him that he could not find any Thing of any Foundation therein and that it was not necessary to Salvation and therefore could not believe it the Priest run upon him in the open Street took him by the neck cloth and dragged him in such manner that he had like to be Strangled. Some Time afterwards the said Lägler was sent to close prison and there detained for Twenty three weeks, During which Time he was Twice examined by the Clergy and Laity, And there he freely gave Glory to God by Professing the Gospel according to the Confession of Augsburg. About that Time many more were convinced of the Truth of the word of God and the Gospel, and united themselves in Devotion by Praying and Singing together. Which alas! altho it is a Service most acceptable to God and even commanded by the Lord himself hath been looked upon as such a Crime that soon after Twenty Peasants were cited to go to Klagenfurth and Seven of them were put into close prison who likewise joyfully made publick Confession of their Faith. As these were carried out of their native Country, our adversaries said to a Bailiff officer, These men shall never come back again to Hear the Ringing of the Bells of that market Town[.] Yet its very observable That soon after a Fire broke out in that Town and consumed many Houses and the Church and melted down the Bells thereof, so that they themselves could no longer hear the Sound of them and yet the said men Confessors returned afterwards to that Town and then there was an order that we should no more meet together in Prayers For that the following Spring there would be an End made of that affair. But since hitherto no alteration hath been made, We meet together in a By-Chappel in Singing and Praying and this year about Two hundred Persons of these parts are not come to Confession. Last year it was publickly ordained That such persons, in whose Custody Lutheran or the like Heretical Books, should be found should forfeit Houses and Lands and the like Penalties have been inserted in the Leases of such as Live in the Jurisdiction of Patria[?]. It was likewise ordered by severe Penalties that no Saltzburger should be admitted into any House! It was also forbidden last year about Martinmas at Nicholas Town to the Roman Catholicks to converse negotiate or to travel with any Heretick! The Schools in such Towns where no Priests reside have been suspended and put down for no other reason but that the Children in Length of Time may be disabled from learning the Reading the word of God; They knowing how powerfully it worketh against their Superstition. In the Jurisdiction of Mühlstadt [Millstatt] the Priests always called our people Hereticks; Upon which at first, Five of them made a free Confession, and as they were asked What they did believe? Answered That they would by the Grace of God believe only according to the Holy Scripture. Then the Priest referred them to the Commandments of the Church, But they replied That they Trusted to be Saved only by the Holy word of God, as many Thousands since the Time of Christ and of the apostles down to this time were Saved. For the Gospel is the Power of God unto Salvation to every one that believeth. Romans 1. 16. As lately the arch Priest came to the Church, at the Visitation, It was read from the Pulpit that such as had any Thing to say in matters of Faith should come forth and appear. On the 6 of June last Four Score persons of the parish of Biberstein made their appearance and professed themselves to be of the Evangelical Faith according to the Confession of Augsburg; and on the 7th Ditto Fifty persons more, which were followed on the 14 June in the Parish of Margaret by above Twenty Persons and according to what we hear many many in other Jurisdictions are by the Grace of God to make open Profession of their Faith. Therefore we the openly Confessors of the Dear Evangelical Truth most humbly entreat for God his Sake Your Excellencies, according to your most laudable Zeal not unknown to the Evangelical Christianity, to be graciously pleased to take our Cause in Hand with the same Vigour as you have already done in behalf of so many Thousand other poor Emigrants and Evangelical Brethren of the Faith and to promote our Welfare and good Religious Cause according to your wise insight and your Influence by a Gracious Intercession. Dated the 4 July 1734.

Extract of a Letter written from Ischel upon the Ens in Austria the 6 July 1734.

As on the 12 June last it was Signified to the Forty Four Heads of Protestant Families by the I. [Imperial] Commissioners of Reformation of Religion That in Three Days they should go on Bord some Ships in order to be carried down to Hungary & Transylvania with Threatnings that there they should receive their Salaries[207] and their rest. such of them as would not readily comply with that order but gave for answer That if his Imperial Majesty would grant them Liberty of Conscience in Hungary or any other Province of his Dominions they did not know why he could not grant them the same Liberty in their native Country; But if they should receive their rest in Hungary it would not be necessary to carry them to so great a Distance. They were ready for the Sake of the Gospel to receive it out of Hand here, were put in Irons and immdiately carried by water to Lintz; Viz. Aster, Thomas Tauber, Matthias Fischer Thomas Kayser Michael Kerfer Andreas Brumpelsberger Jacob Gädler Hans Bäer and Wolfgang Engelleuthner, and they were told this was done not upon account of Religion but for their Disobedience! Besides these Michael Köbel, Hans Kampel, Thomas Feüchtner, Joseph Heuschober, Hans Losserer and Balthasar Eckenreüther etc. are also put in Irons and in prison at Ischel. The others answered that they durst not fight against Power, but must Suffer whatsoever should be done with them; Moreover that they were as ready to die for the Sake of Truth in their native Country as in Hungary etc. which was interpreted by the Commissioners as their Consent to be transported and it was entered into the Register of the Court. All of them insisted on the words and Promises of the Count Von

Serau chief Officer of the Salt office who at the beginning of these motions, before they had made open Profession of the Evangelical Religion and were written down as such, had not only exhorted them all to discover the Bottom of their Heart and Faith and to be no longer Hypocrytes with their Lips but had Promised them faithfully upon his word of a Count to grant necessary Passports and Letters of License to all such as should write themselves down as Protestants, in Order to go where they pleased. They said they had Served his I. Majesty their most gracious Sovereign truly and faithfully to the best of their power and did not know themselves guilty of any Crime committed in Civil matters against his authority, and that they could hardly believe it to be the Emperor's Orders, to be used [treated] in that manner; For some weeks ago Two men were with the Emperor himself and delivered Him a Petition for obtaining Leave of coming out of the Country which Petition his Majesty received in an extraordinary gracious manner and told them he did not know what to do, there were in his Dominions very many such people; That after better peace He would once go into their Country to hunt there the wild Goats [Chamois] with them. Yet for all that, by order of the Commission of Religion, the above mentioned Persons, who spoke chiefly in behalf of the others, were sent to close Prison as above said and on the 4 instant, the remainder of them together with their Wives and Children even the smallest without allowing them the least Time for putting on their Cloaths, were taken at night out of their Beds and houses by some Bailiffs and Soldiers and carried to an uninhabited House belonging to the Government and there altogether locked up. Such of them as did not readily go along were unmercifully beaten and some of them tied by the Hands and Feet, particularly Stephen Hueber who remained Three Hours in that Condition untill the arrival of the Count and Governour and at the same time they were told that on the 12 Instant the Ships should carry them down the River, That such as had some small Things might put them into Small Chests and take them with them; such as had sold their Effects were obliged to deliver the money of the produce thereof to the Governour. Time must Show whether anything and how much thereof at their going off will be returned to them? They have had no work from the Salt office these Six weeks. neither have they obtained any Letters of Leave for withdrawing nor Certificates for shewing in Time to come upon what account they have been in that manner carried out of their native Country.

The Gazette printed in this Town on the 31st July last contained a paragraph dated at Vienna the 17th July last purporting that the 40 Protestants Families consisting of Three Hundred persons coming out of the Land upon the Ens in Austria went by there upon Five Ships, in order to be carried down to Ofen or Buda in Hungary and thence farther to Transylvania where they live among the Saxons Settled in that Country.

A Translation of a Letter from Mr. [J.] Reck dated at Ratisbon 22 July 1734 N.S. No. 12768. N.B.[208] 30 July 1734. To Mr. Henry Newman.

Sir: As I am not so happy as to be Master of English enough, you will excuse me that I answer you in french that I had the honour of your letter of the 25th of June inclosing my Nephew's [Ph. von Reck's] letters from Savannah Town. God be praised that he is happily arrived with the Saltzburg Colonists.

I hope he will stay there sometime and beg Sir you will exhort him not to think of returning so soon to Germany but that he would endeavour to recommend himself to the incorporated Society for Georgia that he may hereafter enjoy their Protection as the principal End of his Voyage. In Germany he has little fortune to hope for.

This is what I have wrote to him in the enclosed letter, which I beg you will put into his hand when he arrives in England.

Be so kind as to present him to our[209] Court and to the Ministry as a Young man zealous for the Service of the King.

I remain with all possible respect
            Sir   Your most humble and Obedient Servant   Reck

Copy of a letter from H. W. Guerdes London August 6, 1734. No. 12771.   Read 6 August 1734.   To the Honourable Society.

Gentlemen: I am desired by three Successive letters of Christopher Ortman your Clerk and Schoolmaster amongst the Saltzburghers of Ebenezer in the Colony of Georgia to acquaint the Honourable Society in the most respectfull manner in his Name (1) That he is not able to take and till any land to his advantage, (2) That it will be absolutely impossible for him to subsist by a Salary of ten pounds Sterling per annum, particularly when the Provisions now granted to the Colonists will come to cease; and therefore to pray in his behalf that the Honourable Society would be graciously pleased, either to let him have his full Salary of twenty pounds per annum, (which he says was promised him at his appointment by the Honouble Trustees of Georgia, and will appear to be so by their Protocoll at their Office) all in money: or to give him such an Equivalent for that one Moiety of his Salary as they think will be of equal Value in Georgia to ten pounds sterling per annum.

Now as this man has been employed by me for six years past as a Schoolmaster and found by experience to be able to instruct Children and make them learn their book and Catachism with the greatest patience and application imaginable and therefore hath been recommended by me upon his repeated requests to the honourable Society for the Office he now enjoys: And whereas this man in all other Stations before has behaved honestly and faithfully, nay [and also] for his Services done to the Crown of England has been rewarded formerly with the Pension of Greenwich Hospital which he quitted because he would not live an idle life but serve God and his Neighbour as long as he was able; for which, if needful, the Testimonies of the Governours and Officers of Greenwich Hospital, and of his Grace the Duke of Montague, may be had.

I hope the Honourble Society will not think it impertinent in me, that I presume to beg that this honest poor Man may have favour and pity shown; and that the faults of his Wife's quarrells may not be imputed to him, but He the rather more be pitied upon that account as he really deserves.

I shall not say too much if I add that any favour shown to him will be deemed by me as done to my Self; for if the Society should think fit to discard him the Charge will fall upon me, thinking my Self in honour and Conscience obliged to support him.

I beg pardon for the trouble I give to the Society by this letter and am with great respect

   Gentlemen Your Honours most obedient humble Servant
             Henry Walther Guerdes

Translation of a letter from Mr. [Ph.] Reck dated Rotterdam 28 October[210] *1734 No. 12806. Read 10 Sept. 1734. To Mr. H. Newman.*

Sir: Being arrived here the 20th Current I set out the 22nd for the Country of Cadzand; many reasons induced me to take this Journey which I shall have the honour to acquaint you with. Being returned but this day, I shall set out for Germany this Evening or if it please GOD without fail tomorrow morning.

From Frankfort or Augsburg I hope to let you know whether and when a Transport of Saltzburghers may be expected.

I beg you to make my Complements to Messrs. Vernon & Oglethorpe and if the latter please to send me a plan [map] of Georgia I shall be glad to receive it at Ratisbon. Mr. Reiche will not fail to put it into the King's Packet. I am without reserve

Sir   Your most humble & Obedient Servant   P. G. de Reck

Copy of a letter from Mr. Ulsperger dated at Augsburg 6th September 1734. No. 12807. Read 10 Sept. 1734. To Mr. H. Newman.

Honoured and Dear Sir: Since my last letter of the 2nd of August I received the favour of yours of the 30 July; and in pursuance of the resolution of the Society, for sending a Transport of between 40 and 50 Saltzburgers to be conducted by Mr. Vat to Rotterdam, and from thence to Gravesend in the River of Thames in the Passage Sloops, consisting of the five Persons left here since last year, and of the friends and relations of such who went first to Georgia. I have ordered a short advertisement to be printed, and sent to several Imperial Towns where these People are dispersed, having fixed the 20th day of this month for giving their Appearance here in this town; and by that time I hope to be fully informed to the Number of such of them as shall be willing and ready to accept of this generous Offer of the Society, so that all possible Care shall be taken, that this Transport shall be in England by the latter end of this Month, Old Stile. Mr. Vat who presents his most humble respects to the Society and to your Self will be as frugal in the Expenses upon the road etc. as possible; and is expecting the Society's full Instructions, how to deport himself upon his Journey and Voyage to England; as also a Passport from his Britannick Majesty for himself and the People of this Transport; which may be sent to Franckfort in Case it be not already sent thither, as it was mentioned in my letter of the 15th of July last. I have likewise received Mr. [Ph.] Van Reck's Journal on the Voyage to and at Georgia, which you sent to the Baron [J.] Van Reck at Ratisbone, and a letter from Mr. [Ph.] Van Reck of the 6th of August acquainting me with his safe Arrival in England. I could have wished to have some letters of the Saltzburgers themselves, in order to print them with the Extracts out of the Journals of the two Reverend Divines and of Mr. Van Reck, I remain

Honoured and Dear Sir   Your most obedient humble Servant

Samuel Urlsperger

P.S.: My Journey to Stutgardt, whence I returned on Saturday last, hath been the Occasion of so much Business to be done that I have not time to answer Mr. [Ph.] Van Reck's letter, but I hope to be able to do it on Thursday next, in the mean time I beg the favour of you, Sir, to give him my humble Service.

Copy of a letter from Mr. Urlsperger dated at Augsburg 16th September 1734. No. 12812. Read 17 September 1734. To Mr. H. Newman.

Honoured and Dear Sir: I received the favour of your letter of the 20th of August and hope you received my last of the 6th Instant. I have the honour to acquaint you that this moment twenty and one Saltzburgers came from the Imperial town of Memmingen, in order to go home [hence?] to Georgia for which Voyage they are very willing. For want of time I must referr the sending a list of them till next post; I dont doubt but Mr. Vat may go hence in eight or ten days. The royal Passport is not yet come to hand, but I shall procure one for him from his Majesty's Envoys at Ratisbone, in such manner as Mr. [Ph.] Van Reck obtained One last year. The Societys Orders of being very frugal upon the Journey and Voyage were communicated to Mr. Vat, and he will certainly do all in his power to answer the expectations of the Society. He is expecting the Instructions of the Society, how to deport himself upon his Journey and Voyage hence to England. If those Instructions should not come to hand before his going, I'll give him such as are in my power. But one thing is, Honoured Sir, which I desire of you Viz. that the Captain of the Ship, which is to transport the people, might behave himself better than the former Captain [Fry], for I find many Complaints in the Ministers Journals against him,[211] There are two Families, One of two, and the Other of three Children, All the Others are single and most of them young People.

Mr. Van Reck is not yet come hither. I cannot omit telling You that the Roman Catholicks made not the least Obstruction for the admitting the said 21 Persons to come from Memmingen to be lodged in this City. I remain

Honoured and Dear Sir   Your most obedient humble Servant

Samuel Urlsperger

Copy of a letter from Mr. Urlsperger Augsburg September 20th 1734. No. 12826. Read 24 September 1734. To Mr. H. Newman.

Honoured and Dear Sir: My last was the 16th Instant. And if the Providence of God was made manifest in regard to the first Transport of Saltzburgers gone to Georgia, it doth much more shine in favour of the Second; for beyond all expectation, forty seven Persons, according to the inclosed list, have actually entered their Names and joyfully intend to undertake the Voyage, and I dont doubt but before Thursday next, their Number will be encreased to Sixty. Moreover they are people that make a good appearance and will be of very good use. As to their true Christianity concerning some of them to the number of 13 the inclosed testimony of the Reverend Minister of Lindau [Riesch] giveth a full Account. In the meantime a great Blessing attends Georgia by these people. And I earnestly beg of You that good care may be taken of them, and that the promises made to them be fully made good; for I have engaged my word for the due performance thereof.[212] I beg particularly that good orders may be given concerning their transport to America. And since by Your letter of the 30th of July you recommended the going of the Transport with great diligence, Its departure is fixed to Thursday next the 23rd Instant. We are in hopes of receiveing by Wednesday next letters from England and the Instructions intended for Mr. Vat, as also his Majesty's Passport. May the Lord of Hosts, who led Israel, also lead these honest people. I remain

Honoured and Dear Sir   Your most obedient and most humble
Servant   Samuel Urlsperger
P.S. My humble respects to the Society and the Trustees.

A List of the Saltzburgers embarked on Board the Prince of Wales for Georgia made out by Mr. Vat 29th October 1734.

5 Bartholomeus Riser, his Wife and 3 Boys[213]
4 Thomas Bacher, his Wife and 2 Girls
4 Ruprecht Schoppacher, his Wife and 2 Girls
2 Ruprecht Kalcher and his Wife
2 Thomas Pichler and his Wife
2 Ruprecht Eischberger and his Wife
2 Matthias Burgsteiner and his Wife
2 Stephen Rothenberger and his Wife
2 Ruprecht Steiner and his Wife

2 Simon Steiner and his Wife
2 Adam Riedelsperger and his Wife
2 Veit Lemenhesser [Lemenhofer] and his Wife
2 Matthias Brandtner and his Wife
1 Ruprecht Zimmerman
1 Sebastian Glantz
1 Georg. Felser
1 Paulus Zietrauer
1 Georg. Bruckner
1 Paulus Lemenhesser [Lemenhofer]
1 Gabriel Bach
1 Car. [Carl] Sigmund Ott
1 Hans Michael Muggizer
1 Bartholomeus Zand
1 Georg. Kogler
1 Nicolaus Riedelsperger
1 Christian Reidelsperger [Riedelsperger]
1 Hans Madreüter
1 Jacob Schortner [Schartner]
1 Ruprecht Zietrauer
1 Sibilla Schawl [Schwab]
1 Veit Landfelder
1 Andreas Rasch
1 Georg. Sanftleben
1 Christian Hassler [Hessler, Hössler]
1 Gabriel Maurer
1 Hans Maurer

A Translation of a letter from Mr. Urlsperger dated Augsburg Sept. 23, 1734. No. 12831. Read 24 Sept. 1734. To Mr. H. Newman.

Honoured and Dear Sir: This day at half an hour past nine Mr. Vat departed hence with a Transport of 54 Persons amongst whom are but seven Children. They set out from my house, whither I had ordered every thing necessary to be brought. That Instant when the march began I received letters from Mr. Commisary [Ph.] de Reck wherein he acquaints me of his arrival at Ratisbon and promised to be with me this day or to morrow and bring the two Royal Passports which are arrived. But every thing being ready for marching I let it go on, and when Mr. de Reck arrives I shall advise with him about what is fit to be done in relation to Himself and other Circumstances. These people will give great Joy to the Trustees. What else has happened in relation to this Transport worth mentioning, I must defer to my next, being hindred to day by a great many other Occurrences in my Pastoral Office.

Mr. Vat begs that some Credit may be procured for Him and his Transports with some Merchants of Rotterdam and if possible with those Mr. Simon formerly recommended him to, the Names of whom I cannot in the hurry I am at present recollect.

I hope that this Transport will soon arrive at Roterdam, it being six weeks sooner than the first went off.

The Moiety of this Sheet I desire you to cut off and deliver it to Mr. Ziegenhagen because it is directed to the Gentlemen Ministers of Ebenezer [Bolzius & Gronau].

I have paid to Mr. Vat about 800 florins for the discharge of the Expences of the Transport between here and Frankfort, having enjoyned him to husband the money every where in the best manner he can and to give a particular account of all Payments.

Besides this Mr. Van Munch has been so good as to procure Credit for Mr. Vat at Frankfort and Cologn.

And whereas I have before advised you in my letter of the 15th February of this Year; that I had then no more than 942 florins 21 creutzers in Cash as a remainder of my Georgia Cash, and whereas I have Since the 4th March last (including the aforesaid 800 florins and all the maintenance of Mr. Vat for 9 months past) disbursed more than 1080 florins; nay on behalf of the Georgians now departed have paid some what more than mentioned.

Therefore you see that not only Georgian Cash is quite exhausted, but likewise that I have been obliged to make use of part of the 1000 florins, which I had else reserved for the Benefit of the Emigrants in general and the maintenance of some Children of theirs here received. I remain

Honoured & Dear Sir   Your most humble Servant   S. Urlsperger

Translation of Mr. [J.] Recks letter dated at Ratisbonne 16 September 1734 N.S.   No. 12833.   Read 24 Sept. 1734.   To Mr. H. Newman.

Sir: I had the Honour of receiving about a Week ago two of your acceptable Letters the 1st and the 15th of August O.S. congratulating me on the happy return of my Nephew [Ph. von Reck], from Georgia and to assure me that my letter was delivered to him at his arrival but that nothing could dissuade him from hazarding a second Transport for that Country of which he had undertaken with the Approbation of the Society.

You do him too much honour Sir in heaping so many Praises upon his Conduct and the happy Success of his Commission. You have likewise the Goodness to add to it the Testimony of his Excellency Governour Belcher to Your Self and his Son too advantageous for the small Services he has done.

I am very much pleased that Mr. Oglethorpe is of the same mind and entirely approve of the resolution that my Nephew has taken for the Glory of the English Nation and of the Society.

Your second letter which I find my Self honourd with is of the 20th of August O.S. accompanying one of mine wrote to him supposing he had been in London.

You repeat to me the agreable News that my Nephew is on his way hither since the beginning of this Month and that he is furnished with a Pass from his Britannick Majesty to return to Georgia with another Colony with which I am very glad because it perfectly answers my Wishes. May GOD bless this glorious design. Since he left London I have no news for him but expect it every moment; as for the rest, I am very much obliged for the advice you give me of your Function which will serve for my future direction and I flatter my Self with your favour and Protection of my Nephew which I recommend to You and remain with all mine heart

Sir   Your most humble and obedient Servant   J. Reck

Translation of a letter from Mr. [J.] Reck dated at Ratisbonne 27 September N.S. 1734. No. 12835. Read 1 October 1734. To Mr. H. Newman.

Sir: It was the 21st of September N.S. before I had the honour of receiving yours of the 29th August O.S. inclosing the two Passports for my Nephew [Ph. von Reck] and Mr. Vat; And as I had then the pleasure of having had my Nephew at my house since the 19th September I had a good Occasion of putting into his hands his Majesty's Passport and the letter accompanying. He only stayed for that to go to Augsburg to concert with Mr. Urlsperger the means and time of undertaking a new Transport of Saltzburgers for which Service he set out the 23rd September where he arrived the day following according to the letters I received yesterday: And as Mr. Urlsperger believes that he has already People enough for that purpose I hope he will soon be in a Condition to set out for Georgia. I find him very zealous to execute that laudable design with the help of GOD of which he himself will more amply satisfy You. I remain

Sir   Your &c.   Reck

Copy of a letter from Mr. John Vat dated Franckfort 4th October 1734. No. 12838. Read 1 October 1734.²¹⁴ To Mr. H. Newman.

Honoured Sir: I don't doubt but the reverend Mr. Senior Urlsperger hath acquainted you with my leaving Augsburg on the 23rd of September last, having under my Care 32 Men, 14 Women; 4 Boys and 4 Girls, under 15 Years of Age; Their Baggage being put upon two Waggons, a Messenger on Horse back went with us thro Donawerth [Donauwörth], Dünckelspiel [Dinkelsbühl], Rothenburg, as far as Wertheim where we took the Water, and thence came hither in two days, GOD be praised, and every one in good health. As I could not get a proper Barge for Holland at Wertheim, I bought one here on Saturday last, which is filling up, and I propose to go down the main [River Main] this afternoon, and hope to be at Holland in Eight or ten days. Mr. [Ph.] Van Reck having sent by an Express which overtook us near Dünckelspiel [Dinkelsbühl] His Britannick Majesty's Passport acquainted me at the same time that Mr. Walters [Wolters] at Rotterdam had Orders from the Society, for providing us with proper Shipping for our Passage from thence to Gravesend and I suppose he is also empowered by the Society for supplying us with Moneys necessary to our Passage. I received at Augsburg from Mr. Urlsperger between seven and eight hundred florins and I intend to take up here of Messrs. Van Münch five hundred Guilders more.

The Prince royal of Prusia and the Russian Auxiliaries passed yesterday thro' this Town coming from the Army on the Rhine and are incamped within Cannon Shot of this Town.²¹⁵

Mr. Degmair Candidatus Ministerii of Augsburg, who some time ago was Preceptor to Mr. Urlspergers Children, and hath been very usefull to him in transcribing and translating English and French into the German Language having been proposed to me by Mr. Urlsperger to be our travelling Chaplain till we should be at Rotterdam, I readily agreed to defray his Expenses on the Voyage, and as he seems enclinable to see England, I humbly submit to the Consideration of the Society whither [whether] it will be proper to invite him for going with us to Gravesend? I believe such an Invitation, as to a Gentleman of fine parts in Divinity,

will be conducive to enlarge the Interest of the Society; And I beg
leave to subscribe my Self

Honoured Sir Your most obedient & most humble Servant
John Vat

*Copy of a letter from Mr. Richard Lowther at Rotterdam 8th
October 1734. No. 12840. Read 1 Oct. 1734. To Mr. H.
Newman.*

Sir: Your favour of the 20 Sept. I received, and observe the
Contents; as soon as the Saltzburgers arrive, I shall be ready to
give them all the Assistance in my power and put them on board
such Sloops where I am sure the Commanders will use [treat]
them well. The price per head to Gravesend will not I believe
exceed five Shillings; they must provide their own Provision, and
sleep between Decks for that price. Here is a poor Family, a Man,
his Wife, and three Children from Philipsburg,[216] lost all they had
and obliged to wander for their bread, they are Lutherans, I fancy
they would be glad to be transported to Georgia if the Trustees
and the Society will give me leave I shall put them on board with
the Saltzburgers. Pray you Sir, present my most humble Service
to Sir John Philipps, Mr. [Erasmus] Philipps, Mr. Vernon and the
other worthy Gentlemen of the Society. You may acquaint them,
whenever the Trustees or the Society think I can be usefull in
forwarding any of their good designs they may freely comand me.
I am, Sir Your most obedient humble Servant Richard Lowther
P.S.: I shall be glad to have your answer as soon as possible to
know whether I may send the poor family from Philipsburg with
the Saltzburgers.

*Translation of a letter from Mr. Urlsperger dated Augsburg Sept.
27 N.S. 1734. No. 12847. Read 8 Oct. 1734. To Mr. H. Newman.*

Sir: By my last letter you have seen that a Transport of 54
Persons did depart from hence the 23rd of this Month. The next
day Mr. [Ph.] de Reck arrived here from Ratisbonne and took
Lodging in my house that we might the better confer with one
another. I sent after Mr. Vat the Royal Passport with some Admonitions hinted by Mr. Reck. I proposed to have sent to you to
day an exact Account of the Transports but the Occurrences of
my Office have taken up all my time. The chiefest of what I must
mention now is that the new Transport may not be sent to the
Indies[217] without a Conductor. Mr. Vat has often been asked by

me if he had a mind to go to Georgia and used to answer he should not refuse going if he should receive a Call for it.

Mr. [Ph.] de Reck is yet a while necessary in Germany partly that we may confer with him, partly that he may see if a new Transport might not offer to be conducted by him. He is else still resolved to return to Georgia. As soon as I shall hear what Mr. Vat took up at Frankfort or Cologn on my Credit I shall let you know it that the Bill of Exchange to be sent from London hither to Mr. Munch may be proportionable to it.

To day the Transport arrives at Marchstehl [Marksteft] where it is to embark for Frankfort. GOD preserve the good People that they may arrive in Georgia in due time as a great Blessing. The Presence of Mr. Van Reck gives me a great deal of Pleasure. This very moment arrives two Saltzburgers here, one a Bricklayer and the other a Carpenter, desirous to go to Georgia.[218] They are good men and have good Testimonies about them and are Brethren, but I doubt whether they can still be sent after[.] I conclude and remain

Sir   Yours &c.   S. Urlsperger

P.S. by Mr. [Ph.] v. Reck: Sir: I take the liberty to add a few lines and a letter to Mr. Vernon which will let you see the State of our Affairs. It goes too hastily so that I cannot conduct this transport and see you again before winter but it must be as god pleases. I am without reserve,

Sir   Your very humble & most obedient Servant   Reck

Translation of a letter P. G. F. de Reck dated Augsburg October 4th N.S. 1734. No. 12851. Read 8 October 1734. To Mr. H. Newman.

Sir: To give you an Account of the State of our Affairs in Germany I have the honour to acquaint you (1) that the Transport arrived the 30th last N.S. at Wertheim where they lay to embark upon the Mayne [Main], and if it please GOD will be in two days at Frankfort. (2) That Mr. Vat will conduct them if the Society please to Georgia. (3) That I am this day going to Saxony and Brunswick. (4) That I am always ready to obey the Orders of the Society, either to stay this Winter in Germany or to follow the Transport cost what it will till they arrive at London; or to conduct towards next Spring a Third Transport and to perform every thing that shall be enjoyned me (5) That Monsieur Count Sinzendorff [Zinzendorf],[219] whose Portrait the Reverend Mr.

Zeigenhagen can better give you than I, wishes to send some of his People to Georgia: (6) That Count Spangenberg will go on his part to London to conferr with the Honourable Society and Trustees for Georgia on their Propositions. God prevent all the [omission][220] which may happen to Georgia[.] We commend you entirely to his Bounty, and for the rest we recommend to you as much as we can the request of Mr. Zwiffler and the Salary of the Preachers which are delivered to Mr. Vat (who I suppose will go to Georgia for he won't leave his Charge) that every thing may be done according to their desire. If you please to communicate this and the other Letter to Mr. Oglethorpe and to the Honourable Society. Procure for me from Mr. Oglethorpe a Plan of Georgia. There is a little general Chart of the coasts of Georgia Carolina and Cape Florida engraved at London. I should be glad [if] you would send me two Copies together with the Plan above mentioned from Mr. Oglethorpe which shall be presented to the King[221] and Mr. Urlsperger. These may be of good Service. If you have any other particulars concerning the Indians be pleased to communicate them. When you favour me with a letter please to send it to Ratisbonne or Augsburg[.] Mr. Urlsperger salutes you respectfully and I beg my Complements may be acceptable to Mr. Zeigenhagen. I am Sir Yours &c. Reck

Copy of a Letter from Mr. John Vat dated at Cologne Oct. 11, 1734. No. 12854. Read 11 Oct. 1734. To Mr. H. Newman.

Honoured Sir: I had the honour of writing to you from Frankfort the 4th of this month and gave you an account of our safe arrival at that town, where two Saltzburgers came from Lindau by the way of Augsburg, and we got on board our Ship on Tuesday the 5th. We were detained by high Winds at Höchst on Wednesday, and at Mayntz [Mainz] by the imperial Commander of that Town from Thursday at Noon till Friday morning at eight of the Clock, and now we are come to this Town, GOD be for ever praised, in very good health. As soon as I shall be dispatched by the military and civil Officers I intend to proceed on our Voyage and I hope to be at Rotterdam in few days and to receive there your Commands I am

Honoured Sir Your most obedient & most humble Servant
John Vat

P.S.: On Monday morning at 10 of the Clock as the Post for Holland goes tomorrow from Cologne I shall send this Letter from

Suntz[222] being 15 Miles English nearer to Rotterdam[.] I cannot sufficiently express the Satisfaction I enjoy with my fellow travellers who are the finest of all Saltzburgers that went this way, if Credit is to be given to the respective Custom House Officers and I am sure the Society will be pleased with the Sight of them. Dusseldorff on Tuesday at 7 of the Clock in the Morning the 12th of October 1734.

Copy of a Letter from Mr. Vat dated Rotterdam Oct. 19, 1734. No. 12856.   Read 15 Oct. 1734.   To Mr. H. Newman.

Honoured Sir: I suppose my last from Dusseldorf is by this time come to your hands, and I beg leave to give you a short account of our Voyage from Franckfort to this time. We left that town the 5th Instant and came to Hoëhet [Höchst] and were there detained by high Winds and heavy rains the 6th all day; the 7th we came to Mayntz [Mainz] by 8 of the Clock in the morning but the Governour of the Emperor would not let us go till the 8th at 8 of the Clock, and the Master of the March Ship going every day at 10 Clock from Francfort to Mayntz and vice versi by virtue of an old Privilege revived within these six months exacted sixteen Creützers per head which I declined paying till I had been with Mr. Stubenrauch, Member of the Privy Council to the Elector, who told me that it was resolved that I should pay the said Shipping which seems to be an unjust Duty altho' the Ship was bought at Francfort. However Ships coming from any Port above Francfort are exempted from paying that Duty. Then having hired a Pilot we came that Night to Caub; the 9th within a League of Coblentz the 10th to Cologne where I dismissed the Pilot of Mayntz, and took another for conducting us hither: the 11th we passed by Santz, the 12 Dusseldorf and Zaysersworth [Kaiserswerth] at which last town the Custom House Officers would not let us proceed till I had paid them one Gold Ducal [ducat?] for the People's Bagage over and above all other Duties, particularly Brandzoll in two Offices, which tho very unjust is required arbitrarily in all Roman Catholick Places, as Perquisites belonging to the respective Custom House Officers; It being to no manner of purpose the contesting with them serving only to prolong the delaying of Passengers, and it cannot be avoided but by having free Passports of the respective Sovereign bordering on the Rhine which perhaps upon succeeding Transports might be obtained by applying for them at Ratisbonne; This would save a good deal of time and

Charges. The 13th we got to Ruhrort where the Prussian Custom house Officers gave us a free ticket for several Places as far as Emmerich but we were detained at Orsoy two nights and one day by high contrary winds. The 14th we made Wesel, and got in the Evening to Emmerich, where at that time my Lord President of Cleve happened to cross the Water in his return to Cleve whilst we went over on the other side, however we overtook him in his Coach and he very graciously ordered the Officers to dispatch us without going to Cleve, by which we saved a Journey by Land and the Charges thereof. The 15th we came to Scheallschans, there the Dutch Officers remitted us Brand Zoll and gave us a free Tickett as far as Arnhem, but the Officers of this last place made us pay Brand Zoll in two Offices. The 16th we came to Wick,[223] the 17th we passed by Schonhoffen [Schoonhoven], and arrived in perfect good health at one of the Clock within half an hours walk of Rotterdam. As it happened to be on Sunday we could not come before the Magistrates till yesterday Noon and having readily obtained leave of our Boat coming into the town, on Condition that our people should not go begging in the Streets, they got into this town by four of the Clock. The Prussian and Dutch Officers every where excepting Arnhem were exceeding kind and obliging particularly Mr. Wrick at Wick[223] requiring only a small matter for the Custom House Servants. We enquired of Capt. Hewers the price of our going in his Sloop and he was pleased to ask fifteen Shillings per head, and Capt. Christian seven Shilings and Six pence but as Capt. Wm. Thompson Master of the Two Brothers North Carolina built for 150 Tuns offered by the mouth of the Reverend Mr. Horner to carry us for 5 Shillings a piece hence to Gravesend we agreed to it and gave him time till Friday to get his Ship in readiness and we have to go on board tomorrow. Messrs. Curtone have offered to comply with the order of Messrs. Simonds concerning the Credit the Society are pleased to give me and I shall make use thereof as I shall be in want. At my coming to this Town I received the favour of your Letters of the 24th September and 1st of this month by the hands of Mr. Walters [Wolters] who has been as well as Mr. Lowther, very obliging, and it is with their Approbation that Capt. Thompson has been pitched [picked] upon. Inclosed is the List of the Emigrants now with me to the Number of 56 and Mr. Degmair is very glad of having an Opportunity of presenting his respects to the Gentlemen of the Society. I hope the Society will get a permit of the Commissioners of the Customs for their Officers at Gravesend that we may be dispatched & kindly used there. I am very glad to hear that the Prince Frederick[224] Capt. Dunbar is appointed to carry the Indian King[225] and his family together with our Saltzburgers directly to Georgia; but the Reverend Mr. Urlsperger has mistaken my declarations concerning my going to Georgia, for as he and several other Gentleman, and some of the Saltzburgers were asking me, whether I would not go to Georgia? My answer was that if it were necessary upon the Society's enquire for a proper Person not to be found I should not scruple to go on the Voyage notwithstanding my weak Constitution of Body, but that I could not give a full answer till I should have the honour of waiting upon the Society in London, and I am still of the same opinion; and could wish the Society had a proper Subject to go along with these honest people. Wherever they went people of the respective Places declared, they never saw so fine a Transport of Saltzburgers.

I am   Honoured Sir: Your most obedient and most humble

Servant   John Vat

Copy of a Letter from Mr. Richard Lowther dated at Rotterdam 19 October 1734.   No. 12858.   Read 15 Oct. 1734.   To Mr. H. Newman.

Sir: Your favour of the 1st October O.S. I received, and as the Society are not at Liberty to send the poor Philipsburg family, I will endeavour to support them here till the Spring and send them to Philadelphia, whither they would rather go.[226]

Last Sunday Mr. Vat and a Clergyman [Degmair] arrived here, they left the Saltzburgers about an hour from this town, but yesterday they all arrived here in perfect health, they are in number 56. Our trading Sloops would not take them to convey them to Gravesend for less than three half Crowns per head, but we have met with an Irish Master bound for London, that we have agreed with to carry them for five Shillings per head, they providing for themselves. I think it cheap, all their Baggage is included: the Masters name is Wm. Thompson the Ship is called the Two Brothers. Mr. Horner who I perceive is known to the Society recommended the Master[.] He promises to use them well; By our agreement he is to sail next Friday if the wind be favourable. He is to be paid for 54 You may assure the worthy Gentlemen of the Society (to who I desire you will present my humble Service) my good Offices shall not be wanting to Mr. Vat and the Emigrants under his care, whilst here. I am   Sir   Your most humble Servant

Richard Lowther

Copy of a Letter from Mr. D. Wolters dated at Rotterdam 19 October 1734. No. 12859. Read 15 Oct. 1734. To Mr. H. Newman.

Sir: The inclosed is a letter which Mr. Vat gave me just now for you by which you will see we had the good luck to meet with a Ship that is going home to England empty. We are agreed at five Shillings a head for the Passage of the Saltzburgers the Cabins of Mr. Vat and the Minister included, which is reckoned very civil for no Sloop would have taken them under seven Shillings. They are to take their own Provisions.

We have been, Mr. Vat and I, to the Magestrates to desire leave for to have the Saltzburgers in town which will be much more convenient for them than being at a mile out of town where all foreign Transports are obliged to stay and after some Considerations I had the pleasure to obtain it.

We thought fit to leave them in their boats till the Ship is ready to take them in, which will be much less expensive and quite as convenient as having them in a private house on shoare. The Capt. has engaged to sail a Friday next if the Wind permits.

I am with a true regard   Sir   Your most humble and most
obedient Servant   D. Wolters
P.S.: If you are so kind to put a direction over my letter *to Mr. Gravius at the Post Office at the Briel* they cost me very little Postage.

No. 12865. Read 15 October 1734. To Mr. H. Newman. A Letter in High Dutch from Mr. Samuel Urlsperger dated at Augsburg in Germany 11 October 1734. [Listed but not copied in the Newman Letterbooks.]

A Letter in High Dutch from Mr. Samuel Urlsperger dated at Augsburg in Germany 18 October 1734. No. 12865. To Mr. H. Newman. [Listed but not copied in the Newman Letterbooks.]

A Letter in High Dutch from Mr. Urlsperger dated at Augsburg in Germany 28 October 1743. No. 12865. To Mr. H. Newman. [Listed but not copied in the Newman Letterbooks.]

Translation of Mr. Urlsperger's letter to Mr. Vat Augsburg Oct. 14, N.S. 1734. No. 12865. Read 15 October 1734.

Dearest Mr. Vat: Having asked last monday Parson Degmayer what he thought of it in case the honourable Society should think fit to offer to his Son at his arrival in London to conduct in Conjunction with Mr. Vat the Transport unto Georgia, and return and he having answered that he could not consent to it for many weighty reasons therefore let it be known to his Son.

Last monday I wrote to Mr. Newman and as none of his letters are come to day I have nothing to acquaint you with except that the Arch Bp. of Saltzburg continues his Persecutions and from time to time some single Persons arrive here and at Ratisbonne, banished out of their native Country and fined a great deal of money for no other Cause or reason than the finding some Evangelick Books about them and their refusing to abjure the Protestant Religion, of whom we have received in our Evangelick Alms House but a few days ago a married woman driven away from her husband and Children for the reasons above mentioned.

What has been promised out of the Evangelick Cash at Ratisbonne is not yet come to hand but the Envoy M. [J.] de Reck has assured me yesterday of its coming by the next Opportunity. I must beg of you Mr. Vat to endeavour to bring it about amongst those Trustees of Georgia who are our friends that the Lands allotted for the Saltzburgers in Georgia may not be restrained to descend only to the Male Heirs[227] but descend a like to the female Sex, and this particularly for this substantial reason, that the Women as well as the men have been exiled for the sake of Religion and therefore ought not to be deemed *deterioris conditionis* less worthy; besides that else the Widows and female orphans of such families as are destitute of male heirs must needs be Beggars.

Mr. [Ph.] Von Reck has brought a long with him several Letters among which there are two happily fallen into my hands written in such a manner as would have given occasion for many Calumnies.[228] One was written by one Braunberger but he is a Saltzburger and the letter was styled by that wicked man *ROTH* the other was a little more moderate, but very imprudent too.

I desire therefore (1) you would speak with Mr. Ziegenhagen that he may send all letters and Packetts from Georgia immediately to me. (2) that you would speak with Messrs. Bolzius & Gronau and desire them partly to reprove Braunberger and Roth for the writing such a Letter; Partly to admonish the people in general that they may raise no clamours against the Country they are in by their imprudent letters; and for that end to look over the letters before they are sent from thence.

The Parents of M. Degmayer not being to give their Consent to his Voyage for Georgia he must content himself with it.

I know God Almighty will work the more amongst you by his

Spirit and excite some amongst you to perform the spiritual office of Deacons to their Children.

But, dear Mr. Vat, you have no Father or Mother, that hinders you from going to Georgia. You, You will go along with my Children: You, You will shew Moses's faithfullness towards them and bring them to that Country GOD almighty has allotted for them. Go but cheerfully and be not afraid, but believe that Jesus who was with his Disciples upon the Sea will be also certainly with you: Go therefore to my Brethren of Eben-Ezer in Georgia, Mr. Bolzius, Gronau, Geschwandel, Mosheimer [Mosshamer], and tell them that I wear them in my heart and pray for them incessantly. All my house saluteth and blesseth You and I remain for prayers for you till death.

S. Urlsperger

P.S.: Pray make my best respects acceptable to the Honourable Trustees and Society and other Friends and Patrons. I dont know whether I told you that Mr. Belcher Governour of New England did send me from Boston a very hearty letter by Mr. [Ph.] de Reck to which I intend to return an answer next monday if possible because I still keep my bed.

To the Congregation of Eben-Ezer in Georgia.　No. 12865

Dearly beloved Children: As a father rejoyceth on account of his Children living in foreign parts, when they do well and are sweet savour to others, and as he likewise heartily grieves at the hearing of their not doing well, or that they become by their wicked life and Conversation an ill Savour to others so have I likewise in part rejoyced heartily when I heard of the wonderfull Providence of God attending you, and how the greatest part of you have been in good earnest working out their Salvation: but in part it has grieved my very heart when I have been told from time to time that some body or other of you not acknowledging the ways of God and his Benefits bestowed upon you but grieving and scandalizing others by their disorderly and sinfull Behaviour.

To you therefore of the latter sort I do call saying. Return O! ye going a stray and let the word the tears and the zeal of these faithfull Servants of God living with you have their Effect upon you for this purpose; for then the Lord will not hide his face from you for he is mercifull and will not be angry eternally; but acknowledge your Crimes that you have sinned against the Lord your God: Make hast and save your Souls that you may not fall into Judgment of GOD and men.

But, to you who have been hitherto the Joy and Crown of your Teachers, I do say, Go on in the good way you have begun with[.] let no body alter your Aim; avoid all deviations as well to the right as to the left. Take heed of bad Christians as well as Heathens that you may not Be ensnared nor fettered by any of them. Do constantly look by faith on the Lord Jesus; pray without intermission, love the word of God; use frequently the Lords Supper for the strengthen your faith; labour willingly; be patient and meek to one another, be afraid of nothing but Sin; let the Helmet of Hope not be taken from your head; have ye care of all false Liberty; let no predominat Vice be told of you; be lightning and burning Lamps a Salt that all the blind neighbouring Heathens may be salted with; Ease your Ministers in their Office as much as you can, for I love your Ministers more than my own Brothers, the manner they have hitherto laboured in with your Souls I approve entirely.

The letters of Moshammer and Huber I have received with Joy but wonder that Gsirandel [Geschwandel] and Riedelsperger who are able to do it did not write to me. If I can serve you in any thing let it be known unto me for I [omission] my self in Love obliged unto you till death; But know ye that I shall believe of no Body that he loves me if he dont before all and above all Love the Lord Jesus. Keeping my bed at this time I have been obliged to dictate this Letter else I should have written it with my own hand. I recommend you all to the infinite mercy of God and remain untill death sincerely praying for you.   S. Urlsperger

Copy of a Letter from Harman Verelst[229] at the Georgia Office Westminister 18 Oct. 1734.   No. 12869.   Read 22 Oct. 1734.   To Mr. Newman.

Sir: Please to speak the following Parcells of Seed to be parcelled in 42 parts each which I shall distribute on board to each Person wanting them which Mr. Oglethorpe thinks the most proper method Viz.

   7 Ounces Cabbage seed of the Sided sort divided into 42 parts
      and bound up together and the Contents indorsed.

14	Pounds	Turnip Seed	"	"
14	Ounces	Parsnip Seed	"	"
14	"	Carrot Seed	"	"
7	"	Brown Dutch Lettice	"	"
7	"	Green Capauchin	"	"
3	Pounds & 1/2	Chard Beet	"	"

7 Pecks Sandwich Beans   "   "
7 Ounces Bore Cole Seed   "   "
7 Quarts large Dutch Kidney Beans "   "
7   "   Spinage Seed   "   "

A little
{ Wheat
  Barley
  Oats &
  Rye   about a handfull of each in a parcell 42 of
        each made up and endorsed.

Sir John Philipps recommended Mr. Pacell a Seedsman near
Bartlet's Buildings in Holborn. Desire him to send them to the
Trustees Office for Georgia and a Bill of Parcells next monday
at 12 to   Sir   Your most humble Servant   Harman Verelst
P.S. If the Medicines and Grocery are not ordered how to go on
board; they may as well be ordered to the Trustees Office at the
same time to go with other things. Let the Medicines be markt
G x C. Medicines and let a Bill of Parcells come with them. I have
one of the Grocery Mr. Palten in Milk Street Market sent me.

*Copy of a Letter from Dr. Guerdes dated London 16 Oct. 1734.
No. 12873. Read 22 Oct. 1734. To Mr. H. Newman.*

Honoured Sir: The Consistory of my Congregation has determined to day (1) That 25 Guineas be distributed to the Saltzburgers in their name viz. 10 Shillings per head for every grown
Person and 5 Shillings per head to the Children. (2) That I shall
distribute this Benefaction in their Name.
Nay they have even dispensed with my Attendance on their Sundays Service, so that I may go down now, and stay even of a Sunday
if required, provided no other hindrance arises. I therefore beg
the favour of you to let me know of the Arrival of the Saltzburgers
at Gravesend as soon as you hear of it, and likewise if or when you
know when and where the Gentlemen are to meet that I may not
miss of the Opportunity of going with them and be no hindrance
to their going but be ready to wait on them. I am with due regard
        Sir   Your very affect. Friend & Servant   H. W. Guerdes

*Copy of a Letter from Mr. John Vat dated at Rotterdam October
22, 1734. No. 12878. Read 25 Oct. 1734. To Mr. H. Newman.*

Honoured Sir: I had the honour, on the 19th Instant of giving
You an Account of our safe arrival at this town on Sunday last
and of our having agreed with Wm. Thompson Commander of
the Two Brothers 150 Tunns, North Carolina built, for carrying to

Gravesend for five Shillings sterling per head, which is the lowest price we could expect to have it. As the weather was very bad all this Week excepting this day the said Thompson's Ship could not be cleared as yet between decks, however we are to go on board of her tomorrow; in the mean time our People dine and lye in the Barge that brought them hither, and we all long for a fair Wind for Conveying us to England being all in very good health. If the wind holds at S.W. till Sunday next Mr. Degmair is to preach a German Sermon to our people, in the English Episcopal Chappel, the Reverend Mr. Lowther having generously proposed that Service. This hath exceedingly rejoyced our Young divines[230] and I hope it will be a great Satisfaction to the Reverend Mr. Senior Urlsperger and to the Body of the Protestant Ministers at Augsburg where the Reverend Mr. Senior Weidner, two days before our setting out thence preached an excellent Sermon to our people whom I headed to the Church of St. Ulrich and after divine Service was over he distributed One florin a piece to all of them that had entered their names in my list being money of the Clergy at Augsburg, and the next day after a most excellent Sermon preached by the Reverend Senior Urlsperger the Protestant Magistrates of Augsburg by the hands of the Clergy gave also one florin to each person old and young and Mr. Burgomaster Morel, who is a zealous and hearty friend to the Saltzburgers, was so kind that morning we set out to give his direction for loading our people's Chests and Baggage, consisting in thirty four Chests Boxes or Trunks. I beg the favour of you Sir to obtain a permit of the Custom house of London, for the Officers at Gravesend, that we may not be at unnecessary trouble at visiting our Baggage, for I can affirm that there is nothing else but what is to be necessary for wearing for our people and some provisions.

The Duke of Richmond got yesterday safe to the Hague and I hear he intends to return to England by the first fair wind.

I also beg the favour of you honoured Sir to make my respects acceptable to all the Gentlemen of the Society and all my old Acquaintance and I beg leave to subscribe my Self

Honoured Sir   Your most obedient & humble Servant   John Vat

Copy of a Letter from Mr. John Vat dated at Rotterdam Oct. 26, 1734.   No. 12878.   Read 22 October 1734.   To Mr. H. Newman.

Honoured Sir: Since my last of the 22nd I received the favour of yours of the 11 Instant, and I received this day six hundred &

sixty Guilders Current money of this Country of Messrs. Courtone & File [Fils] & Denormandie of this Town, which you will be pleased to reimburse to Messieurs Simonds et Freres Merchants in London. We got our people on board on Saturday afternoon, and Capt. Thompson promises to go down to morrow morning with the tide to Helvoet Sluys and if this wind be favourable we hope to be at Gravesend in few days. Messrs. Lowther and Wolters who have been very kind to us present their humble Service to You and I remain

Honoured Sir   Your   John Vat

Copy of a Letter from Mr. Ziegenhagen Kensington Square October 4, 1734. No. 12882. Read 4 October 1734. To Mr. H. Newman.

Dear Sir: Reading over again the Letters from Messrs. Bolzius and Gronau I find they desire some few things which I wonder I have so long forgot viz. 3 Pair of Tongs and 3 Shovels[;] Some quantity of Sugar, Soap & white Starch[;] A Travelling Bed. If it is not too late I should beg the favour of the Society to grant this small request. I am

Dear Sir   Your humble Servant   F. Ziegenhagen

Copy of a Letter from Mr. J. Vat dated at Gravesend 2 November N.S. 1734. No. 12884. Read 24 Oct. 1734. To Mr. H. Newman.

Honoured Sir: We came this Evening at 5 o'the Clock to an Anchor before this town in good health[,] after we sailed the 31st of last month from Helvoet Sluys we got the next day into Oxford Road and did not heave Anchor till this morning the wind being favourable all this day. Mr. Mayor of this town [Haffenden] hath been so kind as to present the Saltzburgers with the Charges of Postage of Letters, and as you order the going of the Saltzburgers to the Red house at Deptford, Capt. Thompson proposes to go up tomorrow and I must postpone the pleasure of waiting on the Society till the Saltzburgers shall be on board the Prince of Wales. I am   Your most obedient & most humble Servant   John Vat

Translation of a Letter from G. Reck dated in the Dutchy of Mahren [Moravia] Oct. 7 O.S. 1734. No. 12890. Read 5 Nov. 1743. To Mr. H. Newman.

Sir: I send you the inclosed Letter for Mr. Vernon to which I refer You; I humbly beseech you to communicate it to both the illustrious Societies. I hope to see you soon (if it please God). I am till death   Sir   Your most humble & obedient Servant

G. Reck

Copy of a Letter from Mr. Vat London Oct. 29, 1734.   No. 12895. Read 5 Nov. 1734.   To Mr. H. Newman.

Sir: I humbly request the Favour of the Society to give my hearty thanks to the Reverend Mr. Senior Urlsperger, Mr. Burgomaster Morel, and the Reverend Clergy of Augsburg, as also to Mr. Von Münch of Augsburg and Messres. Munch of Francfort and likewise to the Reverend Mr. Lowther, Mr. Walter [Wolters] and Messrs. Courtone of Rotterdam for their great Civilities to the Saltzburgers and to my Self.

It is humbly desired to send a Box to the Reverend Mr. Senior Urlsperger at Augsburg by means of Mr. John Herman Zur-Hurst Merchant in London containing Six pewter Dishes and twelve Plates weighing 38½ lb. as a present to Mrs. Urlsperger; and desired; likewise that Mr. Wolters give advice whether Mr. Tervin of Amsterdam hath paid by order of Mr. Horner f 10, 11 [10 florins 11 creutzers] for beer put on board the two brothers not called for on John Vat when he was on board the said Ship but is allowed Mr. Horner as paid by him.   Yours   John Vat

Copy of a Letter from Mr. John Vat dated at Deal in Kent November 6, 1734.   No. 12899.   Read 12 November 1734.   To Mr. H. Newman.

Sir: Having Sailed from Gravesend on Friday last the 1st Instant, we did not come into the downs till this day at Noon[.] God be praised all excepting one of the Saltzburg men in good health, and I received the favour of your two Letters of the 31st of October, and that of the Reverend Mr. Ziegenhagen of the 2nd Instant to whom I sent this day by our Pilot two sermons sent him by Mr. Stockmeyer of Autgard [Stutgard] which I could not find whilst I was in London. The Packet is recommended to Mr. Simond.

Sir Francis Bathurst[231] having got a dangerous wound about fourteen days ago by hitting his Shinbone on board the Ship; I have dressed it every day, and applied Shour's[232] Balsam of Augsburg, and it is in a fair way of doing well provided he takes Care of it. The Ship is so crowded with People & Baggage that I can

hardly find room to write any thing. The Capt. and Tomo Cha-Chi [Tomochichi] return the humble Service to You. We are waiting for a fair Wind and I am Yours &c. John Vat
P.S.: Pray my very humble Service to Mr. & Mrs. and the Reverend Mr. Hales as also the Reverend Mr. Wilson and to all the Gentlemen of my Acquaintance. I desired my Nephew Nicholas Watt[233] who lives with Mr. Lightbourne in Ironmonger Lane near King street in the City, to fetch a Brown Coat with brass buttons at Mrs. Giles's and to carry the same to Mr. Simond and if you send any thing to deal before we sail I beg you will send also the said Coat.

The Lying in Woman [Mrs. Schoppacher] doth extraordinarily well and all the Saltzburgers present their most humble Thanks to their Benefactors, and the Reverend Mr. Degmair to whom I beg the favour of giving my humble Service.

Copy of a Letter from John Martin Bolzius at Ebenezer in Georgia 13 July 1734. No. 12900. Read 12 Nov. 1734. To James Oglethorpe, Esq.

Most honoured Sir: The many favours and benefits you have paid upon me and all the Saltzburgers hath occasioned my writing to you, and I hope your generous good nature will excuse it when these lines cause any hindrance in the urgency of your affairs for that should render us worthy of blame if we dont let you understand that we account our Selves happy in your favour and tender Care towards us and our prayers are daily for your Health and Welfare. We have it already cast in our minds to bring to our father in heaven many Sacrifices of Thanksgiving so soon as we are informed you have finished your Sea Voyage in good health and prosperity. God reward you thousand times for all your Goodness presented to us in the former time and let all your good Councels and weighty affairs redound to the Publick Good and welfare of many poor people. We will make it the future Business of our Lives by the grace and assistance of the Holy Ghost to be no ways behind in Gratitude towards God and all our Benefactors. I cannot but let you know by this that through your fatherly Care and order Mr. Causton has sent for the Saltzburgers very sufficient Provisions and gave me daily several testimonies of his tender regard to us, which is as we see and hear very tedious to the People at Abricon [Abercorn][234] and Savannah wherefore they spread out very many lies and ill things against the Saltzburgers Viz. that they were all given to Laziness Drunkenness and several disorders and were not worthy of so many benefits. I and all persons which are much conversant with these people are obliged to report well of them that they dwell in the fear of God, practise soberness and all other Christian Virtues, and labour so earnestly that some of them have by much troubles and heavy work brought Sickness and Death upon themselves.[235] Five Men and two Women are deceased and some of them have been till now deadly sick. Therefore seeing that we do not find a great abatement in our Congregation we pray you will after your beloved kindness be carefull that some more Saltzburgers may come to our place as soon as possible because a greater number of hands shall ease their burden and very difficult works. Untill this time they are constrained to do several Works which hinder them very often in building their own houses and tilling the Ground. They have put into the Ground some Indian Pease, corn and other Seed which they received from Mr. Causton in abundance, but no more, as the said Pease and some English beans and cucumbers grow up. I believe the seasonable time of sowing was past or the seed is supperannuated[.] As for Mine and Mr. Gronaus health thanks be to God it continues as heretofore and of our livelihood we have no reason to complain. The Indians haunt [visit] us and tell us several words of their Language which we note and learn by heart. So soon as we can quit the Business which are without our vocation we will do our utmost endeavour to learn the said Language after which we have a hearty desire and delight. We wish earnestly that some family might dwell among us in Ebenezer and rest in hopes our wishes shall be by you and Mr. Causton's Care successfull. I have no more to add than that my Colleague Mr. Gronau gives his humble respects to you and so with my heartiest and best wishes I close up this and am evermore with the greatest respects most honoured Sir

Your most humble Servant   John Martin Bolzius

P.S.: Mr. Roll [Roth] gave me the inclosed Letter to send it to you and desire that you Sir grant him leave and license to return to Germany by reason he cannot work in the Ground after the Testimony of all the Saltzburgers.[236]

A Copy of a Letter from Mr. Peter Simonds in Nicolas Lane London.   No. 12912.   Read 23 Nov. 1734.   To Mr. Newman.

Sir: I little expected the Complaints that have been made that the Saltzburgers are ill accommodated on board my Ship the Prince of Wales; I hope that when you'll have heard what I have to say, you'll own they are groundless, and justify me from any such frivolous Assertions.

The Ship's Burthen is 200 Tuns, built on purpose to carry with Conveniency a great number of Passengers; She is 6½ foot high in the Steerage, and 5 feet high between Decks. So large a Ship and of such a Built might contain very easily 150 Common Passengers, whose Passage is paid upon the common footing; if they were to be heaped upon one another as the Palatines[237] it might have contain 300 and these pay however as much as the Saltzburgers. To give more ease and air to the Passengers on board my Ship, and particularly to the Saltzburgers we built Cabins with little Beds along the Sides of the Ship only, and left all along in the middle an empty space 9 feet broad; a thing never practized in the fitting out a Ship for Passengers, where the Owners had their own Interest in view.

Those Cabbins built round the Ship in the Steerage & between Decks, were to contain 118 heads, putting two in the uppermost little Beds and three in the lowermost, I had limited my Self to that number being resolved thro' a principle of honour and humanity to lay aside any Consideration of my own Interests.

We had kept in the Hole sufficient Room as we thought for the Baggage of the Saltzburgers, little imagining that those poor Emigrants would have any considerable quantity; but the truth is that when they came on board they brought such great heaps that they did not know where to put it all; the Hold was soon filld, the rest was put in the space between Decks that had been kept clear for the Conveniency of the Passengers, nor was this sufficient till they had beat down the Cabins that had been built in the Fore Castle which were to hold 12 persons; this Expedient was the sooner resolved upon as Mr. Detzius[238] and Mr. [Peter] Gordon and their families consisting in 11 Persons, said they would not go the Voyage; however they met the Ship at Gravesend, which occasioned some disorder, the Cabins that had been built for them having been beat down; but why had they been beat down, was it not for the Sake of the Saltzburgers and their Baggage, why then do they complain, when they are the occasion of the Inconveniences other people suffer.

I have in that Ship but 5 Tunns of Beer and two Mill Stones that serve for Ballast, so that 'tis filled up entirely with Provisions and Baggage. I had prepared Some Goods for Georgia and Purrysburg, but perceiving there was no room in the Ship I have sent them to Charles Town per Capt. Pick to whom I have paid 50 s per Tun freight.

I dont know why it is intimated that there is in the ship 14 Passengers more than the number I agreed for which they say was 120. I don't know that I ever agreed for any fixed number, I indeed said, that if there were not at least 120 Passengers I should be a Looser, and this is but too true, but I always expected there would be more since there were Cabbins built for 118 in the Steerage and between Decks besides those in the great Cabbin.

According to the muster made at Gravesend by Mr. Verelst there are in all in that ship   Heads 124 5/6[239]
Out of which deduct those that are in the
    Cabbin and State Room viz
8 Indians
2 Mr. Gordon and his Wife   12
1 Mr. Vat   ‾‾‾‾‾‾‾‾
1 Mr. Fullerston [Fullerton]   112 5/6

So that there is but 112 5/6 in the Steerage and between Decks[.] there were places for 118[.] if they were afterwards taken away those must be blamed for whose Baggage the Cabbins were beat down, not I.

The Ship is 200 Tuns, the Water and Provisions take up 80 Tuns, the place for the Passengers 20 Tuns, so that there must be on board 100 Tuns of Baggage or Goods that the Trustees have sent. Had I put up the Ship for Charles Town I should have had 40 s per Tun and 10 per premium, which would have rendred £ 440
Deduct 3 Mo. Wages, Ware Tare & Provisions &c.   240
                                                        ‾‾‾‾‾‾‾
                                        Profits   £ 200

I should have gained £ 200 on the Voyage to Charles Town & my Ship would have been there ready to take in her Loading for London.

I can prove by the Tradesmen's bills, that the Provisions Casks, Cabins etc. for this Imbarkation Cost me   £ 520
    Three month's Wages Ware Tare etc.   210
                                                       ‾‾‾‾‾‾‾
                                                        730
    125 Passengers at £ 5 per head   625
                                                       ‾‾‾‾‾‾‾
                                                       £ 105

That is 105 £ loss to me which added to 200£ that I might have gained makes 305 difference and my Ship will be at Savannah where contrary Winds and other Accidents may detain her a long while, before She can reach Charles Town to take in her Lading, as it has happened to all the Ships that I have hitherto sent there. 'Tis true that if it pleases God to give her a short passage, there will remain some Provisions that may make some amends for a part of my loss but that is very uncertain.[240]

You see then, Sir, that it is not the desire of Gain that makes me undertake these Embarkations, which besides are exceedingly troublesome, and you may easily judge that acting as I do in a view far from mercenary, I must be extremely affected with the reproaches that are groundlessly made me, I hope, Sir, they'll not alienate the friendship you have always honoured me with, which will make me ample amends for the Injustice done me, as I am conscious that I in no ways deserve blame.   I am   Sir

Your most humble and most Obedient Servant   P. Simonds

P.S.: I wish those Gentlemen who complain would inquire if a Ship of 200 Tuns which has 6½ feet high in the Steerage, and 5 foot high between decks, and built purposely for Passengers, may not very easily contain 125 Passengers when there is nothing else in the Ship but their Baggage and provisions and whether, those Passengers paying but 5 £ a head, and nothing at all for the Baggage with which they fill up the Ship and being nourished & accomodated as those are, the Owner can come off without loss; they'll find they are much in the wrong to complain. And indeed I had rather another would undertake it for the future.

A Copy of the Translation of a Letter from the Reverend Mr. Urlsperger dated Augsburg 8th November 1734 N.S.   No. 12915. Read 23 November 1734.   To Mr. Newman.

Honoured & Dear Sir: Two Bohemians who Some years ago left their Country for Religion Sake are arrived here, and by recommendation of an eminent Count of the Empire, applyd to me in order to make their Complaints of what they suffered on Account of their Country men who for Religion Sake are in Prison. What they told me, and I could give Credit to, I have caused to be set down in writing, and send it inclosed to the Honourable Society, since they have often desired me to communicate to them what happened in regard to Sufferers for Religion and Emigrants. It contains Examples that are terrible to read. The two mentioned Bohemians intend to go to Ratisbonne in order to see whether there is any help for them. These two have several times ventured to go into their Country in order to save some, which they also happily performed. God send these Persons some help. I have heard nothing from our Transport since the

26 October from Rotterdam; I hope they are well arrived. I salute Mr. Vat, Degmeyer and the Colonists & remain   Tuus totus S.U. P.S.: in Carinthia 2000 have declared themselves Protestants of which more in my next.

An Account of the Persecution of the Protestants in Bohemia. No. 12915.

In our Country viz. in the Kingdom of Bohemia, especially in the Dominions of Opotsch, Leutomyshel and Neustadt, are many Protestants, but dare not discover themselves for fear of being severely punished. Since two years ago, in the Dominion of Opotsch in the König-Grätres Circle [Kreis Königsgrätz], some discovered themselves by a Petition to their Superiors that they were no Roman Catholicks but Protestants, and therefore humbly begged liberty to edify each other in private, which they also immediately made a Beginning of, in praying, singing and reading the Word of God in private Houses, but they were put in prison, where 60 persons of them are kept to this hour,[241] notwithstanding some of them agreed with the Roman Catholick Clergy in some articles[.] Those who for Conscience sake leave their Country and all what they have, and go to places where Protestants are tolerated are in danger of being taken up by the way and put in prison and beaten till they discover others their Brethren, who are treated in the same manner, and the Prisoners have no hopes of being released, since we have not one Example that any hath been set at liberty unless he hath wronged his Conscience and denied the truth of the Gospel. The Wives of those men imprisoned in the abovementioned Dominion of Opotch were going to petition for their Husbands at Praag or Vienna, but they were hindred from it. Many perish in prison, an Example of which is John Tlaska's Wife, who upon the road leaving her Country, was taken up brought to Neustadt and put in prison, where She being big with Child, brought forth Twins, and for want of Assistance, perished with both Children. Other Examples more this kind, we are willing to give a true account of either by word of mouth or in Writing. When any of those who have left their Country, and actually settled again in other Countries, and for some necessary Business return into their Country and are apprehended, they are likewise unmercifully beaten and put in prison. This puts the rest of the Protestants, thro' out the whole Kingdom, in such fear that none of them declare the religion in their heart, but are continually troubled with Scrupel of Conscience. Because now no Protestant is permitted to go out of the said Kingdom, to seek for help for their Souls and Consciences, and the mean people, who mostly are concerned in these troubles do not understand the German Tongue, nor are capable of taking good Counsel We two Emigrants, Tobias Bernhard and John Sommerbrodt, born in Bohemia, have resolved, since many particular Circumstances, and the great Sufferances of our Brethren is known to us and goes us to heart, to try whether it was not possible upon Earth that some Body would pitty them, and assist them in such a manner as is right before God and Men.

Out of the Dominion Lytomishel, belonging to the Count of Trautmansdorff

Watzlaw Erespivoa, an Inkeeper, and his Wife, have been put in prison about two years ago, and his house and goods taken from him, and are in prison to this hour. The Village where he lived is called Znedossyna.

George Wostrowskj, born in Ezezalsche, hath been beaten very hard and afterwards made a Soldier.

John Anderle, Geo. Bzizschka, Nicolaus Dolezal, born in *Hermannitz,* and *John Penitzschka* from *Syrneden* these four have had the same treatment as *Geo. Wostrowski,* and afterwards been forced to be Soldiers.

Syrezeck, hath got above 200 Stroakes. The Enemies of Truth have a Chest, in which is a Hole so big that a Man can put his head thro' it. Upon this Chest the Sufferers must lay themselves down and put their head into the hole, which afterwards is shut close that the poor peoples crying out when they are beat, may not be heard.

Nicolaus Anderle, a Brother to the abovementioned *Anderle* has likewise got 200 Stroakes and so violently that his Shirt stuck to the Flesh and afterwards he was put into Prison, and left there not cured of his Wounds till Maggots grew in the foul Flesh upon his Back; after that he is set at liberty, but is very weak and miserable.

Nicolaus Butner, hath likewise been beat in the same manner, and these two viz. *Nicolaus Anderle* & *Nicol. Butner* when they were set at liberty, they were obliged to pay 40 Guilders fine and give Security. A certain Countryman's Daughter, namely *Stemadi,* who hath a Sister living in Saxony, resolved to go to her Sister, but She was taken up, and after having also got 200 Strokes, put in prison. She hath afterwoods got her liberty, but is very miserable yet on account of the Stroakes. She hath been set at liberty with this Condition that she should pilgrimage and thereby do penitance for her Sins, but being so unmercifully beaten, that She cannot stirr nor move, her Brother was obliged to take the Pilgrimage upon him in her place. Two other Girls who were resolved to go away with the aforementioned, were also betrayed, and these were likewise beaten and put in prison together with their Parents. The father of one of these two is a Smith, who for fear that his Business should be taken from him, was forced to pay 40 Gilders fine, besides give Security, and promise that never he would let come such thoughts more into his head.

Two Brothers namely *Ezapkya* had also a Brother in Saxony who once visited them and exhorted them earnestly to embrace the Protestant Religion. Afterwards it was known that their Brother from Saxony had been with them, and gone away again but these two were immediately arrested and forced to pay, the one of them 250 and the other 200 Guilders, and give three Men for Security that they never would leave their Country for if they did these three men Security should in their place loose their Estates and be driven out of the Country.

A Countryman namely *Benetsch* from a Village called *Zlubeyso* was put in prison together with his Wife because his Wife's Sister who lives in Saxony had wrote a letter to them, and they should confess the Contents of the Letter, wherefore they both were beaten insomuch that the Child which his Wife was big with died in her Belly.

Penizschka, born in a Village *Syrneden,* and his Son and Daughter were accused of having Protestants Books by them, which they were forced to deliver, and afterwards unmercifully beaten, especially the Daughter, so that her Breast is mortified.

In the Dominion *Opocinzki,* belonging to Count Rudolfa *Skoloredj*[242] a Schoolmaster namely *Dautz,* because he had wrote the Petition above mentioned for 400 Persons that would joyn to edify one another, was unmercifully beaten, insomuch that he sounded [swooned] away, and after that boyling Water was thrown upon his Body and then put into Prison where he is to this hour.

Some of the 400 Persons after they were brought by the Soldiers from the Village to the City in order to put them in prison, and whilst others were carried thither, they were brought into the Walch [watch] houses and put under the tables and Benches; and when the Officers came in they mocked them; among other Games they had with them, they tied their Canes with their hairs and then tore them out again with great force; besides this barbarous Treatment the Officers bid them to bark at them under the table like Dogs, and to sing the Hymn: "Welcome Cross, with all my heart etc."

Sixty of these 400 Persons are still in prison, the rest have abjured the Evangelick Doctrine out of fear; but not [now?] they are troubled in their Conscience and find no rest.

In the Dominion of *Neustadt* belonging to Count *Lessel* a Smith and his Wife living in a Village called *Prowodow* left their Estate worth about 4000 Guilders and went away. When this was known they were persued and taken, but the Husband escaped, and his Wife being with Child could not but was put in prison, where she brought forth Twinns, but being left without Assistance, She together with the Children died in great misery.

Besides, it is to be observed that all those that have been set at liberty have been obliged to give Security.

New Bond Street 22 November 1734. No. 12915. To Mr. Newman.

Sir: You'll find here enclosed the Account sent by Mr. Urlsperger but without any Corrections, for Mr. Wilson and I could not meet about it as yet.

It is my Opinion that this Account should not be handed about in the name of the Society,[243] before we have a Confirmation of it from abroad, because in the very Book Mr. Urlsperger sent to us relating to the History of the Emigration of the Saltzburgers, I find an entire Chapter treating of Impostors that have pretended to great Sufferings for the Sake of Religion till they were found out, so that we cannot be too cautious in the publication of such Accounts.

Be so good to acquaint Sir John Philipps with what I say here and present my most humble respects to his Honour I am Good Sir Your very affectionate Friend and Servant H. W. Guerdes

Augsburg. No. 12920. Read 26 November 1734. To Mr. Hen. Newman

Honoured and Dear Sir: The following Accounts from *Carinthia* and *Saltzburg* show how Affairs stand here. The archBishop of Saltzburg hath lately declared, by his Envoy at *Ratisbonne* to the Envoy of Holland Mr. *Gallieris,* that he will permit the Emigrants who are gone to Prusia Holland or elsewhere, to sell their

Effects left behind them, On which account the Evangelick part of the Magistrate at *Augsburg* have actually wrote to Mr. *Ploto* the Prussian Envoy at Saltzborgh, in behalf of the Saltzburgers who are here, desiring him to take Care of those Effects belonging to them, according to the Letter of Attorney sent him; Query whether something is to be done for our Georgians, who doubtless have also left something behind them! Mr. Goebel Prussian Counselor who is likewise to go to Saltzburg, in order to assist Mr. *Ploto* hath offered himself in a letter to me to take the Georgians Part in case it was required, as I have mentioned some months ago.

The two *Bohemians* namely Tobias Bernhard & John Somerbrod,[244] after they had lodged 10 days incognito in my house, and privately received many Benefactions from the Protestants here, are yesterday gone by water to Ratisbonne. The Envoy Mr. [J.] Von Reck will assist them as much as possible. They are true Confessors, and I was very glad to harbour Christ in them. Pray Sir, if Mr. Vat is with you yet, tell him that they have lodged in his room. In my next I shall send a farther account of them. Having had no letter yet from London I have heard nothing about the Transport since the 26th October N.S. *Sat cito si sat bene.* I remain honoured and dear Sir; *Tuus totius* [sic]

U.

P.S.: For certain Reasons I have this time used another Seal, and caused the Superscription wrote by another hand.

Translation of an Account from Carinthia dated the 7th September 1734. No. 12920.

There being in our Country about Two thousand Persons who publickly have declared themselves Protestants, and perhaps some more will do the same, we suffer very much of the Commissioners, as well Spiritual as Temporal, who continue to throw Protestants in prison, which now amount to 30 or 40 persons, three of which are shut up in a Stable, and young men they starve, and say unto them: *forsake your heretick Religion and then we will give you Victuals.* Many of us have also suffered much of the Penitance—Preachers,[245] notwithstanding that by a Committee, and as they say by the Emperor's Command, it hath been published from the pulpit, that those who declare themselves Protestants might go to their respective places, and have their names set down; Accordingly a great number did, but yet from time to time some were fettered and two Waggons full at once carried from the Court at Spateria to Klagenfurt into prison: besides Seven Countrymen, some unmarried Persons and Jacob Haffel, Matthias Zwang and Windelbaum; and Simon Sigil and Joseph Schwollinger are now above two years in prison. Besides they threaten to banish them and to sell their Effects, which Rigour hath caused that nine Persons are fled into the Woods where they live miserably.

The Fathers of Families which are known to be Protestants and stayed in Cammeryuth [Kammergut] when other Protestants left their home, amount to three hundred and fifty nine Persons their Families included; and Thomas Lasserer, because he went to Vienna to receive the Sacrament in the Protestant Church, lies fettered in prison since a quarter of a year ago, and has been told that he must remain there till his Cause has been heard and made out.

A Declaration of a Saltzburg Woman concerning the hard proceedings against her and her Husband who remained there after her. No. 12920.

Elizabeth Stuchlebner, who came from Berlin hither and expected her Husband to meet her here, but according to his desire went to Oettingen, and from thence to Johannis, arrived here again the 14th September, and declared that after She had been with her Husband at Johannis 14 days she had been betrayed, found out, brought to prison, and kept there three weeks, during which time she had been twice examined, and her Husband pressing her to stay with him in that Country, and She refusing it, the Officers had taken her from her Husband, without speaking one word more with him, and brought her to the Boarders of Bavaria, and then was forced to pay to them 2 Guilders and 30 Krentzers. Moreover she told that her Husband had been obliged to pay 200 Guilders fine for having lodged the Messenger Andreas Pohre, who had brought letters from his Wife and another Woman here, Anna Hirnbacher, two nights and not informed the Magistrate of it and the other Man 100 Guilders.

Translation of a further Account of the Two Bohemians. No. 12920. To Mr. Newman.

A year ago Tobias Bernhard went to his Country Bohemia and took with him from thence a Widow with 6 Children. Two years before that time her Husband was resolved to leave that Country, but he fell sick and died; after his Death his Son took possession of the Estate, but three months after he left it, and went out of the Country with his Mother & the rest of his Brothers and Sisters, and the aforesaid Man conducted them. It was midnight when they set out from the Village, and travelled all night without any hindrance, till daybreak, when they met with some Hostlers, who took them to be Thieves because they travelled by night, and some of them carried Bundles upon their Backs, wherefore they stopped them; But Tobias Bernhard talking with them he made room for his people to escape; the Horslers set after them, but returned soon, and said to Bernhard, it was enough they had him, and He should give an account of the rest. They also sent immediately to the Judge in the next Village, in order to fetch a Man whom they suspected. In the mean time Tobias Bernhard spoke to them saying: God Almighty would call them to an account for what they were going to do to him; if he should be put in prison, his Sighs would not be good for them. This and other such talk more, frightned them, and they said to each other if they had known that before, they would have let the Man pass, and not sent to the Judge. But an old Soldier who, near that place, watched a pease Field, came to them, and this being a wicked man, they were afraid of being betrayed by him, wherefore they asked Tobias Bernhard to give them money to satisfy the Soldier for fear he should betray them. He having no more about him than two Pieces of seventeen Kreutzers gave them to them, wherewith they were satisfied at last, tho' at first they had asked 4 Ducats, and they let him go, and bid him make hast that he might escape the Judges hands. He went away full of joy singing aloud the Hymn, *Let us praise the O Lord* &c whereby he was known to his Companions who stayed for him in a Forest.

Twelve years ago the Enemies took a man up in the fields, namely Christian Sommerbred [Sommerbrodt], a Miller by trade because he had protestant Books, and went to their meeting, whom they put in prison, besides 6 persons more from the Village Moraschitz. In prison he was examined very sharp, and asked about those that were gone with him to the meeting; and because he would not confess, he was punished with 50 Strokes. Three days after he was examined again, and he not confessing then, he got two hundred Strokes. Three days after this he was examined the last time, and he not confessing yet, he suffered 250 Strokes, which made him swoon away, then he was carried into an Office and layd behind an oven where he lay half a day senseless; after which they laid a Cloth dipt in vinegar upon his back which recovered him, but immediately he was brought into prison again, and was forced to remain there 13 Weeks: And since they could get nothing out of him, they sent him, together with 6 other persons, to Praag, two and two tyed together and thrown upon a Waggon, and then forced to be a Soldier for 9 Months, after which time his own Wife purchased his freedom; He then went home, but did not stay there above three days, after which He, his Wife, and four Children left their Country. Two Children they left there with their Grandmother, one of which, viz. a Son, they fetched away in two years after and the other, a Daughter, is to this hour, with the Grandmother and cannot get free, because the Grandmother will then loose all what She hath and be likewise banished.

At the Fustigation their hands were tied upon their backs, and their Heads laid upon their Knees, then a Stick put between the Arms and Knees, so that the Back may be the better exposed to receive the Strokes.

A year and a half ago the following Persons left Bohemia, born in a Village called Hermanitz, Frantz Anderle, Tobias Martineck, and a Widow named Gauckin with four Children; And because they were forced to go out privately, they could take nothing with them of their Goods, but trusted all their Household Goods to one named Martineck, Father to the aforesaid Tobias Martineck, in hopes to get them conveniently when opportunity happened; for which purpose they hired a Waggon last Summer this present year in order to fetch them; but before the Waggon went, the news came that all was betrayed in Bohemia, and the Goods taken away by the Magistracy, and the Man, who had taken them into Custody, was put in prison. Immediately this good man's moveables and immoveables were seized on among which were 200 Guilders Orphan's Money belonging to the four Children of the said Widow Gauckin, and in prison he was told, that if he would have his house and goods again, he must buy them of the Magistracy, else he would never have them again nor come out of prison. Two months ago some persons came from Bohemia and said that this Man was then in prison.

A man, namely Briddel, from a Village called Zworlowist, was last Summer put in Prison, beaten and forced to pay 40 Guilders fine. His Daughter left her Country a year ago, and happily escaped into Lower Lusatia. This present year about 8 Weeks ago She returned thither to see her Father, because he had promised her to go with her, but the misfortune happened that She was catched, and immediately brought before the Magistracy, and put into the same Prison where her father lay, and hath twice been beaten.

Old Palace Yard November 19, 1734.   Read 31 December, 1734.
To the Reverend Mr. Samuel Wesley at Epworth near Gainsboro
in Lincolnshire.

Reverend Sir: I am very glad to find by yours of the 7th Instant that you are upon recovery after your long Illness. I ought to thank you for the kind Concern you express for the Welfare of our Colony. You are very much in the right that an Inundation of Rum would be the destruction of it; to prevent which we have given very strict orders for staving all Rum that shall be landed there, and also to stop the allowance of food and all other Benefits from such Persons as shall venture to sell any Rum. Our people do not go at all upon Sugar Canes; their labour at first was taken up in building Houses, next in raising Provisions; as Corn, Peas, Potatoes and other Garden Stuff Fruit Trees, Poultrey, Hogs and Cattle of which latter they have a large Stock, having now near 500 head in the Colony. The bringing these things to perfection takes up a great deal of time, considering there is a great deal of timber upon the Land which must be cut down. It is true they convert the Timber to use, splitting some for fencing withall, cutting up some to sell for building and pipe Staves and some for fireings. They do not grub the Roots of the Trees but dig between them and let the Stems stand and rot. They have very good Crops in this way which saves a great deal of Labour in grubbing & in a few years time the Roots will rot and come up easily. When they have got every thing round them so as to live comfortably then it will be time enough to think of Exports. Wine, Oyle and Silk will be the Staple Commodities and we have already Mulberry Trees, Olives and Vines in the publick Garden to furnish the people with Plants as soon as they are ready to receive them.

We should have been very happy had a man of your Spirit health enough to have gone thro' the labours of a Missionary, for I can assure you Sir, that we have not found any one who is qualified for that difficult Office willing to undertake it. Though the Indians are mighty desirous to hear Instruction, and have high Notions of a Deity, yet we could not prevail upon any one of a good life and pious Conversation to go and live amongst them. We have one Clergyman[246] in the Town of Savannah and he has so much Business amongst our own people that he has not been able to apply himself to the Indian Language nor can he be of any Service in instructing them, for whosoever would succeed in instructing them, must like them undergo a voluntary Poverty, must live with them in hutts, have all things in Common, abstain not only from Vice and Excess, but from all things that give Scandal; must bear equal hardships with them, learn their Language and teach their Children. They say that when God is pleased to instruct them he will send a man proper to do it. As for their Language they have two kinds, One which is a vulgar Dialect, different in each Town, the other a general Language common to the Creek Nations the Chactaws and the Blew Mouths, which if thoroughly searched into would (I believe) be found to be the radical Language of all America. In this Language are the Songs which contain their History and sacred Ceremonies. In their History they have an Account of deluge, that their Ancestors formerly inhabited great Cities, had Reading and Writing and the use of Shipping. That they came from the Setting Sun, over a great Sea of Salt Water, & that in process of time the Men in their Infancy minding hunting and neglecting Learning they fell into their present Ignorance. It is too long to give a particular Account of their Religion; but in short they believe the Immortality of the Soul and expect an Instructor & Mediator.

I had the Lord's Prayer translated into their general Language of which I send you a Copy. The New England Bible is in the Dialect of a particular Nation that lived where Boston now is and of no use but to that particular Nation.

I am   Reverend Sir   Yours   James Oglethorpe

Wroot December 5th 1734.   To the Reverend Mr. [Samuel] Wesley at Epworth.

Honoured Sir: The Account of Georgia which you were so good as to communicate to me out of Mr. Oglethorpe's Letter gave me a great deal of Pleasure, particularly what relates to the Indians, that they are very desirous of Instruction have high notions of the Deity, believe the Immortality of the Soul and expect a Mediator which I could not but look upon as an excellent Foundation whereon to build the saving Knowledge of Jesus Christ.

That God should throw into our way a people so predisposed for the Reception of the Gospel, looks like an especial call from Providence to our English Clergy to glorify their Master in a more than ordinary manner, by enlarging the Kingdom of Christ, and I cannot but consider it as a very great Talent put into our hands, which it is highly incumbent upon us to improve.

Since that time my heart has been labouring with a Design, which I am almost afraid to disclose, lest it should be thought the Effect of Youthfull Rashness (my Age being only about 26) but being pressed with weighty reasons on the other Side I now resolve not to smother it: It is a design of going over thither, if I can get leave of the Lord Bishop of London, and the honourable Gentlemen abovementioned in order to trye what can be done by the Grace of God towards instructing the Natives in our holy Religion.

Whether I am qualified for such an undertaking I dare not say. I have a Constitution naturally pretty strong and for the first Seventeen Years of my life was enured to as great Hardships of Cold, Heat, Labour, etc. as almost any I have known, which makes me think I should be better able to bear the Difficulties of that kind which a Missionary must expect to undergo, than many others. I think I have an indifferent good Notion of the Christian Religion in general as it is contained in plain and indisputable Texts of Scripture, in our Articles, Creeds, and Church Catechism, and that seems sufficient in the present Case, for nice and controverted points in Divinity I cannot suppose needfull but rather very improper for those with whom I should have to do. And I hope I shall be the firmer in my Principles, as I have examined the most important points, both of natural and revealed Religion with all the freedom of thought and impartiality I was Master of. As for Courage I flatter my self that there are several Instances in my life which show I do not altogether want [lack] it, when I have any matter of Importance upon my hands, and surely nothing can be of greater Importance than the Salvation of so many Souls, to any one, who considers the inestimable worth of them, or the price their Redeemer paid for them: And, as for learning the Language of the natives, having Youth, and a tolerable Memory, I hope I shall not find any great difficulty in that. But what I reckon most considerable of all is, that since the Death of my beloved Wife, I have no Attachment either to the World in general or to my own Countrey in particular.

This is what account I am able to give of my Self, but as every one is an improper Judge in his own Cause, I desire to refer it to your determination who well know my Life, my Abilities and Capacities, whether I be in any degree qualified to be of use in an Affair of this nature.

As to the point, whether my resolutions be the effect of youthfull Heat & Rashness, I can only say I have weighed the matter, with as much deliberation and impartiality as I am able, and that I am at present firmly fixed and heartily willing to leave my native

Countrey, with all the pleasures and advantages of it, in order to bring Glory to God, and promote the Salvation of men, besides which I have an additional reason drawn even' from worldly Prudence (tho' I hope I should have no need of it to determine my Choice) namely that a Life of action and labour would not only much conduce to my health, but would also prevent that drowsiness & Indolence which in an unactive State I am apt to fall into.

If I go over, I believe I can prevail with one or two of my acquaintance, gentlemen of great piety and Spirit, to go over likewise, if there is occasion & assist in the Glorious Enterprize. I rather chose to communicate these things to you because you have been, thro' your whole Life remarkable for your Zeal to promote the Glory of God, and the good of his Church and so I have reason to hope you will encourage a Disposition towards it in others: Wherefore I take the boldness to desire you if you approve of my design to use your Interest for me with proper persons in this Affair, and if it succeed to my Wishes tho' the Glory I would have all given to God, yet the Reward of bringing many to Righteousness will in good measure be yours, and if it does not succeed, you will still have the reward of desiring and endeavouring it. This is what I thought proper to acquaint you with, and may he who is infinite in knowledge direct us to what is most conducive to the Glory of his name and the Good of Mankind.

I am   Honoured Sir   your obedient Son[247]   John Whitelamb P.S.: Should I go into Georgia I should think my Self much obliged to act punctually according to the Instructions I should receive from wise and experienced Men in England, & to live among the Natives & accomodate my Self to their manners as Mr. Oglethorpe proposed in his Letter.

Epworth December 7th 1734.   No. 12972.   Read 7 January 1735. To James Oglethorpe Esq. in Old Palace Yard, Westminster.

Honoured Sir: I can't express how much I'm obliged by your last kind and instructive Letter concerning the Affairs of Georgia, I could not read it over without Sighing (tho' I have read it several times) when I again reflected on my own age and infirmities which make such an Expedition utterly impracticable for me, yet my mind workt hard about it, & 'tis not impossible but providence may have directed me to such an Expedient, as may prove more serviceable to your Colony than I should have ever been.

The thing is thus, there is a young man [John Whitelamb] who has been with me a pretty many years and assisted me in my work of Job, after which I sent him to Oxford to my Son John Wesley Fellow of Lincoln College, who took Care of his Education where he behaved himself very well, and improved in piety and Learning: then I sent for him down, having got him into double Orders, and he was my Curate in my Absence at London where I resigned my small Living at Wroot to him, and he was instituted and inducted there; I likewise consented to his marrying one of my Daughters, there having been a long and intimate friendship between them, but neither he nor I were so happy as to have them live long together, for She dyed in Child bed of her first Child. He was so inconsolable at her Loss, that I was afraid he would soon have followed her, to prevent which I desired his Company here at my own house that he might have some amusement and Business, by assisting me in my Cure during my Illness. 'Twas then Sir I had just received the favour of yours, and let him see it for his diversion, more especially because John Lyndal and he had been fellow Parishioners and School fellows at Wroot and had no little kindness one for the other. I made not great reflection on the thing at first, but soon after when I found that he had thought often on it, was very desirous to go to Georgia himself, and wrote the inclosed Letter to me on that Subject, and I knew not of any person more proper for such an Undertaking, I thought the least I could do was to send the Letter to your honour who would be so very proper a Judge of the Affair, and if you approve of it, I shall not be wanting in my Address to my Lord Bishop of London or any other, since I expect to be in London my Self at Spring to forward the matter as far as it will go.

As for his Character I shall take it upon my self, that he is a good Scholar, and sound Christian and a good Liver, has a very happy memory especially for Languages, and a Judgement and Diligence not inferior. My eldest Son [Samuel, Jr.] at Tiverton has some knowledge of him concerning whom I have writ to him since your last to me: My two other his Tutor at Lincoln [John] and my third [Charles] of Christ Church have been long and intimately acquainted with him, and I doubt not but they will give him at least as just a Character as I have done. And here I shall rest the matter, till I have the Honour of hearing again from you, and shall either drop it or prosecute it as appears most proper to your maturer Judgment. Ever remaining

Your Honours   Most sincere and most obliged   Friend &

Servant   Sam. Wesley, Senior

Old Palace Yard Westminster.   December the 25th 1734.   To the
Reverend Mr. Sam. Wesley at Epworth near Gainsboro in Lin-
colnshire.

Reverend Sir: I received yours with great pleasure and could
not help thinking of good Evander
 sed mihi tarda gelu saeclisque effæta senectus
 Invidet imperium seraeque ad fortia vires
 . . . . . spes et solatia nostri
 Pallanta adjungam[248] . . . . . .
like the old Hero, not able to assist himself, your longing Soul
looks out for one whose youth and Capacity renders him proper
for the undertaking. I can assure you that it was with great Joy I
heard of a man indued with that worthy Spirit and true Christian
Zeal which both his & your Letter expresses him to be. If his resolution be not took up in hast, and as soon laid down again, he may
be the happy Instrument of Salvation to many thousands. He
must consider well before he undertakes so great so glorious but at
the same time so difficult and dangerous a Task. There are not
only the Hazards and Inconveniences of a Sea Passage to be gone
thro, but when he arrives there he must be prepared to meet with
the difference of the Climate and the want of all Luxuries, and
now and then of some necessity. The Air is clearer and more pierc-
ing, the Heats in Summer greater, and the Colds in Winter less
raw, and lasting than in England. There are also Gnats and Flies
which are troublesome to the face, and little red Vermin called
Potatoe Lice which in Summer time crawl up the Legs of those
that lie in the Woods, and if scratched raise Blisters.[249] There are
also Thunders and Lightnings in Summer, but with them there
are violent Showers of Rain which greatly refresh the Earth, and
cause a fragrancy superior to the most blooming Spring in Europe.
Besides these Inconveniences he must also expect much Gainsay-
ing from Men who dare not themselves undertake so great an Enterprize and yet are envious of any one who has Courage and
Virtue enough to attempt it.

With respect to the Indians they are People of an excellent
Temper and will receive with great Joy the glad Tidings of Salvation. They will not only hear with patience but with Pleasure
the Doctrine preached to them: but they will Sift everything that
is alledged with great penetration. They will ask very Shrewd
Questions and will argue with great Strength and Solidity and
always give up their Opinion to Reason. They are greedy of
Knowledge and having plenty of all things meerly necessary, and

desiring nothing more, their Genius's, not being pressed by Poverty nor Clogged by Luxury, exert themselves with great Lustre: their Expressions are high and lofty and their Sentiments noble. In their manner of Living they resemble much the patriarchial age. They often dwell under the Shade of Oak, Laurel or Pine: instead of Beds they lie upon Skins of Beasts; they hunt in the woods for Venison and wild Turkies. Their Houses are in their Corn or maze Fields. The women take Care of the Poultry, dress the Victuals and bake the Cakes upon the hearth. Their houses are Covered with barks of Trees, their Floors of Clay, their Windows are not glazed and their Doors have rarely any Iron Hinges. In these Mansions they live much more contented than our great men in palaces.

> They to themselves do all their wishes grant
> And nothing Coveting do nothing want.

Their King directs them when to hunt, when to Sow, when to fish, and a little Labour thus employed Supplys all the Nations with Food, for whilst one has they never let another want. The rest of the time they spend in conversing together in the Square, in eating together or in playing at Ball or in dancing or Singing. They think the English very unwise who waste Life in Care and Anxiety merely to heap up Wealth, for to raise Discord amongst their Heirs and to build lasting Houses for to make their Children incapable of bearing the Inclemency of their native Air. They do not conceive how People can be in Debt, for they say that if a Man does not want a thing he ought to give it to him who does, and if he doth want it should use it himself and not sell it. They in their little Fields have plenty of Indian Corn, Pumpkins, Potatoes, Mush Mellons, Water-Mellons, Peas, Peaches, Plumbs, Nectarines & Locust Trees. They have Fowls and Hogs in abundance and some Cows and Horses. Besides which their Gunns furnish them with wild Turkeys and Partridges, Ducks, Wild Geese & Teal. They use the Fat of the Bear instead of Oile and the Honey Combs which the Bees Build in the hollow Trees supply the place of Sugar. These are the Men with whom Mr. Whitelamb must converse and whose Language he must learn; this also is the manner in which he must live, in order to convert them to Christian Faith.

I would have him thro'ly well consider these Difficulties, and if upon reflection he thinks he shall want health Courage or Steadiness to go thro' them, I would advise him immediately to lay aside all thoughts of it: but on the Contrary, if he finds that God hath indued him with a Portion of his Spirit sufficient for the great Undertaking (for its only the Grace of God that can support Man under such Labours and Difficulties) I say if he finds his heart warm with the Love of his fellow Creatures and willing and desirous for their Salvation, to sacrifice Pleasure, Youth, Health, nay Life it self; If he finds his Heart glow with the Love of his Saviour, that he is eager to take up the Cross and follow his Crucified Master, I should be far from discouraging him to undertake this Task; for he may be the happy Instrument of converting the Heathen to Christ which God hath so frequently promised in holy Writ and which the Heathen also expect, and of a Truth we know that God in the fullness of time will raise up proper Instruments for their Conversion.

If therefore after a serious Consideration he resolves to offer himself as a Missionary to the Indians, and has a mind to come to Town for that purpose, if he pleases to come to my House in Old Palace Yard Westminster he shall be mighty welcome to a good Bed and share of my Commons that he may be at no expence during his stay here, and I can then acquaint him more particularly with the nature of the thing, of the difficulties he will meet with, and of the methods of obviating them, and then upon a full knowledge he will be able the better to go on with or to quit this Enterprize.

The reason I delayed answering your Letter till now was that I acquainted some of my Friends with it and asked their advice; They were all greatly pleased with the primitive Spirit of Mr. Whitelamb's Letter & very thankfull to you for recommending so worthy a Person. I am.

Reverend Sir   James Oglethorpe

A short letter in French dated Windhausen 1 November 1734 from P. De Reck. Read 26 November 1734 and undertook to be answered by Mr. Vernon. No. 12919. [Listed but not copied in Newman Letterbooks.]

Translation of a Letter from Mr. De Reck dated at Wernigerode 18/29 December 1734. No. 12959. Read 24 December 1734. To Mr. Newman.

Sir: The divine Will has been, and is always in my undertakings as the Pillar which guided the Israelites in their March, and as the Polar Star of which the Brightness encouraged the 3

Eastern Kings to go to Bethlehem. Without a lively Conviction of that, and without the advice of my Superiors, I will not, nor can I stir or imbark in any undertaking whatsoever. Having been tossed till now from Province to Province from City to City from Place to Place, I don't yet see the end of my Pilgrimage. My Intent now is once more to return to Bavaria, and perhaps, if it pleases God, farther. I have learned to resign my Self from day to day to the faithfull and wise providence of my Saviour, hoping he will one day take me out of my troubles, and this miserable Life which I pass for his Glory, and eternal Felicity. If there are any that resolve to go to Georgia this winter, or any other Countrey of America they shall do it at their own expence. As for my Selfe I shall stay at Augsburg and at Ratisbonne for to wait your orders concerning a third Transport. The Map of Georgia which I have so often mentioned, which you have sent by the way of Mr. Reich in the Baggage of a Gentleman is not yet delivered to us. How do the English Nation, that is always glorious, when the Glory of God and the Liberty of Conscience are concerned, regard the Cruel Persecution of the Papists in the hereditary Countries of the Emperor, is it with an Indifferent or an attentive Eye? Give my Service to Mr. Vernon and Mr. Oglethorpe. I am beyond my Expressions
Sir   Your most obedient and humble Servant
Philipp Geo. Fr. De Reck

Rotterdam 30/19 November 1734.   No. 12917.   Read 26 November 1734.   To Mr. Newman.

Sir: I have sent the two letters I have received from England to Mr. Urslperger at Augsburg where I suppose Mr. [Ph.] De Reck is, and I know the Reverend Mr. Urlsperger is in correspondence with him. The Inclosed is a letter for Mr. Vatt I received from Germany. At the beginning of next year I'll have the honour to send you a bill of what I have spent in Postage for account of the Honourable Society.

I am with a due respect Sir   Your most humble & most obedient
Servant   D. Wolters

Translation of a letter from Mr. Urlsperger dated Augsburg 13 December 1734 N.S.   No. 12956.   Read 24 December 1734.   To Mr. Newman.

Honoured and Dear Sir: Your most acceptable Letter of the 1st Nov. O.S. I received in due time, and perceive with great Joy what God has done in London for our beloved Brethren the Saltz-burgers; but of the many letters received from thence, I have made abstracts and caused them to be printed whereof I herewith send you a Specimen: This relation has given here and in the neighbouring imperial Towns, especially at Ratisbonne, such Consolation as to produce Tears of Joy and praises to God, as appears by the many letters I have by me, and can testify the same by my own Experience[.] May God himself be a shield and great reward to the Society Trustees and all others.

As far the rest I am concerned that the Trustees as well as the Society are under some uneasiness occasioned by a Letter wrote by Mr. [Ph.] Von Reck to Mr. Vernon, as I perceive by a Letter from Mr. Agent Wolters enclosing Mr. Von Reck's Letter as also a Letter from Mr. Ziegenhagen. God be preased that every thing was redressed by me before the Letters were sent from London; In the mean I entreat you to excuse my dear Mr. Von Reck who had the Fatality in his Travels to fall into the hands of a Person of distinction who animated him to carry on such a design, but now he, *namely Mr. Von reck*, writes to me that he will for the future enter no more into any design of this kind without communicating of them to me.

I expect to see him with me soon when I shall speak more fully to him. I expect answers soon to my Letters from Mr. Secretary.

Mr. Ziegenhagen's Letter of the 8th of November O.S. I have received, but from Mr. Degmar [Degmair] I have received none Since the 11 Nov. N.S. I recommend you to the Grace of God and remain

Dear Sir   Your most humble Servant   S. Urlsperger

Translation of a Letter from Mr. Urlsperger dated Augsbourg December 20, 1734. No. 12971. Read 31 December 1734. To Mr. Newman.

Honoured & Dear Sir: Your favour of the 1st Nov. O.S. I have received, and that of the 26 of Nov. to day, as I hope mine of the 15 of Nov. and 13 of December N.S. are come to your hands.

The two Bohemian Brethren are gone away from Ratisbonne the first Instant; what passed there with them you may see from the extract of the letter written to me[.] God help these and all others distressed. As I perceive Mr. [Ph.] Von Reck's last Letter refers to the Letter of Mr. Vernon, directed to me some Weeks ago. I hope to see him here short, and thus all the letters to him may be safely addressed to me for the future, because I know best wherever he resides, if he be not here or at Ratisbonne. I could wish also that his letters might be sent open to me provided they contain nothing but the Affairs of the Society so that I may be immediately informed what proper measures to take. The Lord be praised, that our Transport, under the Conduct of Mr. Vat, left the Downs with such a fair Wind. I have composed a particular form of Prayer to implore the Almighty's Assistance and Protection for this present Transport, which by the Approbation of the Magistrate and the Consistory is always read whenever any publick Service is performed in our Church at St. Annes. May the Lord hear our Intercessions. Some days ago I received letters from the two Ministers in Georgia, dated July 16th under Cover of the Reverend Mr. Ziegenhagen, wherein they refer themselves to their Journal, which I dont doubt is still in Mr. Ziegenhagen's Hands. By these letters I learn the particular Tryals they as well as their little Flock have hitherto undergone, so I look upon the present Condition of their Countrymen as a signal Providence. The Lord enable them to overcome all Temptations.

Dear Sir, you owe me still several answers, if you have not time, be pleased to communicate them to Mr. Degmair[.] I am very glad to hear he makes so good a Progress in the English Tongue, which give me hopes he will be usefull to me in my English Correspondence at his return instead of Mr. Vat, to which I shall not forget to encourage him. The Lord bless and strengthen you for the Prosecution of the Interest of the Protestant Mission in the East Indies,[250] and since we are not able here to assist you with our advice, we shall the more earnestly pray to God to bless your pious Endeavours, which you may be always assured of from

Honoured & Dear Sir   Your most humble Servant

Sam. Urlsperger

Extract of a Letter to S. Urlsperger from Ratisbonne December 3d N.S. 1734. No. 12971.

Your kind letter I received with Joy from His Excellency [J.] Von Reck by the two recommended Bohemian Brethren[.] When I saw them wanting an Intercessor, I went with them to all the Protestant Ambassadors, with whom they met not only with a favourable Reception, but also a great deal of Compassion; but the question was, How shall we assist You? In the mean time I procured them a Lodging at one of the Cityzens where they could live incognito, because the Bohemian Ambassador, Count Coloredo,[251] himself has above 60 of such Confessors, Prisoners in his own Territories,[252] and we sent them to Nurenberg December 1st with a safe Opportunity. Mr. ——————— Von[253] composed their memorials, whereof one is to be sent to the Evangelick Body; the other to England, the third to the States General, but they must be subscribed by some of their Brethren, and in Lusatia, before they can be delivered here.

May the Lord bless it with Success; for when he will help, they must lend their Assistance, that have no mind to its present.

Translation of a letter from Mr. Urlsperger dated Augsburg 30 December 1734. No. 12977. Read 7 January 1734/5. To Mr. H. Newman.

Honoured & Dear Sir: Your favour of the 3rd December O.S. I received in due time. What Mr. [Ph.] Von Reck wrote to Mr. Vernon he likewise wrote to his Uncle [J. von Reck] and to me. I have, without delay, writ two letters of his undertaking to Moravia and Bohemia upon the same foundation that the Honourable Society was pleased to use and with the full Consent of the Resident Baron [J.] Von Reck as he had desired me, and to diswade him by all means that he should not undertake any thing in that nature be it from whom it will until he has been here and concerted with me farther thereupon. This is, God be praised, come to the hands of Mr. Von Reck as we were informed by the Letters of yesterday's Post from Professor [G. A.] Frank, that the 21st Instant Mr. Von Reck was arrived at Hall, the 22nd he came to Poltzig to Count Hanckel [Henchler?] and by this means to keep the Hollidays with his Serene Highness the truly pious Duke Regent of Saxon Salfeld [Sachsen Saalfeld], and so from thence will hasten hither thro' Nuremberg and Ratisbonne, that in all probability I may see him here in 4 or 5 days with his dearest Uncle [J. von Reck] as a most Christian, wise experienced, and with regard to the Protestant Interest, an incomparable Statesman with whom I may freely Act and impart to him every thing that is communicated to me on that Subject.

With regard to the people that Mr. Von Reck so highly recommended, most of them are come out of Moravia & Bohemia in a private manner [secretly] above ten years ago for the sake of Religion and with the greatest danger of their lives and loss of Goods and often long Imprisonments. These people have from time to time gone to receive the Protection of the Count of Zinzendorff who has offered them a little part of his Estate in the upper Lusatia to build houses thereupon which is a present called *Herrenhût* or in English the *Lord's Protection*.

I hear they are there narrowly supported but in no respect persecuted. Hitherto they have not declared to be of the Augsburg Confession tho' they do not appear to be any way against it; with regard to their Doctrine they will not be obliged or bound to any Confession, but rather to the former Ecclesiastical Discipline of the Bohemian Brethren but they thro' the visible head of those Herrenhutischen Brethren (as they call themselves) the Count of Zinzendorff has much altered and added to it. The Herrenhutischen Brethren themselves most of them, before they came to Herrenhût were instructed by our Evangelick Ministers in Silesia most of them appearing to be honest and good Christian people, and it should be in no wise be imputed to them what is alledged against their Lord and Count, that upon the whole he seems to have a great Zeal for a reformation, for which reason he has been accused of indiscreet actions and some Errors in Doctrine; from which he cannot intirely acquit himself but that it is given out to have altered many things, but others who know him well will not trust him. But it is not to be denied that since he is a person of distinction & great Gifts and has preached in the pulpit several times publickly in Stralsund, and lastly in Tubengen [Tübingen], and at the last place with his embroidered Star upon a black Velvet Suit, and the little Order of Danningbrook, has occasioned a great Speculation.

The same Gentleman as likewise his Herrenhutischen Brethren have desired me to be concerned with them: but I would not, for the Affair is not sufficiently clear & right to me. But I do not write this to be in any ways prejudicial to the Count, but according to the best of my Knowledge and as the Honourable Society desires to know, I am ready to comply in all Duty when I can have a Testimony of a better Information.

What Count Zinzendorff's project is with regard to Georgia he would send thither 15 of the Herrenhutischen Brethren at his own Expence, or much rather at the Expence of his Brother in Law Count Reuss of Ebursdorff and to the intent that they shall preach the Doctrine of Christ to the Heathen, in which proposal I cannot be any ways concerned since it is not known to me what call they have to entitle them to it. Secondly, since they have not the extraordinary or ordinary Gift of Speaking foreign Languages,[254] while they begin in this manner under Heathens & Christians, it is to be feared many disorders will arise not to mention other things. There is a Rumour goes that they have already begun their Journey from the Voigt Land of Ebersdorff, thereupon 8 days ago I writ to the Count's chief Manager, Mr. Deboni, to exhort the people not to go, and if they were even upon the road, to turn back, and before they begin their march, to have a Correspondence with the Trustees for Georgia; if they will not follow these directions they must impute to themselves any Accident that befalls them. Now it is to be considered that in Case those people do go and entirely upon their own expence, what the Trustees will resolve upon, whether to send them back or to let them go forward. Should it be the last, I with all Submision think it will be proper that Mr. Ziegenhagen should examine them and forewarn them against any singularity of Opinion, and especially that Care be taken that they do not become chargeable to the Saltzburgers Congregation, or introduce any disorders there. Tis not enough to have good Intentions which I do not discommend these people for, but there is still more required.

I have farther to mention that these 15 persons as far as we know are common Artificers, as likewise that Count Zinzendorff gives out upon all Occasions that the Trustees had presented him with some Land. Before I received that Commission, which Donation I hope the Trustees will not retract from. To Mr. Ziegenhagen I wrote the last month as likewise since I sent him an Extract of my whole Correspondence in this Affair from whence a greater light may be gathered. God enlighten and govern the hearts of all, that his Will may be thereby in all things accomplished.

In the last letters from Mr. Resident [J.] Von Reck, he says as follows; That you may give orders at London that your Letters may be directed under Cover to me and be certain that I shall take all due Care of them. But the Resident desires that they should be sent to Mr. Reiche in Pall Mall to be put in his Packet, by this means all letters will go free, and I shall receive mine but one post day later; I request you to make mention of this to Mr. Ziegenhagen.

In Carinthia the numbers of the Confessors of the Gospel grow greater and greater, but the necessity and misery of those that are in prison to the number of above 50 is still heavier; In Ratisbonne they are coming still more & more who forsake their Wives, Children, and Goods,[255] but it is yet unknown where such people shall be supported.

God in mercy cause the fountain to be opened for the relief. All this I write out of a hearty Love for the Promotion of the

Kingdom of Christ. Amidst a multitude of Affairs especially at
this time with intreating you once for all, that when I do not
observe to write accurately you'd be pleased to excuse me.

Under this most hearty wish that God would at the opening of
this new year abundantly bless the Trustees and Society   I remain
Honoured and Dear Sir   Your most humble Servant
Samuel Urlsperger
P.S.: Lord be mercifull to our dear Mr. Vat, to the Transport of
Saltzburgers and even to all that are in the Ship. Mr. Professor
Frank has paid the 302 fl. to Mr. Münch so that they are to be re-
paid to Mr. Ziegenhagen. The Map of Georgia is come to Ratis-
bonne.

Savannah in Georgia January 15, 1734/5.   No. 13154.   Read 13
May 1735.   To Henry Newman Esq. in Bartlet's Buildings Hol-
bonne [Holborn] London [from Samuel Quincy].

I have had the Favour of three Letters from you for which I
esteem my Self much obliged, and should certainly have returned
an answer much Sooner, but that since my Return from New
England I have had no opportunity till very Lately and the Vessel
I now send by has been delayed above a month in our River by
means of a Dispute with a Custom house Officer from Carolina.[256]
Mr. Vat arrived here with the Saltzburgers all in good health
on the 28 of last month, having had a Short and pleasant passage
at Sea, and express themselves well pleased with the Treatment
they met with from Capt. Dunbar who has a very good word from
all his Passengers. Mr. Vat is now gone up with his Charge to their
designated Place of Settlement Ebenezer. He desired me to give his
Humble Service to you and to acquaint you that he would write
very shortly which he will have a good opportunity to do by the
Return of Capt. Dunbar who intends to Load in our River and
Sail directly for London. We hope that he will have his full Freight
in about Six weeks from the Date of this, tho' it may possibly be
Longer. I hear from Ebenezer that the People make very great
complaints of the Badness of their Lands, and that Mr. Vat and
Mr. Bolzius have a great Deal to do, to persuade them to stay on
the Place. This is a very great discouragement and cannot too soon
be Remedyed; for by all accounts the Land is indeed very bad, and
will produce nothing at all. But Mr. Vat I suppose will particularly
acquaint you of this Matter. Mr. Bolzius has merited the best of
Characters for his Prudent Conduct and Singular Care of the

People under his charge whom he endeavours by all means to make easy and contented in their present Situation.

I return a great many Thanks to the Honourable Society in behalf of the poor People here for their Last Benefaction of Books and shall take Care to distribute them where they are necessary and may be most Usefull. I have informed Mr. Bolzius that I am ready to deliver to him so many of the Books as he thinks will be Serviceable to him or his People.

I am Sorry that I cannot Send the Honourable Society a more Agreeable Account of the State of Religion amongst us than I must do if I say any Thing at all upon that head. The Publick Worship of God is very much neglected & Vice and Immorality are Very open and flagrant amongst us. The pretence for not attending Divine Service is the Want of a Church and indeed the place we are at present obliged to make use of as such, is much too Small to contain the People[.] it will not hold above 100 at most, and we might reasonably expect according to our present numbers not less than 300 if the People were at all Zealous to Perform this great Duty towards God. But is very seldom that we have above 30 or 40 People at Church and not near so many who give their Constant attendance. I have heard that the Trustees have money in their hands towards Building a Church and doubt not when the Matter is represented to them, but that they will think it necessary to put it forward, that this may no longer be an objection. They have been so good as to recommend This Duty in particular to the People in a Letter which was by their order publickly read to them and I pray God it may have a good Effect upon their Minds. It is the more necessary to hasten the Building of a Church because we have a great many Dissenters amongst us who talk much of Building a meeting house and getting a Minister. We had last Summer an Embarkation of Scotch all of them Presbyterians and many of them very rigid that way, and I verily believe if we had a Church it would lend very much to keep the English Dissenters especially in our Communion and without their Joyning with the others they will hardly be able of themselves to carry on the Design.

We are much obliged to Mr. Hale for his Friendly Admonition it is very seasonable here; for we have too many addicted to the hatefull Vice of Excessive Drinking which Reigns as much amongst the Women as the Men to the no little Scandal of the Sex. We have very strict Prohibitons against bringing Rum into the Colony but it seems almost impossible wholly to prevent it; for those who have got a habit of it, will by one means or other procure it[.] It is Strange that Vice So apparently destructive, Should Stick so close People that no arguments will prevail with them to Leave it tho their own Misery and Ruin may so plainly be pointed out. But Experience Shows that it is so, and that as soon almost may the Ethiopian change his Skin as those who are accustomed to this Evil forsake it.

Since the Writing of the above Mr. Vat has been down here to acquaint our Magistrates that the Saltzburgers were so discouraged with the Badness of their Lands at Ebenezer that they desired not to be Set down there, but to be removed to Some other place where their Labour in cultivating might not be lost. Upon this one of our Magistrates [Causton] with the Surveyor [Noble Jones] went up, and they found a Small Tract of Land about half a mile distant from the Town which they judged to be very good planting Land, and there will be about 10 Acres of it for each Family, this they will divide amongst them for their present use till they can further know the Pleasure of the Trustees. It would be a great Pity that the People who are come So far and are really deserving for their Industry and Sobriety should not have good Lands; but it Seems the Magistrates here have not a power of removing them with out Leave from the Trustees and therefore what they have done is only to Satisfie them for the present, till they have further orders.

I hear there is a Missionary intended to be Sent to the Indians our Neighbours. This is certainly a very pious and Charitable Design, and I must beg Leave to offer some Informacons I received upon the Subject of Indian Missionaries, from Mr. Quincy our worthy Relation in New England. He assured me that they found by the long Experience of upwards of a 100 years, that is ever Since that Country has been Settled; That the method they had taken to convert the Indians there was intirely wrong, which was to instruct English Missionarys educated amongst them in the Indian Tongue, who were to preach to the Indians and instruct them in their own Tongue without teaching them any English. This they now find has greatly obstructed Christianity amongst them. Whereas if they had been taught English, they would in time have forgot their own Language, and together with it their old Heathenish Customs which are now too much retained amongst their Converts particularly in those Places where this Method only has been pursued. The Indians they have near Boston are a full Confirmation of this; for they not only acknowledge the Christian Religion but have in a great measure forgot their own Language and are so Conform to the measures and manners of the English that they are not to be distinguished but by their Complexions. The Reason of this is very obvious for those who are taught to understand and Speak English can converse universally with the English and therefore have abundantly more opportunity for Instruction than those who only Speak Indian. In the former Case they receive all their Instruction from one Person who perhaps has a whole Nation under his Care and tho' 'tis not to be doubted, but that a good man employed on so excellent a Design himself thoroughly convinced of the Truths and Importance of the Christian Doctrines will carefully instill into them the most essential & fundamental Points of our Holy Religion so far as it in his Power, but it has I think always been found that the greatest difficulty in bringing over the Indians has been to reclaim them from their Wild and Savage Way of Living and Settle them in a regular and orderly manner; and this will be certainly best be attained by Teaching them English whereby they will become more readily acquainted with the Customs and Manners of the English, & Sooner brought over approve of, & imitate them. And it would be to very Little purpose to entertain them with Speculations tho' ever So just and exact without reforming their Behaviour.

The only objection to this I'm Sensible will be, that the generality of Christians themselves are So bad & vicious in their Lives that they will rather expose than recommend Religion to the Infidels, & therefore it will be best to keep them ignorant of the Language that they may be less influenced by the bad Examples of such Christians. It is indeed true that we have too many degenerate Christians who are a Shame & Scandal to their Profession; but I cannot agree that we have any Christians more degenerate and worse than Heathen;[257] This I know is Strongly maintained by the Enemies to Christianity and made use of as an objection against it tho' very unjustly; for if it was true that Christians were indeed So bad as they represent, this would not destroy the Truth and Excellency of the Christian Religion tho' it might eclipse it. But the assertion is in reality very false; for if we take a View of the Shocking & abominable Vices practised by the Indians, without any shame or Remorse, it will Sufficiently Convince any impartial person that Christians, bad as they are, are not arrived to a Pitch of Wickedness and Impiety equal with the heathens. The most abandoned and profligate amongst them if they commit a Crime of a heinous nature as Murder for instance, are generally touched with Some Remorse and Concern for it and cannot conceal the Guilt and Horror of their own Minds. But these wretches are So far from Tenderness and Humanity, that the barbarous Mother frequently dispatches her Sucking Infant, merely to get rid of the Troubles of nursing it. In their Revenge they are most bloody and cruel, not being contented barely with the death of their Enemy, but inflict the most horrid Torments they can invent and nothing occasions greater Joy amongst them than a Spectacle of Inhumane Butcherie, Their Perfidiousness and Treasury [treachery] are notoriously known to all the English in America, particularly in New England where they have often found to their Cost that the Most Solemn Promises and Engagements of Peace & Fidelity would never hold to restrain them from Hostility and Outrage. To Say nothing of their Bestial Vices which they are said to be vilely addicted to, too abominable to mention. And these Crimes which amongst Christians are branded, as they ought, with the utmost Marks of Infamy, are not looked upon by them as Such, but rather gloried in, and committed by them without the least Remorse or Shame. I should be very sorry I have at all misrepresented them, but I verily believe I have not because those things that I have Said out of my own Knowledge I have taken from the Report of Men of Probity who are well acquainted with them. Their War Dances sufficiently show what Sentiments they have of Honour, when the Hero instead of being praised for brave and hazardous Exploits, boasts himself of Some base and insiduous piece of Treachery more worthy of Contempt than applause. Their Religion if they may be Said to have any consists in Songs and Dances Some of them very obscene, these are used upon their going to hunt or to War; and they have likewise a Sort of Jargon consisting of unintelligible Words not understood by themselves which they say over them that are Sick in order to recover them. I might relate many more of their Superstitious and abominable Vices, by all which it will evidently appear with how little Justice it is Said that Heathens by the Light of nature are better than Christians with the help of Revelation. For certainly if we would look for a Sect of People who have been always governed by the meer Dictates of nature, and never have been Corrupted, (as they impiously term it) by Revelation, they are to be met with among the American Indians; for it cannot be proved but that the Heathens of other Countries have had some notices of Revelation and reformed themselves in some measure by it. As for the ordinary Harangues of Deists upon the excellent Morals and Virtues of the Grecian and Roman Sages and Philosophers they are no fair proof at all that Heathens are better Moralists than Christians because these Manifestly received their best Notions from Revelation it self. But if we would See nature in its true Light divested of all Supernatural Assistance, we must turn our Eyes upon these Western Climates, and here I'm Sure the Infidels will have little Cause to extol natural above revealed Religion. Thusly I have endeavoured to answer all the objections that I think can be raised against instructing the Indians in the English Tongue, and I might add to it that tho' this should not be done yet will it not wholly prevent bad Examples before them; but as this cannot be expected I should think it would be very much in the Power of a Missionary to antidote the Mischief of ill Examples to let them know how odious the Characters of bad Men were amongst good Christians.

I heartily pray for the abundant Success of the Society in all their pious undertakings for the Glory of God and Good men and am Dear Sir   Their and your most obliged and most obedient humble
Servant   Samuel Quincy
My most humble Service to Sir John Philipps, Mr. Vernon, Mr. Wilson and Mr. Copping.

No. 12991.   Read 21 January 1735.   Extract of a Letter from the Reverend Mr. Urlsperger to Mr. Newman dated Augsburg 13 January N.S. 1735.

Mr. Urlsperger gives a Copy of a letter of Baron [J.] Von Reck the Envoy to Him in which it is said,

That several hundred of Saltzburgers in the Baillage of Gastein are ready for Emigration, and will have the Liberty for so doing.

He therefore desires to be informed whether he may offer to these people the same Conditions for going to Georgia as before and

Whether he may receive one or two hundred of them, upon accepting of the known Conditions.

Lastly Whether he must allow Mr. [Ph.] Von Reck upon his staying with him at Augsburg his Alimentation money in full or in part, or discharge the Expences of his Journey to and from Augsburg on the Society's Account?

Translation of a Letter from Mr. Urlsperger dated Augsburg 13 Jan. 1735.   No. 13005.   Read 28 Jan. 1734/5.   To Mr. Newman.

Honoured & Dear Sir: My last of Oct. 20th. I hope you have received. It was a fortnight that I did not hear any thing of Mr.

[Ph.] Von Reck where he was but last night I received the following Letter of his dear Uncle [J. von Reck] dated Ratisbonne 11th of this month

"My Cousin being arrived here the 8th of this month in the morning and intending to go with the Stage Coach to morrow for Augsburg to give you a verbal Account of all, I need not to tell you what I heard of [from] him. He will acquaint you among other things.

That there are still remaining in the County of Saltzburg (particularly in the Bailage of *Gastein*) some hundred of Emigrants which will be willingly dismissed; If therefore the English Society would be pleased to receive these people and they should be willing to go thither, it would be a great opportunity to satisfy both.

I leave it principally to the divine Providence and your Reverence's wise direction recommending likewise my Cousin [Nephew] to your farther Instruction, and hoping that you will keep his Religious Fire and Zeal within due bounds, that it may tend only to the Honour of God & the Benefit of His Church.

I remain faithfully   Reverend Sir   Your very affectionate &
obliged Servant   V. R. [J. von Rech]"

I desire therefore an answer by your next upon the following material Questions (1) Whether if these several hundred Saltzburgers of the Baillage of Gastein should actually [now] emigrate, I may offer unto them the old Conditions for going to Georgia, and receive one or two hundred of them if they should accept the Conditions.

2. Whether (as Mr. [Ph.] Von Reck is to conduct the next Transport if any is to be made) I must allow him his Alimentation in full or in part in Case he should stay here, or pay the Expences of his Journey to and from if he should not stay with me?

I do not remember whether I told you in my last that Mr. Professor [G. A.] Frank has paid to Mr. Münch here the 302 fl. and that this Sum consequently is to be repaid by the Society to Mr. Ziegenhagen on account of the Mission at Malabar[.] If I have not do it before I mention it now for this purpose.

Wishing you heartily a happy new year and recommending my Self to the Society   I remain
Honoured & dear Sir   Your most humble Servant   S. U.

*Translation of a letter from Mr. Urlsperger dated at Augsburg
Jan. 14, 1735.   13,005.   To Mr. Newman.*

Honoured & Dear Sir: Here I send you some very remarkable

and certain Accounts which as well as the foregoing can be certified upon Oath. I received these Accounts four days ago from an eminent and pious Person living in a great Town of Silesia[258] who has Spoken partly with these Bohemian Exiles himself and sent me formerly several hundred florins for the benefit of the Saltzburgers, so that they may be depended upon.

I wish that God may move the hearts of some in England to assist these people with some Charitable Gifts. I intend to desire the same here privately from some able and well disposed persons.

Mr. [Ph.] Von Reck is now with me and begins to apprehend the impracticableness of his former design, But it will be necessary when you, dear Mr. Secretary write to him again, that you tell him by order of the Society (and let the Secretary of the Trustees do the like) not to begin any thing absolutely without me, for else it will give me incredible Trouble to adjust and correct things again, and I shall at last not be able to handle all things as required, on account of the multiplicity of the Business of my Office. Of the several hundred Emigrants out of the Baillage of Gastein last mentioned, I have heard nothing farther therefore expecting a convenient answer upon my last I remain unalterably

Tuus totus   S. U.

Copy of a Letter from Messrs. Bolzius and Gronau at Ebenezer dated 6 Feb. 1734/5.   No. 13090.   Read 8 Apr. 1735 and Read 19 Apr. 1735.   To Mr. Newman.

Sir: Being assured that you have kindly received that letter which we took the Liberty to write to you the 10th December last, we now make bold once more to trouble you in the midst of your weighty affairs, by these humble Lines. Whereas you gave us in your last very obliging Letter dated the 29th of October Sufficient marks of the Continuance of your and other great Benefactors favour toward us, and have sent in the name of the Society money, two travelling Beds and other necessary things for our Relief, we should be the most ungratefull Persons in the world if we did use these Benefits without praising Almighty God, and the praise worthy generosity of our great favourers; 'Tis our firm Resolution, which is renewed now by these new Testimonies of divine Blessings, to employ all our Care in beseeching God continually to reward them a thousand times for all Benefits bestowed hitherto upon us and our flock and to grant us his Grace to be answerable to their desires, and extraordinary Intentions. We take the pleasure to acquaint you that the Saltzburgers under the Conduct of Mr. Vat have finished their Sea Voyage so happy and in so short a time, that every one wondered at it. 'Tis impossible to express in words all manner of Joy & pleasure we have had in receiving them in our place Eben-Ezer. We appointed them quickly several Houses and Hutts, builded already here as well as possible we could; And they may live in these Houses and Hutts of their Countrymen till they have prepared some Acres to plant Corn and other things next Spring. They told us with pleasure of mind that by the particular Care of their Benefactors, they have received at London as well as in the Ship a great many kindnesses and benefits, and not only confess themselves highly obliged to them, for the said and many more Benefits, promised for time to come, but are also firmly resolved to make it their Business by the assistance of the holy Ghost to perform to the utmost of their power all that shall become true Christians, to answer the Expectations of the Trustees and the Society. Some of these good people are affected with some Infirmities, which as we hope and wish in our prayers, will go off in a short time.[259] One man [Sebastian Glantz] died at Purisburg before he could be brought up to our place of Abode. The Child [Schoppacher] that was Christened at Gravesend died likewise a few days after their Arrival. For the Young man [Henry Bishop] that was sent to our Service by the providential Care and Goodness of the Society, we return our humble thanks assuring his Soul's and Body's Welfare shall be our special Business, to make him thro' God's Blessing, capable of being serviceable both to God and Men. We accept of with Thankfullness the Salaries the Society are pleased to allow us for the support of our Bodies, moreover relying upon the Blessing of God, who can, and as we hope, will by his fatherly Affection and Care supply all our Wants. The Bill of fifty pounds has been accepted by Mr. Montagut, and the Spanish pieces of Eight, together with the Half pence, are come to our hands, and we could wish that more half pence had been added in lieu of Silver; Copper money being extremely usefull and convenient in this Country; For the English Books the Society are pleased to allow us by Mr. Quincy, we return our humble thanks; We know to take Advantage of them and pray for many more. I am in hopes, You will not take it amiss in troubling you with an account of the manner of the Saltzburgers divine Worship. We are exceedingly glad to perform by God's direction divine Offices among them. Their Love to the holy word of God is very great, and they shew by words and by deeds, that That was the chief and only reason of leaving their native Country. They not only come on Sundays three times to our Assemblies, but their Zeal to Edifying their Souls is so ardent, that at their desires we have appointed in the Evening about half an hour's time for instructing them in Christian Duties and putting up with them, to Almighty God, our prayers in the Weekdays, after they are come from their daily labours and Refreshment so that they may not loose the least time for preparing their ground and they convince us by their sober behaviour, that they make a very good use of the Gospel they have heard, and endeavour to keep strictly, by assistance of the holy Ghost, its precepts, They at all times remember the great many Benefits, they constantly receive from their generous Benefactors in England, particularly for having allowed them Ministers, preaching and administring to them the holy word of God and the holy Sacraments. And, being sensible of those great Blessings, they lift up their hands and Heart to God Almighty for the Prosperity of their noble Benefactors. Their Children being now 12 in number who constantly come to School and give us great hopes of following the footsteps of their pious Parents. And in order that they may be instructed in the Principles of the Christian Religion, and other necessary Qualifications, we both teach them every day, over and above what they are taught by Mr. Orthman [Ortmann] who follows our directions concerning them, wherein he employs at present his best Skill, and we hope he will continue so to do hereafter. Some of the Children begin to be of years fit for assisting their Parents and upon that Account we use our utmost Endeavours to promote their Learning. Lest we weary you with our Scribble, and for fear we steal from you that time which you wholly employ for the Care of the publick, we conclude, professing that we beseech God to prosper the noble designs of the Society, and hope surely he will further mercifully grant a good Issue to all their praiseworthy Enterprises taken for promoting Christian Knowledge, And so wishing you all Happiness imaginable, we take the leave & pleasure to subscribe our Selves

Kind Sir Your most humble Servants John Martin Bolzius[,]
Israel Christian Gronau

P.S.: Mr. Vat presents to the Gentlemen of the Society his most humble Duty and Respect not being yet able to write to them himself by reason of his Infirmities upon his Eyes.

Copy of a Letter from Henry Bishop dated Jan. 2nd 1734. Inclosed in Messrs. Bolzius & Gronau's Letter. No. 13090.

Honoured Father & Mother: I am willing [anxious] to let you know of my safe arrival at Georgia. Thank God we had a fine passage. We were but Six weeks from Lands End to Land's End, only as we had no Sun the Captain did not know it was Georgia, We sailed past it and sailed to Augusteen [St. Augustine] at the Mouth of the [St. John's] River among the Spaniards, but as it happened they did not see us. The Indians being on board knew the place, for some times they are at war with them. The Queen on board lost her Eye with the Spaniards. We had a fair wind and Sun & sailed to Georgia & the pilot boat came and brought us into Georgia River.[260] I don't live at Georgia Town but far up the River. There are three Towns, Georgia Purrysburg and Ebenezer. I live at Ebenezer. I have got a very good Master [Bolzius], he has promised me when I have learned German to furnish me with a great many fine Books[.] Pray remember me to Esq. Oglethorpe and Esq. Newman & Esq. Vernon and return them thanks in particular & also my Master & all the Trustees of the School & also of the Office[.] Pray remember me to my Schoolfellows my Brothers & Sisters & all friends. I delivered Mr. [Noble] Jones his letter being he came on board the Ship when we came in the River. So no more at present I remain

Your loving Son   Henry Bishop

No. 13097.   Read 8 Apr. 1735 and Read 19 Apr. 1735.   To Mr. Henry Newman.

Honoured Sir: I don't doubt but before this cometh to your Hands you'l have heard of our Safe arrival in Georgia and of our Landing at Savannah Town, which was on Saturday the 28th of December last. I should before this Time have given you an Account thereof; But having got a great Cold in my Head and Eyes ever since we cast the Lead for Sounding at sea, I could hardly read or write any Thing; Besides being obliged to attend the Loading and unloading of the Baggage and Provisions for one Quarter of a year; so that I must refer my self for further particulars to the Letters now to be written by the Reverend Messrs. Bolzius and Gronau to James Oglethorpe Esq. to James Vernon Esq. more particularly to the Reverend Mr. Ziegenhagen & to your Self. The substance whereof they have made me acquainted with, as being of the same Opinion with them relating to the Soil of this Place and the great difficulties the People lie under in bringing up their Provisions from Savannah Town and other matters concerning the present settlement of the Saltzburgers.

On the 12th of January we left Savannah Town, and got on Board three Periawgoes,[261] the Smallest of them being gone with the Sick directly for the Landing Place at 4 English miles distance from this Town over Purysbourg and Ebenezer River. We came with the two larger Periawgoes the 12th to Abercorn and the 13th of the same month by Land being 12 or 14 miles to this Town of Ebenezer; At the Sight whereof we were confirmed of what every Body (excepting Mr. Causton and Mr. [Noble] Jones the Land Surveyor) had told us of the Barrenness of this Part of this Province being chiefly Pine barren, a sandy white Ground not above one Fifth or at most one Tenth Part of tolerable mould, Cane Land or Swamps; which swamps seeming to be Good are covered with a black mould about one or Two Inches Deep, but under it appears a White Sand like salt. So that every one who cometh hither saith The People will never be able to get a Livelyhood in this Place be they never so Industrious and Laborious; For upon a rainy Day the Black mould being washed off nothing but white sand is seen in large Places like paths in a Walk so that the poor Saltzburgers were exceedingly struck down[262] and disheartened and begged that according to the Permission given to them by Mr. Causton for looking out for some good Ground and for working it jointly till he and Mr. Jones should come to Set out their Lotts, They might seek out some such Spot; At their Return they reported to have seen some good spotts of Ground near the mouth of Ebenezer River and Savannah River. Upon this Mr. Bolzius and myself went by water the 25 January to the Red Bluff and the Indian Hut, this last being about nine miles distance by Land from this Town in a Straight Line 6 Miles; And meeting in our Passage thither on Ebenezer River the small Perriawgoe, loaden with part of our Baggage and Provisions from Abercorn, I ordered it to go down and to unload its cargo on the Indian Hut and coming out thither, We found that there was on the North side of that Indian Hut up to the Red Bluff and along the South side of Ebenezer River about Two miles high on the Land on the declivity whereof Large oak Trees and others as also large Vines of 3 or 4 Inches Diameter; but at the Top of the high Land mere Pine barren, and judged that were the Lotts there so ordered that one Chain were given on the River side and Three upwards, the People would have one Moiety good Land and the Other Moiety

Pine Barren, with which the People would be exceedingly pleased. The next Day we went to Abercorn, and meeting there by chance Mr. Christie the Recorder, We acquainted him with our Expedition And he carried me in his Boat the 26 to Savannah Town in order to see the Powers given by the Trustees for Georgia to Mr. Gordon Mr. Causton, Mr. Parker and Mr. Christie jointly for Setting out 2500 Acres of Land for our Saltzburgers, Those Writings giving them power to Sett out such Lands, where and in what manner they should think fit. Mr. Christie proposed my Presenting a Petition to the said Four Gentlemen but Mr. Causton and Mr. Parker were of Opinion (Mr. Gordon being absent) Mr. Causton and Mr. Jones should go with me to view the Lands at Ebenezer and the Indian Hut and should make their Report thereof to the other Gentlemen. Accordingly Mr. Causton and Mr. Jones came with me the 29th to Abercorn the 30th to Ebenezer Town. After Dinner we desired to go by Land to the Indian Hut, But missing our way thither We came again to Abercorn. The 31st Mr. Causton designing to return to Savannah Town I desired him to give me his directions for my acting with the People since he did not intend to go to the Indian Hut for want of a Knowing Guide. He told me the People should [pitch?] upon some spott of Ground near Ebenezer Town and clear it and work it jointly till such Time as the Pleasure of the Trustees should be known for which Purpose he would write to them. Then I proposed to him the Permitting the new Settlers under my Care to build a hutt on the Indian hutt Land for working there jointly; but he roundly refused Complying with my request. I returned that day to Ebenezer Town; & some days afterwards I heard Mr. Causton and Mr. Jones had been the 1st instant in the Indian Hut and were gone thence the 2nd in the Morning early. The 3rd I went by Land to the Indian Hut and Ordered such Baggage etc. as was deposited there to be brought up to the Landing Place with the Small Perriawgoe and the Small Boat. The former of which was sometime five days in going from Abercorn to the Landing Place, Which might have been avoided if we had proper Carriages and Horses to be conveyed by Land about 12 or 14 Miles. And this Transport could not be Computed [completed?] before the 5th tho' We have not as yet all our Provisions for one Quarter of [a year from] the store house in Savannah Town. And the People are obliged to bring their Baggage and small Tools and Casks upon their Backs from the Landing Place and the heavy Casks in the small Boat by water, Which will require 8 or 10 Days more provided the Water in the River Ebenezer doth not fall. The Inhabitants of this Province generally compute the distance 20 Miles from Savannah Town to Abercorn Creek. Thence 6 Miles to Purysbourg 10 Miles to the Indian Hutt, 4 Miles to the Red Bluff or the mouth of Ebenezer River, 7 Miles to the landing Place and thence 12 Miles to Ebenezer Town. In all 59 miles by Water. The Town of Abercorn lies 2 Miles from the river Savannah. Indeed our Saltzburgers with a small Boat go down from the Landing Place to Savannah Town, in one Day but cannot come up in Less than Three Days. And that Boat cannot carry above One Thousand Pounds weight and four Men. Neither can it come hither from the Landing in Summer or at low Waters by Reason of some Trees Bushes and sands in the River, By which the People must lose a great deal of Time and Labour. However were the soil of this Place tolerably good these Difficulties might be over-looked[.] But as it is the Opinion of every Body even some of the best Planters in this Country and the Province adjoining, It's humbly hoped the Trustees for Georgia will take the low dejected Condition of these Poor People into their Consideration and grant them the Favour of Removing hence to the Indian Hutt or the Red Bluff or to some other Place higher on Savannah River more likely of being thereby enabled to get a Livelihood. For should they be obliged to remain here in this place, according to the promise made to them in Germany, The Society or the Trustees for Georgia will find themselves under a necessity of Subsisting them with Provisions as long as any of them shall be living And it is to be feared many of them will dye for grief; no less than eleven have died of the first Transport since they came into this Province, and two of the last Transport viz. Sebastian Glantz the 13 January at Purrysbourg, and the Child [Schoppacher] born on board the prince of Wales, whilst the Ship was in the Thames, dyed here the 23 January last; And Some of the first and last Transport are now Sick, but we are in hopes by Mr. Zwifflers Care some of them will do well again, as some others are recovered of their Illness. And here I cannot but observe that indeed 6 pounds of salt Beef per Week for a Man, and 5 pound for a Woman, and as much for 2 Children above two and under 12 years of Age is Sufficient; yet 2 lb of Rice, 2 lb of Flower and 2 lb of Corn or pease per Week, per head is not Sufficient as not being thereby enabled to bake Bread which is the main Support of Health and

Life. Neither is the Allowance of 4 lb of Cheese and 2 lb of Butter per Quarter per head Sufficient if they actually had had Garden Roots or Eatables.[263]

It is our humble Opinion that should the Trustees for Georgia think fit to Settle the People of the new and last Transport near the Indian Hutt, the Town to be built there might go by the name of Ebenezer and would not discourage other Saltzburgers in Germany to come there and settle; but as long as these remain here none are like to be invited either by the first or last Transport.

On the 6th of this month the Reverend Mr. Bolzius proposed by a fine Speech to the people of the last Transport to work jointly on a piece of Land to be pitched upon near this Town, tho' attended with many Difficulties, Some of such as were present seemed to come heartily into it, and such as are gone to Savannah Town for some Provisions, are to be consulted at their Return, and then we shall pitch upon some place for that purpose. My next shall give you an account thereof, and of such other Occurrences, as may happen, I am with great Submission

Honoured Sir   Your most obedient and most humble Servant
John Vat

Ebenezer Feb. 10th 1734

Copy of a Letter from Mr. Chas. Purry London 24 February 1735. No. 13034.   Read 25 Feb. 1734/5.   To Mr. Newman.

Sir: Inclosed is a Duplicate of Capt. Lusk's Receipt for the Box directed to Mr. Quincy[.] I have paid all 8 s. for all Charges viz.

Freight _____	5
Primage to the Capt. _____	1
Portage & Wateridge on Board Ship _____	2
	8 Sh.

I have sent to Mr. Quincy under the Cover of Mr. Thomas Caston [Causton] another receipt for the said Box.

We hear from Carolina that on the 17th December Capt. Geo. Dunbar[264] arrived at the Barr of Charlestown and called there for a Pilot in order to enter Safely in the River Savannah.

I am Sir   Your most humble Servant   Chas. Purry

A Copy of the inclosed Receipt Received of Messrs. Peter Simond & Co.

a large Box directed to the Reverend Mr. Quincy at Savannah which I promise to deliver to Mr. Thos. Causton after my safe

Arrival (the danger of the Seas ex[cepted?] free of freight London february 22nd 1734/5.

James Lusk

A Translation. Copy of a Letter from Mr. Urlsperger dated at Augsburg March 27, 1735. No. 13078. Read 25th March 1735.* To Mr. Newman.

Dear Sir: I long very much to hear what the Court of Great Britain and the Society may have farther resolved in behalf of the Bohemians, having received fresh and undoubted Accounts how barbarously these people are dealt with, which I shall communicate in my next.

That the Carinthians now residing at Ratisbon have resolved to go to Georgia appears by the inclosed if the Society should think fit to receive them on the same foot as the Saltzburgers it would be necessary for the Society to procure and order of his Britannick Majesty to his *Embassy* Ministers at Ratisbon, that by their interceeding with the Imperial Ministers at Ratisbon these Carinthian Colonists for Georgia now residing at Ratisbon may be helped to receive their Wives and Children, as many of them as are left in Carinthia, and likewise if possible their Goods and Chattells.

By the last letters from Mr. Bolzius I see that these Ministers as well as their Congregation need very much the Society's and Trustees Advice Assistance and protection, of which Mr. Ziegenhagen will make ampler remonstration for I must conclude on Account of so many other Occupations and remain

&c. S.U.

P.S.: Mr. [Ph.] Von Reck has left us for Eight days and is gone upon some private Business of his own.

In the Conference of the Evangelick Body held at Ratisbon March 5, 1735. No. 13078.

(1) It has been resolved on account of so many religious Grievances relating to the Country near the Ens laid before them, to make Remonstrances of them to the imperial Ministers here by the Electoral Ministers of Saxony, and recommend this Affair to such Courts as have their Ministers residing at Vienna That Application may be made to the Ministry of the Emperor for the remedying of them.

(2) The Minister of the Elector of Saxony has reported that a Collection has been remitted hither out of Switzerland of

349 fl. 45 creutzers half of which to be given to the Saltzburgers returned from Holland and the other half to such
Emigrants as are most needy.

(3) Because the Emigration of the Saltzburgers for Religion's sake
still continues, and some of them want Assistance It has been
resolved to assist them out of the Cash (at Ratisbon) And
whereas amongst the Saltzburg Emigrants in Prussia and Hanover are a great many disabled, sick and helpless people, and
several of such are found and maintained in the Hospitals of
of the Imperial Cities of Augsburg, Memmingen, Hauffbearen
[Kaufbeuren] and Nordlingen, it is thought meet to take
21000 fl. out of the Cash for their benefit, and regulate the
distribution of it in the next Session.

*Extract of a Letter from a friend at Ratisbon [probably J. von
Reck] to Mr. Urlsperger Mar. 17, 1735.   No. 13078.*

In this letter Mr. Urlsperger is told

(1) That the Carinthians after mature deliberation have resolved
to go to Georgia.
(2) That they desire nothing else than that Application may be
made in their behalf for obtaining their Wives and Children
some of them being left in Carinthia they being willing to
forsake all the rest.
(3) A good Character is given of these people.
(4) 'Tis said that more application will be made in their Behalf
as well at Vienna as at Ratisbon.
(5) The following List of them is subjoyned

A List of the Carinthians now at Ratisbon who have resolved to
go to Georgia.[265]

Bartholomeus Globish his Wife & Seven Children	9
Christian . . . . . and his Wife	2
John Underwald his Wife and five Children	7
Clement Leitner his Wife and two Children	4
Gregory Koffler his Wife and one Child	3
Matthew Egenter his Wife and three Children	5
Nicolaus Neidhart his Wife & three Children	5
Simon Moser his Wife and four Children	6
Franc Santer his Wife & three Children	5
Matthias Auer	1
Lorentz Minchler	1
Rupert Glotz & Belthasar Glotz	2

Jacob Bingel, Geo. Krawol, Geo. Stegoberez 3
Andrew Geiciter, Caspar Walten, Rich. Mitterez 3
George Scheibler, Thos. Scheibler & 1 Child 3
Matthias Edez, Nic. Moser, Helena Nussen 3
Martin Ausserulammer his Wife & 1 Child 3
Vincent Recger [Reiger] 1

In all........ 66

*Copy of a Letter from Messrs. Bolzius & Gronau dated Ebenezer
6 Feb. 1734/5. No. 13112. Read 19 Apr. 1735. To James
Vernon, Esq. in London.*

Most honoured Sir: The great many Affairs you are pleased to
undertake for the Welfare of the Church and State made us formerly afraid to trouble You with our humble Lines, tho' the great
kindness and favour you have laid upon us at Dover, by writing
a very obliging letter to us was at any time fresh in our memory,
however the great many Benefits which are shewed to the Saltzburgers at London, and in their Sea Voyage by Your and other
Benefactors Care, constrain us to let you know by this our Joy
and humble thanks. Almighty and mercifull God who hath moved
your heart to Compassion to this persecuted people will never forget your work and labour of Charity. He will mercifully hear our
prayers for Your and other dear Benefactors Happiness and Welfare, and will give you that what he hath promised Psalm 41 v.
1. 2. 3. *Blessed is he that considereth the poor the Lord will deliver him in the time of trouble; the Lord will preserve and keep
him alive, and he shall be blessed upon the Earth; and thou wilt
not deliver him unto the will of his Enemies;* The Lord will
strengthen him upon the Bed of Anguish. Mr. Vat, under the Conduct of whom the new Saltzburgers came to us, employs now his
whole Care to settle them as happily as possibly he can and to look
very earnestly for a good piece of land in this Country since there
are very few good pieces about Ebenezer. The people have a great
desire to till the Ground, and perform after this manner the Intention of the Honourable the Trustees, if God grant them good
health and his Blessing to their works. Eleven people of the first
Saltzburgers that came over together with us, are dead, and one of
them still afflicted with Sickness and is almost past hope of recovery. Our Doctor Mr. Zwiffler is exceeding glad and we very much
surprized to see the great many Physical herbs and other Medicines and Utensills for distilling allowed to him by our most

Honourable Benefactors; in requital of which he promises to do his utmost Endeavour for the relief and recovery of sick people. We are told the salt Beef and Pork contribute very much to the Sicknesses that are so common in this Country. If the Honourable the Trustees would have the same Goodness as Mr. Oglethorpe in the former time to grant us some Oxen for fresh meat, we surely believe these Benefits would prevent several Infirmities. However we are fully satisfied with whatever the Trustees shall think fit. We intreat you humbly to present our most humble Respects to all Honourable the Trustees wishing them all manner of Prosperity for all Benefits bestowed up us and our flock and so entirely commending You to divine Conduct and Blessings, and us to[gether w]ith our flock to your favour, we subscribe our Selves

Most Honoured Sir   Your most humble Servants
John Martin Bolzius   Israel Christian Gronau

Copy of a letter from Mr. Bolzius dated at Ebenezer in Georgia 7 Feb. 1734/5.   No. 13113.   Read 19 Apr. 1735.   To Mr. Oglethorpe.

Most Honoured Sir: My Duty obliges me to render you my humble Submission and respect, by these lines. I am not ignorant your mighty Affairs give you very little leisure to read them over. I did the same the 16 July & 12 December last, which letters I hope are come to your hands. Eleven people of our small Company are dead, wherefore I am exceeding glad together with the Saltzburgers that You were pleased after your Generosity to send a new Body of their Persecuted Brethren under the Conduct of Mr. Vat, which came safely on shoare the 30 December[.] what Goodness and Benefits they have received by your order, and the good Care of Mr. Dunbar, You will hear from Mr. Newman, to whom I gave a short Account of it. I must return you a thousand thanks for all your favours and concerns for my Self and my Collegue as well as for our flock. Beseeching You to believe we have such a Sense of all your favours that we want [lack] words to express it sufficiently. What thanks and prayers the people put up daily to Almighty God for you and other great Benefactors I need not to tell you since you are ascertained of the Saltzburgers Godliness and tender love to you; Be pleased to assure your Self the longer the more that they fear and love God the more earnestly will they endeavour (as far as lies in their power) to till the Ground according to the Intent and will of their Benefactors. However I cannot forbear to

mention after my humble Duty that at present even so as formerly
the English and other people as often as they come to our place
talk very much from [about] the Settlement of the Saltzburgers;
They call our land Pine barren, where nothing else will grow but
Indian peese and [sweet] potatoes. Hence it is that the poor people
are some times disheartned by such talkings tho' we do our utmost
endeavour to encourage them by the holy word of God. Some
Acres about the [Ebenezer] River seem to be good, but they are
few, and some are covered newly by the high Water of the River
and Swamps[.] notwithstanding they have worked hitherto in the
Ground as much as possibly they could for the great troubles they
have had till this time in fetching their Provisions and other
things from Savannah which by modest Computation is no less
than 44 English Miles by Water. The people in some measure
cleared the River in such manner that if the Water is high they
can come up with a small boat to our Town, but if the water is
low they can come no further than within four miles of the town
to the Landing place; And the Currents of Savannah River from
Abricorn [Abercorn] Creek to the mouth of Ebenezer River are
so strong that the people with a small boat carrying about one
thousand pounds weight cannot perform the Voyage down and up
in less than 4 days time being obliged to land at night in such
places where they can have no Accomodations for refreshing or
resting themselves, unless they make small Hutts, and lie upon the
Ground, which in Summertime weakened their Bodies so much
that they very frequently fall sick especially wanting proper refreshments,[266] and in the Winter and wet Seasons the[y] suffer
very much by the Cold and rainy Nights, but the people is never
out of patience[,] it is a new Testimony of the tender Care of our
Benefactors for our best, that by their order Rolf [Roth] and his
Wife [Mary Barbara] were obliged to quite Ebenezer. There is no
Body in our Congregation that must not suffer several Importunities from them heretofore. This order came just at this time to
Mr. Causton as the said Rolf was willing [intending] according to
his open threatnings to kill treacherously two persons at Ebenezer,
of which wickedness Mr. Causton was informed by my letter to
him. After the aforesaid good order he is hindred to pursue his
wicked purpose. He intended to go by sea for Germany, but since
his Voyage was stopt by my Letter, that I must [had to] send
to Mr. Causton after my Duty and Mr. Causton's desire he shewd
together with his Wife a great Indignation against me with a

scornfull mein. The Behaviour of the Saltzburgers towards God and man gives us a great Satisfaction wherefore I hope no Body should blame me for mine Eagerness to see many more such people in our Congregation. My dear Colleague Mr. Gronau, Mr. Zwiffler and all the Saltzburgers present you their most humble Respects and due Acknowledgements for all your favours and benefits; and so expecting your Commands to do what may please you in all things I remain

Most honoured Sir   Your most humble Servant
John Martin Bolzius

Extrait d'une lettre de Gaisern [Goisern] le 13 Jan. 1735 touchant les Sujets Evangeliques persecutés dans l'Autriche Superieure. No. 13115.   Read 19 April 1735.

Nous confesseurs de la religion Protestante environ 350 personnes enregistes, etant restés depuis la derniere Sortie dans les terres Salinaires[267] de Sa Majesté Impériale vivons dans la plus grande oppression n' ayant ni Service divin ni les Sts. Sacremens, ni la grace ou bien vieillance de nos Superieurs. Par ordre du magistrate quelques uns ont vendue leurs maisons, et biens, avce beaucoup de perte. On ne veut plus batiser nos enfans. La Commission de la reformation nous a aussi defendu toutes les assemblees á prier, et a louer Dieu sou peine d'etre fait Soldat, et envoye aux tranchees et Sosseés [Fossées]. Si nous ne pouvons pas esperer la liberté de conscience dans notre patrie, nous supplions que Dieu veuille toucher les coeurs de Ses princes et de Ses elus, a nous Secourer avec conseils, intercessions, et pouvoirs, et á mettre fin à notre misere. Nous craignons aussi en cas de Sortie qu'on retienne les jeunes gens à les forcer au Papisme, que Dieu veuille empecher On veut debiter de nous beaucoup de demarches, rechutes &c. mais comme il arrive, dans le tens de la tentation quelques uns ne restent pas toujours fideles Pourtant l'eglise Evangelique accroit et s'augmente.

Aussi nous faisons savoir que Thomas Laserer etant parti pour Vienne y recevoir les Saints Sacremens est detenu à Lintz depuis la derniere emigration dans une dure captivité, et jusqu'à cette heure sans esperance de sa liberté[.] son compagnon Balthaser Daubler l'ayant sû s'est retiré peu avant qu'on a visité sa Maison pour le chercher querir. Nous prions pour l'amour de Jesus Christ d'avoir une chrétienne et efficace pitié de ces innocens prisonniers comme aussi de 5 autres nommes Georg Pains, Matthias Reisenbechler, Matthias Hübner, Michael Lohmer, Joseph Lohmer, Au reste nous souhaitions à jouir de la liberté Evangelique dans notre patrie, ou au moins de la permission à partir une fois l'année quelque part a communier; ou á nous rendre chés nos amis et parens en Transylvanie supposé qu'ils ont lá la liberté de conscience, ou enfin aller où Dieu nous conduire, pour qu'on nous donne les passesports necessaires, nous laisse aller, et emporter le peu de bien que nous avons sans en payer le Droit de rabat.

[Translation of the Foregoing letter, by George Fenwick Jones.]

Extract of a letter from Gaisern dated 13 January 1735 concerning the Protestant subjects being persecuted in Upper Austria. No. 13115. Read 19 April 1735.

We adherants of the Protestant Faith, about 350 registered people, having remained since the last emigration in His Imperial Majesty's salt lands [Salzkammergut], live under the direst oppression having neither divine service nor the blessed sacraments nor the understanding and good will of our superiors. By order of the magistrate some have sold their houses and goods at a great loss. We can no longer baptise our children. The Reformation Commission has also forbidden us all prayer meetings for worshiping God under pain of being made a soldier and sent to the trenches. If we cannot hope to have freedom of conscience in our homeland we pray that God might touch the hearts of His Princes and His chosen people, so that they might come to our help with advice, intervention, and power on our behalf and put an end to our misery. We also fear that should anyone depart hence that the young ones will be retained and forced to become Catholic, which may God prevent. People wish to spread many stories about us, our backsliding, etc. but in time of temptation some few do not always remain faithful. Nevertheless, the Protestant Church is growing.

We also make it known that Thomas Laserer, having departed for Vienna to receive the Holy Sacraments there, has been detained at Lintz since the last emigration at hard labor and even to this hour without hope of liberty. His companion, Balthasar Daubler, having known it, withdrew shortly before they visited his house to search for him. We beg you for the love of Jesus Christ to have a Christian and effective pity for these innocent prisoners as also for five others named George Pains, Matthias Reisenbechler, Matthias Hübner, Michael Lohmer, Joseph Lohmer. Moreover, we hope to enjoy Protestant liberty in our homeland, or at least permission to leave once each year to have communion somewhere or to betake ourselves to our friends and relations in Transylvania, assuming that they have liberty of conscience, or finally to go where God leads us, provided that [?] we are given the necessary passports and allowed to go and take the few goods that we have without paying any sales excises.

Relation de la Carinthie de 16. Feb. 1735 touchant les Persecutions et Oppressions de la religion Protestante.

Il arriva qu'on enmena de nouveau 43 garcons de Spaltrian [?] dont quelques unes furent fait soldats, les autres transportés nous ne savons par où. Lorsque nous lisons le Dimanche des livres Evangeliques, le juge est là, qui les arrache de nos mains, lorsque nous nous assemblons de nuit à lire, ou à chanter, les sentinelles sont postés qui nous saississent par voye des chiens. Une femme ayant niée qu'elle avoit les livres evangeliques a eté deshabillee jus qu'au chemise et chargee de 30 coups de baton. Dans le baillage provincial de Biberstein et Mühlstadt [Millstatt] on traine les jeunes à la milice, et les vieux dans une dure captivité comme aussi les femmes; les enfans sont retenus pour les faire Catholiques, et les remettre un jour dans les biens de leurs peres.

Au commencement on nous a permis à lire nos livres, nous ayant confessés à cela; et cette heure on nous les arrache, on les met au feu en sans en etre consumés on les hache, coupe, et taille en pieces. Qu'ils l'ont reiteré trois fois en ayant encore une grande quantité a massée de rester.

[Translation of the foregoing account, by George Fenwick Jones.]

Account from Carinthia, dated 16 Feb. 1735, concerning the persecution and oppressions of the Protestant religion.

They have again taken away 43 young men from Spaltrian, of whom some were made soldiers, the others transported we don't know where. On Sundays, when we read our Protestant books, the judge is there and snatches them from our hands. When we gather at night to read or to sing, the sentries are posted and seize us with their dogs. A woman having denied that she had Protestant books, she was undressed down to her chemise and punished with 30 blows with a stick. In the provincial bailiwick of Biberstein and Mühlstadt they drag the young people to the militia and the old ones to hard labor, as they do the women too. The children are retained to be made Catholic and to be reinstated some day with the goods of their fathers.

At the beginning we were allowed to read our books, having confessed to so doing; now they are snatched from us, put on a fire and before being so much as burnt are hacked at, slashed and cut into pieces. They have repeated this action three times and still have a great pile left.

Extract of the Amsterdam Dutch Gazette translated by Dr. Guerdes viz. No. 13116. Read 19 April 1735.

Ratisbon April 14, 1735.

We are credibly informed and assured by letters that the Roman Catholick Clergy, and at their Instigation the Bailiffs of several great Provinces, and among them particularly in Carinthia, are continually going on in a very hard and astonishing manner to oppress all such as favour the Protestant Religion that they might, if possible, extinguish all light shining unto these people.

And if the unaccountable practices are such in reality as they are related, it must be allowed that they are far beyond any thing that hath hitherto happened any where else to the Oppression of these poor people, which become therefore the most worthy of Compassion as their Sufferings are such as hitherto has not been heard of even amongst Turks and Heathens.

They have renewed their request with the Evangelick Body for an Intercession with his Imperial Majesty that they may have leave to emigrant.

Translation of a Letter from the Reverend Mr. Urlsperger dated Augsburg March 7th 1735. No. 13126. Read 29 April 1735. To Mr. H. Newman.

Dear Sir: Tho' I have nothing particular at present to write to you, still expecting your answers to my Letters since February the 14th yet I would not forbear sending you a Translation of the Inscription of a screwed Medal concerning the East India Mission lately published here, and deserved as well to be seen as those about the Saltzburgers. Value 3½ florins.

An Engraver here being resolved to engrave the Indian King and his Nephew,[268] who lately have been in England, desires me to be informed, whether the Figures upon the Indian Kings Face and Breast, which appear so abundantly in the print published in

England, are natural marks, or made by painting or incision. The Answer to this be pleased only to communicate to Mr. Degmair.

Neither the Plan nor the Map of Georgia is come to hand. My Sickness continuing now 11 days hinders me from writing to Mr. Degmair to wish him a good Voyage. The Accounts continue that many in Saltzburg are resolved to leave their Countrey, but when and in what manner it is unknown. I remain

Dear Sir &c.   Samuel Urlsperger

Translation of a French Letter from the Reverend Mr. Urlsperger dated Augsbourg 28th April 1735 N.S.   No. 13133.   Read 29th April 1735.   To Mr. H. Newman.

Sir: I cannot sufficiently express my Joy with which the Compassion of the Honourable Society, signified in your Letters of the 15 and 18 of the last month, towards the persecuted Bohemians, has filled me; At present we must wait for an answer from the Court of Vienna. Several of the publick Ministers at Ratisbonne desire that a powerfull Intercession may be made in their Behalf at Vienna, by the Crown of Great Britain, and by other powers in Alliance with it, in such a manner that it may be evident that they have this Affair very much at heart.

As soon as I had the honour to be informed by you, that it was designed to extend the same favour to the Carinthians and others, who have quitted their native Countrey for the Sake of Religion, as to the Saltzburgers, I thereupon wrote to Ratisbon to know certainly the Resolution of the Carithians as to the Assistance offered for conducting them to, and settling them in Georgia, and directed them to send to me two Deputies in the name of the rest. The two Deputies are arrived with their answers; They humbly accept the Protection of his Britannick Majesty, and the Charitable Terms granted to them as well as to the Saltzburgers in Georgia, adding only this, that they most humbly intreat His Britannick Majesty will be pleased to obtain that their Wives and Children may be suffered to go with them, without whom they cannot depart from hence. I have conferred thereupon with the Envoy Mr. [J.] de Reck, who is at present here, having assisted at the affiancing of his Son, the Secretary of the Embassy, which was cellebrated here. He is of the same Sentements, that the said Carinthians being engaged as Colonists to his Britannick Majesty, that the Gentlemen who are Trustees will in Compassion procure his Majesty's orders to his Ministers at Vienna and Ratisbon to treat on this Matter with those of his Imperial Majesty, that their Wives and Children may be suffered to come to their Husbands and Parents. I add in order thereto, the number of the Carinthians consisting of 77 Persons; of whom there are 35 Men; married 11; Boys 24; Women 15; married 11; and 5 Daughters, and 27 Small Children. They have amongst them, 6 Masons; 3 Carpenters; 2 Taylors; 6 Linnen Weavers, the others are all Peasants. They confirm the Account, which I have had the honour to send you from time to time; and say their Compatriots at Ratisbon are firmly attached to the Evangelical Religion and were forced to quit their native Countrey by Stealth lest they should have been made Soldiers or Prisoners with their Brethren, Nevertheless they had a free passage and were graciously received at Ratisbon, notwithstanding the Imperial Embassy residing there. The Advertisement published here, I send you. For want of Accounts from Saltzburg I have no news of the motion or Tranquillity of that Archbishoprick. What you are pleased to signify in your kind letter with regard to a separate Account of the Carinthians and Saltzburgers shall be punctually observed. My most humble Complements to the Honourable Society. I am

&c. Samuel Urlsperger

P.S.: Pray acquaint Mr. Ziegenhagen that the worthy Dr. Ramback is dead suddenly at Giessen of a malignant Fever. I have received the letter from Mr. Degmair of the 4 Instant and I have obtained Permission for him to stay 2 or 3 Months longer in England to make himself a better Master of the English Tongue. He has only God to trust in who will take Care of his Subsistence.

Translation of Letter in French from the Reverend Mr. Urlsperger dated Lindau 12 July 1735. No. 13254. Read 22 July 1735. To Mr. H. Newman. (NB This Letter is entr. here by Mistake it should have been in fol. 174 according to the Date.)

Sir: I am now at Lindau upon the Lake of Constance taking the Mineral Waters for restablishing my health. I lodge in the Countrey at Mr. Riesch, who is the Minister who sent from hence to the last Transport of the Colonists of Georgia the best People, of [from] whom I have had the pleasure to understand with great Joy that the Suisse at St. Gal [St. Gallen] have testified to our Emigrants so much Charity, besides giving them good Examples.

Sir You will learn by the enclosed from Mr. [Ph.] de Reck (which I have already answered) what he has wrote to me here, and as the 30 Austrians are ready to be transported, I expect by the first post, a Specification of the persons to regulate the Subsistence of them. I am very sorry that for so long a time, I have not heard whether his Britannick Majesty would most graciously intercede in behalf of the Carinthians, or no, for their being allowed their Wives and Children, because the Transport is thereby very much retarded, and things being thus uncertain I am asked by the Post of this day from two of my Correspondents whether 50 or 60 Bohemians, who have lately left their Country and understand the German Language and who are of the Augsburg Confession may not resolve to go with the Austrians.

Besides that I must acquaint you Sir I have taken up from Mr. de Münch 500 florins to furnish them with Necessaries that come from time to time, and to reimburse what I have taken from the General Cash for the Emigration towards the Expence of the Georgian Colonists, which general Cash for Emigrants has been almost quite exhausted, chiefly by the Young Men who have been bound out Apprentices, and for diverse other Saltzburgers who have suffered by Sickness, and of whom, some of them in working, have broken their Legs and Arms, concerning whom I shall be more particular if it please God, at my return to Augsburg being

Sir   Your most humble Servant   Samuel Urlsperger

Translation of the Extract of a Letter from Monsieur [Ph] de Reck at Ratisbonne of the 5 July N.S. sent from Mr. Urlsperger to H. Newman.   No. 12254.   Read 22 July 1735.

I have the Honour to inform you in answer to your Letter of the 30th of last month that for certain 30 Austrians are always ready to depart hence. They will find here Work and Subsistence till St. James's Day, but after that they must depart or be subsisted by the Society; these honest people and true Christians are as little disposed as I am, to change their resolution for going to Georgia.

Mean time the Grand Bailiff of Carinthia and the Country about Ens has sent a long Writing to the Imperial Embassador here, which being in terms very severe against the Evangelick Religion and the Evangelick Body at Ratisbon, is taken for an Answer to the Griefs proposed, and contains at the same time a fanifest [manifest?] declation [declaration] of the Emperor, that he will not permit[269] any Emigration from his Estates nor even any private Exercise of Religion.

*Translation of a Letter from the Reverend Mr. Urlsperger dated
Augsburg, 9 May 1735. No. 13165. Read 27 May 1735. To.
Mr. Newman.*

Sir: Here I communicate to you what hath been wrote to me
from *Vienna* and what I have wrote to *Ratisbon* on that account.
Likewise a Copy of a Letter which I received this day from Mr.
John Tobler Mathematician in Rehetobel in the Canton of *Appenzell* in *Swisserland,* beseeching you to deliver the same to the
Trustees, whom I most humbly desire to send an Instruction to
Mr. [Ph.] von Reck what Answer he must give to the people who
desire to go to *Georgia* and are no Emigrants, especially to the
above mentioned that have wrote to me, and to those who have
wrote to me from *St. Gal.* It is impossible for me to mind any other
Business, besides that of the Emigrants; but if I am able to give
good advice I will do it with all my heart, and serve Mr. Von Reck
in his Correspondence with the Trustees. Particularly Mr. von
Reck should be informed, whether *Georgia* is really so good and
fruitfull a Countrey as hath been wrote about it 2 years ago to the
end that he may give a good account of it to the People in *Swisserland.*[270] I would fain take the Correspondence with the Trustees
upon me, but am afraid of undertaking more than I am able to
perform.

I remain   Sir   Your most humble Servant   Samuel Urlsperger

*Copy of Mr. Urlsperger's Letter to the Envoy Mr. [J.] von Reck
dated 19 May 1735. No. 13165. Read 27 May 1735.*

Honoured Sir: I take the liberty to acquaint your Excellency
with what hath been wrote to me from *Vienna* the 4th Instant in
the following terms: Concerning the pressed Brethren in
B[ohemia?] a forcible order is come to the Embassy from England
and Hannover, but having annexed no *Speciem Facti* or the like,
whereupon to ground the Representation, nothing hath hitherto
been done in the Case. The Hannoverian Envoy is today gone
from hence to Saxony, and will be some weeks before he returns
hither, Consequently very little will be done in this Affair at present. I have also spoke with the Danish Envoy about the Business,
who declared himself to make it *communem causam* together with
the others; but it is to be wished these Gentleman would earnestly
undertake the Affair, and to be pressing when they make Representations, and not to be put off with equivocal and insignificant

Answers. God Almighty help the miserable People and come to their Assistance for they have little hopes for help from Mankind. I shall not fail to put the others in mind of the Business. I return you hearty thanks for the printed Advertisement sent me. The Valliant and Christian General Mr. von *Wuttgenau* hath also let his light shine here, and I have had the Satisfaction to pass many hours with him with edifying Discourses. God hath been pleased to visit him with a dangerous Sickness here, from which he is not yet quite recovered; however he hopes to be able to set out from hence within 10 or 12 days for *Mantua,* and then he will take his leave of you.

By this you see, how necessary it is, that a Copy of the Memorial from Bohemia and of the Writings annexed from Ratisbonne be sent to the Hannoverian Embassy at Vienna to the End that, at the return of the Hannoverian Envoy at Vienna, the Cause may [be] immediately removed etc.

A Copy of a Letter from Mr. John Tobler, Mathematician in Rehetobel in Appenzel dated the 6th May 1735.   No. 13165.   Read 27th May 1735.   To Mr. Urlsperger.

Having with great pleasure and Satisfaction read several Books and Writings concerning Carolina causes me to write these lines to you, which liberty I would not have taken if Mr. Zublin[271] in St. Gal had not encouraged me by praising your particular Goodness. And some persons here having a desire to go to Carolina I beg the favour of you to let me know whether they can't go thither with the Saltzburgers, and whether they can go from Augsbourg under the same Pasport or not; likewise when or about what time they go thither, to the End that these people here may know what to do in this Case, which is of no little moment; and I should think it a happy Voyage if they could go with so good People. Concerning the people who at present are resolved to go thither, I believe they will be no hindrance to the Colonists in Georgia, but very serviceable, they being well skilled, not only in breeding Cattle, but also manuring the Ground for Flax, especially in manufacturing all sorts of Linnen Cloth, and other arts, and in sowing and planting of Fruits. How many will go thither I cannot specify yet, but will do it next time if required, however 30 or 40 Persons will prepare themselves for it. There would be more, but some cannot afford the Charges and some lead such a life here, that it is better to go from them than with them. For my part I cannot go this year for reason of some Difficulties which happen yet, but my Father and my Son are resolved, God willing, to go, and if they do well, as I hope, next year a great number of brave people will go, and the number would very much encrease, if the King of England would give some of them the free Passage only over Sea, or lend them the money till they can pay it again. Some of them, among which is my Father, take poor people with them and pay their Charges, for which they shall serve them till they have worked it out. In short they want nothing more to undertake this Voyage, than the Ordering and Recommendation by a Gentleman who is well acquainted with this Affair, and I know no person more proper for it than you good Sir, if you would be pleased to do it. I would have wrote also to Mr. [Ph.] von Reck if Mr. Zublin had not told me, he was gone on a Journey. Concerning the Religion, it is here the Reformed, but we will make no Scruple of living with the Lutherans, because we have here many Books of theirs which we read with great Edification. As to the Almanack, we use here, as they do in England, the old Julian Almanack, as the inclosed, which is my Work, sheweth, made from the Year 1722 till 1751 however, as you see, I have had no other Master for my Assistance than God Almighty and what I have found in Books on this Subject.

Extract of a Letter from the Reverend Mr. Urlsperger dated Augsbourg 16 May 1735. No. 13165. Read 27 May 1735. To Mr. Ziegenhagen.

Because the last Suissers had occasioned great Complaints in Holland, their High Mightinesses have ordered their Residents, and amongst these the Resident at Ratisbon, to notify every where, and consequently to the Commissary Mr. [Ph.] von Reck, that they would not permit any more Colonists to pass Their Territories, except they are provided with an authentick Passport, and can travel thro' without giving any trouble to the States: Whereupon I wrote to the Dutch Minister, de Galliers, that the two Transports of the Colonists of Georgia had not given the least trouble to the States, nor was any such thing to be feared of a future Transport; but if His Excellency would be pleased to intercede with His Principals for obtaining a free Passage for the Emigrants, as Colonists of Georgia without paying any Toll or being unnecessarily detained, considering they carry nothing with them but their penury, it would be very thankfully acknowledged. To which

His Excellency was pleased to return this Answer; That as to the free Passage, there would be, in his Opinion, no more required than that the Trustees and the Society did endeavour to represent their Case to Their High Mightinesses in few words by the English Envoy at the Hague not doubting, but it would be taken into Consideration, he himself having humbly recommended it.

Translation of a Letter from the Reverend Mr. Urlsperger dated Augsbourg 19 May 1735. No. 13175. Read 27 May 1735. To Mr. H. Newman [from Ph. von Reck].

Sir: Upon my Commissions given to the Commissary Mr. [Ph.] Von Reck, who is at present in Ratisbonne, he sent me the following Answer, dated the 17th Instant.

I. Concerning the Saxon Envoy Mr. von Schönberg. This Gentleman hath often advised the Carinthians who are here to give a memorial to the Imperial Embassy in behalf of their Wives and Children left behind them, which out of too great fear, they never would do: Wherefore I offered my Self not only to draw up a Memorial, as Lit. A. sheweth and get it signed by the CARINTHIANS but also to deliver it for them, with which his Excellency [von Schönberg] was well pleased, and promises himself a good Effect from it. As to the maintaining the CARINTHIANS who are very poor and have no work to get their Living by, it is thought proper, not to maintain them out of the Emigrants Cash, for fear they should grow idle, and have a mind to stay here; but if they would go to Georgia, the Saxon Envoy would procure them a considerable Viaticum of money.

II. Concerning the Electoral Brunswick Envoy Mr. von Hugo. In respect to the CARINTHIAN Wives and Children left behind he is of the Opinion with the Electoral Saxon Envoy [von Schönberg] and promises to second my Memorial with a Representation by word of Mouth. He, as well as the Electoral Brunswick Envoy at Vienna Mr. von Erff, hath got a Rescript from Court, concerning the B.B. [Bohemian Brothers?] according to which he will do his utmost Endeavour for their best, and send the BOHEMIAN Memorial to Vienna.

In case the Envoy Mr. [J.] von Reck[272] should die, he [von Hugo] will be very glad to correspond with you Sir, as well in Affairs concerning Religion as that of Georgia, likewise His Excellency will have an Opportunity to send your Letters along with the Kings Pacquet to London.[273]

III. With regard to the Envoy from Holland Mr. Gallieres. He assures me that in the *Bohemian* Affair, he entirely concurs with the rest of the Envoys; and that by the last post he had sent to the High and Mighty States General a very forcible and moving Representation in favour of the B.B. which he does not doubt will have a good Effect. He together with some other Envoys doth not only think it proper but highly necessary that the BOHEMIAN Memorial be printed and published in England without delay because it is intended to do the same in Holland. Their High Mightynesses assure the Trustees and the Society of their Assistance in this Affair, and would by the Help of the King of England endeavour that when, as is expected, a Peace shall be concluded with France, a particular Article may be inserted in favour of the B.B. The Crisis of the present time being so favourable that either one must make an advantage of it or by neglecting such an Opportunity renounce his Right almost for ever.[274]

As to the March of the GEORGIAN Transport thro' Holland it would be acceptable to their High-Mightinesses if the Honourable Society or the Trustees would give notice to Mr. Dayrolles,[275] which would contribute very much to a more easy and speedy Journey for us. This Week, God willing, I shall take an opportunity to speak with the Electoral Brandenborgh-Danish-Swedish and other Protestant Envoys.

If anything shall be done in the BOHEMIAN Affair, the Envoys here are of Opinion that the Grievances and the redressing thereof be represented to the Emperor in a particular [private] Audience.

The Privy Counselour Mr. Goebel[276] assures me that in the Berchtolsgaden District are still above *100 Emigrants,* and that he expects to hear within a fortnight the time fixed for their departure.

Memorial of the Carinthians referred to in the foregoing Letter.[277]

(A.)
High and wellborn Free Lords of the Empire.
Gracious Lords: Your Excellencies's praiseworthy Clemency and Commisseration towards all miserable people, causes us also in our Affliction, most humbly to seek our Refuge by you. For whereas we have for the sake of liberty of Conscience, left our Country, Effects, Wives and Children, lived here for a while, and now are

obliged to proceed in our Pilgrimage into other Protestant Countryes, which is very hard as well for our Wives and Children as for us; We most humbly beg your Excellencies graciously to consider our miserable Condition, and to grant that our Wives and Children may follow us, and that we may get somewhat of our Effects left behind us to bring us to our Journey's end. Which act of Compassion the most gracious God will reward, and hear our Prayers for your Excellencies Welfare. In hopes of Your Excellencies granting us our desire, we remain with all Submission,
Your Excellencies,   Most humble and most obedient Emigrants
from CARINTHIA

Extract of a Letter from Ratisbon dated 17 May 1735.[278]

The Dollar your Reverence sent me, to which I have begged another from the Emigrants Cash, which makes in all 3 Guilders, shall certainly be delivered by the first Opportunity into the hands of [Hans] Lerchner the good Saltzburger now in prison in Raab in Hungaria, to whom I sent a while ago some Guilders. I have also procured 8 Guilders for honest Simon Sigel, from K. [Kärnten, Carinthia]) who is likewise at the same place in prison, and as it is said for his life time, which he hath received just in the time when he was in the greatest misery, and as he himself mentions was ready to starve[279] of Cold for want of Cloaths so that no body could know him. I knew him, because by his Letter from K. he was the first that told me of God's wonder [miracle] which happened at the said place, of which more might be said. I pitty him with all my heart, God send him strength and Comfort, and give him grace for his Faithfullness. I have heard good news from Holland, concerning the Emigrants in CADSAND [Cadzand]; those who are there still, thank God and do not desire to go from thence. Several of them have bought themselves necessaries. They have now a Church of their own, and got a house for their Minister Mr. Fisher, God give his Blessing to his Word in their Souls! Many are very well placed in the HANNOVERIAN Countries, some return, the Artists are gone to Nurnberg. Just now the CARINTHIANS were with me and signed a Memorial concerning their Wives and Children, which tomorrow will be delivered to the most excellent Imperial and Austrian Embassy, God grant it a happy Effect! Mr. Von Reck will give you a farther Information. Here follows a Specification of the Age, Names etc. as far as I could be informed by those that are here, especially of those that are married. Several are gone to Anspach to work; In Ratisbon it is impossible that so many Emigrants should get work, however as much as possible.

A further Specification of the names Age and Families of the Emigrants from CARINTHIA, & their Relations.

From the Jurisdiction *Biberstein.*

(i) Frantz Sandler, a Master Linnen Weaver 36 years his wife Brigitta of the same Age, they have 4 Children.

    1. Maria 14   3. Ursula 7

    2. Matthias 10   4. Eva 3 years old.

(ii) Nicolaus Neidhart, a Master Taylor 42 years, his Wife Maria 26 years, and 4 Children.

    1. Matthias - 8   3. Balthasar 3

    2. Simon - 6   4. Caspar 1 year

(iii) Christian Steinacher, a Bricklayer 52 years his Wife Margaretha 43 years and one Child.

    1. Elizabeth 6 years.

From the same Jurisdiction *Biberstein,* three Single Women, namely

    1. Magdalena Anna Weinin 25 years

    2. Maria Sublin - 21 years

    3. Catherina Sieblin - 16 years

From the Jurisdiction *Mühlstadt* [Millstatt]

(I) Matthias Egarter, a Countreyman 34 Years, his Wife Susanne 27 years, and one Child.

    1. Christiana 6 years.

(II) Gregory Rochler, a Countryman 32 years, his Wife Lucia 27 years and one Child.

    1. Maria 1 year.

(III) Clement Leidter, a Countryman 48 Years, his Wife ＿＿＿＿ 44 years, and 2 Children.

    1. Maria 13 years & 2. Maria 6 years.

(IV) Simon Moser, a Master Lennen Weaver & Bricklayer 43 Years his Wife 39 years, and 3 Children.

    1. Maria 18 years. 2. Christina 12 & 3dly Afra 6 years.

(V) John Unterwald a Countryman 49 years, his Wife Maria 47 Years and 5 Children whose names are yet unknown.

(VI) Bartholomäus Globischeig a Countryman 49 years his Wife Christina 45 Years and 7 Children, whose names are unknown, the Father being in Anspach at work.

(VII) Johann Egger, a Countreyman 53 years, whose Wife and Children never owned themselves Protestants.

NB. The Wives and Children are yet in Carinthia as I have signified in my former Letter.

Sir I expect an answer upon this as well as my former Letters, as soon as possible, because they contain weighty Affairs, and hereby acquaint you that at last I have got leave for Mr. Degmair from his father and the protestant privy Counselor here, that he may stay 2 or 3 Months longer in England, which I have done for reason that he may make himself more perfect in the English Tongue and afterwards be usefull to me in the Correspondence to England; But because his father's Circumstances, he being but poor and since some years sick and having many Children, will not permit to maintain him any longer, and the Protestant privy Counselor, upon my Request having giving him 200 Guilders I am obliged to acquaint the honourable Society with the Circumstances of this young Mr. Degmair, of whom one may promise ones Self much Good, and therefore recommend him to the Charity of the Honourable Society, humbly beging Pardon for taking that Liberty.

I remember very well that upon the desire of the Society, I have promised to give a further Account of the Demarches and Views of Count Zinz; but because it cannot be done now, it shall perhaps be done in my next. Since 3 Weeks ago things have happened which are not to be allowed. I most heartily recommend the Saltzburgers in Eben Ezer and remain

Sir   Your most humble Servant   S.U.

Copy of a Letter from the Reverend Mr. Urlsperger dated Augsburg 2 Jun. 1735. No. 13188. Read 3 Jun. 1735. To Mr. H. Newman.

Sir: I write and give notice to the Honourable Society in answer to theirs of the 29th April and 6th May by the hands of a Merchants Clerke, whom God hath truly converted through the Service of my Son in law [Wenndrich] when he was still Preacher in Venice, and who being now on his Journey from Venice to Hamburgh, where he has an English Mother and German Father, has during these Whitsun-holdays lodged in my house.

(1) That I shall send over the 25 Screw Medals of the Protestant Mission in the East Indies, as soon as possible

(2) 'Tis impossible to write whether any *Saltzburger* out of that Archbishoprick may come this Summer or not, altho' 'tis known for certain that more than a thousand have declared themselves Protestants, but that it is

(3) Rather to be expected that one hundred of Berchtolsgaders will go out

(4) That altho' this were not, yet will a Transport to Georgia of an hundred Persons meet together mostly out of CARINTHIA with some SALTZBURGERS and OSTERREICHERS [Austrians], provided those of Carinthia, as I have already mentioned, can get their Wives & Children

(5) That the Dutch Resident at Ratisbon Mr. de Gallieris let me know again by a third person, that it would be very good for a new Transport to Georgia if the Society or Trustees would by means of the British minister at the Hague Mr. de Ayrolles[280] procure a Pass for them from the General States.

As for the rest I have read the Society's printed Circular Letter with great Joy, since I see thereby that God is with this Society, and because God is with it, all future Undertakings towards the Promotion of the great Gospel, whereof I preached last Monday out of St. John Chap. 3 v. 16 must for certain have good Success, which I likewise wish and pray for, and am also heartily willing to sacrifice all that is in my power in behalf thereof.

You find here Extracts of two Letters, the first one, of the Privy Counsellor Mr. Van Herold, and the other of the Commissary Mr. [Ph.] von Reck to which I add that after the Sermon on Whitsunday, where in I had made mention of our *Protestant East India Mission*, and as I was going out of the Church, a topping Person, whom two days before, I had communicated the 37th Continuation of that Mission,[281] squeezed five and twenty guilders with many tears of Joy into my hand, with the proviso to keep the name concealed. The Lord be praised who works in England and Germany, that the hearts are moved to administer both great and small Gifts, and that, because God hath given them his Son. I remain

Your etc.   S. Urlsperger

Extract of a Letter from the Prussia Privy Counsellor Mr. Van Herold to Mr. Urlsperger, dated Berlin the 17 May 1735. Viz.

In the interim I have not omitted directly to consign unto the Elders of this poor Church of the BOHEMIAN Brethren, the fifty Guilders that were sent me, and the adjoyned receipt which themselves have set to paper, will more particularly shew, with what hearty thankfullness they received this unexpected Gift. The said Church is already encreased to upwards of 600 Souls, and continually still more appear in secret; the most of them are pious and labourious Persons, and as his Royal Majesty [of Prussia] has in general most graciously committed the Direction of the Established Collonists in the new Frederick Town to me, so hath he likewise, even since last year, and also when His Majesty was in the Campagne, particularly enjoyned me to take care of these people in the very best manner. Now because their former Preacher Liberta has been arrested in Saxony, and likewise this poor Church with their many Children that are not yet brought up, going astray, I with the help of Professor [G. A.] Frank in Hall, have in the intrim got a pious Student out of Hungary for them, who is also pretty well versed in the *Bohemish* Language, and takes Care of the poor Children, the number whereof almost amounts to Eighty, that they are instructed in Christianity, and the German Language, where in 2 grown Bohemish Scholars assist him, and since the Clearing of the aforesaid Liberta seems yet so far off, I am in hopes to obtain the permission of his Majesty to get them for a Preacher a certain fit and pious man, whose name is Macher and is at present in the School in Cottbus, he understands also Bohemian well. The good people were much rejoyced when they heard by the Relation given them by you Reverend Sir what has been resolved in England in regard to their Affair, and I assure you my most gracious King and Lord will withall efficaciously officiate in this affair. With the SALTZBURGERS that are settled in Prussia thank God it goes better than has perhaps been reported with you; they are reputed as pious and diligent people in their work, seeing their Preachers that have been given them so well as those they serve, testify the same of them; This only is to be pitied that they have been brought, I know not by whom, to a Notion as if the Archbishoprick of Saltzburg must pay them for what was left behind them according to their Tax, and as in the present Conjunctures it proves very difficult in the Sale of the Goods, to obtain hardly half so much as what they have vallued them at, most of them seem ill pleased therewith, and refuse to give up the Documents they have in hand to this purpose, wherefore one hath much to do with them, tho' many indeed submit themselves, seeing that his Majesty the King useth all pains & diligence for them and that the Case cannot possibly be pushed farther whereas for want of inspection their good would only decline more and more in price.

Extract of a Letter from the Comissary [Ph.] von Reck to Mr. Urlsperger, dated at Ratisbon the 24th May 1735 etc.

The 21st Instant two CARINTHIANS delivered their Memorial to the Embassy of Austria[.] Mr. Concommis Saxy Baron of Joddci has not let them come before him, but sent however for the Memorial and returned it with the answer "He would not interpose for them, they were perhaps gone astray, they must have recourse to the Baron de Palm as Embassador of Austria," which accordingly they did, who let them come before him, took the Memorial, read it over, kept it, and gave for answer. "He would write to the Captain of the Land thereabouts,[282] that still an Emigration of some few would not be hindred, but by no means permitted or allowed to many, because the Land would Suffer a very great damage thereby."

Translation of a Letter from the Reverend Mr. Urlsperger dated Augsburg 2 May 1735. No. 13191. Read June 10th 1735. To Mr. H. Newman.

Sir: I hope my last letter, dated April 28th You have received. I desire only to deliver those three pieces to the Honoured Gentlemen, Mr. Oglethorpe and Mr. Vernon, as also to acquaint the Honourable Society of the Contents of them.

I have lately mentioned to Mr. Labhart Merchant in St. Gallen that he might directly send his Project to the Honourable Trustees or to one of them Mr. Oglethorpe in the French Tongue; but I wrote to him at the same time, preliminary

(1) That the Honourable Society and the Trustees had promised to take Care of the Saltzburgers that they might always live together by themselves, consequently never mixed with other People.

(2) That I hardly could believe that the Honourable Society would grant all that to the 30 Men of St. Gallen which they have done to the Saltzburgers and such like because it is quite an other thing with this people.

(3) They should not think Eben-Ezer to be so great a Town as to set up now already Contoirs [business establishments] for great Merchants.

(4) That it methinks would be much more convenient to settle their first Transport on a particular Place, tho' it might be in the Neighbourhood of Eben-Ezer, because they designed to send some hundred Persons of St. Gallen after.

(5) That, if the sending over of the Swissar Colonists should meet with good Success, I might assure them that they would not be hindred in their Exercise of the reformed Religion common amongst the Swissers.

(6) That I besides had to acquaint them that as much as I know, only such Fabriques and Manufactures would be allowed to be erected in Georgia which may not hinder those in Great Britain.[283]

Now you will see what the abovementioned Mr. Labhart with his Companions will write to Mr. Oglethorpe. I only would mention this at this time to the Honourable Trustees a little before, that they may have time enough to consider of this matter.

Mr. Commissary [Ph.] von Reck is sent by me to Regenspurg to perform some Business there according to my order. To conclude I commend you to the divine Grace and Protection giving my humble Service to the Honourable Society remaining.

Sir   Your most humble Servant   S. Urlsperger

Translation.   The following three Pieces are desired to be communicated to the two Honourable Gentlemen Mr. Oglethorpe and Mr. Vernon.   No. 13191.   Read 10 June 1735.

An Enquiry and Proposal of the Senior Minister Mr. Urlsperger in behalf of Commissary [Ph.] Von Reck.

My last letter directed to the Society takes more particular Notice of the Stability of the Resolution, the Carinthians (who are at Ratisbon) have taken of going to Georgia and that as soon as they shall by the means of England have their Wives and Children (still remaining in Carinthia) Returned to them. And as there is a great likelihood that this Summer a fresh Transport will set out for that Colony under the Conduct of Mr. [Ph.] Von Reck, because the number of the Carinthians alone amounts to 78 persons, I had a mind to sound that Gentleman whether he might not be disposed to remain in Georgia for good, in case the Trustees should think fit to allot him a certain quantity of Land, and to give him a place with a Salary or only a Pension sufficient for his Subsistence untill he could put himself in a way to do without?

To which he answered: That in Case he had a Call from the Trustees and necessary Provision was made for his Subsistence, he could gladly resolve for the sake of his fellow Creatures to stay during Life in Georgia to continue in the Service of the Trustees, and to employ what little Fortune he has of his own in that Countrey.[284] Wherefore I beg the two Honourable Gentlemen Mr. Oglethorpe and Mr. Vernon would be pleased to take this Affair into Consideration, and if they should think it may be brought to bear, lay before the Trustees at their Meeting in order for their speedy resolution that the said Mr. Von Reck may be able to govern himself accordingly, and get such of his Effects in readiness against the departure of a Transport as may be proper to be carried with him from Germany. For my part, should this able, brisk, couragious, disinterested, serviceable and pious Man be employed by the Trustees and engaged to continue there, it is my opinion it would be of very great Advantage to the Colony. For he is equally qualified to be of use in Spiritual as well as in temporal Concerns. His Uncle the Publick Minister at Ratisbon [J. von Reck], who was lately here, and with whom I had some Discourse about this Affair spoke to this Effect: "I know not which way my Nephew can serve both God and man better, and if the Trustees should require him, I would gladly resign him for ever on so good an account."

Translation of a Copy of a Letter from Mr. John Henry Labhart Merchant, dated St. Gall April 25th 1735. No. 13191. Read 10 June 1735. To Mr. Senior Urlsperger.

Very Reverend Sir: I take the Liberty to acquaint you with my humble Thoughts concerning the intended Transportation of some People for Eben-Ezer in Georgia. A Set of Gentlemen sent last year 3 of our Citizens upon their Charges to Purisburg, who were joyned by two others. This was all entirely done pursuant to the Invitation, and promises of Colonel Pury sent to me. He promised each man should have 50 Acres of good Land given him for ever, besides 1 Acre to build upon in the Town for the habitation of Self and family, together with the necessary Subsistence for one year. The sending those Persons was with a view of becoming able to represent this year to our gracious Governours with good Grounds the Possibility, and Facility to find out ways and means of providing in a more plentifull manner for the poorer sort of our Inhabitants, and at the same time ease the publick Treasury of very considerable Burthens thence arising. But neither of these

3 Men (contrary to expectation notwithstanding the repeated Charge and Instruction they had given them) having hitherto sent the least Account, it is to be feared the Success has not answered the Prospect either thro' their own or Mr. Pury's Fault. Thus the proposed End remains unanswered and some of those Gentlemen who have disbursed money towards it are quite disheartned, others on the Contrary cheered up by the report of Mr. Zubly my Countryman from what he heard of [from] the highly honoured Commissary [Ph.] von Reck during his full stay at Augsburg begin to take new Courage, which induces me to do my Self the Honour to send to you inclosed my thoughts on that Subject humbly requesting you would be pleased after perusal to correct them where it may appear necessary and to favour me with your Opinion, whether there be room under your Patronage for me to flatter my Self of bringing about such an establishment, and moreover to give me some Instruction what farther steps you think ought to be taken etc.

The inclosed Thoughts referred to in the foresaid Letter.

The desire of this City as well as others in our Confederacy of erecting a Colony in Carolina has induced last year some Gentlemen to send 3 Burghers to Purisburg, in order to gain by that method a previous, certain, and impartial Information of the Condition and nature of that Country the better to be able to lay the Case before our High Magistrate. But having to this very hour not received any such account, we cannot possibly think of any Colony this year, nay people have in a manner lost all inclination towards it.

However since we are assured by diverse Accounts that the Establishment of the *Saltzburgers* at Eben Ezer in Georgia has met with Success, and that the land is better, and clearer from Woods,[285] than Carolina, we heartily wish from a peculiar and just Confidence we bear to the Saltzburgers, we might make a Settlement among them at Eben Ezer and that in the following manner.

We would pick out about 30 good tempered labourious and ingenious men fit for Agriculture, planting of Vineyards, skilled in Handicraft Trades and Sciences, as also understanding the Silk and Linnen Manufactures. For these people we should desire a district of Land answering to 50 Acres per head, the Enjoyment of which they should have on the same Condition with the same

Prerogatives and Privileges which the Saltzburgers are endowed with. They desire to continue in the free Exercise of the Reformed Religion as established in Switzerland; in all civil Affairs they subject themselves to the same terms with the Saltzburgers. It is farther judged necessary that these men should have gratis allowed them in the Town of Eben Ezer a certain spot where they might build a spacious house, with yard, gardens & stabling. They must also be sure of the same quantity of necessaries of Life at their arrival, as was given to the 5 Men who went from hence to Purisburg last year, for one whole year viz.

Provisions for 5 Men one year.

Three hundred pound weight of Beef 50 of pork 20 quarter of Flower, 2 quarter of Salt, 1 Cow, 1 Calf, 1 Hogg, an Axe, two Hatchets. It is calculated that to compleat their Voyage from hence to Carolina they had given them 2000 florins which summe was judged sufficient for their support during their Journey along the Rhein by land, and so afterwards to London, and from thence to Georgia by Sea, and the purchasing of some Horses, Cattle etc.

Of the number of these 30 Men 20 should be Husbandmen the which should immediately after their Arrival in the month of November, sow one hundred Acres with Corn, Rice or other Grain, which with the Blessing of God, would yield about 200 Bushells of Indian Corn per Acre,[286] or about 6 hundred Weight of Rice to be ripped about May. The 20th part of which should be laid up for their provision of the year ensuing. This, together with the Cattle reared the first year is judged to be sufficient for their Subsistence, in so much that they have an opportunity of selling the rest of the produce of their land either Northward or any where else where it turns to the best account, the profit of which they may employ either for building, materials or other incumbent Charges. If after such a Tryal, the end of this current year, or the beginning of the next, should furnish us, with the News of the good Success of these people, and in the mean time a convenient tract of Land should be pointed out proper to establish a Colony on, our gracious Magistracy will not be wanting [will not fail] to send a Colony of some hundred persons, taking proper time for and more mature deliberation about it, as also making necessary Provision & Regulation. In the mean time assure your self that for this purpose none but warlike,[287] labourious, peaceable and experienced Men will be employed, provided with Pastors and prudent Leaders and what else they shall have occasion for.

No. 13191.   Read 24 June 1735.

The Humble Petition of the Subjects of his Imperial Majesty, Professors of the Gospel and Inhabitants of Gaisern [Goisern], in upper Austria, being three hundred and fifty persons, concerning the Persecutions that they Suffer from the Papists, dated at Gaisern May 25th 1735.

Grace and Blessing from GOD be to all those who love the Lord Jesus Christ.

We the undermentioned Protestants of his Imperial Majesty's Salt Office do humble and prostrate our Selves chiefly in the presence of Almighty God, seeking with the greatest Contrition to move his Compassion towards us, in humble Confidence that our Submissive Complaints and Sighs will obtain of him Deliverance from the Miseries we endure, and next to the hope we have in him, we rely upon and intreat to be favoured with the Assistance, and Succours, and good Offices, which our deplorable Condition requires, from the Compassion of those temporal Princes, and Ecclesiastical Dignitaries in the Church of Christ, whose Glory and Duty it is in a more eminent manner powerfully to forward every good Work, and to succour and intercede for the oppressed, according to the Will of God, beging they will pardon the liberty we take of throwing our selves under their Protection, and entertaining a comfortable Expectation of every thing proper for us to hope for, and necessary in our Condition, from their well known Piety & Christian Affection.

Since the last Transport of our Brethren removed into Transilvania is gone, We who are all the registered Protestants that are left here, are 350 Persons, and are left here as Sheep without a Sheppard, being tormented and persecuted, on all sides, experiencing in some measure what our Lord Jesus Christ says Mark 13.9 And ye shall be brought before Rulers and Kings, for my sake.

Six of our people are in close imprisonment at Lintz; one of them Thomas Laseres was taken up a year ago and imprisoned because he went to Vienna to receive the Sacraments. The other five, whose names are Matthias Chiesenbuhler, Matthias Humber, Michael & Joseph Leimber & George Gimps, were fettered and chained 16 Weeks ago, and thrown into the water Tower at Lintz, where they lie in a most sordid Dungeon because they were found praying to God, and visiting some of their sick Brethren who had desired it of them.

We earnestly request you in the name of our Saviour Jesus Christ to remember their Chains; to intercede and pray for them, that God would be pleased to endue them with Christian Patience, and Strength to persevere in his Truth, notwithstanding the Examinations our Persecutors compel 'em to undergo thrice a Week in order to pervert them to Popery.

Another great hardship we undergo is their taking away our Evangelical Books of Instruction, and our hymns and Catechisms etc. which is the more grievous because we cannot get others. Among other hardships we have suffered it happened that some friends being met on a Saturday at the house of Matthias Humber (whose Son is one of the Prisoners abovementioned) to sing Psalms, and to read a Sermon; the Bailiff of the Place, with a Priest, a Register and two Archers [constables], came and surprized them at ten o Clock at night, and took away all the Books they could find. As God is just and incomprehensible in all his ways, we are not much surprized at their showing so little regard for the word of God, and submit to him in all these Sufferings, adoring his Justice, acknowledging Our Unworthiness, and not doubting, but that every thing will end to his Glory and our Salvation.

Notwithstanding how much they deprive us of all that may yield any Consolation or Comfort to us in this miserable Life, they have gone so far as even to forbid us under very severe Penalties to sing or read.

Upon the death of one of our Brethren, some friends went to comfort the Parents of the deceased, and to strengthen them to bear the loss, by joyning in prayer and singing some Hymns, which they continued to do every day till the day of his Interment; They have thereupon forbid us this pious Exercise, the great duty of Christians, and which was a great Consolation to us; and since this rigorous Prohibition, a Patrole of Priests and Archers go often about upon such Occasions to molest the afflicted.

When the Popish Priests understand that any of us are sick, they come every day to us, forcing themselves upon us in that weak Condition to try if they can make us stagger in the faith; at first they appear to us in Sheeps Cloathing, making use of soft insinuating Language, and if we do not give Ear to that, they fly out into a violent rage, and terrible Excreations, which would make one tremble if possible. They also refuse Christian Burial to us allowing it only to Children, and have forbade the Bells to be rung at our Interments on pain of Imprisonment, and will scarcely suffer the nearest Relations to attend our Funerals; and yet their Priests and other Clergy demand their Fees to be given them by us, as if every thing were done, as the law appoints, and in good order.

We most humbly intreat from our hearts the divine mercy and the pity and Compassion of the Protestant Princes and Lords, that they would be pleased to deliver us from our Sufferings, that we might according to the Principles of the Evangelical Religion which we profess, go to Church and receive the holy Sacraments, and if this cannot be obtained in our native Country, that we may be allowed to go from thence to some neighbouring place where we may receive the Communion, and freely joyn in all other Acts of Religion with the Protestants, for a remote Emigration, is what we have no prospect of at present, neither do we desire to leave our Country, and so good a Prince as his Imperial Majesty is, for whom we are ready, and better disposed than formerly to sacrifice our Goods, Blood, life and all things, might we but enjoy liberty of Conscience, and the freedom of publick Worship and not to be left any longer as a Flock abandoned of their Shepherd and destitute of Pasture.

We made mention above, that part of our People were removed into Transilvania, some of us being forced to depart thither, and others to remain here, it so happened that Part of the Baggage belonging to those here was carried to Transylvania, and our Persecutors, on that Occasion separated even married Persons; in Particular Thomas Deiller was carried thither in Chains, and his wife left here; so little regard they pay to the Sacrament of Marriage, as they call it, when they resolve to distress us.

As to what remains. We suffer great hardships from our Superiours, with regard to temporal Things, They conceal part of our Goods, and in effect will not suffer us to live.

We commit all into the hands of the Almighty and recommend our deplorable Condition to the Counsel, Succour, Assistance & Intercession of all those who love our Lord Jesus Christ and his word; Saluting them a thousand and a thousand times, most humbly; By us

The whole remaining Evangelical Body of registered Protestants of the Salt Office, who since the last Removal are 350 Persons.

Translation of a Letter out of French from Mr. Urlsperger dated Augsburg 30 June 1735. No. 13238. Read 8th July 1735. To Mr. H. Newman.

Sir: I have the honour to communicate to You the Extract of Letters which Mr. [J.] De Reck sent last to me from Ratisbonne. I beg you Sir to let me know as soon as may be the resolution of the Honourable Society thereon, If they agree that Mr. [Ph.] De Reck, having assembled a number of 50 Persons, may transport them to England, that you would procure for him a Passport to this Effect on the part of Her Majesty the Queen without naming the precise Number, because that cannot be yet known.

The letter here inclosed is with their permision to recall Mr. Degmair who has lost his father some days since whose funeral Sermon I preached from the 8th Chap. & 27th v. of the Epistle of St. Paul to the Romans.

I beg you Sir to let me know by the first Opportunity an Account of his behaviour, and how the Gentlemen of the Society have found him because that may be of good use and you will very much oblige

Sir Your most humble Servant Urlsperger

P.S.: Since it is impossible for me to send You a Translation of Mr. [Ph.] De Reck's Letter being on the point of going hence to use the mineral Waters in the Imperial Town of Lindau because of the much impaired State of my Health, I take the liberty to communicate it to you as I received it in High Dutch.

Translation of an Extract of a Letter from Mr. [Ph.] Von Reck to Mr. Urlsperger.

The Austrians, who are about 30 in number, declared the day before yesterday to me that they could no longer subsist here till St. James's day, being obliged then to give their Landlords Warning, and to leave their Work to others, that they already had prepared themselves for their Journey, and sold their Things that were of no use to them.

The Carinthians, who some weeks ago went to their Country for their Wives, are returned back without them, assuring that it is impossible to get their Wives and Children privately [secretly] out of the Country. Mr. von Palm hath sent for them and told them that the answer he had got upon His letter was, that the Emperor would not let their Wives and Children depart, and the Case being carried on too far, they could not be permitted to depart privately; and that the Carinthians might return back into their Country, and worship God in private, and avoid Assemblies. Notwithstanding this Threatening and Flattery the Carinthians insist upon going to Georgia, and expect some time or other, help from England for their Wives and Children; But this Negotiation going on very slowly, the unmarried Carinthians resolving perhaps to depart immediately; The Austrians being not able to stay any longer than till St. James's day the best time for travelling passing away, and as soon as good news from this Transport comes to Germany, a considerable number of Austrian Emigrants, besides the Carinthians, will doubtless follow their Country People. I expect your Opinion whether the Society would not take it amiss, if I within four weeks time the farthest, set out with these few people; for which purpose I humbly beg them to procure a pass from Her Majesty the Queen, and to send it to me about that time.

Postscript by Mr. Urlsperger to the Society.

My Cash being almost quite emptied I have borrowed 500 fl. from Mr. van Münch, which I desire may be reimbursed. As soon as I have occasion again to write, I shall let you know to what the rest of the money which was in my hands hath been employed.

Farewell

A Translation of the Receipt accompanying the Origin of the foregoing Letter.

A Receit for 500 florins, which at my request was paid me in ready money by Mr. Christian Von Münch for him to be reimbursed by the highly Honourable Society for Promoting Christian Knowledge at London likewise for a double receit of the same tenure tho' but for one and the same Summ Attested in Augsburg June 30th 1735.

Samuel Urlsperger   Senior Minister   Pastor of St. Ann

£ 56   12.0 St.

(at 8 5/6 of a Florin for one pound Sterling)

Copy of a Letter from the Reverend Mr. Urlsperger dated Lindau July 12th 1735.   No. 13254 vid. Entered fol. 147.   To Mr. Newman. [Entered on page 553 above.]

Translation of a Letter from Baron [J.] von Reck dated Ratisbon 8th August N.S. 1735.   To Mr. Newman.   No. 13278.   Read 12 August 1735.

Sir: Your kind Letter of the 11th July O.S. advises me of a new Transport designed to be sent to Georgia and at the same time covers the Royal Passport for my Nephew [Ph. von Reck] that he may conduct them.

As to my Nephew he only waited the Arrival of this order to give them Satisfaction but as their numbers cannot be yet determined for want of the Emperor's Resolution whether the Wives and Children of the Carinthians who desire to follow their Husbands and parents may be permitted to go out of their Countrey, concerning which Mr. Robinson hath undertaken to sollicit, and therefore we must have a little patience.

Mr. Urlsperger is on his return to Augsburg, and will do all that he possibly can with my Nephew to execute the orders of the Society as soon as it shall be practicable. My Nephew will tell you the rest and I am

Sir   Your most humble Servant   J. De Reck[288]

Translation of a Letter from Mr. Ph. Geo. Fred. De Reck at Ratisbon 8th August N.S. 1735.

Sir: In hopes of sending you news of my departure with a Transport I have deferred writing to you. Being provided now with a Passport from Her Majesty which you have been so good to send under Cover to my dear Uncle [J. von Reck] and being provided with Letters of Credit I wish and hope that Mr. Robinson may procure us the Colonists, who tho' they have long since promised to go to Georgia neverthless stay for their Wives and Children from Carinthia.

Messrs. the Honourable Trustees for Georgia have presented me with 500 Acres of Land in Georgia, the Passage & support for one year for 10 Persons beside a Gratuity which Mr. Oglethorpe will give me. I gratefully acknowledge this particular favour as a Gift from God, of which may his Grace render me worthy. I joyn the freshest news from Carinthia and am with great esteem

Your most obedient humble Servant   Philip Geo. Fred De Reck
P.S. I beg you'l forward the inclosed as directed.

Translation of an Account from Carinthia 14th June 1735.

They refuse the Labourers the necessary work the only means whereby they can subsist in their Countrey; they force us to send our Children to the Roman Catholick Catechism under the penalty of a Ducat to be paid every time they miss going there; therefore they condemned Steven Engelmeyer and Peter Walter in the penalty of 3 Ducats each. Many are beat because they will not go to their Churches. The 12th June, being assembled to sing Psalms and pray, the Bailiff and Serjeant surprized us, and fired upon us, and encouraged a great Dog to fall upon us. Four of us were beaten with Sticks and drove to Church and to the Procession. They have kept back 44 small Children and sent their Fathers and Mothers to Transylvania. They brought hither from Clagenfurt the 9th May 3 Men with their Wives and 3 Boys. Elizabetha Sateleekeim[289] is still in prison with her Daughter. The Prisoners live at their own Expences and pay 30 to 40 florins, their Goods are appraised two or three times to increase the Expences, and they seek by forcible means and by threats to oblige the people to the Belief of their Doctrine, of which we could hereto add a very great and ample Account.

Translation of Letter from Mr. Urlsperger dated Augsburg 11 August 1735. To Mr. Newman. No. 13299. Read 19 August 1735.

Sir: My last of the 12th July I hope you have received, as I have your most agreable Letters of 20th June and 8th & 15th July all in due time. Some few days ago I returned from Lindau in a better State of Health after having spent a Week at St. Gall in Swisserland, where I received from the Magistrates of the reformed Ministry, as also from several other persons, most uncommon marks of Love and Friendship. I have met with there many worthy people of all Ranks and Conditions, who were greatly rejoyced when I told them of the uncommon Care the honourable Society have taken of the poor Emigrants.

Mr. [Ph.] Von Reck hath received the Royal Pass and sent me a Copy thereof. It is great Pity Mr. Robinson has not yet writ neither to this place nor to Ratisbonne any Account of the Imperial Court's permitting the Wives and Children of the Carinthians to follow them, notwithstanding I immediately, on the Reception of yours of the 20th June, writ from Lindau to him at Vienne and sent him an accurate list of the Carinthians, and this

delay of the Imperial Resolution is the Occasion that the Transport cannot yet set out. I have therfore writ by the last post to
the Envoy [J.] Von Reck to intreat Mr. Robinson to be very pressing in this Affair. I have taken Care that some of the Bohemian
Emigrants that understand the German Tongue and who are
united to our Augsburg Confession, and in a Condition to labour,
to get some of them that they should go and be united to the
Colony of Ebenezer. My Correspondent to whom I have committed this Affair, is not only capable but has in him the fear of
God, is therfore already gone from *Silesia* to *Cottbus* and *Gross-
herrnersdorff* [Grosshennersdorf] where the greatest part of those
Bohemian Emigrants have resided for these two years past and I
expect [await] by the first Opportunity what Success he has had
therein. Should 50, 60 or 80 of these Bohemians resolve to go, of
which I can scarce make the least doubt, they should march under
the Conduct of my Correspondent to Frankfort, where Mr. Commissary [J.] Von Reck might joyn them with his 30 Austrichians
[Austrians] and Saltzburgers that are at Ratisbonne, and so proceed on their Journey to Rotterdam England and Georgia.

Should it so happen that after the departure of these Transports
consisting of Austrichians Bohemians and Saltzburgers that the
Carinthians Wives and Children should have leave to come to
them, these Carinthians might make a second Transport; What
may occur in this Affair I shall give you a regular Account thereof
from time to time.

The good Testimony you my dearest Secretary gives of Mr.
Degmayer has greatly rejoyced both me & others. I likewise believe
that upon the filling the ministerial Vacancies a due Consideration
will be had of him. I have learnt with great Satisfaction not only
that the honourable Society has been pleased to reimburse the 500
florins to Messrs. Meyer and Jansen that were advanced to me by
Mr. Von Münch, but also that the Trustees have granted such
great Advantages to Mr. Commissary [Ph.] Von Reck towards his
Settlement in Georgia, of which Mr. Oglethorpe has writ to Mr.
Von Reck who immediately thereupon sent me Copies thereof.
Mr. Commissary Von Reck, as likewise his Uncle [J. von Reck] are
highly rejoyced therewith. And since the Trustees are pleased to
have so tender a regard to my recommendation, and that Mr.
Oglethorpe and Mr. Vernon have so generously seconded the same,
therefore I beg my most humble thanks may be made to them for
the same as likewise to the honourable Society that they have lately
been pleased to grant so generous an Assistance to Mr. Degmayer.

The Lord bless the undertakings of the Society in a most abundant manner I remain

Sir   Your most obedient humble Servant   Samuel Urlsperger
P.S.: I hear for some time past that the Bohemians inclining to Protestantism are not so severely treated as formerly; they have declared from the Pulpit that whoever would become Bohemian Emigrants, have only to declare themselves as such, but this the people with good reason have no Confidence in.

Copy of a Letter from Mr. John Vat dated 30 May at Ebenezer in Georgia 1735. No. 13300. Read 26th August 1735. To Mr. Newman.

Sir: Inclosed is a Copy of a Letter I had the honour of writing to you the 10th February last, which I suppose to be in your hands long before this time. Few days afterwards I was so ill, and brought so low, that on the 15th February when Mr. Causton, Mr. [Noble] Jones, and Capt. Dunbar were here I could hardly stand upon my legs, which Weakness continued several Weeks, but upon taking proper Medicines of Mr. Zwiffler, and recovering some Strength, altho' the defluction upon my Eyes held on, I resolved about Easter to leave this place, and to return to England; which Resolution however I since altered upon a Rumour of some motions of the Indians in the Spanish Interest; and I am thinking of continuing to be here, till I shall hear of the final Resolution of the Trustees for Georgia concerning the present Settlement of the Saltzburgers; for should the People be obliged to settle in this barren place, I could not but with the greatest Grief behold the Misery which must evidently attend them. The Experiments we have made this Spring evidently confirm the Opinion, and the dismall Accounts given by every body concerning Pine-barren Lands! For all the Seeds we had from the Trustees were sown in due time, and most of them came forth plentifully to the great Joy of the people, but there being no Substance in the Ground of Sand, hardly any of them are coming to any Seed. The Kidney and Sandwich Beans will scarely yield one pint for sowing next Spring, and the Indian Corn, which the people have planted here and there every week since the beginning of February last, gives but a very small Prospect, there being only here and there a fine plant coming up, yet Mr. Causton, who, together with several Persons hath been here thrice within these three months magnifies a small spot of Ground of about twenty feet Square in the midst of an

Inclosure of half an Acre near Guhwandel's [Geschwandel's] House, where some Cattle stood for some Months, and thence concludeth that by a small Stock of Cattle this Soil may be improved, so as to produce plentifully hereafter, and cries this place up as the best land in all the province. It even seems by his Actings that he is fully bent upon the Continuance of the Settlement in this place, for on the 14th of February last he sent hither two men for sawing of Boards for finishing the Six Houses ordered by Mr. Oglethorpe to be built here One of them viz. Mr. Bolzius's being finished. Two others viz. the Store House and the Schoolmasters, joining together are so wretchedly slight, that by making one single step, both Houses shake so, as to be in danger of falling to the ground, and upon any rain I am forced to shift my Bed, in one of rooms therein, occupied by my Self and three families besides. And the frames of the three others, now standing naked, are so bad, that I wonder how any one shall be prevailed on to occupy them when they shall be covered and boarded. These Sawyers have sawed no more than 159 boards, and have left of [off] working, but Mr. Causton saith he will send other Sawyers to finish the said Houses. Upon the Return of some of our men from Savannah, the women resolved to clear some Ground by themselves for Gardens; the single men took thereby Occasion to do the like, and then on the 3rd of March the men began joyntly to clear a Spot of Ground, which hath been since fenced in, and planted with Indian Corn and Pease. On the 8th of March, hearing of some disturbances at Savannah, we began the Building of a Block Watch House 28 Feet in length and 18 feet in Breadth, which is now made use of as a Church and School, as also as a Warehouse; and we afterwards built a Bridge over the River in this town, and another Bridge over a small Swamp in going hence to the landing place, in order to bring thence our Provisions, partly upon our backs or by a small Waggon, whilst the Waters are so low that no boat could be of any service since the 16th of April last to this time; and this is like to be so all this Summer. On the 15th April last I went to Savannah, and prevailed with Mr. Causton for sending us Provisions for six months. He agreed with Mr. Mamour to bring part of them in his Periawgoe [periagua, pirogue] to the landing, but Mamour could come no higher than within two miles of the mouth of Ebenezer River; from thence we carried them in our own small Boat to the Landing. As about 40 Bushells of Indian Corn were scattered loose in the Periawgoe, and some Hogs and pigs lying amongst it, which our people cannot eat for its nastiness, and as there were Six Inches of a Cask of Wine of twenty Six Inches deep sent by another Boat wanting, I desired Mr. Causton to buy a proper boat for our carrying our own provisions. That Periawgoe employed 12 days in coming from Savannah town to the mouth of our River, one Mr. Guthry hath since made two trips in five days each, with part of our provisions from Savannah town to our landing place, for which he is to have Forty pounds Currency[290] and Mamour 30 £ besides the Wages of five men at Eighteen pence a day, each. This Shews that the Charges of carrying down, or bringing up anything will for ever keep our people in a very low state, even were the Soil as good as some people would have it. And I cannot see how the poor people will get any thing for procuring linnen and Shoes, of which they are now in great want, not to mention many other necessaries, as Earthen Ware and Utensels for the Kitchen. I could wish the Society had given orders for one hand Saw to Each Freeholder instead of but Eight for all, also some large Coppers for boiling of beer, the River water being very bad especially in Summer. We indeed very lately upon searching for a vein of water, and fixing a Rice Cask, found a very fine and strong spring, but in this great drought it runs so very small, that it's apprehended it will soon be dried up. It is very observable that hitherto all our Childbearing Women are delivered of their Children before their full Growth, and that most of the Women died, and the former happened to Mrs. Smith an English Woman, who lately miscarried in this town, so that such of our Women as are now pregnant are in deadly apprehensions that the present Soil is pernicious both to the growth of Children and Seeds. The list of such as died since we have been here is as follows—

Jan. 23 Margaretha Schoppacher
    26 Christian Steiner
Feb. 13 Maria Hueber
Apr. 2 Margaretha Guhwandel [Geschwandel]
    4 Maria Schoppacher
    8 Anna Schwaigger
    16 Ruprecht Schoppacher
    30 Hans Madreuiter

As the Bread kind Provisions of 6 pounds a Week per head will not admit baking bread, it's supposed the want of bread is of no small detriment to the health of our people, who were used thereto from their tender years. We have no Ovens but some of us bake Bread in our large iron Pots, which is very tedious as well as our grinding of Corn with small iron hand Mills almost good for nothing; some Mill Stones of a midling Size would be very usefull, as also some fishing Tackle for catching Fish. As Sebastian Glantz, who died at Purrysburg without a Will and no Relation here, left some small matter, I desire to know the pleasure of the Trustees concerning his Effects; Mr. Bolzius being of opinion for dividing of them to the poorer sort of his Congregation, which may be attended with some difficulties; but my opinion is for selling them to the highest Bidder, and for lodging the produce thereof in Mr. Boltzius's hands. Capt. Dunbar hath, as I doubt not, by this time, given you an Account of our Voyage, and of the Scituation of the place, and I hope a faithfull Representation of the nature of the Soil thereof. As in the Account of the Stores put on board the prince of Wales mention is made of a Box of medicines shipped by Mr. Newman, but could find no such Box, we suppose a Trunk of Medicines B.G. No. 5, which Mr. Zwiffler has received, is meant thereby. The Barrell of Molassus was so slight that it was broke at Sea, and tho' we shifted it into another Cask, yet we saved but fifty four quarts thereof when we received it here. Capt. Thompson as also Mr. [John?] West[291] will doubtlessly give you a faithfull Account of the Barrenness of this place as having both been here. And I submit it to the Consideration of the Society, Whether, in Case the people are removed hence to a better Soil, they will be so kind to send the following Particulars Viz. Some Sand-hour-Glasses, Tinn-Funnels, Pewter-Quarts, Pints and half Pints, for measuring Wine & Beer; Half Bushells, Pecks etc. for measuring Corn etc. Divers sorts of Ropes; Small Scales of about 18 Inches Diameter and Brass weights, Gold Scales & Weights for weighing small things; Divers sorts of Iron Wires; One middling sized Bell for ringing to Church and publick Work. Some Joyner's Glue; Flannel for night Wast Coats, Some great Gunns for alarm and defence; Black Smiths Tools; Some Coopers Tools for making pails & Casks; and large Bellows; Tinn Plates; Sives of several Sizes, and Turners Tools. But every thing must be packed up, for the Sailors broke three of the four Lantherns sent on board the Prince of Wales; and they took several of the Bedding Blankets which we could never see again. When I was last at Savannah, I went several times to Mr. Spangenberg's five Acres Lot to see his men who seem to be very industrious at work. I wish the Soil of Ogeechee which is designed for Count Zinzendorf, may be as good as That within two miles of Savannah Town; for I look upon the beginning of a Settlement, as upon the Foundations of Houses, unless these be solid in themselves the Superstructure must in Course be affected; and that a Soil to be made good by dung is an under-taking too precarious for poor Husbandmen, and not so easily to be done as some persons would persuade others, who know the Consequences thereof by Experience; and it is no small Discour-agement to our Saltzburgers, that they do not yet know their respective Lots, altho' in the printed Accounts published in Ger-many, they had promises of having immediately at their Arrival here, their Portions of Land assigned to them. And in this the Reputation of the Reverend Mr. Senior Urlsperger and some of the chief Magistrates at Augsburg is highly concerned, for it was upon their publickly appearing in the Affair, that these innocent people ventured their all, in leaving their Services in good Fami-lies;[292] and the Roman Catholicks in that Town will not be want-ing in insults for sending people into such a desart, where in two years they cannot reap the Corn or Seed they sowed. Some people here indeed lay the fault upon the unseasonable Coming of the first Saltzburgers into this Countrey, and upon their not knowing the manner of sowing and planting therein, likewise the extraordi-nary heats of this Spring almost without rain. Next year perhaps the Fault will be laid upon the Rainy Season; but I am fully per-suaded and convinced, that the real Cause of its not producing is in the Foundation of the Soil, as being sandy without any Cohesion of Particles. This seems to be the reason that Pine Trees wholly consumed and burnt, leave no manner of Ashes, but only sandy Particles, and Pine Trees cut, or thrown down and laying on the Ground one year, are generally decayed and rotten as if they had been twenty years before on the Ground. As to the Cattle, they do indeed at present look very well, but it is quite the reverse in the Heats of the Summer and dead of the Winter, being obliged to range a great way off for getting their Subsistence in Swamps or Cane lands, the Grass being too rank and sour. The old Saltz-burgers[293] did not see their Cattle all last Winter, and of thirty head of Cattle given them last year, they have now but five Cowes that casted their Calves this Spring, the others being either wild or lost. At the latter end of February last, Mr. Causton sent twelve Hogs to Abercorn for the new Saltzburgers, one of them died in bringing hither, two of them broke loose from the ropes and are lost, and after having kept the others several Weeks in a Stye, three of them likewise run away and have not since been seen. Mr. Caus-ton hath given us Six Bushells and a half of rough rice for sewing in our Swamps which are now quite dry, we are very glad of try-ing every thing, and are now preparing a proper place for that purpose, but I apprehend the Success thereof will be no better than with the Indian Corn, for it is to be observed that in Carolina the Negroes, as the only proper Planters thereof, are made use of, and that whenever White People are employed in that way of working, they die like flies, as being unable to endure the Waters in such Swamps, much less the Heats of the greatest part of the day in Summer time, the clearing such Swamps being more difficult and labourious, than the dry land, be it never so much over-run with large or small wood of any kind; And considering these people were born and bred in high and rocky lands, which are as different from the nature of these here as the day is from the night, some knowing people say, It would be better to shoot the people at once than to put them into such a way of planting. Mr. Boltzius hath taken a memorandum of such demands as some of our People have to make in the ArchBishoprick of Saltzburg, or of some of their Countrymen who are gone into Prussia. The attempting of getting them in, is, as I humbly conceive, a work to be recommended by the Crown of Great Britain!

Moreover should the Trustees think fit to remove hence these Industrious and worthy people, I should notwithstanding the Indisposition upon my Eyes, be inclined to assist them in a New Settlement till next Spring, and in that Case I hope the Society will give proper Instructions for my Support for I receive from the Stores no more than any other man, unless when I am at Savannah. Mr. Causton and his Lady overheap me with Goodness and Civilities, and I am constantly troublesome to them.

    I am,   Sir   Your most obedient and most humble Servant

                                John Vat

Ebenezer May 30th 1735.

Copy of a Letter from Messrs. Bolzius and Gronau dated Ebenezer April 2nd, 1735. No. 13301. Read 26 August 1735. To Mr. Newman.

Dear Sir: We doubt not but that our letters, dated the 10th of December and 8th of February last, are come to your hands, and that our Journals have been laid by the Reverend Mr. Zeigenhagen before the Society. We shall be no ways wanting to give from time to time to our Benefactors an exact account of the State and Condition of our Hearers as well as of our Selves, and we are in hopes they will take our freedom in writing in good part. We are much pleased to understand by Your letter of 21st Jan. that the Society are in hopes our endeavours in performing divine Offices will redound to the honour of our Blessed Saviour, and to the happiness of our people, as well as of the poor Indians in this Colony, which sit in darkness without any knowledge of our Lord Jesus Christ. It grieves us to the very bottom of our hearts to see so many people walk in many ways that lead to eternal misery. We therefore endeavour as much as lies in our power to be able to shew them by and by the way of their Happiness according to the holy Gospel.

The progress, made in learning the Indian Language, is yet very small, by reason from the beginning of our Settlement here divers affairs have prevented our applying our Selves thereto, so that we have had very little time to acquire more knowledge in the English Tongue. Having no Indian Linguist[294] is likewise a great hindrance to our undertaking. We indeed hear frequently many words spoken by the Indians, in shewing them the things, whereof we want to know their Signification, but these are only Substantives, and we cannot learn from them the Signification of Verbs and Particals, which are necessary to have the Construction thereof. We are apt to flatter our selves that the proposals we lately made to Mr. Ziegenhagen will be approved of by the Society and Trustees for Georgia, and we believe that then this Undertaking will go on with more Success. Our Conversation with the Indians for some months since hath been but very small because our Settlement is too far distant from Savannah River, and from the places of their ordinary Dwellings. If we should be so happy to have some sober Indian Families at Ebenezer according to Mr. Oglethorpe's Direction, their dwelling among us might be by God's Blessing a great advantage in learning this Language. Your letter informs us of Mr. Spangenberg's Intention of settling in this Colony. We wish with all our heart, he may approve himself in such a manner as you are pleased to mention. He made some time ago many disturbances at Hall being of the same Principles of Count Zinzendorff, who seems to be endeavouring to reform the Christian Church. His new notions were then very much disliked to every one that loves truth, peace and the Conservation of our holy doctrine, which unhappy divisions are yet fresh in our memory as living at the same time in that town. If his Behaviour in time to come is such as to avoid the intended Separation from the Communion of our Church, we shall be exceeding glad to live with him in the same cordial manner as we did before we knew fully his Principles, to exalt every where the Glory of our great

Redeemer, and his glorious Kingdom, for this is and ought to be thro' God's Grace the chief and only Business of our whole life. And the more fellow-labourers there are for this purpose, the greater is our Comfort and joy thereof. We have frequent opportunities of conversing with English people, and being persuaded that some of them take great delight to borrow or to have English Books for their Edification, we beg the Society will be pleased to send us some for this purpose. Mr. Quincy hath favoured us with a good many of them, but we wish we might have some more of divers sorts. Mr. Ortman our Schoolmaster and his Wife go on in the path of Christian Virtues, and he is more and more serviceable to the Children of our School. We pray therefore the Society would forgive them their former Offences, for they repent heartily of the same. The present State of our Congregation, as well as of our Selves, will be, reported by Mr. Vat, particularly the nature of the Soil of this place, which is almost nothing else but sand. We humbly thank the Society for all their favours and bounties bestowed upon us and our flock, wishing that God Almighty may grant them, all manner of Blessings to their own persons, and to their noble Undertakings they persue. Be pleased to present our most humble respects to all the Gentlemen of the Society begging the favour of them to continue their paternal Care for us and our flock, and to give us leave to subscribe our Selves their and Sir
Your obedient humble Servants   John Martin Bolzius,
Israel Christian Gronau

Ebenezer 2 April 1735
Postscript: Mr. Vat who intended to leave our Countrey hath altered his Resolutions for such reasons as are mentioned in our Letters to Mr. Ziegenhagen. He presents his humble Respects to the Gentlemen of the Society, and intends to give a short Account of the present State and Settlement of the poor Saltzburgers, as soon as possibly he can. We received the 1st of May the second Bill of Exchange, together with the Copies of Letters and Invoice of the Goods we received by Capt. Dunbar. That the said Bill was accepted and since paid by Mr. Montaiguh [Montaigut] is already mentioned in the letter, which we took the liberty to write to you the 8th of Feb. last. God bless you with happiness & good health and reward you a thousand times for all your kindness and favours laid upon us. And so we remain   Dear Sir
Yours etc.   J. M. Bolzius, I. C. Gronau
Ebenezer May 5, 1735

Copy of a Letter from Messrs. Bolzius and Gronau dated Ebenezer 19 May 1735.   To Mr. H. Newman.

Sir: We received the favour of your letter dated the 19th of February last, after the parcel of our letters and Journal for London was already delivered to Mr. Savey, who is going by the way of England to Geneve. We have likewise sent some letters the 2nd of April to Charles Town to be conveyed by Capt. Dunbar which we hope will come safe to your hands. By the said letters and Journals you will be informed of the State and Condition of the Saltzburgers. It is a very great Comfort to us, that the Society, as your letters tell us, go on to be very carefull in promoting the Spiritual and temporal Welfare of our flock, which encourageth them to put up constantly their fervent and united prayers to God for rewarding with his divine Blessings, such great Kindnesses and favours. For the Books you were pleased to send us by the hands of the Reverend Mr. Quincy in the name of the Society, we are highly obliged, being very desirous to read them for our own use, as well as to distribute them to several people, we have some times occasion for. Those we have already distributed formerly, have been accepted with great pleasure & thankfulness, and we hope God will accompany the reading of them with his blessing to every Readers Welfare. Some persons in this Colony have no Opportunity to hear any Sermon preached by the Reverend Mr. Quincy, because they are constrained by their Business to live far off from Savannah, and therefore they account it a great Benefit to receive good Books for private Edification of their Souls. We firmly believe it would be not a small help for converting the poor Indians, if the Christian People in this Colony should be brought by the holy Word of God and other wholesome remedies to such a way as to set every time a good Example to these heathens. We long very much for the coming of Capt. Meckperson [MacPherson] to our place, and want to speak with him to let us have for a short time one of his Servants, who speaks the Indian Language, and was lately willing to give us some Intimation thereto. We are by Gods Grace so desirous to be serviceable to every Body in this Colony for their eternal Happiness that we want words to express it, and are in hopes God Almighty will himself lead us by his holy Spirit in that path we must go in performing his blessed Will and Intention, concerning the Souls of men. If there should be any Opportunity to teach the Indian Children, it is supposed it would be easier to instill the principles of Christian Knowledge into their minds than into those of the grown people. We dare not trouble

You with a farther Account of the Settlement of the Saltzburgers, being very well persuaded that Mr. Vat will do it with this Opportunity better than it may be expected by us. We recommend our Selves and our Flock to the continuance of the fatherly care and favour of the Society, and wishing You and them all manner of divine Blessings to Your & their praise worthy Undertaking We remain

Dear Sir   Your most humble Servants   John Martin Bolzius, Israel Christian Gronau

P.S. Mr. Vat presents his humble Service to you & intends to write by this Opportunity unless his receiving of the Provisions for half a year & delivering some part thereof doth prevent him.

Copy of a Letter from Mr. Quincy dated Savannah 4th July 1735. To Mr. Newman.

Dear Sir: I received your favour by Capt. Lusk sometime since together with a Box of small Tracts from the Honourable Society. I sent to Mr. Bolzius according to order 100 Journals and some of the Tracts against Popery which he desired and shall be ready to deliver to him as many more of them as he thinks will be serviceable to his People.

Mr. Bolzius desires me to acquaint you that he humbly begs the Assistance and direction of the Honourable Society in an Affair that has lately happened to him. Mr. Montagute having an order to pay Mr. Bolzius a Sum of money and being well acquainted with the Carolina Currency had received counterfeit Bills; which false Bills, not knowing them to be such, he paid to Mr. Bolzius to the number of 8 or 9 £ 15 Bills which amounts to £ 16 or 18 Sterling money. This Loss is like to fall upon Mr. Bolzius; for he having kept the Bills by him 3 Months, not suspecting that they were bad; when he would have returned them, Mr. Montagute absolutely refused to take them, because of the distance of time, alledging that if he should do it, all the counterfeit Bills in the province might be brought to him. Our Magistrates have given it as their Opinion that Mr. Montagute ought to make the Bills good, but he refuses to stand to their determination and they cannot oblige him, because he belongs to Purrysbourg and is not under the Jurisdiction of their Court. Mr. Bolzius begs directions how to proceed in this Affair; Such a loss would be heavy upon him in his present Circumstances.

I have endeavoured to inform my Self as much as possible concerning what you write that I might acquaint the Honourable

Society therewith Viz. Whether we have any Romish Missionaries from Home that keep a Correspondence here; but I cannot find that there are any. We have several Persons supposed to be Roman Catholicks and some known to be such, but if they carry on any designs of proselyting others, it is extremely private [secret], and I rather believe there is no such design, because Religion seems to be the least minded of any thing in the place; and if there were any such thing a foot I apprehend there would at least be more of the face of it. Since I have received your Letter I have put my Clerke who is a sober young man, upon getting a Society of other young men, to meet every Sunday night, which they have done for about 6 Weeks past to the number of 7 or 8 of them, after the Example of Some Societies in London. There method is to read the Epistle and Gospel of the day, with Comments upon them, to say the Evening Service with a Collect composed for the Occasion, and confer on what they have heard. I look upon this by the Blessing of God, to be one likely means to preserve them from being tainted with Errours. And if there are any designing Persons of the Romish Communion it may be a means of discovering them because as I have heard it observed, they frequently mix themselves with such young Societies. I shall therefore narrowly watch over them, and often visit them to give them Instructions and directions.

You desire in one of your letters to know whether the Jews amongst us seem inclined to embrace Christianity. We have here two sorts of Jews, Portugueze and Germans. The first, having professed Christianity in Portugal or the Brazils, are more lax in their way, and dispense with a great many of their Jewish Rites; and two young men, the Sons of a Jew Doctor [Nunez],[295] sometimes come to Church, and for these reasons are thought by some People to be inclined to be Christian but I cannot find they really are so, only that their Education in those Countries, where they were obliged to appear Christians, makes them less rigid and stiff in their way. The German Jews,[296] who are thought the better sort of them, are a great deal more strict in their way, and rigid Observers of their Laws. Their kindness shewed to Mr. Bolzius and the Saltzburgers, was owing to the good temper and humanity of the people, and not to any Inclination to change their Religion, as I understand. They all in general behave themselves very well, and are industrious in their Business.

I have by this Opportunity conveyed Letters from Mr. Bolzius and Mr. Vat, which I believe are chiefly on the Subject of their

Lands. I mentioned in a letter, which I hope you have long since received, something of the same matter. There Dependence is very much on the Honourable Society to use their Interest with the Trustees, to get them removed to a more fertile Soil, without which they have no prospect of ever subsisting themselves.

I am   Dear Sir   Your most obedient humble Servant   S. Quincy

Translation of a Letter from Mr. P. G. F. de Reck dated at Ratisbon 15 August 1735.   To Mr. Newman.

Sir: I hope that my letters of the 8th Instant are come safe. Herewith I send you his Excellency Mr. Robinson's Answer which I have just received, whereby you may perceive that he has not obtained the Imperial Resolution, and it is much to be doubted whether ever he will receive any the Imperial Court being accustomed for many years to give not the least answer to the Grievances that have been represented in behalf of the Evangelical Body. I expect your order that in Case of the Refusal or Silence, or in case there do not come any more Emigrants, I might engage other poor Protestants who deserve Charity as well as the Emigrants. I am with all imaginable Respect

Sir   Your most humble and most obedient Servant
Philip George Frederick de Reck

Translation of the Answer from Mr. Robinson dated Vienna the 10th of August to Mr. de Reck.

'Twas impossible for me to answer sooner the Letter you wrote me the 19th of last month in which you sent me the List of the Carinthians who are to go to the Colony of Georgia. Immediately upon the reception of that List, I presented a new Memorial to this Court in the most pressing Terms, and the most proper for obtaining immediately the Liberty of the Wives and Children to come to their Husbands and Fathers. I shall take Care to inform you of the Success of my reiterated Instances; and I pray you to believe that no one can be etc.   J.[297] Robinson

Translation of a Letter from the Reverend Mr. Urlsperger dated 21st August at Augsburg 1735.   To Mr. Newman.

Sir: Your most acceptable Letter of the 22nd of July I received by the way of Ratisbonne, and hope that mine of the 15th of August will be delivered to you in due time.

Last Thursday I writ to the Reverend Mr. Waneck at Gross-hennersdorf being not far from Zittau, where the greatest part of the Bohemian Emigrants are assembled, and sent him a full power to deliver to my Correspondent that is gone thither from Silesia, which power is confirmed by our Evangelical Privy Council here; That when he had or could bring together Sixty Bohemians he should without delay march with them directly to Frankfurt. I have likewise writ to Mr. Commissary [Ph.] von Reck that since Mr. Vernon in his last presses Mr. von Reck's speedy departure, that he would without delay come from Ratisbonne with the few Austrians and joyn the few Saltzburgers that are here, and so hasten with them to Frankfurt, where I wish my appointed Commissary in Lussatia may get a sufficient number of Bohemians and meet them in due time in Frankfurt that Mr. von Reck may proceed in his March with the United Transport.

I am not able to give an exact Account of the number of those that will go from hence and from Ratisbonne; partly, since many at Ratisbonne, who were resolved to Go, are dissauded from it; partly, since many will not determine to go till the others are ready to set out; but I shall inform You as soon as possible, and time enough. I only wish that Mr. Oglethorpe would stay for Dear Mr. von Reck, that He with his Transport may have the happiness to go with him to Georgia.

Mr. Robinson writes to me from Vienna on the 10th of August 1735 as follows

"I received under Cover of Mr. Von Reck the letter you writ to me on the 20th of last Month. As I had already received from Ratisbonne some days before a list of the Carinthians who are to go to Georgia, I had likewise already presented a new Memorial to this Court in the most proper terms to procure immediately liberty to the Wives and Children of these Carinthians to come to their husbands & Fathers. I shall inform you of the Success of my Representations and I am perfectly etc."   Robinson

Since we know not when the Imperial Court will think fit to notify their Resolution, tho' both at Ratisbon and here we have little hopes of its being favourable, and tho' it were favourable, the Romish Priests have been known in many Instances in Carinthia and Saltzburg to detain them notwithstanding. Therefore I think it convenient to take the first Opportunity to dispatch them and that Mr. [Ph.] Von Reck should have the advantage of going with Mr. Oglethorpe. If it should happen as I mentioned in my last, that they should obtain liberty to leave the Countrey, I hope the good Providence of God will furnish me with a fit person as Commissary to conduct those that are left behind to England.

I expect Mr. Von Reck here in two days in order to consult with him fully on this Affair. We will loose no time, if God gives us but Strength and Wisdom in this weighty Affair, and bless all our Endeavours for the honour of his Holy Name.

I am   Sir   Your most humble Servant   S. Urlsperger
P.S. I wish Mr. Ziegenhagen a better State of health and shall write to him by the first Opportunity. I hope Mr. Degmair is set out on his Journey, for his presence here is very necessary.

Copy of a Letter from Mr. Urlsperger a Augsbourg le 1 Septembre N.S. 1735.   To Mr. Newman.

Sir: Mr. le Commissaire [Ph.] de Reck arriva ici hier au soir avec 16 Autrichiens, auxquels se joindront 20 Saltzbourgeois qui partiront ensemble d'ici lundi prochain pour Francfort. Voila la liste de ces gens, mais qui selon toutes apparences sera augmentee de quelques uns qu'on attend encore de jour à autre; la premiere poste en portera une liste plus accurate. Il est à plaindre que la cour imperiale n'a pas donnée la moindre resolution touchant les Carinthiens; De sorte que Mr. de Reck est necessite a se mettre en marche sans les attendre. Aus si vois je par les dernieres lettres de Lusatie que les Bohemians ne se peuvent pas resoudre à aller en Georgie sans un Ministre qui leur preche L'Evangile dans leur langue martenelle [maternelle] à cause qu'ils 'nentendent point l'Allemand; Vous verres tout par la copie ci 'jointe de la lettre de Lusatie. Il s'agit à ce qu il me semble, Si Messrs. les Trusteés agreeront a prendre quelques cens Bohemians aux mêmes conditions qui ont ete recu les Saltzbourgeois agreeront à leur donner un ministre qui entend la langue Boheme; et enfin Si s'interesseront à la cour de Saxe pour le Lipperda dont la sudite lettre fait mention Je vois dans cette Affaire de tous cotés beaucoup de difficultes et autant de depenses. Le Transport qui part se montera à 40 personnes, y contant Mr. le Commissaire et son frere[298] qui l'accompagne. Dieu se Souvienne des siens Souffrans et je Suis
      Monsieur   Votre tres humble et tres obeissant Servant
                                                    Sam. Urlsperger
P.S. N'ayant rien à ajouter que Vous assurer de mes respects comme aussi Mr. Vernon et Mr. Oglethorpe pe je Vous prie de le retardes,[299] et de leur communiques comme aussi a Mr. Simond la liste des Emigrans.   Reck

[Translation of foregoing letter, by George Fenwick Jones.]

Sir, Mr. [Ph.] de Reck the Commissary arrived here yesterday evening with 16 Austrians, who will be joined by 20 Saltzburgers and will leave here together for Frankfurt next Monday. Here is the list of these people, but one which will to all appearances be increased by a few more whom we expect any day now; the first post will bring a more accurate list. It is to be regretted that the Imperial Court has not made the slightest resolution concerning the Carinthians so that Mr. de Reck is obliged to set out without waiting for them. I also see, from the latest letters from Lusatia, that the Bohemians cannot bring themselves to go into Georgia without a minister to preach the Gospel to them in their mother tongue as they do not understand German; you will see it all in the copy of the letter from Lusatia enclosed herewith. In my opinion, it is a question of whether the Lord Trustees will agree to take a few hundred Bohemians on the same terms as the Saltzburgers and to grant them a minister who speaks Bohemian; and finally whether the Court of Saxony will take an interest in the Lipperda of whom mention is made in the letter mentioned above. I foresee on all sides of this affair many difficulties and as many expenses. The transport which is leaving will amount to 40 people, including the Commissary and his brother[298] who will accompany him. May God remember his suffering children, I remain, Sir,

Your very humble and obedient servant   Sam. Urlsperger
P.S. having no more to add but to assure you of my respects to you as well as to Mr. Vernon and Mr. Oglethorpe, whom I beg you to detain,[299] and to communicate the list of emigrants to them as well as to Mr. Simon.

The list of emigrants who are to be found here at this time and who will leave next Monday. All peasants:

La liste des Emigrants qui se trouvent à cette heure ici et qui partiront lundi prochain. Tous paisans

1.	Frantz Haverfahner	Age 47 ans
2.	Marie Sa femme	46
3.	Susanne enfant	15
4.	Magdalene le Second	12
5.	Johann Schmidt	27
6.	Catharine Sa femme	30
7.	Jacques enfant	1-1/2
8.	Andreas Bauer	27

9. Johann Pletter	30	
10. Susanne Hartbergerin	48	
11. Catherine Enfant	12	
12. Joseph Ernst	27	
13. Marie Sa femme	30	
14. Sabine enfant	1-1/2	
15. Andreas Grimminger	27	
16. Sabine Sa femme	28	
17. Frantz Sigmund Herrenberger	37	
18. Johann Troel [Floerl]	23	
19. Carolus Troel [Floerl]	30	
20. Thos. Oseneker	24	
21. Martin Lackner	28	
22. Peter Reuter	20	
23. Joseph Leidtner	23	
24. Rosina Spiebieglerinn	50	
25. Gertraud Einacherin	27	
26. Barbara Einacherin	31	
27. Barbara Maurin	23	
28. Maria Theresia Hammerinn	42	
29. Joh. Michel Lieser [Rieser]	31	
30. Anna Marie Sa femme	26	
31. Gottlieb enfant		1/2
32. Leonhard Kraus	20	
33. Anna Cath. Meierin	20	
34. Anna Marie Hopflingerin	20	

Translation of a Letter from the Reverend Mr. Samuel Urlsperger dated Augsburg September 8, 1735. To Mr. Newman.

Honoured & Dear Sir: I have received your agreable Letter of 15th of August O.S. by the way of Ratisbonne which came to my hand but yesterday the 7th of September N.S. I must acquaint you that You would for the future direct your Letters by the way of Frankfort on the main which is a more Expeditious way for if they come by the way of Ratisbonne they will be three or four days later.

The Transport set out from hence the 6th Instant[.] I accompanied them half a days Journey with some persons of distinction as likewise good Christians when Mr. [Ph.] Von Reck's Brother and my parting with him was very Affecting and moving. I have given to Mr. Commissary Von Reck all necessary Instructions, particularly that he would pursue his Journey with all speed that he may have the great Advantage of overtaking Mr. Oglethorpe and I make not the least doubt but that he will pursue my directions. I here recommend him with his Brother in the best manner possible to the worthy Trustees and Honourable Society.

The letter that my dear Mr. Secretary wrote to him in the name of the Society I shall send him this Instant and hope he will receive it in three days time. With regard to the proper Expedient for the Carinthians to *leave three or four grave and elderly persons to wait the release of their Wives and Children which are now pressed at the Imperial Court by the British Minister there, and that the Trustees for Georgia will at any time receive and transport those Women & Children after their husbands and parents,* it has been already proposed to them, but they would not accept of it, Since scarce any one even of the Evangelical Envoys at Ratisbonne will believe that the Imperial Court will give a favourable answer, or should the Court even give one the Romish Clergy in Carinthia would raise many difficulties to oppose it, as we have hitherto found by many Examples.

I for my part am in hopes that the Imperial Court will give a favourable Resolution, and in that Case I shall employ my utmost endeavours that the 4th Transport should proceed. But for this purpose I shall want a proper Commissary to conduct those people to Rotterdam or London, it would be therefore necessary that I should have a Pass from Her Majesty the Queen in such a manner that a Blank therein should be left to put in the Commissary's name which to me hitherto is unknown. I write this day to Mr. Commissary Von Reck to send me, by every post, the Route he takes, as likewise An Account of all other Circumstances that in case a new Commissary should conduct the Carinthians, he might have the use and benefit of the same.

Be pleased to assure the honourable Society that I have hitherto omitted nothing that could be done towards the dispatch of this Affair, and they may be assured that for the future every thing shall be done according to the power that God is pleased to give me.

Mr. Von Reck is witness that this Transport has occasioned my whole family to be employed from Morning till late at night & has created more trouble and Care than any other Transport has hitherto done.

I praise God that he hath supported me in it, and that I brought together to the number of 37 persons. I write again this day to Vienna and Ratisbonne with regard to the Carinthians[.] the Lord in mercy grant his Blessing to all   I remain

Honoured & Dear Sir   Your most humble Servant

Samuel Urlsperger

P.S.: I hope that our most worthy Mr. Oglethorpe will wait for the Transport under Mr. Von Reck's Conduct, for upon that Account even I and my Family have done almost beyond our power.

Translation of a letter from the Reverend Mr. Urlsperger dated Augsburg 22 September 1735.   To Mr. Newman.

Dear Sir: Yesterday I received yours dated the 26 Aug. upon which I know nothing to answer but what I have mentioned in my former. I have received three letters from Ebenezer. The Journals are still with Mr. Ziegenhagen. By the letters I see that it will be very necessary as well for the Saltzburgers who are there and for their Posterity as for those who are to come to them that another place and land be given to them, because as Experience shews the land about Ebenezer is not so good as it hath been taken to be at first. I therefore most humbly beseech the honourable Society to entreat the Trustees especially Mr. Oglethorpe that our Saltzburgers may get a better land and that the representations made by the Ministers who have shewed themselves hitherto as very good men in every respect, likewise by Mr. Vat who certainly is impartial, may be heard[.] It is certainly a matter of great moment that these people may soon be comforted in this point, and that they may be assisted in their temporal Business so that their Souls my remain in enjoying daily the word of God. These good people acknowledge the great Benefactions they have received, but they would also fain see themselves in such a Condition that they might eat their own bread. It is a particular Temptation for the Saltzburgers that whereas Georgia being a fruitfull Country they should see just the contrary of it[.] However it must also be said in this case per *Crucem ad Lucern.*[300]

The Ministers praise Mr. Zwiffler very much for his Diligence and faithfullness in assisting the sick People, wherefore I wish he also might be provided for that he may have something certain for the future having as yet nothing certain besides his Maintenance and the extraordinary Presents of the Society. I remain

Honoured & Dear Sir   Your most humble Servant   Urlsperger

Translation of a Letter from the Reverend Mr. S. Urlsperger dated Augsburg 15th September 1735. To Mr. Newman.

Dear Sir: I hope my Letters dated the 15th and 22nd of August and the 1st and 8th Instant are come to your hands. Yesterday I received yours dated the 19th August upon which I have nothing to answer but that Mr. [Ph.] Von Reck is arrived at Marcksteft the 10th Instant and hopes to be at Frankfort the 16th and will not tarry; and to the end that he may the more easily pass by Mayntz [Mainz], I write by this post to the General Count of Seckendorff who is Commander in chief there and have the honour to be particularly acquainted with him. I am very glad that this Transport will come in time to go with Mr. Oglethorpe. Concerning the Carinthians no Resolution is come yet, and if it was, I doubt whether they could go this year. However on this head I expect an Instruction as well from the Society as from the Trustees. The day before the departure of Mr. Von Reck I received the full Account of the Saltzburg Emigrants who are settled in America, which I have caused to be printed at Hall and given to Mr. Von Reck in order to deliver it to the Secretaries of the Trustees and the Society that they might produce the same in due places because I have ventured to dedicate it to the Trustees and the Society which I hope they will kindly accept of.

Having mentioned in my former that the last Colonists, especially the Austrians, were very poor I now acquaint you that the 500 florins which Mr. Van Münch paid me are spent and more too. I have paid out of it Mr. Von Reck's Maintenance from the 1st April to the 9th September weekly

	Creutz.
3 florins and 30 Creutzers makes	f. 80.30
for Matlasses [?]	60.
Cloths Stockings & Shoes	150.
P sick and Surgeons	20.
Books Bibles, John Arnds Christianity &c.	30.
2 Waggons from hence to Marcksteft	92.
Their Lodgings & Victuals	50.
Linnen for Shirts & necessary Bedding	40.
	522.30

Besides other trifling things. These charges will be divided into two, so that the Society pay one half and the Trustees the other half. For tho' the Saltzburgers are the most in number yet they have not wanted so much Clothing as the others.

And because all the money for the Colonists being thus spent

I have kept the 262 florins which I was to remit to Mr. Ziegen-hagen for the Congregations in Pensylvania Virginia etc. desiring that they may be paid again to Mr. Ziegenhagen by the honour-able Society. I remain

Dear Sir   Your most humble Servant   Urlsperger

Translation of a letter from Mr. P. G. F. de Reck on Bord the London Merchant in the Downs dated 27 October 1735.   To Mr. Newman.

Sir: 'Tis by God's Providence we just now arrived here where I received the letter which you was pleased to send me I humbly thank you and pray God to keep you always in good health and to reward you both here and hereafter for all the favours you have bestowed on me. Thank God we are now all in good health. I have had the happiness to see my dear Friend Mr. Oglethorpe on bord the Symonds and our ship in good health. we are now at Anchor waiting for a fair wind. I recommend to you the enclosed and beg you'd forward it to his Excellency.

My most humble respects to Mr. Vernon and Mr. Ziegenhagen. I am always ready to be

Sir   Your most humble & most obedient Servant   Reck

P.S. My Brother gives his humble Service to you.

 *Notes*

OUTWARD CORRESPONDENCE

1. Ziegenhagen's secretary who translated most of the German letters for the S.P.C.K.
2. The S.P.C.K. was then contributing to the Danish mission at Fort St. George near Madras in India, which was staffed with German missionaries from Halle.
3. See intro., p. 5.
4. John Deane, Archbishop of Rochester, treasurer of the S.P.C.K.
5. Particularly the Treaty of Westphalia, which guaranteed religious dissenters the right to leave their country. See intro. p. 4.
6. Probably Benjamin Martyn's account of Georgia: *Reasons for Establishing the Colony of Georgia,* (London, 1732) .
7. This promise was not kept for two years.
8. Thomas Hollis, merchant of London, Trustee for Georgia and benefactor of Salzburgers and Harvard College.
9. Purysburg (Purrysburgh) on the Savannah River.
10. After comparing Newman's hand with that of his clerk, it is easy to see that Urlsperger's translator must have been delighted.
11. In the 18th century the word "curiosity" included art works and other objects of value.
12. Urlsperger published these in his *Ausführliche Nachrichten* . . . (Halle, 1735) , I, 2-5.
13. Because some Calvinists in Germany had been inhospitable to the Salzburgers, Urlsperger expected the same of the Presbyterians.
14. Also written Jennison, Jenison, & Jennings.
15. Because this stipulation of inheritance only by tail male caused much discontent, it was finally dropped.
16. Newman was unaware that many more had died by the time he wrote this letter.
17. Despite its name, the Swedish Church in Trinity Lane was actually German Lutheran.
18. The Waldensians.
19. Later William IV of the Netherlands. Married Anne, daughter of George II.
20. Probably the copyist's error. See letter of 13 November.
21. War of the Polish Succession, 1733-35.
22. Another chalice was bequeathed by a dying youth in Germany.
23. This is one of the few times that Boltzius is written with a *t* in these letters. The t-sound should be pronounced whether written Boltzius or Bolzius.
24. He had studied law at Helmstedt and he later studied administration at Halle.
25. Consort of Queen Anne.
26. The name Fischer was perhaps omitted inadvertently.
27. See note 2.
28. To try to collect moneys owed them there.

29. But only during droughts. In the rainy season it was often submerged.
30. In 1687 King James expelled the fellows of Magdalen College, Oxford, for refusing to accept Farmer, a Catholic appointed as their president.
31. Franco (prepaid) to Cologne.
32. See note 23.
33. The first two lines of this letter are in Newman's hand.
34. A good example of the English sense of due process of law.
35. Chief among these was Benjamin Sheftal, who was from Germany and spoke good German.
36. The Prince Frederick was soon renamed the Prince of Wales.
37. Yesterday was sennight—a week ago.
38. Weisiger, a German returning to Pennsylvania after solliciting money in Germany for church building. Also written Weisinger and Wisiger.
39. Mrs. Schoppacher, whose baby died soon after reaching Georgia, as did nearly all those born there during the first few years.
40. His home town.
41. *An Extract of the Journals of Mr. Commissary Von Reck, . . . and of the Reverend Mr. Bolzius, . . .* (London, 1734).
42. Urlsperger's German script, preserved among the Miscellaneous letters in the S.P.C.K. archives, is truly difficult to read.
43. The Moravians claimed to be closer to Luther than were the Pietists of Halle.
44. Comparison of this copy with the original draft in Newman's hand, which is preserved in the miscellaneous letters of the S.P.C.K., reveals the great care taken by the copyist.
45. This engraving appears as frontispiece of Vol. I of the *Ausführliche Nachrichten.*
46. Michael Jastram, who later joined Schalkhauser, Flügel, & Jastram, a German firm in Venice that donated to the Georgia Salzburgers.
47. The States General, government of the Netherlands.
48. This ending is most unusual for the Puritan Newman. Perhaps he felt it suitable in writing to a Continental nobleman.
49. This description of the early Georgia colonists is certainly more frank than tactful.
50. This description of Georgia is unlike those then being circulated in Germany.
51. Gronau to Catherina and Bolzius to Gertraut Kröer, the daughters of Barbara Rohrmoser of the first transport.
52. To stimulate missionary zeal, Oglethorpe had previously reported that the Indians were awaiting a "White Man" who would instruct them in wisdom.
53. See note 15 above.
54. Probably coconut shells given him by a travelling companion in New England who had just come from Jamaica.
55. Having received no further grant from the Trustees, Zwiffler departed for Europe but seems to have gotten no further than to Pennsylvania.
56. See note 55 above.
57. Was this the gentleman who refused to refund Bolzius for the counterfeit bills, or was it his son?
58. Georg Sanftleben, a carpenter from Silesia, who returned home to fetch his sister and five other marriageable women.
59. The War of Jenkins' Ear.

INWARD CORRESPONDENCE

1. Urlsperger seems to have received 8.63 Florins per £ Sterling.
2. Reference to a miracle purported to have occurred near Radstadt in Salzburg.
3. As personal chaplain to the King, Ziegenhagen was mostly at the Court of St. James.

4. Matthew 18:20.
5. James Vernon, member of S.P.C.K. and Trustee of Georgia.
6. Hochdeutsch, High German. Both the Germans and the German Swiss were called "Dutch". Cf. Dutchtown, near Savannah.
7. This letter must have been dated 17 July, as indicated in Urlsperger's letter of 1 August.
8. The designation after 345 looks like R 1, perhaps for Reichstaler (rixdollars). Later in the letter (p. 6) Urlsperger explains that 50 £ equals 437 Florins 30 Creutzers, so perhaps the 345 RT is its equivalent.
9. This suggests the over-head clapping so popular in later revival meetings.
10. Schiessgraben, a target range outside of Augsburg.
11. Johann Caspar Schaur (Schauer, Schaurer), a distiller and inventor of Schauer's Balsam, benefactor of the Salzburgers.
12. Johann Georg Morell, mayor of the Protestant part of Augsburg.
13. Frederick William I, Elector of Brandenburg and King of Prussia.
14. The northeast part of East Prussia, particularly in Memel, where they remained until ejected by the Russians in 1944.
15. Possibly J. von Reck, but more probably Göbel.
16. Protestant mission in India, operated by German missionaries and supported by Danish government and S.P.C.K.
17. Probably rixdollars (Reichsthalers).
18. This was also true of Urlsperger, whose forefathers had been expelled from Austria because of their Protestant religion.
19. See p. 252. Delay in recording was probably due to delay in translation.
20. Probably Göbel.
21. Possibly Esterlin.
22. See Intro. p. 5.
23. Almost an enclave in Salzburg. The S.P.C.K. included the exiles from Berchtesgaden with those from Salzburg.
24. This is the beginning of the Tirnberger debacle, which caused such problems for the Protestant authorities. These Salzburgers, from Tirnberg (Dürnberg near Hallein) were settled in an unfavorable location at Cadzand in Holland.
25. The Treaty of Westphalia, which ended the Thirty Years War in 1648, guaranteed the prince's right to determine the religion of his subjects.
26. Apparently serfs and therefore without civil rights. As hereditary ruler of Bohemia, the Emperor could determine the religion of his subjects.
27. Pfyffer, the Upper German spelling of Pfeifer.
28. Usually written Schaitberger. He was the author of the "Exile's Song" (Exulantenlied).
29. Göbel and Plotho.
30. Twenty German miles equal about 100 English miles.
31. Probably from Göbel.
32. Bussprediger, in this case Capuchins.
33. That he was not entirely wrong in this regard is shown by Thomas Geschwandel's illegitimate child. See page 368.
34. Allusion to the "Oath of Salt" by which the Salzburg Protestants swore to be loyal to their faith.
35. Upper Hungary is now in Slovakia. Schemnitz was then predominantly German.
36. Members of the Commission for Reforming Religion [Religionskommission], a kind of inquisitorial commission.
37. Thomas Wilson, Bishop of Soder and Man.
38. Probably Göbel, but possibly J. von Reck.
39. Cajetan Anthon, Freiherr von Nottaffel, Prince Abbot of Berchtesgaden.
40. Göbel (Goebel, Gobel) seems to have been the anonymous correspondent most often mentioned by Urlsperger. Being accredited to the court of Salz-

burg, he had to conceal his activities in luring Saltzburg subjects for his master, the King of Prussia.

41. This may be the letter of 16 October N.S.

42. Hieronymus Christiani von Rell (Roll), the real villain of the expulsion.

43. A supreme court of the Holy Roman Empire, established by Maximilian I in 1501.

44. Eugene of Savoy, hero of the Turkish wars, under whom Oglethorpe had fought.

45. Now Elblag in Poland. These were the exiles en route to East Prussia.

46. The term Austria then included Lower Austria (Niederösterreich), the area immediately around Vienna, and sometimes upper Austria, or the region immediately adjacent to Salzburg.

47. The original translation (apparently in Martini's hand) had Tirnberg & Berchtolsgaden. The town of Dürnberg is on the boundary between Berchtolsgaden and Salzburg.

48. The original translation said "residing here". The copyist did not realize that that meant at Ratisbon.

49. They had done so in 1618 and subsequently but had been suppressed.

50. S.P.C.K. and Georgia Company (Trustees).

51. Georgia was generally included in the term "West Indies", especially by the S.P.C.K. and the Francke Stiftungen, who also supported missions in the East Indies.

52. Johann Arnd's *Vom Wahren Christenthum* was perhaps the most influential religious treatise in the 18th century. Also popular in Pietistic circles was his *Paradiesgärtlein*.

53. See note above.

54. Donau or Danube.

55. This was probably von Hugo, envoy of the Duke of Brunswick (George II of England).

56. Probably an error for 25 Sept., since no letter of 25 Dec. is recorded. See memorandum at close of letter.

57. Possibly the Eisenburg mentioned on p. 252, which is a county next to Austria.

58. Zell was one of George II's German territories, usually associated with Brunswick.

59. The Protestant cause was greatly weakened in parts of Germany by the rivalry between the Lutherans and the Calvinists, which surpassed their common hate of Popery.

60. The name was Puckler. The *in* was merely a feminine ending added to the maiden or married names of a woman.

61. A mistranslation of *Schütz*, literally a marksman but used of constables and militia.

62. This is a typical list of edifying Protestant literature.

63. It was already 1733 in Germany (New Style) but still 1732 in England (Old Style).

64. *Bussprediger,* Capuchins or Jesuits. See letter of 3 March 1733, item 6.

65. This reveals that the anonymous correspondent was an agent of the King of Prussia entrusted with procuring Salzburger exiles and suggests that it was Göbel. See Urlsperger's letter of 6 Oct. 1732 to Newman and of 16 March to Ziegenhagen.

66. All references to Prof. Francke as living refer to Gotthilf August (sometimes written August Gotthilf) Francke, the son and Successor of August Hermann Francke, who had died in 1727. Gotthilf August was one of the "Reverend Fathers" of the Georgia Salzburgers.

67. Augustus II, Elector of Saxony, had become officially Catholic in order to become King of Poland in 1697, but he remained Lutheran in sentiment. When

he died, a Polish nobleman ruled until ousted by Augustus III in the War of the Polish Succession of 1733-35.

68. Kaschau, now in Slovakia.
69. The M. could be for Monsieur or for Marines.
70. Urlsperger was a devotee of the Halle school of Pietism.
71. These were soon rejected as being tainted by Zinzendorf.
72. Esterlin had been too zealous in trying to proselytize Catholics in Ratisbon.
73. Charles Albertus later succeeded in having himself crowned as Charles VII, Emperor of the Holy Roman Empire, from 1742 to 1745.
74. The German term was *Gericht* (jurisdiction, court).
75. There seems to be an omission at this point in this badly copied passage.
76. The stipulation in the Treaty of Westphalia that one should not lure away foreign subjects.
77. The ending *in* is the feminine ending, not part of the name.
78. See pp. 309-312.
79. According to the Treaty of Westphalia, the British could accept only those Salzburgers who had voluntarily left their country and had not been persuaded to do so by promises from the British. Subjects were looked upon as a natural resource and the property of the ruler.
80. The copyist erroneously added "with us".
81. The prefix "dis" was added, no doubt in error.
82. As Duke of Hanover, George II of England was an elector of the Holy Roman Empire.
83. Evangelical Body.
84. Although his uncle gives his first name as George, the nephew nearly always signed himself Philip Georg Friedrich (the only exception in this collection being his letter of 7 Oct. 1734), and that is how all legal documents list him. Nevertheless, perhaps because of his uncle's letter, he often appears as George in contemporary English records.
85. Feeling mistreated, some of the emigrants to Brunswick returned to Ratisbon.
86. They were discovered to be disciples of Zinzendorf.
87. Probably a confusion for "has not been willing to permit".
88. See note 66 above.
89. See note 86 above.
90. This is not clear, but it seems to mean that Urlsperger cannot employ a minister and catechist until the transport is complete.
91. Urlsperger's letters were, of course, 7 & 14 Sept. *N.S.*, or 25 Aug. and 2 Sept. O.S.
92. Johannes Mosshamer and Maria Kröer.
93. Protestants of Savoy, followers of Peter Waldo.
94. Stephen Hales, M.A., of Teddington, 1677-1761, Georgia Trustee and member of S.P.C.K. Author of *Vegetable Statiks*, London, 1727.
95. This long letter, badly copied in a minute hand, has been omitted as irrelevant to the Georgia Salzburgers.
96. The Lutheran articles of faith.
97. A verb has been omitted.
98. Johann Martin Bolzius and Israel Christian Gronau.
99. City in Livonia, now in Esthonia.
100. Possibly an error for Mischke, who was inspector at Halle and Bolzius' immediate superior.
101. See note 52 above.
102. Secretary to Prince George of Denmark, consort of Queen Anne.
103. See note 52 above.
104. Johann Anastasius Freylinghausen, Pietistic theologian at Halle.
105. Now Kronshtadt, city on island in Gulf of Finland.

106. This is evidence of J. von Reck's active, even if anonymous, efforts to encourage dissenters to emigrate.

107. Error for Meuse (Maas). They would have already passed the confluence of the Moselle at Koblenz.

108. Copyist repeated *of* by error, the German word having been *Bruderliebe* (brotherly love).

109. Johannes Mosshamer and Maria Kröer.

110. Should be compatriots (from *Landsleute*).

111. See note 82 above.

112. Hoff (now Bad Hofgastein) is a village in the district of Gastein in southern Salzburg.

113. Premstall, an estate in Anger Hundsdorf near Bad Hofgastein.

114. Probably error for *Oberknecht* (foreman).

115. Probably Lannthal, on road between Saalfelden and Zell.

116. This name is variously spelled Kräer, Kröher, Barbara, with her two older daughters, left her unconverted husband and their smaller children in Salzburg.

117. Probably by Kellbach in Saalfelden. Several centuries after the Germanic invaders settled in Salzburg, a population increase caused many new settlements to be founded, and these often took the name of the parent settlement, with the designation Ober, Unter, Nieder, Mitte, etc. added.

118. Probably an anglicization of some name like Stockheim auf der Heide.

119. This was the marriage performed in Augsburg.

120. The rich peasant who changed his mind.

121. Martin Herzog, a miller.

122. St. Ulrich is nearly 10 miles from Salzburg, a long way for a peasant woman to travel in those days.

123. Now Dorfgastein.

124. Now Unterladerting, formerly Underlaidratting, between Bad Hofgastein and Dorfgastein. For this and other information I am indebted to Sebastian Hinterseer of Bad Hofgastein in Salzburg.

125. Röstermeister is probably not a name, but a profession (smelting-master, master-smelter).

126. Mühlbach is a few miles up the Salzach from Mittersill.

127. In the Bavarian dialect of Salzburg, unaccented *e* is generally dropped. This name was therefore correctly spelled Gschwändl, a fact that explains why our scribes and copyists always misspelled it. In America Gschwändl used the standard German form Geschwandel.

128. This shows that the Priest in Augsburg was not entirely wrong about the Salzburgers illegitimacy rate. Common-law marriage was common among the peasants, Catholic as well as Protestant, because of the many ecclesiastical restrictions upon marriage, particularly with regard to consanguinity and divorce, since peasants could not afford the luxury of dispensations.

129. Gut Niederberg in Gadaunern, near Hoff in Gastein.

130. Mühlberg near Bad Bruck at the foot of Bad Gastein.

131. Faschingberg, above Gadaunern near Bad Hofgastein.

132. The name Riste seems to have been interchangeable with Rieser (Risser).

133. Between Saalfelden and Weissbach on the road to Lofer.

134. Immediately next to Zell am See.

135. Lehm Grube (clay pit), in local dialect Loamgrubn.

136. Roth, not a Salzburger, later caused difficulty.

137. Rieser and his family remained behind because his son George broke his leg, but they joined the second transport.

138. This man seems to have used the names Bartholomeus and Balthazar indiscriminately, the latter being also the name of his nine-year-old son. He like-

wise vacillated between the last name Riste and Rieser (Risser), but the latter survived in America.

139. Apparently a term for their Catholic persecutors.
140. *Vogel friss oder stirb*—"Bird, eat or die".
141. For Schaitberger, see Intro. p.
142. *Vriss Vogel*—"Eat, Bird". See note 140 above.
143. Good German word order. This anonymous letter would seem to be from John Sartorius (see p. 437).
144. Urlsperger gives this as Evermercken.
145. Karl Hildebrand, Freiherr von und zu Canstein (1667-1719), financed Bible printing at the Orphans House in Halle.
146. The copyist seems to have missed a line.
147. Von Reck's loyalty to his confession in preference to his Empire is symptomatic of German politics before Bismarck.
148. The Waldensians.
149. Faulty translation of Herr Kaufmann Minet, merchant being Isaac Minet's profession, not his name. Butjenter always wrote German.
150. The famous Waysenhaus, or orphanage, at Glaucha near Halle.
151. Like his English colleagues, Martini did not distinguish between minister (of state) and minister (of the Gospel). Here the context implies the latter, in this case probably a Huguenot.
152. Compare von Reck's youthful enthusiasm and rhetoric with Vat's maturer realism in his letter of 26 August 1735.
153. This was bequeathed by a dying youth in Germany.
154. Apparently an omission.
155. Probably renders *merkwürdig*.
156. J. A. Sartorius and J. E. Geister were German Lutheran missionaries from Halle sent by the S.P.C.K. to the Danish mission at Fort St. George near Madras, India. This Sartorius should not be confused with John Sartorius, the persecuted pastor from Hungary. (See pp. 379, 437).
157. To serve in the Danish mission founded by the Rev. Benjamin Schultze in Madras.
158. *Ludowic's Grammar.*
159. Capt. Coram was still resentful that the Trustees would not expel the Jews who arrived in Georgia on 11 July 1733.
160. The War of the Polish Succession.
161. The Anabaptists of Holland, followers of Melchior Hofmann, were influential in Eastern Germany, as also in Russia and America.
162. Urlsperger' account of this event gives Fry's name as Joseph instead of Tobias.
163. Von Reck seems to have begun using the English (Julian) calendar as soon as he reached Dover.
164. This is the only reference to this questionable person.
165. Von Reck gives the date according to the Georgian calendar, which was normal for him, but also to the Julian calendar, which he henceforth uses in letters to British recipients.
166. The S.P.C.K. soon published extracts of this journal in English under the title of *Extracts from the Journals of Mr. Commissary Von Reck . . . and the Reverend Mr. Bolzius* (London, 1734), which subsequently furnished the chief source for many Georgia historians.
167. In the War of the Polish Succession.
168. Apparently an omission.
169. Probably from Urlsperger at Augsburg to Newman at London, who had written him on 18 & 28 December.
170. This is the end of the extract.
171. This explains the anonymity of so many of the propagandistic letters.

172. A marginal note explains: "The remaining Cash will be mentioned in my next."
173. Urlsperger continued to suppress all unfavorable reports for some time.
174. An apparent omission.
175. Austria and Kärnten (Carinthia)?
176. See letters of 30 Oct. and 18 Dec. 1733.
177. Dr. Octaviano Plossen.
178. See note 2, p. 599.
179. Probably a transliteration of *Ihre Hochgeehrten*.
180. Von Reck was young and impressionable. No one else noticed any storm.
181. *Exulantenlied*: "Ich bin ein armer Exulant", by Joseph Schaitberger.
182. Having been sick during most of his second sojourn in Georgia, von Reck renounced this grant, returned to Germany, studied administration, married an heiress, and lived to a ripe old age.
183. Article 4 of the Treaty of Ryswick (1697).
184. Lutheran Senior Pastor at Frankfort.
185. The Protestant Privy Council.
186. See pp. 376-377.
187. A poor copy of this Latin letter follows.
188. "that he admit as colonists any exiles or people emigrating from their native land for the sake of the Protestant religion who might wish to join the afore-mentioned colony and set out for America."
189. Old Ebenezer was several feet above sea level, but much of it was below the water table in the rainy season!
190. William Stanhope (Lord Harrington), Secretary of State.
191. There seems to be an omission between folios 29 & 30 .
192. See note 194 below.
193. Probably an error for or imitation of *Lutherthum* (Lutherdom, Lutheran lands).
194. Despite his protests to the contrary, the Catholic authorities believed that Lerchner was instigating dissent and proselytizing their subjects against the terms of the Treaty of Westphalia.
195. Chief of the Imperial Salt Office in Salzkammergut.
196. Being apochryphal, the book of Syrach is omitted from the King James version of the Bible.
197. Either from Styria (Steiermark) or from the town of Steyer.
198. Such steps to make Lutheran services in the colonies conform to Anglican usage helps explain how some Lutheran congregations became Anglican.
199. Faulty translation of *Gericht*, meaning here "district".
200. The name Savannah Town was soon transferred to a settlement high up the Savannah River.
201. This is New Style, because the Salzburgers reached Savannah on 12 March, O.S.
202. Father etc. is in apposition with Oglethorpe, not with Gov. Johnson.
203. In the 18th century, particularly among the Pietists, sentiments were freely expressed. This is probably a translation of *Liebesbrief*, or affectionate letter.
204. Pury had returned to Switzerland for more colonists.
205. This rate was nearly as high as that of the first group of Salzburgers in Ebenezer.
206. As a redemptioner.
207. Poor translation of *Lohn* (rewards, pay) .
208. No reason is given for writing *Nota Bene* instead of the usual "read".
209. The court of George II, who was both king of England and also Duke of Hanover.
210. An error for August, as is shown by von Reck's journal.

211. The chief complaint was that the ministers were excluded from the great cabin. Urlsperger deleted these complaints from his *Ausführliche Nachrichten*.
212. This promise was to cause Urlsperger much anxiety. See Vat's letter of 30 May 1735.
213. One of these was the child that broke his leg.
214. The apparent confusion in dates was due to the use of both calendars. The 4th of Oct. N.S. was the 22 Sept. O.S., so the letter took nine days to arrive and be read.
215. To oppose the French in the War of the Polish Succession.
216. This family, from Philippsburg on the Rhine in the Palatinate, was among the numerous impoverished Palatines then seeking passage to America.
217. Georgia was usually included in the term West Indies.
218. Nicolaus and Christian Riedelsperger.
219. Count Nicolaus Ludwig von Zinzendorf, a lay minister of the Bohemian Brothers, sent a group of co-religionists to Georgia under Count Spangenberg.
220. This may have been deleted as too unkind. Like all the other Pietists, von Reck did not care for Zinzendorf's innovations.
221. It is not clear what king or why.
222. An error. Urlsperger gives it as Zuntz.
223. Wijk bij Duurstede.
224. This ship was soon renamed the Prince of Wales.
225. Tomochichi, the Mico of Yamacraw, who was returning with his family from a state visit in England.
226. Despite many unfavorable letters from there, Pennsylvania remained the El Dorado for impecunious Germans.
227. Constant complaints finally persuaded the Trustees to rescind their law of allowing inheritance only by tail male, a policy adopted for military reasons.
228. This is the first illustration of the strict censorship clapped on the Georgia Salzburgers.
229. Trustee for Georgia.
230. Von Reck does not indicate which divines. Degmaier was the only one listed.
231. He settled between Savannah and Ebenezer but died soon after.
232. Schauer's Balsam. See note 11 on p. 601 above.
233. Illustrative of the way in which foreign names like Vat were naturalized into English.
234. Old Ebenezer's landing place on the Savannah River. The Salzburgers usually Germanized it as Habercorn (Oat-corn), perhaps influenced by memories of the Lutheran theologian Peter Habercorn (1604-1676).
235. Most of these died of dysentery.
236. Bolzius did not yet know about the "calumnious" letter intercepted by Urlsperger (see Urlsperger's letter of 14 Oct. 1734 to Vat). Being neither a Salzburger nor a religious exile, Roth did not fit at Ebenezer.
237. Like the African slavers, the captains who transported Palatines had to determine at what point the law of diminishing returns would begin as they crowded their ships. The tighter they were packed, the higher the death rate, and therefore the smaller the cargo to be sold.
238. Andreas Gottfried Dietzius, a German who had lived in Batavia. He was offered a grant in Georgia but preferred to go to Purysburg.
239. Children counted only as a fraction.
240. Here he lets the cat out of the bag!
241. See letter of 3 Dec. 1734 N.S. to Urlsperger.
242. This seems to be a Czechish form or corruption of Rudolf Colloredo, the Bohemian Ambassador. See note 251 below.
243. See introduction, p. 8
244. See account beginning p. 511 below.

245. Mostly Capuchins and Jesuits.

246. Quincy had actually been in New England for most of the year.

247. His son-in-law. He had married Wesley's daughter, recently deceased.

248. *Aeneid* VIII, 508-09, 514-15.

249. Anyone who has camped on the coast of Georgia in the summer will agree that Oglethorpe is more candid in this letter than in any of his others.

250. At Madras and Tranquebar.

251. Rudolf Joseph, Prince of Colloredo-Mels und Waldsee.

252. One of these is mentioned in the account of the persecution of the Protestants of Bohemia enclosed in Urlsperger's letter of 8 Nov. 1734 to Newman. Cf. the name Rudolfa Skoloredj (p. 508).

253. Possibly J. von Reck, who always attempted to conceal his religious activities.

254. Their successors were the first to render the Bible into a Georgia Indian language (at Springdale).

255. This echoes Luther's hymn "Ein' feste Burg".

256. The South Carolina and Georgia authorities soon began squabbling over trade in the Savannah River, particularly with regard to Georgia's prohibition of rum. For reference to staving of rum barrels, see Oglethorpe's letter of 19 November 1734 to Samuel Wesley, Sr.

257. Compare this realistic account with the romantic views in Oglethorpe's letter of 25 December to Samuel Wesley, Sr.

258. Silesia, while predominantly Protestant, was still subject to Catholic Austria.

259. The chief among these was scurvy.

260. Bishop consistently confused Savannah and Georgia.

261. Periagua, pirogue (large dugout).

262. Depressed (probably from *niedergeschlagen*).

263. This non-sequitur suggests a copyist's error.

264. Carrying Vat and the second transport.

265. None of these people reached Georgia. Many of the names are inaccurate, those ending in *z* should probably end in *g*. See copy of this list on p. 561.

266. By this time all the first transport were infected with malaria.

267. A faulty translation of Salzkammergut.

268. See note 45 above.

269. Not as emperor, but as duke of Carinthia.

270. When Tobler's Swiss finally passed through Ebenezer on their way to New Windsor, they complained that they had been deceived by the good reports sent back by the Salzburgers.

271. Zublin soon removed to Purysburg. He was a close kinsman of Joachim Zubly.

272. Ph. von Reck seems to have acted for his uncle who was often, although anonymously, engaged in such projects.

273. As envoy from the Court of Brunswick-Lüneburg, von Hugo was also a subject of George II and von Reck's natural successor as champion of the persecuted Protestants.

274. Von Reck advised the Protestant States to exact certain guarantees from the Emperor in return for their aid against France.

275. Minister at the Hague, usually written de Ayrolles.

276. The King of Prussia's commissioner for the Salzburg exiles. Usually written Göbel or Gobel.

277. Apparently composed by Ph. von Reck. See first paragraph of preceding letter.

278. Probably Ph. von Reck.

279. Transliteration of *sterben* (to die).

280. See note 275 above.

281. Urlsperger's publications about the charitable works in India and Georgia were a major factor in eliciting German donations.

282. I.e. Landeshauptmann, or governor.

283. This mercantilistic policy was the greatest handicap to Georgia's development.
284. His hardships and nearly fatal diseases caused him to relinquish his claim. Having no funds of his own, he married an heiress.
285. The land was clearer only where it was too poor to support trees.
286. Labhart had been misled by von Reck's optimism. Seven or eight bushels of corn per acre would have been more realistic.
287. Able to bear arms. One purpose for founding Purysburg had been to supply a regiment of militia, of which Pury was to be the colonel.
288. This is one of the two times that von Reck signs with his first initial.
289. This name, which was badly copied, probably began with *Schl.*
290. This may mean in colonial currency, worth about a seventh or eighth of the same value sterling. See Quincy's letter of 4 July 1735 below.
291. John West was a baillif of Georgia.
292. Urlsperger had feared wisely. See his letter of 20 September 1734.
293. The first transport.
294. In his original letter, which is preserved among the autographed manuscripts of S.P.C.K., Bolzius wrote "linguister".
295. Samuel Nunez, who arrived on 11 July 1733 in a party of 41 Jews, was credited with ending an epidemic. Nevertheless, while praising his Latin, Bolzius doubted his medical knowledge.
296. Chief among these were Benjamin Sheftall and his wife Perla.
297. Despite the initial "J", this must have been Thomas Robinson (Baron Grantham), British ambassador to Vienna. Perhaps the scribe misread his "T" as "J".
298. In his letters and journal von Reck seldom mentions and never names his brother.
299. Urlsperger was hoping that the third transport would be able to sail with Oglethorpe (See p. 594).
300. Through the cross (tribulations) to the light.

Index of Authors

Index of Proper Names

Names on lists of persecuted Protestants who did not go to Georgia are omitted. Asterisks indicate names which are further explained. 1st, 2nd, and 3rd tr. are abbreviations for first, second, and third Salzburger transports.